T0174238

HANDBOOK ON STANDARDS AND GUIDELINES IN ERGONOMICS AND HUMAN FACTORS

Human Factors and Ergonomics

Gavriel Salvendy, Series Editor

Bullinger, H.-J., and Ziegler, J. (Eds.) : Human–Computer Interaction: Ergonomics and User Interfaces, *Volume 1 of the Proceedings of the 8th International Conference on Human–Computer Interaction*

Bullinger, H.-J., and Ziegler, J. (Eds.) : Human–Computer Interaction: Communication, Cooperation, and Application Design, *Volume 2 of the Proceedings of the 8th International Conference on Human–Computer Interaction*

Karwowski, W. (Ed.) : *Handbook of Standards and Guidelines in Ergonomics and Human Factors*

Stephanidis, C. (Ed.) : *User Interfaces for All: Concepts, Methods, and Tools*

Smith, M. J., Salvendy, F., Harris, D., and Koubeck, R. J. (Eds.) : *Usability Evaluation and Interface Design: Cognitive Engineering, Intelligent Agents and Virtual Reality*

Smith, M. J., and Salvendy, G. (Eds.) : *Systems, Social and Internationalization Design of Human–Computer Interaction*

Stephanidis, C. (Ed) : *Universal Access in HCI: Towards an Information Society for All*

Stanney, K. (Ed.) : *Handbook of Virtual Environments Technology: Design, Implementation, and Applications*

HANDBOOK ON STANDARDS AND GUIDELINES IN ERGONOMICS AND HUMAN FACTORS

Edited by

Waldemar Karwowski
University of Louisville

CRC Press
Taylor & Francis Group
Boca Raton London New York

CRC Press is an imprint of the
Taylor & Francis Group, an **informa** business

Senior Acquisitions Editor:	Anne Duffy
Editorial Assistant:	Rebecca Larsen
Cover Design:	Kathryn Houghtaling Lacey
Full-Service Compositor:	TechBooks
Text and Cover Printer:	Hamilton Printing Company

This book was typeset in 10/12 pt. Times Roman, Italic, Bold, and Bold Italic.
The heads were typeset in Helvetica, Helvetica Italic, Helvetica Bold, Helvetica Bold Italic.

CRC Press
Taylor & Francis Group
6000 Broken Sound Parkway NW, Suite 300
Boca Raton, FL 33487-2742

First issued in paperback 2019

© 2006 by Taylor and Francis Group, LLC
CRC Press is an imprint of Taylor & Francis Group, an Informa business

No claim to original U.S. Government works

ISBN-13: 978-0-8058-4129-9 (hbk)
ISBN-13: 978-0-367-39153-9 (pbk)

This book contains information obtained from authentic and highly regarded sources. Reasonable efforts have been made to publish reliable data and information, but the author and publisher cannot assume responsibility for the validity of all materials or the consequences of their use. The authors and publishers have attempted to trace the copyright holders of all material reproduced in this publication and apologize to copyright holders if permission to publish in this form has not been obtained. If any copyright material has not been acknowledged please write and let us know so we may rectify in any future reprint.

Except as permitted under U.S. Copyright Law, no part of this book may be reprinted, reproduced, transmitted, or utilized in any form by any electronic, mechanical, or other means, now known or hereafter invented, including photocopying, microfilming, and recording, or in any information storage or retrieval system, without written permission from the publishers.

For permission to photocopy or use material electronically from this work, please access www.copyright.com (http://www.copyright.com/) or contact the Copyright Clearance Center, Inc. (CCC), 222 Rosewood Drive, Danvers, MA 01923, 978-750-8400. CCC is a not-for-profit organization that provides licenses and registration for a variety of users. For organizations that have been granted a photocopy license by the CCC, a separate system of payment has been arranged.

Trademark Notice: Product or corporate names may be trademarks or registered trademarks, and are used only for identification and explanation without intent to infringe.

Library of Congress Cataloging-in-Publication Data

Handbook on standards and guidelines in ergonomics and human factors / edited by Waldemar Karwowski.
    p. cm. – (Human factors and ergonomics)
  Includes bibliographical references and index.
  ISBN 0-8058-4129-6 (casebound : alk. paper)
  1. Human engineering—Handbooks, manuals, etc. 2. Human engineering—Standards.
  I. Karwowski, Waldemar, 1953– II. Series.
TA166.H2775 2005
620.8′20218–dc22
                                       2005030798

**Visit the Taylor & Francis Web site at
http://www.taylorandfrancis.com**

**and the CRC Press Web site at
http://www.crcpress.com**

Contents

Foreword

With the rapid introduction of highly sophisticated computers, (tele)communication, service, and manufacturing systems, a major shift has occurred in the way people use technology and work with it. The objective of this Handbook is to provide researchers and practitioners a platform where important issues related to these changes can be discussed, and methods and recommendations can be presented for ensuring that emerging technologies provide increased productivity, quality, satisfaction, safety, and health in the new workplace and the Information Society.

When designing products, services, and workstations, the first thing the ergonomics designer and engineers want to ensure is that their design meets the pertaining guidelines and standards associated with the design. When processes are designed for the operation of enterprises one wants to ensure this process meets the ergonomics guidelines and standards. The various ergonomic guidelines and standards were scattered in a large number of diverse documents around the world; hence, the practioner has great difficulty accessing them. This problem has been totally eliminated now with the publication of this *Handbook of Standards and Guidelines in Ergonomics and Human Factors*.

The 32 chapters of this handbook written by authorities from Germany, India, Italy, Japan, P.R. China, Poland, Portugal, Sweden, The Netherlands, United Kingdom, and the United States provide a wealth of international know-how, guidelines, and standards which are crucial as corporations are operating, manufacturing, and marketing their products and services worldwide. These guidelines and standards presented in this Handbook cover every conceivable situation from material handling to human–computer interaction. This Handbook should be in the toolbox of each and every practicing ergonomics and human factors practitioner.

Gavriel Salvendy
Purdue University and Tshinghua University

P. R. China
Series Editor

Preface

Human factors and ergonomics (HFE) standards and guidelines play an important role in facilitating the design of optimal working conditions with regard to human safety, health and general well-being, as well as system performance in the context of technological advances and opportunities for economic development worldwide. Such standards and guidelines offer guidance on the design of work systems, including tasks, equipment and workplaces, as well as working conditions in relation to human capacities and limitations.

Standardization efforts help to recognize the significance of human factors and ergonomics discipline by focusing on the requirements that need to be taken into account during the design, development, testing and evaluation of workplaces and systems. This process involves active participation of a wide range of organizations and other institutions at the national and international levels. HFE standards and guidelines are developed through a process that aims to ensure that all interested and potentially affected parties, including representatives of industrial and commercial entities, government agencies, professional and consumer associations, and general public, have an opportunity to represent their interest and participate in the development process.

HFE standards and guidelines often represent the best available knowledge and practices, both in their explicit and tacit forms that can be used in system design, testing, and evaluation processes. Although the application of relevant HFE standards and guidelines by itself cannot always guarantee optimal workplace design, it can provide clear and well defined requirements for ergonomics design process. For example, by ensuring consistency of the human-system interface and improving quality of the human-system interfaces and their components, HFE standards and guidelines contribute to the enhanced usability and overall system performance. The application of standards and guidelines also facilitates dissemination and promotion of human factors and ergonomics knowledge and the HFE discipline across the society at large.

This handbook offers a comprehensive review of and knowledge about the selected international and national standards and guidelines in the broadly defined area of ergonomics and human factors (including relevant safety and health issues). The Handbook consists of 32 chapters divided into nine sections, including:

1. The Standardization Efforts in Human Factors and Ergonomics (3 chapters)
2. Nature of HF/E Standards and Guidelines (4 chapters)
3. Engineering Anthropometry and Working Postures (5 chapters)
4. Design and Evaluation of Manual Material Handling Task (6 chapters)

5. Human-Computer Interaction (5 chapters)
6. Management of Occupational Safety and Health (4 chapters)
7. Safety and Legal Protection Standards (3 chapters)
8. Military Human Factors Standards (1 chapter)
9. Sources of Human Factors and Ergonomics Standards (1 chapter).

The intended audience for this Handbook are professionals in the HFE and other related fields, including ergonomics specialists, consultants and practitioners; researchers and graduate students from a great variety of interrelated fields, including engineering, safety, manufacturing, and product design; industrial and system designers; managers and supervisors responsible for management of corporate ergonomics and safety and health; architects; quality specialists; occupational health & safety and environmental protection managers; and government officials at all levels.

I would like to thank all the authors for their contributions and patience with the formidable editing process. I would like to express my appreciation to Laura Abell, Editorial Assistant, and Bohdana Sherehiy and David Rodrick, graduate students from the Department of Industrial Engineering, University of Louisville, for their help with the logistics of this demanding editing project. I would also like to acknowledge Anne Duffy, Lawrence Erlbaum Associates and Gavriel Salvendy, Series Editor, for their trust, encouragement and guidance in bringing this challenging project to its fruition.

Waldemar Karwowski
Louisville, Kentucky

About the Editor

Waldemar Karwowski, CPE, P.E. is Professor of Industrial Engineering and Director of the Center for Industrial Ergonomics at the University of Louisville, Louisville, Kentucky, USA. He holds an M.S. (1978) in Production Engineering and Management from the Technical University of Wroclaw, Poland, and a Ph.D. (1982) in Industrial Engineering from Texas Tech University. He received a D.Sc. (dr hab.) degree in Management Science, by the Institute for Organization and Management in Industry (ORGMASZ), Warsaw, Poland (June 2004). He was also awarded an Honorary Doctorate in Science (Doctor Honoris Causa) from the South Ukrainian State K.D. Ushynsky Odessa Pedagogical University of Ukraine (May 2004) for his outstanding contributions to the fields of human factors engineering and management of work systems.

Dr. Karwowski is the author or co-author of over 300 scientific publications (including over 100 peer-reviewed journal papers) in the areas of: work systems design, organization and management; macroergonomics; human-system integration in advanced manufacturing; industrial ergonomics; fuzzy systems and neuro-fuzzy modeling in human factors; and forensics. Dr. Karwowski currently serves as Editor of the *Human Factors and Ergonomics in Manufacturing*, an international journal published by John Wiley & Sons, New York, and the Editor-in-Chief of *Theoretical Issues in Ergonomics Science (TIES)*, a new journal designed to stimulate and develop a theoretical basis for the unique science of ergonomics (Taylor & Francis, Ltd., London). Dr. W. Karwowski also serves as Co-Editor of the *International Journal of Occupational Safety and Ergonomics*, and Consulting Editor of the *Ergonomics* journal. He is also a member of editorial boards for several peer-review journals, including: *Human Factors, Applied Ergonomics, International Journal of Human-Computer Interaction, Universal Access to the Information Society: An International Interdisciplinary Journal, Occupational Ergonomics*, and *Industrial Engineering Research: An International Journal of IE Theory and Application* (Hong Kong).

Dr. Karwowski was named the Alumni Scholar for Research (2004–2006) by the J. B. Speed School of Engineering of the University of Louisville. He also received the University of Louisville Presidential Award for Outstanding Scholarship, Research and Creative Activity in the Category of Basic and Applied Science (1995), Presidential Award for Outstanding International Service (2000), and the W. Jastrzebowski Medal for Lifetime Achievements from the Polish Ergonomics Society (1995). Dr. W. Karwowski was elected an Academician of the International Academy of Human Problems in Aviation and Astronautics (Moscow, Russia, 2004). He is Fellow of the International Ergonomics Association (IEA), Fellow of the Human

Factors and Ergonomics Society (HFES, USA), Fellow of the Institute of Industrial Engineers (IIE, USA), and Fellow of the Ergonomics Society (United Kingdom). He is a recipient of the highest recognition in occupational safety and health in Poland, Pro Labore Securo (2000).

Dr. W. Karwowski served as Secretary-General (1997–2000) and President (2000–2003) of the International Ergonomics Association (IEA). He is past President of the International Foundation for Industrial Ergonomics and Safety Research, as well past Chair of the US TAG to the ISO TC1S9: Ergonomics/SC3 Anthropometry and Biomechanics. He served as Fulbright Scholar and Visiting Professor at Tampere University of Technology, Finland (1990–1991). His research, teaching, and consulting activities focus on human system integration and safety aspects of advanced manufacturing enterprises, human-computer interaction, prevention of work-related musculoskeletal disorders, workplace and equipment design, and theoretical aspects of ergonomics science. He can be reached at: karwowski@louisville.edu.

List of Contributors

Thomas J. Albin
Auburn Engineers, Inc.
Minnesota, United States

Heiner Bubb
Technical University of Munich
Germany

Charles A. Cacha
Ergonix, Inc.
California, United States

Gustav Caffier
Federal Institute for Occupational Safety
   and Health
Dortmund, Germany

Gerald Chaikin
United States

Denis A. Coelho
University of Beira Interior
Portugal

Daniela Colombini
Research Unit, Ergonomics of Posture
   and Movement (EPM)
Milan, Italy

João Carlos de Olveira Matias
University of Beira Interior
Portugal

Henk J. de Vries
Erasmus University Rotterdam
The Netherlands

Nico J. Delleman
TNO Human Factors, Soesterberg
The Netherlands

Jan Dul
Erasmus University Rotterdam
The Netherlands

Anne Ferguson
British Standards Institution
London, United Kingdom

Kaj Frick
National Institute for Working Life
Stockholm, Sweden

David A. Graeber
The Boeing Company
Washington, United States

Antonio Grieco
Research Unit, Ergonomics of Posture
   and Movement (EPM)
Milan, Italy

Sadao Horino
Kanagawa University
Japan

Waldemar Karwowski
University of Louisville
Kentucky, United States

Robert S. Kennedy
RSK Assessments, Inc.
Florida, United States

Falk Liebers
Federal Institute for Occupational Safety
  and Health
Dortmund, Germany

Joe W. McDaniel
Air Force Research Lab
Ohio, United States

Omar Merhi
University of Minnesota
United States

Pranab K. Nag
National Institute of Occupational Health
Ahmedabad, India

Anjali Nag
National Institute of Occupational Health
Ahmedabad, India

Masatoshi Nomura
NEC Corporation
Japan

Enrico Occhipinti
Research Unit, Ergonomics of Posture
  and Movement (EPM)
Milan, Italy

Akira Okada
Osaka City University
Japan

Daniel Podgórski
Central Institute for Labour
  Protection—National Research Institute
Warsaw, Poland

Heather A. Priest
University of Central Florida
United States

Robert W. Proctor
Perdue University
Indiana, United States

Pei-Luen Patrick Rau
Tsinghua University
Beijing, China

David Rodrick
University of Louisville
Kentucky, United States

Sohsuke Saitoh
Human Factor Co., Ltd.
Japan

Eduardo Salas
University of Central Florida
United States

Peter Schaefer
München University of Technology
Germany

Karlheinz G. Schaub
Darmstadt University of Technology
Germany

Andreas Seidl
Human Solutions
Kaiserslautern, Germany

Bohdana Sherehiy
University of Louisville
Kentucky, United States

Thomas J. Smith
Human Factors Research Laboratory School
  of Kinesiology
Minnesota, United States

Kay M. Stanney
University of Central Florida
United States

Ulf Steinberg
Federal Institute for Occupational Safety
  and Health
Dortmund, Germany

Tom Stewart
System Concepts, Ltd.
London, England

Carol Stuart-Buttle
Stuart-Buttle Ergonomics
Pennsylvania, United States

Henk F. van der Molen
Arbouw Foundation
Amsterdam, Netherlands

Kim-Phuong L. Vu
California State University–Long Beach
California, United States

Harmen Willemse
Erasmus University Rotterdam
The Netherlands

Katherine A. Wilson
University of Central Florida
United States

Toshiki Yamaoka
Wakayama University
Japan

Kazuhiko Yamazaki
IBM Japan Ltd.
Tokyo, Japan

Koji Yanagida
SANYO Design Center Co., Ltd.
Japan

HANDBOOK ON STANDARDS AND GUIDELINES IN ERGONOMICS AND HUMAN FACTORS

I

Standardization Efforts in Human Factors and Ergonomics

1

An Overview of International Standardization Efforts in Human Factors and Ergonomics

Bohdana Sherehiy
David Rodrick
Waldemar Karwowski
University of Louisville

INTRODUCTION

This chapter provides an overview of the international and U.S.-based standards and guidelines in human factors and ergonomics (HFE). In general, standardization is the means by which society gathers and disseminates technical information (Spivak & Brenner, 2001). Standards provide quality control and support legislation and regulations to ensure equal opportunity and fairly operating international markets. One of the main purposes of standardization is to assure uniformity and interchangeability in a given area of application. For example, standards may limit the diversity of sizes, shapes, or component designs, and prevent generation of unneeded variation of products, which do not provide unique service. Harmonization of international standards reduces trade barriers; promotes safety; allows interoperability of products, systems, and services; and promotes common technical understanding (Wettig, 2002).

Standard is defined as *a documented agreement containing technical specifications or other precise criteria to be used consistently as rules, guidelines, or definitions of characteristics, to ensure that materials, products, processes, and services are fit for the purpose served by those making reference to the standard* (International Organization for Standardization [ISO], 2004). The standardization process can be performed at the national, regional, and international levels (see Figure 1.1). The basis for worldwide standardization in all areas is provided mainly by the following three organizations: the ISO, the International Electrotechnical Commission (IEC), and the International Telecommunications Union (ITU). Standards related to the human factors and ergonomics are mainly developed by the ISO and the European Committee for Standardization (CEN).

In addition to the CEN, in Europe there are two other standardization organizations, that is, the European Committee for Electrotechnical Standardization (CENELEC) and the European Telecommunications Standards Institute (ETSI). Their mission is to develop a coherent set of voluntary standards that can serve as a basis for a Single European Market/European Economic Area (Wettig, 2002). At the national level, almost every nation has its own national body for standards development. Examples of the national standardization organizations

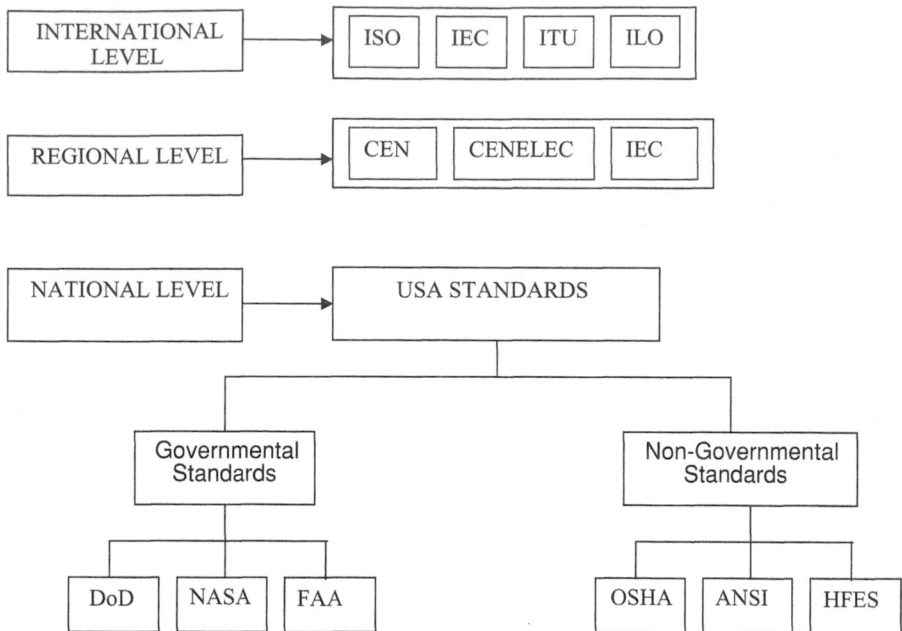

FIG. 1.1.   Hierarchy of Standards levels

include the American National Standards Institute (ANSI), British Standards Institution (BSI), the Deutsches Institut für Normung (DIN), and the Association Française de Normalisation (AFNOR). In general, standards can also be prepared by technical societies, labor organizations, consumer organizations, trade associations, and governmental agencies.

International, regional, and national standards are distinguished by documented development procedures. These procedures have been designed to ensure that all interested parties that can be affected by a particular standard will have an opportunity to represent their interest and participate in the standards development process. For example, ISO standards are developed by technical committees consisting of experts from the industrial, technical, and business sectors that are in need of the standards. Many national members of the ISO apply public review procedures to consult draft standards with the interested parties, including representatives of government agencies, industrial and commercial organizations, professional and consumer associations, and the general public (ISO, 2004). The ISO national bodies are expected to take into account any feedback they receive and present a consensus position to appropriate technical committees.

ISO STANDARDS FOR ERGONOMICS

The ISO is a worldwide federation of national standardization bodies from 146 countries. ISO is a nongovernmental organization, and it considers the diverse interests of users, producers, consumers, governments, and the scientific community. The mission of ISO is to promote the development of standardization and related activities in the world to facilitate the international exchange of goods and services and to enhance cooperation in the areas of intellectual, scientific, technological, and economic activity (ISO, 2004).

TABLE 1.1
Organizational Structure of ISO TC 159 "Ergonomics"

Committee	Title
TC 159/SC 1: Ergonomic guiding principles	
TC 159/SC 1/WG 1	Principles of the design of work systems
TC 159/SC 1/WG 2	Ergonomic principles related to mental work
TC 159/SC 1/WG 4	Usability of every day products
TC 159/SC 3: Anthropometry and biomechanics	
TC 159/SC 3/WG 1	Anthropometry
TC 159/SC 3/WG 2	Evaluation of working postures
TC 159/SC 3/WG 4	Human physical strength: manual handling and force limits
TC 159/SC 3/WG 5	Ergonomic procedures for applying anthropometry and biomechanics standards
TC 159/SC 4: Ergonomics of human-system interaction	
TC 159/SC 4/WG 1	Fundamentals of controls and signaling methods
TC 159/SC 4/WG 2	Visual display requirements
TC 159/SC 4/WG 3	Control, workplace and environmental requirements
TC 159/SC 4/WG 5	Software ergonomics and human–computer dialogues
TC 159/SC 4/WG 6	Human-centered design processes for interactive systems
TC 159/SC 4/WG 8	Ergonomic design of control centers
TC 159/SC 5: Ergonomics of the physical environment	
TC 159/SC 5/WG 1	Thermal environments
TC 159/SC 5/WG 2	Lighting environments
TC 159/SC 5/WG 3	Danger signals and communication in noisy environments

In 1975, ISO formed the technical committee (TC) 159 to develop standards in the field of ergonomics (Parsons, Shackel, & Metz, 1995). The scope of the ISO/TC 159 activity has been described as standardization in the field of ergonomics, including terminology, methodology, and human factors data. The ISO TC159 promotes the adaptation of working and living conditions to the anatomical, psychological, and physiological characteristics of man in relation to the physical, sociological, and technological environment. The main objectives of such standardization efforts are safety, health, well-being, and effectiveness (Parsons, 1995c).

All ISO standards can be obtained at the ISO Web site (http://www.iso.org). At present, the ISO TC159 organizational structure is administrated by the German Standards Association (DIN). The ergonomics standardization group consists of four subcommittees: SC1, SC3, SC4, and SC5. The areas of work of subcommittees and their organizational structure are presented in the Table 1.1.

ISO Standards for Ergonomics Guiding Principles

The TC159/SC1 subcommittee focuses on the standards related to basic principles of ergonomics (see Table 1.2). The ISO 6385:2004 standard specifies the objectives for ergonomics system design and provides definitions of basic terms and concepts in ergonomics. This standard establishes ergonomics principles of the work system design as basic guidelines that should be applied for designing optimal working conditions with regard to human well-being, safety and health, and due consideration of technological and economic efficiency (Parsons, 1995a).

TABLE 1.2
ISO Standards for Ergonomic Guiding Principles

Reference Number	Title
ISO 6385:2004	Ergonomic principles in the design of work systems
ISO 10075:1991	Ergonomic principles related to mental workload—General terms and definitions
ISO 10075-2:1996	Ergonomic principles related to mental workload—Part 2: Design principles
ISO/FDIS 10075-3	Ergonomic principles related to mental workload—Part 3: Principles and requirements concerning methods for measuring and assessing mental workload
ISO/CD 20282-1	Ease of operation of everyday products—Part 1: Context of use and user characteristics
ISO/CD TS 20282-2	Ease of operation of everyday products—Part 2: Test method

The ISO 10075 standard that deals with mental workload comprises three parts. The first part *General Terms and Definitions* (1991) present-terminology and main concepts. The Part 2 *Design Principles* (1996) covers guidelines on the design of work systems, including task, equipment, workspace, and work conditions with references to the mental workload. Part 3, *Measurement and Assessment of Mental Workload*, which provides guidelines on measurement and assessment of mental workload, is currently at the stage of a *Final Draft International Standard*. The third part specifies the requirements for the measurement instruments to be met at different levels of precision in measuring mental workload. Because any human activity includes mental workload, the described standards on mental workload are relevant to all kinds of work design (Nachreiner, 1995).

ISO Standards for Anthropometry and Biomechanics

The TC159/SC3 subcommittee focuses on the standards related to anthropometry and biomechanics. This subcommittee consists of four working groups: WG1, "Anthropometry"; WG2, "Evaluation of Working Postures"; WG3, "Human Physical Strength"; and WG4, "Manual Handling and Heavy Weights." A list of published standards and standards in this area is presented in the Table 1.3. The description of anthropometric measurements, which can be used as a basis for definition and comparison of population groups, is provided by the ISO 7250:1996 standard. In addition to the lists of the basic anthropometric measurements, this document contains definitions and measuring conditions.

The three-part standard for the Safety of Machinery (ISO 15534) provides guidelines for determining the dimensions required for openings for human access for machinery. The first part of this standard (ISO 15534-1:2000) presents principles for determining the dimensions for opening whole body access to machinery; the second part (ISO 15534-2:2000) specifies dimensions for the access openings. The third part of the safety of machinery standard (ISO 15534-3:2000) provides the requirements for the human body measurements (anthropometric data) that are needed for the calculations of access-opening dimensions for machinery specified in the two previous parts of these standards (Parsons, 1995c). The anthropometric data are based on the static measurements of nude people and is representative of the European population of males and females.

The ISO 14738:2002 standard describes principles for deriving dimensions from anthropometric measurements and applying them to the design of workstations at nonmobile machinery. This standard specifies also the body space requirements for equipment during normal operation in sitting and standing positions. The ISO 15535:2003 standard specifies general requirements

TABLE 1.3
Published ISO Standards and Standards Under Development for Anthropometry
and Biomechanics

Reference Number	Title
ISO 7250:1996	Basic human body measurements for technological design
ISO 11226:2000	Ergonomics—Evaluation of static working postures
ISO 11228-1:2003	Ergonomics—Manual handling—Part 1: Lifting and carrying
ISO 14738:2002	Safety of machinery—Anthropometric requirements for the design of workstations at machinery
ISO 14738:2002/Cor 1:2003	
ISO 15534-1:2000	Ergonomic design for the safety of machinery—Part 1: Principles for determining the dimensions required for openings for whole-body access into machinery
ISO 15534-2:2000	Ergonomic design for the safety of machinery—Part 2: Principles for determining the dimensions required for access openings
ISO 15534-3:2000	Ergonomic design for the safety of machinery—Part 3: Anthropometric data
ISO 15535:2003	General requirements for establishing anthropometric databases
ISO/TS 20646-1:2004	Ergonomic procedures for the improvement of local muscular workloads—Part 1: Guidelines for reducing local muscular workloads
ISO/CD 11228-2	Ergonomics—Manual handling—Part 2: Pushing and pulling
ISO/CD 11228-3	Ergonomics—Manual handling—Part 3: Handling of low loads at high frequency
ISO/DIS 15536-1	Ergonomics—Computer manikins and body templates—Part 1: General requirements
ISO/CD 15536-2	Ergonomics—Computer manikins, body templates—Part 2: Structures and dimensions
ISO/FDIS 15537	Principles for selecting and using test persons for testing anthropometric aspects of industrial products and designs
ISO/DIS 20685	3D scanning methodologies for internationally compatible anthropometric databases

for anthropometric databases and their associated reports that contain measurements taken in accordance with ISO 7250. This standard presents information, such as characteristics of the user population, sampling methods, and measurement items and statistics, to make international comparison possible among various population segments.

The ISO 11228-1:2003 standard describes limits for manual lifting and carrying with consideration, respectively, of the intensity, frequency, and duration of the task. The recommended limits can be used in the assessment of several task variables and the health risks evaluation for the working population (Dickinson, 1995). This standard does not include holding of objects (without walking), pushing or pulling of objects, lifting with one hand, manual handling while seated, and lifting by two or more people. Holding, pushing, and pulling objects are included in other parts of the ISO 11228 standard, which are currently at the stage of committee drafts. The ISO/TS 20646-1:2004 standard presents guidelines for application of various ergonomics standards related to local muscular workload (LMWL) and specifies activities to reduce the level of LMWL.

ISO Standards for Ergonomics of Human–System Interaction

The TC159/SC4 subcommittee develops standards related to ergonomics of human–system interaction. The subcommittee is divided into six working groups, which consider the following topics: controls and signaling methods, visual display requirements, control, workplace and

TABLE 1.4
ISO Standards for Controls and Signaling Methods

Reference Number	Title
ISO 9355-1:1999	Ergonomic requirements for the design of displays and control actuators—Part 1: Human interactions with displays and control actuators
ISO 9355-2:1999	Ergonomic requirements for the design of displays and control actuators—Part 2: Displays
ISO/DIS 9355-3	Safety of machinery—Ergonomic requirements for the design of signals and control actuators—Part 3: Control actuators
ISO/DIS 9355-4	Safety of machinery—Ergonomic requirements for the design of displays and control actuators—Part 4: Location and arrangement of displays and control actuators

environmental requirements, software ergonomics and human–computer dialog, human-centered design processes for interactive systems, and ergonomics design of control centers.

ISO Standards for Controls and Signaling Methods

The ISO 9355 standard on ergonomic requirements for the design of displays and control actuators provides guidelines for the design of displays and control actuators on work equipment, especially machines (see Table 1.4). A list of the parts of the ISO 9355 standard is presented in Table 1.4. Part 1 describes general principles of human interactions with display and controls. The other two parts provide recommendations for the selection, design, and location of information displays (Part 2), and control actuators (Part 3). Part 4 covers general principles for the location and arrangement of display and actuators.

ISO Standards for Visual Display Requirements

The ISO 9241 *Ergonomic Requirements for Office Work with Visual Display Terminals* (VDTs) standard is one of the most important standards for ergonomic design (Eibl, 2005; Stewart, 1995). This standard presents general guidance and specific principles that need to be considered in the design of equipment, software, and tasks for office work with VDTs (see Table 1.5).

Part 1 of the ISO 9241 standard describes the underlining principles of the user performance approach. Part 2 describes how task requirements may be identified and specified in organizations and how task requirements can be incorporated into the system design and implementation process. Parts 3 through 9 provide assistance in the procurement and specification of the hardware and environmental components. Three parts present image quality requirements (performance specification) for different types of displays: white and black displays (Part 3), color displays (Part 8), and display with reflections (Part 7). Part 4 provides criteria for the keyboard; and Part 9, for non-keyboard-input devices. Parts 5 and 6 concern ergonomic principles of the design and procurement of computer workstations, workstation equipment, and work environment for office work with VDTs. Those two parts include issues such as technical design of furniture and equipment for the workplace, space organization and workplace layout, and physical characteristics of office work environment, such as lighting, noise, and vibrations. Part 10 presents the core ergonomics principles that should be applied to the design of dialogues between humans and information systems. These principles are intended for in the specifications, design, and evaluation of dialogues for office work with VDTs. Part 11 defines usability and specifies the usability evaluation in terms of the user performance and

TABLE 1.5
ISO 9241 Ergonomic Requirements for Office Work With Visual Display
Terminals (VDTs)

Reference Number	Title
ISO 9241-1:1997	Part 1: General introduction
ISO 9241-2:1992	Part 2: Guidance on task requirements
ISO 9241-3:1992	Part 3: Visual display requirements
ISO 9241-4:1998	Part 4: Keyboard requirements
ISO 9241-5:1998	Part 5: Workstation layout and postural requirements
ISO 9241-6:1999	Part 6: Guidance on the work environment
ISO 9241-7:1998	Part 7: Requirements for display with reflections
ISO 9241-8:1997	Part 8: Requirements for displayed colors
ISO 9241-9:2000	Part 9: Requirements for non-keyboard-input devices
ISO 9241-10:1996	Part 10: Dialogue principles
ISO 9241-11:199	Part 11: Guidance on usability
ISO 9241-12:1998	Part 12: Presentation of information
ISO 9241-13:1998	Part 13: User guidance
ISO 9241-14:1997	Part 14: Menu dialogues
ISO 9241-15:1997	Part 15: Command dialogues
ISO 9241-16:1999	Part 16: Direct manipulation dialogues
ISO 9241-17:1998	Part 17: Form-filling dialogues

satisfaction measures. Part 12 provides recommendations for information presentation on the text-based displays and graphical user interfaces. Part 13 presents recommendations for different types of user guidance attributes of software interfaces such as feedback, status, help, and error handling. Parts 14 to 17 deal with different types of dialog styles: menus, commands, direct manipulation, and form filling.

The ISO 13406 standard provides additional recommendations to ISO 9241 with respect to visual displays based on flat panels. Two parts of this standard cover image quality requirements for the ergonomic design and evaluation of flat-panel displays. ISO 14915 provides additional recommendations to ISO 9241 concerning multimedia presentations.

ISO Standards for Software Ergonomics

ISO 14915: Software Ergonomics for Multimedia User Interfaces, specifies recommendations and principles for the design of interactive multimedia user interfaces that integrate different media such as static text, graphics, and images; and dynamic media such as audio, animation, and video. This standard focuses on issues related to integration of different media, whereas hardware issues and multimodal input are not considered. The standard consists of three parts (see Table 1.6), which address general design principles (Part 1), multimedia navigation and control (Part 2), and media selection and combination (Part 3). The Committee draft ISO/CD 23973 also considers ergonomics design principles for World Wide Web user interfaces.

ISO Standards for Ergonomic Design of Control Centers

ISO 11064 *Ergonomic Design of Control Centers* specifies requirements and presents principles for the ergonomics design of control centers (see Table 1.7). This standard is concerned with the following issues: principles for the design of control centers, principles of control suite

TABLE 1.6
ISO Standards for Software Ergonomics

Reference Number	Title
ISO 14915-1:2002	Software ergonomics for multimedia user interfaces—Part 1: Design principles and framework
ISO 14915-2:2003	Software ergonomics for multimedia user interfaces—Part 2: Multimedia navigation and control
ISO 14915-3:2002	Software ergonomics for multimedia user interfaces—Part 3: Media selection and combination
ISO/CD 23973	Software ergonomics for WWW-user interfaces

TABLE 1.7
ISO 11064 Ergonomic Design of Control Centers

Reference Number	Title
ISO 11064-1:2000	Part 1: Principles for the design of control canters
ISO 11064-2:2000	Part 2: Principles for the arrangement of control suites
ISO 11064-3:1999	Part 3: Control room layout
ISO 11064-4:2004	Part 4: Layout and dimensions of workstations
ISO/DIS 11064-6	Part 6: Environmental requirements for control canters
ISO/CD 11064-7	Part 7: Principles for the evaluation of control canters

arrangements, control room layout, workstation layout and dimensions, displays and controls, environmental requirements, evaluation of control rooms, and ergonomic requirements for specific applications.

ISO Standards for Human–System Interaction

Two ISO standards focus on accessibility issues in the design of usable systems (see Table 1.8). The ISO/AWI 16071 provides guidance on accessibility in reference to the software, whereas ISO/TS 16071:2003 addresses accessibility in reference to the human–computer interfaces. The guidelines on human-centered design process throughout the life cycle of the computer-based interactive systems are described in the ISO 13407:1999 and ISO/TR 18529:2000n standards.

Usability methods supporting the human-centered design are described in the ISO/TR 16982:2002 standard. Further standards concerned with the human–system interaction address issues such as development and design of icons (ISO/IEC 11581), design of typical controls for multimedia functions (ISO 18035), icons for typical WWW-browsers (ISO 18036), and definitions and metrics concerning software quality (ISO 9126).

ISO Standards for Ergonomics of the Physical Environment

The ISO TC159 SC5 document describes an international standard in the area of ergonomics of the physical environment. The subcommittee in charge of this standard development is divided

TABLE 1.8

Published ISO Standards and Standards in Development for Human–System Interaction

Reference Number	Title
ISO 13407:1999	Human-centered design processes for interactive systems
ISO 1503:1977	Geometrical orientation and directions of movements
ISO/AWI 1503	Ergonomic requirements for design on spatial orientation and directions of movements
ISO/AWI 16071	Ergonomics of human–system interaction—Guidance on software accessibility
ISO/TS 16071:2003	Ergonomics of human–system interaction—Guidance on accessibility for human–computer interfaces
ISO/TR 16982:2002	Ergonomics of human–system interaction—Usability methods supporting human-centered design
ISO/PAS 18152:2003	Ergonomics of human–system interaction—Specification for the process assessment of human–system issues
ISO/TR 18529:2000	Ergonomics—Ergonomics of human–system interaction—Human-centered lifecycle process descriptions
ISO 13406-1:1999	Ergonomic requirements for work with visual displays based on flat panels—Part 1: Introduction
ISO 13406-2:2001	Ergonomic requirements for work with visual displays based on flat panels—Part 2: Ergonomic requirements for flat panel displays
ISO/CD 9241-301	Ergonomic requirements and measurement techniques for electronic visual displays—Part 301: Introduction
ISO/CD 9241-302	Ergonomic requirements and measurement techniques for electronic visual displays—Part 302: Terminology
ISO/CD 9241-303	Ergonomic requirements and measurement techniques for electronic visual displays—Part 303: Ergonomic requirements
ISO/AWI 9241-304	Ergonomic requirements and measurement techniques for electronic visual displays—Part 304: User performance test method
ISO/CD 9241-305	Ergonomic requirements and measurement techniques for electronic visual displays—Part 305: Optical laboratory test methods
ISO/CD 9241-306	Ergonomic requirements and measurement techniques for electronic visual displays—Part 306: Field assessment methods
ISO/CD 9241-307	Ergonomic requirements and measurement techniques for electronic visual displays—Part 307: Analysis and compliance test methods
ISO/CD 9241-110	Ergonomics of human system interaction—Part 110: Dialogue principles
ISO/CD 9241-400	Physical input devices—Ergonomic principles
ISO/AWI 9241-410	Physical input devices—Design criteria for products
ISO/AWI 9241-420	Physical input devices—Part 420: Ergonomic selection procedures
ISO/IEC 11581-1:2000	Information technology—User system interfaces and symbols—Icon symbols and functions Part 1: Icons—General
ISO/IEC 11581-2:2000	Information technology—User system interfaces and symbols—Icon symbols and functions Part 2: Object icons
ISO/IEC 11581-3:2000	Information technology—User system interfaces and symbols—Icon symbols and functions Part 3: Pointer icons
ISO/IEC 11581-5:2004	Information technology—User system interfaces and symbols—Icon symbols and functions Part 5: Tool icons
ISO/IEC 11581-6:1999	Information technology—User system interfaces and symbols—Icon symbols and functions Part 6: Action icons
ISO/IEC TR 9126-1:2001	Software engineering—Product quality—Part 1: Quality model
ISO/IEC TR 9126-2:2003	Software engineering—Product quality—Part 2: External metrics
ISO/IEC TR 9126-3:2003	Software engineering—Product quality—Part 3: Internal metrics
ISO/IEC TR 9126-4:2004	Software engineering—Product quality—Part 4: Quality in use metrics

TABLE 1.9
Published ISO Standards on the Ergonomics of the Thermal Environment

Reference Number	Title
ISO 7243:1989	Hot environments—Estimation of the heat stress on working man, based on the WBGT-index (wet bulb globe temperature)
ISO 7726:1998	Ergonomics of the thermal environment—Instruments for measuring physical quantities
ISO 7730:1994	Moderate thermal environments—Determination of the PMV and PPD indices and specification of the conditions for thermal comfort
ISO 7933:1989	Hot environments—Analytical determination and interpretation of thermal stress using calculation of required sweat rate
ISO 8996:1990	Ergonomics—Determination of metabolic heat production
ISO 9886:1992	Evaluation of thermal strain by physiological measurements
ISO 9920:1995	Ergonomics of the thermal environment—Estimation of the thermal insulation and evaporative resistance of a clothing ensemble
ISO 10551:1995	Ergonomics of the thermal environment—Assessment of the influence of the thermal environment using subjective judgment scales
ISO/TR 11079:1993	Evaluation of cold environments—Determination of requisite clothing insulation (IREC)
ISO 11399:1995	Ergonomics of the thermal environment—Principles and application of relevant International Standards
ISO 12894:2001	Ergonomics of the thermal environment—Medical supervision of individuals exposed to extreme hot or cold environments
ISO 13731:2001	Ergonomics of the thermal environment—Vocabulary and symbols
ISO/TS 13732-2:2001	Ergonomics of the thermal environment—Methods for the assessment of human responses to contact with surfaces—Part 2: Human contact with surfaces at moderate temperature

into three working groups (WGs): thermal environments (WG1), lighting (WG2), and danger signals and communication in noisy environments (WG3).

Standards on Ergonomics of Thermal Environment

The standards on ergonomics of thermal environments are concerned with the issues of heat stress, cold stress, and thermal comfort, as well as with the thermal properties of clothing and metabolic heat production due to work activity (Olesen, 1995). Physiological measures, such as the skin reaction due to a contact with hot, moderate, and cold surfaces, and thermal comfort requirements for people with special requirements are also considered (see Tables 1.9 and 1.10). The comfort standard ISO 7730 provides a method for predicting human thermal sensations and the degree of discomfort, which can also be used to specify acceptable environmental conditions for comfort. This method is based on the predicted mean vote (PMV) and predicted percentage of dissatisfied (PPD) thermal comfort indices (Olesen & Parsons, 2002). It also provides methods for the assessment of local discomfort caused by drafts, asymmetric radiation, and temperature gradients. Other thermal environment standards address issues such as thermal comfort for people with special requirements (ISO TS 14415), responses on contact with surfaces at moderate temperatures (ISO 13732, Part 2), and thermal comfort in vehicles (ISO 14505, Parts 1–4). Standards concerned with thermal comfort assessment specify measuring instruments (ISO 7726), methods for estimation of metabolic heat production

TABLE 1.10

ISO Drafts and Standards in Development on the Ergonomics of the Thermal Environment

Reference Number	Title
ISO/DIS 7730	Ergonomics of the thermal environment—Analytical determination and interpretation of thermal comfort using calculation of the PMV and PPD indices and local thermal comfort
ISO/FDIS 7933	Ergonomics of the thermal environment—Analytical determination and interpretation of heat stress using calculation of the predicted heat strain
ISO/FDIS 8996	Ergonomics of the thermal environment—Determination of metabolic rate
ISO/CD 9920	Ergonomics of the thermal environment—Estimation of the thermal insulation and evaporative resistance of a clothing ensemble
ISO/CD 11079	Ergonomics of the thermal environment—Determination and interpretation of cold stress when using required clothing insulation (IREQ) and local cooling effects
ISO/DIS 13732-1	Ergonomics of the thermal environment—Methods for the assessment of human responses to contact with surfaces—Part 1: Hot surfaces
ISO/DIS 13732-3	Ergonomics of the thermal environment—Touching of cold surfaces—Part 3: Ergonomics data and guidance for application
ISO/CD TS 14415	Ergonomics of the thermal environment—Application of International Standards to the disabled, the aged, and other handicapped persons
ISO/DIS 14505-1	Ergonomics of the thermal environment—Evaluation of thermal environment in vehicles—Part 1: Principles and methods for assessment of thermal stress
ISO/DIS 14505-2	Ergonomics of the thermal environment—Evaluation of thermal environment in vehicles—Part 2: Determination of equivalent temperature
ISO/CD 14505-3	Ergonomics of the thermal environment—Thermal environments in vehicles—Part 3: Evaluation of thermal comfort using human subjects
ISO 15265	Ergonomics of the thermal environment—Risk-assessment strategy for the prevention of stress or discomfort in thermal working conditions
ISO/CD 15743	Ergonomics of the thermal environment—Working practices in cold: Strategy for risk assessment and management

(ISO 8996), estimation of clothing properties (ISO 9920), and subjective assessment methods (ISO 10551). ISO 11399:1995 provides information needed for the correct and effective application of international standards concerned with ergonomics of the thermal environment (Table 1.10).

Standards on Communication in Noisy Environments

This set of standards considers communication in noisy environments including warning, danger signals and speech (see Table 1.11). The ISO 7731:1986 document specifies the requirements and test methods for auditory danger signals and provides guidelines for the design of the signals in the public and work spaces. This document also provides definitions that guide in the use of the standards concerned with noisy environment. The criteria for perception of the visual danger signals are provided in ISO 11428:1996, which addresses safety and ergonomic requirements and corresponding physical measurements.

ISO 11429:1996 specifies a system of danger and information signals in reference to different degrees of urgency. This standard applies to all danger signals that have to be clearly perceived and differentiated: from extreme urgency to "all clear." Guidance on detectability is provided in terms of luminance, illuminance, and contrast, considering both surface

TABLE 1.11
ISO Standards for Danger Signals and Communication in Noisy Environments

Reference Number	Title
ISO 7731:1986	Ergonomics—Danger signals for public and work areas—Auditory danger signals
ISO 11428:1996	Ergonomics—Visual danger signals—General requirements, design and testing
ISO 11429:1996	Ergonomics—System of auditory and visual danger and information signals
ISO 9921-1:1996	Ergonomic assessment of speech communication—Part 1: Speech interference level (SIL) and communication distances for persons with normal hearing capacity in direct communication (SIL method)
ISO/TR 19358:2002	Ergonomics—Construction and application of tests for speech technology

TABLE 1.12
Organizational Structure of CEN/TC 122

Working Group	Title
CEN/TC 122/WG 1	Anthropometry
CEN/TC 122/WG 2	Ergonomic design principles
CEN/TC 122/WG 3	Surface temperatures
CEN/TC 122/WG 4	Biomechanics
CEN/TC 122/WG 5	Ergonomics of human–computer interaction
CEN/TC 122/WG 6	Signals and controls
CEN/TC 122/WG 8	Danger signals and speech communication in noisy environments
CEN/TC 122/WG 9	Ergonomics of personal protective equipment (PPE)
CEN/TC 122/WG 10	Ergonomic design principles for the operability of mobile machinery
CEN/TC 122/WG 11	Ergonomics of the thermal environment
CEN/TC 122/WG 12	Integrating ergonomic principles for machinery design

and point sources. ISO 9921-1:1996 describes a method for prediction of effectiveness of speech communication in the presence of noise generated by machinery, as well as in other noisy environments. The following parameters are taken into account in this standard: an ambient noise at the speaker's position, an ambient noise at the listener's position, a distance between the communication partners, and a variety of physical and personal conditions. The ISO/TR 19358:2002 standard deals with testing and assessment of speech-related products and services.

Standards on Lighting of Indoor Work Systems

ISO 8995 (1989): *Principles of Visual Ergonomics—The Lighting of Indoor Work* systems was developed by the ISO 159 SC5 WG2 "Lighting" group in collaboration with the International Commission on Illumination (CIE; Parsons, 1995b). This standard describes principles of visual ergonomics, identifies factors that influence human visual performance, and presents criteria for acceptable visual environments.

CEN STANDARDS FOR ERGONOMICS

The main aim of European standardization efforts is the development of a coherent set of voluntary standards that can provide a basis for a Single European Market/European Economic Area. The work of European standardization organizations is carried out in cooperation with worldwide bodies and the national standards bodies in Europe (Wettig, 2002). Members of the European Union (EU) and the European Fair Trade Association (EFTA) have agreed to implement CEN standards in their national system and withdraw any conflicting national standards.

In 1987, the CEN established the CEN TC122 "Ergonomics," as a body responsible for the development of European standards in ergonomics (Dul, et al. 1996). The scope of the CEN TC122 "Ergonomics" has been defined as *standardization in the field of ergonomics principles and requirements for the design of work systems and work environments, including machinery and personal protective equipment, to promote the health, safety, and well-being of the human operator and the effectiveness of the work* (CEN, 2004). The organizational structure of the CEN TC122 "Ergonomics" is presented in Table 1.12.

The ISO and CEN have signed a formal *agreement on technical cooperation between ISO and CEN* (The Vienna Agreement) that established a close cooperation between these two standardization bodies. ISO and CEN decided to harmonize the development of their standards and to cooperate regarding exchange of information and standards drafting. According to this agreement, the ISO standards can be adopted by CEN and vice versa. Table 1.13 presents published CEN ergonomics standards. The CEN ergonomic standards in development are shown in the Appendix 1.1.

Other International Standards Related to Ergonomics

Because of historical and organizational factors, many ISO and CEN standards in the field of ergonomics have not been developed by technical committees ISO TC159 and CEN TC122. Some ergonomics areas covered by other ISO and CEN technical committees are presented in the Table 1.14. A list of published ISO standards related to ergonomics area, but developed by other than TC159 committee, is provided in Appendix 2.

ILO GUIDELINES FOR OCCUPATIONAL SAFETY AND HEALTH (OSH) MANAGEMENT

Implementation of the ISO-standardized approach to management systems led to the view that this type of approach can also improve management of occupational safety and health. Following this idea, the International Labor Organization (ILO) developed voluntary guidelines on OSH management systems, which reflect ILO values and ensure protection of workers' safety and health (ILO-OSH, 2001). The main objective ILO is the promotion of social justice and internationally recognized human and labor rights (ILO, 2004). ILO represents interests of the three parties treated equally: employers and employees organizations and government parties.

The ILO-OSH (2001) provides recommendations concerning design and implementation of OSH management systems (MS) in the way allowing for integration of OSH with the general enterprise management system. The ILO guidelines state that these recommendations are addressed to all who are responsible for the occupational safety and health management. These guidelines are nonmandatory and are not intended to replace the national laws and regulations. The ILO-OSH document distinguishes two levels of the guidelines application: national and organizational. At the national level, ILO-OSH provides recommendations for

TABLE 1.13
Published CEN Standards for Ergonomics

CEN Reference	Title	ISO Standard
Ergonomics principles		
EN ISO 10075-1:2000	Ergonomic principles related to mental workload—Part 1: General terms and definitions	ISO 10075:1991
EN ISO 10075-2:2000	Ergonomic principles related to mental workload—Part 2: Design principles	ISO 10075-2:1996
ENV 26385:1990	Ergonomic principles of the design of work systems	ISO 6385:1981
EN ISO 6385:2004	Ergonomic principles in the design of work systems	ISO 6385:2004
Anthropometrics and biomechanics		
EN 1005-1:2001	Safety of machinery—Human physical performance—Part 1: Terms and definitions	
EN 1005-2:2003	Safety of machinery—Human physical performance—Part 2: Manual handling of machinery and component parts of machinery	
EN 1005-3:2002	Safety of machinery—Human physical performance—Part 3: Recommended force limits for machinery operation	
EN 13861:2002	Safety of machinery—Guidance for the application of ergonomics standards in the design of machinery	
EN 547-1:1996	Safety of machinery—Human body measurements—Part 1: Principles for determining the dimensions required for openings for whole-body access into machinery	
EN 547-2:1996	Safety of machinery—Human body measurements—Part 2: Principles for determining the dimensions required for access openings	
EN 547-3:1996	Safety of machinery—Human body measurements—Part 3: Anthropometric data	
EN 614-1:1995	Safety of machinery—Ergonomic design principles—Part 1: Terminology and general principles	
EN 614-2:2000	Safety of machinery—Ergonomic design principles—Part 2: Interactions between the design of machinery and work tasks	
EN ISO 7250:1997	Basic human body measurements for technological design	ISO 7250:1996
EN ISO 14738:2002	Safety of machinery—Anthropometric requirements for the design of workstations at machinery	ISO 14738:2002
EN ISO 15535:2003	General requirements for establishing anthropometric databases	ISO 15535:2003
Ergonomics design of control centers		
EN ISO 11064-1:2000	Ergonomic design of control centers—Part 1: Principles for the design of control centers	ISO 11064-1:2000
EN ISO 11064-2:2000	Ergonomic design of control centers—Part 2: Principles for the arrangement of control suites	ISO 11064-2:2000

(Continued)

TABLE 1.13
(Continued)

CEN Reference	Title	ISO Standard
EN ISO 11064-3:1999	Ergonomic design of control centers—Part 3: Control room layout	ISO 11064-3:1999
EN ISO 11064-3:1999/AC:2002	Ergonomic design of control centers—Part 3: Control room layout	ISO 11064-3:1999/Cor.1:2002
Human–system interaction		
EN ISO 13406-1:1999	Ergonomic requirements for work with visual display based on flat panels—Part 1: Introduction	ISO 13406-1:1999
EN ISO 13406-2:2001	Ergonomic requirements for work with visual displays based on flat panels—Part 2: Ergonomic requirements for flat panel displays	ISO 13406-2:2001
EN ISO 13407:1999	Human-centered design processes for interactive systems	ISO 13407:1999
EN ISO 13731:2001	Ergonomics of the thermal environment—Vocabulary and symbols	ISO 13731:2001
EN ISO 14915-1:2002	Software ergonomics for multimedia user interfaces—Part 1: Design principles and framework	ISO 14915-1:2002
EN ISO 14915-2:2003	Software ergonomics for multimedia user interfaces—Part 2: Multimedia navigation and control	ISO 14915-2:2003
EN ISO 14915-3:2002	Software ergonomics for multimedia user interfaces—Part 3: Media selection and combination	ISO 14915-3:2002
EN ISO 9921:2003	Ergonomics—Assessment of speech communication	ISO 9921:2003
Danger signals		
EN 457:1992	Safety of machinery—Auditory danger signals—General requirements, design and testing	ISO 7731:1986, modified
EN 842:1996	Safety of machinery—Visual danger signals—General requirements, design and testing	
EN 981:1996	Safety of machinery—System of auditory and visual danger and information signals	
Thermal environments		
EN 12515:1997	Hot environments—Analytical determination and interpretation of thermal stress using calculation of required sweat rate	ISO 7933:1989 modified
EN 27243:1993	Hot environments—Estimation of the heat stress on working man, based on the WBGT-index (wet bulb globe temperature)	ISO 7243:1989
EN 28996:1993	Ergonomics—Determination of metabolic heat production	ISO 8996:1990
EN ISO 10551:2001	Ergonomics of the thermal environment—Assessment of the influence of the thermal environment using subjective judgement scales	ISO 10551:1995
EN ISO 11399:2000	Ergonomics of the thermal environment—Principles and application of relevant International Standards	ISO 11399:1995

(Continued)

TABLE 1.13
(Continued)

CEN Reference	Title	ISO Standard
EN ISO 12894:2001	Ergonomics of the thermal environment—Medical supervision of individuals exposed to extreme hot or cold environments	ISO 12894:2001
EN ISO 7726:2001	Ergonomics of the thermal environment—Instruments for measuring physical quantities	ISO 7726:1998
EN ISO 7730:1995	Moderate thermal environments—Determination of the PMV and PPD indices and specification of the conditions for thermal comfort	ISO 7730:1994
EN ISO 9886:2001	Evaluation of thermal strain by physiological measurements	ISO 9886:1992
EN ISO 9886:2004	Ergonomics—Evaluation of thermal strain by physiological measurements	ISO 9886:2004
EN ISO 9920:2003	Ergonomics of the thermal environment—Estimation of the thermal insulation and evaporative resistance of a clothing ensemble	ISO 9920:1995
ENV ISO 11079:1998	Evaluation of cold environments—Determination of required clothing insulation (REQ)	ISO/TR 11079:1993
EN 13202:2000	Ergonomics of the thermal environment—Temperatures of touchable hot surfaces—Guidance for establishing surface temperature limit values in production standards with the aid of EN 563	
EN 563:1994	Safety of machinery—Temperatures of touchable surfaces—Ergonomics data to establish temperature limit values for hot surfaces	
EN 563:1994/A1:1999	Safety of machinery—Temperatures of touchable surfaces—Ergonomics data to establish temperature limit values for hot surfaces	
EN 563:1994/A1:1999/AC:2000	Safety of machinery—Temperatures of touchable surfaces—Ergonomics data to establish temperature limit values for hot surfaces	
EN 563:1994/AC:1994	Safety of machinery—Temperatures of touchable surfaces—Ergonomics data to establish temperature limit values for hot surfaces	
Displays and control actuators		
EN 894-1:1997	Safety of machinery—Ergonomics requirements for the design of displays and control actuators—Part 1: General principles for human interactions with displays and control actuators	
EN 894-2:1997	Safety of machinery—Ergonomics requirements for the design of displays and control actuators—Part 2: Displays	
EN 894-3:2000	Safety of machinery—Ergonomics requirements for the design of displays and control actuators—Part 3: Control actuators	

TABLE 1.14

Ergonomic Areas Covered in Standards Developed by the Other ISO
and CEN Technical Committees (Dul, de Vlaming, & Munnik, 1996)

Topic	Technical Committee	
	ISO	*CEN*
Safety of machines	TC 199	TC 114
Vibration and shock	TC 108	TC 211
Noise and Acoustics	TC 43	TC 211
Lighting		TC 169
Respiratory protective devices		TC 79
Eye protection		TC 85
Head protection		TC 158
Hearing protection		TC 159
Protection against falls	TC 94	TC 160
Foot & leg protection		TC 161
Protective clothing		TC 162
Radiation protection	TC 85	
Air quality	TC 146	
Assessment and workplace exposure		TC 137
Office machines	TC 95	
Information procession	TC 97	
Road vehicles	TC 22	
Safety color and signs	TC 80	
Graphical symbols	TC 145	

the establishment of a national framework for OSH-MS. The guidelines suggest that this process should be supported by the provision of the relevant national laws and regulations. The establishment of the national framework for OSH-MS (Figure 1.2) includes the following actions (ILO-OSH):

1. Nomination of competent institution(s) for OSH-MS
2. Formulation of a coherent national policy
3. Development of the national and tailored guidelines

At the organizational level, the ILO-OSH (2001) guidelines establish responsibilities of the employers regarding occupational safety and health management, and emphasize the importance of compliance with the applicable national laws and regulations. The ILO-OSH suggests that OSH-MS elements should be integrated into an overall organizational policy and management strategies actions. The OSH-MS in the organization consist of five main sections: policy, organizing, planning and implementation, evaluation, and action for improvement. These elements correspond to Demming's cycle of a Plan-Do-Check-Act, internationally accepted as the basis for the "system" approach to management. The OSH-MS main sections and their elements are listed in Table 1.15.

The ILO-OSH (2001) guidelines require establishment by the employer of the OSH policy in consultation with workers and their representatives and defines the content of such a policy. These guidelines also indicate the importance of OSH policy integration and compatibility with

TABLE 1.15

ILO–OSH-MS Main Sections and Their Elements (ILO-OSH, 2001)

Section	Elements
Policy	3.1. Occupational safety and health policy
	3.2. Worker participation
Organizing	3.3. Responsibility and accountability
	3.4. Competence and training
	3.5. OSH management system documentation
	3.5. Communication
Planning and implementation	3.6. Initial review
	3.7. System planning and implementation
	3.8. Occupational safety and health objectives
	3.9. Hazard prevention
Evaluation	3.10. Performance monitoring and measurement
	3.11. Investigation of work-related incidents, and their impact on BHP
	3.12. Audit
	3.13. Management review
Action for improvement	3.15. Preventive and corrective action
	3.16. Continual improvement

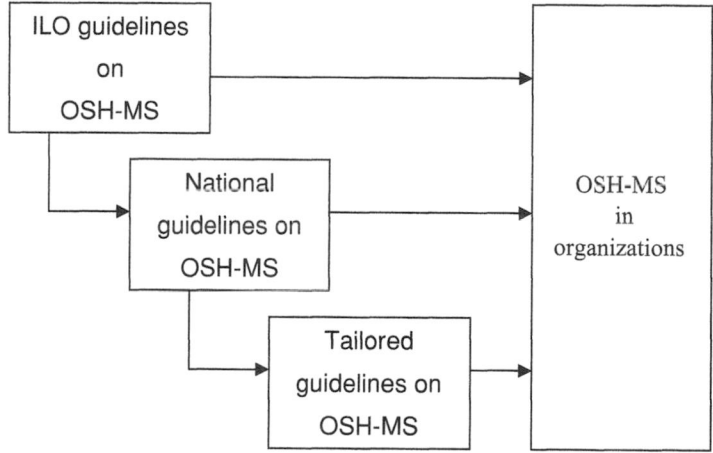

FIG. 1.2.   Establishment of national framework for the OSH-MS (ILO-OSH, 2001)

other management systems in the organization. The guidelines also emphasize the necessity of worker participation in the OSH management system in the organization. Workers should be consulted regarding OSH activities and should be encouraged to participate in OSH-MS, including a safety and health committee. The organizing section of the guidelines underlines the need for allocation of responsibility and accountability for the implementation and performance of the OSH management system to the senior management. This section also includes requirements related to competence and training in the field of OSH and defines the needed documentation and communications activities.

The planning and implementation section of the ILO-OSH (2001) guidelines includes the elements of initial review, system planning, development and implementation, OSH objectives, and hazard prevention. The initial review identifies the actual states of the organization in regards to the OSH and creates the baseline for the OSH policy implementation. The evaluation section consists of performance monitoring and measurement, investigation of work-related diseases and incidents, audit, and management review. The guidelines require carrying out internal audits of the OSH-MS according to the established policies. Action for improvement includes the elements of preventive and corrective action and continual improvement. The last section underlines the need for continual improvement of OSH performance through the development of policies, systems, and techniques to prevent and control work-related injuries and diseases.

HUMAN FACTORS AND ERGONOMICS/STANDARDS IN THE UNITED STATES

U.S. Government Standards for Human Factors and Ergonomics

Among the human factors and ergonomics (HFE) U.S. government standards, two documents are usually mentioned as basic: a military standard providing human engineering design criteria (MIL-STD-1472) and man–system integration standards (NASA-STD-300; Chapanis, 1996; McDaniel, 1996). In addition, there are more specific standards that have been developed by U.S. government agencies such as Department of Defense, Department of Transportation, Department of Energy, and U.S. Nuclear Regulatory Commission. Additionally, a large number of handbooks that contain more detailed and descriptive information concerning human factor and ergonomics guidelines, preferred practices, methodology, and references data that can be needed during design of equipment and systems have also been developed. These handbooks provide assistance in the use and application of relevant government standards during the design process. A list of human factors standards used by the U.S. government agencies and a description of their scope can be found in the *Index of Government Standards* (Human Factors Standardization Sub TAG [HFS SubTAG], 2004).

Human Factors Military Standards (Department of Defense)

A set of consensus-type military standards was developed by human factors engineers from the U.S. military's three services (Army, Navy, and Air Force), industry, and technical societies. As a result of the standardization reform of the late 1990s, most of the single-service standards were cancelled and were integrated into a few Department of Defense (DoD) standards and handbooks. However, the distinction between the two main categories of human factors military standards—general (MIL-STD-1472 and related handbooks) and aircraft (JSSG 2010 and related handbooks)—remain unchanged, which reflects the criticality of aircraft design. A list of the main military standards and handbooks is presented in Table 1.16.

The basic human engineering principles, design criteria, and practices required for integration of humans with systems and facilities are established in the MIL-STD-1472 F *Human Engineering Design Criteria for Military Systems, Equipment and Facilities*. This standard document can be applied to not only the military but also the commercial design of all systems, subsystems, equipment, and facilities. The MIL-STD-1472 F includes requirements for displays, controls, control–display integration, anthropometry, ground work space design, environment, design for maintainability, design of equipment for remote handling, small systems and equipment, operational and maintenance ground and shipboard vehicles, hazards and

TABLE 1.16
Military Standards and Handbooks for Human Factors and Ergonomics
(Based on HFS SubTAG, 2004)

Document Number	Title	Date	Source
Standards			
MIL-STD-882D	Standard Practice for System Safety	2000	http://assist.daps.dla.mil/docimages/ 0001/95/78/std882d.pd8
MIL-STD-1472F	Human Engineering	1999	http://assist.daps.dla.mil/docimages/ 0001/87/31/milstd14.pd1
MIL-STD-1474D	Noise Limits	1997	http://assist.daps.dla.mil/docimages/ 0000/31/59/1474d.pd1
MIL-STD-1477C	Symbols for Army Systems Displays	1996	http://assist.daps.dla.mil/docimages/ 0000/42/03/69268.pd9
MIL-STD-1787C	Aircraft Display Symbology	2001	This is a controlled distribution document
Handbooks			
DOD-HDBK-743A	Anthropometry of U.S. Military Personnel	1991	http://assist.daps.dla.mil/docimages/ 0000/40/29/54083.pd0
MIL-HDBK-759C	Human Engineering Design Guidelines	1995	http://assist.daps.dla.mil/docimages/ 0000/40/04/mh759c.pd8
MIL-HDBK-767	Design Guidance for Interior Noise Reduction in Light-Armored Tracked Vehicles	1993	http://assist.daps.dla.mil/docimages/ 0000/13/24/767.pd1
MIL-HDBK-1473A	Color and Marking of Army Materiel	1997	http://assist.daps.dla.mil/docimages/ 0000/85/40/hdbk1473.pd6
MIL-HDBK-1908B	Definitions of Human Factors Terms	1999	http://assist.daps.dla.mil/docimages/ 0001/81/33/1908hdbk.pd9
Mil-HDBK-46855	Human Engineering Requirements for Military Systems Equipment and Facilities		

safety, aerospace vehicle compartment design requirements, and man–computer interface. The MIL-STD-1472 also includes a nongovernmental standard ANSI/HFS 100 on VDT workstations. After standardization reform, the design data and information part of MIL-STD-1472F were removed and inserted into the MIL-HDBK-759.

Another important military standardization document is the MIL-HDBK-46855 A *Human Engineering Requirements for Military Systems Equipment and Facilities*. This handbook presents human engineering program tasks, procedures, and preferred practices. MIL-HDBK-46855 covers topics such as analysis functions including human performance parameters, equipment capabilities, and task environments design; test and evaluation; and workload analysis, dynamic simulation, and data requirements. This handbook also adopted materials from DoD-HDBK-763 *Human Engineering Procedures Guide* concerned with human engineering methods and tools, which remained stable over time. The newest rapidly evolving automated human engineering tools are not described in the MIL-HDBK-46855, but can be found at Directory of Design Support Methods (DSSM) on the MATRIS Web site (http://dtica.dtic.mil/ddsm/).

Other military standards cover topics such as standard practice for conducting system safety (MIL-STD-882D); acoustical noise limits, testing requirements, and measurement techniques (MIL-STD-1474D); physical characteristics of symbols for army systems displays

TABLE 1.17
USA Governmental Human Factors/Ergonomics standards (Based on HFS SubTAG, 2004)

Document Number	Title	Date	Source
National Aeronautics and Space Administration			
NASA-STD-3000B	Man–Systems Integration Standards	1995	http://msis.jsc.nasa.gov
Department of Transportation, Federal Aviation Administration			
HF-STD-001	Human Factors Design Standard	2003	http://www.hf.faa.gov/docs/508/docs/wjhtc/hfds.zip
DOT-VNTSC-FAA-95-3	Human Factors in the Design and Evaluation of Air Traffic Control Systems	1995	http://www.hf.faa.gov/docs/volpehndk.zip
FAA-HF-001	Human Engineering Program Plan	1999	http://www.hf.faa.gov/docs/did_001.htm
FAA-HF-002	Human Engineering Design Approach Document—Operator	1999	http://www.hf.faa.gov/docs/did_002.htm
FAA-HF-003	Human Engineering Design Approach Document—Maintainer	1999	http://www.hf.faa.gov/docs/did_003.htm
FAA-HF-004	Critical Task Analysis Report	2000	http://hfetag.dtic.mil/docs-hfs/faa-hf-004_critical_task_analysis_report.doc
FAA-HF-005	Human Engineering Simulation Concept	2000	http://hfetag.dtic.mil/docs-hfs/faa-hf-005_human-engineering_simulation.doc
Department of Transportation, Federal Highway Agency			
FHWA-JPO-99-042	Preliminary Human Factors Guidelines for Traffic Management Centers	1999	http://plan2op.fhwa.dot.gov/pdfs/pdf2/edl10303.pdf
FHWA-RD-98-057	Human Factors Design Guidelines for Advanced Traveler Information Systems (ATIS) and Commercial Vehicle Operations (CVO)	1998	http://www.fhwa.dot.gov/tfhrc/safety/pubs/atis/index.html
FHWA-RD-01-051	Guidelines and Recommendations to Accommodate Older Drivers and Pedestrians	2001	http://www.tfhrc.gov/humanfac/01105/cover.htm
FHWA-RD-01-103	Highway Design Handbook for Older Drivers and Pedestrians	2001	http://www.tfhrc.gov/humanfac/01103/coverfront.htm
Department of Energy			
DOE-HDBK-1140-2001	Human Factors/Ergonomics Handbook for the Design for Ease of Maintenance	2001	http://tis.eh.doe.gov/techstds/standard/hdbk1140/hdbk11402001_part1.pdf
Multiple Departments			
FED-STD-795	Uniform Federal Accessibility Standards	1988	http://assist.daps.dla.mil/docimages/0000/46/05/53835.pd5

(MIL-STD-1477C); and symbology requirements for aircraft displays (MIL-STD-1787C). The definitions for all human factors standard documents are provided in the MIL-HDBK-1908B *Department of Defense Handbook: Definitions of Human Factors Terms.*

Other U.S. Governmental Standards: NASA and FAA

A list of other relevant government standards is provided in the Table 1.17. The NASA-STD-3000 provides generic requirements for space facilities and related equipment, important for proper human–system integration. This document is integrated with the Web site that also offers video images from space missions and illustrates the human factors design issues. This standard document is not limited to any specific NASA, military, or commercial program, and can be applied to almost all kinds of equipment. The NASA-STD-3000 consists of two volumes: *Volume I—Man-Systems Integration Standards* presents all of the design standards and requirements, and Volume II—Appendices contains the background information related to standards. The NASA-STD-3000 the covers following areas of human factors: anthropometry and biomechanics, human performance capabilities, natural and induced environments, health management, workstations, activity centers, hardware and equipment, design for maintainability, and facility management.

The standards of the Federal Aviation Administration (FAA) are concerned with following topics: human factors design criteria oriented to the FAA mission and systems (HF-STD-001), design and evaluation of air traffic control systems (DOT-VNTSC-FAA-95-3), elements of the human engineering program (FAA-HF-001), evaluation of human factor criteria conformance of equipment that have interface with both operator (FAA-HF-002) and maintainer (FAA-HF-003).

The Department of Energy (DOE) in their standard DOE-HDBK-1140-2001 provides the system maintainability design criteria for DOE systems, equipment, and facilities. The Federal Highway Administration (FHA) establishes standards concerning development and operation of traffic management centers (FHWA-JPO-99-042). The FHA also describes human factors guidelines and recommendations for design of Advanced Traveler Information Systems (ATIS), Commercial Vehicle Operations (CVO), and concerning accommodation of the older drivers and pedestrians. The Nuclear Regulatory Commission provides guidelines of HFE conformance evaluation of the interface design of the nuclear power plants systems (NUREG-0700 & NUREG-0711). For use in federal and federally funded facilities FED-STD-795 has been developed, which establishes standards for facility accessibility by physically handicapped persons.

Relevant OSHA Standards

Development of occupational safety and health standards in the United States is mandated by the general duty clause, Section 5(a)(1), the *Occupational Safety and Health Act of 1970*, which states that "Each employer shall furnish to each of his employees, employment and a place of employment which is free from recognized hazards that are causing or are likely to cause death or serious harm to his employees." In general, penalties related to deficient and unsafe working conditions have been issued under this general duty clause. The general duty clause has also been supplemented by the *Americans with Disabilities Act* (Public Law 101-336, 1990). The disabilities act has an important bearing on ergonomics design of workplaces. The ADA prohibits disability-based discrimination in hiring practices and requires that all employers make reasonable accommodations to working conditions in order to allow qualified disabled workers to perform their job functions.

In 1990, OSHA issued a set of voluntary guidelines entitled "Ergonomics Program Management Guidelines for Meatpacking Plants" (OSHA 3123), which have been successfully used by many types of industries, including those from outside the food production business. In 2000, the U.S. federal government proposed the *Ergonomics Program Rule* (OSHA, 2000). The main elements of the standard included (a) training in basic ergonomics awareness, (b) providing medical management of work-related musculoskeletal disorders, (c) implementing a quick fix or going to a full program, and (d) implementing a full ergonomic program when indicated, including such elements as management leadership, employee participation, job hazard analysis, hazard reduction and control, training, and program evaluation. However, the regulation was repealed in March 2001.

Recently, OSHA has published three voluntary guidelines to assist employers of the specific types of industries in recognizing and controlling hazards. These guidelines are:

- Nursing Home Guideline (issued on March 13, 2003)
- Draft Guideline For Poultry Processing (issued on June 3, 2003)
- Guideline For The Retail Grocery Industry (issued on May 28, 2004)

In addition, OSHA plans to develop additional voluntary guidelines with the use of a standard protocol (OSHA, 2004). The objective of this standard protocol is to establish a fair and transparent process for developing industry- and task-specific guidelines that will assist employers and employees in recognizing and controlling potential problems. By using this protocol, each set of guidelines will address a particular industry or task. It is intended that the industry- and task-specific guidelines will generally be presented in three major parts: (a) program management recommendations for management practices addressing ergonomic hazards in the industry or task; (b) worksite analysis recommendations for worksite and workstation analysis techniques geared to the specific operations that are present in the industry or task; and (c) hazard control recommendations that contain descriptions of specific jobs and that detail the hazards associated with the operation, possible approaches to controlling the hazard, and the effectiveness of each control approach. Because there are many different types of work-related hazards, injuries, and controls vary from industry to industry and task to task, OSHA expects that the scope and content of the guidelines will vary.

Other Relevant Standards for Occupational Safety and Health

In 2000, the National Safety Council (NSC), acting on behalf of the Accredited Standards Committee (ASC), issued a draft document (known as Z-365) entitled *Management of Work-Related Musculoskeletal Disorders* (MSD). The Draft defines the following areas of importance for preventing work-related injuries: (a) management responsibility, (b) employee involvement, (c) training, (d) surveillance, (e) evaluation and management of work-related MSD cases, (f) job analysis and design, and (g) follow-up.

Independently of the aforementioned efforts, in 2001 another ANSI committee, ASC Z-10 *Occupational Health Safety Systems* was formed under the auspices of the American Industrial Hygiene Association (AIHA). The main objective of ASC Z-10 is to develop a standard of management principles and systems for improving the occupational safety and health in companies.

ANSI Standards

A list of the HFE-relevant standards developed by the ANSI is presented in Table 1.18.

American National Standard for Human Factors Engineering of Visual Display Terminals—ANSI/HFS 100-1988

This standard presents ergonomics principles related to visual display terminals. The standard has been updated by BSR/HFES 100 Draft Standard dated March 31, 2002.

TABLE 1.18

List of ANSI Standards for Human Factor and Ergonomics

Reference Number	Title
ANSI/HFS 100-1988	American National Standard for Human Factors Engineering of Visual Display Terminals
BSR/HFES 100 DS	Human Factors Engineering of Computer Workstations
HFES 200–1998	Ergonomic Requirements for Software User Interfaces
ASC Z-365	Management of Work-Related Musculoskeletal Disorders
ASC Z-10	Occupational Health Safety Systems
ANSI B11	Ergonomic Guidelines for the Design, Installation and Use of Machine Tools

TABLE 1.19

Main Chapters and Described Topics of the Human Factors Engineering
of Computer Workstations—BSR/HFES 100 Draft Standard

Chapter	Issues
Installed systems	Hardware components, noise, thermal comfort, and lighting
Input devices	Keyboards, mouse and puck devices, trackballs, joysticks, styli and light pens, tablets and overlays, touch-sensitive panels
Visual displays	Monochrome and color CRT, and flat-panel displays (viewing characteristics, contrast, legibility, etc)
Furniture	Specifications for workstation components (chairs, desks, etc.); postures (reference postures, reclined sitting, upright sitting, declined sitting and standing); anthropometry

Human Factors Engineering of Computer Workstations—BSR/HFES 100 Draft Standard

The BSR/HFES 100 *Human Factors Engineering of Computer Workstations* (HFES 100) is *a specification of the recommended human factors and ergonomic principles related to the design of the computer workstation* (Albin, 2004). This standard is organized into four major chapters: (1) Installed Systems, (2) Input Devices, (3) Visual Displays, and (4) Furniture. The major topics described in each of these chapters are listed in Table 1.19.

Ergonomic Requirements for Software User Interfaces (HFES 200—1998)

The HFES/HCI 200 Committee that operates under the auspices of the Human Factors and Ergonomics Society's Technical Standards Committee has been working on the development of a proposed American national standard for software user interfaces. This standard will provide requirements and recommendations for software interfaces with a primary focus on business and personal computing applications. The standard is related to the ISO 9241 series

TABLE 1.20
Topics addressed in the Ergonomic Requirements for Software User Interfaces
(HFES 200—1998)

Chapter	Issues
Accessibility	Keyboard input; multiple keystrokes
	Customization, repeat rates, acceptance delays
	Pointer alternative; accelerators; remapping, navigation
	Display fonts: size, legibility, styles, colors
	Audio output: volume and frequencies, customization, content and alerts; graphic
	Color: palettes, background–foreground, customization, coding
	Errors and persistence; online documentation and help
	Customization: cursor, button presses, click interval, pointer speed, chording
	Window appearance and behavior: navigation and location, window focus, titles
	Input focus: navigation, behavior, order, location
Color	Color selection: chromostereopsis, blending and depth effects, use of blue and red, identification and contrast
	Color assignments: conventions, uniqueness and reuse, naming, cultural assignments
	General use consideration: number of colors, highlighting, positioning, and separation
	Special uses: warnings, coding, state indications, pointers, area identification
Voice and telephony	Speech recognition (input): commands, vocabularies, prompts, consistency, feedback, error handling, dictation
	Speech output: vocabularies message format, speech characteristics, dialog techniques, physical properties, alerting tones, stereophonic presentation
	Nonspeech auditory output: consistency, tone format, critical messages, frequency, amplitude
	Interactive voice response
Technical sections	Presentation of information, user guidance, menu dialogs, command dialogs, direct manipulation, dialogue boxes and form-filling dialogues windows

of user interface standards. The topics described in each section of the HFES 200 standard are listed in Table 1.20. These topics are (a) accessibility, (b) color, and (c) voice and telephony.

ANSI B11 Ergonomic Guidelines for the Design, Installation, and Use of Machine Tools

ANSI B11 *Technical Report: Ergonomic Guidelines for the Design, Installation and Use of Machine Tools* is a consensual ergonomic guidelines developed by the *Machine Tool Safety Standards Committee* (B11) of the American National Standards Institute. The subcommittee responsible for the preparation of these guidelines consisted of the representatives from manufacturing, higher education, safety, design, and ergonomics. The document specifies the ergonomic guidelines to assist the design, installation, and use of individual and integrated machine tools and auxiliary components in manufacturing systems.

The guidelines document underlines the importance of three basic ideas for achievement of the effective and safe design, installation, and use of the machine tools: (a) communication among all people that are involved with the machine tools (users, installers, manufacturers, and designers), (b) dissemination of the knowledge concerning ergonomics concepts and principles among all individuals, and (c) ability to apply ergonomics concepts and principle knowledge effectively to machine tools and auxiliary components. The guidelines document states that the provision of worker safety, work efficiency, and optimization of the whole production system requires consideration of the following ergonomics issues:

1. The variation of the employee's physiological and psychological characteristics such as strength and capacity
2. Incorporation of ergonomics concepts and principles into new project, tool, machine, and work process at the beginning of this process.
3. Routine tasks that are done precisely, rapidly, and continuously, and especially tasks in hazardous environments which should be performed by machines
4. Tasks that require judgment and integration of information; that is, the tasks the humans do the best should be assigned to workers.
5. A system that does not consider human limits such as information handling, perception, reach, clearance, posture, or strength exertion can predispose an accident or injury.

The aforementioned documents also recommend matching design of the tools and processes with the physical characteristics and capability of workers in order to ensure accommodation, compatibility, operability, and maintainability of the machine tools and auxiliary components.

ISO 9000:2000—QUALITY MANAGEMENT STANDARDS

Quality standards can also play an important role in assuring safety and health at the workplace. ISO stipulates that if a quality management system is appropriately implemented utilizing the eight *Quality Management Principles* (see following), and in accordance with ISO 9004, all of an organization's interested parties should benefit. For instance, people in the organization will benefit from the (a) improved working conditions, (b) increased job satisfaction, (c) improved health and safety, (d) improved morale, and (e) improved stability of employment, whereas the society at large will benefit from the (a) fulfillment of legal and regulatory requirements, (b) improved health and safety, (c) reduced environmental impact, and (d) increased security.

As discussed by Hoyle (2001), the term *ISO 9000* refers to a set of quality management standards. ISO 9000 currently includes three quality standards: ISO 9000:2000, ISO 9001:2000, and ISO 9004:2000. ISO 9001:2000 presents requirements, whereas ISO 9000:2000 and ISO 9004:2000 present guidelines. ISO first published its quality standards in 1987, revised them in 1994, and then republished an updated version in 2000. These new standards are referred to as the "ISO 9000:2000 Standards."

It is recommended that the ISO 9001:2000 standard be used if an organization is seeking to establish a management system that provides confidence in the conformance of its product to established requirements. The standard recognizes that the word "product" applies to services, processed material, and hardware and software intended for, or required, by the customer (Hoyle, 2001).

The ISO 9000:2000 Standards apply to all kinds of organizations such as manufacturing, service, government, and education. The standards are based on eight quality management principles:

- Principle 1:   Customer focus
- Principle 2:   Leadership
- Principle 3:   Involvement of people
- Principle 4:   Process approach
- Principle 5:   System approach to management

- Principle 6:   Continual improvement
- Principle 7:   Factual approach to decision making
- Principle 8:   Mutually beneficial supplier relationships

There are five sections in the standard that specify activities that need to be considered when implemented to the quality management system. According to Hoyle (2001), following a description of the activities that are used to supply products, the organization may exclude the parts of the product realization section that are not applicable to its operations. The requirements in the other four sections such as quality management system, management responsibility, resource management, and measurement analysis and improvement apply to all organizations, and the organization needs to demonstrate how it applies them to the organization's quality manual or other documentation. These five sections of ISO 9001:2000 define what the organization should do consistently to provide products that meet customer and applicable statutory or regulatory requirements, and enhance customer satisfaction by improving its quality management system. ISO 9004:2000 can be used to extend the benefits obtained from ISO 9001:2000 to the employees, owners, suppliers, and society, in general.

ISO 9001:2000 and ISO 9004:2000 are harmonized in structure and terminology to assist the organization to move smoothly from one to the other. Both standards apply a process approach. Processes are recognized as consisting of one or more linked activities that require resources and must be managed to achieve predetermined output. The output of one process may directly form the input to the next process and the final product is often the result of a network or system of processes. The eight Quality Management Principles stated in ISO 9000:2000 and ISO 9004:2000 provide the basis for the performance improvement outlined in ISO 9004:2000. ISO 9000 standards cluster also includes other 10000 series standards. Table 1.21 shows a list of the relevant standards and their purposes.

As discussed by Hoyle (2001), ISO requires that the organization determine what it needs to do to satisfy its customers; establish a system to accomplish its objectives; and measure, review, and continually improve its performance. More specifically, the ISO 9001 and 9004 requirements stipulate that an organization shall:

1. Determine the needs and expectations of customers and other interested parties
2. Establish policies, objectives, and a work environment necessary to motivate the organization to satisfy these needs
3. Design, resource, and manage a system of interconnected processes necessary to implement the policy and attain the objectives
4. Measure and analyze the adequacy, efficiency, and effectiveness of each process in fulfilling its purpose and objectives
5. Pursue the continual improvement of the system from an objective evaluation of its performance

ISO identified several potential benefits of using the Quality Management Standards. These benefits may include the connection of quality management systems to organizational processes, encouragement of a natural progression toward improved organizational performance, and consideration of the needs of all interested parties.

EMPIRICAL STUDIES ON SELECTED HF/E STANDARDS

The existing body of literature shows that not many empirical studies were conducted on human factors and ergonomics standards. This section provides a review of empirical

TABLE 1.21
The ISO 9000 Quality Management Standards and Guidelines

Standards and Guidelines	Purpose
ISO 9000:2000, Quality management systems—Fundamentals and vocabulary	Establishes a starting point for understanding the standards and defines the fundamental terms and definitions used in the ISO 9000 family to avoid misunderstandings in their use.
ISO 9001:2000, Quality management systems—Requirements	This is the requirement standard to be used to assess the organization's ability to meet customer and applicable regulatory requirements and thereby address customer satisfaction. It is now the only standard in the ISO 9000 family against which third-party certification can be carried.
ISO 9004:2000, Quality management systems—Guidelines for performance improvements	This guideline standard provides guidance for continual improvement of the organization's quality management system to benefit all parties through sustained customer satisfaction.
ISO 19011, Guidelines on Quality and/or Environmental Management Systems Auditing (currently under development)	Provides the organization with guidelines for verifying the system's ability to achieve defined quality objectives. You can use this standard internally or for auditing your suppliers.
ISO 10005:1995, Quality management—Guidelines for quality plans	Provides guidelines to assist in the preparation, review, acceptance and revision of quality plans.
ISO 10006:1997, Quality management—Guidelines to quality in project management	Guidelines to help the organization to ensure the quality of both the project processes and the project products.
ISO 10007:1995, Quality management—Guidelines for configuration management	Gives the organization the guidelines to ensure that a complex product continues to function when components are changed individually.
ISO/DIS 10012, Quality assurance requirements for measuring equipment—Part 1: Metrological confirmation system for measuring equipment	Gives the organization guidelines on the main features of a calibration system to ensure that measurements are made with the intended accuracy.
ISO 10012-2:1997, Quality assurance for measuring equipment—Part 2: Guidelines for control of measurement of processes	Provides supplementary guidance on the application of statistical process control when this is appropriate for achieving the objectives of Part 1.
ISO 10013:1995, Guidelines for developing quality manuals	Provides guidelines for the development, and maintenance of quality manuals, tailored to your specific needs.
ISO/TR 10014:1998, Guidelines for managing the economics of quality	Provides guidance on how to achieve economic benefits from the application of quality management.
ISO 10015:1999, Quality management—Guidelines for training	Provides guidance on the development, implementation, maintenance, and improvement of strategies and systems for training that affects the quality of products.
ISO/TS 16949:1999, Quality systems—Automotive suppliers—Particular requirements for the application of ISO 9001:1994	Sector specific guidance to the application of ISO 9001 in the automotive industry.

studies on the human factors and ergonomics standards and evaluation of these standards. Following the emergence of any standard, it is beneficial to investigate the standard with respect to (a) evaluation of the success of the standard and (b) estimation of the extent of standard's usage. There are a few basic researches in the existing body of literature that primarily focused on the standard's validity, reliability, usability, and scope of application. Ideally, a standard should be evaluated before it formally becomes available to the practitioners.

Software Engineering and Human-Centered Design

Though numerous standards currently exist for software engineering, it is argued that there is a poor adoption of the standards (Fenton, Littlewood, & Page, 1993) and that the extent to which software engineering standards are actually used is unknown (El Emam & Garro, 2000). El Emam and Garro conducted their study on ISO/IEC 15504, which is an international standard on software process assessment. El Emam and Jung (2001) also empirically evaluated the ISO/IEC 15504 assessment model.

As reported in El Emam and Jung (2001), ISO/IEC 15504 consists of nine parts: (a) concepts and introductory guide, (b) a reference model for processes and process capability, (c) performing an assessment, (d) guide to performing assessments, (e) an assessment model and indicator guidance, (f) guide to qualification of assessors, (g) guide for use in process improvement, (h) guide for use in determining supplier process capability, and (i) vocabulary. The architecture of the standard is two-dimensional. One dimension is the "process" dimension that consists of five categories of processes. The other dimension is the "capability" dimension that consists of five levels and nine attributes. A 4-point achievement scale constitutes a rating scheme that rates each of the attributes. In this study, a questionnaire method was used to obtain data on (a) use of the assessment model, (b) usefulness and ease of use, (c) meaningfulness of rating aggregation scheme, (d) usability of the rating scale, (e) usefulness of the indicators, and (f) understanding of the process and capability dimensions.

A total of 70 worldwide assessments were collected from Asia Pacific, Europe, and the United States. About 33% of the assessments were performed in such organizations as software production, other IT products, and services. The remaining assessments were done by distribution and logistics, business, and defense. Most of the assessors were either managers or senior technical personnel. Approximately 98% of the assessors have received at least one training on software process assessment. Most assessors positively responded to the usefulness and ease of use of the standard model. However, the assessors indicated problems while rating the process attributes on the process dimension. They also had difficulty in rating achievement. In general, the study found that most assessors positively responded to the meaningfulness of aggregated rating scheme, usability of the rating scale, usefulness of the indicators, and comprehension of the process and capability dimensions.

In their study, Earthy, Jones, and Bevan (2001) reviewed ISO 13407 and associated ISO TR 18529, which are the standards for human-centered design processes for interactive systems. Earthy et al. argued that Human Factors has processes that can be managed and integrated with existing project processes and that this internationally agreed set of human-centered design processes provides a definition of the capability that an organization must possess in order to implement user-centered design effectively. Earth et al. further speculated that the standard can also be used to assess the extent to which a particular development project employs user-centered design.

Usability and Human–Computer Interaction

The body of literature indicates that the most research on ergonomics-related standards was done on different parts of ISO 9241 standard series, which can be termed as usability or ergonomics of human–computer interaction standard. As reported in Bastien et al. (1999), user-interface design guidelines appear in different types such as design guides, style guides, guidelines compilations, principles, and general guides. For years, guidelines compilations have been given considerable emphasis in writing design guides. In this context, the most cited work is the compilation of guidelines by Smith and Mosier (1984), which contains 944 guidelines for the design of user interfaces in single sentence forms with illustrated examples.

Subsequently, Mosier and Smith (1986) conducted an empirical survey on the utilization of compiled guidelines. In their analysis of questionnaire responses, it was found that most respondents considered the compilation useful. However, the respondents also reported problems in its use. For instance, users had difficulty in (a) locating relevant guidelines within the compilation, (b) choosing which guidelines to use, and (c) translating general guidelines into specific design rules.

Bastien et al. (1999) conducted a pilot evaluation of the ISO/DIS 9241-10 "dialogue principles" that compared the part with other guidelines (termed *Ergonomic Criteria*) and a control group that did not use any guidelines. Each group evaluated a musical database application. The study found that the median time spent evaluating the application was lowest for the control group followed by the ISO and Ergonomic Criteria groups. Multiple comparisons between groups showed that only the control and the Ergonomic Criteria groups differed significantly. The study also found that the highest numbers of usability problems were identified by the Ergonomic Criteria group followed by the ISO and control groups. As in the first case, multiple comparisons between groups showed that only the control and the Ergonomic Criteria groups differed significantly. The study concluded that Ergonomic Criteria group spent significantly more time evaluating the application than did the Control group. The Ergonomic Criteria group also uncovered significantly more usability problems, though no significant differences were found between the Control and the ISO groups as well as between the ISO and the Ergonomic Criteria groups.

Dzida and Freitag (1998) investigated the use of scenarios to validate analysis and design according to ISO 9241-11 "Guidance on usability." To validate scenarios and requirements, Dzida and Freitag argued that "the process of requirement construction can be conceived of a task domain (represented in terms of a context scenario) into the initial model of an artifact (represented in terms of a use scenario or prototype)." In the study, the semantic structure of scenarios consisted of three parts: (a) context scenario, (b) concept of use, and (c) use scenario. A subsequent case study resulted in the development of a new context scenario, which suggests a number of requirements that may induce an additional iteration step in the prototyping process. Dzida and Freitag concluded that by consensus, the analyst, the designer, and the user can decide not to consider the obsolete prototype version as a valid solution of the problem.

Several studies evaluated computer input devices with respect to ISO 9241-9 standard on "Requirements for non-keyboard input devices" by utilizing the Fitt's law (e.g., Douglas, Kirkpatrick, & MacKenzie, 1999; Isokoski and Raisamo, 2002; Keates, Hwang, Langdon, Clarkson, & Robinson, 2002; MacKenzie & Jusoh, 2001; MacKenzie et al., 2001; MacKenzie & Oniszczak, 1998; Oh & Stuerzlinger, 2002; Poupyrev et al., 2004; Silfverberg, MacKenzie, & Kauppiner, 2001, Sohn & Lee, 2004, Zhai, 2004). These studies utilized different types of input devices (such as different types of mice, touch pads, trackballs, and joysticks). Soukoreff and MacKenzie (2004) reviewed several studies and compared the results of those studies that utilized only Fitt's law with those that utilized Fitt's law with ISO 9241-9 standard. Soukoreff and MacKenzie (2004) throughput values lower for those studies that utilized both Fitt's law and ISO 9241-9. Furthermore, Soukoreff and MacKenize provided seven specific recommendations that were argued to support and in some instances supplement the ISO 9241-9 standard on the evaluation of pointing devices.

In a recent study, Sohn and Lee (2004) introduced an ultrasonic pointing device system that was developed for an interactive TV system. The study evaluated the pointing device and compared it to other conventional pointing devices using ISO 9241-9. The results showed that the throughput of the new pointing device was slightly lower than the touchpad and trackball. The average movement time of the new pointing device was higher than that of the trackball and the touchpad. Further, the average error rate of the new pointing device was higher than that of the touchpad and the trackball.

Besuijen and Spenkelink (1998) reviewed ISO 9241-3 "Visual display requirements". According to the standard (ISO 9241-3), visual display quality was defined as sufficient if a total of 25 requirements were met. The requirements were defined based on physical measurements in luminance and spatial domains. Besuijen and Spenkelink (1998) identified several problems with the standard, such as (a) modeling problem—concerns with inability of the standard to estimate a meaningful overall quality index and (b) compliance to the individual physical parameters for image quality (because there are many characteristics that have interaction effects). Following the review of the standard, Besuijen and Spenkelink (1998) proposed comparative user performance test methods that included both qualitative and quantitative performance tests, visual measurement, subjective evaluation of the image quality, and utilization of the Display Evaluation Scale (DES), which is already established as a valid and reliable measurement.

Becker (1998) reviewed ISO/DIS 13406-2, "Ergonomic requirements for flat panel displays," and ISO 9241-7, "Requirements for display with reflections," and stated that the rating and classification of the ergonomic performance of visual display terminals in the presence of ambient light sources requires assessment of the reflective properties of the display unit. In the study, Becker introduced a novel measurement of the two-dimensional "Bidirectional Reflectance Distribution Function" (BRDF). By utilizing this new approach and instrument, the study evaluated several properties of visual display performance (such as haze, gloss, and distinctness of image).

Another study by Umezu et al. (1998) investigated the measurement of specular and diffuse reflection (ISO 9241-7) of two types of video display units such as CRT (Cathode Ray Tube) and LCD (Liquid Crystal Display). The study utilized typical commercially available photometers to verify the certainty of specular reflection coefficient measurement from an ergonomic standard. The results showed that the measured value difference among the three different photometers was 20%, which the study concluded to be high. Umezu et al. (1998) argued that because the specular reflection coefficient itself was very small, a slight difference would create a high error ratio. The study also utilized inspectors to change working distance and site. The results showed that such changes did not significantly affect ISO 9241-7 requirements.

Lindfors (1998) examined the accuracy and repeatability of the ISO 9241-7 for testing CRT-type video display units. The display units were tested in various laboratory settings of Europe and Japan. One of the two hypotheses tested in the study was to investigate whether "the defined test methods are repeatable and reliable." The results showed that with respect to specular and diffuse reflection and interlaboratory variation, the hypothesis was found to be false. However, Lindfors (1998) argued that the problems associated with the rejection of the hypothesis could be covered and rectified either by interlaboratory agreement or by a future minor update of ISO 9241-7.

In a recent study, Marmaras and Papadopoulos (2003) investigated the extent to which ergonomic requirements for work on computers are met in Greek office workstations. The study assessed 593 office workstations using an assessment tool consisting of 70 assessment points. The study results showed that the ergonomic requirements that are independent of the specific characteristics of individual work spaces and environments, such as design standards for seats, monitors, and input devices, are adequately met. Ergonomic requirements that should take into consideration the specific characteristics and constraints of individual work content, work spaces, and environments (e.g., requirements dealing with workplace layout, environmental conditions, software, and work organization) are inadequately met. Based on these results, the study recommended for (a) enhancement of efforts for the application of ergonomic principles in the enterprises, (b) involvement of ergonomists in decision making, design, and selection of computerized office components, (c) spreading of ergonomics awareness and knowledge through training, and (d) development of office workplace design methods and tools and facilitation of the application of ergonomics principles.

Physical Environment

In their study, Griefahn and Brode (1999) investigated the sensitivity of lateral motions relative to vertical motions. The sensitivity was then compared to predictions provided by ISO 2631. Two experiments were conducted where lateral and vertical motions were applied consecutively or simultaneously and where the magnitude of a single- or dual-axis test signal was adjusted until it was judged as equivalent to a preceding single-axis reference motion of the same frequency. From the results of those two experiments, it was substantiated that ISO 2631 was qualitatively valid.

Another study by Griefahn (2000) examined the validity of cold stress model of ISO/TR 11079. The study investigated the possible limitations of the applicability of the IREQ model and necessities and possibilities to improve the model. In the study, 16 female and 59 male workers (16–56 years) were monitored during their work. According to their cold stress at the workplace they were allocated to three groups. First group of 33 participants were exposed to constant temperatures of more than 103°C, second group of 32 participants were exposed to less than 103°C, and a third group of 10 participants experienced frequent temperature changes of 133°C. The study also manipulated the degree of physiological workload. First group of 8-participants worked at metabolic rates of less than 100 W/m^2. Second group of 50 participants worked between 101 and 164 W/m^2, and the third group of 17 worked at more than 165 W/m^2. The results revealed that the IREQ model applies for air temperatures up to 15°C and for temperature changes of 13°C (at least) but needs to be improved with respect to gender. The study also found that the IREQ model did not apply sufficiently for high and largely varying workloads (165 W/m^2 and more). However, the study concluded that these situations were beyond the currently available possibilities to protect workers adequately with conventional clothing material.

A recent theoretical study by Olesen and Persons (2002) discussed the existing ISO standards related to thermal comfort. The existing thermal comfort standard (ISO 7730), metabolic rate (ISO 8996), and clothing (ISO 9920) were critically appraised with respect to validity, reliability, usability, and scope for practical application.

Ishitake et al. (2002) investigated the effects of exposure to whole-body vibration (WBV) and the ISO 2631/1-1997 frequency weighting on gastric motility. The gastric motility was measured by electrogastrography (EGG) in nine healthy volunteers. In the study, sinusoidal vertical vibration at a frequency of 4, 6.3, 8, 12, 16, 31.5, or 63 Hz was given to the subjects for 10 min. The magnitude of exposure at 4 Hz was 1.0 m/s^2 (RMS). The magnitudes of the other frequencies gave the same frequency-weighted acceleration according to ISO 2631-1. The pattern of the dominant frequency histogram (DFH) was changed to a broad distribution pattern by vibration exposure. The results showed that vibration exposure had the effect of significantly reducing the percentage of time for which the dominant component had a normal rhythm and increasing the percentage of time for which there was tachygastria ($p < 0.05$). It was also found out that vibration exposure generally reduced the mean percentage of time with the dominant frequency in normal rhythm component. There was a significant difference between the condition of no vibration and exposure to 4 and 6.3 Hz of vibration frequency ($p < 05$). From the findings, it was concluded that the frequency weighting curve given in ISO 2631/1-1997 was not adequate for use in evaluating the physiological effects of WBV exposure on gastric motility.

Enterprise Engineering and Integration

Kosanke and Nell (1999), in their study, reviewed the existing ISO standards for enterprise engineering and integration. In their review, Kosanke and Nell (1999) provided a road map for standardization in enterprise engineering and integration. This road map essentially incorporated the existing ISO 15704 "Requirements for Enterprise Reference Architecture" with ISO 14258, 9000, 14000, and ENV 12204. Kosanke and Nell also advocated for future research needs with respect to process representation, human role representation, infrastructure integration, facilitation of terminology, and standards landscaping.

CONCLUSIONS

Although human factor and ergonomics standards cannot guarantee appropriate workplaces design, they can provide clear and well-defined requirements and guidelines, and therefore basis for good ergonomics design. Standards for workstation design and work environment can ensure the safety and comfort of working people through establishing requirements for the optimal working condition. By providing consistency of the human–system interface and improving ergonomics quality of the interface components, ergonomics standards can also contribute to the enhanced systems usability and overall system performance. This benefit is based on the general requirement of harmonization across different tools and systems, in order to support user performance and avoid unnecessary human errors.

One of the most important benefits from standardization efforts is a formal recognition of the significance of ergonomics requirements and guidelines for system design on the national and international levels (Harker, 1995). The consensus procedure applied to standards development demands consultation with a wide range of the commercial, professional, and industrial organizations. Therefore, the decision to develop standards and consensus of diverse organizations concerning the need for standards reflects the formal recognition that there are important human factors and ergonomics issues that need to be taken into account during design and development of workplaces and systems.

Standards represents the essence of the best available knowledge and practice extracted from variety of academic sources and presented in the way that is easy to use by professional designers and to implement this knowledge into the design process. The consensus procedure makes the standards under development known and available to the interested parties and general public. Such a procedure also facilitates dissemination and promotion of the human factors and ergonomics knowledge across the nonexperts.

APPENDIX 1.1

CEN Standards for Ergonomics Under Development

Reference	Title	DAV
prEN ISO 13732-3	Ergonomic of the thermal environment—Methods for the assessment of human responses to contact with surfaces—Part 3: Cold surfaces (ISO/FDIS 13732-3:2004)	2005-01
prEN 1005-4	Safety of machinery—Human physical performance—Part 4: Evaluation of working postures and movements in relation to machinery	2003-09
CEN/TC 122 N 291	Personal protective equipment—Ergonomic principles—Part 2: Application of anthropometric measurements in design and specification	2002-09
prEN 13921-3	Personal protective equipment—Ergonomic principles—Part 3: Biomechanical characteristics	2003-10
prEN 13921-4	Personal protective equipment—Ergonomic principles—Part 4: Thermal characteristics	2003-10
prEN 13921-6	Personal protective equipment—Ergonomic principles—Part 6: Sensory factors	2003-10
prEN 14386	Safety of machinery—Ergonomic design principles for the operability of mobile machinery	2005-02
prEN ISO 15537	Principles for selecting and using test persons for testing anthropometric aspects of industrial products and designs (ISO/DIS 15537:2002)	2004-05
prEN 894-4	Safeguarding crushing points by means of limitation of the active forces	1998-11
	Safety of machinery—Ergonomic requirements for the design of displays and control actuators—Part 4: Location and arrangement of displays and control actuators	2003-04
EN ISO 11064-4:2004	Ergonomic design of control centers—Part 4: Layout and dimensions of workstations (ISO 11064-4:2004)	2004-07
ISO/CD 11064-5	Ergonomic design of control centers—Part 5: Displays and controls	2003-10
prEN ISO 11064-6	Ergonomic design of control centers—Part 6: Environmental requirements for control centers (ISO/DIS 11064-6:2003)	2003-10
ISO/CD 11064-7	Ergonomic design of control centers—Part 7: Principles for the evaluation of control centers	2003-10
prEN 13921-1	Personal protective equipment—Ergonomic principles—Part 1: General guidance	2003-10
ISO/NP 12892	Reach envelopes	2003-01
prEN 1005-5	Safety of machinery—Human physical performance—Part 5: Risk assessment for repetitive handling at high frequency	2005-04
prEN ISO 15536-1	Ergonomics—Computer manikins and body templates—Part 1: General requirements (ISO/DIS 15536-1:2002)	2003-12
prEN ISO 15536-2	Ergonomics—Computer manikins and body templates—Part 2: Verification of function and validation of dimensions for computer manikin systems	2002-10
prEN ISO 10075-3	Ergonomic principles related to mental workload—Part 3: Principles and requirements concerning methods for measuring and assessing mental workload	2004-10
prEN 614-1 rev	Safety of machinery—Ergonomic design principles—Part 1: Terminology and general principles	2004-01
prEN ISO 8996 rev	Ergonomics of the thermal environment—Determination of metabolic rate	2004-10
prEN ISO 7933	Ergonomics of the thermal environment—Analytical determination and interpretation of heat stress using calculation of the predicted heat strain	2004-10
ISO/NP 15743	Ergonomics of the thermal environment—Working practices for cold indoor environments	2003-07

(Continued)

APPENDIX 1.1
(Continued)

Reference	Title	DAV
prEN ISO 7730 rev	Ergonomics of the thermal environment—Analytical determination and interpretation of thermal comfort using calculation of the PMV and PPD indices and local thermal comfort (ISO/DIS 7730:2003)	2005-05
prEN ISO 11079	Evaluation of cold environments—Determination of required clothing insulation (IREQ) (will replace ENV ISO 11079:1998)	2003-03
prEN ISO 20685	3D scanning methodologies for internationally compatible anthropometric databases (ISO/DIS 20685:2004)	2005-06
prEN ISO 15265	Ergonomics of the thermal environment—Risk assessment strategy for the prevention of stress or discomfort in thermal working conditions	2004-09
prEN ISO 13732-1	Ergonomics of the thermal environment—Methods for the assessment of human responses to contact with surfaces—Part 1: Hot surfaces (ISO/DIS 13732-1:2004)	2006-01
prEN ISO 23973	Software ergonomics for WWW user interfaces	2006-01
prEN ISO 14505-1	Ergonomics of the thermal environment—Evaluation of thermal environment in vehicles—Part 1: Principles and methods for assessment of thermal stress	2006-02
prEN ISO 14505-2	Ergonomics of the thermal environment—Evaluation of thermal environment in vehicles—Part 2: Determination of equivalent temperature	2006-02
	Ergonomics of the thermal environment—Application of International Standards to the disabled, the aged, and other handicapped persons	2004-12
prEN ISO 9920 rev	Ergonomics of the thermal environment—Estimation of the thermal insulation and evaporative resistance of a clothing ensemble	2006-05
	Ergonomics of the thermal environment—Thermal environments in vehicles—Part 3: Evaluation of thermal comfort using human subjects (ISO/CD 14505-3)	2007-01
	Ergonomics of human–system interaction—Ergonomic requirements and measurement techniques for electronic visual displays—Introduction	2007-04
	Ergonomics of human–system interaction—Ergonomic requirements and measurement techniques for electronic visual displays—Terminology	2007-04
	Ergonomics of human–system-interaction—Ergonomic requirements and measurement techniques for electronic visual displays—Ergonomic requirements	2007-04
	Ergonomics of human–system interaction—Ergonomic requirements and measurement techniques for electronic visual displays—User performance test methods	2007-04
	Ergonomics of human–system interaction—Ergonomic requirements and measurement techniques for electronic visual displays—Optical laboratory test methods	2007-04
	Ergonomics of human–system interaction—Ergonomic requirements and measurement techniques for electronic visual displays—Field assessment methods	2007-04
	Ergonomics of human–system interaction—Ergonomic requirements and measurement techniques for electronic visual displays—Analysis and compliance test methods	2007-04
EN ISO 14738:2002/prAC	Safety of machinery—Anthropometric requirements for the design of workstations at machinery (ISO 14738:2002)	2003-05

APPENDIX 1.2

HFE Standards Published by Other Than TC 159 ISO Technical Committees

---

CIE—International Commission on Illumination
ISO/CIE 8995:2002	Lighting of indoor work places

JTC 1/SC 6—Telecommunications and information exchange between systems
ISO/IEC 10021-2:2003	Information technology—Message Handling Systems (MHS): Overall architecture

JTC 1/SC 7—Software and system engineering
ISO/IEC TR 9126-4:2004	Software engineering—Product quality—Part 4: Quality in use metrics
ISO/IEC 12119:1994	Information technology—Software packages—Quality requirements and testing
ISO/IEC 12207:1995	Information technology—Software life-cycle processes
ISO/IEC 14598-1:1999	Information technology—Software product evaluation—Part 1: General overview
ISO/IEC 14598-4:1999	Software engineering—Product evaluation—Part 4: Process for acquirers
ISO/IEC 14598-6:2001	Software engineering—Product evaluation—Part 6: Documentation of evaluation modules
ISO/IEC 15288:2002	Systems engineering—System life-cycle processes
ISO/IEC TR 15504-5:1999	Information technology—Software Process Assessment—Part 5: An assessment model and indicator guidance
ISO/IEC 15910:1999	Information technology—Software user documentation process
ISO/IEC 18019:2004	Software and system engineering—Guidelines for the design and preparation of user documentation for application software
ISO/IEC TR 19760:2003	Systems engineering—A guide for the application of ISO/IEC 15288 (System life-cycle processes)
ISO/IEC 20926:2003	Software engineering—IFPUG 4.1 Unadjusted functional size measurement method—Counting practices manual
ISO/IEC 20968:2002	Software engineering—Mk II Function Point Analysis—Counting Practices Manual

JTC 1/SC 22—Programming languages, their environments, and system software interfaces
ISO/IEC TR 11017:1998	Information technology—Framework for internationalization
ISO/IEC TR 14252:1996	Information technology—Guide to the POSIX Open System Environment
ISO/IEC TR 15942:2000	Information technology—Programming languages—Guide for the use of the Ada programming language in high-integrity systems

JTC 1/SC 27—IT Security techniques
ISO/IEC TR 13335-4:2000	Information technology—Guidelines for the management of IT Security—Part 4: Selection of safeguards
ISO/IEC 21827:2002	Information technology—Systems Security Engineering—Capability Maturity Model

JTC 1/SC 35—User interfaces
ISO/IEC 15411:1999	Information technology—Segmented keyboard layouts
ISO/IEC 18035:2003	Information technology—Icon symbols and functions for controlling multimedia software applications

TC 8/SC 5—Ships' bridge layout
ISO 8468:1990	Ship's bridge layout and associated equipment—Requirements and guidelines
ISO 14612:2004	Ships and marine technology—Ship's bridge layout and associated equipment—Additional requirements and guidelines for centralized and integrated bridge functions

TC 8/SC 6—Navigation
ISO 16273:2003	Ships and marine technology—Night vision equipment for high-speed craft—Operational and performance requirements, methods of testing, and required test results

TC 20—Aircraft and space vehicles
ISO/TR 10201:2001	Aerospace—Standards for electronic instruments and systems

TC 20/SC 1—Aerospace electrical requirements
ISO 6858:1982	Aircraft—Ground support electrical supplies—General requirements

---

(Continued)

APPENDIX 1.2
(Continued)

TC 20/SC 14—Space systems and operations
ISO 16091:2002 — Space systems—Integrated logistic support
ISO 17399:2003 — Space systems—Man–systems integration

TC 21/SC 3—Fire detection and alarm systems
ISO 12239:2003 — Fire detection and fire alarm systems—Smoke alarms

TC 22/SC 3—Electrical and electronic equipment
ISO 11748-2:2001 — Road vehicles—Technical documentation of electrical and electronic systems—Part 2: Documentation agreement
ISO/TR 15497:2000 — Road vehicles—Development guidelines for vehicle based software

TC 22/SC 13—Ergonomics applicable to road vehicles
ISO 2575:2004 — Road vehicles—Symbols for controls, indicators, and tell-tales
ISO 3958:1996 — Passenger cars—Driver hand-control reach
ISO 4040:2001 — Road vehicles—Location of hand controls, indicators, and tell-tales in motor vehicles
ISO 6549:1999 — Road vehicles—Procedure for H- and R-point determination
ISO/TR 9511:1991 — Road vehicles—Driver hand-control reach—In-vehicle checking procedure
ISO/TS 12104:2003 — Road vehicles—Gearshift patterns—Manual transmissions with power-assisted gear change and automatic transmissions with manual-gearshift mode
ISO 12214:2002 — Road vehicles—Direction-of-motion stereotypes for automotive hand controls
ISO 15005:2002 — Road vehicles—Ergonomic aspects of transport information and control systems—Dialogue management principles and compliance procedures
ISO 15007-1:2002 — Road vehicles—Measurement of driver visual behavior with respect to transport information and control systems—Part 1: Definitions and parameters
ISO/TS 15007-2:2001 — Road vehicles—Measurement of driver visual behavior with respect to transport information and control systems—Part 2: Equipment and procedures
ISO 15008:2003 — Road vehicles—Ergonomic aspects of transport information and control systems—Specifications and compliance procedures for in-vehicle visual presentation
ISO/TS 16951:2004 — Road vehicles—Ergonomic aspects of transport information and control systems (TICS)—Procedures for determining priority of on-board messages presented to drivers
ISO 17287:2003 — Road vehicles—Ergonomic aspects of transport information and control systems—Procedure for assessing suitability for use while driving

TC 22/SC 17—Visibility
ISO 7397-1:1993 — Passenger cars—Verification of driver's direct field of view—Part 1: Vehicle positioning for static measurement
ISO 7397-2:1993 — Passenger cars—Verification of driver's direct field of view—Part 2: Test method

TC 23/SC 3—Safety and comfort of the operator
ISO 4254-1:1989 — Tractors and machinery for agriculture and forestry—Technical means for ensuring safety—Part 1: General
ISO/TS 15077:2002 — Tractors and self-propelled machinery for agriculture and forestry—Operator controls—Actuating forces, displacement, location and method of operation

TC 23/SC 4—Tractors
ISO 4253:1993 — Agricultural tractors—Operator's seating accommodation—Dimensions
ISO 5721:1989 — Tractors for agriculture—Operator's field of vision

TC 23/SC 7—Equipment for harvesting and conservation
ISO 8210:1989 — Equipment for harvesting—Combine harvesters—Test procedure

TC 23/SC 14—Operator controls, operator symbols and other displays, operator manuals
ISO 3767-1:1998 — Tractors, machinery for agriculture and forestry, powered lawn and garden equipment—Symbols for operator controls and other displays—Part 1: Common symbols
ISO 3767-2:1991 — Tractors, machinery for agriculture and forestry, powered lawn and garden equipment—Symbols for operator controls and other displays—Part 2: Symbols for agricultural tractors and machinery

(Continued)

APPENDIX 1.2
(Continued)

ISO 3767-3:1995	Tractors, machinery for agriculture and forestry, powered lawn and garden equipment—Symbols for operator controls and other displays—Part 3: Symbols for powered lawn and garden equipment
ISO 3767-5:1992	Tractors, machinery for agriculture and forestry, powered lawn and garden equipment—Symbols for operator controls and other displays—Part 5: Symbols for manual portable forestry machinery

TC 23/SC 15—Machinery for forestry

ISO 11850:2003	Machinery for forestry—Self-propelled machinery—Safety requirements

TC 23/SC 17—Manually portable forest machinery

ISO 8334:1985	Forestry machinery—Portable chain-saws—Determination of balance
ISO 11680-1:2000	Machinery for forestry—Safety requirements and testing for pole-mounted powered pruners—Part 1: Units fitted with an integral combustion engine
ISO 11680-2:2000	Machinery for forestry—Safety requirements and testing for pole-mounted powered pruners—Part 2: Units for use with a back-pack power source
ISO 11681-1:2004	Machinery for forestry—Portable chain-saw safety requirements and testing—Part 1: Chain-saws for forest service
ISO 11681-2:1998	Machinery for forestry—Portable chain-saws—Safety requirements and testing—Part 2: Chain-saws for tree service
ISO 11806:1997	Agricultural and forestry machinery—Portable hand-held combustion engine driven brush cutters and grass trimmers—Safety
ISO 14740:1998	Forest machinery—Backpack power units for brush-cutters, grass-trimmers, pole-cutters and similar appliances—Safety requirements and testing

TC 23/SC 18—Irrigation and drainage equipment and systems

ISO/TR 8059:1986	Irrigation equipment—Automatic irrigation systems—Hydraulic control

TC 38—Textiles

ISO 15831:2004	Clothing—Physiological effects—Measurement of thermal insulation by means of a thermal manikin

TC 43/SC 1—Noise

ISO 11690-1:1996	Acoustics—Recommended practice for the design of low-noise workplaces containing machinery—Part 1: Noise control strategies
ISO 15667:2000	Acoustics—Guidelines for noise control by enclosures and cabins

TC 46—Information and documentation

ISO 7220:1996	Information and documentation—Presentation of catalogues of standards

TC 59/SC 3—Functional/user requirements and performance in building construction

ISO 6242-1:1992	Building construction—Expression of users' requirements—Part 1: Thermal requirements
ISO 6242-2:1992	Building construction—Expression of users' requirements—Part 2: Air purity requirements
ISO 6242-3:1992	Building construction—Expression of users' requirements—Part 3: Acoustical requirements

TC 67—Materials, equipment and offshore structures for petroleum, petrochemical, and natural gas industries

ISO 13879:1999	Petroleum and natural gas industries—Content and drafting of a functional specification
ISO 13880:1999	Petroleum and natural gas industries—Content and drafting of a technical specification

TC 67/SC 6—Processing equipment and systems

ISO 13702:1999	Petroleum and natural gas industries—Control and mitigation of fires and explosions on offshore production installations—Requirements and guidelines
ISO 15544:2000	Petroleum and natural gas industries—Offshore production installations—Requirements and guidelines for emergency response
ISO 17776:2000	Petroleum and natural gas industries—Offshore production installations—Guidelines on tools and techniques for hazard identification and risk assessment

TC 69—Applications of statistical methods

ISO 10725:2000	Acceptance sampling plans and procedures for the inspection of bulk materials

(Continued)

APPENDIX 1.2
(Continued)

TC 72/SC 5—Industrial laundry and dry-cleaning machinery and accessories

ISO 8230:1997 — Safety requirements for dry-cleaning machines using perchloroethylene

ISO 10472-1:1997 — Safety requirements for industrial laundry machinery—Part 1: Common requirements

TC 72/SC 8—Safety requirements for textile machinery

ISO 11111:1995 — Safety requirements for textile machinery

TC 85/SC 2—Radiation protection

ISO 17874-1:2004 — Remote handling devices for radioactive materials—Part 1: General requirements

TC 92/SC 3—Fire threat to people and environment

ISO/TS 13571:2002 — Life-threatening components of fire—Guidelines for the estimation of time available for escape using fire data

TC 92/SC 4—Fire safety engineering

ISO/TR 13387-1:1999 — Fire safety engineering—Part 1: Application of fire performance concepts to design objectives

TC 94/SC 4—Personal equipment for protection against falls

ISO 10333-6:2004 — Personal fall-arrest systems—Part 6: System performance tests

TC 94/SC 13—Protective clothing

ISO 11393-4:2003 — Protective clothing for users of hand-held chainsaws—Part 4: Test methods and performance requirements for protective gloves

ISO 13688:1998 — Protective clothing—General requirements

ISO 16603:2004 — Clothing for protection against contact with blood and body fluids—Determination of the resistance of protective clothing materials to penetration by blood and body fluids—Test method using synthetic blood

ISO 16604:2004 — Clothing for protection against contact with blood and body fluids—Determination of resistance of protective clothing materials to penetration by blood-borne pathogens—Test method using Phi X 174 bacteriophage

TC 101—Continuous mechanical handling equipment

ISO/TR 5045:1979 — Continuous mechanical handling equipment—Safety code for belt conveyors—Examples for guarding of nip points

TC 108/SC 2—Measurement and evaluation of mechanical vibration and shock as applied to machines, vehicles, and structures

ISO 14964:2000 — Mechanical vibration and shock—Vibration of stationary structures—Specific requirements for quality management in measurement and evaluation of vibration

TC 108/SC 4—Human exposure to mechanical vibration and shock

ISO 2631-1:1997 — Mechanical vibration and shock—Evaluation of human exposure to whole-body vibration—Part 1: General requirements

ISO 2631-2:2003 — Mechanical vibration and shock—Evaluation of human exposure to whole-body vibration—Part 2: Vibration in buildings (1 Hz to 80 Hz)

ISO 2631-4:2001 — Mechanical vibration and shock—Evaluation of human exposure to whole-body vibration—Part 4: Guidelines for the evaluation of the effects of vibration and rotational motion on passenger and crew comfort in fixed-guideway transport systems

ISO 2631-5:2004 — Mechanical vibration and shock—Evaluation of human exposure to whole-body vibration—Part 5: Method for evaluation of vibration containing multiple shocks

ISO 5349-1:2001 — Mechanical vibration—Measurement and evaluation of human exposure to hand-transmitted vibration—Part 1: General requirements

TC 108/SC 4—Human exposure to mechanical vibration and shock

ISO 5982:2001 — Mechanical vibration and shock—Range of idealized values to characterize seated-body biodynamic response under vertical vibration

ISO 6897:1984 — Guidelines for the evaluation of the response of occupants of fixed structures, especially buildings and off-shore structures, to low-frequency horizontal motion (0.063 to 1 Hz)

ISO 8727:1997 — Mechanical vibration and shock—Human exposure—Biodynamic coordinate systems

(Continued)

APPENDIX 1.2
(Continued)

ISO 9996:1996	Mechanical vibration and shock—Disturbance to human activity and performance—Classification
ISO 10068:1998	Mechanical vibration and shock—Free, mechanical impedance of the human hand-arm system at the driving point
ISO 13090-1:1998	Mechanical vibration and shock—Guidance on safety aspects of tests and experiments with people—Part 1: Exposure to whole-body mechanical vibration and repeated shock
ISO 13091-1:2001	Mechanical vibration—Vibrotactile perception thresholds for the assessment of nerve dysfunction—Part 1: Methods of measurement at the fingertips
ISO 13091-2:2003	Mechanical vibration—Vibrotactile perception thresholds for the assessment of nerve dysfunction—Part 2: Analysis and interpretation of measurements at the fingertips

TC 121/SC 1—Breathing attachments and anaesthetic machines

ISO 7767:1997	Oxygen monitors for monitoring patient breathing mixtures—Safety requirements

TC 121/SC 3—Lung ventilators and related equipment

ISO 8185:1997	Humidifiers for medical use—General requirements for humidification systems
IEC 60601-1-8:2003	Medical electrical equipment—Part 1-8: General requirements for safety—Collateral standard: General requirements, tests, and guidance for alarm systems in medical electrical equipment and medical electrical systems
IEC 60601-2-12:2001	Medical electrical equipment—Part 2-12: Particular requirements for the safety of lung ventilators—Critical care ventilators

TC 123/SC 5—Quality analysis and assurance

ISO 12307-1:1994	Plain bearings—Wrapped bushes—Part 1: Checking the outside diameter

TC 127/SC 1—Test methods relating to machine performance

ISO 8813:1992	Earth-moving machinery—Lift capacity of pipelayers and wheeled tractors or loaders equipped with side boom

TC 127/SC 2—Safety requirements and human factors

ISO 2860:1992	Earth-moving machinery—Minimum access dimensions
ISO 2867:1994	Earth-moving machinery—Access systems
ISO 3164:1995	Earth-moving machinery—Laboratory evaluations of protective structures—Specifications for deflection-limiting volume
ISO 3411:1995	Earth-moving machinery—Human physical dimensions of operators and minimum operator space envelope
ISO 3449:1992	Earth-moving machinery—Falling-object protective structures—Laboratory tests and performance requirements
ISO 3450:1996	Earth-moving machinery—Braking systems of rubber-tired machines—Systems and performance requirements and test procedures
ISO 3457:2003	Earth-moving machinery—Guards—Definitions and requirements
ISO 3471:1994	Earth-moving machinery—Roll-over protective structures—Laboratory tests and performance requirements
ISO 3471:1994/Amd 1:1997	Laboratory tests and performance requirements
ISO 5006-2:1993	Earth-moving machinery—Operator's field of view—Part 2: Evaluation method
ISO 5006-3:1993	Earth-moving machinery—Operator's field of view—Part 3: Criteria
ISO 5010:1992	Earth-moving machinery—Rubber-tired machines—Steering requirements
ISO 5353:1995	Earth-moving machinery, and tractors and machinery for agriculture and forestry—Seat index point
ISO 6682:1986	Earth-moving machinery—Zones of comfort and reach for controls
ISO 7096:2000	Earth-moving machinery—Laboratory evaluation of operator seat vibration
ISO 8643:1997	Earth-moving machinery—Hydraulic excavator and backhoe loader boom-lowering control device—Requirements and tests
ISO 9244:1995	Earth-moving machinery—Safety signs and hazard pictorials—General principles
ISO/TR 9953:1996	Earth-moving machinery—Warning devices for slow-moving machines—Ultrasonic and other systems

(Continued)

APPENDIX 1.2
(Continued)

ISO 10262:1998 Earth-moving machinery—Hydraulic excavators—Laboratory tests and performance requirements for operator protective guards

ISO 10263-1:1994 Earth-moving machinery—Operator enclosure environment—Part 1: General and definitions

ISO 10263-2:1994 Earth-moving machinery—Operator enclosure environment—Part 2: Air filter test

ISO 10263-3:1994 Earth-moving machinery—Operator enclosure environment—Part 3: Operator enclosure pressurization test method

ISO 10263-4:1994 Earth-moving machinery—Operator enclosure environment—Part 4: Operator enclosure ventilation, heating, and/or air-conditioning test method

ISO 10263-5:1994 Earth-moving machinery—Operator enclosure environment—Part 5: Windscreen defrosting system test method

ISO 10263-6:1994 Earth-moving machinery—Operator enclosure environment—Part 6: Determination of effect of solar heating on operator enclosure

ISO 10533:1993 Earth-moving machinery—Lift-arm support devices

ISO 10567:1992 Earth-moving machinery—Hydraulic excavators—Lift capacity

ISO 10570:2004 Earth-moving machinery—Articulated frame lock—Performance requirements

ISO 10968:1995 Earth-moving machinery—Operator's controls

ISO 11112:1995 Earth-moving machinery—Operator's seat—Dimensions and requirements

ISO 12117:1997 Earth-moving machinery—Tip-over protection structure (TOPS) for compact excavators—Laboratory tests and performance requirements

ISO 12508:1994 Earth-moving machinery—Operator station and maintenance areas—Bluntness of edges

ISO 13333:1994 Earth-moving machinery—Dumper body support and operator's cab tilt support devices

ISO 13459:1997 Earth-moving machinery—Dumpers—Trainer seat/enclosure

ISO 17063:2003 Earth-moving machinery—Braking systems of pedestrian-controlled machines—Performance requirements and test procedures

TC 130—Graphic technology

ISO 12648:2003 Graphic technology—Safety requirements for printing press systems

ISO 12649:2004 Graphic technology—Safety requirements for binding and finishing systems and equipment

TC 131/SC 9—Installations and systems

ISO 4413:1998 Hydraulic fluid power—General rules relating to systems

ISO 4414:1998 Pneumatic fluid power—General rules relating to systems

TC 136—Furniture

ISO 5970:1979 Furniture—Chairs and tables for educational institutions—Functional sizes

TC 163/SC 2—Calculation methods

ISO 13790:2004 Thermal performance of buildings—Calculation of energy use for space heating

TC 171/SC 2—Application issues

ISO/TR 14105:2001 Electronic imaging—Human and organizational issues for successful Electronic Image Management (EIM) implementation

TC 172/SC 9—Electro-optical systems

ISO 11553:1996 Safety of machinery—Laser processing machines—Safety requirements

TC 173—Assistive products for persons with disability

ISO 11199-1:1999 Walking aids manipulated by both arms—Requirements and test methods—Part 1: Walking frames

ISO 11199-2:1999 Walking aids manipulated by both arms—Requirements and test methods—Part 2: Rollators

ISO 11334-1:1994 Walking aids manipulated by one arm—Requirements and test methods—Part 1: Elbow crutches

ISO 11334-4:1999 Walking aids manipulated by one arm—Requirements and test methods—Part 4: Walking sticks with three or more legs

TC 173/SC 3—Aids for ostomy and incontinence

ISO 15621:1999 Urine-absorbing aids—General guidance on evaluation

TC 173/SC 6—Hoists for transfer of persons

ISO 10535:1998 Hoists for the transfer of disabled persons—Requirements and test methods

(Continued)

APPENDIX 1.2
(Continued)

TC 176/SC 1—Concepts and terminology
ISO 9000:2000   Quality management systems—Fundamentals and vocabulary

TC 176/SC 2—Quality systems
ISO 9004:2000   Quality management systems—Guidelines for performance improvements

TC 178—Lifts, escalators and moving walks
ISO/TS 14798:2000   Lifts (elevators), escalators, and passenger conveyors—Risk analysis methodology

TC 184—Industrial automation systems and integration
ISO 11161:1994   Industrial automation systems—Safety of integrated manufacturing systems—Basic
                        requirements

TC 184/SC 4—Industrial data
ISO 10303-214:2003   Industrial automation systems and integration—Product data representation and
                        exchange—Part 214: Application protocol: Core data for automotive mechanical
                        design processes

TC 184/SC 5—Architecture, communications, and integration frameworks
ISO 15704:2000   Industrial automation systems—Requirements for enterprise-reference architectures
                        and methodologies
ISO 16100-1:2002   Industrial automation systems and integration—Manufacturing software capability
                        profiling for interoperability—Part 1: Framework

TC 188—Small craft
ISO 15027-3:2002   Immersion suits—Part 3: Test methods

TC 199—Safety of machinery
ISO 12100-2:2003   Safety of machinery—Basic concepts, general principles for design—Part 2: Technical
                        principles
ISO 13849-1:1999   Safety of machinery—Safety-related parts of control systems—Part 1: General
                        principles for design
ISO 13851:2002   Safety of machinery—Two-hand control devices—Functional aspects and design
                        principles
ISO 13856-1:2001   Safety of machinery—Pressure-sensitive protective devices—Part 1: General principles
                        for design and testing of pressure-sensitive mats and pressure-sensitive floors
ISO 14121:1999   Safety of machinery—Principles of risk assessment
ISO 14123-2:1998   Safety of machinery—Reduction of risks to health from hazardous substances emitted
                        by machinery—Part 2: Methodology leading to verification procedures
ISO/TR 18569:2004   Safety of machinery—Guidelines for the understanding and use of safety of machinery
                        standards

TC 204—Intelligent transport systems
ISO 15623:2002   Transport information and control systems—Forward vehicle collision warning
                        systems—Performance requirements and test procedures

TC 210—Quality management and corresponding general aspects for medical devices
ISO 14969:1999   Quality systems—Medical devices—Guidance on the application of ISO 13485 and
                        ISO 13488
ISO 14971:2000   Medical devices—Application of risk management to medical devices

TC 212—Clinical laboratory testing and *in vitro* diagnostic test systems
ISO 15190:2003   Medical laboratories—Requirements for safety
ISO 15197:2003   In vitro diagnostic test systems—Requirements for blood-glucose monitoring systems
                        for self-testing in managing diabetes mellitus

TMB—Technical Management Board
IWA 1:2001   Quality management systems—Guidelines for process improvements in health service
                        organizations
ISO/IEC Guide 50:2002   Safety aspects—Guidelines for child safety
ISO/IEC Guide 71:2001   Guidelines for standards developers to address the needs of older persons and persons
                        with disabilities

CASCO—Committee on conformity assessment
ISO/IEC 17025:1999   General requirements for the competence of testing and calibration laboratories

REFERENCES

Albin, T. J. (2005). Board of Standards Review/Human Factor and Ergonomics Society 100—Human Factors Engineering of Computer Workstations—Draft Standard for Trial Use. In Karwowski, W. (Ed.), 2005, Handbook of Human Factors and Ergonomics Standards and Guideliness, Lawrence Erlbaum Publishers (in press).

Bastien, J. M. C., Scapin, D. L., & Leulier, C. (1999). The ergonomic criteria and the ISO/DIS 9241-10 dialogue principles: A pilot comparison in an evaluation task. *Interacting With Computers, 11*(3), 299–322.

Becker, M. (1998). Evaluation and characterization of display reflectance, *Displays, 19*, 35–54.

Besuijen, K., & Spenkelink, G. P. J. (1998). Standardizing visual display quality, *Displays, 19*, 67–76.

CEN. (2004). European Standardization Committee Web site: http://www.cenorm.be/cenorm/index.htm

Chapanis, A. (1996). *Human factors in systems engineering.* New York: Wiley.

Dickinson, C. E. (1995). Proposed manual handling international and European Standards. *Applied Ergonomics, 26*(4), 265–270.

Department of Industrial Relation. (2004). California Department of Industrial Relation homepage. http://www.dir.ca.gov/

Douglas, S. A., Kirkpatrick, A. E., & MacKenzie, I. S. (1999). Testing pointing device performance and user assessment with the ISO 9241, Part 9 standard. In *Proceedings of the ACM Conference on Human Factors in Computing Systems—CHI '99* (pp. 215–222). ACM, New York.

Dul, J., de Vlaming, P. M., & Munnik, M. J. (1996). A review of ISO and CEN standards on ergonomics. *International Journal of Industrial Ergonomics, 17*(3), 291–297.

Dzida, W. (1995). Standards for user-interfaces. *Computer Standards & Interfaces, 17*(1), 89–97.

Dzida, W., & Freitag, R. (1998), Making use of scenarios for validating analysis and design, *IEEE Transactions on Software Engineering, 24*(12), 1182–1196.

Earthy, J., Jones, B. S., & Bevan, N. (2001). The improvement of human-centered processes—Facing the challenge and reaping the benefits of ISO 13407. *International Journal of Human-Computer Studies, 55*, 553–585.

Eibl, M. (in press). *International Standards of Interface Design.* In W. Karwowski (Ed.), 2005 *Handbook of Human Factors and Ergonomics Standards and Guideliness.* Lawrence Erlbaum Associates.

El Emam, K. & Garro, I. (2000). Estimating the extent of standards use: The case of ISO/IEC 15504. *The Journal of Systems and Software, 53*, 137–143.

El Emam, K., & Jung, H.-W. (2001). An empirical evaluation of the ISO/IEC 15504 assessment model. *The Journal of Systems and Software, 59*, 23–41.

Fenton, N., Littlewood, B., & Page, S. (1993). Evaluating software engineering standards and methods. In R. Thayer & R. McGettrick (Eds.), *Software engineering: A European perspective*, IEEE Computer Society Press: Silver Spring, MD, (pp. 463–470).

Griefahn, B. (2000). Limits of and possibilities to improve the IREQ cold stress model (ISO/TR 11079): A validation study in the field. *Applied Ergonomics, 31*, 423–431.

Griefahn, B., & Brode, P. (1999). The significance of lateral whole-body vibrations related to separately and simultaneously applied vertical motions: A validation study of ISO 2631. *Applied Ergonomics, 30*, 505–513.

Harker, S. (1995). The development of ergonomics standards for software. *Applied Ergonomics, 26*(4), 275–279.

Hoyle, D. (2001). *ISO 9000: Quality systems handbook.* Oxford: Butterworth Heinemann.

Human Factors and Ergonomics Society. (2002). Board of Standards Review/Human Factors and Ergonomics Society 100—Human factors engineering of computer workstations—Draft Standard for Trial Use. Human Factors and Ergonomics Society, Santa Monica, CA.

Human Factors Standardization SubTAG. (HFS SubTAG). (2004). Index of Government Standards on Human Engineering Design Criteria, Processes, and Procedures. Version 1 (Draft), March 15, 2004. Department of Defense, *Human Factors Engineering Technical Advisory Group.* Web site. http://hfetag.dtic.mil/docs/index_govt_std.doc

ILO (2004). International Labor Organization Web site. http://www.ilo.org/public/english/index.htm

ILO-OSH (2001).*Guidelines on occupational safety and health management systems, ILO-OSH 2001.* Geneva, International Labour Office. http://www.ilo.org/public/english/protection/safework/managmnt/guide.htm

Ishitake, T., Miyazaki, Y., Noguchi, R., Ando, H., & Matoba, T. (2002). Evaluation of frequency weighting (ISO 2631-1) for acute effects of whole-body vibration on gastric motility, *Journal of Sound and Vibration, 253*(1), 31–36).

Isokoski, P., & Raisamo, R. (2002). Speed-accuracy measures in a population of six mice. In *Proceedings of the Fifth Asia Pacific Conference on Human-Computer Interaction—APCHI 2002* (pp. 765–777). Bejing, China: Science Press.

ISO (2004). International Standardization Organization Web site. http://www.iso.org/iso/en/ISOOnline.openerpage

Keates, S., Hwang, F., Langdon, P., Clarkson, P. J., & Robinson, P. (2002). Cursor measures for motion impaired computer users. In *Proceedings of the Fifth ACM Conference on Assistive Technology—ASSETS 2002.* (pp. 135–142). New York: ACM.

Kosanke, K., & Nell, J. G. (1999). Standardization in ISO for enterprise engineering and integration, Computers in Industry, *40*, 311–319.

Lindfors, M. (1998). Accuracy and repeatability of the ISO 9241-7 test method. *Displays, 19*, 3–16.

MacKenzie, I. S., & Jusoh, S. (2001). An evalution of two input devices for remote pointing. In *Proceedings of the Eighth IFIP Working Conference on Engineering for Human-Computer Ineraction—EHCI 2001.* (pp. 235–249). Springer-Verlag, Heidelberg, Germany.

MacKenzie, I. S., Kauppinen, T., & Silfverberg, M. (2001). Accuracy measures for evaluating computer pointing devices. In *Proceedings of the Human Factors in Computing Systems*, CHI Letters, *3*(1), 9–16.

Marmaras, N., & Papadopoulus, S. (2003). A study of computerized offices in Greece: Are ergonomic design requirements met? *International Journal of Human-Computer Interaction, 16*(2), 261–281.

MATRIS. (2004). Directory of Design Support Methods (DSSM). Web site: http://dtica.dtic.mil/ddsm/

McDaniel, J. W. (1996). The demise of military standards may affect ergonomics. *International Journal of Industrial Ergonomics, 18*(5-6), 339–348.

Nachreiner, F. (1995). Standards for ergonomics principles relating to the design of work systems and to mental workload. *Applied Ergonomics, 26*(4), 259–263.

Oh, J.-Y., & Stuerzlinger, W. (2002). Laser pointers as collaboratve pointing devices. In *Proceedings of Graphics Interface* (pp. 141–149). AK Peters and CHCCS.

Olesen, B. W. (1995). International standards and the ergonomics of the thermal environment. *Applied Ergonomics, 26*(4), 293–302.

Olesen, B. W., & Parsons, K. C. (2002). Introduction to thermal comfort standards and to the proposed new version of EN ISO 7730. *Energy and Buildings, 34*(6), 537–548.

Occupational Safety and Health Administration. (OSHA) (2000). *Ergonomics Program Rule.* Federal Register, Vol 65, No 220.

OSHA (2004). Occupational Safety and Health Administration Web site. http://www.osha-slc.gov

Parsons, K. (1995a). Ergonomics and international standards. Applied Ergonomics, *26*(4), 237–238.

Parsons, K. C. (1995b). Ergonomics of the physical environment: International ergonomics standards concerning speech communication, danger signals, lighting, vibration and surface temperatures. *Applied Ergonomics, 26*(4), 281–292.

Parsons, K. C. (1995c). Ergonomics and international standards: Introduction, brief review of standards for anthropometry and control room design and useful information. *Applied Ergonomics, 26*(4), 239–247.

Parsons, K. C., Shackel, B., and Metz, B. (1995). Ergonomics and international standards: History, organizational structure and method of development. *Applied Ergonomics, 26*(4), 249–258.

Poupyrev, I., Okabe, M., & Maruyama, S. (2004). Haptic feedback for pen computing: Directions and strategies. In *CHI '04 extended abstracts on Human factors in computing systems* (pp. 1309–1312). Vienna Austria.

Public Law 101-336 (1990). Americans with Disabilities Act. Public Law 336 of the 101st Congress, enacted July 26, 1990.

Poupyrev, I., Okabe, M., & Maruyama, S., (2004). Haptic feedback for pen computing: directions and strategies. *Extended Abstracts of the ACM Conference on Human Factors in Computing Systems—CHI 2004* (pp. 1309–1312). New York: ACM.

Reed, P., Holdaway, K., Isensee, S., Buie, E., Fox, J., Williams, J., & Lund, A. (1999). User interface guidelines and standards: Progress, issues, and prospects. *Interacting with Computers, 12*(2), 119–142.

Seabrook, K. A. (2001). International Standards Update: Occupational Safety and Health Management Systems. In *Proceedings of the American Society of Safety Engineers' 2001 Professional Development Conference*, Anaheim, CA.

Silfverberg, M., MacKenzie, I. S., & Kauppinen, T. (2001). An isometric joystick as a pointing device for handheld information terminals. In *Proceedings of Graphics Interface of Canadian Information Processing Society* (pp. 119–126). Toronto, Canada.

Smith, S. L., & Mosier, J. N. (1984). The user interface to computer-based information systems: A survey of current software design practice. *Behaviour and Information Technology, 3*, 195–203.

Sohn, M., & Lee, G. (2004). SonarPen: An ultrasonic pointing device for and interactive TV. *IEEE Transactions on Consumer Electronics, 50*(2), 413–419.

Spivak, S. M., & Brenner, F. C. (2001). *Standardization Essentials: Principles and Practice.* New York: Dekker.

Stewart, T. (1995). Ergonomics standards concerning human-system interaction: Visual displays, controls and environmental requirements. *Applied Ergonomics, 26*(4), 271–274.

Umezu, N., Nakano, Y., Sakai, T., Yoshitake, R., Herlitschke, W., & Kubota, S. (1998). Specular and diffuse reflection measurement feasibility study of ISO 9241 Part 7 method. *Displays*, 19, 17–25.

Wettig, J. (2002). New developments in standardization in the past 15 years—product versus process related standards. *Safety Science, 40* (1–4), 51–56.

Zhai, S. (in press). Characterizing computer input with Fitt's law parameters—The information and non-information aspects of pointing. *International Journal of Human-Computer Studies.*

2

Positioning Ergonomics Standards and Standardization

Jan Dul
Henk J. de Vries
Erasmus University Rotterdam

INTRODUCTION

Ergonomics is just one of the fields where standards have been developed. In comparison to other fields, ergonomics standardization is relatively young: only some decades old, whereas technical fields like telecommunication, electrotechnology, mechanical engineering, and civil engineering have a standardization tradition of more than a century, and the origins of standards can be found many centuries before Christ in China, Mesopotamia, and Egypt. In those times, it was often the emperor who took the initiative. For example, the Chinese emperor Qing-Shihuang set compulsory standards on measurement, seeds, cloth sizes, and weaponry (Wen, 2004).

Is ergonomics standardization an area with exceptional characteristics, or is it just one of the many areas of standardization? In the first case, ergonomic standardization can learn lessons from the experiences in other fields. In the second case, the same applies; but then, we should reckon with the typical differences.

In this chapter, we distinguish between standards (the result of standardization, usually a document) and standardization (the process of making standards). We draw some general lines on standards and standardization, describe ergonomics standards and standardization, and subsequently examine to which extent the situation in ergonomics differs from the situation of standards and standardization in general.

In Part I, we start with a general description of standards and standardization. In Part II, we describe ergonomics standards and standardization. We conclude that the goals of ergonomic standards and standardization differ from those that generally apply, and discuss the possible lessons.

PART I: GENERAL STANDARDS AND STANDARDIZATION

Standards

Standards are very common in everyday life. The A4 series of paper sizes, specifications of credit cards, ISO 9000 requirements for quality management systems, the SI system of units (SI = *Système International d'Unités*), McDonald's product and service specifications, and the specifications of the GSM mobile phone system have in common that they are used repeatedly by a large number of people and, therefore, are laid down in standards. Standards can be considered as a lubricant for the modern industrial society.

Definition

The definition of a standard used by formal standardization organizations is *a document, established by consensus and approved by a recognized body, that provides, for common and repeated use, rules, guidelines or characteristics for activities or their results, aimed at the achievement of the optimum degree of order in a given context* (ISO/IEC, 1996). However, not all of the aforementioned examples fit in this definition: Not all standards are consensus based or approved by a recognized body, and standards may have another format than a document, for example, software. Therefore, De Vries (1999) has developed another definition: *A standard is an approved specification of a limited set of solutions to actual or potential matching problems, prepared for the benefits of the party or parties involved, balancing their needs, and intended and expected to be used repeatedly or continuously, during a certain period, by a substantial number of the parties for whom they are meant.*

Users

Standards can be used by companies, but also by other groups in society, such as governments and testing organizations. We distinguish between direct users—parties that read the standard to apply it—and indirect users—parties that have a stake in the application of the standard. A party can be a person, a group of people, and an organization.

Direct Users. In general, the category of direct users includes the following:

- Designers: parties that design products, services, and processes in order to make a design that meets criteria laid down in a standard
- Testers: parties that test products, services, or processes against the requirements in a standard or that use a test method specified in a standard
- Advisors: parties that provide advice concerning designing or testing according to a standard
- Regulators: parties that develop regulations, other requirements, standards, rules, or laws based on a standard

The activities of these direct users of standards are carried out by, or for, organizations like companies, testing laboratories, certification bodies, consultancy firms, and governmental agencies. Standardization can be important for these organizations: Standards may apply to, or can be developed for, the products or services delivered or purchased by these organizations, or may apply to the business processes in these organizations.

Indirect Users. Indirect users can be end-users of a standard: consumers, workers, and the general public (e.g., safety and environmental issues). These groups normally do not read standards, but the results of applying standards in the product and processes are important for these groups. Other indirect users can be special interest groups, such as branch of business organizations (for the common interests of their members), trade unions (for the interests of workers), and environmental pressure groups. In the next sections we discuss the interests of three main direct and indirect users of standards: companies, governments, and consumers.

Companies.   Companies can be both direct and indirect users of a standard. In company practice, the main aim of using standards is to contribute to business results and to the effectiveness and efficiency of the organization. Standards can reduce the costs of products and services. Meeting or not meeting certain standards can be the difference between success and failure in the market. General aims of standardization include (after Sanders, 1972):

- Reduction of the growing variety of products and procedures
- Enabling communication
- Contributing to the functioning of the overall economy
- Contributing to safety, health, and protection of life
- Protection of consumer and community interests
- Eliminating trade barriers although standards at the national or regional level can also create trade barriers, according to Hesser et al. (1995).

More specifically, the following issues illustrate the importance of standards for companies (Schaap & De Vries, 2004):
Within the company:

- Not re-inventing the wheel, but using existing solutions (laid down in standards) which have already been well thought out
- Bringing procedures into line with what is normal elsewhere, so that cooperation is simpler and purchasing cheaper
- Efficient working, by repeatedly using the same solution
- Using recognized requirements in the field of, for example, quality and safety
- Cheaper purchasing, due to economies of scale, and more transparency in the market which results in an increase of price competition
- Less cost of stock and logistics by making use of standard solutions
- More possibilities for outsourcing
- Fewer problems in the field of occupational health, safety, and environmental issues

Outside the company:

- Being able to demonstrate the quality of products and services (using test methods laid down in standards)
- Giving clients confidence: the product (the production method) meets accepted requirements
- Being allowed to bring products onto the market, because conformity to standards may be a means to demonstrate conformity to legal requirements
- Being able to bring products onto the market, because they meet requirements (laid down in standards) that are important for customers
- Being successful with products, because they meet customers' wishes and are compatible with other products.

Standards may also have disadvantages for companies.

Within the company:

- Too much formalization may cause too much routine for employees, resulting in dissatisfaction.
- Standards for processes may hinder process improvement.
- Standards for products or services may hinder product innovation.

Outside the company:

- In the case the company is not able to meet the requirements in a standard, it will lose market share.
- In the case a competitor is better able to meet the requirements, the company faces a competitive disadvantage.
- Standards may make the market more transparent and in this way cause an increase in price competition, at the cost of profit margins.

De Vries (1999) suggested that the importance of standards is growing, because:

1. Companies, in general, can no longer be regarded as isolated organizations, not only in trade transactions but also in their technical operations. Especially in the area of information and communication technology (ICT), companies are connected to other companies. Also in other areas, technical specifications chosen by the company have to fit specifications of the company's environment. For example, ICT without standardization is impossible, and the chemical composition of petrol should not differ per country.
2. The tendency to concentrate on core business and to contract out other activities makes it necessary to agree with suppliers on, for instance, product specifications, product data, communication protocols, and the quality of the production and delivery processes. Because the company usually has several suppliers, each with several customers, the most profitable way to solve these matching problems is by using widely accepted standards.
3. There is a tendency to pay more attention to quality and environmental issues in a systematic way. This has increased the need for management systems standards, such as the ISO 9000 series of standards for quality management and the ISO 14000 series for environmental management. It has also increased the need for other standards, because management systems cause companies to perform activities in a structured way; standards for products, production means, and information systems contribute to the structure needed.
4. There is an increasing need to provide confidence to customers and other stakeholders, which can be partly achieved through certification. Certification is a procedure by which a third party gives written assurance that a product, process, or service conforms to specified requirements (ISO/IEC, 1996). In general, these requirements are laid down in standards.

Governments.   Most of the benefits and risks of standards for companies largely also apply for governments when governmental agencies use standards for improving business processes and performance. Additionally, the following issues seem to be important for governments (De Vries, 1999):

- Governments may support standardization as a part of the government's general role in stimulating business performance and international trade.
- Governments may supplement, simplify, or improve their legal system with standardization by making references to standards in laws.

- By standardization, the market may govern itself with less regulation from the government.
- International standards may help to remove trade barriers.

Also for governments, the importance of standards seems to be growing, because (De Vries, 1999):

1. Globalization of trade increases the need for international standardization.
2. Within the *European Union* and the *European Free Trade Association*, the choice for one single market without barriers to trade causes replacement of different national standards by European ones. This makes export for companies that export to several countries easier. There is no longer a need to produce different variants of products to meet different standards in different countries. However, companies that mainly serve national markets, especially in smaller countries, have been confronted with a substantial increase in the number of standards that are used.
3. Both at the European and at the national level there is a general tendency to link standards to legislation, in a way that standards provide detailed requirements that correspond to global requirements laid down in laws. This causes an increase in the number of standards and an increase in obligations to use them.

Consumers.   For consumers, standards may contribute to (ANEC, 2003):

- Accessibility/design for all
- Adaptability
- Consistent user interfaces
- Ease of use
- Functionality of solutions
- Service quality
- System reliability and durability
- Health and safety
- Environmental issues
- Information (product information, directions for use, and system status information)
- Reliability of information
- Privacy and security of information
- Interopability and compatibility
- Multicultural and multilingual aspects
- More transparency in the market, enabling consumers to better compare products and services from different suppliers on price, quality, and other characteristics

Additionally one can argue that standards also may lead to lower prices, due to economies of scale and increased price competition.

Classification

De Vries (1998) has provided several classifications of standards related to the contents of the standards, the intended users, or the process of developing the standards. For the purpose of this book we concentrate on content-related classifications.

Entities.   The content of a standard concerns entities or relations between entities. Thus, standards can be classified according to these entities. An entity may be: a person or group of persons; a "thing" such as an object, an event, an idea or a process. "Things" include

plants and animals. Gaillard (1933, p. 33) provides a rather complete list of possible entities; a combination of the first two types of entity (e.g., a car with a driver).
Matching problems can concern:

- Matching thing—thing (e.g., bolts and nuts)
- Matching man—thing (e.g., safety requirements)
- Matching man—man (e.g., procedures and management systems).

Horizontal, Vertical. Standardization organizations often distinguish between *horizontal* and *vertical* standards. Horizontal standards set general requirements for a collection of different entities, for instance, biocompatibility criteria for medical devices. Vertical standards set several requirements for one kind of entity, for instance, medical gloves.

Basic, Requiring, and Measurement. Distinctions can be made among basic standards, requiring standards, and measurement standards (De Vries, 1998).

- *Basic standards* provide structured descriptions of (aspects of) interrelated entities to facilitate human communication about these entities, or to be used in other standards. Examples are terminology standards, standards providing quantities and units, standards providing classifications or codes, and standards providing systematic data or reference models.
- *Requiring standards* set requirements for entities or relations between entities. These can include specifications of the extent to which deviations from the basic requirements are allowed. There are two subcategories: performance standards and standards that describe solutions.
  - *Performance standards* set performance criteria for the solution of matching problems. They do not prescribe solutions. Performance standards can include specifications of the extent to which deviations from the basic requirements are permissible.
  - *Solution-describing standards* describe solutions for matching problems.
    In general, companies and other stakeholders in standardization prefer performance standards rather than standards that describe solutions (Le Lourd, 1992, p. 14), but most developing countries prefer descriptive standards with a large number of technical details (Hesser & Inklaar, 1997, p. 38). The percentage of performance standards is growing, at the expense of standards that prescribe certain solutions.
- *Measurement standards* provide methods to be used to check whether requiring standards criteria have been met.

Interference, Compatibility, and Quality. Simons (1994) distinguishes among interference standards, compatibility standards, and quality standards:

- *Interference standards* set requirements concerning the influence of an entity on other entities. Examples are standards on safety, health, environmental issues, and EMC (electromagnetic compatibility with respect to electrical disturbances). Companies often have to use interference standards because of governmental requirements. They, therefore, have no choice: They must use them.
- *Compatibility standards* concern fitting of interrelated entities to one another, in order to enable them to function together, for example, specifications for films and cameras and cell phone specifications. Choices regarding compatibility standards are often commercial decisions that can have direct impact on market share. Thus, although the choices are up to the company, it is often the market situation that strongly influences these choices.

- *Quality standards* set requirements for entity properties to assure a certain level of quality. Examples of quality standards include ISO 9000 quality management standards, film sensitivity standards (to enable standard film processing), measurement standards, and standards for company procedures. Quality standards are often related to the company's business operations.

Standardization

Definition

The activity of making standards may be called standardization. *Standardization concerns establishing and recording a limited set of solutions to actual or potential matching problems directed at benefits for the party or parties involved and intending and expecting that these solutions will be repeatedly or continuously used during a certain period by a substantial number of the parties for whom they are meant. A matching problem should be understood as a problem of interrelated entities that do not harmonize with each other; solving it means determining one or more features of entities in a way that they harmonize with one other or of determining one or more features of an entity because of its relation with other entities* (De Vries, 1997).

Developers

The main stakeholders in most standardization projects are manufacturers of products and services, the professionals that support the development of these products and services, and the customers that buy the products and services. Other stakeholders include the organizations that represent these stakeholders (e.g., branch of business organizations, professional societies, and consumer organizations), governmental agencies, organizations for testing and certification, consultancy firms, research institutes, universities, and special interest groups, such as trade unions and environmental pressure groups. Many times, the manufacturers dominate the standardization.

Standards can be developed by companies, industrial consortia, branch of business organizations, professional associations, governmental agencies or, last but not least, formal standardization organizations. In the remaining part of this chapter, we concentrate on formal standardization. Three levels of formal standardization can be distinguished: national, regional, and global.

National Standardization.   More than 150 countries have a national standardization organization (NSO). Well-known examples are Association Française de Normalisation (AFNOR), American National Standards Institute (ANSI), British Standards Institution (BSI), Deutsches Institut für Normung, Germany (DIN), Japanese Industrial Standards Committee (JISC), Standardization Administration of China (SAC), and Standards Australia. In Northwest and Central Europe and in Italy, Australia, the United States and Japan, such organizations were founded in the first decades of the 20th Century. Before and shortly after the Second World War, India, South Africa, and several South American and other European countries followed. More than 100 other NSOs are less than 30 years old (Toth, 1997). Most of the first and second generation of NSOs were founded by people from industry with the aim to establish national standards. The NSOs provided the platform for national stakeholders to agree on standards and published the standards.

NSOs usually distinguish the following steps for the development of national standards: (a) request, (b) assignment to a committee, (c) drafting, (d) public comment, (e) review of comments, (f) approval, (g) publishing, (h) publicity, (i) implementation, and (j) evaluation.

The following description is mainly based on a Dutch publication (NEN, 2004a, pp. 12–13), with additions from British Standard 0-2 (BSI, 1997, pp. 16–25). Terms have been taken from BS 0-2; other NSOs may use different terms. Procedures for steps 1 to 8 may slightly differ per NSO but are similar to NEN and BSI procedures. Steps 9 and 10 are up to standards users, though NSO procedures require evaluation of standards after a certain period.

1. *Request.* Any company, organization, person, or the NSO itself can indicate a need for new standards, or for improvements in existing standards. Most NSOs have Sector Boards that are responsible for dealing with such requests. They may ask advice from a standardization committee or have a feasibility study carried out. To be able to decide on the proposal for a new work item, they need justification, including:
   ○ Reasons for standardization: What advantages can be expected for whom?
   ○ The topic of standardization: Is it (technically) convenient for standardization?
   ○ The amount of support in the market, including willingness to finance the project
   ○ Reasons, if any, to standardize at the national level, instead of at the—preferred—international or regional level.
   The title and scope of the standard need to be clear; when necessary, further specifications may be added, as well as a schedule for development.

2. *Assignment to a committee.* The Sector Board will decide on the proposal. When positive, the new work item will be assigned to a technical committee (TC). When no TC exists, a new one may be set up. A TC may establish a subcommittee (SC) or working group (WG) to handle the topic. A WG exists only for the time necessary to draft one or more standards. When the standards are ready, the WG is discontinued. An SC is a more permanent committee, responsible for a field of activities. This, however, does not exclude the SC from being disbanded. Interested parties are invited to get involved. Additionally, the new work item and the establishment of a new TC, SC, or WG, if any, is publicly announced, so that representatives of all interested parties have the opportunity to join the committees. Often, in the first meeting its composition is discussed, and organizations that were overlooked still get an invitation to participate. In general, the committee one step higher in the hierarchy has to agree on this.

3. *Drafting.* The responsible TC, SC, or WG prepares a first draft, based on professional expertise, deliberations, and consensus. Sometimes, research is carried out to obtain data to be used in the standard. Often, the committee initiates the discussion making use of an initial document brought forward by one of the participants. The standard may require testing, to be carried out either by participants or, as part of a special program set up for the project, by a testing organization. When the TC has delegated the work to an SC or WG, their approval is followed by the TC's approval. After this, NSO employees check the draft on conformity to standards in adjacent areas, and to the NSOs rules for the drafting and presentation of standards. Some NSOs have laid down these rules in national "standards for standards." Examples include the American national standard SES 1 (SES, 1995), British standard BS 0-3 (BSI, 1997b), the French standard X 00-001 (AFNOR, 1993), and the German standard DIN 820-2 (DIN, 1996).

4. *Public comment.* The draft standard is published for comments. It is announced in the NSOs regular media, such as its monthly magazine and its Web site, and in selected media, such as specialists' journals. Sometimes, copies are sent to experts and to interested parties not represented in the committees. Other NSOs are notified and get the opportunity to comment. The party that requested the standard, if not a committee member, also gets a copy. Parties get a certain amount of time, mostly a few months, to give their comments, if any.

5. *Review of comments.* The TC, or its SC or WG, discusses the comments and uses them to improve the standard. Sometimes, major contributors are invited in a meeting

to discuss their comments with the committee. All contributors are informed about the committee's decision on their comments.

6. *Approval.* The TC decides on the proposed standard. NSO officers check it again for conformity to standards in adjacent areas and to rules for drafting and presentation of standards. If it is decided that the standard is still appropriate but the content or structure of the document is changed significantly as a result of comments received, a second draft for public comment may be issued.
7. *Publishing.* The NSO does the final editing and publishes the standard.
8. *Publicity.* The NSO uses its own media, such as its journal or Web site, to announce the new standard. Additionally, press releases are sent, media events organized, seminars or courses held, and so forth.
9. *Implementation.* Of course it is up to the companies and other organizations to implement the standard. Some NSOs provide support for this by means of written guidelines, courses, or advice.
10. *Evaluation.* Most NSOs review their standards after 5 years. The responsible committee then decides to withdraw, revise, or to maintain them unchanged. Of course, market developments may be a cause for earlier revision. Evaluation of standards is a problem when the responsible TC has been disbanded. Then the NSO Bureau has to decide whether or not the standards should remain in the collection.

Nowadays, the role of NSOs has shifted to the preparation of the national standpoint in international standardization. In the field of electrotechnology, there has been international standardization from the outset, in 1906. Often, international standards are adopted as national standards, whether or not translated into the national language. The NSO provides the national selling point for such standards. Additionally to their role in standards development and standards selling, many NSOs serve their national market by providing information related to standards in the form of, for instance, informative books or courses. More than 50% of the NSOs provide activities in the field of certification. In many developing countries, the institutes also play a role in accreditation, testing, and metrology (De Vries, 1999).

International Standardization: Global.   Formal standardization at the global level comprises the International Telecommunication Union (ITU), the International Electrotechnical Commission (IEC) and the International Organization for Standardization (ISO). ISO and IEC have been established by NSOs and, in turn, recognize only one NSO per country. This is a main reason why countries with lots of sectoral standardization organizations, such as Japan, Norway, and the United States, have also an NSO. The NSO has to be intermediary between the national stakeholders and the global and regional standardization arena. Therefore, NSOs tend to form national TCs that more or less reflect the committee structure of ISO/IEC.

ISO and IEC have a hierarchy of committees for standards development and approval. The "lowest" level are working groups (WG), where experts develop standards. A WG develops one or a few standards and can then be disbanded. Decision making on standards is done in Technical Committees (TCs). Many TCs have delegated a part of their activities to an in-between level: Subcommittees (SCs). Participants in TCs and SCs are national delegates, whereas NSO committee members represent national stakeholder groups. Delegates are expected to speak and act on behalf of their country, not on behalf of their company or stakeholder group. NSO committees formulate the national standpoints for voting or for input by participation in committees at the international level. In national standardization, decisions are based on consensus. ISO and IEC use the consensus principle within committees that draft standards, but voting also takes place at several crucial stages of standards development. NSOs vote on

TABLE 2.1
Regional Formal Standardization Organizations

Region	Organization
Africa	African Regional Organization for Standardization (ARSO)
	Common Market for Eastern and Southern Africa (COMESA)
Arab countries	Arab Industrial Development and Mining Organization (AIDMO)
	Standardization and metrology organization for the Gulf Cooperation Council (GSMO)
Europe	Comité Européen de Normalisation (CEN)
	Comité Européen de Normalisation Electrotechnique (CENELEC)
	European Telecommunication Standards Institute (ETSI)
Most countries of the former Soviet Union	Euro-Asian Council for Standardization, Metrology and Certification (EASC)
South Asia	South Asian Association for Regional Cooperation (SAARC)
Southeast Asia	ASEAN Consultative Committee for Standards and Quality (ACCSQ)
Asia/Pacific	Pacific Area Standards Congress (PASC)
North America	NAFTA Committee on Standards-Related Measures (CSRM)
Caribbean	Caribbean Common Market Standards Council (CCMSC)
South America	Comisión Panamerican de Normas Técnicas (COPANT)
	Comité MERCOSUR de Normalización (CMN)

behalf of their country. The main stages in developing an ISO standard can be found in ISO (2004).

International Standardization: Regional. In many regions of the world, special standardization organizations have been established (see Table 2.1).

Most of these organizations do not develop their own standards but provide a forum for discussing standards matters of common interest. And the list is not complete—other, more informal, forms of cooperation exist, for example, among the Baltic countries, among Turkey and its neighboring countries, and among China, Japan, and South Korea. In one region there is a huge amount of standards development: Europe. CEN, CENELEC (electrotechnical standardization), and ETSI (standardization in the field of telecommunication) have developed thousands of standards. The standards development processes of CEN are almost identical to those of ISO. A main difference is that in ISO each member country had one vote, no matter its size, whereas CEN operates a system of weighted voting. A second difference is that once a European standard has been adopted, CEN members must implement it by giving it the status of a national standard, either by publication of an identical text or by endorsement, and by withdrawing any conflicting preexisting national standards, no matter whether they agree to the contents of this European standard. In the case of ISO standards, they are free whether or not to implement a standard. Though in the case of European standards it is obligatory to adopt them as national standards, this does not imply that the use of this standard is obligatory as well. In principle, ISO as well as CEN standards and national standards are all voluntary standards. Only through reference in legislation or through market forces they may become more or less compulsory.

Combined Development of European and International Standards. The political wish to have one common European market without barriers to trade has caused an enormous growth in European standardization. Most European standards are identical to the international ones, but they are developed for the simple reason that they have to be implemented into the national standards systems of all EU and EFTA member countries, whereas this does not apply for ISO standards. So in fact, the same standards are developed within two committees, one at the global and one at the European level. Of course, this is not efficient and, therefore, CEN and ISO have agreed to establish the possibility of a tight cooperation in standards development. In this "Vienna Agreement" (IEC and CENELEC have a comparable agreement called "Dresden Agreement"), CEN and ISO have agreed on four main methods of achieving common standards (ISO/CEN, 1995; Smith, 1995):

1. Adoption of existing ISO standards by CEN, using a short procedure within CEN. If it seems likely that an ISO standard will need to be revised to meet Europe's requirements, the ISO committee will be given the opportunity to consider whether it is prepared to revise the standard.
2. Submission of work items developed by ISO to parallel approval procedures in ISO and CEN
3. Submission of work items developed by CEN to parallel approval procedures in CEN and ISO
4. Submission of European standards for adoption by ISO, using a short procedure within ISO

One might argue that such cooperation strengthens Europe's position within ISO. Because of the need to arrive at common European standards, European countries might join forces in ISO. Moreover, they have the majority in more than 50% of all ISO committees. In this way, Europe could impose its standards to the rest of the world. In some exceptional cases this has occurred indeed, but almost always it appears that the Europeans do not agree with one another and that the controversies in international standardization are not between Europe and the rest of the world.

European Legislation.   In some cases, there is a close relation between standards and legislation. At the European level this applies especially to the so-called *New Approach*. In the "Old Approach," member states imposed their own technical specifications and conformity controls for manufactured products. Any technical harmonization across the European Union relied on agreeing directives for individual products. The requirements in these directives then had to be implemented into the legal systems of the member states. European directives set detailed requirements, mainly on product safety. It took many years to agree on the safety requirements, and often the directive was outdated already once it was ready. Therefore, it was decided to replace this by a New Approach in which the directives set only the essential requirements on, for instance, safety, health, or environment, and these requirements are formulated globally. Linked to these directives, European standards are developed in which detailed requirements or test methods are laid down. A company that meets the relevant standards is assumed to meet the general requirements set in the directives. Thus, implementing the standards is an efficient way to meet the legal requirements. The company, however, is allowed to meet these requirements in another way. Though principally voluntary, in practice, these standards are almost obligatory. Conformity to requirements in the directives is indicated by means of the CE mark (CE = *Conformité Européenne*; see Leibrock, 2002, and Huigen, Inklaar, & Paterson, 1997).

Stakeholder Participation

An important issue in standardization is the involvement of stakeholders in the process of developing standards. In practice, not all relevant stakeholders are involved in standardization, and there can even be one-sided representation (De Vries et al., 2003). In most TCs, industry is overrepresented. As a result, the developed standard may not be acceptable for other relevant potential users. De Vries, Verheul, and Willemse (2004) have listed 27 barriers for participation of stakeholders in formal standardization.

In order to avoid underrepresentation of a relevant stakeholder, a model has been developed to identify and select relevant stakeholders to be involved in standardization (Willemse, 2003). This model has been applied to ergonomics standardization (Dul, Willemse, & De Vries, 2003). In Chapter 20 of this book, the model and the application to ergonomics standardization are presented (Willemse, Dul, & De Vries, 2005).

PART II. ERGONOMICS STANDARDS AND STANDARDIZATION

During 2 million years of human history, implicit knowledge has been used to fit natural tools and artifacts to the needs of the individual human being. After the Industrial Revolution, which introduced mass production, a need arose for explicit knowledge about the interactions between the human and the complex man-made environment (Dul, 2003a). From this need, the discipline of ergonomics emerged. Product ergonomics developed as a field that can fit mass products to the individual consumer needs and characteristics. Production ergonomics developed as a field that can fit complex technical-organizational systems to the needs and characteristics of workers. In 1949, the first ever national ergonomics society was established in the United Kingdom, at which moment the name "ergonomics" was coined to name the field. This was followed in 1961 by the creation of the International Ergonomics Association (IEA), which nowadays represents some 19,000 ergonomics scientists and practitioners world-wide. In the early 1970 s, the IEA decided to initiate the development of ergonomic standards (Parsons & Shackel, 1995) and requested the ISO to start ergonomics standardization. The aim was to design products and (production) processes to optimize well-being and performance of consumers and workers interaction with the products and production systems. Since that time, a large number of ergonomics standards have been developed for different types of products, production systems, and work environments.

Ergonomics Standards

Definition

Ergonomics is a broad scientific and professional field, which is described by the IEA as follows: *Ergonomics (or human factors) is the scientific discipline concerned with the understanding of interactions among humans and other elements of a system, and the profession that applies theory, principles, data and methods to design in order to optimize human well-being and overall system performance.* In this chapter, we limit ourselves to two types of ergonomics standards: global standards from ISO and regional (European) standards from CEN. The reason is that it appears that most ergonomics standards are developed by these organizations. Figure 2.1 shows the growth of the number of ISO and CEN standards over the years.

Ergonomics standards can be also developed by companies, consortia of companies or industry branch organizations, but these standards, if available, are not considered here. We estimate that in other fields of standards (e.g., electrotechnology), the number of standards from companies or consortia is considerably larger than that in the ergonomics field. We also estimate that several standards from companies and branches of industry organizations are, at least partly, based on formal standards from ISO and CEN.

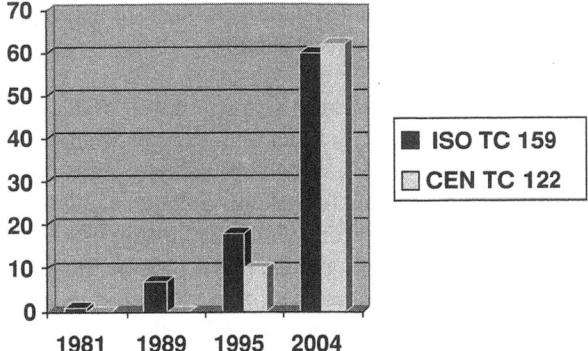

FIG. 2.1. The number of ergonomics standards developed by ISO TC159 "Ergonomics" and CEN TC122 "Ergonomics" (After: Dul et al., 1996, 2004; Metz, 1991).

Within ISO and CEN, we consider only standards that have been developed by the technical committees ISO TC 159 "Ergonomics", and CEN TC 122 "Ergonomics," although ergonomics standards have also been developed by other technical committees (Dul, De Vlaming, & Munnik, 1996).

Hence, in the remainder of this chapter, the definition of ergonomics standards is *formal standards developed by ISO TC 159 or CEN TC 122.*

Users

In some cases, potential users of ergonomics standards can be identified by looking at the field of application mentioned in these standards. In the "mother standard" of all ergonomics standards, many potential users are mentioned. The "mother standard" is the first ergonomics standard. It was published in 1981 as ISO 6385 "Ergonomic principles in the design of work systems," and revised in 2003 (Eveleens, 2003). This standard describes the design steps for a (work) system and the required ergonomics inputs in order to design an ergonomic system:

1. Formulation of goals of the system
2. Analysis and allocation of functions between human and technology
3. Design of the concept of the system
4. Detailed design (the term *work* is used in a broad sense)
   ◦ Design of work organization
   ◦ Design of work tasks
   ◦ Design of jobs
   ◦ Design of work environment
   ◦ Design of work equipment, hardware, and software
   ◦ Design of workspace and workstation
5. Realization, implementation, and validation.

This standard mentions the following users: "users of this standard will include managers, workers (or representatives), professionals such as ergonomists, project managers and designers who are involved in the design of work systems."

Direct Users. A party (person, group of people, organization) that reads an ergonomics standard, and uses it for the professional activities, is considered as a direct user of the ergonomics standard. In particular product and production system designers, testers and advisors are potential direct users of ergonomics standards.

Designers.   The objective of ergonomics is to contribute to the *design* of the technical and organizational environment. Hence, product and process designers are important direct users. Many ergonomics standards are formulated in general terms because the specific requirements depend on the organizational context in which the standard is used. For example, the requirements for the maximum mass for safe manual lifting of a load also depend on the lifting frequency and duration, and other organizational factors. Some ergonomics scientists emphasize that the body of scientific knowledge in certain areas is not yet sufficient to formulate specific standards to limit the exposure to risk factors. For example, Fallentin et al. (2001) evaluated standards that limit the physical workload and concluded that "the scientific coherency of specific quantitative criteria was limited, whereas general process-type standards were more favorable."

However, such generally formulated standards cannot be readily applied by designers (Dul, De Vries, Verschoof, Eveleens, & Feilzer, 2004). Wulff, Westgaard, and Rasmussen (1999) found, for instance, that designers may not understand the general ergonomic recommendations, do not know how to make them concrete in specific situations, or do not consider them important enough if they are in conflict with other design requirements. Then, ergonomic advisors can assist the designers to apply ergonomics standards in specific situations. Most standards other that ergonomics standards, are product standards with specific requirements on technical issues, and these can be more simply implemented by designers.

Testers.   Several parties use ergonomic standards to test whether products or processes meet ergonomics requirements. For example, testers can judge whether certain products (e.g., hand tools, VDU-screens, and chairs) meet the ergonomics requirements formulated in a standard. Also, product standards can be used by consumer organizations to judge the ergonomic quality of a consumer product. Occupational Health Services or the Labor Inspectorate can use ergonomics standards (e.g., on physical and mental workload) to evaluate production systems and work environments for performing a risk assessment of the work situation.

Recently, the International Ergonomics Association has taken an initiative to stimulate testing of the ergonomics quality of products and processes. The aims of this so-called Ergonomics QUality In Design (EQUID) initiative are:

- To define process criteria and requirements for the ergonomic design of products, work systems, and services
- To define a system for accrediting certifying bodies that will assess the ergonomics quality in design, using the relevant criteria and requirements
- To design, implement, and manage a system for regularly assessing and updating the process requirements for the ergonomic design of products, work systems, and services
- To design, implement, and manage a system for regularly evaluating and improving the accreditation program.

In this initiative, the product or process itself is not tested, but whether the ergonomic input during the development process of the product was appropriate. This test plan in ergonomics deviates from the testing practice in most other fields, where many standards provide methods for testing products, services, or processes.

Advisors.   The initiative to develop ergonomic standards came from the International Ergonomics Association, and ergonomics experts have been heavily involved in the development of ergonomics standards. Hence, in contrast to general standardization that is dominated by manufacturers of products and services, ergonomics standards seem to be dominated by ergonomics experts and may primarily reflect the professional insights about what can be considered as "good ergonomics." Putting ergonomics knowledge into standards may have supported the transfer of ergonomics knowledge and may have given status to the ergonomics discipline. Ergonomics advisors are probably the most important direct users of ergonomics standards. Ergonomists can use the standards when contributing to the design or testing of products and (production) processes within companies and other organizations. A review among certified European ergonomists (Breedveld, 2005) showed that 69% of these ergonomists regularly or always use international ergonomics standards in their work during the last year, whereas only 3% never used such standards during the last year. Most of the ergonomists were internal or external consultants, giving ergonomics advice to companies.

Regulators.   Certain organizations may use ergonomics standards for regulatory purposes. Industrial branch organizations may adopt or adapt ergonomics standards for their branch of industry. Trade unions can use ergonomics standards to develop own guidelines to be used for the interest of the workers. Governments can be a direct user of ergonomics standards, when using the standard in for example health and safety regulation. In Europe, several ergonomics CEN standards are related to European legislation to stimulate free trade within Europe, by setting the same minimum health and safety requirements for machinery produced in European countries. In particular, the Safety of Machinery Directive of 1998 (European Union, 2003) formulates general requirements for health and safety related to machinery, and refers to CEN standards for specific requirements (see subsection List of CEN Standards Related to Legislation). Other European directives that have a relationship to ergonomics are, for example, Use of Work Equipment Directive (European Union, 1989), Manual Handling Directive (European Union, 1990a), and the Work With Display Screen Equipment Directive (European Union, 1990b). These directives are part of the European social policy on worker protection, but no reference is made to specific CEN standards.

Indirect Users.   In ergonomics standards, by definition, end users of products (e.g., consumers) and end users of (production) processes (e.g., workers) are essential entities: One of the objectives of ergonomics is to contribute to the well-being of the end users (see previous definition of ergonomics). End users normally do not read standards, but the result of applying standards is important for them. In many design teams, the ergonomist can be considered as the "voice" of the end user.

Other indirect users of ergonomics standards can be branch of business organizations (for the common interests of their members) and trade unions (for the interests of workers), when not using standards in a direct way as described earlier.

In the next sections we discuss the interests of the main direct and indirect users of ergonomics standards: companies, governments, and workers and consumers.

Companies.   Companies can have interest in applying ergonomics, because integration of ergonomics in the design of products and processes can contribute to social and economic business goals, and strengthen the company's competitive advantage (Dul, 2003a, 2003b). When ergonomics standards are used to integrate ergonomics in the design of (production) processes, worker satisfaction and worker motivation may improve, sickness absence may reduce, and the worker performance can increase in terms of reduction of human error (quality) and increase of human output (productivity; see, for example, Breedveld, 2004, and Dul et al., 2004).

Ergonomics standards can also contribute to the development of user-friendly products, which better meet customer demands and improve customer satisfaction. Other reasons for companies to apply ergonomics standards are that standards can help to prevent occupational health and safety problems, or are useful means to obey to Occupational Health and Safety regulations. Also, companies may want to meet ergonomics standards in order to demonstrate to customers and other stakeholders "good social practice" and social responsibility, both in the products that are sold, as well as in the (production) processes.

In contrast to most other standards, where economic goals are the major driver, ergonomics standards may be primarily used for the social dimension in terms of worker's satisfaction, health and safety, without disregarding the economic dimension (Dul et al., 2004).

Governments Other Than Regulator.   Governments can be users of ergonomics standards in another role than regulator (see subsection List of CEN Standards Related to Legislation). Similarly to companies, governments have also the role of employer, and can use ergonomics standards to improve employee satisfaction and improve service quality (Cook et al., 2003). Also similarly to companies, governmental agencies deliver products or services that may benefit from the use of ergonomics standards in order to improve the satisfaction of citizens. Governments can use other instruments than legislation, to simulate the development and application of ergonomics standards as a contribution to better health and safety of the population. For example in Germany, the government supports the Commission for Occupational Health and Safety and Standardization (KAN, 2004), which observes the standardization process, and ensures that standards makers devote sufficient attention to Occupational Health and Safety. Members of KAN are employer and employee organizations, the State, the statutory accident insurance institutions, and the German national standardization institute DIN.

Within a government, ministries of social affairs, health or labour may be interested to stimulate the development and use of ergonomics standards. However, in case of standards in general, in particular ministries of economic affairs may be interested to stimulate standardization.

Workers and Consumers.   When ergonomics standards are applied to production systems, workers can expect that working conditions and occupational health and safety criteria are taken into account. Unions and work councils within organizations can represent workers when working conditions are evaluated with ergonomics standards.

When products are developed according to ergonomics standards, customers can expect that besides functionality, also health and safety and user-friendliness are assured. Consumer organizations represent consumers when testing products based on ergonomics standards.

With ergonomics standards, both workers and consumers can benefit from the use of ergonomics standards in production systems and products. Other standards are primarily beneficial for suppliers and their customers. The latter may be consumers or professional users.

List of ISO Standards

Since the first ergonomics ISO standard was published in 1981, the number of ISO ergonomics standards has increased rapidly (see Figure 2.1 and Table 2.2). Tables 2.3 and 2.4 show ergonomics ISO standards that have been published, or are in preparation by ISO TC 159 (Dul et al., 2004). Certain standards show up both as a published standard and as a standard in preparation. Then, the standard in preparation is a revision of the published standard. After publication, it will replace the existing standard. In Table 2.5, the ISO standards are organized according to ergonomics topics. This table shows that most standards are developed for the topics "Physical Environment" and "Visual information, VDTs and software."

TABLE 2.2
ISO Standards on Ergonomics

	1989 (Metz, 1991)	*1995 (Dul et al., 1996)*	*2004 (Dul et al., 2004)*
ISO published	7	18	60
ISO in preparation	14	31	25
Total	21	49	85

List of CEN Standards

Since the first ergonomics CEN standard was published in 1990 as ENV 26385, which was an adoption of the above ISO 6385, the number of CEN standards increased considerably (see Figure 2.1 and Table 2.6). Tables 2.7 and 2.8 show ergonomic CEN standards that have been published or are in preparation by CEN TC 122 (Dul et al., 2004). It turns out that several CEN and ISO standards are identical. This is a result of a policy of CEN and ISO to harmonize the development of their standards, according to the *Agreement on technical cooperation between ISO and CEN (Vienna Agreement)* of 1991 (CEN, 2004; ISO/CEN, 1991). The topics covered in CEN standards are presented in Table 2.9. Most standards are developed for the topics "Safety of Machinery," "Workplace and Equipment Design," "Physical Environment" and "Visual Information, VDTs, and Software."

List of CEN Standards Related to Legislation.   As stated earlier, certain CEN standards have a legal status. In the so-called *New Approach*, several CEN standards are related to legislation formulated in *European Directives*. For the ergonomics field, certain standards are related to the Machinery Directive 98/37/EC (EU, 2003). This Directive puts generally formulated essential requirements on safety and health when using machinery. Linked to this directive the CEN standards shown in Table 2.10 give detailed requirements. A company that meets these standards is assumed to meet the general requirements set in the Directive. Apart from ergonomics standards that are related to legislation, ergonomics standards can also be useful for governments to stimulate voluntary actions in the field of Occupational Health and Safety. Because ergonomics in general, and ergonomics standards in particular, have (by definition, see subsection Definition) both a social AND an economic goal (Dul, 2003b), governments have two reasons to stimulate the use of ergonomics standards. By emphasizing this dual goal, ergonomics standards may be a positive incentive for companies for designing both healthy and efficient production systems.

The obligatory implementation of European standards may also lead to problems. An example is the European standard EN 1335-1 "Office furniture—Office work chair—Part 1: Dimensions—determination of dimensions" (CEN, 2000). Although this standard was not developed by CEN TC 122 and is therefore not listed in one of the tables presented here, the standard includes some important ergonomics issues, such as specification for the adjustment of the seat height. Because Dutch people are on average the tallest people in the world, and the population also includes small people due to its cultural diversity, the range of motion for height adjustments specified in the European standard is too small for the Dutch population. The Dutch TC on office furniture has developed a deviating Dutch guideline for a Dutch work chair, which provides guidance on how to tackle this situation (NEN, 2004b).

TABLE 2.3
Published Standards From ISO TC 159

ISO 6385:2004 Ergonomic principles in the design of work systems

ISO 7243:1989 Hot environments—Estimation of the heat stress on working man, based on the WBGT index (wet bulb globe temperature)

ISO 7250:1996 Basic human body measurements for technological design

ISO 7726:1998 Ergonomics of the thermal environment—Instruments for measuring physical quantities

ISO 7730:1994 Moderate thermal environments—Determination of the PMV and PPD indices and specification of the conditions for thermal comfort

ISO 7731:2003 Danger signals for work places—Auditory danger signals

ISO 7933:1989 Hot environments—Analytical determination and interpretation of thermal stress using calculation of required sweat rate

ISO 8996:1990 Ergonomics—Determination of metabolic heat production

ISO 9241-1:1997 Ergonomic requirements for office work with VDTs—Part 1: General introduction

ISO 9241-2:1992 Ergonomic requirements for office work with VDTs—Part 2: Guidance on task requirements

ISO 9241-3:1992 Ergonomic requirements for office work with VDTs—Part 3: Visual display requirements

ISO 9241-4:1998 Ergonomic requirements for office work with VDTs—Part 4: Keyboard requirements

ISO 9241-5:1998 Ergonomic requirements for office work with VDTs—Part 5: Workstation layout and postural requirements

ISO 9241-6:1999 Ergonomic requirements for office work with VDTs—Part 6: Guidance on the work environment

ISO 9241-7:1998 Ergonomic requirements for office work with VDTs—Part 7: Requirements for display with reflections

ISO 9241-8:1997 Ergonomic requirements for office work with VDTs—Part 8: Requirements for displayed colors

ISO 9241-9:2000 Ergonomic requirements for office work with VDTs—Part 9: Requirements for non-keyboard-input devices

ISO 9241-10:1996 Ergonomic requirements for office work with VDTs—Part 10: Dialogue principles

ISO 9241-11:1998 Ergonomic requirements for office work with VDTs—Part 11: Guidance on usability

ISO 9241-12:1998 Ergonomic requirements for office work with VDTs—Part 12: Presentation of information

ISO 9241-13:1998 Ergonomic requirements for office work with VDTs—Part 13: User guidance

ISO 9241-14:1997 Ergonomic requirements for office work with VDTs—Part 14: Menu dialogues

ISO 9241-15:1997 Ergonomic requirements for office work with VDTs—Part 15: Command dialogues

ISO 9241-16:1999 Ergonomic requirements for office work with VDTs—Part 16: Direct manipulation dialogues

ISO 9241-17:1998 Ergonomic requirements for office work with VDTs—Part 17: Form-filling dialogues

ISO 9355-1:1999 Ergonomic requirements for the design of displays and control actuators—Part 1: Human interactions with displays and control actuators

ISO 9355-2:1999 Ergonomic requirements for the design of displays and control actuators—Part 2: Displays

ISO 9886:2000 Evaluation of thermal strain by physiological measurements

ISO 9920:1995 Ergonomics of the thermal environment—Estimation of the thermal insulation and evaporative resistance of a clothing ensemble

ISO 9921:2003 Ergonomic assessment of speech communication—Part 1: Speech interference level and communication distance for persons with normal hearing capacity in direct communication (SIL method)

ISO 10075:1991 Ergonomic principles related to mental workload—General terms and definitions

ISO 10075-2:1996 Ergonomic principles related to mental workload—Part 2: Design principles

ISO 10551:1995 Ergonomics of the thermal environment—Assessment of the influence of the thermal environment using subjective judgment scales

ISO 11064-1:2000 Ergonomic design of control centers—Part 1: Principles for the design of control centres

ISO 11064-2:2000 Ergonomic design of control centers—Part 2: Principles for the arrangement of control suites

ISO 11064-3:1999 Ergonomic design of control centers—Part 3: Control room layout

ISO/TR[a] 11079:1993 Evaluation of cold environments—Determination of requisite clothing insulation (IREC)

ISO 11226:2000 Ergonomics—Evaluation of static working postures

ISO 11228-1:2003 Ergonomics—Manual Handling—Part 1: Lifting and carrying

ISO 11399:1995 Ergonomics of the thermal environment—Principles and application of relevant International Standards

ISO 11428:1996 Ergonomics—Visual danger signals—General requirements, design, and testing

ISO 11429:1996 Ergonomics—System of auditory and visual danger and information signals

(Continued)

TABLE 2.3

(Continued)

ISO 12894:2001 Ergonomics of the thermal environment—Medical supervision of individuals exposed to extreme hot or cold environments

ISO 13406-1:1999 Ergonomic requirements for work with visual displays based on flat panels—Part 1: Introduction

ISO 13406-2:2001 Ergonomic requirements for work with visual displays based on flat panels—Part 2: Ergonomic requirements for flat panel displays

ISO 13407:1999 Human-centered design processes for interactive systems

ISO 13731:2001 Ergonomics of the thermal environment—Vocabulary and symbols

ISO/TS[a] 13732-2:2001 Ergonomics of the thermal environment—Methods for the assessment of human responses to contact with surfaces—Part 2: Human contact with surfaces at moderate temperature

ISO 14738:2002 Safety of machinery—Anthropometric requirements for the design of workstations at machinery

ISO 14915-1:2003 Software ergonomics for multimedia user interfaces—Part 1: Design principles and framework

ISO 14915-2:2003 Software ergonomics for multimedia user interfaces—Part 2: Multimedia navigation and control

ISO 14915-3:2003 Software ergonomics for multimedia user interfaces—Part 3: Media selection and combination

ISO 15534-1:2000 Ergonomic design for the safety of machinery—Part 1: Principles for determining the dimensions required for openings for whole-body access into machinery

ISO 15534-2:2000 Ergonomic design for the safety of machinery—Part 2: Principles for determining the dimensions required for access openings

ISO 15534-3:2000 Ergonomic design for the safety of machinery—Part 3: Anthropometric data

ISO 15535:2003 General requirement for establishing anthropometric databases

ISO/TS 16071:2003 Ergonomics of human-system interaction—Guidance on accessibility for human-computer interfaces

ISO/TR 16982:2002 Ergonomics of human–system interaction—Usability methods supporting human-centered design

ISO/TR 18529:2000 Ergonomics—Ergonomics of human–system interaction—Human–centered life-cycle process descriptions

ISO/TR 19358:2002 Ergonomics—Construction and application tests for speech technology

---

[a]TR, Technical Report; TS, Technical Specification.

Classification

Entity. Because ergonomics deals with the interaction between a person and the environment, a major characteristic of all ergonomics standards is that the person is always one of the entities concerned. Some standards only deal with the entity person (e.g., human body measurements), other standards deal with both the person and the technical or organizational environment, in particular when matching person and "thing." In most other standards the matching entities concern "things," for example, a product or a part thereof.

Horizontal, Vertical. The current set of ergonomics standards contains both "horizontal" (general) and "vertical" (specific) standards. ISO 6385 on ergonomic design principles is an example of a horizontal standard. Examples of "vertical standards" are ISO 9241 on ergonomic requirements for visual display terminals, and ISO 11064-3 on control room layout. Also in the set of other standards, both vertical and horizontal standards can be found.

In the environmental field, additional to (horizontal) environmental standards (the ISO 14000 series of standards for environmental management and hundreds of standards that specify methods to measure pollution), initiatives have been taken to get more attention for

TABLE 2.4
ISO Standards from ISO TC 159, in preparation

ISO/DIS[a] 7730 Ergonomics of the thermal environment—analytical determination and interpretation of thermal comfort using calculation of the PMV and PPD indices and local thermal comfort
ISO/FDIS[a] 7933 Ergonomics of the thermal environment—analytical determination and interpretation of heat stress using calculation of the predicted heat strain
ISO/FDIS 8996 Ergonomics of the thermal environment—Determination of metabolic heat rate
ISO/CD[a] 9241-10 Ergonomics of human system interaction—Part 10: Dialogue principles
ISO/9920 Ergonomics of the thermal environment—Estimation of the thermal insulation and evaporative resistance of a clothing ensemble
ISO/CD 10075-3 Ergonomic principles related to mental workload—Part 3: Measurement and assessment of mental workload
ISO/CD 11064-4 Ergonomic design of control centres—Part 4: Layout and dimensions of workstations
ISO/CD 11064-6 Ergonomic design of control centres—Part 6: Environmental requirements
ISO/CD 11064-7 Ergonomic design of control centres—Part 7: Principles for the evaluation of control centers
ISO/CD 11079 Evaluation of the thermal environment—Determination and interpretation of cold stress when using required clothing insulation (IREQ) and local cooling effects
ISO/CD 11228-2 Ergonomics—Manual handling—Part 2: Pushing and pulling
ISO/CD 11228-3 Ergonomics—Manual handling—Part 3: Handling of low loads at high frequency
ISO/DIS 13732-3 Ergonomics of the thermal environment—Touching of cold surfaces—Part 3: Ergonomics data and guidance for application
ISO/CD 14505-1 Ergonomics of the thermal environment: Thermal environment in vehicles—Part 1: Principles and method for assessment for thermal stress
ISO/CD 14505-2 Ergonomics of the thermal environment: Thermal environment in vehicles—Part 2: Determination of equivalent temperature
ISO/CD 14505-3 Ergonomics of the thermal environment: Thermal environment in vehicles—Part 3: Evaluation of thermal comfort using human subjects
ISO/FDIS 15265 Ergonomics of the thermal environment—Risk assessment strategy for the prevention of stress or discomfort in thermal working conditions
ISO/DIS 15536-1 Ergonomics—Computer manikins and body templates—Part 1: General requirements
ISO/DIS 15536-2 Ergonomics—Computer manikins and body templates—Part 2: Structures and dimensions
ISO/DIS 15537 Principles for selecting and using test persons for testing anthropometric aspects of industrial products and designs
ISO/CD 15743 Ergonomics of the thermal environment—Working practices in cold: Strategy for risk assessment and management
ISO/CD 20282-1 Ease of operations of everyday products—Part 1: Context of use and user characteristics
ISO/CD 20282-2 Ease of operations of everyday products—Part 2: Test method
ISO/PRF TS 20646 Ergonomic procedures for the improvement of local muscular loads
ISO/CD 20685 3D scanning methodologies for internationally compatible anthropometric databases

---

[a]CD, Committee Draft, registered draft standard; DIS, Draft International Standard, registered draft standard; FDIS, Final Draft International Standard, registered for formal approval.

environmental aspects in (vertical) product standards (CEN, 1998; Commission of the European Communities, 2004; ISO, 1997). In this way a specific field attempts to get more impact in other fields. In ergonomics we have not observed a similar strategy.

Basic, Requiring, Measurement. "Basic," "requiring," and "measurement" standards show up in the tables presented here. Almost all ergonomics standards belong to the category of "Requiring standards." For most of these standards it is difficult to make the difference between performance standards and solution-describing standards. Some of the standards are basic standard, such as ISO 10075, which includes terms and definitions on mental workload, and data standards such as ISO 15534-3, with data on human body dimensions.

TABLE 2.5
ISO Standards from Tables 2.3 and 2.4, Organized According to Ergonomics Topics[a]

---

1. General design principles
   ISO 6385, ISO 13407
2. Safety of machinery
   ISO/FDIS 14738, ISO 15534-1, ISO 15534-2, ISO 15534-3
3. Physical environment
   Noise/speech: ISO 7731, ISO 9921, ISO 11428, ISO 11429 ISO/TR 19358
   Climate: ISO 7243 ISO 7726, ISO 7730, ISO 7933, ISO 8996, ISO 9241-6, ISO 9886, ISO 9920,
   ISO 10551, ISO/TR 11079, ISO 11399, ISO 12894, ISO 13731, ISO/TS 13732-2, ISO/DIS 13732-3
4. Physical workload
   ISO 11226, ISO 11228-1
5. Mental workload
   ISO 9241-2, ISO 10075, ISO 10075-2, ISO/CD 10075-3
6. Workplace and equipment design
   General: ISO 9241-5, ISO 9241-6, ISO 11064-1, ISO 11064-2, ISO 11064-3
   Anthropometry: ISO 7250, ISO 14738, ISO 15534-1, ISO 15534-2, ISO 15534-3, ISO 15535, ISO/DIS 15536-1,
   ISO/DIS 15537
7. Visual information, VDTs and software
   General: ISO 9241-1, ISO 9241-2, ISO 9241-3, ISO 9241-4, ISO 9241-5, ISO 9241-6, ISO 9241-7, ISO 9241-8,
   ISO 9241-9, ISO 13406-1, ISO 13406-2, ISO 16071
   Software: ISO 9241-10 ISO 9241-11 ISO 9241-12 ISO 9241-13 ISO 9241-14 ISO 9241-15 ISO 9241-16,
   ISO 9241-17, ISO 13407, ISO 14915-1, ISO 14915-2, ISO 14915-3, ISO/TR 16982, ISO/TR 18529
8. Displays and controls
   ISO 9241-4, ISO 9355-1, ISO 9355-2, ISO 11428, ISO 11429
9. Personal protection equipment

---

[a] Standards can be listed under more than one topic.

TABLE 2.6
CEN Standards on Ergonomics

	1989 (Metz, 1991)	1995 (Dul et al., 1996)	2004 (Dul et al., 2004)
CEN published	0	10	62
CEN in preparation	0	38	27
Total	0	48	89

Some standards are "Measurement standards." Examples are ISO 7726, on methods for measuring physical quantities of the thermal environment, and ISO/TS 13732-2, on methods for measuring human responses to contact with cold or hot surfaces.

Compared with other fields of standardization, the number of measurement standards in ergonomics is relatively small.

Interference, Compatibility, Quality. Because ergonomics deals with the interaction between a person and the environment, most ergonomics standards are interference standards, setting requirements on the influence of the entity "environment" on the entity "person." Examples are standards on human–system interaction, or standards on manual handling of loads.

TABLE 2.7
Published Standards from CEN TC 122

EN 457:1992 Safety of machinery—Auditory danger signals—General requirements, design, and testing (ISO 7731:1986 modified)

EN 547-1:1996 Safety of machinery—Human body measurements—Part 1: Principles for determining the dimensions required for openings for whole-body access into machinery

EN 547-2:1996 Safety of machinery—Human body measurements—Part 2: Principles for determining the dimensions required for access openings

EN 547-3:1996 Safety of machinery—Human body measurements—Part 3: Anthropometric data

EN-563 1994 Safety of machinery—Temperature of touchable surfaces—Ergonomics data to establish temperature limit values for hot surfaces

EN 614-1:1995 Safety of machinery—Ergonomic design principles—Part 1: Terminology and general principles

EN 614-2:2000 Safety of machinery—Ergonomic design principles—Part 2: Interactions between the design of machinery and work tasks

EN 842:1996 Safety of machinery—Visual danger signals—General design requirements, design, and testing

EN 894-1:1997 Safety of machinery—Ergonomics requirements for the design of displays and control actuators—Part 1: General principles for human interactions with displays and control actuators

EN 894-2:1997 Safety of machinery—Ergonomics requirements for the design of displays and control actuators—Part 2: Displays

EN 894-3:2000 Safety of machinery—Ergonomics requirements for the design of displays and control actuators—Part 3: Control actuators

EN 981:1997 Safety of machinery—System of auditory and visual danger and information signals

EN 1005-1:2001 Safety of machinery—Human physical performance—Part 1: Terms and definitions

EN 1005-2 Safety of machinery—Human physical performance—Part 2: Manual handling of machinery and component parts of machinery

EN 1005-3:2002 Safety of machinery—Human physical performance—Part 3: Recommended force limits for machinery operation

EN ISO 7250:1997 Basic human body measurements for technological design (ISO 7250:1996)

EN ISO 7726:2001 Ergonomics of the thermal environment—Instruments for measuring physical quantities (ISO 7726:1998)

EN ISO 7730:1995 Moderate thermal environments—Determination of the PMV and PPE indices and specification of the conditions for thermal comfort (ISO 7730: 1994)

EN ISO 9241-1:1997 Ergonomic requirements for office work with VDTs—Part 1: General introduction (ISO 9241-1:1997)

EN ISO 9241-2:1993 Ergonomic requirements for office work with VDTs—Part 2: Guidance on task requirements (ISO 9241-2:1992)

EN ISO 9241-3 :1993 Ergonomic requirements for office work with VDTs—Part 3: Visual display requirements (ISO 9241-3:1992)

EN ISO 9241-4:1998 Ergonomic requirements for office work with VDTs—Part 4: Keyboard requirements (ISO 9241-4:1998)

EN ISO 9241-5:1999 Ergonomic requirements for office work with VDTs—Part 5: Workstation layout and postural requirements (ISO 9241-5:1998)

EN ISO 9241-6:1999 Ergonomic requirements for office work with VDTs—Part 6: Guidance on the work environment (ISO 9241-6:1999)

EN ISO 9241-7:1998 Ergonomic requirements for office work with VDTs—Part 7: Requirements for display with reflections (ISO 9241-7:1998)

EN ISO 9241-8:1997 Ergonomic requirements for office work with VDTs—Part 8: Requirements for displayed colours (ISO 9241-8:1997)

EN ISO 9241-9:2000 Ergonomic requirements for office work with VDTs—Part 9: Requirements for non-keyboard-input devices (ISO 9241-9:2000)

EN ISO 9241-10:1996 Ergonomic requirements for office work with VDTs—Part 10: Dialogue principles (ISO 9241-10:1996)

EN ISO 9241-11:1998 Ergonomic requirements for office work with VDTs—Part 11: Guidance on usability (ISO 9241-11:1998)

EN ISO 9241-12:1998 Ergonomic requirements for office work with VDTs—Part 12: Presentation of information (ISO 9241-12:1998)

(Continued)

TABLE 2.7

(Continued)

EN ISO 9241-13:1998 Ergonomic requirements for office work with VDTs—Part 13: User guidance
(ISO 9241-13:1998)

EN ISO 9241-14:1999 Ergonomic requirements for office work with VDTs—Part 14: Menu dialogues
(ISO 9241-14:1995)

EN ISO 9241-15:1997 Ergonomic requirements for office work with VDTs—Part 15: Command dialogues
(ISO 9241-15:1997)

EN ISO 9241-16:1999 Ergonomic requirements for office work with VDTs—Part 16: Direct manipulation dialogues
(ISO 9241-16:1999)

EN ISO 9241-17:1998 Ergonomic requirements for office work with VDTs—Part 17: Form filling dialogues
(ISO 9241-17:1998)

EN ISO 9886:2001 Evaluation of thermal strain by physiological measurements (ISO 9886:1992)

EN ISO 9920 Ergonomics of the thermal environment—Estimation of the thermal insulation and evaporative resistance
of a clothing ensemble (ISO 9920:1995)

EN ISO 9921:2003 Ergonomics—Assessment of speech communication (ISO 9921:2003)

EN ISO 10075-1:2000 Ergonomic principles related to mental workload—Part 1: General terms and definitions
(ISO 10075:1991)

EN ISO 10075-2:2000 Ergonomic principles related to mental workload—Part 2: Design principles
(ISO 10075-2:1996)

EN ISO 10551:2001 Ergonomics of the thermal environment—Assessment of the influence of the thermal environment
using subjective judgment scales (ISO 10551:1995)

EN ISO 11064-1:2000 Ergonomic design of control centers—Part 1: Principles for the design of control centres
(ISO 11064-1:2000)

EN ISO 11064-2:2000 Ergonomic design of control centers—Part 2: Principles for the arrangement of control suites
(ISO 11064-2:2000)

EN ISO 11064-3:1999 Ergonomic design of control centres—Part 3: Control room layout (ISO 11064-3:1999)

ENV[a] ISO 11079:1998 Evaluation of cold environments—Determination of required clothing insulation (IREC)
(ISO/TR 11079:1993)

EN ISO 11399:2000 Ergonomics of the thermal environment—Principles and application of relevant International
Standards (ISO 11399:1995)

EN 12515:1997 Hot environments—Analytical determination and interpretation of thermal stress using calculation of
required sweat rate (ISO 7933:1989 modified)

EN ISO 12894:2001 Ergonomics of the thermal environment—Medical supervision of individuals exposed to extreme
hot or cold environments (ISO 12894:2001)

EN 13202:2000 Ergonomics of the thermal environment—Temperatures of touchable hot surfaces—Guidance for
establishing surface temperature limit values in production standards with the aid of EN 563

EN ISO 13406-1:1999 Ergonomic requirements for work with visual display based on flat panels—Part 1: Introduction
(ISO 13406-1:1999)

EN ISO 13406-2:2001 Ergonomic requirements for work with visual displays based on flat panels—Part 2: Ergonomic
requirements for flat panel displays (ISO 13406-2:2001)

EN ISO 13407:1999 Human-centered design processes for interactive systems (ISO 13407:1999)

EN ISO 13731:2001 Ergonomics of the thermal environment—Vocabulary and symbols (ISO 13731:2001)

EN 13861:2002 Safety of machinery—Guidance for the application of ergonomics standards in the design of machinery

EN ISO 14738:2002 Safety of machinery—Anthropometric requirements for the design of workstations at machinery
(ISO 14738:2002)

EN ISO 14915-1:2002 Software ergonomics for multimedia user interfaces—Part 1: Design principles and framework
(ISO 14915-1:2002)

EN ISO 14915-2:2003 Software ergonomics for multimedia user interfaces—Part 2: Multimedia control and navigation
(ISO 14915-2:2003)

EN ISO 14915-3:2002 Software ergonomics for multimedia user interfaces—Part 3: Media selection and combination
(ISO 14915-3:2002)

EN ISO 15535:2003 General requirements for establishing an anthropometric database (ISO 15535:2003)

ENV 26385:1990 Ergonomic principles of the design of work systems (ISO 6385: 1981)

EN ISO 27243:1993 Hot environments—Estimation of the heat stress on working man, based on the WBGT-index (wet
bulb globe temperature) (ISO 7243: 1989)

EN 28996:1993 Ergonomics—Determination of metabolic heat production (ISO 8996:1990)

[a]ENV, Preliminary European Standard.

TABLE 2.8
Standards From CEN TC 122, In preparation

prEN 614-1rev[a] Safety of machinery—Ergonomic design principles—Part 1: Terminology and general principles. Under approval[a]

prEN 894-4 Safety of machinery—Ergonomics requirements for the design of displays and control actuators—Part 4: Location and arrangement of displays and control actuators. Under development[a]

prEN 1005-4 Safety of machinery—Human physical performance—Part 4: Evaluation of working postures and movements in relation to machinery. Under approval

prEN 1005-5 Safety of machinery—Human physical performance—Part 5: Risk assessment for repetitive handling at high frequency. Under approval

prEN ISO 6385 rev Ergonomic principles in the design of work systems (ISO/FDIS 6385:2003). Ratified[a]

prEN ISO 7730 rev Ergonomics of the thermal environment—Analytical determination and interpretation of thermal comfort using calculation of the PMV and PPD indices and local thermal comfort (ISO/DIS 7730:2003). Under approval

prEN ISO 7933 Ergonomics of the thermal environment—Analytical determination and interpretation of heat stress using calculation of the predicted heat strain (ISO/DIS 7933:2003). Under approval

prEN ISO 8996 rev Ergonomics—Determination of metabolic heat production (ISO/DIS 8996:2003). Under approval

prEN ISO 9886 rev Ergonomics—Evaluation of thermal strain by physiological measurements (ISO FDIS 9886: 2003). Under approval

prEN ISO 9920 rev Ergonomics of the thermal environment—Estimation of the thermal insulation and evaporative resistance of a clothing ensemble. Under development

prEN ISO 10075-3 Ergonomic principles related to mental workload—Part 3: Measurement and assessment of mental workload (ISO/DIS 10075-3:2002). Under approval

prEN ISO 11064-4 Ergonomic design of control centers—Part 4: Layout and dimensions of workstations (ISO/DIS 11064-4:2002). Under development

prEN ISO 11064-6 Ergonomic design of control centers—Part 6: Environmental requirements for control centers (ISO/DIS 11064-6:2003). Under approval

prEN ISO 11079 Evaluation of cold environments—Determination of required clothing insulation (IREQ) (will replace ENV ISO 11079:1998). Under development

prEN ISO 13732-1 Ergonomics of the thermal environment—Methods for the assessment of human responses to contact with surfaces—Part 1: Hot surfaces (ISO/DIS 13732-1:2003). Under approval

prEN ISO 13732-3 Ergonomics of the thermal environment—Touching of cold surfaces—Part 3: Ergonomics data and guidance for application (ISO/DIS 13732-3:2002). Under approval

prEN 13921-1 Personal protective equipment—Ergonomic principles—Part 1: General requirements for the design and the specification. Under approval

prEN 13921-3 Personal protective equipment—Ergonomic principles—Part 3: Biomechanical characteristics. Under approval

prEN 13921-4 Personal protective equipment—Ergonomic principles—Part 4: Thermal characteristics. Under approval

prEN 13921-6 Personal protective equipment—Ergonomic principles—Part 6: Sensory factors. Under approval

prEN 14386 Safety of machinery—Ergonomic design principles for the operability of mobile machinery. Under approval

prEN ISO 14505-1 Ergonomics of the thermal environment: Thermal environment in vehicles—Part 1: Principles and method for assessment for thermal stress. Under development

prEN ISO 14505-2 Ergonomics of the thermal environment: Thermal environment in vehicles—Part 2: Determination of equivalent temperature. Under development

prEN ISO 15536-1 Ergonomics—Computer manekins and body templates—Part 1: General requirements (ISO/DIS 15536-1:2002). Under approval

prEN ISO 15537 Principles for selecting and using test persons for testing anthropometric aspects of industrial products and designs (ISO/DIS 15537:2002). Under approval.

prEN ISO 20685 3D scanning methodologies for internationally compatible anthropometric databases. Under development

prEN ISO 23973 Software ergonomics for World Wide Web user interfaces. Under development

[a]Under development, active work item which has not yet reached the stage of enquiry; Under approval, active work item at a stage between the beginning of the enquiry and the end of formal vote; Ratified, work item at a stage between ratification and publication; rev, standard under revision.

TABLE 2.9
CEN Standards From Tables 2.7 and 2.8, Organized According to Ergonomics Topics[a]

---

1. General design principles
   EN 614-1, prEN 614-1, EN 614-2, prEN ISO 6385, EN 13407, ENV 26385
2. Safety of machinery
   EN 457, EN 547-1, EN 547-2, EN 547-3, EN 563, EN 574 EN 614-1, prEN 614-1, EN 641-2, EN 842,
   EN 894-1, EN 894-2, EN 894-3, EN 894-4 EN 981, EN 1005-1, EN 1005-2, EN 1005-3, prEN 1005-4,
   prEN 1005-5, EN 13861, prEN 14386, prEN ISO14738
3. Physical environment
   Noise/speech: EN 457, EN 981, EN ISO 9921
   Climate: EN 563, EN ISO 7243, EN ISO 7726. EN ISO 7730, prEN ISO 7730, prEN ISO 7933, EN ISO 8996,
   prEN ISO 8996, EN ISO 9241-6, EN ISO 9886, prEN ISO 9886, EN ISO 9920, prEN ISO 9920, EN ISO 10551,
   ENV ISO 11079, prEN ISO 11079, EN ISO 11399, EN 12515, EN ISO 12894, EN 13202, EN ISO 13731,
   prEN ISO 13732-1, prEN ISO 13732-3, prEN ISO 14505-1, prEN ISO 14505-2, EN 27243, EN 28996
4. Physical workload
   EN 1005-1, EN 1005-2, EN 1005-3, prEN 1005-4, prEN 1005-5
5. Mental workload
   EN 614-2, EN 9241-2, EN ISO 10075-1, EN ISO 10075-2, prEN ISO 10075-3
6. Workplace and equipment design
   General: EN ISO 9241-5, EN ISO 9241-6, EN ISO 11064-1, EN ISO 11064-2, EN ISO 11064-3,
   prEN ISO 11064-4, prEN ISO 11064-6, prEN 14386
   Anthropometry: EN 547-1, EN 547-2, EN 547-3, EN ISO 7250, EN 14738, EN ISO 15535, prEN ISO 15536-1,
   prEN ISO 15537, prEN ISO 20685
7. Visual information, VDTs, and software
   General: EN ISO 9241-1, EN ISO 9241-2, EN ISO 9241-3, EN ISO 9241-4, EN ISO 9241-5, EN ISO 9241-6,
   EN ISO 9241-7, EN ISO 9241-8, EN ISO 9241-9, EN ISO 13406-1, EN ISO 13406-2
   Software: EN ISO 9241-10, EN ISO 9241-11, EN ISO 9241-12, EN ISO 9241-13, EN ISO 9241-14,
   EN ISO 9241-15, EN ISO 9241-16, EN ISO 9241-17, EN ISO 13407, EN ISO 14915-1, EN ISO 14915-2,
   EN ISO 14915-3, prEN ISO 23973
8. Displays and controls
   EN ISO 9241-4
9. Personal protection equipment
   prEN 13921-1, prEN 13921-3, prEN 13921-4, prEN 13921-6

---

[a] Standards can be listed under more than one topic.

In most other areas of standardization, compatibility standards (e.g., in information technology) or quality standards (e.g., in management systems) are dominant.

Ergonomics Standardization

Definition

Ergonomics standardization is the activity of making ergonomics standards. As mentioned before, in this chapter we concentrate on ergonomics standards and hence ergonomics standardization from ISO and CEN because, in contrast to other fields of standardization, most ergonomics standards are developed by ISO and CEN and not by companies or consortia of companies. Therefore our definition of ergonomics standardization is *ergonomics standards making by ISO and CEN.*

Developers

National Standardization. It is common practice that national standardization organizations adopt ISO and—in Europe—CEN standards as national standards. This is an obligation

TABLE 2.10
Ergonomics CEN Standards Linked to the European Machinery Directive

---

EN 457:1992 Safety of machinery—Auditory danger signals—General requirements, design and testing (ISO 7731:1986 modified)

EN 547-1:1996 Safety of machinery—Human body measurements—Part 1: Principles for determining the dimensions required for openings for whole-body access into machinery

EN 547-2:1996 Safety of machinery—Human body measurements—Part 2: Principles for determining the dimensions required for access openings

EN 547-3:1996 Safety of machinery—Human body measurements—Part 3: Anthropometric data

EN-563 1994 Safety of machinery—Temperature of touchable surfaces—Ergonomics data to establish temperature limit values for hot surfaces

EN 614-1:1995 Safety of machinery—Ergonomic design principles—Part 1: Terminology and general principles

EN 614-2:2000 Safety of machinery—Ergonomic design principles—Part 2: Interactions between the design of machinery and work tasks

EN 842:1996 Safety of machinery—Visual danger signals—General design requirements, design and testing

EN 894-1:1997 Safety of machinery—Ergonomics requirements for the design of displays and control actuators—Part 1: General principles for human interactions with displays and control actuators

EN 894-2:1997 Safety of machinery—Ergonomics requirements for the design of displays and control actuators—Part 2: Displays

EN 894-3:2000 Safety of machinery—Ergonomics requirements for the design of displays and control actuators—Part 3: Control actuators

EN 981:1997 Safety of machinery—System of auditory and visual danger and information signals

EN 1005-1:2001 Safety of machinery—Human physical performance—Part 1: Terms and definitions

EN 1005-2 Safety of machinery—Human physical performance—Part 2: Manual handling of machinery and component parts of machinery

EN 1005-3:2002 Safety of machinery—Human physical performance—Part 3: Recommended force limits for machinery operation

EN ISO 7250:1997 Basic human body measurements for technological design (ISO 7250:1996)

EN ISO 14738:2002 Safety of machinery—Anthropometric requirements for the design of workstations at machinery (ISO 14738:2002)

---

for CEN standards, but it also occurs with many ISO standards. For example, the "mother ergonomics standards" ISO 6385 "ergonomic principles in the design of work systems" has been adopted by national standardization organizations in many countries.

Additionally, national standardization organizations may develop own national standards on ergonomics. In Chapter 20 of this book, Stuart-Buttle (2005) gives an overview of national ergonomics standards in the United States, Australia, and Japan.

International Standardization: Global.   In the early 1970s, the International Ergonomics Association decided to initiate the development of ergonomic standards (Parsons & Shackel, 1995) and requested the ISO to set up a technical committee on ergonomics. In 1974, the ISO established TC 159. The scope of this TC is *Standardization in the field of ergonomics, including terminology, methodology, and human factors data.* Since then until 2004, 60 standards have been developed and published (Dul et al., 2004), and new standards are being developed or revised within four subgroups on the following topics:

- Ergonomic guiding principles
- Anthropometry and biomechanics
- Ergonomics of human–system interaction
- Ergonomics of the physical environment

In 2004, the secretariat of TC 159 is with the German Standardization Institute DIN. ISO TC 159 has 24 countries that participate in developing the standards, and 29 observing countries.

International Standardization: Regional.   Since the 1980s, Europe has developed toward a free internal market without barriers to trade. For this common market, common standards were desired, for example, for the safety requirements of machinery. In 1989, CEN established the Technical Committee TC 122 "Ergonomics" to address ergonomics requirements in relation to safety of machinery. The first European ergonomics standard was published in 1990 as ENV 26385, which was an adoption of the above ISO 6385. Afterwards CEN has published ergonomics standards on safety of machinery and other ergonomics issues. The scope of CEN TC 122 is *Standardization in the field of ergonomics principles and requirements for the design of work systems and work environments, including machinery and personal protective equipment to promote the health, safety, and well-being of the human operator and the effectiveness of the work systems.* Until 2004, CEN TC 122 has developed and published 62 standards (Dul et al., 2004). New standards are being developed or revised within nine working groups on the following topics:

- Anthropometry
- Ergonomic design principles
- Surface temperatures
- Biomechanics
- Ergonomics of human–computer interaction
- Signals and controls
- Danger signals and speech communication in noisy environments
- Ergonomics of personal protective equipment
- Ergonomics of the thermal environment

In 2004, the secretariat of CEN TC 122 is with the German Standardization Institute DIN.

Cooperation ISO-CEN.   Many ergonomics standards have been developed in cooperation between ISO and CEN, within the framework of the Vienna Agreement. Since 1989, CEN has adopted many ergonomics ISO standards developed by ISO TC 159. Also, ISO and CEN have used parallel approval procedures for work items developed by ISO or CEN. Common standards are listed as "EN ISO" standards in Tables 2.7 and 2.8.

Stakeholder Participation

A model, that has been developed to identify and select relevant stakeholders to be involved in standardization (Willemse, 2003), has been applied to ergonomics standardization (Dul et al., 2003). The model and the application to ergonomics standardization are presented in Chapter 6 of this book (Willemse et al., 2005). One conclusion is that ergonomics experts seem to be the major participants in ergonomics standardization, and that there is an underrepresentation of other relevant stakeholders, for example, designers.

CONCLUSIONS

Ergonomics Standards

During the past 30 years, more than 150 ergonomics standards have been published by ISO and CEN, or are in preparation. Because of this high production of new standards in the last three decades, it seems that some duplication of work, inconsistencies, and contradictions have occurred (Nachreiner, 1995). In future standards development, a reduction of overlap, and more clear relationships among the standards are desirable. Feedback on the existing collection of standards, for example, answers to questions like "are the standards known?",

TABLE 2.11
Summary of Major Differences Between the Majority of Ergonomics Standards and the Majority
of Other Standards

Aspect	Other Standards	Ergonomics Standards
Standardization		
Age	Thousands years	30 years
Initiator/developer	Manufacturer	Ergonomics expert
Organization of the standardization	Standardization organization, company, consortium	Standardization organization
Standards		
Aim	Economic	Social
Government interest	Economic affairs	Social affairs
User groups	Designer, Tester	Ergonomics expert
End users	Consumer	Worker
Application	Specific, technical	General
Object/Entity	Product, test method	Human, process

"are the standards used?", and "are the standards considered to be useful?" could help further development of ergonomics standardization as well. Also the principle of *"user participation,"* advocated by most ergonomists, could be better applied to the ergonomics standardization: It is suggested to strengthen the involvement of relevant stakeholders in future ergonomics standardization, in particular, designers of products and production processes.

Ergonomics Standards Versus Other Standards

The large number of ergonomics standards is only a small fraction of all available ISO and CEN standards. In the introduction we raise the question: *"Is ergonomics standardization an area with exceptional characteristics, or is it just one of the many areas of standardization?"* By comparing ergonomics standards and standardization with other standards and standardization, we conclude that there are major differences (Table 2.11). When considering Table 2.11, we stress that also many similarities exist between ergonomics standards and other standards, that this table is only a rough estimate of the differences that we observed, and that we did not take into account the existing nuances; hence, the table is open for discussion. Nevertheless, with respect to *standardization* (making of standards), we observed the following differences:

- Ergonomics standardization started only recently, whereas standardization in most other fields is about one century old, with roots thousands of years ago.
- Ergonomics experts are the major initiators and developers of ergonomics standards, in contrast to manufacturers in most other fields of standardization.
- Ergonomics standards are mainly formal standards developed by formal standardization organizations, whereas many other standards are also developed by companies and consortia of companies.

With respect to *standards* (the results of standardization) we observed the following major differences:

- The main aim of ergonomics standards is a social aim (well-being of the workers or the consumers), whereas the major aims of other standards are economic.

- Ministries of social affairs, health, or labor may have governmental interest in ergonomic standards, for example, for support of regulation on occupational health and safety. In other fields, ministries of economic affairs may support standardization in general as a way to stimulate the economy.
- The major user group of ergonomics standards is the group of ergonomics experts. For other standards, designers and testers are the major user groups.
- The end users of ergonomics standards are primarily workers, whereas the end users of other standards are primarily consumers.
- Most ergonomics standards apply to technical and organizational aspects of (work) situations. Other standards usually apply to specific technical aspect of products.
- Ergonomics standards usually concern humans or the interaction of human with (production or work) processes, whereas other standards concern primarily products or test methods.

Lessons

What can we learn from these differences? We feel that the following questions can be raised to contribute to the discussion on the future of ergonomics standards:

- Should we attempt to integrate more economic goals ("total system performance") in ergonomics standards?
- Should we then look for governmental support from ministries of economics affairs, apart from existing support from ministries of social affairs, health, and labor?
- Should we more involve the designers, manufacturers, and consumers in the development or ergonomics standards, according to the ergonomic principle of user participation?
- Should we stimulate that more ergonomics standardization is organized by companies and consortia of companies?
- Should we develop more (technical) standards that can be applied to specific situations?
- Should we develop more standards for products and test methods?

It is not easy to give answers to these questions. But even if answers can be given, one main question remains: *"Is it better to develop ergonomic standards as a separate field* (as was done during the past 30 years), *or is it better* (similar to the standardization policy in the environmental field) *to integrate ergonomics in standardization activities of other fields?"* Then the ergonomics field may gain more impact, but may also lose identity.

REFERENCES

AFNOR. (1993). *X 00-001 Normes françaises—Règles pour la rédaction et la présentation—Conseils pratiques.* Paris, Association Française de Normalisation.

ANEC. (2003). *Consumer requirements in standardisation relating to the information society.* European Association for the Co-ordination of Consumer Representation in Standardisation, Brussels.

Baraton, P., & Hutzler, B. (1985). *Magnetically induced currents in the human body.* Technology Trend Assessment No.1, Geneva, International Electrotechnical Commission.

Breedveld. P. (2005). *Factors influencing perceived acceptance and success of ergonomists within European organizations.* MSc thesis, Rotterdam School of Management, Erasmus University Rotterdam, The Netherlands.

BSI. (1997a). *BS-01 A standard for standards—Part 2: Recommendations for committee procedures.* British Standards Institution, London.

BSI. (1997b). *BS-03 A standard for standards—Part 3: Specification for structure, drafting and presentation.* London: British Standards Institution.

CEN. (1998). *CEN Guide 4 Guide for the inclusion of environmental aspects in product standards.* Brussels: Comité Européen de Normalisation.

CEN. (2000). *EN 1335-1 Office furniture—Office work chair—Part 1: Dimensions—Determination of dimensions.* Brussels: Comité Européen de Normalisation.

CEN. (2004). *European standardization in a global context.* Brussels: European Committee for Standardization (CEN).

Commission of the European Communities. (2004). *Integration of environmental aspects into european standardisation.* Brussels: COM(2004)130 final. Commission of the European Communities.

Cook, L. S., Bowen, D. E., Chase, R. B., Dasu, S., Stewart, D. M., & Tansik, D. A. I. (2003). Human issues in service design. *Journal of Operations Management 20*(2), 159–174.

De Vries, H. J. (1997). Standardization—What's in a name? *Terminology—International Journal of Theoretical and Applied Issues in Specialized Communication* 4 (1), (rectification in 55–83).

De Vries, H. J. (1998). The classification of Standards. *Knowledge Organization, 25*(3), 79–89.

De Vries, H. J. (1999). *Standardization—A business approach to the role of National Standardization Organizations.* Boston/Dordrecht/London: Kluwer Academic Publishers.

De Vries, H. J., Verheul, H., & Willemse, H. (2003). *Stakeholder identification in IT standardization processes.* In John L. King & Kalle Lyytinen (Eds.), *Proceedings of the workshop on standard making: A critical research frontier for information systems* (pp. 92–107). Seattle, WA, December 12–14, 2003.

De Vries, H. J., Feilzer, A., & Verheul, H. (2004). *Removing barriers for participation in formal standardization.* In Françoise Bousquet, Yves Buntzly, Heide Coenen, & Kai Jakobs (Eds.), *EURAS Proceedings 2004.* Aachener Beiträge zur Informatik, Band 36, Wissenschaftsverlag Mainz in Aachen, Aachen (pp. 171–176).

DIN. (1996). *DIN 820 Teil 2 Normungsarbeit—Teil 2: Gestaltung von Normen.* Berlin: Beuth Verlag GmbH.

Dul, J., De Vlaming, P. M., & Munnik, M. J. (1996). A review of ISO and CEN standards on ergonomics. *International Journal of Industrial Ergonomics, 17*(3), 291–297.

Dul, J. (2003a). *De mens is de maat van alle dingen—Over mensgericht ontwerpen van producten en processen.* (Man is the measure of all things—On human-centered design of products and processes). Inaugural address. Erasmus Research Institute of Management, Erasmus University Rotterdam.

Dul, J. (2003b). *The strategic value of ergonomics for companies.* In H. Luczak & K. J. Zink. (Eds.), *Human factors in organizational design and management VII.* (pp. 765–769). Santa Monica: IEA Press.

Dul, J., Willemse, H., & De Vries, H. J. (2003). Ergonomics standards: Identifying stakeholders and encouraging participation. *ISO-Bulletin* (September), 19–23.

Dul, J., De Vries, H., Verschoof, S., Eveleens, W., & Feilzer, A. (2004). Combining economic and social goals in the design of production systems by using ergonomics standards. *Computers and Industrial Engineering 47*(2–3), 207–222.

European Union. (2003). Commission communication in the framework of the implementation of Directive 98/37/EC of the European Parliament and of the Council of 22 June 1998 in relation to machinery amended by directive 98/79/EC, *Official Journal of the European Union, C192,* 2–29.

European Union. (1989). Council Directive 89/655/EEC of 30 November 1989 concerning the minimum safety and health requirements for the use of work equipment by workers at work (second individual Directive within the meaning of Article 16 (1) of Directive 89/391/EEC, *Official Journal of the European Union* L 393, 30.12.1989, p. 13.

European Union (1990a). Council Directive 90/269/EEC of 29 May 1990 on the minimum health and safety requirements for the manual handling of loads where there is a risk particularly of back injury to workers (fourth individual Directive within the meaning of Article 16 (1) of Directive 89/391/EEC), *Official Journal of the European Union* L 156, 21.06.1990, p. 9.

European Union. (1990b). Council Directive 90/270/EEC of 29 May 1990 on the minimum safety and health requirements for work with display screen equipment (fifth individual Directive within the meaning of Article 16 (1) of Directive 89/391/EEC), *Official Journal of the European Union* L 156 of 21.06.1990, p. 14.

Eveleens, W. (2003). A basic ergonomic standard. How to provide optimal working conditions for personnel. *ISO-Bulletin* (June), 3–6.

Fallentin, N., Viikari-Juntura, E., Waersted, M., & Kilbom, A. (2001). Evaluation of physical workload standards and guidelines from a Nordic perspective. *Scandinavian Journal of work environment and health,* 27(Suppl. 2), 1–52.

Gaillard, J. (1933). *A study of the fundamentals of industrial standardization and its practical application, especially in the mechanical field.* Delft: NV W.D. Meinema.

Hesser, W., & Inklaar, A. (1997) *Aims and functions of standardization.* In Wilfried Hesser & Alex Inklaar (Eds.), *An introduction to standards and standardization* (pp. 39–45). *DIN Normungskunde Band 36.* Berlin/Vienna/Zürich: Beuth Verlag. pp. 33–45.

Huigen, H. W., Inklaar, A., & Paterson E. (1997). *Standardization and certification in Europe.* In Wilfried Hesser & Alex Inklaar (Eds.), *An introduction to standards and standardization* (pp. 230–251). Berlin: Beuth Verlag.

ISO. (1997). *ISO Guide 64 Guide for the inclusion of environmental aspects in product standards.* Geneva: International Organization for Standardization.

ISO. (2004). http://www.iso.org/iso/en/stdsdevelopment/whowhenhow/proc/proc.html

ISO/CEN. (1991). *Agreement on technical cooperation between ISO and CEN (Vienna Agreement)* First Revision. Geneva: ISO.

ISO/IEC. (1996). *ISO/IEC Guide 2 Standardization and related activities—General vocabulary* (7th ed.). Geneva: International Organization for Standardization/International Electrotechnical Commission.

KAN. (2004). Kommission Arbeitsschutz und Normung. http://www.kan.de

Leibrock, G. (2002). *Methods of referencing standards in legislation with an emphasis on European legislation.* Enterprise Guides. Brussels: European Commission, Enterprise Directorate-General, Standardisation unit.

Le, Lourd, Ph. (1992) *La normalisation et l'Europe—Secteurs de l'agro-alimentaire, du bois, et de l'eau.* Paris: Editions Romillat/AFNOR.

Metz, B. G. (1991). Outcomes of international standardization work in the field of ergonomics between 1995 and 1989. In Y. Quéinnec & F. Daniellou (Eds.), *Designing for everyone* (pp. 987–980). London: Taylor and Francis.

NEN. (2004a). *Handleiding Commissieleden* (3rd ed.). Delft: NEN.

NEN. (2004b). De Nederlandse Werkstoel. *NormalisatieNieuws, 13*(2), 1.

Nachreiner, F. (1995). Standards for ergonomics principles relating to design of work systems and to mental work load. *Applied Ergonomics, 26*(4), 259–263.

Parsons, K. C., & Shackel, B. (1995). Ergonomics and international standards. History, organisational structure and method of development. *Applied Ergonomics, 26*(4), 249–258.

Sanders, T. B. R. (Ed.), (1972). *The aims and principles of standardization.* Geneva: International Organization for Standardization.

Schaap, A., & De Vries, H. J. (2004). *Evaluatie van normalisatie-investeringen—Hoe MKB-bedrijven kunnen profiteren van deelname aan normalisatie.* Zoetermeer: FME-CWM.

SES (1995). *SES 1 Recommended practice for standards designation and organization—An American national standard.* Dayton, OH: Standards Engineering Society.

Simons, C. A. J. (1994). *Kiezen tussen verscheidenheid en uniformiteit.* Inaugural address. Erasmus University Rotterdam, Rotterdam.

Smith, M. A. (1995). *Vienna Agreement of Technical Cooperation Between ISO and CEN.* ISO/IEC Directives seminar 1995, Geneva: ISO.

Stuart-Buttle, C. (2006). Overview of National and International Standards and Guidelines. In W. Karwowski (Ed.), *Handbook on standards and guidelines in ergonomics and human factors* (pp. 133–147). Mahwah, New Jersey: Lawrence Erlbaum Associates.

Susanto, A. (1988). *Methodik zur Entwicklung von Normen.* Berlin/Cologne: DIN-Normungskunde Band 23, Berlin, DIN Deutsches Institut für Normung e.V., Beuth Verlag GmbH.

Tamm Hallström, K. (2004). *Organizing international standardization—ISO and the IASC in quest of authority.* Cheltenham, UK/Northampton, MA: Edward Elgar Publishing.

Toth, R. B. (Ed.). (1997). *Profiles of national standards-related activities.* Gaithersburg, MD: NIST Special Publication 912, National Institute of Standards and Technology.

Wen, Z. (2004). *Reform and change—An introduction to China standardization.* Presentation at the 11th International Conference of Standards Users IFAN 2004, 2004-11-11-12, Amsterdam. Delft: NEN.

Willemse, H. (2003). *Management van normalisatie. Een model voor het managen van normalisatie* (In Dutch: *Management of standardization. A model for managing standardization*). MSc Thesis, Rotterdam School of Management, Erasmus University Rotterdam.

Willemse. H., De, Vries, H. J., Dul, J. (2006). Balancing Stakeholder Representation: An Example of Stakeholder Involvement in Ergonomics Standardization. In W. Karwowski (Ed.), *Handbook on standards and guidelines in ergonomics and human factors* (pp. 149–156). Mahwah, New Jersey: Lawrence Erlbaum Associates.

Wulff, I. A., Westgaard, R. H., and Rasmussen, B. (1999). Ergonomic criteria in large-scale engineering design-II Evaluating and applying requirements in the real work of design. *Applied Ergonomics, 30* (3), 207–221.

3

Ergonomic Performance Standards and Regulations—Their Scientific and Operational Basis

Thomas J. Smith
Human Factors Research Laboratory School of Kinesiology

Omar Merhi
University of Minnesota

INTRODUCTION AND BACKGROUND

This chapter addresses the use of performance standards in occupational health and safety (OHS) regulations, with particular reference to ergonomic performance standards. The sections in the following deal with (a) some relevant definitions, and an historical perspective on OHS performance standards; (b) how performance standards differ from prescriptive standards; (c) an introduction to ergonomic performance standards, and an analysis of their scientific and operational basis plus evidence regarding their criterion validity; and (d) recommendations regarding the application of ergonomic performance standards.

Definitions

No consensus agreement has emerged regarding the descriptive terminology applied to OHS standards and regulations generally, and to ergonomic standards in particular. Because there are instances of uncertainty, redundancy, and ambiguity in terminology employed by different authors in this area, Table 3.1 is provided to clarify the meanings of various terms in this chapter. It is not assumed that the definitions in Table 3.1 necessarily apply to other chapters in this handbook.

It is first important to clarify the distinctions among a law or regulation, a standard, and a guideline or recommendation. The terms *law* and *regulation*, considered synonymous here, both apply to a statement that requires legal compliance. In the OHS field, the meaning associated with the term *standard* is ambiguous. For example, usage by the U.S. Department of Labor Occupational Safety and Health Administration (OSHA) indicates that a regulation and a standard are synonymous (OSHA, 1980). However, Webster's dictionary definition—"anything recognized as correct by common consent, by approved custom, or by those most competent to decide"—does not imbue the term with legal status. This is the meaning adopted in this chapter, based on the definition provided (Table 3.1) by the American Academy of Pediatrics, American Public Health Association, and National Resource Center for Health and Safety in

TABLE 3.1
Definitions of Terms Relevant to OHS/Ergonomic Standards

Term	Definition	Reference[a]
Behavioral hazard	A manifestation of worker or organizational system behavior that creates an OHS hazard. An unsafe act.	1–3
Code	An entire system of rules or laws.	Webster
Engineering hazard	An engineering design factor that creates an OHS hazard. An unsafe condition.	1–3
Environmental hazard	An environmental design factor that creates an OHS hazard. An unsafe condition.	1–3
Ergonomic standard/regulation	A standard/regulation directed at OHS hazard mitigation through ergonomic intervention targeting correction of a specified workplace design factor or condition. Ergonomic standards address a broad range of workplace design issues, such as physical workload, workplace, equipment, work station, job, or warning design factors and conditions.	9–12
Generic performance standard or regulation	A performance standard/regulation that specifies the person or persons whose performance is responsible for meeting an OHS hazard mitigation objective, but does not specify how that objective should be achieved	6–8
Guideline	A statement of advice or instruction pertaining to practice. Like a recommendation, it originates in an organization with acknowledged professional standing. Although it may be unsolicited, a guideline is developed in response to a stated request or perceived need for such advice or instruction.	4
Hazard	An engineering, behavioral, or operational factor that elevates the risk of detrimental performance by one or more workers (employees or managers), or by an organizational system.	5
Law	That which is laid down or fixed. A rule laid down or established, whether by custom or as an expression of the will of a person or power able to enforce its demands.	Webster
Operational hazard	An OHS hazard arising as a consequence of interaction between behavioral factors and physical design factors.	3, 5
Performance standard/regulation	A standard/regulation that explicitly identifies the person or persons whose performance is responsible for meeting an OHS hazard mitigation objective specified in the standard (Definition 1).	3, 6–8
	A standard/regulation directed at mitigation of an OHS hazard associated with work demands that give rise to decrements in behavioral performance (Definition 2).	9–11
Physical hazard	An engineering or environmental design factor that creates an OHS hazard. An unsafe condition.	1–3
Physical standard	A standard directed at mitigation of a physical OHS hazard.	1–3
Physical workload standard	An ergonomic standard directed at mitigation of work performance and/or musculoskeletal health hazards arising out of worker exposure to work demands related to physical exertion and/or biomechanical forces.	9–11
Prescriptive standard/regulation	A standard/regulation that prescribes how a specific OHS hazard is to be mitigated, without any indication as to whose performance is responsible for mitigating the hazard. Typically directed at mitigation of physical OHS hazards.	3, 6–8
Process standard/regulation	A qualitative performance or prescriptive standard/regulation that defines a program approach for purposes of mitigating one or more specified OHS hazards.	10
Qualitative standard/regulation	A process standard/regulation that lacks precise, quantitative, numerical compliance criteria.	9–11
Quantitative standard/regulation	A performance or prescriptive standard/regulation that specifies precise, quantitative, numerical compliance criteria.	9–11

(Continued)

TABLE 3.1
(Continued)

Term	Definition	Reference[a]
Recommendation	A statement of practice aimed at providing an OHS benefit to the population served, usually initiated by an organization or a group of individuals with expertise or broad experience in the subject matter. A recommendation is not binding on the practitioner; that is, there is no obligation to carry it out.	
	A statement may be issued as a recommendation because it addresses a fairly new topic or issue, because scientific supporting evidence may not yet exist, or because the practice may not yet enjoy widespread acceptance by the members of the organization or by the intended audience for the recommendation.	4
Regulation	A regulation originates in an agency with either governmental or official authority, has the power of law, and is usually accompanied by an enforcement activity. A regulation often imbues a previous recommendation or guideline with legal authority. The components of a regulation vary by topic addressed and by area and level of jurisdiction. Because a regulation prescribes a practice that every agency or program must comply with, it usually is the minimum or the floor below which no agency or program should operate.	4
Rule	(1) a regular established method of procedure or action; or (2) a code.	Webster
Specification standard/regulation	A prescriptive standard/regulation.	3, 6–8
Standard	A statement that defines a goal of practice. It differs from a recommendation or a guideline in that it carries great incentive for universal compliance. It differs from a regulation in that compliance is not necessarily required for legal operation. It usually is legitimized or validated based on scientific or epidemiological data, or when this evidence is lacking, it represents the widely agreed on, state-of-the-art, high-quality level of practice. An entity that does not meet a standard may incur disapproval or sanctions from within or outside the organization. Thus, a standard is the strongest criteria for practice set by an organization or association.	4

[a]Numbers refer to references cited in the following footnotes.
[1]Heinrich (1931, 1959); [3]Smith (1973); [4]American Academy of Pediatrics, American Public Health Association, and National Resource Center for Health and Safety in Child Care (2002); [5]Smith (2002a); [6–8]Bryce (1983, 1985a, 1985b); [9]Fallentin et al. (2001); [10]Westgaard and Winkel (1996); [11]Viikari-Juntura (1997); [12]Dul, de Vlaming, and Munnik (1996).

Child Care (2002). Because it typically emanates from an entity with professional standing and credentials, a standard is a statement that carries a strong, but not regulatory, incentive for compliance. A *guideline* and *recommendation*, considered synonymous here, are statements that carry less incentive for compliance than a standard.

The basic objective of an OHS or ergonomic regulation, standard, guideline, or recommendation is to mitigate a hazard that poses a risk to the health and safety of the worker. There are essentially two distinct ways of formulating a requirement of this kind—as a performance standard or as a prescriptive standard. This chapter focuses on the former type of formulation, but it is important to define both approaches to OHS or ergonomic standard writing.

Typically, the term *performance regulation* rather than *performance standard* is used by authors concerned with a performance-based approach to formulation of OHS regulations (Bryce,

1983, 1985a, 1985b). The term *performance standard* is ambiguous, because it has been used in the OHS and ergonomic literature to refer to two different concepts. As noted by Fallentin, Viikari-Juntura, Waersted, and Kilbom, (2001), and by Westgaard and Winkel (1996), physical workload standards (Table 3.1) first were systematically delineated about 100 years ago as performance criteria concerned with mitigation of OHS hazards associated with work demands that give rise to decrements in behavioral performance (Definition 2 in Table 3.1). Extending the earlier work of Lavoisier in the 18th century, Taylor, Gilbreth, and others early in the 20th century developed performance guidelines for the physical workload capacity of workers that were directed at optimizing production efficiency (Fallentin et al., 2001).

On the other hand, as outlined in the next subsection, performance regulations and standards that specify who is responsible for meeting a specified requirement have been known since ancient times. As applied to OHS, this sense of the term 'performance standard' refers to a formulation that explicitly identifies the person or persons whose performance is responsible for meeting an OHS hazard mitigation objective specified in the standard (Table 3.1, Definition 1). In the remainder of this chapter, use of the terms *performance standard* or *performance regulation* is understood to refer to Definition 1 in Table 3.1.

There also is some uncertainty regarding the terminology to apply to an OHS standard or regulation that is not performance based. In this case, the formulation for the standard or regulation refers to what the OHS requirement is, but not to who is responsible for meeting the requirement. To this type of formulation, the terms *prescriptive, specification,* or *physical standard* or *regulation* (among others) have been applied (Table 3.1; Bryce, 1983, 1985a, 1985b; Fallentin et al., 2001; Smith, 1973; Westgaard & Winkel, 1996).

It is assumed here that the meanings of each of these terms are synonymous. However, in the opinion of these authors, none of these terms are entirely satisfactory. After all, an OHS requirement formulated as a performance standard or regulation obviously offers a prescription or specification about mitigation of a specified hazard. However, in line with customary usage, the terms *prescriptive standard* or *prescriptive regulation* are used in this chapter to refer to a formulation for an OHS or ergonomic standard or regulation that specifies what the requirement is, without any specification as to responsibility for compliance.

Let us close this section by addressing the distinction between a *process* or *qualitative standard* and a *quantitative standard* (Table 3.1). Whereas a quantitative standard specifies precise, quantitative, numerical compliance criteria for meeting an OHS or ergonomic requirement, a qualitative standard lacks such criteria and instead specifies a process or program approach for purposes of mitigating one or more OHS hazards (Fallentin et al., 2001; Viikari-Juntura, 1997; Westgaard & Winkel, 1996). It is assumed here that the terms *process standard* and *qualitative standard* have synonymous meanings. It also should be emphasized that qualitative and quantitative standards each can be formulated as either prescriptive or performance standards, although quantitative standards typically are formulated as prescriptive standards.

History of Performance Standards

The formulation of regulations, standards, and guidelines that explicitly state whose performance is responsible for meeting specified compliance requirements have an ancient history. Table 3.2 provides some examples relevant to health and safety, dating back to ancient Mesopotamia. The most widely known of these, likely familiar to hundreds of millions of people worldwide, is the sixth commandment. Its literal formulation (scanning right to left) is "no murder," but this is inaccurate. The first letter of the second Hebrew word is the imperfect inflection of "to murder" meaning "you will murder" (Harrison, 1955, p. 69) hence the translation.

TABLE 3.2
Some ancient performance standards relevant to health and safety.

Standard		Origin	Approximate Date[a]	Reference
Original Text	English Translation			
Not available	When a lion kills [animals in an enclosed yard], his keeper shall pay all damages, and the owner of the yard shall receive the killed animals.	Ancient Mesopotamia (Akkadian through Neo-Babylonian periods)	2250–550 BCE	www.fordham.edu/ halsall/ancient/ 2550mesolaws.html
Not available	If anyone opens his ditches to water his crop, but is careless, and the water flood the field of his neighbor, then he shall pay his neighbor for his loss.	Code of Hammurabi, Article 55	1955–1913 BCE	www.yale.edu/ lawweb/avalon/ medieval/hammint. htm
Not available	The builder has built a house for a man and his work is not strong and if the house he has built falls in and kills a householder, that builder shall be slain.	Code of Hammurabi, Article 229	1955–1913 BCE	Foliente (2000, p. 13)
לֹא תִּרְצָח	You shall not murder.	6th Commandment	1300 BCE	Plaut (1981, p. 554)
Not available	No person shall hold meetings in the City at night.	The Twelve Tables (original Roman Code of Law)	451–450 BCE	www.csun.edu/ ~hcfl1004/ 12tables.html

[a]BCE, before current era.

Six of the remaining 10 commandments—numbers 2, 3, and 7 to 10—are similarly formulated as performance standards. Thus, well over 3,000 years ago, ancient Jews understood that God's fundamental words, passed on to them through Moses, not only prescribed basic rules of behavior but also carried the connotation of personal responsibility for performance in complying with these rules—the essence of a performance standard formulation.

It also is noteworthy that the term *law*, with an Old Norse etymology dating back to development by the Vikings of a code of common law first written down in the Icelandic Jonsbok Manuscript in the 12th century (www.mnh.si.edu/vikings/, Room 6), refers to a rule laid down or established (Table 3.1). A *rule*, in turn, is defined as a regular established method of procedure or action. Thus, there are connotations of both prescription and performance in our modern understanding of what a law means.

Recent Trends in Performance Standards and Regulations

Despite their ancient origins, performance standards or regulations have received little analytical attention from those in the HF/E or legal communities concerned with the formulation of OHS or ergonomic rules. In their recent reviews of ergonomic standards, no reference to the Definition 1 sense of performance standards (Table 3.1) is made by Dul, de Vlaming, and Munnik (1996), Fallentin and colleagues (2001), Viikari-Juntura (1997), or Westgaard and Winkel (1996). A February 2003 search of the World Wide Web (using Google) found over 500,000 Web pages with the exact phrases "performance standards" or "performance standard," and over 1,600 with the exact phrases "performance regulations" or "performance regulation." In contrast, searches for all combinations of "occupational health and safety"/"occupational safety and health"/"OHS" or "OSH" with "performance," followed by "standards"/"standard"/ "regulations" or "regulation" revealed only 35 sites. Searches for all combinations of "ergonomic performance" followed by "standards"/"standard" or "regulations"/"regulation" found one site containing the first author's treatment of the topic (Smith, 2002b; http://www.doli.state.mn.us/pdf/overview.pdf).

Nevertheless, over the past 3 decades, a number of jurisdictions have adopted performance regulations as a routine part of their OHS regulatory code. In Canada, the OHS Acts of every province and territory feature routine and extensive use of performance regulations, prompted in no small part by a 1981 report ("Reforming Regulations") from the Economic Council of Canada encouraging Canadian governments to establish a national set of OHS regulatory provisions that would provide uniform and consistent OHS legislation for all Canadian workers directed at reducing occupational injuries and diseases (Bryce, 1983). This report referenced the developing use of performance regulations and pointed out that a performance-based approach allowed employers to make cost-efficient choices in complying with OHS regulatory provisions.

In a similar vein, the OHS Acts for every state and territory in Australia also make extensive use of performance regulations. Web site links to OHS Acts for governmental jurisdictions in both Canada and Australia may be found at http://www.eu-ccohs.org and http://www.nohsc-eu.gov.au/, respectively.

There also are U.S. examples of deployment of OHS performance regulations. Thus, although they are built largely around prescriptive regulations, OSHA Occupational Safety and Health Standards (OSHA, 1980) nevertheless feature 218 performance-based "employer shall," and 90 "employee shall" provisions (http://www.osha-slc.gov/pls/oshaweb/owastand. display_standard_group?p_toc_level=1&p_part_number=1910&p_text_version=FALSE). In Minnesota, the AWAIR (A Workplace Accident and Injury Reduction) Act makes extensive use performance regulations in order to target workplace accident and injury reduction (Minnesota Department of Labor and Industry, 1993).

Another health and safety domain in which the performance-based approach to standard setting is receiving growing emphasis is that of building performance (Foliente, 2000; International Council for Building Research Studies and Documentation [CIB], 1982; Prior & Szigeti, 2003). The application of performance-based building regulations dates back to the code of Hammurabi, in which Article 229 mandated the death penalty for any builder whose building collapsed and killed an occupant (Table 3.2). Prior and Szigeti (IBID) summarize a record of growing application of performance-based building codes and regulations in the United States, Europe, and Australasia over the past 5 decades. More broadly, in 1997, the World Trade Organization, recognizing that prescriptive codes and standards represent major nontariff barriers to trade, adopted the following clause in its Agreement on Technical Barriers to Trade (Foliente, 2000): "Wherever appropriate, members shall specify technical regulations based on product requirements in terms of performance rather than design or descriptive characteristics."

The basic philosophy underlying this approach is summarized by CIB (1982): "the performance approach is, first and foremost, the practice of thinking and working in terms of ends rather than means." Foliente (2000) notes the advantage of this approach for building construction: "The most serious problem with the prescriptive approach is that it serves as a barrier to innovation. Improved and/or cheaper products may be developed, yet their use might not be allowed if construction is governed by prescriptive codes and standards."

A prominent example of the application of performance standards outside the realm of OHS are the ISO 9001 quality management and ISO 14000 environmental management standards, to which many thousands of companies worldwide are certified, and which are formulated exclusively as performance standards (International Organization for Standardization, 1994).

The perspective on OHS/ergonomic performance regulations offered in this chapter is informed by two primary background sources. The first is the seminal chapter by K. U. Smith (1973), the first publication (to our knowledge) to provide a systematic human factors rationale for a performance-based approach to the formulation of OHS regulations. The publications of Bryce (1983, 1985a, 1985b) are the second source, generated in the course of his work with the Province of Alberta, Canada, on systematic incorporation of performance regulations into their occupational health and safety act. Examples in the following sections of how performance regulations are used in OHS legislation are drawn from the OHS Regulation for the Province of British Columbia (B.C.), Canada, which features extensive use of performance regulations (Workers' Compensation Board of British Columbia, 1999c).

PERFORMANCE VERSUS PRESCRIPTIVE REGULATIONS IN OCCUPATIONAL HEALTH AND SAFETY LEGISLATION

A performance standard may contain both performance and prescriptive criteria. However, a *generic performance standard or regulation* (Table 3.1) describes what the OHS objective is (the prescriptive criterion), and who is responsible for meeting the objective (the performance criterion), but does not prescribe explicitly how the objective should be achieved (Bryce, 1983). This section compares and contrasts the performance versus the prescriptive approach to formulation of regulations promulgated in actual OHS legislation.

To illustrate some key differences between these two approaches, consider the performance-based regulation dealing with general requirements for machine guarding adopted by B.C., versus the prescriptive regulation adopted by OSHA.

Example 1

The B.C. regulation reads as follows (Workers' Compensation Board of British Columbia, 1999a, Section 12.2, Safeguarding requirement, p. 12-1):

"12.2 Safeguarding requirement
The employer must ensure that machinery and equipment is fitted with adequate safeguards which

- a. protect a worker from contact with hazardous power transmission parts,
- b. ensure that a worker cannot access a hazardous point of operations, and
- c. safely contain any material ejected by the work process which could be hazardous to a worker.'

In contrast, the comparable OSHA regulation (OSHA, 1980, 1910 Subpart O—Machinery and Machine Guarding, 1910.212, General requirements for all machines) reads:

"1910.212 General requirements for all machines
1910.212(a)
Machine guarding.

1910.212(a)(1)

Types of guarding. One or more methods of machine guarding shall be provided to protect the operator and other employees in the machine area from hazards such as those created by point of operation, ingoing nip points, rotating parts, flying chips and sparks. Examples of guarding methods are barrier guards, two-hand tripping devices, electronic safety devices, etc.

1910.212(a)(2)

General requirements for machine guards. Guards shall be affixed to the machine where possible and secured elsewhere if for any reason attachment to the machine is not possible. The guard shall be such that it does not offer an accident hazard in itself.

1910.212(a)(3)

Point of operation guarding.

1910.212(a)(3)(i)

Point of operation is the area on a machine where work is actually performed upon the material being processed.

1910.212(a)(3)(ii)

The point of operation of machines whose operation exposes an employee to injury, shall be guarded. The guarding device shall be in conformity with any appropriate standards therefor, or, in the absence of applicable specific standards, shall be so designed and constructed as to prevent the operator from having any part of his body in the danger zone during the operating cycle.

1910.212(a)(3)(iii)

Special handtools for placing and removing material shall be such as to permit easy handling of material without the operator placing a hand in the danger zone. Such tools shall not be in lieu of other guarding required by

this section, but can only be used to supplement protection provided.

1910.212(a)(3)(iv)

The following are some of the machines which usually require point of operation guarding:

1910.212(a)(3)(iv)(a)

Guillotine cutters.

1910.212(a)(3)(iv)(b)

Shears.

1910.212(a)(3)(iv)(c)

Alligator shears.

1910.212(a)(3)(iv)(d)

Power presses.

1910.212(a)(3)(iv)(e)

Milling machines.

1910.212(a)(3)(iv)(f)

Power saws.

1910.212(a)(3)(iv)(g)

Jointers.

1910.212(a)(3)(iv)(h)

Portable power tools.

1910.212(a)(3)(iv)(i)

Forming rolls and calenders.

1910.212(a)(4)

Barrels, containers, and drums. Revolving drums, barrels, and containers shall be guarded by an enclosure which is interlocked with the drive mechanism, so that the barrel, drum, or container cannot revolve unless the guard enclosure is in place.

1910.212(a)(5)

Exposure of blades. When the periphery of the blades of a fan is less than seven (7) feet above the floor or working level, the blades shall be guarded. The guard shall have openings no larger than one-half (1/2) inch.

1910.212(b)

Anchoring fixed machinery. Machines designed for a fixed location shall be securely anchored to prevent walking or moving."

END EXAMPLE 1.

Example 1 provides a framework for addressing in the following subsections: (a) generic versus narrowly focused OHS performance regulations; (b) formulation of performance and prescriptive OHS regulations; (c) advantages of performance relative to prescriptive OHS regulations; and (d) disadvantages of performance relative to prescriptive OHS regulations.

Generic Versus Narrowly Focused OHS Performance Regulations

There are two general types of OHS performance regulations, *broadly focused or generic* (Example 1) and *narrowly focused* (Bryce, 1983, 1985a). An OHS *generic* performance regulation embodies the following major elements:

- It identifies a particular class of participant in an OHS system whose performance is responsible for meeting a specified OHS objective.
- It describes an OHS objective but does not explicitly dictate or prescribe how the objective should be achieved.
- It represents a single, complete regulatory statement whose meaning and intent are clearly stated without being accompanied by numerous qualifying provisions.
- It is conceptually broader in its application than a prescriptive regulation, and as such, its regulatory scope encompasses a broader range of hazards and hazard abatement possibilities.
- It is neither hazard specific nor industry specific in its formulation and, therefore, is potentially applicable to any workplace.

Referring to Example 1, the B.C. generic performance regulation dealing with general requirements for machine guarding differs from the parallel OSHA prescriptive regulation with respect to all of these major elements. Noteworthy differences that merit emphasis are that, relative to the B.C. regulation, formulation of the OSHA regulation (a) is longer and more complicated, (b) contains a number of narrow qualifying provisions, and (c) is both hazard specific and industry specific.

In contrast to a generic OHS performance regulation, a *narrowly focused* OHS performance regulation is hazard specific and may also be industry specific. Consequently, relative to the former, the formulation of the latter more closely resembles that of a prescriptive regulation. However, as the following example illustrates, a narrowly focused OHS performance regulation still exhibits key differences from a prescriptive regulation. Example 2 contrasts a narrowly focused OHS performance regulation promulgated by the Province of British Columbia, Canada, for control of toxic and hazardous substances, with a prescriptive regulation promulgated by OSHA for the same purpose.

Example 2

The B.C. regulation (Workers' Compensation Board of British Columbia, 1999b, Sections 5.48 & 5.55, pp. 5-10 to 5-12) reads:

"5.48 Exposure limits
The employer must ensure that a worker's exposure to a substance does not exceed the exposure limits listed in Table 5-4.
5.49 through 5.54
[not included here]
5.55 Type of controls

(1) If there is a risk to a worker from exposure to a hazardous substance by any route of exposure, the employer must eliminate the exposure, or otherwise control it below harmful levels and below the applicable exposure limit listed in Table 5-4 by
  a. substitution,
  b. engineering control,
  c. administrative control, or
  d. personal protective equipment
(2) When selecting a suitable substitute, the employer must ensure that the hazards of the substitute are known, and that the risk to workers is reduced by its use.
(3) The use of personal protective equipment as the primary means to control exposure is permitted only when
  a. substitution, or engineering or administrative controls are not practicable, or
  b. additional protection is required because engineering or administrative controls are insufficient to reduce exposure below the applicable exposure limits, or
  c. the exposure results from temporary or emergency conditions only."

The comparable OSHA regulation (OSHA, 1980; 1910 Subpart Z—Toxic and Hazardous Substances, 1910.1000, Air contaminants) reads:

"1910.1000   Air contaminants
An employee's exposure to any substance listed in Tables Z-1, Z-2, or Z-3 of this section shall be limited in accordance with the requirements of the following paragraphs of this section.
1910.1000(a) through **1910.1000(d)**
[not included here]
1910.1000(e)
To achieve compliance with paragraphs (a) through (d) of this section, administrative or engineering controls must first be determined and implemented whenever feasible. When such controls are not feasible to achieve full compliance, protective equipment or any other protective measures shall be used to keep the exposure of employees to air contaminants within the limits prescribed in this section. Any equipment and/or technical measures used for this purpose must be approved for each particular use by a competent industrial hygienist or other technically qualified person. Whenever respirators are used, their use shall comply with 1910.134.

END EXAMPLE 2.

Table 5-4 referenced by the B.C. regulation and Tables Z-1 through Z-3 referenced by the OSHA regulation each contains prescriptive exposure limits for a long list of chemicals that largely are identical for the two regulations.

Example 2 illustrates the point that even though a narrowly focused performance regulation may contain prescriptive elements, it still may differ in important ways from a purely prescriptive regulation. Thus, even though both the B.C. and the OSHA regulations establish specific, prescriptive exposure limits for each chemical listed, only the B.C. regulation:

- Explicitly specifies that it is the employer's responsibility to ensure that a worker's exposure to a listed chemical does not exceed the exposure limits
- Allocates to the employer the responsibility for determining exactly *how* the prescribed exposure limit for each chemical will be achieved
- Does not explicitly specify the particular means the employer must use to satisfy the objective of controlling worker exposure to any of the listed chemicals

Formulation of Performance and Prescriptive OHS Regulations

This section compares and contrasts basic considerations underlying the formulation, or drafting, of performance and prescriptive OHS regulations (Bryce, 1985a). A well-drafted prescriptive OHS regulation typically exhibits the following features, or satisfies the following criteria, in its formulation.

1. It clearly states the specific means that must be applied or satisfied for purposes of compliance with the objective stated in the regulation. That is, it provides explicit instructions about what to do and how to do it, for purposes of regulatory compliance. Little or no latitude is provided for pursuing alternative strategies for compliance.
2. The class of personnel with performance responsibility for compliance with the stated regulatory objective is not explicitly identified in the regulation. However, virtually without exception, the implication is that the employer has responsibility for compliance.
3. A prescriptive regulation is most often promulgated when:
    ◦ The nature of the hazard, whose mitigation is targeted by the regulation, is such that at least one specific hazard mitigation strategy can be identified;
    ◦ One defined method, approach, or design feature, prescribed in the regulation, will definitely accomplish the hazard mitigation objective of the regulation (a prescriptive provision may lay out a series of alternatives for hazard mitigation, but each alternative typically is formulated as a prescriptive option).
    ◦ A particular, defined safety standard needs to be adopted and applied in a regulatory framework.
    ◦ Certification of regulatory compliance is required by a certifying body or individual.
4. It may be difficult to ascertain the underlying rationale for the regulatory prescription (the "why" behind the stated objective) from the stated prescriptive provisions.

In contrast, a well-drafted OHS performance regulation typically satisfies the following criteria, or exhibits the following features, in its formulation.

1. It should clearly specify the condition(s) under which the provision will apply.
2. It should identify a class of personnel whose work performance is responsible for meeting the stated regulatory objective.
3. It should clearly indicate the regulatory objective to be achieved, through use of appropriate descriptive language or definitions pertaining to mitigation of a specified hazard or hazards.
4. Criteria for gauging whether compliance with the stated regulatory objective has been achieved should be readily apparent from the formulation of the regulation, without need for recourse to administrative or legal interpretation. That anyone referencing the

regulation will be able to readily determine that compliance with its regulatory objective has been satisfied represents a key test of a meaningful performance regulation.

5. The performance objective should be phrased so as to be potentially applicable to a range of possible contexts or situations involving the hazard addressed by the stated objective.
6. Use of general qualifying terms, such as *suitable* or *sufficient,* should be avoided, unless such terms are accompanied by additional guidance in the regulation that clearly indicates what will be considered to be suitable, sufficient, and so forth.
7. It can be generic or narrowly focused (previous section).
8. The specified objectives can include or be accompanied by qualifying provisions. A regulation with such "qualified performance objectives" combines a performance objective with a description of provisions that need to be considered for purposes of compliance with the objective.
9. It can be combined with prescriptive provisions in a given section of OHS regulations (see previous section).
10. Its promulgation generally is advisable under the following circumstances: (a) when the hazard of concern is chronic (i.e., long-standing) in nature; or (2) when the primary deficiency in the regulatory code is lack of information for the employer about mitigation of a particular hazard—hazards whose mitigation may best be served by such "informational" regulations typically are of a low-risk nature.
11. The performance requirements specified in the regulation should be clearly compatible with the performance capabilities and limitations of the class of personnel specified in the regulation.

To illustrate the rationale for criterion 11, it would be reasonable to require an employer, but not an employee or even a supervisor, to establish OHS policies and programs that comply with a promulgated OHS code.

To illustrate some of the above criteria regarding performance regulation formulation, the following example contrasts a more loosely with a more carefully drafted performance regulation. The more loosely drafted regulation, the so-called OSHA 'general duty' clause (OSHA, 1970, Section 5—Duties), reads as follows.

Example 3

"5. Duties

(a) Each employer
   (1) shall furnish to each of his employees employment and a place of employment which are free from recognized hazards that are causing or are likely to cause death or serious physical harm to his employees;
   (2) shall comply with occupational safety and health standards promulgated under this Act.
(b) Each employee shall comply with occupational safety and health standards and all rules, regulations, and orders issued pursuant to this Act which are applicable to his own actions and conduct."

In contrast, the B.C. general duty regulations for employers and employees (Workers' Compensation Board of British Columbia, 1999c, Sections 115-124, pp. xiii–xv) read as follows.

"Section 115 General duties of employers

(1) Every employer must
   a. ensure the health and safety of
      (i) all workers working for that employer, and

(ii) any other workers present at a workplace at which that employer's work is being carried out, and

b. comply with this Part, the regulations and any applicable orders.

(2) Without limiting subsection (1), an employer must

a. remedy any workplace conditions that are hazardous to the health or safety of the employer's workers,

b. ensure that the employer's workers

(i) are made aware of all known or reasonable foreseeable health or safety hazards to which they are likely to be exposed by their work,

(ii) comply with this Part, the regulations and any applicable orders, and

(iii) are made aware of their rights and duties under this Part and the regulations,

c. establish occupational health and safety policies and programs in accordance with the regulations,

d. provide and maintain in good condition protective equipment, devices and clothing as required by regulation and ensure that these are used by the employer's workers,

e. provide to the employer's workers the information, instruction, training and supervision necessary to ensure the health and safety of those workers in carrying out their work and to ensure the health and safety of other workers at the workplace,

f. make a copy of this Act and the regulations readily available for review by the employer's workers and, at each workplace where workers of the employer are regularly employed, post and keep posted a notice advising where the copy is available for review,

g. consult and cooperate with the joint committees and worker health and safety representatives for workplaces of the employer, and

h. cooperate with the board, officers of the board and any other person carrying out a duty under this Part or the regulations.

Section 116: General Duties of Workers

(1) Every worker must

a. take reasonable care to protect the worker's health and safety and the health and safety of other persons who may be affected by the worker's acts or omissions at work, and

b. comply with this Part, the regulations and any applicable orders

(2) Without limiting subsection (1), a worker must

a. carry out his or her work in accordance with established safe work procedures as required by this Part and the regulations,

b. use or wear protective equipment, devices and clothing as required by the regulations,

c. not engage in horseplay or similar conduct that may endanger the worker or any other persons

d. ensure that the worker's ability to work without risk to his or her health or safety, or the health or safety of any other person, is not impaired by alcohol, drugs or other causes,

e. report to the supervisor or employer

(i) any contravention of this Part, the regulations or an applicable order of which the worker is aware, and

(ii) the absence of or defect in any protective equipment, device or clothing, or the existence of any other hazard, that the worker considers is likely to endanger the worker or any other person,

       f. cooperate with the joint committee or worker health and safety representative for the workplace, and

       g. cooperate with the board, officers of the board and any other person carrying out a duty under this Part or the regulation."

END EXAMPLE 3.

The B.C. general duty regulation also goes on to specify general duties for supervisors, owners, suppliers, and directors and officers of a corporation.

Referring to criteria 1–11 listed previously for a well-drafted performance regulation, the B.C. general duty regulation largely or entirely conforms to criteria 1 to 8, and 11, and thus can be considered to represent a more carefully drafted generic performance regulation. The OSHA general duty regulation largely or entirely conforms to criteria 1 to 3, 5, and 7, and thus can be considered to represent a more loosely drafted generic performance regulation. For example, in contrast to the qualifying provisions in the B.C. regulation (Part 2 in Sections 115 and 116), there are no qualifying provisions in the OSHA regulation that make it readily apparent whether or not compliance has been achieved (criteria 4 and 8).

Furthermore, in the OSHA regulation, the precise meanings of the terms "free of recognized hazards" and "serious physical harm" are unclear (criterion 6), in the sense that the regulation itself provides no guidance either as to what constitutes serious harm or as to how an employer should perform to recognize hazards and remove them from the workplace. The B.C. regulation also refers to health and safety hazards, but the mandate for employer performance is "to remedy any workplace conditions that are hazardous" (Section 115-(2)-a), the implication being that the nature of a particular hazardous workplace condition has become known to the employer through prior effect.

Finally, it is not clear from the OSHA general duty regulation for employees how a U.S. worker should go about fulfilling the performance requirement that mandates employee awareness and understanding of the full range and scope of "occupational safety and health standards and all rules, regulations, and orders issued pursuant to this Act which are applicable to his own actions and conduct." It turns out that sprinkled throughout the OSHA Act are information provision and training requirements aimed at assisting employees in meeting this performance requirement. Nevertheless, no guidance is provided in the Section 5 of the act itself as to how an employee should satisfy the compliance performance requirement (criterion 11).

In contrast, although the B.C. regulation has a similar performance provision mandating employee compliance with OHS regulations (Section 116-(1)-b), this is accompanied by explicit qualifying provisions that proscribe employee behaviors (such as horseplay and substance abuse) that may compromise compliance. Furthermore, there are two qualifying provisions for employer performance designed to assist with worker compliance, one dealing with training (Section 115-(2)-e), the other with posting of information (Section 115-(2)-f). These explicit qualifying provisions in the B.C. general duty regulations for both employees and employers provide reasonably clear guidance for B.C. workers regarding the manner in which their compliance performance requirement is to be achieved.

Advantages of Performance Relative to Prescriptive OHS Regulations

Arguments for building OHS legislation around performance rather than prescriptive regulations may be categorized into three classes, relating to regulatory efficiency, regulatory effectiveness, and the goal of regulatory intervention. Subsections in the following sections deal with each of these areas.

Regulatory Efficiency of OHS Performance Regulations

The term *regulatory efficiency* as used here refers to the volume and detail of regulatory provisions that may be required to achieve a particular OHS regulatory objective. In contrast to prescriptive OHS regulations, generic OHS performance regulations typically are more efficient, in the sense that their formulation is designed to address a broad range of hazards in a manner that lacks excessive detail and complexity (Bryce, 1983). Example 1 dealing with machine guarding in the preceding section illustrates this point. That is, relative to the more efficient B.C. machine guarding generic performance regulation, the less efficient prescriptive OSHA machine-guarding regulation (a) is longer and more complicated; (b) contains a number of narrow qualifying provisions, and (c) is both hazard specific and industry specific.

When it comes to OHS regulation, is regulatory efficiency a desirable objective? An affirmative answer to this question is suggested by considerations that, relative to a longer, more detailed regulation, a regulation with moderate length, detail, and complexity (a) is easier to understand; (b) is likely to more effectively support OHS information transmittal and training; (c) may be more readily interpretable by workers, employers, OHS specialists, and regulatory bodies; and (d) thereby may encourage a greater dedication to compliance. Because the appropriate research is lacking, these predictions remain speculative at this point.

Regulatory Effectiveness of OHS Performance Regulations

As used here, the term *regulatory effectiveness* refers to the degree to which the stated objective of an OHS regulation—mitigation of a specified hazard—is in fact realized. It is the apparent lack of regulatory effectiveness that represents one of the strongest arguments against the use of prescriptive OHS regulations. Documentary evidence for such a lack is derived largely from evaluations of the effectiveness of U.S. OSHA regulations, regulations that are largely prescriptive in nature. A rather consistent finding is that OSHA prescriptive regulations directed at mitigating specified hazards do not reliably and predictably result in accident and injury prevention, the putative goal of the regulations (Smith, 2002a).

For example, Jones in 1973 documented the limitations of dependence on prescriptive engineering control regulations in bringing down occupational injury rates to acceptable levels. Gill and Martin (1976) echo this finding with the claim that the prescriptive engineering approach to safety management is not sufficient to prevent all accidents, and that innovative hazard control strategies such as performance standards and worker participation should be considered to achieve further improvement in accident prevention. M.J. Smith and colleagues (1971) and Gottlieb and Coleman (1977) report that inspections carried out by OSHA are relatively ineffective in preventing accidents, because most hazards cannot be identified by traditional workplace inspections that rely on prescriptive regulations. Results from these and other studies suggest that, in general, only 5% to 25% of accidents can be avoided by rigorous compliance with conventional safety standards (Ellis, 1975; Smith, 1979). OSHA itself has noted that, "OSHA's own statistics appear to indicate that 70% to 80% of all deaths and injuries each year are not attributable to a violation of any OSHA specification standard" (Occupational Safety and Health Reporter, 1976, p. 684).

What is the basis for this apparent lack of effectiveness of prescriptive OHS regulations in preventing occupational accidents and injuries? The answer offered here is that a prescriptive regulation only addresses one part of what a hazard actually represents. In operational terms, a hazard manifests itself as a human factors consequence of performance–design interaction, the fundamental concern of human factors/ergonomic (HF/E) science (T. J. Smith, 1993, 1994, 1998, 2002a; T. J. Smith, Henning, & Smith, 1994). That is, realization of the effect of an OHS hazard (degradation in operational performance, accident, injury, or death) depends on the interaction of worker behavior and performance with a physical workplace condition. The assumption is that because their formulation ignores this operational reality of the actual nature of OHS hazards, prescriptive OHS regulations are inherently limited in their impact on the consequences of such hazards.

It is from this perspective that K.U. Smith (1973) advocates use of OHS performance standards as the most effective strategy for achieving the hazard mitigation and accident and injury prevention goals of OHS regulation. This author introduces and defines the term *operational hazards* (IBID; K.U. Smith, 1979, 1988, 1990) to refer to hazards that arise as a consequence of highly variable (relative to both within- and between-worker performance) and largely unpredictable behavioral interactions of individuals or groups with existing physical (microergonomic) or organizational (macroergonomic) workplace design conditions.

From this human factors systems perspective therefore, most occupational hazards represent incidents or conditions which exist or arise through the variable, sometimes transitory, and typically synergistic convergence of organizational, behavioral, operational, and physical-environmental factors specific to particular tasks, job operations, and sociotechnical systems. The complexity of this interaction mocks simplistic attempts to pigeonhole hazards into a limited number of categories (e.g., physical vs. behavioral); defeats efforts to attribute occupational accidents, injuries, and safety problems to one type of hazard or another; and stymies abatement strategies based on dissective, prescriptive assumptions about the nature of hazards.

The basic assumption underlying this analysis is that in today's workplace, operational hazards represent the most prevalent source of work-related performance decrements, accidents, or injuries. Typically, these are not addressed by prescriptive OHS regulations, are not detected by outside inspectors, and are not abated by remedial approaches prompted by standardized accident and injury investigations. Further, operational hazards are context specific to particular workers, tasks, jobs, and operations, in a manner that mirrors the context specificity observed generally with human behavior and performance (T. J. Smith, 1998; T. J. Smith, Henning, & Smith, 1994). These considerations suggest that, guided by dictates of the Pareto principle (Juran, 1964, 1995), OHS regulations should focus on operational hazards as a major priority for hazard mitigation and accident and injury prevention.

The effectiveness argument, therefore, posits that in contrast to prescriptive regulations, OHS performance regulations are better designed to provide this focus because their formulation embodies the closed-loop relationship between worker behavior and specified workplace conditions characteristic of operational hazards. What evidence is available to support this prediction?

No systematic research is yet available to provide a conclusive answer to this question. Negative evidence cited previously regarding the limited effectiveness of prescriptive OHS regulations is suggestive. More persuasive, in a narrow sense, is the demonstrated effectiveness of B.C. ergonomics performance regulations in reducing musculoskeletal injuries (MSIs), presented in Section 3. Yet a third line of evidence stems from research on the ability of workers to recognize and control workplace hazards.

The basic premise of this research—grounded in the concept of worker self-regulation of hazard management (T. J. Smith, 2002a)—is that compared with managers, safety professionals, or government regulators, workers often know as much or more about their jobs, about job-related hazards that they encounter on a daily basis, and about how to reduce or eliminate those hazards. That is, whereas prescriptive OHS regulations tend to focus narrowly on physical–environmental hazards, workers tend to identify a broader range of hazards, often related to job performance, to dynamic operational circumstances, or to intermittent or transitory work situations or conditions, for which no explicit prescriptive regulations or standard operating procedures exist.

Studies on hazard self-regulation by workers support the conclusion that workers represent a relatively untapped resource for hazard control. Thus, Swain (1973, 1974) refers to workers as subject-matter experts on their work situations, whose input to management can serve as a key contributor to safety performance. He notes that worker hazard data represent a rare but effective source of information about near accidents and potentially unsafe situations. Hammer (1972, 1976) likewise presents evidence showing that workers have high awareness of job-related hazards, and that asking them directly represents the best way of obtaining this information. Both authors note marked discrepancies between worker-provided hazard data and that provided by more conventional specialist reports and postaccident investigations.

That workers can recognize hazards, and have knowledge and skill in self-controlling them to improve safety performance, likewise has been tested with positive results in a series of cross-industry studies conducted through the University of Wisconsin Behavioral Cybernetics Laboratory (Cleveland, 1976; Coleman & Sauter, 1978; Coleman & Smith, 1976; Gottlieb, 1976; Kaplan & Coleman, 1976; Kaplan, Knutson, & Coleman, 1976; Richardson, 1973; Smith, 1994). Findings from this research reveal that employees are one of the best sources of information about day-to-day hazards of their jobs, and support the following major conclusions:

- Workers generally identify hazards that most directly affect their personal safety. Unsafe conditions and operating procedures often are specified, but it is not uncommon for possibly hazardous actions or behaviors to be candidly cited.
- Worker-identified hazards largely comprise those of immediate self-concern.
- Studies of six Wisconsin metal processing plants (Cleveland, 1976; Coleman & Sauter, 1978) found that employee-identified hazards bear a significantly closer resemblance to actual causes of accidents than do those specified in prescriptive OHS regulations or found by state inspections.
- Workers typically can clearly delineate their actions in avoiding, reducing, or controlling hazards.
- Workers tend to identify health hazards in terms of their acute toxic effects, rather than of their long-term potential for causing illness or disease. There is substantial awareness of engineering defects in ventilation or personal protection which may cause increased risk of toxic exposure.
- Employers provided with worker hazard data generally conclude that (a) employees can provide more and better information about job hazards than other available sources, (b) workers often identify hazards that management is totally unaware of, and (c) survey results are of value in establishing priorities for safety and HM procedures.

What do these findings regarding the ability of workers to understand job-related hazards and their mitigation have to do with the effectiveness of OHS performance regulations? First, the findings tend to validate performance regulations dealing with employee duties, featured in the OHS acts for B.C. and other Canadian provinces, that call for the employee to "carry out his or her work in accordance with established safe work procedures as required by this Part and the regulations" (Example 3). Second, unlike prescriptive OHS regulations, both workers and generic OHS performance regulations are more attuned to the nature and significance of operational hazards, the former in terms of awareness and understanding, the latter in terms of design and scope. The argument therefore is that the consequences of a regulation mandating performance that will accommodate mitigation of operational hazards is likely to mirror those of efforts to call on worker expertise for the same purpose.

Performance Regulations and the Purpose of OHS
Regulatory Intervention

This brings us to perhaps the most provocative argument regarding the putative benefits of OHS performance regulations, an argument that gets at the fundamental purpose of OHS regulatory intervention. The two basic questions are: Is regulatory intervention necessary, and if so, how should regulatory provisions be formulated to best accomplish their objectives? The first question represents a matter of active debate with regard to ergonomics regulations (Fallentin et al., 2001), and will not be considered here. The second question has been considered in depth by Bryce (1983), with the conclusion that there is an inherent contradiction between the avowed purpose of OHS regulatory intervention, and use of prescriptive regulations to achieve this purpose.

To understand the logic of this argument, consider the OSHA regulations. They comprise many hundreds of prescriptive provisions, almost all narrowly focused, many of them task specific or industry specific (OSHA, 1980). The OSHA general duty clause (Example 3) mandates that each employer shall furnish "a place of employment . . . free from recognized hazards that are causing or are likely to cause death or serious physical harm to his employees." The presumption of this mandate is that it is the employer's responsibility to ensure that his or her workplace is free from hazards that otherwise might violate the regulation. The code then goes on to describe hundreds of hazards that the employer should consider for compliance with the mandate. Yet the prescriptive formulation for many of the regulations not only describes *what* the hazard is, it goes on to prescribe *how* the hazard should be mitigated. In other words, as a consequence of the manner in which the regulations are drafted, it is not the employer but the regulatory body—the government—that has assumed major responsibility for ensuring that the workplace is free of dangerous hazards, by essentially dictating to the employer, under force of law, both the need for action and the course of action.

Clearly, there are political, socioeconomic, and legal implications of OHS regulations generally, and this perspective on the inherently contradictory nature of prescriptive OHS regulations particularly, that go beyond the scope of this chapter. From a human factors perspective, however, it is probably fair to say that neither the employer nor the government find entirely satisfactory a situation, arising primarily as a consequence of the prescriptive regulatory approach, in which the former abrogates substantial responsibility to the latter for prescriptive guidance on how a workplace should be operated to protect worker health and safety.

As the analysis of Bryce (1983) suggests, there have been three primary influences behind the accretion of today's voluminous code of largely prescriptive OHS regulations in the United States. The first is the influence and tradition of English common law, with its emphasis on a prescriptive approach to legislation drafting. A second, related influence is that an early, major impetus for OHS regulations was to control worker exposure to hazardous substances through formulation of prescriptive exposure limits. A third influence is lack of rationalization over the years of the OHS regulatory code, related in no small part to lack of appreciation on the part of both legislators and drafting bodies that primarily informational regulations targeting (typically lower risk) hazards need not necessarily contain the detailed guidance that regulations targeting higher risk hazards might require.

Because they inform the employer of the existence and nature of a hazard that requires mitigation, but leave it up to the employer to determine how the mitigation might be accomplished, generic OHS performance regulations avoid the inherently contradictory design of prescriptive OHS regulations as regards regulatory intervention. This may be deemed a benefit for both employers and regulatory bodies alike. Moreover, as noted earlier, if mitigation of a particular hazard (such as a hazardous substance) is best served by the prescriptive approach (e.g., exposure limits), then a narrowly focused performance regulation can be drafted that is accompanied by prescriptive provisions (Example 2).

Disadvantages of Performance Relative to Prescriptive OHS Regulations

Some of the major advantages of performance regulations cited in the previous section also may be construed as major disadvantages in terms of translating the abstract aims of OHS regulatory intervention into practical reality. That is, relative to prescriptive regulations, the brevity, breadth, and prudent avoidance of detailed prescriptive guidance on the part of generic performance regulations creates the potential for a more efficient, effective, and flexible regulatory approach. Yet these same qualities conceivably can make the approach more difficult to implement.

One such difficulty is that reliance on generic performance regulations may make the OHS inspector's job more difficult. That is, when it comes to citing employers for violations of OHS regulations, government inspectors tend to find it easier to cite violations of a specific regulation, as opposed to citing violations of a generic performance regulation that may apply to a specific situation. Referring to Example 1, an OSHA inspector may find it relatively straightforward to conclude that a fan less than 7 feet above the floor with a guard opening of 1 inch represents a clear violation of the prescriptive regulation (OSHA, 1980; 1910 Subpart O-Machinery and Machine Guarding, 1910.212, General requirements for all machines, Section 1910.212(a)(5)) mandating that, "... when the periphery of the blades of a fan is less than seven (7) feet above the floor or working level, the blades shall be guarded. The guard shall have openings no larger than one-half (1/2) inch." On the other hand, the inspector may find it more difficult to decide whether or not this condition represents a clear and unequivocal violation of a generic performance standard that states, "The employer must ensure that machinery and equipment is fitted with adequate safeguards which...ensure that a worker cannot access a hazardous point of operations" (Workers' Compensation Board of British Columbia, 1999a, Section 12.2, Safeguarding requirement, p. 12-1, Section 12.2). In other words, relative to reliance on the prescriptive regulation, reliance on the generic performance regulation for purposes of regulatory intervention regarding machine guarding may call for a more judgmental, and therefore less definitive and predictable, decision on the part of the inspector.

A second, related disadvantage of generic performance regulations is that, relative to prescriptive standards, the former may provide less explicit guidance to employers about how to control specific workplace hazards that may pose a risk for job-related injury or illness. This concern can be considered particularly applicable to small business operations, that typically feature less formalized and less informed safety management systems.

Yet a third disadvantage is that generic performance regulations pose a greater challenge for worker safety training in that, again, such regulations provide less explicit guidance regarding the job-related injury or illness risk associated with specific workplace hazards.

Given these disadvantages, it is not surprising that OHS performance regulations may not enjoy the trust and support of the labor community (Hansen, 1985; Walker, 1984, personal communication). The fundamental source of labor concern in this regard is that, by their very nature, performance regulations place an undeniable emphasis on employer self-reliance and self-regulation when it comes to mitigation of OHS hazards. Labor may not be prepared to accept that employers will in fact follow through with satisfactory performance in self-responsibility for OHS hazard mitigation.

More surprisingly perhaps, some employers may also view the installation of performance regulations with suspicion, for much the same reasons. For example, as Bryce (1983) points out, employers in industries that are regulated by relatively separate, well-defined codes of prescriptive OHS regulations (such as construction, mining, etc.) may have structured their OHS management programs around such codes, and therefore may feel more comfortable in maintaining the regulatory status quo.

ERGONOMIC PERFORMANCE REGULATIONS

Ergonomic regulations represent one type of OHS regulation. Foregoing sections have dealt with the latter, because of limited and relatively recent experience with the former. However, available evidence suggests that observations and conclusions summarized above regarding performance regulations are specifically applicable to ergonomics regulations.

In their comprehensive review of ergonomics standards and regulations, Fallentin and colleagues (2001) summarize the major characteristics of a total of nine government-promulgated general ergonomics standards and regulations. All but one deal with the mitigation of hazards predisposing to work-related MSIs (upper limb or low back). As of the date of this handbook, only four of these have established regulatory enforcement status, namely general ergonomics rules promulgated by the states of California and Washington in the U.S. (State of California, 1997; Washington State Department of Labor and Industries, 2000), by the Province of British Columbia in Canada (Workers' Compensation Board of British Columbia, 1999d, Sections 4.46 to 4.53, pp. 4-7 to 4-8), and by Sweden (National Board of Occupational Health and Safety, 1998). In 2004, the Washington ergonomic rule was rescinded as the result of a statewide referendum.

The 9 government-promulgated general ergonomics standards and regulations reviewed by Fallentin and colleagues (2001) feature a mix of prescriptive, narrowly focused performance plus prescriptive, and generic performance regulations. As examples of prescriptive ergonomics standards dealing with musculoskeletal load, Westgaard and Winkel (1996) cite the following early formulations along with more recent European Committee for Standardization (CEN) formulations, the latter possibly influenced by the former.

Barnes (1937):
"The two hands should begin as well as complete their motions at the same time."
"Work should be arranged to permit an easy and natural rhythm wherever possible."

Grandjean (1969):
"Arm movements should be in opposition to each other or otherwise symmetrical."

CEN-standard prEN 614-1, December, 1991 (Westgaard & Winkel, 1996):

"Repetitive movement that causes illness or injury shall be avoided."
"Work equipment shall be designed to allow the body or parts of the body to move in accordance with their natural paths and rhythms of motion."

The ergonomics rules promulgated by Washington State (Washington State Department of Labor and Industries, 2000) and B.C. (Workers' Compensation Board of British Columbia, 1999d, Sections 4.46–4.53, pp. 4-7–4-8) offer an instructive contrast between reliance on primarily prescriptive versus exclusively generic performance regulations for purposes of mitigating MSI hazards. The Washington State regulation mandated that employers with 'caution zone jobs' first must determine whether these jobs have work-related musculoskeletal disorder (WMSD) hazards, and if so, must reduce WMSD hazards so identified. As summarized in Example 4 below, this approach features narrowly focused performance combined with detailed prescriptive regulations.

Example 4

Guidance as to what constitutes a 'caution zone job' first is provided with a narrowly focused performance regulation (Washington State Department of Labor and Industries, 2000, Section WAC 296-62-05103):

> **"WAC 296-62-05105 What is a 'caution zone job"?**
> Employers having one or more "caution zone jobs" must comply with Part 2 of this rule. "Caution zone jobs' may not be hazardous, but do require further evaluation."

This clause is accompanied first by a series of qualifying provisions, and then by 12 detailed prescriptive provisions that define a "caution zone job" in relation to postural, upper limb, and lifting risk factors.

Part 2 of the regulation (IBID, Section WAC 296-62-05130) offers employers what are termed general or specific performance options.

> **"WAC 296-62-05130 What options do employers have for analyzing and reducing WMSD hazards?**
> **General Performance Approach**
> (1) The employer must analyze "caution zone jobs" to identify those with WMSD hazards that must be reduced. A WMSD hazard is a physical risk factor that by itself or in combination with other physical risk factors has a sufficient level of intensity, duration or frequency to cause a substantial risk of WMSDs. The employer must use hazard control levels as effective as the recommended levels in widely used methods such as . . . [8 different methods prescribed here].
> (2)–(4) [qualifying performance provisions not included here]
> (5) Employers must reduce WMSD hazards as described below . . . [design change, work practice, and personal protective equipment hazard control prescriptions specified here]."

This formulation actually represents a narrowly focused performance regulation with qualifying performance and prescriptive provisions.

The formulation for what the regulation terms the specific performance option for analyzing and reducing WMSD hazards (IBID, Section WAC 296-62-05130) reads:

> **"Specific Performance Approach**
> The employer must analyze "caution zone jobs" to identify those with WMSD hazards that must be reduced. A WMSD hazard is a physical risk factor that exceeds the criteria in Appendix B of this rule."

Appendix B that this clause refers to contains carefully defined prescriptive WMSD hazard exposure criteria—for awkward postures (shoulders, neck, back, and knees), high hand force (pinching or gripping), highly repetitive motion of the neck, shoulders, elbows, wrists, or hands, and repeated impact of the hands or knees—that if exceeded qualify a job as a caution zone job. This formulation represents a narrowly focused performance regulation with qualifying prescriptive provisions only.

The regulation also contains provisions pertaining to employee education and involvement, plus hazard mitigation evaluation.

END EXAMPLE 4.

In contrast, the B.C. regulatory approach targets the same objective as the Washington State approach—mitigation of WMSD hazards—through use of generic performance regulations only, as illustrated in Example 5.

Example 5

Relevant portions of the B.C. ergonomics rule (Workers' Compensation Board of British Columbia, 1999d, Sections 4.46 to 4.53, pp. 4-7–4-8) read:

"ERGONOMICS (MSI) REQUIREMENTS
Section 4.47   Risk identification
The employer must identify factors in the workplace that may expose workers to a risk of musculoskeletal injury (MSI).
Section 4.48   Risk assessment
When factors that may expose worker to a risk of MSI have been identified, the employer must ensure that the risk to workers is assessed.
Section 4.49   Risk factors
The following factors must be considered, where applicable, in the identification and assessment of risk of MSI:

a. the physical demands of work activities, including: (i) force required, (ii) repetition, (iii) duration, (iv) work postures, and (v) local contact stress;
b. aspects of the layout and condition of the workplace or workstation, including: (i) working reaches, (ii) working heights, (iii) seating, and (iv) floor surfaces;
c. the characteristics of objects handled, including: (i) size and shape, (ii) load condition and weight distribution, and (iii) container, tool and equipment handles;
d. the environmental conditions, including cold temperature;
e. the following characteristics of the organization of work: (i) work-recovery cycles, (ii) task variability, (iii) work rate.

Section 4.50   Risk control

(1) The employer must eliminate or, if that is not practicable, minimize the risk of MSI to workers.
(2) Personal protective equipment may only be used as a substitute for engineering or administrative controls if its is used in circumstances in which those controls are not practicable.
(3) The employer must, without delay, implement interim control measures when the introduction of permanent control measures will be delayed.

Section 4.51   Education and training; Section 4.52 Evaluation
Section 4.53   Consultation
Not included here."

END EXAMPLE 5.

Although the Washington State and B.C. ergonomics rules share the same regulatory purpose, their respective approaches differ markedly. Both rules are quite similar in informing employers what the specific regulatory objectives are—hazard identification, analysis and control, education, employee involvement, and hazard mitigation evaluation. However, with its prescriptions regarding exposure criteria for hazard identification, hazard analysis methods, hazard reduction measures, and so forth, only the Washington State rule goes on to carefully define how the employer must go about meeting these objectives. On the other hand, with the exception of one prescriptive provision pertaining to personal protective equipment (Section 4.50), the B.C. generic performance rule leaves it up to the employer to determine how the objectives will be met.

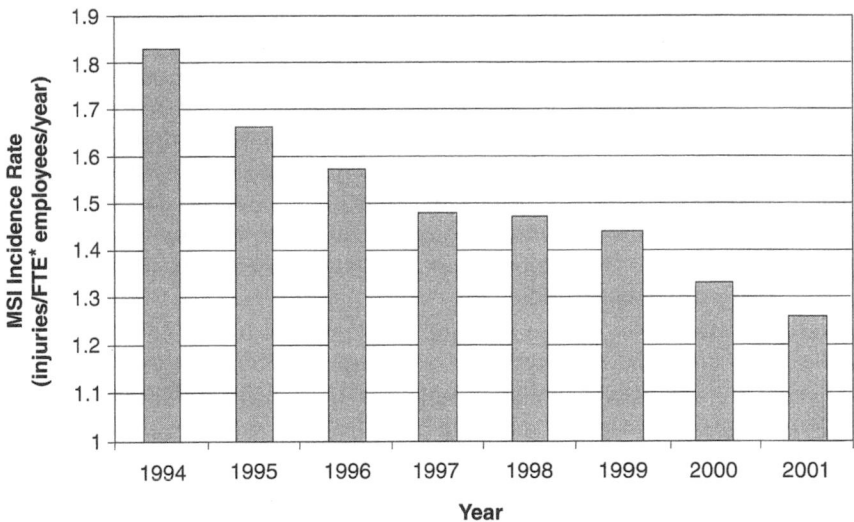

FIG. 3.1.   MSI incidence rate in B.C., 1994–2001 (*FTE = full-time equivalent).

Arguably, there are advantages to both approaches outlined previously Nothing in the B.C. rule matches the useful prescriptive guidance offered to employers by the Washington State rule about what methods to consider for purposes of MSI hazard analysis. Alternatively, the B.C. rule is more informative regarding the range of potential MSI risk factors that an employer should consider for purposes of MSI hazard identification and analysis.

In the last analysis, regardless of the formulation approach, it is the effectiveness and the scientific legitimacy of an ergonomics rule that matters most. The B.C. rule, as outlined in the next two subsections, appears to satisfy the former criterion, and arguably is superior to the Washington State rule regarding the latter.

Regulatory Effectiveness of Generic Ergonomic Performance Regulation

In their review of ergonomic standards and regulations, Fallentin and colleagues (2001, Table 12) report that no data are yet available from any jurisdiction regarding the jurisdiction-wide effectiveness of an ergonomics rule in reducing adverse health effects. However, information provided to the first author by the Senior Ergonomist for B.C. (K. Behiel, 2002, personal communication) indicates that the proposal and ultimate promulgation of the B.C. ergonomics rule has been associated with a steady decline in the province-wide MSI incidence rate between 1994 and 2001.

Data illustrating this conclusion are presented in Figure 3.1 which shows that the MSI incidence rate in B.C. decreased more or less steadily by about one-third, from 1.83 to 1.26 injuries per full-time-equivalent employees per year, between 1994 and 2001. The B.C. ergonomics rule (Workers' Compensation Board of British Columbia, 1999d, Sections 4.46 to 4.53, pp. 4-7–4-8) was first proposed to the provincial legislature in 1994, and subsequently promulgated in 1998.

The significance of the observations in Figure 3.1 should be treated cautiously. The data are from only one jurisdiction—whether or not a comparable impact of ergonomics rules will occur in other jurisdictions remains to be established. Moreover, it is possible that the observed decline in MSI incidence rate would have occurred even if the ergonomics rule had not been introduced.

Nevertheless, the association documented in Figure 3.1 between consideration and promulgation of an ergonomics rule in B.C., and reduction of the province-wide MSI incidence rate, is suggestive and points to two plausible conclusions. The first is that regulatory attention to ergonomics may have a positive impact on MSI injury reduction, even in the absence of actual regulatory intervention (note decline in Figure 3.1 in incidence rate between 1994 and 1998). One interpretation is that introduction of the rule in 1994 prompted greater attention to ergonomics and to MSI reduction among provincial employers after 1994, even before the rule actually was promulgated in 1998.

The second, and more important, conclusion is that an ergonomics rule built around generic performance regulations can be effective in achieving a positive health impact by reducing MSIs. This in turn implies that possible advantages of formulating OHS regulations using the generic performance approach may outweigh possible disadvantages (previous section). The B.C. experience thus joins that documented for dozens of other non-governmental operations in demonstrating the effectiveness of ergonomic intervention in reducing MSIs (Karsh, Moro, & Smith, 2001; Westgaard & Winkel, 1997).

There is case evidence from B.C. supporting this second conclusion. Prompted by the impetus of the B.C. ergonomics rule, an Industrial Musculoskeletal Injury Reduction Program was introduced into the B.C. sawmill industry in 1998 (McHugh, 2002). Industry-wide since that time, MSI claims for B.C. sawmills have dropped 44 percent, and the cost per MSI claim is estimated to have declined over 20%.

Scientific Legitimacy of Generic Ergonomic Performance Regulations

Fallentin and colleagues (2001) evaluate what they term the scientific coherency of the ergonomic standards and regulations addressed in their review. Two of the three North American jurisdictions with promulgated ergonomics rules—B.C. and Washington State—both are judged to have reasonable scientific legitimacy, in relation to sufficiency of the factual basis of the regulations (both rules), and to adequacy of the program elements (B.C. rule). The scientific legitimacy of the California rule is not similarly evaluated by these authors.

These authors, as well as Radwin, Marras, and Lavender (2002), Viikari-Juntura (1997), and Westgaard and Winkel (1996), also cite extensive epidemiological and biomechanical evidence pointing to the conclusion that excessive work-related external physical loading is related to an elevated risk for both back pain and upper limb paint, discomfort, or impairment or disability. Thus, for purposes of mitigating work-related MSIs, the scientific justification for establishing an ergonomics standard or regulation aimed at controlling worker exposure to excessive physical loading of the low back and/or upper limb seems unequivocal (Bernard, 1997; National Institute of Occupational Safety and Health [NIOSH], 2000).

Nevertheless, all of the authors cited above call attention to a significant gap in this body of evidence—the lack of empirical support for definitive exposure–dose and dose–response relationships between exposure to different levels of external physical loading and consequent proportional decremental effects on musculoskeletal function of the low back or upper limb. Without firm evidence for such relationships, the scientific legitimacy of establishing prescriptive ergonomics rules that mandate regulatory intervention when external physical loads exceed explicit, fixed levels remains open to question. For example, Westgaard and Winkel (1996) conclude their review of occupational musculoskeletal load guidelines as a basis for intervention by noting, "at present guidelines to prevent musculoskeletal disorders can only give directions, not absolute limits." Yet both the Washington State ergonomics rule (2000), and the repealed U.S. federal OSHA ergonomics rule (OSHA, 2000), prescribe a series of exact external physical loading limits, related to occupational exposure durations for different upper limb and low back postures and movements, above which ergonomic intervention for at-risk jobs is mandated.

Fallentin and colleagues (2001), Radwin and colleagues (2002), and Westgaard and Winkel (1996) all present conceptual models that depict the roles and influences of multiple factors in the development of musculoskeletal disorders of both the low back and upper limb. The models underscore the multiplicity of both biological (person) factors and workplace design factors whose interaction, in relation to exposure magnitude, duration, and frequency, may influence the etiology of MSIs. The general tenor of analysis in these reviews is that it is insufficient scientific knowledge and understanding of the complexity of this interaction that has forestalled clear delineation of the relationships between relative dose of external physical loading and relative early and late musculoskeletal response. Thus, Fallentin and colleagues note:

> Insufficient knowledge of the mechanisms involved implies that there are no criteria available to define whether the responses studied are relevant intermediate variables between an assumed target tissue dose and disease or disorder. In a risk analysis approach this lack of knowledge regarding exposure-dose relationship and dose-response relationship introduces a large element of doubt.

In the remainder of this section I call attention to an additional source of variability influencing the musculoskeletal loading-response relationship, not considered in the models referenced above, that raises the distinct possibility that delineating a well-defined exposure-dose and dose-response calibration for this relationship represents a scientifically unobtainable goal (T. J. Smith, 2002b). Given that definitive research on this problem area currently is lacking (above), this prediction obviously remains speculative at this point. The major purpose in raising this possibility is to suggest that putative problems with scientific legitimacy inherent to the prescriptive approach to ergonomics regulations may be avoided by adopting generic performance regulations for purposes of ergonomics rule-making.

The additional source of variability alluded to previously arises as a consequence of the basic properties of motor behavioral control—that is, we are concerned with kinesiological variability in musculoskeletal performance. Scientific justification for prescriptive limits for movement patterns associated with MSI risk must deal with three kinesiological sources of variability inherent to movement behavior: those related to the degrees-of-freedom (DOF) problem, the context specificity problem, and the imperfect control system problem.

The degrees-of-freedom problem pertains to the fact that the number of DOFs in three-dimensional (3-D) space inherent to movements of either the upper limb or the trunk is greater than the number of degrees of freedom of a target with which these movements may interact. Specifically, relative to 3 DOF of a fixed target in space, reaching and grasping a target by the upper limb features 7 DOF (mediated by the shoulder, elbow, and wrist), and lifting of a target by the trunk and knees features 4 DOF. As a consequence, there are an infinite number of different 3-D trajectories which an individual may deploy to move the upper limb to reach and grasp a target, or to move the trunk to lift a target (Rosenbaum, 1991).

There are two operational behavioral consequences of the DOF problem. One is that movement behavior is highly context specific—most of the variability observed in movement performance is attributable to the design of the environment in which performance occurs (Smith, 1998; T. J. Smith, Henning, and Smith, 1994). Context specificity arises because during development and maturation, to deal with the DOF problem, an individual develops a constellation of neural models of movement patterns that enable the individual to move effectors in a relatively reliable and reproducible manner (but see following) to interact with targets. The term commonly applied to this process is motor learning.

However, because each individual interacts with performance environments of different designs during the course of development and maturation, context specificity in movement behavior is different from individual to individual. In other words, confronted with common tasks with the same design (such as eviscerating a turkey or assembling a small part), different workers are very likely to deploy different patterns of movement behavior to interact with the design. This is because during development and maturation, each individual has established a distinctive repertoire of neural models to guide movement behavior.

A third source of kinesiological variability in movement behavior is related to the fact that humans are imperfect control systems (Schmidt & Lee, 1999). Empirical evidence shows that when interacting with a target in a repetitive fashion, successive movement patterns executed by an individual may be very similar but never are exactly identical. This is a feature of all effector movements. Such disparity in fidelity of movement patterning is exacerbated under conditions of fatigue or emotional stress. In other words, confronted with common tasks with the same design, the movement patterns of any given worker in interacting with the design may vary over time.

These considerations suggest that when it comes to delineating possible sources of between- and within-worker response variability that may underlie differential susceptibility, and lack of consistent exposure–dose and dose–response relationships, observed in the etiology of MSIs, kinesiological variability in movement behavior is superimposed on variability attributable to other factors specified in conceptual models referenced above (Fallentin et al., 2001; Radwin et al., 2002; Westgaard & Winkel, 1996). Indeed, given that within- and between-subject timing and accuracy of motor behavioral responses can vary over a two- to fivefold range as a consequence of context specific influences (Smith et al., 1994), it is possible that kinesiological factors represent the most prominent source of variance in on-the-job musculoskeletal performance.

The foregoing analysis suggests that the goal of developing reliable and reproducible exposure–dose and dose–response relationships relating musculoskeletal effects of exposure to the design attributes of a given on-the-job task may be scientifically unrealistic. Within- and between-worker variability intrinsic to behavioral guidance of movements, outlined previously, means that the degree of musculoskeletal damage (the response) engendered by a given "dose" of task effort (such as lifting or upper limb repetition at a particular frequency) is likely to be inherently unpredictable for a given worker over time, and from worker to worker under identical task demand conditions.

If further research supports this prediction, the implication is that there may be fundamental questions of scientific legitimacy associated with prescriptive ergonomic regulations that reference MSI risk to explicit duration limits for postures or movements of the low back and/or upper limb. The further implication is that relying on generic performance regulations in the formulation of ergonomic rules for mitigating MSI hazards avoids this problem.

CONCLUSIONS

HF/E science is concerned with designing performance environments to effectively accommodate the capabilities and limitations of performers in those environments. This chapter deals with the question of how best to design OHS/ergonomic regulations and standards to accommodate the capabilities and limitations both of those who must enforce and of those who must abide with such codes.

The first thing to be said about this question is how little attention it has received on the part of both the HF/E and legal communities. Two to 3 decades ago, in the course of advocating for a performance-based approach to formulating OHS regulations, K. U. Smith (1973) and Bryce (1983, 1985a, 1985b) defined a series of basic HF/E and regulatory issues and concerns associated with designing regulations for mitigating OHS hazards. Since then, however, there has been no follow-up to this work and no systematic evaluation of HF/E design of OHS regulatory code.

The analyses of these authors, coupled with subsequent regulatory experience in the United States and Canada, points to a number of distinct advantages associated with the application of performance-based OHS regulations, based on considerations of efficiency, effectiveness, and the purpose of regulatory intervention. Because they focus on assigning performance responsibilities for meeting specified regulatory objectives, without specifying in detail exactly how the objectives are to be achieved, performance regulations are undeniably more efficient than prescriptive regulations. If regulatory attention to specific hazards or specific industries is required, narrowly focused performance regulations combined if needs be with qualifying prescriptive provisions can be applied. Examples 1 and 2 in Section 2 illustrate these points.

Adoption of an ergonomics rule in B.C. based on generic performance regulations (Example 5), associated with a decline in the MSI incidence rate in the province (Figure 3.1), provides putative, case-based evidence for the effectiveness of OHS performance regulations. However, a broader, default argument for performance regulations along these lines may be mounted based on evidence regarding the relative lack of effectiveness of prescriptive regulations. The U.S. OSHA code is built largely around prescriptive regulations. However, various authors plus OSHA itself cite evidence indicating that strict adherence on the part of U.S. employers to OSHA regulations would prevent only a fraction of occupational accidents and injuries that actually occur (Smith, 2002a). This evidence implies that the design of prescriptive regulations is deeply flawed, in the sense that the OSHA code lacks criterion validity—the regulatory outcome of the prescriptive approach does not meet the regulatory intention.

K. U. Smith (1973, 1979, 1988, 1990) calls attention to the likely basis of this design defect by pointing out that most job-related hazards are operational hazards, involving a systems interaction between the behavioral performance of a worker and some design attribute of the work environment. From this perspective, the low criterion validity of OHS prescriptive regulations may be attributed to the fact that they are concerned only with the latter component of the interaction. In contrast, because they address how performance is supposed to interact with a specified design objective, performance regulations arguably are better designed to deal with the nature and pervasiveness of operational hazards.

The purported intention of OHS regulatory intervention represents a third argument favoring performance regulations. In specifying both the regulatory objective and how the objective is to be achieved, the prescriptive approach arguably fosters a babysitting or handholding relationship between the regulatory body and those being regulated. It strains credulity to believe that, in a general sense, either party finds such a relationship entirely satisfactory. Performance regulations avoid this shortcoming by specifying the objective and the party responsible, but not the method, for hazard mitigation.

When it comes to the specific case of ergonomic rules aimed at mitigation of MSI hazards, the prescriptive approach confronts the fundamental question of scientific legitimacy. The lack of empirical dose–response evidence linking hazard exposure levels and musculoskeletal performance decrements is attributed here to inherent within- and between-worker variability in the control of movement behavior. This variability makes it unlikely that it will ever be possible to document any consistent, graded relationships between exposure to job-related movement demands, and consequent musculoskeletal effects. Yet prescriptive ergonomic rules aimed at MSI hazard mitigation assume that such relationships exist (Example 4), a scientific conundrum that a performance-based approach avoids.

The foregoing arguments favoring OHS performance regulations are not meant to suggest that this regulatory approach represents a panacea for dealing with OHS hazards. Major potential disadvantages of the performance-based approach include (a) the lack of explicit guidance they provide regarding the need for regulatory intervention (b) the consequent difficulty they pose for the regulatory process in terms of unequivocally establishing whether or not compliance or a violation has occurred, and (c) their reliance on self-reliance and self-responsibility on the part of the employer for purposes of hazard mitigation implies a trust in employer performance that may not always be justified.

Beyond putative advantages and disadvantages associated with the performance-based approach to OHS regulations, a broader difficulty is lack of systematic, multijurisdictional, comparative research on operational experience with performance versus prescriptive OHS regulations. Consequently, the analysis presented here must be viewed as partially conceptual in nature.

In the context of this handbook, the underlying message of this chapter is that those concerned with promulgating HF/E and OHS standards and regulations should pay as much attention to the design of their formulation as to their actual content. Those favoring prescriptive regulations must confront shortcomings in efficiency, effectiveness, and intervention philosophy inherent to this approach. Those favoring performance regulations must confront potential difficulties with application inherent to this approach. It is assumed here that ensuring the ultimate success of a given OHS standard or regulation must start with settling on an appropriate HF/E design for its formulation. It is our judgment that, relative to the prescriptive approach, the performance-based approach to the latter is more likely to give rise to the former, particularly as regards ergonomics regulations.

REFERENCES

American Academy of Pediatrics, American Public Health Association, & National Resource Center for Health and Safety in Child Care (University of Colorado Health Sciences Center at Fitzsimons). (2002). *Caring for our children. National health and safety performance standards: Guidelines for out-of-home child care* (2nd ed). Elk Grove Village, IL: American Academy of Pediatrics.

Barnes, R. M. (1937). *Motion and time study.* New York: Wiley.

Bernard, B. P. (Ed.). (1997). *Workplace factors. A critical review of epidemiologic evidence for work-related musculoskeletal disorders of the neck, upper extremity, and low back.* Cincinnati, OH: National Institute for Occupational Safety and Health.

Bryce, G. K. (1983). *Some comments and observations on the application of generic performance regulations in occupational health and safety legislation.* Paper presented to the 42nd Annual Meeting of the Canadian Association of Administrators of Labour Legislation, Sept. 28, 1983.

Bryce, G. K. (1985a). Performance regulations and beyond. *British Columbia Workers' Health Newsletter, 11,* 1–6.

Bryce, G. K. (1985b). *The concept and implications of performance regulations in occupational health and safety.* Presentation to British Columbia Worklife Forum, Oct. 30, 1985.

Cleveland, R. J. (1976). *Behavioral Safety Codes in Select Industries.* Madison: Wisconsin Department of Industry, Labor and Human Relations.

Coleman, P. J., & Sauter, S. L. (1978). *The worker as a key control component in accident prevention systems.* Presentation to the 1978 Convention of the American Psychological Association, Toronto, Ontario, Canada.

Coleman, P. J., & Smith, K. U. (1976). *Hazard management: Preventive approaches to industrial injuries and illnesses.* Madison: Wisconsin Department of Industry, Labor and Human Relations.

Dul, J., de Vlaming, P. M., & Munnik, M. J. (1996). Guidelines. A review of ISO and CEN standards on ergonomics. *International Journal of Industrial Ergonomics, 17,* 291–297.

Ellis, L. (1975). A review of research on efforts to promote occupational safety. *Journal of Safety Research, 7*(4), 180–189.

Fallentin, N., Viikari-Juntura, E., Waersted, M., & Kilbom, Å. (2001). Evaluation of physical workload standards and guidelines from a Nordic perspective. *Scandinavian Journal of Work Environment & Health, 27*(Suppl 2), 1–52.

Foliente, G. C. (2000). Developments in performance-based building codes and standards. *Forest Products Journal, 50,* 12–21.

Gill, J., & Martin, K. (1976). Safety management: Reconciling rules with reality. *Personnel Management, 8* (6), 36–39.

Gottlieb, M.S. (1976). *Worker's awareness of industrial hazards: An analysis of hazard survey results from the paper mill industry.* Madison: Wisconsin Department of Industry, Labor and Human Relations.

Gottlieb, M. S., & Coleman, P. J. (1977). *Inspection impact on injury and illness totals*. Madison: Wisconsin Department of Industry, Labor and Human Relations.

Grandjean, E. (1969). *Fitting the task to the man*. London: Taylor & Francis.

Hammer, W. (1972). *Handbook of systems and product safety*. Englewood Cliffs, NJ: Prentice-Hall.

Hammer, W. (1976). *Occupational safety management and engineering*. Englewood Cliffs, NJ: Prentice-Hall.

Hansen, K. (1985). Performance regulations—A naive hope for workplaces without conflict. *British Columbia Workers' Health Newsletter, 10*, 1–6.

Harrison, R. K. (1955). *Biblical hebrew*. Bungay, Suffolk, England: Hodder & Stoughton.

Heinrich, H. W. (1931). *Industrial accident prevention. A scientific approach* (1st ed). New York: McGraw-Hill.

Heinrich, H. W. (1959). *Industrial accident prevention. A scientific approach* (4th ed). New York: McGraw-Hill.

International Council for Building Research Studies and Documentation (1982). *Working with the performance approach in building (CIB report, publication 64)*. Rotterdam, Netherlands: International Council for Building Research Studies and Documentation.

International Organization for Standardization. (1994). *International standard ISO 9001. Quality systems—model for quality assurance in design, development, production, installation and servicing*. Geneva, Switzerland: International Organization for Standardization.

Jones, D. F. (1973). *Occupational Safety Programs—Are They Worth It?* Toronto, Ontario: Labour Safety Council of Ontario, Ontario Ministry of Labour.

Juran, J. M. (1964). *Managerial breakthrough. A new concept of the manager's job*. New York: McGraw-Hill.

Juran, J. M. (1995). *Managerial breakthrough. The classic book on improving management performance* (revised ed). New York: McGraw-Hill.

Kaplan, M. C., & Coleman, P. J. (1976). *County highway department hazards: A comparative analysis of inspection and worker detected hazards*. Madison: Wisconsin Department of Industry, Labor and Human Relations.

Kaplan, M. C., Knutson, S., & Coleman, P. J. (1976). *A new approach to hazard management in a highway department*. Madison, WI: Wisconsin Department of Industry, Labor and Human Relations.

Karsh, B.-T., Moro, F. B. P., & Smith, M. J. (2001). The efficacy of workplace ergonomic interventions to control musculoskeletal disorders: a critical analysis of the peer-reviewed literature. *Theoretical Issues in Ergonomics Science, 2*(1), 23–96.

McHugh, A.-R. (2002). Ergonomic tool kits give sawmills bang for their buck. *Worksafe. The WCB Prevention Magazine on Occupational Health and Safety Issues, 3*(6, November–December), 10–11.

Minnesota Department of Labor and Industry. (1993). *An employer's guide to developing a workplace accident and injury reduction (AWAIR) program*. Saint Paul: Minnesota Department of Labor and Industry, Occupational Safety and Health Division.

National Board of Occupational Safety and Health (NBOSH) (1998). *Ergonomics for the prevention of musculoskeletal disorders* (AFS 1998:1). Stockholm, Sweden: NBOSH.

National Institute of Occupational Safety and Health. (2000). *NIOSH testimony to OSHA. Comments on the proposed ergonomics program* (29 CFR Part 1910, Docket No. S-777). Cincinnati, OH: National Institute for Occupational Safety and Health.

Occupational Safety and Health Reporter. (1976). Washington, DC: Bureau of National Affairs.

OSHA (1970). *Occupational safety and health act of 1970* (Public Law 91-596, 91st Congress, S.2193, December 29, 1970). Des Plaines, IL: U.S. Department of Labor, Occupational Safety and Health Administration, Office of Training and Education.

OSHA (1980). *Code of federal regulations. 29, Labor* (Part 1910, Occupational Safety and Health Standards). Washington, DC: Office of the Federal Register, National Archives and Records Service, General Services Administration.

OSHA (2000). Ergonomics program standard: final rule. *Federal register 2000* (65(220), 68262–870). Washington, DC: Office of the Federal Register, National Archives and Records Service, General Services Administration.

Plaut, W. G. (1981). *The Torah. A modern commentary*. New York: Union of American Hebrew Congregations.

Prior, J. J., & Szigeti, F. (2003). *Why all the fuss about performance based building. http://www.auspebbu.com/files/Why%20all%20the%20fuss%20-%20PBB.pdf.*

Radwin, R. G., Marras, W. S., & Lavender, S. A. (2002). Biomechanical aspects of work-related musculoskeletal disorders. *Theoretical Issues in Ergonomics Science, 2*(2), 153–217.

Rosenbaum, D. A. (1991). *Human motor control*. San Diego: Academic Press.

Richardson, V. L. (1973). *Hazard surveys at select employers*. Madison, WI: Wisconsin Department of Industry, Labor and Human Relations.

Schmidt, R. A., & Lee, T. D. (1999). *Motor control and learning. A behavioral emphasis* (3rd Ed). Champaign, IL: Human Kinetics.

Smith, K. U. (1973). Performance safety codes and standards for industry: The cybernetic basis of the systems approach to accident prevention. In J. T. Widner (Ed.), *Selected readings in safety* (pp. 356–370). Macon, GA: Academy Press.

Smith, K. U. (1979). *Human-factors and systems principles for occupational safety and health.* Cincinnati, OH: NIOSH, Division of Training and Manpower Development.

Smith, K. U. (1988). Human factors in hazard control. In P. Rentos (Ed.), *Evaluation and control of the occupational environment* (pp. 1–7). Cincinnati, OH: NIOSH, Division of Training and Manpower Development.

Smith, K. U. (1990). Hazard management: Principles, applications and evaluation. In *Proceedings of the Human Factors and Ergonomics Society 38th annual meeting* (pp. 1020–1024). Santa Monica, CA: Human Factors and Ergonomics Society.

Smith, M. J. (1994). Employee participation and preventing occupational diseases caused by new technologies. In G.E. Bradley & H.W. Hendrick (Eds.), *Human factors in organizational design and management–IV* (pp. 719–724). Amsterdam: North-Holland.

Smith, M. J., Bauman, R. D., Kaplan, R. P., Cleveland, R., Derks, S., Sydow, M., et al. (1971). *Inspection effectiveness.* Washington, DC: OSHA.

Smith, T. J. (1993). The scientific basis of human factors–A behavioral cybernetic perspective. In *Proceedings of the Human Factors and Ergonomics Society 37th annual meeting* (pp. 534–538). Santa Monica, CA: Human Factors and Ergonomics Society.

Smith, T. J. (1994). Core principles of human factors science. In *Proceedings of the Human Factors and Ergonomics Society 38th annual meeting* (pp. 536–540). Santa Monica, CA: Human Factors and Ergonomics Society.

Smith, T. J. (1998). Context specificity in performance - the defining problem for human factors/ergonomics. In *Proceedings of the Human Factors/ Ergonomics Society 42nd annual meeting* (pp. 692–696). Santa Monica, CA: Human Factors and Ergonomics Society.

Smith, T. J. (2002a). Macroergonomics of hazard management. In H.W. Hendrick, and B. Kleiner, (Eds.), *Macroergonomics* (pp. 199–221). Mahwah, NJ: Lawrence Erlbaum Associates.

Smith, T. J. (2002b). Ergonomics task-force recommendations. In Minnesota Department of Labor and Industry (Ed.), *The state of ergonomics in Minnesota. A summary of the Minnesota Department of Labor and Industry's Ergonomics Task-Force activities and recommendations* (pp. 12–17). Saint Paul, MN: Minnesota Department of Labor and Industry.

Smith, T. J., Henning, R. H., & Smith, K. U. (1994). Sources of performance variability. In G. Salvendy & W. Karwowski (Eds.), *Design of work and development of personnel in advanced manufacturing* (Chap. 11, pp. 273–330). New York: Wiley.

State of California. (1997). Repetitive motion injuries. *Cal/OSHA standards, California code of regulations* (Title 8, Division 1, Department of Industrial Relations, Chapter 4, Division of Industrial Safety, Subchapter 7, General Industry Safety Orders, Group 15, Noise and Ergonomics, Article 106, Ergonomics, Section 5110). Sacramento, CA: Cal/OSHA.

Swain, A. D. (1973). An error-cause removal program for industry. *Human Factors, 15*(3), 207–221.

Swain, A. D. (1974). *The human element in systems safety: A guide for modern management.* London: Industrial and Commercial Techniques.

Viikari-Juntura, E. R. A. (1997). The scientific basis for making guidelines and standards to prevent work-related musculoskeletal disorders. *Ergonomics, 40*(10), 1097–1117.

Washington State Department of Labor and Industries. (2000). Ergonomics. *Washington Industrial Safety and Health Act* (Section WAC 296-62-051). Olympia: Washington State Department of Labor and Industries.

Westgaard, R. H., & Winkel, J. (1996). Guidelines for occupational musculoskeletal load as a basis for intervention: A critical review. *Applied Ergonomics, 27*(2), 79–88.

Westgaard, R. H., & Winkel, J. (1997). Ergonomic intervention research for improved musculoskeletal health: a critical review. *International Journal of Industrial Ergonomics, 20*, 463–500.

Workers' Compensation Board of British Columbia. (1999a). General requirements. *Occupational Health & Safety Regulation. B.C. Regulation 296/97, as amended by B.C. Regulation 185/99. Book 2, General Hazard Requirements, Parts 5-19. Part 12—Tools, Machinery and Equipment.* Vancouver, BC: Workers' Compensation Board of British Columbia.

Workers' Compensation Board of British Columbia. (1999b). Controlling exposure. *Occupational Health & Safety Regulation. B.C. Regulation 296/97, as amended by B.C. Regulation 185/99. Book 2, General Hazard Requirements, Parts 5-19. Part 5 - Chemical and Biological Substances.* Vancouver, BC: Workers' Compensation Board of British Columbia.

Workers' Compensation Board of British Columbia. (1999c). Division 3—General duties of employers, workers and others. *Occupational Health & Safety Regulation. B.C. Regulation 196/97, as amended by B.C. Regulation 185/99. Book 1, Core Requirements, Parts 1-4. Part 3—Occupational Health and Safety.* Vancouver, BC: Workers' Compensation Board of British Columbia.

Workers' Compensation Board of British Columbia. (1999d). Ergonomics (MSI) requirements. *Occupational Health & Safety Regulation. B.C. Regulation 196/97, as amended by B.C. Regulation 185/99. Book 1, Core Requirements, Parts 1-4. Part 4: General Conditions.* Vancouver, BC: Workers' Compensation Board of British Columbia.

II

Nature of HF/E Standards
and Guidelines

4

National Standardization Efforts in Ergonomics and Human Factors

Heather A. Priest
Katherine A. Wilson
Eduardo Salas
University of Central Florida

INTRODUCTION

As technologies are being introduced as a way to support human performance in complex tasks, the fact remains that improper design or implementation of these technologies can have a negative effect on performance (e.g., errors, stress, and increased workload; Alhstrom, Longo, & Truitt, 2002). Therefore, standards have been developed as a way to help ensure its proper design and implementation. However, standards not only influence technology but also impact everything surrounding us in our daily lives (e.g., roads, clothing, and housing). Evidence of the impact of standards on economical and health issues in a variety of industries speaks to the importance of the application of standards. For example, new intensive care unit standards are expected to save 30,000 lives and $1.5 billion per year (Appleby, 2001). New training standards developed by the Occupational Safety and Health Administration (OSHA) are expected to prevent 94,000 injuries, and save up to 11 lives and $135 million in employer costs each year (U.S. Department of Labor Office of Public Affairs, 1998). Since 1968, federal vehicle safety standards have saved thousand of lives, evidenced by the decrease in fatalities per million vehicle miles traveled (i.e., has dropped from 5.7 to 1.6) and overall fatalities (i.e., has decreased from 53,041 to 43,000) despite an increase in cars on the road (National Highway Traffic Safety Administration [NHTSA], 1999).

Ultimately, standards are a way for designers, manufacturers, trainers, and evaluators to ensure a level of safety, quality, and performance across products, organizations, disciplines, and, in some cases, nationalities. Standards for areas relevant to human factors and ergonomics are provided by a number of agencies and diverse arenas, such as government, military, national, and international organizations, and are developed for professionals in a variety of disciplines (e.g., human–computer interaction, engineering, manufacturing, and health care). For human factors and ergonomics professionals, designers, researchers, and practitioners, standards provide a blueprint for industry and organizational wants, needs, and expectations.

Objective and Scope of Chapter

The purpose of this chapter is to provide an overview of the standards available to human factors and ergonomics researchers and practitioners. We will focus on the standards development system in the United States, specifically private organizations and several government organizations, which have adopted private standards. Although it is recognized that international organizations and the military support numerous human factors and ergonomics standards relevant to the field, a discussion of these is beyond the scope of this chapter. Therefore, we encourage readers to look to other chapters contained in this handbook, which provide a more in-depth discussion of international standards (chap. 1.4) and military standards (chap. 20). As such, the structure of the presentation of information is twofold. First, we define the term *standard* and discuss the benefits of implementing standards. Second, we focus on the standards development system itself, specifically four areas: (a) its history, (b) those organizations that serve as liaisons between organizations in the United States and abroad, (c) standard developing organizations in the United States and the standards that they develop, and (d) organizations which provide searchable databases available to researchers and practitioners. As space restrictions for this chapter limit the number of the development organizations and standards and the depth of information that may be included from the field of human factors and ergonomics, we have selected those that are strongly associated with the field and that appeared to be the most widely referenced.

Standards Defined

The National Standards Policy Advisory Committee (1978, p. 6) defines a standard as "a prescribed set of rules, conditions, or requirements concerning definitions of terms; classification of components; specification of materials, performance, or operations; delineation of procedures; or measurement of quantity and quality in describing materials, products, systems, services, or practices." Others define a standard as a uniform technical and engineering rule established by authority which is used to assess the quantity, weight, extent, value, or quality of a discipline's processes, procedures, practices, and methods (National Aeronautics and Space Administration [NASA], n.d.). In general, the term standard is often used to guide a course of action within a specific domain. Privately sponsored standards, as opposed to government standards, are traditionally voluntary. However, voluntary standards that are bound by a legal document (e.g., contract) can be used in legal proceedings (National Institute for Standards and Technology [NIST], n.d.), as well as to demonstrate industry standards that are deemed "acceptable" (Breitenberg, 1987).

There are at least five *kinds* of standards available for organizations today (see American Society for Testing and Materials [ASTM], n.d.): (a) company standards (i.e., agreements among an organization's employees); (b) consortium standards (i.e., agreements among a small group of "like-minded" organizations); (c) industry standards (i.e., agreement among many organizations within a "society"); (d) government standards (i.e., mandatory standards adopted by the government and under its jurisdiction); and (e) voluntary consensus standards (i.e., agreement from representatives from all sectors interested in the standard). Voluntary consensus standards, the main focus of the current chapter, are developed by private organizations and seek to establish rules and guidelines addressing public health and safety, manufacturing products, exporting to foreign markets, and procurement, areas which are addressed by human factors and ergonomics professionals (National Technology Transfer and Advancement Act [NTTAA], 1995). We recognize that guidelines are also used in conjunction with standards, however, because guidelines can refer to a wider variety of recommendations more applicable in other arenas (e.g., the implementation of training programs) and are often embedded within the standards. In this chapter we refer to the "rules" of human factors and ergonomics as standards.

Benefits of Standards

The benefits of implementing voluntary consensus standards can be found in a number of areas, discussed by organizations such as the American National Standards Institute (ANSI), National Standards Systems Network (NSSN), and ASTM International. Standards are valuable to the design, development, implementation, and evaluation of all products for many reasons. First, standards offer a means of communication between buyers and sellers by reinforcing the quality of their products, which is often marketed on packages and labels (ASTM, n.d.). As such, standards provide improved confidence in products, services, quality control, and public safety. Moreover, standards may enable companies to improve their image as a positive corporate entity, through the application of acceptable practices and guidelines that the public desires and expects. In turn, the standardization of products and systems can reduce product liability exposure for companies by meeting an industry minimum for product and system requirements. Through standards, consumers and organizations gain a better understanding of the process of commerce and of each other (Breitenberg, 1987). The issuance and utilization of standards also help industries to establish new markets and expand existing global markets while making technologies more available through the regulation of the industry and emerging technologies (Breitenberg). By having a national consensus for the standards of U.S. products and systems, industry will be able to enforce a united front while opening themselves up to obtain new technology and network with others in their own and related fields. Furthermore, monetary expenses are reduced as unnecessary grades of products are eliminated and thus reduce the inventories for producers and users, reduce the need for negotiations, provide more efficient inspections and testing, reduce the cost of acquiring materials and components for production through multisourcing, and pass on lower prices to users due to competitive bidding (ASTM, n.d.; Breitenberg, 1987). Standards also facilitate faster product development and implementation due to the ability to stock standard items. In addition, and perhaps most importantly, the adherence to national standards helps ensure a safer environment for workers and protection against fraud or some other harm for consumers. Finally, most companies can find additional benefits by complying with consensus national standards within their own organization regarding employee satisfaction, cost and productivity, and image (based on NSSN and ASNI http://www.nssn.org/about.html).

U.S. STANDARDS DEVELOPMENT SYSTEM

History of U.S. Standards

Voluntary standards have historically been used to satisfy citizens' needs, foster industrial competition, and ensure safe, quality production. Standards have existed as long as recorded history. In fact, most publications regarding the history of standards (ANSI, n.d.; Breitenberg, 1987) trace their origination back to 1120, when King Henry I of England decreed that the ell (i.e., the ancient yard) should be as long as his forearm and should be used as the standard unit of length for his kingdom. However, a majority of standards in the United States originated much later than that with the industrial revolution.

Standardization began in the United States in the late 1700s when Eli Whitney was contracted by Vice President Thomas Jefferson to produce 10,000 muskets (ASTM, n.d.; Breitenberg, 1987). Whitney developed the idea of mass production of products to be identical, and therefore interchangeable, making the production of goods simpler and more efficient. After demonstrating the benefits of mass production to the Secretary of War, Whitney was deemed the "father of standardization," and others quickly began to follow suit (e.g., Henry Ford).

Around the turn of the century, U.S. organizations began to coordinate their standard development efforts in order to improve safety (ASTM, n.d.). With the industrial revolution came the need to transport goods from coast to coast, leading to an alternate mode of transportation: the railroad. As the United States began to expand economically and therefore geographically, the need for a standardized transportation system was more evident as rail systems from region to region (and even state to state) varied. For example, railways in the Midwest differed from those in the east. As such, a train traveling from Pennsylvania to California would be forced to stop in Chicago and transfer cargo to another train that could travel the railways through the Midwest. Therefore, one of the earliest and most famous standards was the railroads' standard track gauge, which standardized the space between rails in the tracks throughout the United States (American National Standards Institute [ANSI], n.d.; Breitenberg, 1987). Other safety concerns also propagated as it became apparent that the steel used to construct the railways was of inferior quality. As such, a group of engineers developed the ASTM in 1898 specifically to develop standards to improve railway safety (ASTM, n.d.).

In addition to increasing railroad transportation, traveling from state to state by automobile increased. At that time, each state determined the color lights to be used for traffic signals. For example, in New York, a red light meant caution, whereas in other states, a red light meant stop. In some states, a green light (as it is today) meant safe to go, whereas in others, a green light meant to stop. As the amount of traveling was increasing, it was recognized that without a standardized national code of colors for traffic signals, chaos would ensue each time visitors would make their way through a new city. As such, in 1927 the American Associations of State Highway Officials, the National Bureau of Standards (now NIST), and the National Safety Council introduced standards to regulate colors used for traffic signals nationwide, which are still in use today (Breitenberg, 1987).

Throughout the 20th century, both government and private organizations worked together to develop new and improved standards as times changed. For example, as World War II approached, the need for international standards to increase compatibility across international boundaries due to the coordination of Allied forces was recognized and addressed. As time has passed, development and oversight of standards (e.g., ASTM and ANSI, respectively) has moved away from the government to private organizations (ANSI, 2000). Today, as the concern for environmental and health issues, combined with a growing global economy and competition from the international community, the government relies heavily on the standards developed by private organizations to be adopted for their own purposes.

Standards Today

As one would probably expect, the number of standards that have been developed over the years to be applied in industry and organizations is staggering. Over a decade ago, NIST estimated there were over 30,000 voluntary standards being used at that time that were developed by over 400 organizations in the United States alone (Breitenberg, 1987). Although some organizations specific to human factors and ergonomics (e.g., HFES) have attempted to accumulate resources specifically geared to the disciplines encompassed by human factors and ergonomics, a closer examination of the available resources exhibits the difficulty of parsing out the relevant standards. The diversity of the field is partly to blame, but even more importantly, standards and guidelines often address several issues, some related to human factors and ergonomics, and some not. Furthermore, standards available from some organizations (e.g., Society of Automotive Engineers [SAE]) may apply to several domains, for several different products, intended for several different users. Therefore, a great deal of the organizations and sources discussed in this chapter are not human factors and ergonomics specific, but are useful and relevant to those of us who practice this science.

Unlike many countries abroad, the standard development system in the United States consists of multiple organizations (both government and nongovernment organizations), which develop standards (ASTM, n.d.). The organizations within the standards system each have their own responsibilities. More specifically, the system has three key players. Some organizations play the part of liaison between organizations within the United States and between the United States and abroad. Others focus on the development of standards and are referred to as standards development organizations (SDO). The third key player makes the standards developed accessible through online searchable databases. We discuss each of these next.

Liaison Organizations

ANSI

ANSI (http://www.ansi.org/) is a nonprofit, private organization that works in conjunction with a number of private and public industry organizations. ANSI has served as the administrator for national voluntary consensus standards for the past 80 years (ANSI, n.d.). The purpose of ANSI is to coordinate the U.S. standards system by bringing together public and private sectors. In addition, ANSI seeks to promote U.S. business by protecting, assessing, and confirming national consensus standards from a number of organizations. This has, for the most part, been successful with 270 accredited organizations, both private and public, from over 200 disciplines amassing over 11,000 documents regarding voluntary consensus standards.

No single standards organization is more influential in the area of national private sector standards today than ANSI. Therefore, ANSI warrants a closer examination than some of the other organizations contained in this chapter. The extra attention is credited for several reasons. First, ANSI is the primary standards coordination organization in the United States, as well as internationally. Second, ANSI accreditation is recognized by business, government, users, and consumers. Further validation for the credibility of ANSI and the importance to human factors and ergonomics is that ANSI standards are also relied on in the courts and by government agencies.

The principles of the standardization process set forth by ANSI are designed to be fair and impartial, with the intent of allowing all members to participate. These principles for the development of standards include (a) decisions are based on consensus; (b) all those affected by standards are welcome to participate; (c) the process must be transparent with information available to all; (d) due process is a requirement; (e) the process must be flexible and timely; and (f) standards should be coherent, while avoiding conflict or overlap. When based on these principles, ANSI ensures the standards developed through this process will be relevant, responsive to the real world, and performance based (ANSI, 2000).

When questions or issues surrounding standards arise, there are four boards contained within ANSI that help determine solutions to standards issues: Healthcare Informatics Standard Board (HISB), Information Systems Standards Board (ISSB), Medical Devices Standards Board (MDSB), and Information Infrastructure Standards Planning Panel (IISP). These forums provide industry members with the means to discuss, coordinate, identify problems, and come up with solutions with others who are in similar businesses regarding standards. The four boards are divided based on specific industry sectors and needs within certain domains. We discuss these boards next.

Health Care Informatics Standard Board (HISB)

ANSI HISB consists of 27 members who vote and over 100 participants who discuss standards issues related to health care. This board provides a forum for coordination, on a voluntary basis, of all health care informatics for all U.S. organizations that develop standards. The board's objectives include the creation of common reference information, common terminology, common methods of implementation of the exchange of healthcare information, common trust frameworks, and common approaches for coordination and conflict resolution within the health care community. The ultimate goal is to create interoperability, harmony, and communication among organizations involved in health care informatics.

Information Systems Standards Board (ISSB)

Similar to the HISB, the ISSB acts as a forum for professionals with similar interests and as a tool in selecting and evaluating standards for a specific application, often of concern to human factors and ergonomic professionals and students. ISSB is concerned with the interfaces between system or machine components and the interfaces between the user and the system. The scope of this board is any discussion or debate concerning all information system interfaces outside the domain of the other boards, across multiple fields. For example, the ISSB may act as the standards forum within the fields of aviation and manufacturing, as long as the standards are addressing interfaces between machine components or systems and humans.

Medical Devices Standards Board (MDSB)

Along those same lines, MDSB serves to act as a coordination and information exchange agent for professionals who deal with medical devices, either manufacturing, designing, or using them. The scope of the MDSB is to discuss and approve those standards concerned with devices designed for diagnosis, monitoring, or treatment in medical fields (e.g., neurosurgery, anesthesiology, dentistry, biological analysis, and orthopedics). Any device used within medical professions and not covered by other ANSI boards is within the domain of the MDSB. In addition to the regulation of standards, the MTSB seeks to monitor and maintain the U.S. position on medical devices internationally.

Information Infrastructure Standards Planning Panel (IISP)

The IISP was formed in 1994 to determine the needs and standards associated with the implementation of the Global Information Infrastructure (GII)/Global Information Society (GIS). The GII/GIS refers to the implementation of a global information superhighway. The IISP is a multidisciplinary board that acts to facilitate the coordination of standards and information exchange with respect to the components of the GII/GIS, including telecommunications, wireless networks, computers, and broadcast organizations. The focus of the IISP is the standards needed to implement and maintain the GII/GIS.

In addition to the creation of these boards, ANSI works in conjunction with over 1,000 businesses, standards developers, institutes, professional societies and trade associations, consumer and labor interest groups, and government agencies. In order to promote the interaction of all of these organizations, ANSI also has a number of forums directed at all of its members and participants. There are currently four major forums: (a) Company Member Forum, (b) Government Member Forum, (c) Organizational Member Forum, and (d) Consumer Interest Forum. The first three were formed to provide a forum for discussions and networking, to provide "early warnings" of emerging industry trends, to address issues of interest, to identify issues to be addressed by boards and developers, to play active roles in recruitment and retention of members, and to give members an avenue to represent their organizations. The Consumer Interest Forum is slightly different. This forum consists of consumer producers, retailers, organizations, distributors, government, and industry councils. The goal of this forum is to promote consumer understanding of ANSI, the standards development organization, and the process of standardization.

ANSI appears to make great efforts to provide an open, fair process that is vital in today's environment for the safety and health of U.S. workers and consumers. Furthermore, the benefits and opportunities provided to developers and organizations make it a valuable asset. By all accounts, ANSI seems to be steadfast in their goal to be fully involved in its support of the quality of life for all global citizens and of U.S. and global standardization.

NIST

The NIST (www.nist.gov) serves a dual role of coordinator between organizations in the U.S. and abroad, as well as assists U.S. industries and government (i.e., Department of Commerce) in the development and promotion of standards (discussed later). The history of the NIST can be traced back to the early 20th century. In 1901, the National Bureau of Standards was developed in order to improve the reliability of U.S. products, which had otherwise become known as unreliable, and to improve the accuracy of product measurement. As a part of the Omnibus Trade and Competitiveness Act in 1988, the National Bureau of Standards was changed to NIST.

NIST assists the Department of Commerce (DOC) by providing editorial assistance to assure the technical soundness of standards and that all valid disagreements stated throughout the development process were addressed. In addition, NIST offers DOC secretarial functions for standards committees, determines compliance with procedures, and publishes developed standards as public documents. NIST also assists private sector organizations by facilitating agreement among organizations and providing technical expertise. Furthermore, trade agreements are supported by NIST through the development of conformity-assessment infrastructures. Finally, NIST coordinates the use of voluntary standards by other federal agencies.

NIST also assists U.S. organizations abroad by providing support in overcoming barriers relating to standards faced in foreign markets. One way that this is accomplished is by placing and supporting standards experts in key foreign markets at U.S. embassies. In addition, NIST contracts nongovernment experts in some foreign markets (i.e., India and Saudi Arabia). The purpose of NIST's role abroad is to identify any barriers and to remove them so as to foster export and competitiveness of U.S. products.

Standard Developing Organizations (SDO)

Society and Association SDOs

Human Factors and Ergonomics Society (HFES). The HFES (www.hfes.org), the largest single human factors association, is an ANSI-accredited organization whose goal is to develop, share, and apply knowledge concerning the interaction of humans and machines or systems. HFES has historically partnered with different organizations to develop and accredit standards that can be applied to human-factors-related fields. The subjects covered by current and ongoing standards developed by HFES and accredited by ANSI include human–computer interaction, color-coding of displays, user interface design, voting machine standardization, and workstation design. Some of the organizations that have paired up with HFES to partner in the development of standards include Institute of Electrical and Electronics Engineers (IEEE), NIST, Federal Election Commission, Association of the Advancement of Medical Instrumentation (AAMI), ANSI, and International Organization for Standards (ISO). The following are a few examples of the work being done by HFES in conjunction with other standards organizations.

TABLE 4.1

The Sections of the Current Version of ANSI/HFES 200[a]

Section	Title
1.	Introduction
2.	Accessibility
3.	Presentation of Information
4.	User Guidance
5.	Direct Manipulation
6.	Color
7.	Forms Fill-In
8.	Command Languages
9.	Voice Input/Output,
	Voice Recognition
	Nonspeech Auditory Output
	Interactive Voice Response
10.	Visually Displayed Menu

[a] Adapted from Blanchard (1997).

ANSI/HFES 200.   An update on ANSI/HFES 100 standards for visual display technology (VDT) by the HFES Human Computer Interaction Standards Committee (HFES/HCI), which was a subcommittee created to monitor and develop standards for the design of software and for user–machine interaction will be released soon (Blanchard, 2000). This committee's aim was to create an HCI standards document, which will become an ANSI standard in the United States. The ANSI/HFES 200 is the primary source for the design of software user interface. The multisection document covers a number of HCI design issues and incorporates sections from ISO document ISO 9241 Parts 10–17 and expands on ANSI/HFES 100 chapter 6. The initial five categories have been recently expanded to 10. These parts, shown in Table 4.1, are in varying stages of development and all pertain to HCI.

IEEE-SA Project 1583.   A more recently initiated project has HFES teaming up with the IEEE-Standards Association (SA) on an initiative to develop advanced electronic voting standards (Business Wire, 2002). The goal of this initiative is to make the voting process more reliable, secure, and accessible and help guide states and others in replacing existing voting equipment. The standards will utilize the latest engineering, quality, usability, accessibility, information, and security technologies. The initiative has only been active since July 2002 and, therefore, is still in the beginning stages at the time of publication.

IEEE.   The IEEE (www.standards.ieee.org) is the world's largest technical professional society and an important influence in a number of industries and organizations. The IEEE has an organizational goal of promoting the engineering process through the creation, sharing, development, and application of knowledge, specifically in the domain of electrical and information technologies. The IEEE Standards Association is a membership organization and a division of IEEE that develops standards that consist of numerous operating procedures and guidelines that aim to facilitate trade and commerce, safeguard against hazards, increase competitiveness in industry, foster quality design and manufacturing, and create commerce and trade. The IEEE standards consist of 900 active standards with 700 under development. The

TABLE 4.2
Examples of IEEE Standards

Category	Standard
All-Inclusive Standards	Communication
	Computer Simulation
	Test Technology
Power and Energy	Electric Machinery
	Power Electronics
	Roadway Lighting
Electromagnetics	Electromagnetic Compatibility[a]
Telecommunications	Aerospace Electronics
	Internet Best Practices
	Power Electronics
Transportation	All-Inclusive Transportation
	Intelligent Transportation
	Vehicular Technology
Other Topics	Electronic Devices
	Learning Technology
	Nuclear Engineering

[a]Only standard in category.

standards are divided into six distinct categories: (a) all-inclusive standards, (b) power and energy, (c) electromagnetics, (d) telecommunications, (e) transportation, and (f) other topics. These categories and examples of some of the standards available through the IEEE Web site are shown in Table 4.2.

Society of Automotive Engineers (SAE).   The SAE (www.sae.org/servlets/index) whose purpose is to promote the engineering of advancing mobility in sea, air, and space through standards development, liaisons among a number of organizations relating to the air and space industry, including the Aerospace Industries Association (AIA), the General Aviation Manufacturing Association (GAMA), and the Department of Defense (DOD). The goal of standards produced through SAE is to provide procedures and guidelines to promote and ensure that the organizations within and related to the industry produce quality products and save on valuable engineering time. The SAE standards database contains over 4,000 documents in such fields as aerospace material specifications. A major focus of the standards for SAE organizations is the health and safety of employees of the industry. Furthermore, SAE is becoming increasingly aware of environmental concerns and has recently formed a new committee to ensure established and future standards are environmentally conscious. The standards developed and maintained by SAE cover all fields of aerospace and automotive engineering, such as aircraft, automation, human factors, modeling, and vehicle systems. A few examples of some standards relevant to the field of human factors and ergonomics are shown in Table 4.3.

American Society of Mechanical Engineers (ASME).   The American Society of Mechanical Engineers (ASME; www.asme.org) founded in 1880, has a membership base of 125,000 and conducts over 30 technical conferences each year. Due to the importance of society within the field of engineering and the shear size of the organization, ASME produces many

TABLE 4.3
Examples of SAE Standards for Human Factors Applications

Document Number	Title	Brief Description
ISO6456	Accident Reconstruction	Specifies information necessary for the study of accidents to vehicle occupants wearing seat belts.
ISO2575	Road Vehicle—Symbols of Controls	Establishes the symbols, that is, conventional signs, for use on controls, indicators, and tell-tales of a road vehicle to ensure identification and facilitate use
ISO9790/1	Anthropomorphic Side-Impact Dummy	One of six reports that describe laboratory test procedures and impact response requirements suitable for assessing the impact biofidelity of side-impact dummies
J2094	Vehicle and Control Modifications for Drivers with Physical Disabilities Terminology	Includes only those terms that are pertinent to the adaptive devices discipline
J1050	Describing and Measuring the Driver's Field of View	Describes three methods for measuring the direct and indirect fields of view and the extent of obstructions within those fields
J1757/1	Standard Metrology for Vehicular Displays	Provides methods to determine display optical performance in all typical automotive ambient light illumination
J2395	Its In-Vehicle Message Priority	Recommended practice describes the method for prioritizing its in-vehicle messages or displayed information based on a defined set of criteria
J4001	Implementation of Lean Operation User Manual	Tests lean implementation within a manufacturing organization and include those areas of direct overlap with the organizations suppliers and customers
J284	Safety Alert Symbol for Agricultural, Construction, and Industrial Equipment	Presents the general uses, limitations on use, and appearance of the safety alert symbol.
J1727	Injury Calculations Guidelines	Describes numerical methods used to process impact test data

of the industry and manufacturing standards. The goal of the standards developed by ASME is to both facilitate the technical competencies of their members and the safety of its members. ASME attempts to promote such ideals through the development and collection of codes and standards to be applied to mechanical engineering and related fields. Many of the wide variety of standard divisions are applicable to human factors and ergonomics applications, including controls, elevators and escalator design, nuclear guidelines, and surface quality standards. An examination of some of these standards is shown in Table 4.4.

National Fire Protection Association (NFPA). The National Fire Protection Association (NFPA; www.nfpa.org) established in 1896, is an international nonprofit organization whose purpose is to reduce the problems of fire and other hazards, which affect safety and quality of life. NFPA does this in part by providing and advocating standards and codes that are scientifically based. Codes and standards are developed, published, and disseminated worldwide by NFPA, which impact nearly all buildings, services, designs, processes, and instillations today. NFPA supports 230 technical standard and code development committees that consist of 6,000 plus professional volunteers. Standards and codes are proposed by committees and

TABLE 4.4
Example of ASME Issued Standards

Document Number	Title	Brief Description
CSD-1	Control and Safety Devices for Automatically Fired Boilers	This code cover requirements for the assembly, maintenance, and operation of controls and safety devices
RA-S	Probabilistic Risk Assessment for Nuclear Power Plant Applications	Sets forth requirements for probabilistic risk assessments (PRAs) used to support risk-informed decisions for commercial nuclear power plants
HPS	High Pressure Systems	Presents requirements for high pressure systems, excluding nuclear power applications
A17.1-2000	Handbook on Safety Code for Elevators and Escalators	Contains the rationale for code requirements; explanations, examples, and illustrations of the implementations of the requirements
QRO-1	Standards for the Qualification and Certification of Resource Recovery Facility Operators	Provides requirements to be used in certifying operators of resource recovery facilities
B46.1	Surface Texture, Surface Roughness, Waviness, and Lay	Defines surface texture and its constituents: roughness, waviness, and lay
B107.8M	Adjustable Wrenches	Covers adjustable wrenches open-end, heavy-duty, rack and worm adjustment generally used on both hexagonal and square bolts and nuts

then voted on by all NFPA members before being published. Due to NFPA's dedication to "true consensus," it has earned accreditation from ANSI for its code development process. See Table 4.5 for examples of NFPA standards and codes.

Association for the Advancement of Medical Instrumentation (AAMI).   The Association for the Advancement of Medical Instrumentation (AAMI; www.aami.org) was established in 1967 with the purpose of increasing the understanding of medical instrumentation as well as making its use more beneficial. AAMI has almost 6,000 members with a range of interests, such as nursing, hospital administration, clinical and biomedical engineering, manufacturing and distributing. AAMI's diverse interests and its ANSI accreditation allow the organization to be a leading resource in the medical domain. Today, AAMI is the primary source of information regarding national and international standards for the medical community by assisting with the development, management, and use of medical instrumentation and related technologies in a safe and effective manner. Standards development has been made in such areas as sterilization, dialysis equipment, biological evaluation, electromedical equipment, and symbols. Table 4.6 lists several of the standards developed by AAMI.

ASTM International.   ASTM International (formerly American Society for Testing and Materials; www.astm.org) is a not-for-profit organization that develops and publishes voluntary consensus standards. ASTM International, was founded in 1898 and since then has developed over 11,000 standards, which are used not only in the United States, but worldwide. ASTM standards are used in a variety of areas such as research and development, product testing, and quality systems. Standards are developed when a need or an interested party is identified by one of 130 plus ASTM technical committees. Committee members include producers, consumers, users, and government and academia representatives. The appropriate committee

TABLE 4.5
Sample of NFPA Standards and Codes

NFPA Number	Title	Brief Description
NFPA 1[a]	Fire Protection Code	Regulation of conditions to prevent fire and explosions
NFPA 10	Standard for Portable Fire Extinguishers	Regulates the selection, installation, inspection, maintenance, and testing of equipment
NFPA 54[a]	National Fuel Gas Code	Covers the installation of fuel gas utilization equipment, fuel gas piping systems, and related accessories
NFPA 80	Standard for Fire Doors and Fire Windows	Applies to the installation and maintenance of doors and windows to restrict fire and smoke from spreading
NFPA 101[a]	Life Safety Code	Seeks to minimize dangers to life before buildings are vacated by covering construction, protection, and occupancy features
NFPA 110	Standard for Emergency and standby Power Systems	Regulates performance requirements for systems
NFPA 403	Standard for Aircraft Rescue and Fire Fighting Services at Airports	Specifies minimum fire safety requirements
NFPA 1001	Standard for Fire Fighter Professional Qualifications	Identifies the professional qualifications required of fire department members
NFPA 1670	Standard on operations and training for Technical Rescue Incidents	Specifies performance levels at technical rescue incidents required for safe effective operations
NFPA 5000[a]	NFPA Building Code	Regulates and controls issues such as permitting, design, construction, material quality, use and occupancy, location and maintenance of structures and buildings by providing minimum design regulations to ensure safety

[a]Most widely referenced and respected NFPA codes.

TABLE 4.6
Sample of Standards Developed by Association for the Advancement of Medical Instrumentation

Standard Number	Title
ANSI/AAMI EC13:2002	Cardiac monitors, heart rate meters, and alarms
ANSI/AAMI ST46:2002	Steam sterilization and sterility assurance in health care facilities
ANSI/AAMI EC71:2001	Standard communications protocol—Computer-assisted electrocardiography
ANSI/AAMI ST65:2000	Processing of reusable surgical textiles for use in health care facilities, 1ed
ANSI/AAMI RD16:1996	Hemodialyzers, 2ed
ANSI/AAMI ST35:1996	Safe handling and biological decontamination of medical devices in health care facilities and in nonclinical settings, 2ed
ANSI/AAMI ST37:1996	Flash sterilization—Steam sterilization of patient care items for immediate use, 3ed
ANSI/AAMI/ISO 14155:1996	Clinical investigation of medical devices, 1ed
ANSI/AAMI NS14:1995	Implantable spinal cord stimulators, 2ed
ANSI/AAMI HE48:1993	Human factors engineering guidelines and preferred practices for the design of medical devices, 2ed

will draft the standard, which must be approved before being passed on to the main committee. For a standard to be approved and published, full concurrence must be met by the ASTM Standing Committee on Standards, which ensures that the proper procedures and processes were followed. These standards are then made available to manufacturers who may choose to test their products based on these standards and provide this information to customers.

NIST. In addition to being a liaison between organizations in the United States and abroad (discussed previously), NIST (www.nist.gov) assists in the development of standards. NIST focuses on developing increasingly more precise ways to measure physical quantities (e.g., temperature, time, and mass) of all types of technology leading to the development of measurement standards. These measurement standards are in turn used to improve process and product standards developed by other organizations. NIST depends on its eight laboratories to develop the appropriate standards: (a) Building and Fire Research Laboratory, (b) Electronics and Electrical Engineering Laboratory, (c) Information Technology Laboratory, (d) Manufacturing Engineering Laboratory, (e) Materials Science and Engineering Laboratory, (f) Materials Science and Engineering Laboratory, (g) Physics Laboratory, and (h) Technology Services. Table 4.7 presents each of these laboratories, their respective divisions and offices, and their missions.

Federal and State SDOs

OSHA. The Occupational Health and Safety Administration (OSHA; www.osha.gov) a division of the U.S. Department of Labor, has developed an ergonomics program standard (29 CFR 1910.900) to reduce the risk of musculoskeletal disorders in workplace environments where employees' tasks involve frequent repetitive motion, awkward positions, contact vibration and stress, or force. The standards provide guidance to organizations to help develop a program that fits the task to the worker through such methods as workstation adjustment, use of mechanical assists, or rotation between tasks. There are several elements of the program that must be included: hazard information (i.e., common musculoskeletal disorders, signs and symptoms of disorders, and importance of reporting these disorders as soon as signs and symptoms are noticed) and reporting, training, musculoskeletal disorder management, management leadership and employee participation, job hazard analysis and control, and program evaluation. This standard covers all employees who are covered by the OSHA Act of 1970 with the exception of those employees involved in railroad operations or those covered by OSHA's construction, maritime, and agricultural standards. It is estimated that approximately 4.6 million musculoskeletal disorders relating to work tasks will be prevented over the next decade and $9.1 billion saved each year as a result of this standard.

Washington Industrial Safety and Health Act (WISHA). Similar to that proposed by OSHA, the state of Washington has developed its own state ergonomics standards that organizations must abide by. WISHA (www.lni.wa.gov/wisha) enforces two general standards: the General Safety and Health Standards and the General Occupational Health Standards. These two standards are a part of the Washington Administrative Code (Chapter 296-24 WAC and Chapter 296-62 WAC). The scope and purpose of chapter 296-24, General Safety and Health Standards, is to provide safety requirements that control safety hazards in order to protect employees' safety and health in their work environment. Similarly, the purpose of chapter 296-62, General Occupational Health Standards, is to create a healthy work environment and to protect employees' health by instituting requirements that control hazards affecting employees' health. Tables 4.8 and 4.9 provide a breakdown of the specific sections that each chapter entails.

TABLE 4.7
List of NIST Laboratories, Their Offices, and Mission

NIST Laboratory	Divisions/Offices	Purpose/Mission
Building and fire research laboratory	• Applied economics • Building materials • Fire research • Structures • Building environment	To improve construction quality and productivity in the United States; to reduce losses (i.e., economic, human) due to hazards (e.g., fire, wind)
Chemical science and technology laboratory	• Biotechnology • Surface and microanalysis Science • Analytical chemistry • Process Measurements • Physical and Chemical Properties	To conduct measurement science research; to develop measurements, data, models, and reference standards for chemicals, biochemicals, and chemical engineering; to improve health and safety of public, and environmental quality.
Electronics and Electrical Engineering Laboratory	• Microelectronics Programs • Electricity • RadioFrequency Technology • Optoelectronics • Law Enforcement Standards • Semiconductor Electronics • Electromagnetic Technology • Magnetic Technology	To improve quality of life and U.S. economy through standard advancement and providing measurement science and technology; to provide basis for all U.S. electrical measurements and metrology support to some agencies
Information Technology Laboratory	• Mathematical and Computational Sciences • Computer Security • Convergent Information Systems • Software Diagnostics and Conference Testing • Advanced Network Technologies • Information Access • Information Services and Computing • Statistical Engineering	To conduct research and develop methods for testing and standards to improve the usability, reliability and security of technology (computers, computer networks).
Manufacturing Engineering Laboratory	• Precision Engineering • Intelligent Systems • Fabrication Technology • Manufacturing Technology • Manufacturing Systems Integration	To develop standards, methods for measurement, and technologies; to maintain the U.S. basic units for measuring mass and length
Materials Science and Engineering Laboratory	• Center for Theoretical and Computational Materials Science • Materials Reliability • Metallurgy • Ceramics • Polymers • NIST Center for Neutron Research	To provide expertise and technical leadership for the U.S. materials measurement and standards infrastructure
Physics Laboratory	• Electron and Optic Physics • Optical Technology • Time and Frequency • Electronic Commerce in Scientific and Engineering Data • Atomic Physics • Ionizing Radiation • Quantum Physics	To conduct research and provide support by providing measurement services and developing new standards, methods, and data
Technology Services	• Standards Services • Measurement Services • Technology Partnerships • Information Services	To provide products and services to U.S. industries and public; to collaborate with other NIST labs, and U.S. organizations and agencies

TABLE 4.8
General Safety and Health Standards Chapter 296-24 WAC

Part	Title	Sections
Part A-1	Purpose and Scope	296-24-003–296-24-012
Part A-2 (Reserved)	Personal Protective Equipment	*Has been moved to WAC 296-800-160 and WAC 296-24-980*
Part A-4	Safety Procedures	296-24-110–296-24-119
Part B-1	Sanitation, Temporary Labor Camps and Nonwater Carriage Disposal Systems	269-24-120–296-24-12523
Part B-2	Safety Color Code for Marking Physical Hazards, etc., Window Washing	296-24-135–296-24-14519
Part C	Machinery and Machine Guarding	296-24-150–296-24-20730
Part D	Material Handling and Storage, Including Cranes, Derriks, etc., and Rigging	296-24-215–296-24-29431
Part E	Hazardous Materials, Flammable and Combustible Liquids, Spray Finishing, Dip Tanks	296-24-295–296-24-450
Part F-1	Storage and Handling of Liquefied Petroleum Gases	296-24-475–296-24-47517
Part F-2	Storage and Handling of Anhydrous Ammonia	296-24-510–296-24-51099
Part G-1	Means of Egress	296-24-55001–296-56701
Part G-2	Fire Protection	296-24-58503–296-24-58517
Part G-3	Fire Suppression equipment	296-24-592–296-24-63599
Part H-1	Hand and Portable Powered Tools and Other Hand-Held Equipment	296-24-650–296-24-67005
Part H-2	Safe Practices of Abrasive Blasting Operations, Ventilation	296-24-675–296-24-67521
Part I	Welding, Cutting and Brazing	296-24-680–296-24-722
Part J-1	Working Surfaces, Guarding Floors and Wall	296-24-735–296-24-85505
Part J-2	Scaffolds	296-24-860–296-24-862
Part J-3	Powered Platforms, etc.	296-24-875–296-24-90009
Part K	Compressed Gas and Compressed Gas Equipment	296-24-920–296-24-94003
Part L	Electrical	296-24-956–296-24-985

Federal Aviation Administration Technical Standard Orders (FAA TSO). Due to the rapid growth of the aviation industry during the 1930s and 1940s, the Civil Aeronautics Administration (CAA, now known as the FAA; http://av-info.faa.gov/tso/) was overburdened with the necessary inspections that had to be conducted on the aircraft parts and products, resulting in flight delays and high costs. As one way to minimize these problems, the CAA delegated responsibility to the manufacturers of aircraft by proposing the Technical Standard Orders (TSO) System. The proposal suggested that TSOs would be developed for specified processes, materials, parts, and appliances to be used on civil aircraft stating the minimum

TABLE 4.9
General Occupational Health Standards Chapter 296–62 WAC

Part	Title	Sections
Part A	General	296-62-005–296-62-050
Part A-1	Ergonomics	296-62-05101–296-62-05176
Part B	Access to Records	296-62-052–296-62-05223
Part B-1	Trade Secrets	296-62-05305–296-62-05325
Part C	Hazard Communication	296-62-054–296-62-05412
Part D	Controls and Definitions	296-62-060–296-62-07005
Part E	Respiratory Protection	296-62-071–296-62-07295
Part F	Carcinogens	296-62-073–296-62-07316
Part G	Carcinogens (Specific)	296-62-07329–296-62-07477
Part H	Air Contaminants	296-62-075–296-62-07515
Part I	Air Contaminants (Specific)	296-62-07517–296-62-07660
Part I-1	Asbestos, Tremolite, Anthophyllite, and Actinolite	296-62-077–296-62-07755
Part J	Biological Agents	296-62-080–296-62-08005
Part J-1	Physical Agents	296-62-090–296-62-09013
Part K	Hearing Conservation	296-62-09015–296-62-09055
Part L	Atmospheres, Ventilation, Emergency Washing	296-62-100–296-62-130
Part M	Confined Spaces	296-62-141–296-62-14176
Part N	Cotton Dust	296-62-14533–296-62-14543
Part O	Coke Ovens	296-62-200–296-62-20029
Part P	Hazardous Waste Operations and Treatment, Storage, and Disposal Facilities	296-62-300–296-62-3195
Part Q	Hazardous Chemicals in Laboratories	296-62-400–296-62-40025

performance that must be met. Although the CAA would maintain its responsibility for certifying airplanes, engines, and propellers, manufacturers would now be responsible for guaranteeing that their products met the TSO specifications. This proposal was instituted in 1947 and is still in use today. At the time of publication there are 126 current TSOs (see Table 4.10 for a sample) and 4 being proposed (see Table 4.11).

Uniform Federal Accessibility Standards (UFAS) . As the age of the U.S. population grows older and are more likely to use mobility devices (e.g., wheelchairs), it is appropriate that national standards be provided to ensure the proper design, construction, and alteration of buildings to accommodate individuals who are mobility challenged. These standards (www.access-board.gov/ufas/ufas-tml/ufas.htm) ensure that accessibility and use of all buildings will be available to all individuals in accordance with the Architectural Barriers Act (i.e., 42 U.S.C. 4151-4157). There are four regulatory agencies which enforce these standards: (a) Department of Defense (DoD), (b) Department of Housing and Urban Development (HUD), (c) U.S. Postal Service (USPS), and (d) General Services Administration (GSA). The first three agencies set standards for the buildings that they control (e.g., military facilities, residential properties, and postal facilities, respectively), whereas GSA sets standards for all other buildings, which must comply with the Architectural Barriers Act. Compliance is ensured by the Architectural and Transportation Barriers Compliance Board established by Congress as a part of the Rehabilitation Act of 1973 (29 USC 792 Section 502). Table 4.12 provides examples of the UFAS standards developed.

TABLE 4.10

Sample of Current FAA Technical Standard Orders

TSO Number	Title
TSO-C1c	Cargo Compartment Fire Detection Instruments
TSO-C2d	Airspeed Instruments
TSO-C3d	Turn and Slip Instrument
TSO C4c	Bank and Pitch Instruments
TSO C5e	Direction Instrument, Nonmagnetic (Gyroscopically Stabilized)
TSO-C8d	Vertical Velocity Instruments (Rate of Climb)
TSO-C9c	Automatic Pilots
TSO-C10b	Altimeter, Pressure Actuated, Sensitive Type
TSO-C11e	Powerplant Fire Detection Instruments (Thermal and Flame Contact Types)
TSO-C13f	Life Preservers
TSO-C14b	Aircraft Fabric, Intermediate Grade, External Covering Material
TSO-C19b	Portable Water-Solution-Type Fire Extinguishers
TSO-C41d	Airborne Automatic Direction Finding (ADF) Equipment
TSO-C54	Stall Warning Instruments
TSO-C57a	Headsets and speakers
TSO-C69c	Emergency Evacuation Slides, Ramps, Ramp/Slides, and Slide/Rafts
TSO C72c	Individual Flotation Devices
TSO-C89	Oxygen Regulators, Demand
TSO-C92c	Airborne Ground Proximity Warning Equipment
TSO-C101	Over Speed Warning Instruments
TSO-C116	Crewmember Protective Breathing Equipment
TSO-C126	406 MHz Emergency Locator Transmitter (ELT)
TSO-C135	Transport Airplane Wheels and Wheel and Brake Assemblies
TSO-C141	Aircraft Fluorescent Lighting Ballast/Fixture Equipment
TSO-C153	Integrated Modular Avionics Hardware Elements

TABLE 4.11

Proposed FAA Technical Standard Orders

TSO	Title
TSO-C39c	Aircraft Seats and Berths
TSO-CRIPS	Recorder Independent Power Supply
TSO-C138	Miscellaneous Nonrequired Equipment
TSO-C152	Flammability Test Method for Aircraft Blankets

Private Industry Standards Databases

National Standards Systems Network (NSSN).   The NSSN (www.nssn.org) is an international database that is readily available to a wide variety of audiences and may be particularly useful for students, researchers, and practitioners in human factors and ergonomics. The NSSN is a product of ANSI and consists of standards from U.S. private organization standards developers, international standards organizations, and government agencies. The NSSN provides a database network for developing and approved national, foreign, regional, and international standards and regulatory documents (NSSN, n.d.). This database contains

TABLE 4.12
Examples of UFAS Standards

Accessible Elements and Spaces	Example of Specifications
Accessible Housing	Elevators, entrances, and common areas
Space Allowance and Reach Ranges	Wheelchair passage width (single and dual wheelchairs), turning space, size and approach, forward and side reach
Accessible Route	Location, width, passing space, egress, head room, slope
Parking and Passenger Loading Zones	Location, number of spaces, width of spaces, signage, vertical clearance, and loading zone areas
Curb Ramps	Width, slope, surface, sides, warning textures
Elevators	Automatic operation, hall lanterns, hall call buttons, door protective and reopening device
Doors	Width, maneuvering clearance, thresholds, hardware
Toilet Stalls	Size and arrangement, toe clearance, hold bars, doors
Alarms	Audible, auxiliary, and visual alarms
Signage	Character proportion, color contrast, symbols of accessibility
Restaurants and Cafeterias	Food service lines, tableware areas, vending machines
Healthcare	Entrances, patient bedrooms, patient toilet rooms
Libraries	Reading and study areas, check-out areas, card catalogs, stacks

standards from more than 600 national, international, foreign, and governmental organizations. Because of the growing global community, NSSN adopted international standards as well and now refer to themselves as "a national resource for global standards" (NSSN, n.d.).

IEEE Xplore. IEEE (www. ieeexplore.ieee.org/) provides a searchable database containing full-text versions of all of their published standards dating back to 1980, in addition to their publications and conference proceedings. Table of contents for all standards, in addition to conference proceedings, papers, and publications, is free to the public. Members of IEEE have slightly more access, but for the most part, corporate, government, and university subscribers are the only users who have access to the full range of full-text standards.

In addition to NSSN and IEEE Xplore, there are a number of searchable databases on the World Wide Web. Many are a service provided by a private company (e.g., Systems-Concepts), a government organization (e.g., FDA and FAA), a society (HFES) or an academic institution (e.g., University of Washington). The amount of information provided for free and the price of membership on these Web sites vary, but can be very useful for initial information gathering efforts. A list of some of the searchable databases is shown in Table 4.13.

In addition, a great deal of information regarding U.S. government standards is available as well. Although this chapter is primarily concerned with private voluntary consensus standards, the standards issued by the government offer additional guidelines for the U.S. private sector. As such, the following are a sample of the availability of these standards on the Internet.

U.S. Government Standards Databases

ASSIST Quick Search. This database (www.assist.daps.dla.mil) is available to private citizens and provides access to nearly 10,000 standards issued and enforced by the U.S. Department of Defense (DoD). This Web site is useful because it provides full-text PDF versions of their standards to registered users free of charge. In order to access the full text documents, the user must register for a password and provide minimal information about themselves

TABLE 4.13
Searchable Standards Databases on the Web

Web Site Address	Database	Description
http://standards.ieee.org/	IEEE Standards Online	Provides online access of IEEE standards in PDF format with IP access or Password-Protected Access to Enterprise level subscribers
http://stneasy.cas.org	International Standards Database	A bibliographic database containing citations to German, Austrian, Swiss, French, U.S., Japanese, European, and international standards, guidelines, and other regulations
http://www.system-concepts.com/stds/	Ergonomics Standards & Legislation	A database provided by Systems-Concepts, an ergonomics firm out of the UK
http://www.fda.gov/search/databases.html	Recognized Consensus Standards	Consists of those national and international standards recognized by FDA
http://www2.faa.gov/asd/standards/std-db.cfm	FAA Standards Database	Contains descriptions of more than 5,000 FAA, military, and commercial standards and specifications used in existing FAA contracts and projects
http://db.lib.washington.edu/standards/	Engineering Standards Database	A searchable standards database aimed at engineers through the University of Washington Library

(e.g., e-mail, phone, and organization). In addition, the DoD has STINET (www.stinet.dtic.mil), which provides links to full-text documents that does not require registration.

DOE Technical Standards. This Internet database is provided by the Department of Energy (DOE; www.tis.eh.doe.gov/techstds/standard/standfrm.html). The DOE provides approved technical standards, recently approved standards, and as drafts. The documents can be viewed in PDF or as HTML through your Web browser.

OSHA Regulations. In addition to being available through the NSSN Web site, OSHA (www.osha.gov/OCIS/stand_dev.html), discussed in depth earlier, has its own searchable database. OSHA provides full-text standards, which are contained within the Code of Federal Regulations.

Federal Vehicle Standards. The General Services Administration provides this database (www.apps.fss.gsa.gov/pub/vehicle-standards.cfm) for vehicle standards. The available full-text documents provide standards for sedans; station wagons; and light, medium, and heavy trucks.

Index of Federal Specifications, Standards, and Commercial Item Descriptions. This database (www. http://apps.fss.gsa.gov/pub/fedspecs/index.cfm) covers federal specifications and standards issued through the Federal Property Management Regulations. Documents are listed on the main Web page and may be ordered by calling, e-mailing, or faxing the U.S. General Services Administration (GSA), with specific instructions on the Web site. However, the standards provided are not full text, although some of these standards can be accessed through ASSIST, mentioned earlier.

CONCLUSION

The information provided in this chapter is meant to be used as a tool to assist human factors and ergonomics professionals in their search for the standards needed to design, maintain, and evaluate systems, machines, and products. The history of the development of standards clearly shows how the need drives the development. When a need for a standard measurement, requirement, or system was expressed by citizens or observed in the public, government, or industry created a standard. This is important for the field because, as technology is being developed at an alarming rate, those of us who are trained in human factors and ergonomics must take up the challenge to make technology and human–machine systems safer, more reliable, easier to use, and more productive. Standards are a way for us, as a community of researchers and practitioners, to do so across the board. As humans increasingly continue to interact with machines, emerging technologies will challenge us as scientists, and we must meet the challenge in a systematic, scientific way. Standardization is one tool we can use to consistently and effectively make products and systems more useful and more reliable. The goal of a chapter of this scope is to provide professionals and students with the knowledge of what has come before, what exists now, and what to expect in the future with regard to human factors and ergonomics standards in private industry and related sectors (i.e., government standards that may be enforced in private industry). In addition to the specific information provided in this chapter, we attempted to provide you with tools to further explore the standards development system that is available and to use the tools provided by that system. We hope that the information provided will help guide those interested in accessing national standards being developed and that exist today and to contribute to the growing body of knowledge established by human factors and ergonomics professionals.

REFERENCES

Alhstrom, V., Longo, K., & Truitt, T. (2002). *Human factors design guide update (Report Number DOT/FAA/CT-96/01): A revision to chapter 5: Automation guidelines* (Technical Report Number DOT/FAA/CT-02/11). Washington D.C.: Department of Transportation.

American National Standards Institute (ANSI). (n.d.). *About ANSI: Introduction.* Retrieved January 2, 2003, from http://www.ansi.org/about_ansi/introduction/introduction.aspx?menuid=1

American National Standards Institute (ANSI). (n.d.). *ANSI: A historical view.* Retrieved January 2, 2003, from http://www.ansi.org/about_ansi/introduction/history.aspx?menuid=1

American National Standards Institute (ANSI). (2000). *National standards strategy for the United States.* Retrieved on December 3, 2002, from http://www.ansi.org/standards_activities/nss/nss.aspx?menuid=3

American Society for Testing and Materials (ASTM). (n.d.). *The handbook of standardization: A guide to understanding today's global standards development system.* Retrieved on December 1, 2002, from http://www.astm.org/cgi-bin/SoftCart.exe/NEWS/handbook02/index.html

Appleby, J. (2001, February 4). Study: ICU standards could save lives, money. *USA Today,* D2.

Blanchard, H. E. (1997). International standards on human-computer interaction: What is out there and how will it be implemented? In G. Salvendy, M. J. Smith, & R. J. Koubeck (Eds.), *Design of computing Systems: Cognitive considerations* (pp. 599–602). Amsterdam: Elsevier.

Blanchard, H. E. (2000, April). New year's update on human computer interaction standards: Part 1: The U.S. *SIGCHI Bulletin, 32*(3), 11–12.

Breitenberg, M. A. (1987, May). *The abc's of standards-related activities in the United States* (NBSIR 87-3576). Gaithersberg, MD: National Institute of Standards and Technology.

Business Wire (2002). *Human factors and ergonomics society joins IEEE effort to create advanced voting equipment standard*. Retrieved on December 2, 2002, from http://www.securepoll.com/Archives/Archive83.htm#Human

National Aeronautics and Space Administration (NASA). (n.d.). *NASA technical standards program Overview*. Retrieved on December 3, 2002, from http://standards.nasa.gov/Npts/public_login.taf#

National Highway Traffic Safety Administration (NHTSA). (1999). *Traffic safety facts of 1998*. Retrieved on December 3, 2002, from http://www-nrd.nhtsa.dot.gov/pdf/nrd-30/NCSA/TSFAnn/TSF1998.pdf

National Institute for Standards and Technology (NIST). (n.d.). *NIST products and services: Standards*. Retrieved on December 3, 2002, from http://www.nist.gov/public_affairs/standards.htm

National Standards Policy Advisory Committee. (1978, December). *National policy on standards for the United States and a recommended implementation plan*. National Standards Policy Advisory Committee: Washington, DC.

National Standards Systems Network (NSSN). (n.d.). *About NSSN*. Retrieved on December 1, 2002, from http://www.nssn.org/about.html

National Technology Transfer and Advancement Act (NTTAA). (1995). *National Technology Transfer and Advancement Act of 1995*, PL 104-113.

U.S. Department of Labor Office of Public Affairs. (1998). *New OSHA Training Standard To Save Lives, Reduce Injuries* (OSHA 98-4). Retrieved on August 12, 2002, from http://www.osha.gov/media/oshnews/ dec98/trade98-4.html

5

Overview of National and International Standards and Guidelines

Carol Stuart-Buttle
Stuart-Buttle Ergonomics

INTRODUCTION

The Human Factors and Ergonomics profession has multiple domains and specialty areas for which there are specific standards, and not all standards can be addressed in this handbook. So the reader may have to go searching. Likewise, some nations may not have standards at the time of this publication, but the guidance in this section will assist the reader in finding standards and guidelines in the future.

Within a country or group of countries, such as the European Union (EU), there are only a few standards that are legislated as mandatory. Typically, at the mandatory level, the standards are in general terms so that they do not go out of date. There are numerous nonmandatory ergonomics standards that are desirable to follow as they often provide the substantive guidance to those that are mandatory. Standard setting bodies mostly oversee the development of nonmandatory standards and guidelines, and they develop them on professional consensus or by using experts in the topic area. Such nonmandatory documents are guidelines, although they may be entitled as either standards or guidelines. On occasion, for example, in the United States, a nonmandatory standard may be cited and enforced by legislation.

A specialty group or individuals from a specific practice domain may be involved in developing a standard that then becomes widely adopted by users in other areas. Such standards can be hard to find by someone not affiliated with the standard-developing group. An effective approach to seek specific standards is to find related societies or associations through which to make inquiries.

Many countries adopt international standards, and European standards are enfolded into most European Union members' laws. There is a rising need for standardization as business becomes more global. Therefore standards are always being developed or updated, so standards should be periodically checked to ensure they are current.

SCOPE

This chapter provides general guidance to finding international, European, and national standards and guidelines on human factors and ergonomics. Also specific guidance is given on the primary standards of the United States, Canada, UK, Australia, and Japan. Some of the more widely known ergonomics standards and guidelines are introduced, and resources are shared for the main standard developing bodies.

SEARCHING FOR STANDARDS

There are several groups that provide European and international standards. In addition, European countries and many other countries each has their own standards-developing group. Individual country standardization companies can be found through central groups such as the European Committee for Standardization (CEN). The individual standardization committees or corporations of each country usually serve as the local point of contact for obtaining international standards. The following groups are resources for obtaining European and international standards, and in some cases, standards of individual countries. There is a fee for most standards, especially those that are non-mandatory.

Perinorm

Perinorm (www.perinorm.com) provides a subscriber-based service that offers a database of international, European, and national standards. Standards of 21 countries are available which, apart from European countries, includes the United States, Japan, Australia, Turkey, and South Africa.

International Organization for Standardization (ISO)

There are 40 main groups of ISO (www.iso.org) standards organized by general topic area (international classification for standards [ICS]). There are also listings of Standards by Technical Committee. The TC 159 is the Ergonomics Technical Committee. Apart from specific ergonomics related standards, some of the categories may be pertinent to a particular industry, such as electronics or material handling equipment. Many of the industry-based or equipment-based standards pertain to manufacturing issues, for example, equipment dimensions or stability tests. Human factors and ergonomics ISO standards are discussed in detail in this handbook.

International Ergonomics Association (IEA)

Another possible resource for finding standards in a country is the IEA (www.iea.cc). The association may provide a point of contact in a member country, who could assist with providing information on standards and guidelines of that country.

Worldwide Governments on the World Wide Web (WWW)

If a country is not available through the aforementioned contacts, then another method to determine whether there are mandatory regulations is to go through the government bodies. Web sites for governments throughout the world can be found at Worldwide Governments on the WWW (www.gksoft.com/govt/en/world.html).

Other International Standards Groups

There are other international groups that have worked to standardize pertinent aspects of a group's area of interest. A few additional groups are listed in the following as resources.

International Telecommunication Union (ITU)   www.itu.org
International Civil Aviation Organization (ICAO)   www.icao.org
World Wide Web Consortium (W3C)   www.w3.org

A couple of comprehensive resources for links to the numerous standards organizations are:

NASA Technical Standards Program (NASA requires
   the public to register to log on to the standards page)   http://standards.nasa.gov
World Standards Services Network (WSSN)   www.wssn.net

European Standards

European standards are addressed in detail in this handbook. The following websites are sources for European standards and related information.

European Union (EU)   www.europa.eu.int

This website provides background on the European Union. The EU has several institutions one of which sets Directives that provide a minimum standard for health and safety at work.

CEN   www.cenorm.be

The European Committee for Standardization (CEN) develops European standards. Through the CEN web page there are links to each member country's main standards setting group. Note, these groups are not the legislative groups of the countries.

OSHA/EU Cooperation

The OSHA (www.osha-slc.gov/us-eu) of the United States and the EU have a joint Web page to facilitate communication. Both European and U.S. legislation can be accessed through this site. In addition there are links to each member country and Switzerland, Iceland, Norway, Canada, and Australia.

Other European Standards Organizations

Some additional European standards-setting organizations, other than the main ones related to each country, are:

European Committee for Electrotechnical Standardization (CENELEC)   (www.cenelec.org)
European Telecommunications Standards Organization (ETSI)   (www.etsi.org)

CANADIAN STANDARDS

British Columbia (BC)

The Secretariat for Regulation Review, Board of Governors, Workers' Compensation Board of British Columbia issued a draft ergonomics regulation in the fall of 1994 (www.worksafebc.com). The regulation failed to be adopted by the BC Legislature in 1995. However, since the issue of a draft regulation, there is a two-page section on ergonomics in Part 4, General Conditions of the Occupational Health and Safety Regulations of the Workers' Compensation Board of BC. Sections 4.46–4.53, Ergonomics Musculoskeletal Injury (MSI) Requirements, require employers to identify factors that might expose workers to the risk of MSI, to assess the identified risks, and to eliminate or minimize the risks. Employees are to receive education and training and be consulted by the employers. The requirements also include evaluations of effectiveness.

Ontario (ON)

Draft legislation of Physical Ergonomics Allowable Limits were prepared for the Ministry of Labour, Government of Ontario (www.gov.on.ca). The report was rescinded and shelved in 1995 to 1996. The Occupational Health and Safety Act of 1979 was changed in 1990 with some significant additions. All employers have to have a health and safety policy and program, and the Officers of corporations have direct responsibility. A joint labor and management Health and Safety Committee that is responsible for health and safety in the workplace is required in facilities with greater than 20 workers. The committee is to meet regularly to discuss health and safety concerns, review progress, and make recommendations. Workplaces of fewer than 20 workers must have a Health and Safety Representative. By 1995, employers had to certify that the members of their joint Health and Safety Committees were properly trained.

In February of 2005, The Government of Ontario set a goal to reduce workplace injuries by 20 percent by 2008. In order to meet this goal the Ministry of Labour established an ergonomics working panel that will make recommendations on how to reduce ergonomics-related injuries, plans to coordinate 14 health and safety associations to develop prevention strategies, provide research monies, allocate funds for bed lifts to reduce back injuries to nurses and hire more inspectors. The government's approach does not include an ergonomics standard.

Canadian Standards Association (CSA)

The (CSA) (www.csa.ca) produces voluntary standards pertaining to many areas. One guideline that is widely used is "Guideline on Office Ergonomics", CSA—Z412. The latest version of 2000 is produced as interactive CDROM, Adobe.pdf file or hardcopy.

U.S. STANDARDS

Occupational Safety and Health (OSH) Act

The primary mandatory standard of the United States is the OSH Act of 1970. (www.osha.gov) The section pertinent to ergonomics is the general duty clause, Section 5 (a) (1), that states:

> Each employer shall furnish to each of his employees, employment and a place of employment which is free from recognized hazards that are causing or are likely to cause death or serious harm to his employees.

Citations for ergonomics have been under the general duty clause.

Americans with Disabilities Act (ADA)

The mandatory act (Public Law 101-336) was effective in 1990 (www.access.gpo.gov). There is some bearing on ergonomics. One part of the act addresses accessibility for the disabled, and a second part pertains to employment. Two main points of the act under the employment section are:

- The ADA prohibits disability-based discrimination in hiring practices and working conditions.
- Employers are obligated to make reasonable accommodations to qualified disabled applicants and workers, unless doing so would impose undue hardship on the employer. The accommodations should allow the employee to perform the essential functions of the job.

Often modifications to a job to accommodate someone disabled can benefit all workers. Defining essential functions of a job may involve those responsible for ergonomics.

California Ergonomics Standard

This is a mandatory state law that was effective 1997 and addresses formally diagnosed work-related repetitive motion injuries that have occurred to more than one employee. The employer has to implement a program to minimize the repetitive motion injuries through worksite evaluation, control of the exposures, and training (www.dir.ca.gov/title8/5110/html).

Washington State Ergonomics Standard

In 2000, Washington State adopted a mandatory ergonomics rule (www.lni.wa.gov/wisha). The rule states that employers have to look at their jobs to determine if there are specific risk factors that make a job a "caution zone job" as defined by the standard. All caution zone jobs must be analyzed, employees of those jobs are to participate and be educated, and the identified hazards must be reduced. The rule was repealed in November 2003, effective December 2003, after state residents voted against a standard. Much of the information of the rule, such as evaluation tools, remain on the Washington State Labor and Industries web site as resources in ergonomics.

OSHA Standards

Ergonomics Program Standard—Repealed

OSHA of the federal government released an Ergonomics Program rule in November 2000 that was repealed in March 2001 by a new administration. The standard was issued in the Federal Register November 14, 2000, Vol. 65, No. 220 (www.osha.gov). (29 CFR 1910.900) and is still available.

The main elements of the standard were:

a. Train in basic ergonomics awareness
b. Provide medical management of work-related musculoskeletal disorders
c. Implement a quick fix or go to a full program
d. Implement a full ergonomic program when indicated:
   - Management leadership
   - Employee participation
   - Job hazard analysis

- ○ Hazard reduction and control
- ○ Training
- ○ Program evaluation

The withdrawn standard was similar to long-standing voluntary guidelines issued by OSHA "Ergonomics Program Management Guidelines for Meatpacking Plants" (OSHA 3123) of 1990. These guidelines have been used successfully for many years by industries other than meatpacking.

OSHA Activity Since the Standard

Four-pronged Approach. The government has a four-pronged approach to address occupational musculoskeletal disorders (MSDs). The four parts are:

1) *Guidelines.* Industry specific, voluntary guidelines are being developed to provide effective and feasible solutions to ergonomics-related problems. The industries picked for which to develop guidelines are driven by the injury and illness incidence rates of the industry.

2) *Enforcement.* OSHA continues to inspect facilities for ergonomics-related issues and will issue citations or hazard alert letters based on the general duty clause of the OSH Act (see above). In addition, OSHA conducts a Site Specific Targeting (SST) Inspection Program that identifies the workplaces with high incident rates. These facilities will receive a notice that their incidence rates are high and that they could be inspected. In 2004, rates greater than 15 or more injuries or illnesses resulting in days away from work, restricted work activity or job transfer for every 100 full-time workers (known as the DART rate) were sent letters. The primary list will also include sites that have a days away from work injury and illness (DAFWII) rate or 10 or higher. The average national DART rate in 2002 for private industry was 2.8, while the national average DAFWII rate was 1.6. The program stems from OSHA's Data Initiative for 2003 that surveyed approximately 80,000 employers to attain injury and illness data for 2002.

3) *Outreach and Assistance.* Many compliance tools and information are provided through the OSHA web page including "etools" on specific topics and industries, case studies and training and education resources. The administration has also developed Cooperative Programs, some existing prior to the four-pronged initiative such as the Voluntary Protection Program (VPP) and Safety and Health Achievement Recognition Program (SHARP) that guides and encourages high standards of safety and health processes that include ergonomics. Alliances and Strategic Partnerships have been forged between OSHA and many industries and associations. Alliances are formal agreements with groups and companies to cooperate on developing resources to address workplace safety and health. Strategic Partnership are entered with companies that want help to address specific safety and health issues. Ergonomics is frequently part of all these cooperative programs.

4) *National Advisory Committee on Ergonomics (NACE).* The Committee was formed at its initial meeting in January 2003 and was chartered until November 2004. The committee accomplished the following:
   - provided information related to various industry or task-specific guidelines
   - identified gaps in the existing research base related to applying ergonomic principles to the workplace
   - identified current and projected research needs and efforts

- determined methods of providing outreach and assistance that will communicate the value of ergonomics to employers and employees, and
- provided ways to increase communication among stakeholders on the issue of ergonomics.

Guidelines. The development of voluntary industry guidelines is part of the four-pronged approach. There are three released in final versions and one being developed. OSHA is expected to develop more guidelines but there has been no announcement to date of the next industry guideline. Those that have been developed are anticipated to be useful to industries other than the target audience. The format is heavily illustrated and emphasizes identifying and controlling hazards that are common in the industry. The guidelines in order of release are:

- Guidelines for nursing homes
- Guidelines for retail grocery stores
- Guidelines for poultry processing
- Guidelines for shipyard industry (in process).

ANSI Standards

The American National Standards Institute (ANSI; www.ansi.org) is a voluntary censensus standard body that issues many standards in a year. The following are just a few of the main ergonomics standards that exist or are being developed. There are many others that relate to safety in the workplace. Information about ANSI documents may be obtained through ANSI, but often they are purchased directly from the group responsible for developing the standard in coordination with ANSI. Although voluntary, at times ANSI standards are cited in OSHA regulations.

There are laws, such as the National Technology Transfer and Advancement Act (Public Law 104-113), 1996, that said all Federal agencies and departments shall use technical standards published by voluntary consensus standards bodies. Exceptions to this are allowed but have to be justified. There is also a movement in the government to go away from military standards and to adopt private sector standards as much as possible.

The following ANSI standards are some of the general ergonomics ones that are less industry specific.

a. Human Factors Engineering of Computer Workstations—BSR/HFES 100 Draft Standard.

   This standard, dated 3/31/02, is an update of ANSI/HFES 100-1988 (www.hfes.org) standard and it is issued as a draft standard for trial use until the end of 2004, at which time the standard can be considered for acceptance as an ANSI standard. The draft provides basics on ergonomics of computer workstations including workstation design specifications for the anthropometric range of 5th to 95th percentile and for four reference positions: reclined, upright, and declined sitting and standing. Several input devices are covered, and there is a section on how to integrate all the workstation components for an effective system.

b. ASC Z-365 Management of Work-Related Musculoskeletal Disorders

   The Accredited Standards Committee Z-365 was formed in 1991 (www.nsc.org). The secretariat, National Safety Council (NSC) issued the most recent working draft that was near completion, in October 2000. The draft contained similar elements to that of the repealed federal standard and the OSHA meatpackers' guideline. The document is programmatic rather than specific, in that it does not provide details on how to conduct analyses or on interventions. In the fall of 2003 the secretariat withdrew from the process citing extraordinary costs involved with this particular standard. After twelve years and with an almost completed standard, the committee was disbanded as no alternative secretariat came forward. This status report is provided due to the many years of development.

c. ASC Z-10 Occupational Health Safety Systems

The committee was established in 2001 under the ANSI secretariat of the American Industrial Hygiene Association (AIHA) (www.aiha.org). The objective is to develop a standard of management principles and systems to allow organizations to design and implement approaches to improve occupational safety and health. The document has now been released as ANSI/AIHA Standard Z-10, Occupational Health and Safety Management Systems.

d. HFES 200 Software User Interface Standard

This is a five-part standard (www.institute.hfes.org) being developed by The Human Factors and Ergonomics Society (HFES) under the auspices of ANSI. It will closely mirror the ISO 9241 standard on visual display terminals, except for original parts that will be on color, accessibility, and voice input and output.

ACGIH TLVs

The American Conference of Governmental Industrial Hygienists (ACGIH; www.acgih.org) has developed Threshold Limit Values (TLVs) for chemical substances and physical agents. There are TLVs for hand–arm vibration and whole-body vibration as well as for thermal stress. Two new TLVs are for Hand Activity Level, which is intended for "monotask" jobs performed for 4 hours or more, and a Lifting TLV, which provides weight limits based on frequency and duration of lift, horizontal distance and height at the start of the lift. The three lifting tables are based on current journal literature that includes dynamic biomechanics. Trunk twisting is not included in the tables.

National Institute of Standards and Technology (NIST)

The NIST (www.nist.gov) helps to develop standards that address measurement accuracy, documentary methods, conformity assessment and accreditation, and information technology standards. A current initiative of NIST is to develop industry usability reporting guidelines that directly affect software ergonomics. NIST can also be a source to link to military standards.

Miscellaneous Standards Setting Groups

There are many other sources of standards that may be important to certain domains or specialties. A few of the other organizations that develop standards are:

American Society of Mechanical Engineers (ASME)	www.asme.org
American Society for Testing and Materials (ASTM)	www.astm.org
Institute of Electrical and Electronics Engineers (IEEE)	www.ieee.org
Society of Automotive Engineers (SAE)	www.sae.org

UNITED KINGDOM STANDARDS

Mandatory Regulations

As with most member countries of the European Commission, the United Kingdom (UK) laws originate with the proposals from the EC. There is a statutory Health and Safety at Work

Act under which there are regulations that are law (www.hse.gov.uk). These regulations are general but are interpreted in "Approved Codes of Practice" documents developed by the Health and Safety Executive (HSE). Approved Codes of Practice have special legal status and can be used in prosecutions. Guidance documents interpret the law and provide further detail for compliance that can be especially useful, although they are not legally binding. There are several useful regulations known as the "Six Pack" that were issued in 1992. These regulations apply across all industries.

- Manual Handling Operations
- Display Screen Equipment
- Workplace (Health, Safety, and Welfare)
- Provision and Use of Work Equipment
- Personal Protective Equipment at Work
- Management of Health and Safety at Work

Very practical Guidance documents and approved Codes of Practice correspond to the regulations and could be useful outside of the UK. Additional HSE materials include employers' guides and manual handling solutions. There is a nominal charge for the HSE documents. UK Standards are discussed in more detail elsewhere in this handbook.

Nonmandatory Standards

As in most countries there are many standards that are nonmandatory. The British Standards Institute (BSI) is the primary standards-setting body in the UK (www.bsi-global.com). Two notable publications of BSI are the BS 8800 "Guide to Occupational Health and Safety Management Systems" (1996) and OHSAS 18001 "Occupational Health and Safety Management Systems-Specification" (1999). The BS 8800 is based on the ISO 14001 Environmental Management Systems model. The OHSAS 18001 is part of the Occupational Health and Safety Assessment Series, and the 18001 document provides guidance to a management system that can be assessed and certified (Seabrook, 2001).

AUSTRALIAN STANDARDS

The Commonwealth Government has federal jurisdiction and in addition, there are six states and two mainland territories that have each their own laws. The approach overall is similar to the European model that relies on a general duty of care by employers and employees. The focus is risk management that is based on risk assessment and control. Most of the states and territories have their own general health and safety act, but they vary in the degree of development of their regulations and codes of practice.

National Occupational Health and Safety Commission (NOHSC)

The NOHSC (www.nohsc.gov.au) is the federal body of the Commonwealth of Australia. The Commission is the primary source of National Standards, Regulations and Codes of Practice although the role of developing new standards has diminished. The Commonwealth standards are very general, and individual states and territories either adopt them or go above and beyond these standards. The NOHSC Web page links to all other states and territories.

Specific to ergonomics, there is one Standard and two Codes of Practice that are widely used. They stem from a National Occupational Health and Safety Commission Act of 1985 (Cwlth).

a. Manual Handling, National Standard NOHSC:1001(1990) In very general terms, this standard delineates employers to conduct a risk assessment and control any issues. The approach acknowledges that the risk is multifactorial and does not recommend weight control alone. The Code of Practice that is referenced provides greater guidance.
b. Manual Handling, National Code of Practice NOHSC:2005(1990) The Code of Practice provides considerable detail of risk assessment, criteria of risk, and examples of potential control methods. There are many illustrations and checklists.
c. National Code of Practice for the Prevention of Occupational Overuse Syndrome (NOHSC:2013(1994)

   The approach of the Code of Practice is one of risk identification, assessment and control. Specific risk factors are discussed but not quantified, and include work organization and design issues. A checklist approach is used and controls are presented in the form of principles. The document includes screen-based workstations (office environments).

Comcare (www.comcare.gov.au) is an informational branch of the commonwealth government that publishes some useful booklets and reports on pilot programs to assist with compliance with the law. One example is a booklet "A Guide to Health and Safety in the Office."

New South Wales (NSW) WorkCover Authority

NSW (www.workcover.nsw.gov.au) has recently (2000 and 2001) revised their Occupational Health and Safety Act and regulations and they are quite comprehensive in their legal coverage.

a. Occupational Health and Safety Act 2000 (No 40)

   The Act is general and delineates the duties of employers and employees. It refers to related regulations and codes of practice that are more specific to comply with the law. Inspections and legalities related to noncompliance are also provided in the Act.
b. Occupational Health and Safety (OHS) Regulation 2001

   Consistent with the national approach, the regulation delineates the steps required for risk management, namely, identification, assessment, and control of hazards. Employers are to identify potential manual handling and occupational overuse hazards and assess lighting and workstation design, among many other safety and health issues that are listed. Specifics are given on workplace consultation. An occupational health and safety committee or representative or consultant is to contribute to risk management. Training of those involved is also specified.

Victorian WorkCover Authority

a. Occupational Health and Safety Act 1985 (www.workcover.vic.gov.au; includes amendments up to 2001)
b. Occupational Health and Safety (Manual Handling) Regulations 1999
   This regulation is more expansive than the Commonwealth manual handling standard, although with a similar approach of risk management.

South Australian WorkCover Authority

a. Occupational Health, Safety and Welfare Act 1986
b. Occupational Health, Safety and Welfare Regulations 1995 (Updated 1999).
   The Regulation (www.workcover.com) is nonspecific, but employers are obliged to follow approved Codes of Practice that are listed in the Regulation, one of which is for manual handling.

WorkSafe Western Australia

a. Occupational Safety and Health Act 1984
b. Occupational Safety and Health Regulations 1996
   The Regulation, based on a risk management approach, is less specific than other states, such as NSW.
c. Code of Practice Manual Handling 2000
   This Code of Practice is a simpler and updated version of the national Code of Practice (www.safetyline.wa.gov.au).

Queensland Division of Workplace Health and Safety

a. Workplace Health and Safety Act 1995 (includes amendments to 2000)
b. Workplace Health and Safety Regulations 1997
c. Advisory Standards
   ◦ Manual Handling (Building Industry) 1999
   ◦ Manual Tasks 2000
   ◦ Manual Tasks Involving People 2000

   The Manual Tasks Advisory Standard 2000 provides specific guidance in risk identification, assessment, and control. Checklists and discomfort surveys are provided, and a task analysis is described with example task analysis forms. The risk control section includes an implementation plan and evaluation step (www.whs.qld.gov.au).

Workplace Standards Tasmania

a. Workplace Health and Safety Act 1995
b. Workplace Health and Safety Regulations 1998
   The Regulations are very general (www.wsa.tas.gov.au).

Australian Capital Territory (ACT)

a. ACT Occupational Health and Safety Act 1989
b. ACT Occupational Health and Safety Regulations 1991
c. ACT Occupational Health and Safety (Manual Handling) Regulations 1997
d. ACT Manual Handling Code of Practice 1999
   ACT bases the material handling Regulation and Code of Practice on the national (commonwealth) ones. Similar checklists and illustrations are used. ACT also supports use of the National Code of Practice for the Prevention of Occupational Overuse Syndrome (www.workcover.act.gov.au).

Northern Territory Work Health Authority

There are no laws pertaining to ergonomics beyond the scope of the national ones for the Northern Territory (www.nt.gov.au).

Standards Australia

Standards Australia (www.standards.com.au) is an organization that is recognized as a main source for developing nonmandatory technical and business standards and the dissemination of Australian and international standards. Their principle is to adopt or closely align their standards with international standards whenever possible. The group developed two standards related to ergonomics that has been adopted by both Australia and New Zealand.

1. Occupational health and safety management systems—specification with guidance for use. AS/NZS 4801:2001
2. Occupational health and safety management systems—General guidelines on principles, systems and supporting techniques. AS/NZS 4804:2001

JAPANESE STANDARDS

Japan has a general Labour Standards Law, revised in 1998, which states that employers should take measures to ensure reasonable working conditions and to improve working conditions. Additional laws supplement the general Labour Standards Law, including the Industrial Safety and Health Law. There is a national system that ensures the law is followed through guidance as well as through inspection.

Ministry of Health, Labour, and Welfare

This national ministry (www.mhlw.go.jp) oversees the Industrial Safety and Health Law that was passed in 1972. To support the law, there is an Enforcement Order and many Ordinances that describe the minimum actions required of employers to comply with the law. Of particular note are:

- Ordinance on Industrial Safety and Health
- Ordinance on Safety and Health of Work under High Pressure
- Ordinance on Health Standards in the Office
- Guideline for Occupational Safety and Health Management Systems

Although there is more information in an Ordinance, the guidance remains general. The expectation is to prevent disease and to actively maintain and enhance health. This is to be accomplished through having an occupational safety and health management system under which to identify and control risks and hazards. Japan also has a national initiative to reduce working hours as part of improving working conditions.

Additional English synopses of the national laws and supporting Ordinances and Guidelines are available through the Japan International Center for Occupational Safety and Health (JICOSH). See below for more information.

National Institute of Industrial Safety (NIIS)

The NIIS (www. anken.go.jp) is a research branch of the Ministry of Health, Labour, and Welfare that focuses on safety issues. Ergonomics is a main research topic of this group.

National Institute of Industrial Health (NIIH)

This (www.niih.go.jp) is a multidisciplinary research limb of the Ministry of Health, Labour, and Welfare that focuses on occupational diseases. The Institute provides the government scientific and technical information related to industrial health. There are several main activities that are industrial hygiene oriented, as well as:

- Work management and human factor engineering in response to changes in working conditions
- Working capacity and fitness of women and the elderly
- Assessment of physical hazards.

Japanese Standards Association (JSA)

The JSA (www.jsa.or.jp) is the main resource for purchasing voluntary standards. The Association supports the standards development group, the Japanese Industrial Standards Committee (JISC), which is the primary producer of national voluntary Japanese Industrial Standards (JIS). These JIS are numerous and mostly very technical. ISO standards are also available through the JSA and are adopted by Japan in keeping with the ISO policy for a contributing country.

Japan International Center for Occupational Safety and Health (JICOSH)

JICOSH (www. jicosh.gr.jp) is a useful resource as their mission is outreach to industry of other nations. Therefore, their Web page is in English and they have some overviews of the industrial laws of Japan. There are also some useful links to other Japanese Web sites.

REFERENCES

Seabrook, K. A. (2001). International standards update: Occupational safety and health management systems. *Proceedings of American Society of Safety Engineers' 2001 Professional Development Conference,* Anaheim, CA, Session 507.

WEB PAGE REFERENCES

www.acgih.org	American Conference of Governmental Industrial Hygienists (ACGIH), USA
www.aiha.org	American Industrial Hygiene Association, USA
www.ansi.org	American National Standards Institute, USA
www.asme.org	American Society of Mechanical Engineers (ASME), USA
www.astm.org	American Society for Testing and Materials (ASTM), USA
www.workcover.act.gov.au	Australian Capital Territory (ACT)
www.worksafebc.com	British Columbia (BC) Government, Canada
www.bsi-global.com	British Standards Institute, UK
www.dir.ca.gov	California Ergonomics Standard, USA
www.csa.ca	Canadian Standards Association, Canada
www.comcare.gov.au	Comcare Australia
www.lni.wa.gov/wisha	Department of Labor and Industries, Washington State, USA

www.cenelec.org	European Committee for Electrotechnical Standardization (CENELEC)
www.cenorm.be	European Committee for Standardization (CEN)
www.etsi.org	European Telecommunications Standards Organization (ETSI)
www.europa.eu.int	European Union (EU)
www.access.gpo.gov	General Printing Office (GPO)—USA
www.hse.gov.uk	Health and Safety Executive (HSE)—UK
www.hfes.org	Human Factors and Ergonomics Society (HFES), USA
www.icao.org	International Civil Aviation Organization (ICAO)
www.iea.cc	International Ergonomics Association (IEA)
www.iso.org	International Organization for Standardization (ISO)
www.itu.org	International Telecommunication Union (ITU)
www.ieee.org	Institute of Electrical and Electronics Engineers (IEEE), USA
www. jicosh.gr.jp	Japan International Center for Occupational Safety and Health (JICOSH)
www.jsa.or.jp	Japanese Standards Association (JSA)
www.mhlw.go.jp	Ministry of Health, Labour, and Welfare, Japan
http://standards.nasa.gov	NASA Technical Standards Program
www.niih.go.jp	National Institute of Industrial Health (NIIH), Japan
www. anken.go.jp	National Institute of Industrial Safety (NIIS), Japan
www.nist.gov	National Institute of Standards and Technology (NIST), USA
www.nohsc.gov.au	National Occupational Health and Safety Commission (NOHSC), Australia
www.nsc.org	National Safety Council, USA
www.workcover.nsw.gov.au	New South Wales (NSW) WorkCover Authority, Australia
www.nt.gov.au	Northern Territory Work Health Authority, Australia
www.osha.gov	Occupational Safety and Health Administration (OSHA)—USA
www.useuosh.org	Occupational Safety and Health Administration (USA) and European Union (EU)
www.gov.on.ca	Ontario Government, Canada
www.perinorm.com	Perinorm, private database of international, European and national standards
www.whs.qld.gov.au	Queensland Division of Workplace Health and Safety, Australia

www.sae.org	Society of Automotive Engineers (SAE), USA
www.workcover.com	South Australian WorkCover Authority
www.standards.com.au	Standards Australia
www.workcover.vic.gov.au	Victorian WorkCover Authority, Australia
www.worksafebc.com	Workers' Compensation Board of British Columbia, Canada
www.wsa.tas.gov.au	Workplace Standards Tasmania, Australia
www.safetyline.wa.gov.au	WorkSafe Western Australia
www.gksoft.com/govt/en/world.html	Worldwide Governments on the WWW
www.wssn.net	World Standards Services Network (WSSN)
www.w3.org	World Wide Web Consortium (W3C)

6

Balancing Stakeholder Representation: An Example of Stakeholder Involvement in Ergonomics Standardization

Harmen Willemse
Henk J. de Vries
Jan Dul
Erasmus University Rotterdam

INTRODUCTION

During the last 50 years, ergonomics has developed rapidly as an independent science and profession, with its own knowledge, methods, and networks. Scientists have published their knowledge in increasing numbers of ergonomics journals. However, much of the ergonomics knowledge has not yet reached engineers and managers. Engineers could use the large number of ergonomics standards that have become available during the last two decades, to integrate ergonomics knowledge in the design process. But somehow, most of these standards have not yet found their way into the design process.

This chapter states that an important reason for the limited use of ergonomics standards is insufficient stakeholder involvement in standardization. Because one of the main principles in ergonomics is user involvement, intended users should be involved in the process of designing an environment in order to assure a fit between this environment and its user. In the development of ergonomics standards, however, this principle seems to be ignored. Ergonomics standardization is considered to be the concern of "professional ergonomists" only. Lack of involvement of users and other stakeholders may result in standards that fail to meet users' needs.

CASE STUDY

The example of stakeholder participation in ergonomics standardization presented here is based on a case study performed at the Dutch national standardization body, NEN. The subject of the case study is the revision of the Dutch national standard, NEN 1824:1995, on the requirements for the space of office workplaces. The composition of the working group that revised the standard is analyzed in order to find out whether the working group was a well-balanced representation of all of the stakeholders. An analysis will be made of what stakeholders have been involved, what stakeholders have not been involved, and for what

149

reason. This will offer the possibility to answer the question: "Was the working group that was established for the revision of NEN 1824:1995 a well balanced representation of all of the stakeholders?"

In the following sections, this question is answered. First, a short description of the development of the standard and its content is given. This is followed by a description of the methods that have been used to analyze the (un)balance of stakeholder representation.

The subsequent section describes the results of the stakeholder analysis: which stakeholders were actively involved, and which were not. This answers the research question of the case study. In addition, the situation concerning ergonomics standardization on Dutch national and international levels in general is considered to provide a first impulse to a discussion concerning stakeholder participation in ergonomics standardization.

HISTORY OF THE STANDARD

The standard presented in this case study is the Dutch national standard, NEN 1824, on the requirements for the space of office workplaces. The first version of this standard was published in 1990 and was developed at the request of the Dutch Department of Social Affairs and Employment as part of a policy to stimulate the improvement of working conditions in the Netherlands. The first version's title was NEN 1824:1990 "Ergonomics—Ergonomic Recommendations for the Dimensions of Office Rooms." It described the size of the floor area that was preferred from an ergonomics point of view.

In 1995, this standard was revised in order to reflect amendments in the Dutch Labor Law and other legislation. For this version, ergonomics was no longer used as the starting point, meaning that the standard would no longer classify as an ergonomics standard. The new version solely described the minimal requirements for the width, height, and amount of floor area of office workplaces. Its title was NEN 1824:1995 "Requirements for the Space and Height of Office Workplaces." The 1995 version confronted the users with difficulties concerning the interpretation of the standard for application, especially on innovative office concepts. The standard left too much room for disparity in interpretation, resulting in a lot of questions and even disputes between the Dutch Labor Inspectorate and facility managers on the appropriateness for, in most cases, call-center workplaces.

In 1999, it was decided that the standard should be revised again in order to take into account not only the traditional office rooms but also other workstations like call-center workplaces.

In the beginning of the year 2000, the committee that coordinates the Dutch standardization activities on the field of ergonomics composed a working group. The committee selected a group of members that in their opinion had to be involved in the revision process. This working group commenced by bringing up additional working group members, and when it was decided that the working group contained sufficient representation, the revision of NEN 1824:1995 started. The total process, including the two startup meetings took seven meetings, within a period of 15 months.

In 2001, the revision was published as NEN 1824:2001 "Ergonomics—Ergonomic Requirements for the Space of Office Workplaces." This revised version consists of a total of 10 pages of which the main content lists the factors that have to be kept in mind when equipping an office. A calculation method is given for the minimum floor space, taking into account the employee, the desk, the drawers, and other factors. For the office desk, a distinction is made between a workplace with a flat monitor and one with a regular CRT monitor. In addition, minimum dimensions for the circulation space and dimensions for visual separation of workplaces are given. Furthermore, physical factors of workplaces, like the indoor climate, ventilation, and daylight are discussed. In three appendices, explanations and examples of calculations for the minimum space of common workplaces are given.

The last revision is the subject of this case study. In the following sections, the composition of the working group will be compared with the stakeholders that need to be involved in standardization according to a theoretical stakeholder selection.

METHODS

This section describes the methods that were used to judge whether the representation of the working group that revised NEN 1824:1995 was well balanced. First, a stakeholder classification theory was used to define which stakeholders should be involved in ergonomics standardization in general. Second, the involvement of these stakeholders in the working group that revised NEN 1824:1995 was analyzed.

Stakeholder Classification

The theory of Mitchell, Agle, and Wood (1997) on stakeholder classification for organizations was applied to the standardization process by Willemse, Verheul, and De Vries (2003).

The stakeholders are classified according to the following attributes: power, legitimacy, and urgency with respect to the standardization process. Power (P) entails the capability of an actor to bring about the outcomes one desires. A stakeholder possesses legitimacy (L) if the members of a working group, subcommittee, or committee recognize a certain party as a stakeholder. Urgency (U) is defined as the degree to which the stakeholder demands immediate attention for its interest. These three stakeholder attributes (P, L, and U) allow the classification of stakeholders in eight categories (Figure 6.1).

From this list of eight categories of stakeholders, four categories are selected to be relevant for participation in standardization processes in general. These are the "definitive stakeholders"

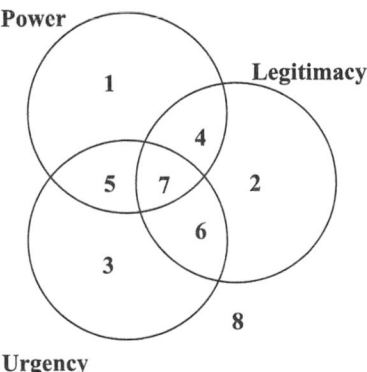

FIG. 6.1.   Stakeholder classification (adapted from Mitchell, Agle & Wood (1997)). These different stakeholder groups are labeled:
1, Dormant Stakeholder (P); 2, Discretionary Stakeholder (L); 3, Demanding Stakeholder (U);
4, Dominant Stakeholder (P, L); 5, Dangerous Stakeholder (P, U); 6, Dependent Stakeholder (U, L);
7, Definitive Stakeholder (P, U, L); 8, Nonstakeholder (−)

(P, U, L), the "dependent stakeholders" (U, L), the "dominant stakeholders" (P, L), and the "discretionary stakeholders" (L).

The "**definitive stakeholders**" have the resources to work on the development of the standard (P), consider the standard being developed as important (U), and their involvement is indisputable (L). These stakeholders can be the driving force of the standardization process. In this case example, ergonomics consultants and occupational health and Safety services can be considered as "definitive stakeholders." In addition, large companies and the Dutch Department of Spatial Planning and the Environment, being large and important users of the standard, can be considered as "definitive stakeholders."

The "**dependent stakeholders**" are important for general support of a standard (L) and see the importance of participating in the standardization process (U). These stakeholders, however, have limited (financial) resources to participate in the development of standards. Stakeholders for the revision of NEN 1824 are labor unions, architects, and office furnishers. Sector organizations that represent small- and medium-sized enterprises fall into this category as well. Through their representation of a large group of small- and medium-sized enterprises, the sector organizations become valuable working group members, whereas individual representatives from a small- or medium-sized enterprise are worthless.

The "**dominant stakeholders**" do not see immediate interest in joining the working group. In some way, however, they are of great importance for the standardization process. This can result from a certain position the stakeholder has, which, for example, makes the stakeholder important for generation of support for the standard. This power base gives the stakeholder legitimacy to join the working group. For a standard for the space of office workplaces, the Dutch Department of Social Affairs and Employment is an important addition to the working group, as it is the legislative power in the field of working environment. The supervising power, the Dutch Labor Inspectorate, is also a party of importance.

The "**discretionary stakeholders**" are perceived to be important for the development of the standard (L), although they are not interested in participating themselves. A research institute is an example of a "discretionary stakeholder." Researchers can serve as an independent party to offer solutions to conflicts of interest between other stakeholders.

For the revision of NEN 1824:1995, involvement of the following stakeholders is considered essential (Table 6.1).

TABLE 6.1
Essential Stakeholders of Ergonomics Standardization

Stakeholder category	*Stakeholder*
"Definitive Stakeholder" (P, U, L)	• Ergonomics Consultants • Occupational Health and Safety Services • Large Employers • Department of Housing, Spatial Planning, and the Environment
"Dependent stakeholder" (U, L)	• Labor Unions • Small Companies (representation) • Architects • Office Furnishers
"Dominant Stakeholders" (P, L)	• Department of Social Affairs and Employment • Labor Inspectorate
"Discretionary stakeholder" (L)	• Research Institutes

Representation

In order to answer the question: "Was the working group that was established for the revision of NEN 1824:1995 a well-balanced representation of all of the stakeholders?," the parties involved in the working group were compared to the list of stakeholders presented in Table 6.1.

Meeting minutes were analyzed to identify the parties that were involved in the revision of this standard, and which parties were not. We assume that each of the working group members represented the line of business of the company that sent them. This was never explicitly discussed during working group meetings. Interviews were conducted with three working group members and the secretary to analyze the presence and role of the stakeholders in the working group. Representatives of the stakeholders that were not represented were interviewed in order to find out if they were informed about the revision of the standard, and if they would have been interested in participation. In addition, the working group members were interviewed to learn why some stakeholders were not represented.

RESULTS

Table 6.2 shows the stakeholders that were represented in the working group and how many meetings the representatives have attended. It turns out that the standard producing core of the group consisted of one large employer, one ergonomics consultant, one representative from the Dutch Department of Housing, Spatial Planning, and the Environment, two representatives from Institutes for Occupational Health and Safety Service and one representative from a research institute.

TABLE 6.2
Representation and Attendance of Stakeholders of Ergonomics Standardization

| | | Attendance in 7 Meetings | |
| | | *Representative* | |
Stakeholder Category	*Stakeholder*	*1*	*2*
"Definitive stakeholder" (P, U, L)	• Ergonomics Consultants	• 0	• 5
	• Occupational Health and Safety Services	• 3	• 2
	• Large Employers	• 0 (2 times written feedback)	• 7
	• Department of Housing, Spatial Planning, and the Environment	• 6	
"Dependent stakeholder" (U, L)	• Labor unions	• Not Represented	
	• Small and Medium-sized Companies (representation)	• Not Represented	
	• Architects	• Not Represented	
	• Office Furnishers	• Not Represented	
"Dominant Stakeholders" (P, L)	• Department of Social Affairs and Employment	• Not Represented	
	• Labor Inspectorate	• Not Represented	
"Discretionary stakeholder" (L)	• Research Institutes	• 3	

Furthermore, the results show that that six out of eleven essential stakeholders were not represented and three stakeholders were represented by two representatives. In two of these cases, however, only one of the two representatives attended the meetings.

STAKEHOLDERS NOT INVOLVED

Three stakeholder groups, namely, the labor union, the office furnishers, and the architects or its sector organization have not been approached for participation in the working group, because the working group thought they would not be interested. However, interviews with some of these parties taught that they do use the standards in their operations and that they would have been interested in joining the workgroup. Most of the users interviewed indicate that they would have been able to offer some input in the working group from their practical experiences with the old version of the standard.

The Department of Social Affairs and Employment has repeatedly been approached by the working groups secretary to join the group, but for the following reason did not show any interest. The Dutch Occupational Health and Safety Act contains references to the standard NEN 1824:1995. This standard is seen as a possible elaboration of the legal framework for working conditions that is laid down in the act. The Department states that because it is a legislative institute, it cannot be involved in the development of a possible elaboration of the act, because this might lead to a conflict of interest. This same story goes for the Dutch Labor Inspectorate, which is the supervising institute in the field of working conditions.

DISCUSSION

In this Dutch case, the ergonomics profession was overrepresented and several other important stakeholders did not participate. The situation of overly represented professional ergonomists, presented in our case study, seems no exception in working groups and committees in the field of ergonomic standardization.

For example, the Dutch committee 302002 called "Physical load," shadows the activities of ISO/TC 159/SC 3 "Anthropometry and Biomechanics" and CEN/TC 122. "Ergonomics." The Dutch committee gives feedback on standards being developed by these international committees. This committee consists of five representatives of research institutes and universities and two representatives from occupational health and safety institutes.

Another example is the Dutch committee 302015 called "Ergonomic Design Principles" that follows the activities of three international working groups. These are ISO/TC 159/SC1/WG1 "Ergonomic Guiding Principles," ISO/TC 159/SC1/WG2 "Ergonomic Principles Related to Mental Workload," and CEN/TC 122/WG2 "Ergonomic Design Principles." This committee consists of representatives from two ergonomic consultancy companies, one research institute and one occupational health and safety institute. In both examples, the working groups show an unbalanced stakeholder representation.

The problem of unbalanced stakeholder representation can also be considered on an international level. Most ergonomics standards are international ones and, as far as Western and Central Europe are concerned, European standards as well. The standardization bodies at these levels, ISO and CEN, operate the "country model," in which participants in technical committees represent their own country.

This raises the question of which people the ISO and CEN members should delegate. Let us take the example of the working group ISO/TC 159/SC3/WG2 "working postures." This working group consists of about two thirds of ergonomic researchers, five standardization organization employees, a small number of ergonomic consultants, one government representative, and one industry representative.

Often, CEN and ISO members send employees of the national standards body. They are experienced in expressing the voice of their national standardization committee, but, in general, lack ergonomics expertise. The other option is to send ergonomics experts, but these may have the disadvantage of being less familiar with the standardization profession. The representation of countries by either an ergonomics expert or a standardization expert leads to an unbalanced stakeholder representation on European and international level as well.

User involvement in standardization is thus even more neglected on European and international level than on the Dutch national level. This seems inevitable, but it is, according to De Vries (1999, p. 68), who provides a listing of advantages and disadvantages of the country model, at least partly compensated thanks to the country model. In this model, participants in ISO are backed by a national standardization committee, to which "weak" stakeholders have easier accesses.

CONCLUSION

In this chapter, we present an example of stakeholder involvement in ergonomic standardization. This example has been selected in order to illustrate the situation of stakeholder involvement in ergonomics standardization.

It appears that many working groups responsible for developing and revising ergonomics standards are dominated by ergonomics professionals and that many other important stakeholders are not involved. We therefore state that the key to increasing the usage of ergonomic standards is to involve users, other than ergonomists, in the development of the standards. User involvement will allow the adoption of new points of view in ergonomic standardization, which will increase the usability and comprehensibility of ergonomic standards.

When ergonomics professionals desire that parties adopt ergonomic standards, they should also adopt these parties as stakeholders in ergonomic standardization.

ACKNOWLEDGMENTS

A number of persons, who contributed to the realization of the case study and this chapter, deserve a special word of thanks.

Hugo Verheul, assistant to professor at the Faculty of Technology, Policy and Management at Delft University of Technology, was the project leader of a joint research of the Delft University of Technology and the Dutch standardization institute. For this research, case studies have been performed in order to improve the standardization process of the Dutch Standardization Institute. The improvements are aimed at increasing stakeholder involvement in standardization processes and to make standards more widely known.

Sandra Verschoof, standardization consultant at the Dutch standardization institute, has provided valuable information about standardization in the field of ergonomics.

Machiel van Dalen, head of the Taskforce Business Development of the Dutch standardization institute, has helped critically reviewing the standardization process and stakeholder involvement in the revision of the standard.

Michiel Kerstens has been very helpful correcting the English and offering suggestions for textual improvements.

REFERENCES

De Vries, H. J. (1999). *Standardization—A business approach to the role of national standardization organizations*, Boston/Dordrecht/London: Kluwer Academic Publishers.

Mitchell, R. K., Agle, R. B., & Wood, D. J. (1997). Toward a theory of stakeholder identification and salience: Defining the principle of who and what really counts. *Academy of Management Review, 22*, 853–886.

Willemse, H., Verheul, H. H. M., & De Vries, H. J. (2003). *Stakeholderparticipatie bij de herziening van de norm NEN 1824*, Delft: C.P.S.

7

Some Ergonomics Standards in Ancient China

Pei-Luen Patrick Rau
Tsinghua University

INTRODUCTION

The Song dynasty (960–1279) in Chinese history has been considered successful in the development of military technology. To defend attacks from its Northern enemies, such as the Liao dynasty (916–1125), the Jin dynasty (1115–1234), and the Yuan dynasty (1206–1368), the central government of the Song dynasty established an army with millions of soldiers equipped with various kinds of weapons. The majority of the Song army was infantry because the Song, as an agricultural society, could recruit many infantry soldiers quickly and easily. The Song's enemies were mainly nomadic people with strong cavalry. For example, the Jin dynasty, the ancestor of the Manchurians, wiped out the first Song central government (Northern Song) and forced the Song to move their capital to the South. The Manchurians that established the last dynasty in China, the Ching dynasty, were famous for their cavalry skill. Another Song enemy, the Mongolian descendants of Genjis Khan who were victorious in the 13th century on the Euro-Asian continent, were also famous for Mongolian cavalry.

To fight against cavalry, the Song had to design weapons, armor, and a strategy for their infantry troops. Long spears and heavy armor can help the infantry to attack and repel cavalry. It was an important issue for the Song to design spears long enough for this purpose. Furthermore, if the armor to protect the soldiers was too heavy, it could hamper soldier mobility on the battlefield. Selecting and training infantry soldiers to fight against cavalry was a significant issue. The Song expended great effort on their military system, strategy, and military technology during their 319-year reign. The Song selected, organized, and paid soldiers according to their height, body size, age, and fighting skills. The Song central government built armories and developed national standards for manufacturing weapons and armor. The purpose of some of the national standards was to limit the weight of the equipment to match the physical capabilities of the soldiers. One national standard regulated the upper limit for the total weight for Song infantry armor and the weight of each part of the armor. This standard was issued by the 10th emperor (Gaozong, 1127–1162) of the Song dynasty.

OBJECTIVE AND SCOPE OF THE CHAPTER

The purpose of this chapter is to introduce three ergonomic standards of the Song dynasty: the height standard for the organization and remuneration of soldiers, the body size standard for soldier selection, and the weight standard for Song infantry armor. Since 960, all military volunteers had a height requirement to be soldiers of the Song imperial army. In 1057, the remuneration system was linked to the anthropometric results and all recruited soldiers were grouped into five levels according to their height. The records in or after 1127 indicated that the height of the soldiers was used as the standard for organizing new troops. Based on the number of the troops, the height distribution of soldiers that were adult Chinese males in the 12th century could be estimated. The Song dynasty measured the body size of soldiers for armor. In 1062, it was recoded that soldiers that could not fit into the armor were retired. In 1134, an order was issued by the 10th emperor (Gao Zong) that regulated the upper limit for a Song infantry soldier's weight.

DISCUSSION OF THE SPECIFIC GUIDELINES

Height Standards

The Song dynasty military was composed of six troop levels. The first level, the imperial army (Jin Jun), was the standard army for fighting. The role of the imperial army was replaced by the central army (Tun Zhu Da Bing) and the new army (Xin Jun) in the Southern Song (Toktoghan, 1345; Wang & Yang, 2001; Zhou, Yang, & Wang, 1998). Height was one of the major criteria for the soldier selection in the Song imperial army, as recorded in the military section (Bin Zhi) of the official imperial history of the Song dynasty (Toktoghan). The unit of measurement was the ancient Chinese length unit (Chi). The ancient Chinese length unit (Chi) was getting bigger as years passed by. One Chi equaled about 29.6 centimeter officially recorded in Sui dynasty (581–618) and Tang dynasty (618–907) (Qiu, 1994; Wei, 636; Zheng & Tang, 2000). But the excavated Song rulers were about 31.2 centimeters (Zheng & Tang). In this chapter, both are used. The ancient Chinese inch (Cun) was one tenth of one ancient Chinese meter.

The Song central government developed two approaches to measure the height of soldiers. The first approach used human models or soldier models (Bin Yang; Toktoghan, 1345). The first Song emperor, Tai Zu (960–976) chose strong soldiers as references and sent them to recruiting locations in 960 (Toktoghan, in vol. 187, record [Zhi] 140, military [Bin] 1, imperial army [Jin Jun Shang], and vol. 193, record 146, military 7, recruiting [Zhao Mu Zhi Zhi]). The human model approach was replaced by the wood stick or sticks of equal length (Deng Zhang or Deng Chang Zhang) approach. The wood sticks were used to measure the height of soldiers later on (Toktoghan; in vol. 187, record 140, military 1, imperial army, and vol. 193, record 146, military 7, recruiting).

Between 1008 and 1016, under the third Song emperor (Zhen Zong), volunteers had to be tall enough to qualify as soldiers in the imperial army. Short volunteers could only be recruited into the second troop level, the local army (Xiang Jun). The local army was primarily responsible for local security and public service rather than fighting (Toktoghan 1345; in vol. 193, record 146, military 7, recruiting). The volunteers that passed the lower height limit for the imperial army were distinguished by their height into five groups:

1. Under 5 Chi and 5 Cun (162.8 centimeter in Sui Chi; 171.6 centimeter in Song Chi)
2. Between 5 Chi and 5 Cun to 5 Chi and 6 Cun (162.8 centimeter to 165.8 centimeter in Sui Chi; 171.6 centimeter to 174.7 centimeter in Song Chi)
3. Between 5 Chi and 6 Cun to 5 Chi and 7 Cun (165.8 centimeter to 168.7 centimeter in Sui Chi; 174.7 centimeter to 177.8 centimeter in Song Chi)

TABLE 7.1
The Height Standards in the Song Dynasty and Jin Dynasty

Year	Emperor	Approach	Standards	Reference (Toktoghan 1345)
960	Tai Zu	Human Model		• Vol. 187, record 140, military 1, imperial army A • Vol. 193, record 146, military 7, recruiting
960–976	Tai Zu	Sticks of Equal Length		• Vol. 187, record 140, military 1, imperial army A • Vol. 193, record 146, military 7, recruiting
1008–1016	Zheng Zong	Sticks of Equal Length	5 Chi and 5 Cun to 5 Chi and 8 Cun (assignment)	• Vol. 193, record 146, military 7, recruiting
1047	Ren Zong	Sticks of Equal Length	5 Chi and 7 Cun (lower limits)	• Vol. 193, record 146, military 7, recruiting
1051	Ren Zong	Sticks of Equal Length	5 Chi and 6 Cun (lower limit)	• Vol. 194, record 147, military 8, selection criteria
1057	Ren Zong	Sticks of Equal Length	5 Chi and 4 Cun to 5 Chi and 7 cun (linked with remuneration)	• Vol. 193, record 146, military 7, recruiting
1060	Ren Zong	Sticks of Equal Length	5 Chi and 3 Cun (lower limit)	• Vol. 194, record 147, military 8, selection criteria
1081	Shen Zong	Sticks of Equal Length		• Vol. 194, record 147, military 8, selection criteria
1127 (?)	Qin Zong	Sticks of Equal Length	5 Chi and 3 Cun to 5 Chi and 8 Cun (assignment)	• Vol. 194, record 147, military 8, selection criteria
1199	Zhang Zong (Jin)	Sticks of equal length	5 Chi and 5 Cun to 5 Chi and 6 Cun	• Toktoghan 1344, Vol. 4, record 25, military, imperial army

4. Between 5 Chi and 7 Cun to 5 Chi and 8 Cun (168.7 centimeter to 171.7 centimeter in Sui Chi; 177.8 centimeter to 180.9 centimeter in Song Chi)
5. Above 5 Chi and 8 Cun (171.7 centimeter in Sui Chi; 180.9 centimeter in Song Chi)

The soldiers were assigned into different units or bases according to the height group that they were assigned (Toktoghan; in vol. 193, record 146, military 7, recruiting). The fourth Song dynasty emperor, Ren Zong (1022–1063) established a complete recruiting and remuneration military system (Table 7.1). The standard had been changed several times in the Song dynasty:

1. In 1047, all imperial army volunteers had to be 5 Chi and 7 Cun or taller (168.7 centimeter in Sui Chi; 177.8 centimeter in Song Chi) (Toktoghan; in vol. 193, record 146, military 7, recruiting).
2. In 1051, the limit was reduced as 5 Chi and 6 Cun (165.8 centimeter in Sui Chi; 174.7 centimeter in Song Chi) (Toktoghan; in vol. 194, record 147, military 8, selection criteria [Jian Xuan Zhi Zhi]).
3. The limit was again reduced as 5 Chi and 3 Cun (156.9 centimeter in Sui Chi; 165.4 centimeter in Song Chi) in 1057 (Toktoghan; in vol. 194, record 147, military 8, selection criteria).

FIG. 7.1. Height standards.

In 1057, the imperial army remuneration was linked to anthropometric data (Toktoghan 1345; in vol. 193, record 146, military 7, recruiting).

- Soldiers taller than 5 Chi and 7 Cun tall or taller (168.7 centimeter in Sui Chi; 177.8 centimeter in Song Chi) could be paid 1,000 Song dollars (Qian).
- Soldiers that were from 5 Chi and 5 Cun to 5 Chi and 7 Cun (162.8 centimeter to 168.7 centimeter in Sui Chi; 171.6 centimeter to 177.8 centimeter in Song Chi) were paid 500 or 700 Song dollars.
- Soldiers that were about 5 Chi and 5 Cun (162.8 centimeter in Sui Chi; 171.6 centimeter in Song Chi) were paid 300 or 400 Song dollars.
- Soldiers that were from 5 Chi and 2 Cun to 5 Chi and 4 Cun (153.9 centimeter to 159.8 centimeter in Sui Chi; 162.2 centimeter to 168.5 centimeter in Song Chi) were paid 200 or 300 Song dollars.

The records in or after 1127 indicated that the height of the soldiers was used as the standard for organizing new troops (Toktoghan 1345; in vol. 194, record 147, military 8, selection criteria).

- There were only enough soldiers taller than 5 Chi and 8 Cun (171.7 centimeter in Sui Chi; 181 centimeter in Song Chi) to form one troop (Table 7.2).
- There were enough soldiers about 5 Chi and 7 Cun (168.7 centimeter in Sui Chi; 177.8 centimeter in Song Chi) in height to form four troops.

TABLE 7.2
The Height Distribution of Soldiers in the Song Dynasty

Ancient Chinese	Height		Number of Troops	Troops
	Sui Chi (cm)	Song Chi (cm)		
5 Chi and 8 Cun	171.7	181	1	Tian Wu Di Yi Jun
5 Chi and 7 Cun	168.7	177.8	4	Peng Ri, Tian Wu Di Er Jun, Shen Wei, Long Wei
5 Chi and 6.5 Cun	167.2	176.3	6	Gong Sheng, Shen Yong, Sheng Jie, Xiao Jie, Long Meng, Qing Shuo
5 Chi and 6 Cun	165.8	174.7	6	Xiao Ji, Yun Ji, Xiao Sheng, Xuan Wu, Dian Qian Si Hu Yi, Dian Qian Si Hu Yi Shui Jun
5 Chi and 5 Cun	162.8	171.6	33	Wu Ji, Ning Shuo, Bu Jun Si Hu Yi Shui Jun, Jian Zhong, Long Wei, Shen Ji, Guang Yong, Long Ji, Xiao Meng, Xiong Yong, Tu Hun, Qin Rong, Xin Li Xiao Jie, Xiao Wu, Guang Rui, Yun Yi, You Ma Jin Yong, Bu Wu, Wei Jie, Wu Wei, Chuang Zinu Xiong Wu, Fei Shan Xiong Wu, Shen Rui, Zhen Wu, Xin Zhao Zhen Wu, Xin Zhi Zhen Wu, Zhen Hua Jun, Xiong Wu Nu Shou, Shang Wei Meng, Ting Zi, Wu Di, Shang Zhao Shou, Ji Zhou Xiong Sheng, Cheng Hai Shui Jun Nu Shou
5 Chi and 4.5 Cun	161.3	170	17	Guang Jie, Wei Sheng, Guang De, Ke Sheng, Shan Fu Xiong Sheng, Xiao Xiong, Xiong Wei, Shen Hu, Bao Jie, Qing Bian Nu Shou, Zhi Sheng, Qing Jian, Ping Hai, Xiong Wu, Long De Gong Qing Wei, Ning Yuan, An Yuan
5 Chi and 4 Cun	159.8	168.5	19	Ke Rong, Wan Jie, Yun Jie, Heng Sai, Zhuo Sheng, You Ma Xiong Lue, Xiao Zhong, Xuan Yi, Jian An, Wei Guo, Quan Jie, Zhou Xiao Zhong, Jian Zhong Xiong Yong, Huai Shun, Zhong Yong, Jiao Yue Zhong Jie, Shen Wei, Xiong Lue, Xia Wei Meng
5 Chi and 3.5 Cun	158.4	166.9	14	Buo Zhou Xiong Sheng, Fei Ji, Wei Yuan, Fan Luo, Huai En, Yong Jie, Shang Wei Wu, Xia Wei Wu, Zhong Jie, Jing An, Chuan Zhong Jie, Gui Yuan, Zhuang Yong, Xuan Xiao
5 Chi and 3 Cun	156.9	165.4	21	Ji Zhou Xiong Sheng, Qi She, Qiao Dao, Qing Sai, Feng Xian, Feng Guo, Wu Ning, Wei Yong, Zhong Guo, Jin Yong, Xia Zhao Shou, Zhuang Wu, Xiong Jie, Jing Jiang, Wu Xiong, Guang Jie, Cheng Hai, Huai Yuan, Ning Hai, Dao Pai Shou, Bi Sheng

- Soldiers that were about 5 Chi and 6.5 Cun (167.2 centimeter in Sui Chi; 176.3 centimeter in Song Chi) formed six troops.
- Soldiers that were about 5 Chi and 6 Cun (165.8 centimeter in Sui Chi; 174.7 centimeter in Song Chi) formed six troops.
- Soldiers that were about 5 Chi and 5 Cun (162.8 centimeter in Sui Chi; 171.6 centimeter in Song Chi) formed 33 troops.
- Soldiers that were about 5 Chi and 4.5 Cun (161.3 centimeter in Sui Chi; 170 centimeter in Song Chi) were assigned into 17 troops.
- Soldiers that were about 5 Chi and 4 Cun (159.8 centimeter in Sui Chi; 168.5 centimeter in Song Chi) were assigned into 19 troops.
- Soldiers that were about 5 Chi and 3.5 Cun (158.4 centimeter in Sui Chi; 166.9 centimeter in Song Chi) were assigned into 14 troops.
- There were 21 troops formed by soldiers that were about 5 Chi and 3 Cun (156.9 centimeter in Sui Chi; 165.4 centimeter in Song Chi).

The numbers of soldiers in each troop may not be the same, but usually close. Based on the number of the troops, the height distribution of soldiers that were adult Chinese males in the 12th century could be estimated. Most soldiers were about or shorter than 5 Chi and 5 Cun (162.8 centimeter in Sui Chi; 171.6 centimeter in Song Chi). There were only a few taller than 5 Chi and 7 Cun (168.7 centimeter in Sui Chi; 177.8 centimeter in Song Chi).

The Song dynasty was not the only dynasty that used height to select soldiers. One of the enemies of the Song, the Jin dynasty also established height standards. The fifth emperor of the Jin dynasty, Zhang Zong (1189–1208) recruited volunteers from 5 Chi and 5 Cun to 5 Chi and 6 Cun (162.8 centimeter to 165.8 centimeter in Sui Chi; 171.6 centimeter to 174.7 centimeter in Song Chi) as soldiers in the Jin imperial army in 1199 (Toktoghan 1344; in vol. 4, record 25, military, imperial army).

Body Size Standards

Using height to select soldiers was not enough for the Song dynasty because height does not necessarily mean strength. The Song dynasty used armor to measure the body size of soldiers (Table 7.3). In 1062, it was recoded that soldiers that could not fit into the armor be retired (Toktoghan 1345; in vol. 193, record 146, military section 7, recruiting). Two years later (AD 1064), the same issue was brought up for the soldiers in the imperial army (Toktoghan; in vol. 194, record 147, military section 8, selection criteria). In 1068, the armor was officially considered a standard for selecting soldiers. Soldiers had to be strong enough to wear the armor to be recruited (Toktoghan; in vol. 194, record 147, military section 8, selection criteria).

TABLE 7.3
The Body Size Standards in the Song Dynasty

Year	Emperor	Approach	Standard	Reference
1062	Ren Zong	The Song Infantry Armor	Fitness	• Vol. 193, record 146, military section 7, recruiting
1064	Ying Zong	The Song Infantry Armor	Fitness	• Vol. 194, record 147, military section 8, selection criteria
1068	Shen Zong	The Song Infantry Armor	Fitness	• Vol. 194, record 147, military section 8, selection criteria

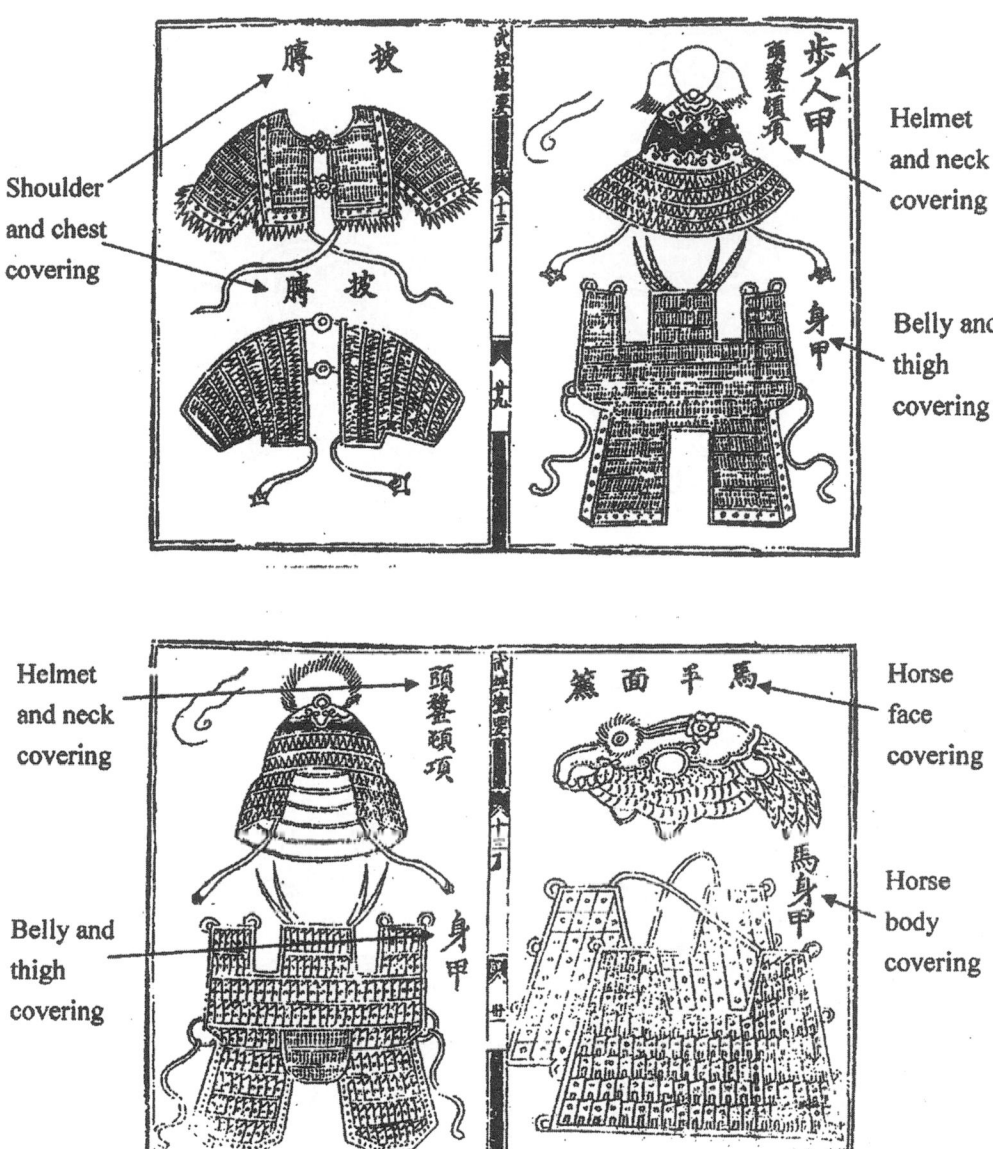

FIG. 7.2. The Song Infantry Armor (Zeng & Ding, 1044, reprinted by Zheng, 1998)

The Weight Standard for the Song Infantry Armor

Song armors had to be standardized before it could be used as a measurement tool. The standard for Song infantry armor was originally published in 1044 (Zeng & Ding, 1044; in vol. 13). The third emperor, Ren Zong, ordered two scholars to edit a military encyclopedia (Wu Jing Zong Yao). This encyclopedia covers famous campaigns, military organizations, strategy, weaponry, and so forth. The design, manufacture, and function of weapons were described and standardized in this book. There are five parts of the Song infantry armor (Figure 7.2): helmet, neck covering, shoulder covering, chest covering, and belly and thigh covering.

In 1134, an order was issued by the 12th emperor (Gao Zong) that regulated the upper limit for the weight of Song infantry armor and the weight of each part of the armor (Toktoghan 1345;

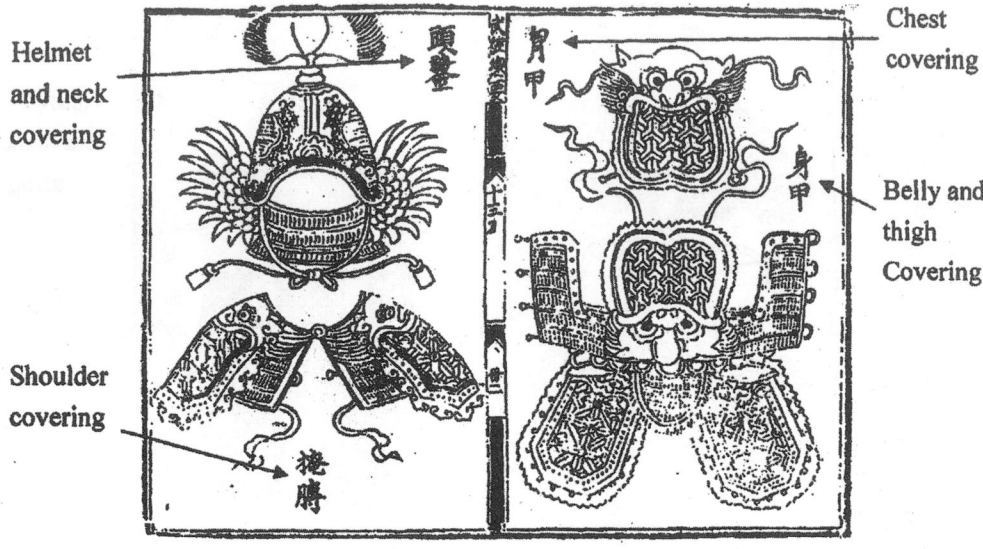

Helmet and neck covering — 頭盔

Chest covering — 身甲

Belly and thigh Covering — 身甲

Shoulder covering — 掩膊

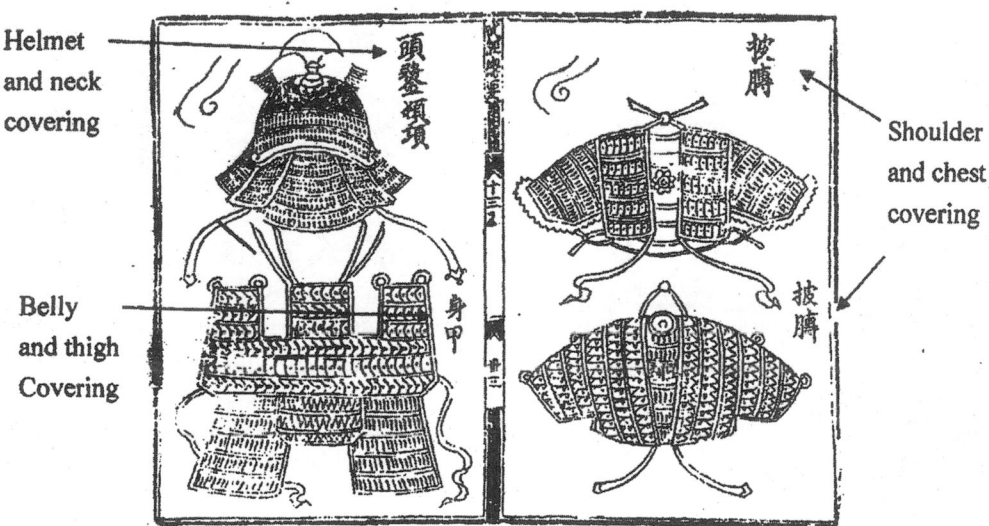

Helmet and neck covering — 頭盔頭項

Shoulder and chest covering — 披膊

Belly and thigh Covering — 身甲

Shoulder and chest covering — 披膊

FIG. 7.2.   (*continued*)

in vol. 197, record 150, military section 11, weapons and armors). The armor was composed of 1,825 pieces of steel. The total weight was from about 27 to about 30 kilograms. The upper limit was about 30 kilograms. The standards for each component are listed in the following (Table 7.4).

- The helmet weighed about 637.5 grams.
- The neck covering consisted of 311 pieces of steel. Each piece weighed about 9.375 grams.
- The shoulder covering consisted of 540 pieces of steel. Each piece weighed about 9.75 grams.
- The chest covering consisted of 332 pieces of steel. Each piece weighed about 17.625 grams.

TABLE 7.4
The Weight Standards for the Song Infantry Armor.

Component	Number of Pieces of Steel	Weight for Each Piece (grams)	Net Weight (grams)
Helmet			637.5
Neck covering	311	9.375	2915.625
Shoulder covering	540	9.75	5265
Chest covering	332	17.625	5851.5
Belly and thigh covering	679	16.875	11458.13
Leather string			3468.75
Total			29596.51

- The belly and chest covering consisted of 679 pieces of steel. Each piece weighed about 16.875 grams.
- The leather string for fixing all of the pieces together weighed about 3468.75 grams.

It was stressed in the order that every piece of steel be weighted according to the standard. If the weight did not meet the requirement, that piece of steel would not be used. The armor upper weight limit was about 30 kilograms. The U.S. military load-carrying standard is 30% of the body weight for combat missions, and 45% of the body weight for marching conditions (MIL-STD-1472E, 1999). The total weight of the Song infantry armor is about 2 more kilograms than the U.S. military load-carrying standard for marching. Considering personnel with fifth percentile body weight, the total load for close combat operations should not exceed 18.5 kilograms (41 pounds) and, for marching, 27.7 kilograms (61 pounds) (MIL-STD-1472E, 1999).

CONCLUSIONS

The ergonomics standards introduced in this chapter include the height standard for the organization and remuneration of soldiers, the body size standard for soldier selection, and the weight standard for Song infantry armor. Some standards were issued from the central government and the emperor, indicating that human capabilities were considered an important factor from the 10th to the 12th century in China. These standards indicated that the awareness of ergonomics could be traced to hundreds of years ago in human history.

REFERENCES

Kuodo, K. (1999). *Ancient weapons in China.* Wan Li Book: Hongkong.
MIL-STD-1472E (1999). Department of defense design criteria standard: Human engineering. Retrieved September 29, 2003, from http://www.r6.gsa.gov/fss/hss/1472F.htm.
Qiu, G. (1994). *Length, volume, and weight measures in ancient China.* The Commercial Press: Taipei.
Toktoghan (1344). *The official imperial histories of the Jin (Jin Shi).* Reprinted by Chung Hua Book: Beijing, 1975.
Toktoghan (1345). *The official imperial histories of the Song (Song Shi).* Reprinted by Chung Hua Book: Beijing, 1977.
Wang, X., & Yang L. (2001). *Military system in China* (Hu Jun Jian Bang: Li Dai Bin Zhi), Wan Juan: Taipei.
Wei, Z. (636). *The official imperial histories of the Sui (Shuishu).* Reprinted by Chung Hua Book: Beijing, 2000.

Zeng, G., & Ding, D. (1044). Military Encyclopedia (Wu Jing Zong Yao), cited in Ji, (1783). *The complete books in the four parts of the Imperial Library* (Si Ku Chuan Shu).

Zhao, H., & Mao, X. (2001). *Ancient military in China*. Wen Chin: Taipei.

Zheng, T., & Tang, Q. (2000). *Chinese history dictionary*. Shanghai Dictionary Publisher: Shanghai.

Zheng, Z. (1998). *Chinese ancient print series, Vol. 1*. Shanghai Chinese Classics Publishing House: Shanghai.

Zhou, B, Yang, M., & Wang, Z. (1998). *The history of northern Song and southern Song*. Chung Hwa: Hongkong.

III

Engineering Anthropometry and Working Postures

8

Standards in Anthropometry

Andreas Seidl
Human Solutions

Heiner Bubb
Technical University of Münich

WHAT IS ANTHROPOMETRY

Anthropometric design is one of the most important disciplines within the area of ergonomic approaches With the correct anthropometry layout, products and working tools can be used safely, comfortably, and in a healthy manner. As in today's modern tailored production, an immediate and direct adaptation of individual dimensions and proportions cannot realistically be realized. Data are needed which can describe the diversity of human body dimensions and which can also be used to adapt the general size of a technical device and that device's adaptability to the human body dimensions. Here the main objective is to design these products in such a way that they can be used effectively and in a healthy and fatigue-free manner. In the case of highly developed products, a design that focuses on comfort can play a major role. The definition of anthropometry is derived from this: *Anthropometry* is the scientific measurement and statistical analysis of data about human physical characteristics and the application (engineering anthropometry) of these data in the design and evaluation of system, equipment, manufactured products, human-made environments and facilities.

The most important objective of anthropometry is the *standardization* of data procurement and therewith the reproducibility of the results (Martin & Knussmann, 1988). Standardization is the means whereby anthropometric data from different fact-finding and analysis sources can be compared and exchanged. Product sales on a global scale have now made the transferability of survey data from different parts of the world directly to the ergonomic design departments of companies significantly more important. This necessary requirement has been fulfilled by the international standardization of anthropometric data and processes. The extensive national standards were analyzed and summarized by scientific and industrial advisory boards and committees. The result is a collection of international standards in which the measurement of human beings and the use of their body measurement data is described (see Table 8.1). The most important objective of anthropometry is its application to product development. However, to apply it properly, one must first understand the creation of anthropometric tables. There are three relevant steps involved.

TABLE 8.1
ISO Standards for Anthropometry

ISO 7250	Basic human body measurements for technological design
ISO 14738	Safety of machinery—Anthropometric requirements for the design of workstations at machinery
ISO 15534-1	Ergonomic design for the safety of machinery—Part 1: Principles for determining the dimensions required for openings for whole-body access into machinery
ISO 15534-2	Ergonomic design for the safety of machinery—Part 2: Principles for determining the dimensions required for access openings
ISO 15534-3	Ergonomic design for the safety of machinery— Part 3: Anthropometric data
ISO 15535	General requirements for establishing anthropometric databases
ISO 15537	Principles for selecting and using test persons for testing anthropometric aspects of industrial products and designs

For the *measurement of anthropometric data*, intensive knowledge about the construction and biomechanical function of the human body is an absolute prerequisite. Since the 1920s, it has been accepted that the measurement of the human body must be carried out with the help of defined measuring points and distances, using the skeletal bone structure as a reference (Martin, 1914). Resulting from this definition, mechanical measurement tools were created and these tools have hardly changed at all in the last 80 years. It is only in the last 10 years that the development of contact-free body scanners has enabled the measurement of soft body parts to enhance and supplement the traditional procurement of data using the skeletal points method.

A further essential prerequisite for the quality of anthropometric data is the selection of a *random sample* that exactly matches the objective of the study or survey. A particular body measurement, for example, arm length, can only be correctly applied if the random sample concerned also matches the future end user of a product. It is logical that data from men and women should not be mixed, but the differentiability of the measurements can be extremely complex and extensive. Sex, age, region, or social environment are only a few of the many. Dynamic changes in the population also play an important role. For over 100 years, the height of young people has been seen to be steadily increasing. This effect, described as *secular growth*, amounts to roughly 1 to 2 cm every 10 years, depending on the region involved. Since the life cycle of high-quality products, for example, cars, can amount to 25 years, this effect must be taken into consideration in the design of such products.

The measurement of anthropometric data takes place with the individual test person who thus provides a random sample. In contrast, the future end customer is taken into account when the data are used for product creation and design. This transition takes place through *statistical analysis* of the body measurement data. In the last few decades, the most important specific value has become the percentile. The declaration of a percentile x states that x percent of the surveyed basic population falls below the given measuring value. A developer can thus guarantee, relatively simply, that a defined number of customers can use the product. Here is an example: A vehicle designer must take into account 95% of male American customers during design of a vehicle interior. To do this, he takes from the tables the value of the 95th percentile of sitting height. This value amounts to 935 mm. If he now ensures that the distance between the driver's seat and the interior roof contour is at least 935 mm he can guarantee that 95% of male Americans will be able to sit in this vehicle without their heads touching the roof.

The engineer or designer uses available anthropometric data when developing a product to match human requirements. The application of anthropometric data is thereby characterized, in that the dimensions of machines and workplaces are developed both with the help of statistical body measurement data and with knowledge about the morphology and the mechanics of the human body. The application typically takes place on the basis of the percentile values, because the statistical data of the future user group is integrated by definition. Typically, percentile length, width, breadth, and circumference are used, with percentile grasp and pedal space also used for more complex tasks. The direct use of anthropometric data from literature is limited to problem areas where either individual body measurements or the envelope contour of a human being is sufficient to provide a representative answer. Direct applications can be resolved with sufficient accuracy by means of tables.

In the second case, statistical body measurements are assembled to synthetic models of the human being created. This can be realized either through templates or through computer-supported manikins. Here, care must be exercised in assembling body data correctly, because correlations of different strengths typically exist within such data. For instance, when you add the 95th percentile leg length to the 95th percentile torso and 95th percentile head length, you will not get a 95th percentile body length—a body length that is much too long will be created instead. The explanation for this is that the 95th percentile body height is, on the one hand, typically represented by relatively long legs, but by a more normal torso on the other. A correct assembly of individual anthropometric data can only be guaranteed if the relevant correlation tables are taken into account.

THE MEASUREMENT AND ANALYSIS OF ANTHROPOMETRIC DATA

In order to use anthropometric tables as a basis for designing products that match human requirements, the data must be acquired accurately and reproducibly. Here, it is not only the selection and accuracy of the measuring tools that have a role to play but also the selection of the random sample to be analyzed that is especially critical.

The Selection of the Test Persons

It would really be necessary to survey the whole population in order to create anthropometric tables. However, this only occurs very rarely—for instance, all potential astronauts can be measured for the development of a space capsule. For the creation of generally valid anthropometric tables, a method must be sought by means of which the correct test can be selected from a very large population group. To create an anthropometric table of the American population, for instance, a correct narrowing-down of the collective to be measured is necessary. It is well known that anthropometry data vary greatly, depending on population, human race, social group, specialized vocation, and so on. In order to get reliable values, certain demands must be made on the population of the sample. Data may never be taken from so-called "lump samples" as is the case, for instance, when university students are the main source of a survey. The sample should be representative. For example, in Europe, the population of the Southern countries is on average smaller than that of the Northern ones.

The correct selection of test persons takes place by means of a statistical narrowing-down of the population. In doing so, the population's physiological differentiation characteristics and its demography must be determined. This process is typically carried out based on earlier anthropometric surveys.

Screening Test

A screening test must be carried out (if none are available). Here the test persons are selected by genuine random selection, or a preselection is made based on statistical information about the population. A screening test should comprise roughly 0.5% of the entire population, if the

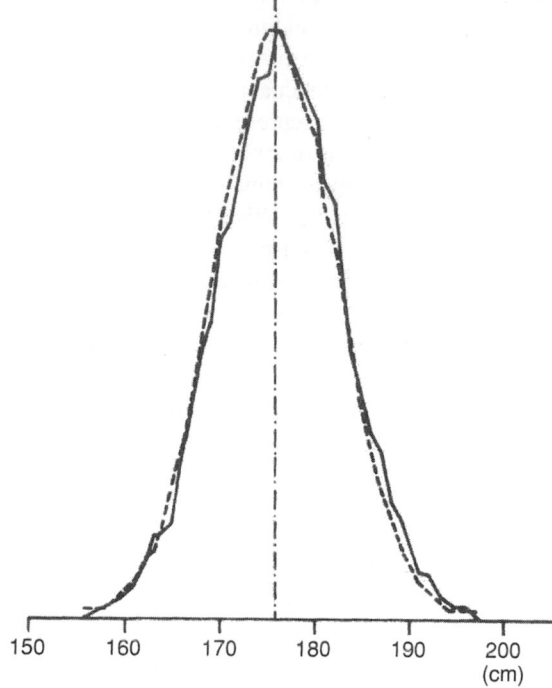

FIG. 8.1.   Comparison of random samples and collectivity.

results are to be analyzed in accordance with the traditional statistical process and partition/subdivision (Juergens, 1989).

In Figure 8.1 you can see the distribution of the heights of 7,144 German military personnel (solid line) born in 1949 and taken in a random sample, compared to the heights of all German military personnel born in 1949 (356,000). The diagram shows that a good dovetailing of both surveys does indeed exist.

Physiological Differentiation Characteristics

In the second step following the screening test, the test person's collective must also be narrowed down in accordance with the application-relevant criteria (physiological differentiation characteristics). Just how important the correct limitation of the test person's collective can be is clearly shown in the following example. When manual workplaces in a car factory in Detroit have to be set up, it doesn't make any sense to include Japanese men or Indian women in the test collective. However, when the driver's cockpit for that particular vehicle is being designed, then the Japanese and Indians *must* be taken into account because the vehicle is going to be exported to those two countries.

As mentioned at the start, there are many physiological differentiation characteristics. The influence of these characteristics on the later application of the data must be very carefully checked during selection of the random sample. The most important four characteristics are examined in the following sections.

Race and Birth Nationality

The influence that a test person's race has on anthropometric measurements is clearly apparent. Typically, Asians usually have a less robust physique and are not as tall as other races. In

TABLE 8.2

Percentile Body Height of German, Italian, Jugoslavian People (DIN 33402)

Nationality	Age	Gender	5th Percentile Body height (mm)
Germany	16–60	Male	1,629
Italy	16–60	Male	1,587
Former Jugoslavia	16–60	Male	1,632
Germany	16–60	Female	1,510
Italy	16–60	Female	1,439
Former Jugoslavla	16–60	Female	1,473

Table 8.2 the difference of German, Italian, and former Jugoslavian people, living in Germany, are shown. The 5th percentile body height for males differs 45 mm (male) and 71 mm (female). The same effect can be observed inside one country, too. In Europe, it is well known that in Italy, for instance, heights differs strongly, Southern Italians are significantly smaller than those born in the north of the country. Juergens, Aune, and Pieper (1989) compared and categorized the results, based on various international serial measurement surveys. Race-typical differentiation characteristics were established, and demographic distribution was also taken into account. Based on the statistical analyses of internationally valid survey results, Juergens suggests a distribution of the world population into 20 regions—North America, Latin America-Indo, Latin Americeuropean, Northern Europe, Central Europe, Eastern Europe, South-East Europe, France, Iberian Peninsula, North Africa, West Africa, Southwest Africa, Near East, Northern India, Southern India, Northern Asia, South China, Southeast Asia, Australia, and Japan.

Gender

Basically, anthropometric studies are gender-differentiated, because the dependency of the results on gender is very apparent. In literature, however, the anthropometric data of women are taken into account much less than are those of their male counterparts. This is because the great majority of surveys were carried out in connection with military studies. Thanks to increasing female participation in the consumption of consumer goods, female data are playing an increasingly important role, and in product development in particular.

Age

The role a test person's age plays is also a parameter that obviously influences anthropometric data. Although young people can continue to grow until their 25th year, the bodies of those who attain 45 years of age begin to change shape due to the process of involution.

Social Differentiation

As has been established by various surveys (e.g., Juergens, Habicht-Benthin, and Lengsfield in 1971 or Backwin and McLaughlin in 1964), there are also significant anthropometric data variances within a nationality, race, and age group. Analyses show that social pedigree (school education and income of the test person as well as the school education and income of the parents) correlates to an increase in body height.

Overlapping Effect, Secular Growth

Secular growth is a very important phenomenon. It can be seen (over a long period of time) that average anthropometric data, especially body height, increase depending on the year when the data were acquired. This observation has been carried out because the first anthropometric measurements were taken for the medical examination of soldiers (in general only body height), as was effected in all national states by the introduction of conscription. Average body height can be seen to increase by between 10 and 20 mm per decade. Until now, no indication of an end to this development has been found (Greil, 2001). Many theories have been put forward attempting to explain these phenomena. It is mostly assumed that the improvement of nutrition is the cause. However, in spite of the poor nutrition available for most people in Europe in the years during and after the Second World War, no collapse of the constant increase of average body height could be identified.

Therefore the date source and the field of application are two factors that must be carefully considered in the application of data. For successful standardization, the validity and reliability of data must be guaranteed, but data acquisition is very costly in terms of money and time, with the result that existing data are often quite old. One example here is that out-of-date data should never be used in the design of long-life industrial products (e.g., cars, aircraft, and railway carriages) for obvious reasons.

Overlapping Effect, Industrialization

A further overlapping effect is created by increasing internationalization and economically related population movement. It has been known for centuries that the inhabitants of towns and cities have larger body height than do their rural cousins. Worldwide industrialization is causing population relocation, country dwellers are moving into the proximity of the industrial centers. Italy is an excellent example of this, the somewhat smaller Southern Italians are emigrating to the industrial north of the country. Although an increase in body height is to be expected from the effects of the secular growth, statistical mean body height values are reduced, thanks to the mixing of large and small inhabitants. Similar effects can be seen in industrial countries which have a significant proportion of foreign residents. In Germany, for example, the secular growth effect is also slowed by the integration of Italians and Turks (see DIN 33402). In the United States too, the influx of Asian and Latin American immigrants causes statistical displacements.

ISO 15537—Principles for Selecting and Using Test Persons

For the correct selection of test persons for anthropometric tests, ISO has defined its own standard, ISO 15537. It has been developed by the CEN/C 122 Work Group (Ergonomics) in cooperation with the SC3 Subcommittee of the ISO/TC 159 Work Group.

ISO 15537 was created for testing the anthropometric aspects of industrial products and designs that have direct contact with the human body. This standard is also for testing the safety aspects of products.

The procedure of test person selection is fully described in ISO 15537. Here the selected procedure is primarily oriented on the development of machines and not so much on the creation of anthropometric tables.

The following tasks are also described in ISO 15537—dimensions of products; critical anthropometry measurements; the combination of both of these; body types, screening, and detailed tests; selection of test persons for screening and detailed tests; and documentation with results. To help you understand this, a descriptive example of anthropometric tests for

TABLE 8.3

Nineteen Body Measurements of Worldwide Anthropometric Data in ISO 15537 Worldwide
Human Body Measurements for Persons Between 25 and 45 Years of Age, Divided Into Two
Categories, that is, "Smaller Type" and "Larger Type"

Human Body Measurement[a]	Smaller type[b]		Value, mm	Larger type[b]	
	P5	P50	P95/P5	P50	P95
Stature (body height)	1,390	1,520	1,650	1,780	1,910
Sitting height (erect)	740	800	870	935	1,000
Eye height, sitting	620	690	750	815	880
Forward reach (fingertips)	670	740	810	880	950
Shoulder (bideltoid) breadth	320	365	410	455	500
Shoulder (biacrominal) breadth	285	325	360	395	430
Hip breath, standing	260	300	335	375	410
Knee height	405	455	505	550	600
Lower leg length (popliteal height)	320	365	410	460	505
Elbow-grip length	270	305	340	375	410
Buttock-knee length	450	505	560	615	670
Buttock-heel length	830	920	1,010	1,100	1,190
Hip breadth, sitting	260	305	350	395	440
Hand length	140	155	170	185	200
Hand breadth at metacarpals	65	75	90	100	110
Foot length	200	225	250	275	300
Head circumference	475	505	540	570	600
Head length	160	175	185	195	205
Head breadth	120	135	145	160	175

Note: For children and elderly populations, separate datasets sometimes are needed.

[a] Source: Hans W. Juergens, Ivar A. Aune, Ursula Pieper, 1989.

[b] Both types should be considered when testing products designed for the whole world. "Smaller-type" and "Lager-type" categories are given if it is not possible to create a product design for the whole world. "Smaller-type" data are based on females from "Smaller type" populations. "Larger type" data are based on males from "larger-type" populations.

an elevator has been included in Annex A. Besides detailed, step-by-step instructions, the test procedure is also graphically summarized in a table.

The ISO 15537 is supplemented by two tables. Table 1 of ISO 15537 contains 20 measurements of European persons (18 to 60 years old) with the 5th, 50th, and 95th percentile values. Table 2 of ISO 15537 shows 19 body measurements taken in worldwide anthropometric surveys. These have been analyzed according to physique types (small and large) and the percentile type has been entered. For the smaller type, the 5th, 50th, and 95th percentiles were calculated. For the larger type the same procedure was used, whereby the 95th percentile of the smaller type corresponds to the 5th percentile of the larger type (see Table 8.3).

The Selection of Anthropometric Data

In the last few decades, measurement performed on live persons (somatology) for anthropometric tables has become firmly established. In order to define anthropometric data, extensive measurements on skeletons (osteology) had to be carried out in the preliminary stages (Bräuer, 1988; Karoly, 1971). Juergens further developed these studies for ergonomic design in industrial

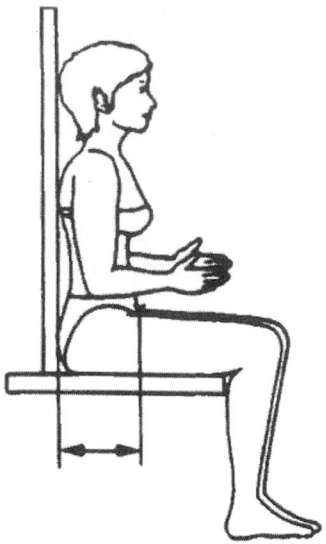

FIG. 8.2.   Measurement with vertical panel (ISO 7250—4.2.17).

anthropology (Juergens & Matzdorff, 1988). The studies describe the objective of industrial anthropology thus . . . "to design the environment of the human being in accordance with his physical needs, in the broadest sense. The acquisition of the morphology and biomechanics of the human beings is thus permanently oriented on a technical precondition, how they arise from the design of workplaces, utility objects, etc."

Although purely anthropometric measurements are derived only from the body and clearly identifiable measuring points, with industrial anthropological data, the demands of the products to be designed are also integrated. This means that for the definition and selection of measurements, it is not only body posture or movement that plays an important role, but also clothing as an influencing factor must be taken into account. It may also be necessary to re-enact complex workplace situations for the test.

Axes and Reference Systems

In order to be able to describe and compare measuring points according to position and orientation, standardization of the axes and reference systems is necessary. In industrial anthropology, the reference to standing and seating areas has established itself. Prerequisite here is that these areas be flat and nondeformable. In various earlier studies (e.g., Lewin & Juergens, 1969), further reference points (e.g., heel point, contact point with the ischium protuberance) are defined, but these should be avoided because the positions involved are difficult to reproduce. As an alternative, contact points or contact areas (e.g., the pelvis or shoulder, see Figure 8.2) or functional, product-related reference points are used (e.g., seat reference point SRP).

The human head plays a special role due to its degree of mobility. As early as 1884, the *Frankfurt Horizontal* was established; it was named after the conference location. A test person must take up the following head posture: The connecting line between the lowest point of the bony eye socket and the highest point on the upper edge of the auditory canal must run horizontally, parallel to the standing area (see Figure 8.3). These definition is used for all measurements where the head is involved (Helbig, 1987).

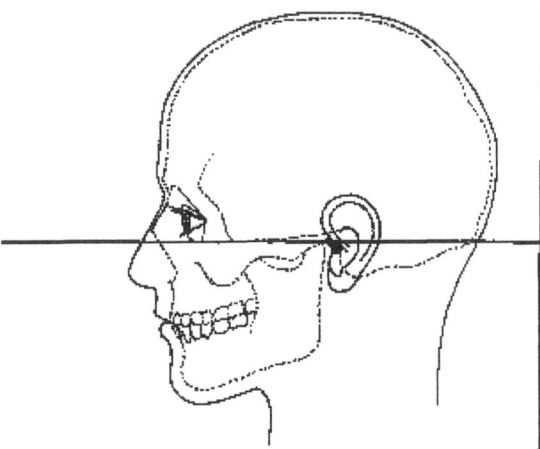

FIG. 8.3. Frankfurt Horizontal.

The Influence of Postures

Because the human being typically takes up different postures while working or using products, very representative measurement approaches must be defined for anthropometric studies. In all serial measurements, standing and sitting postures with various arm positions have established themselves as being the most valid.

Studies of standing postures are typically made of test persons standing ramrod-stiff. The feet are closed and parallel to one another and head posture is in accordance with the Frankfurt Horizontal. Today it is recognized that a stiff posture does not correspond to the physiological working posture—a more relaxed posture is preferred. Nevertheless, studies (Lewin & Juergens, 1969) have shown that reproducibility in the case of physiological postures is at best only inadequately attainable.

A similar effect was observed with sitting postures. A rotation of the pelvis occurs due to the curvature of the spinal column (kyphosis). This leads to deviations of up to 50 mm between stiff and relaxed sitting postures. These deviations are overlapped by two further influencing factors, the asymmetry of the human being and the increasing shortening of certain measurements as the day goes on.

Several studies (Gaupp, 1909; Ludwig, 1932) substantiate that the majority of human beings display *asymmetry*. Although around 75% of all adults' right arms are longer than their left arms, the left leg is longer than the right in over 50% of the population. This asymmetry continues throughout the entire body—the different leg lengths lead to a tilt in the pelvis, which in turn leads to scoliosis (lateral curvature of the spine)—and this in turn is followed by an asymmetry of the acromion (shoulder line). The average difference in asymmetry is between 10 and 20 mm in the case of the acromion (shoulders, right lower than the left), and 10 to 30 mm difference in arm lengths.

The cause of this asymmetry is probably to be found in the one-sided actions that we humans so often carry out. A close relationship can be seen between being right-handed (or left-handed) and the excess length of the more active right (or left) arm. The crossed asymmetry in the leg can be explained by the crossing of right- or left-handed dexterity and the anklebone. A right-handed person will typically jump with the left leg, while a left-handed person will move the right leg first.

The effect of the *daytime shortening* of torso and leg lengths can amount to a body height difference of between 20 and 40 mm. The reason for this is the compression of the intervertebral discs of the spinal column, the articular cartilage of the legs, and the sinking of the foot arch.

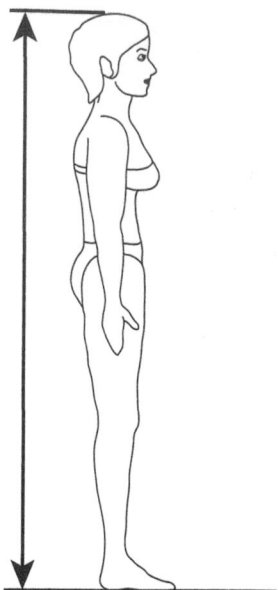

FIG. 8.4. Description of body height (4.1.2) in ISO 7250.

ISO 7250—Basic Human Body Measurements for Technological Design

The ISO 7250:1996 provides a description of anthropometric measurements which can be used as a base for comparison of population groups. It describes the most important body measurements, used in anthropometric surveys. As the authors remark, the body measurements should neither be seen as a binding definition, nor as instructions for taking measurements—they should be used as a recommendation for tests, studies, and surveys. The ISO 7250 was also created by the SC3 Subcommittee of the ISO/TC 159.

The ISO 7250:1996 is currently one of the most important standards, because it introduced an international harmonization of the body measurements to be acquired. Fifty-five body measurements are defined, based on detailed postures. For each of the body measurements there is a clear-cut description (name and number), a sketch, a notation of the measurements, and the recommended measuring method and the measuring instrument. In Figure 8.4 the description of the ISO 7250 is shown with an example of body height.

Further details of the ISO 7250 are given in a detailed definition of all technical terms as well as the measuring conditions (clothing of subject, support surfaces, body symmetry, and the measuring tools).

ISO 15535—General Requirements for Establishing Anthropometric Databases

In closed conjunction to ISO 7250 the ISO 15535 was defined. The goal of the standard is to formulating a database structure and their associated reports. This standard was prepared by the same Subcommittee like ISO 7250. Similar to discussed ISO standards, ISO 15535 defines terms, references, the collection design, and data-collection requirements like sample size, type of clothing, or accuracy of measuring instruments. In a detailed description, a database format and the minimum content of the database is defined; In Annex A, the method for estimating

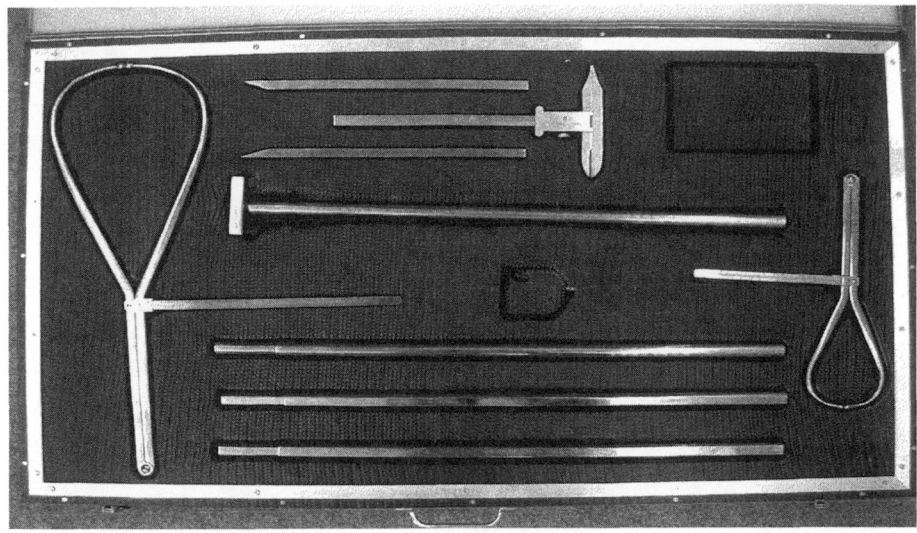

FIG. 8.5. Measurement setting by Martin.

the number of subjects is described; in Annex B and C a definition and an example of an anthropometric data sheet is printed. Annex D, E, and F focus statistical analysis of data.

The Standard Manual Measuring Tools

Martin (1914) is seen as the founder of a catalog of anthropometric measurement prescriptions and methods. He also developed the measurement setting that is named after him. Essentially it consists of different calipers, specific vernier calipers, and a calibrated measuring tape, by means of which individual body dimensions can be taken (see Figure 8.5). The essential idea is to confine measuring (as far as possible) to measurements of bone to bone distances. In every case, bone areas are measured, which, even in the case of corpulent subjects, are immediately under the skin surface and can be clearly felt by hand. In general, anthropometric data are taken from a subject who either stands on a table or sits on it with legs dangling. The body posture is totally 'stretched' and the head is kept in Frankfurt Horizontal. Measurements are generally taken from nude subjects (Bradtmiller, Gordon, Kouchi, Juergens, & Lee, 2004).

Circumference measurements, on the other hand, show much more inner deviation. Very important is the actual use of the measuring tape—it should be taut, with no loosely hanging section. The result depends on the "elastic" properties of the subject and the intention of the person carrying out the test.

Advantages and Disadvantages of Manual Measurement

Anthropometric measurement with manual measuring tools is a proven and standardized methodology. The true advantage in manual measurement is the determination of skeletal characteristics, because they must be actually felt and measured.

One disadvantage of manual measurement can be seen in measurement accuracy and reproducibility. Kirchner et al. (1990) show that in manual measurement interindividual and intraindividual measurement value differences of up to 100 mm occur for body height or sitting height. A further disadvantage is that body shape and posture measurements of soft body parts can only be carried out with a great deal of effort. A lot of this disadvantages can be solved by 3D Body Scanner.

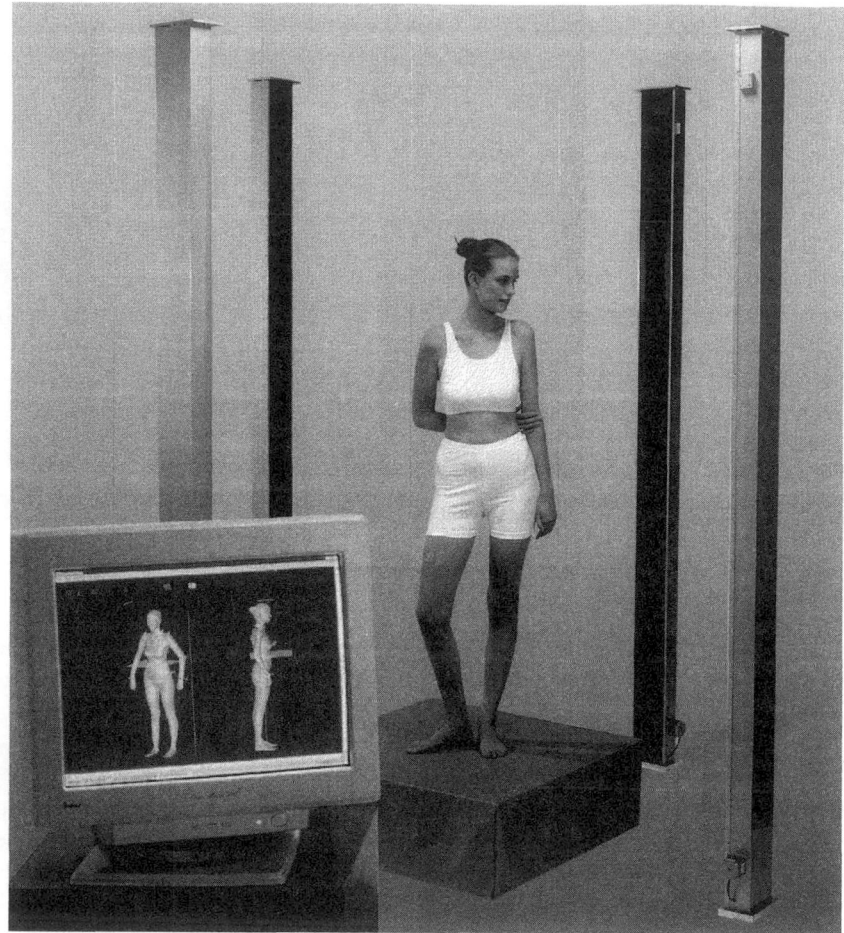

FIG. 8.6.   Body Scanner

Contact-Free Measurement with Body Scanning

For the past few years, the rapid development of laser and camera technologies has enabled the contact-free, accurate and fast scanning of body surfaces by means of body scanners. Within a few seconds, a three-dimensional image of the test person can be created.

Nowadays (Robinette et al. 2004) there are four different principles for three-dimensional measuring tools in use. *Contact devices* are biomechanical systems, by means of which a measuring point on the subject beings is contacted and the three-dimensional coordinates defined. This genre includes mechanical measuring machines (Wenzel, Steinbiehler, etc.) or mechanical measuring arms (e.g., Faro-Arm). *Tracking systems* follow markers attached to the subject with several detectors and calculate the corresponding three-dimensional coordinates of the markers by the triangulation method. This type includes tracking systems from a wide range of manufacturers (Vicon, Polhemus, Peak, Qualisys, etc.) based on the ultrasonic principle, and also systems based on optical and electromagnetic principles. *Stereophotogrammetry systems* work with several cameras, which photograph the test person from different angles. The allocation of the skin point to be measured takes place by means of manual allocation or automatically with the help of projected patterns.

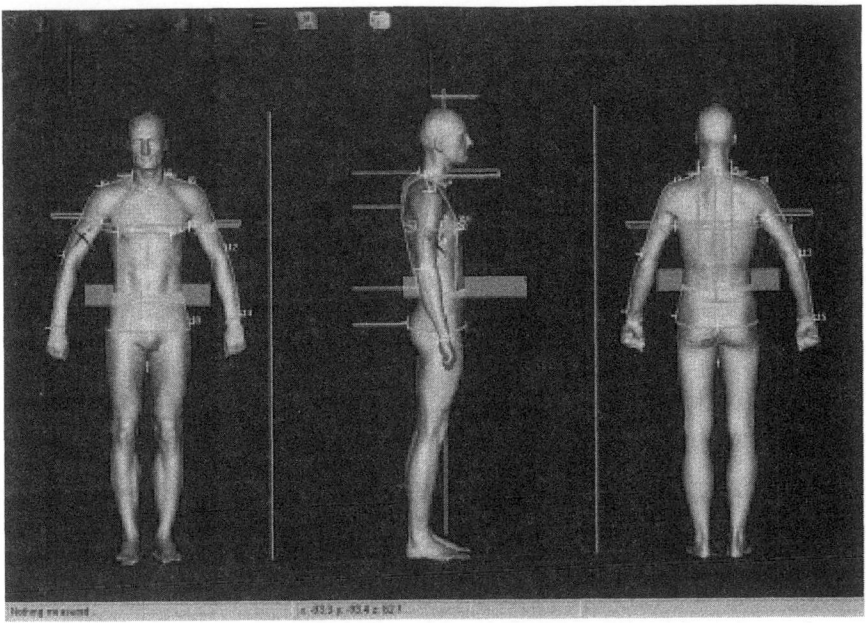

FIG. 8.7. Automatic measurement of body dimensions.

3D Body Scanners (see Figure 8.6) detect the overall outer surface of the human being in one single pass. Using one or more projection units or lasers, lines or patterns are projected on to the subject. Various fixed or mobile CCD cameras detect the deformation of a line or pattern (triangulation method). Three-dimensional surface information is then calculated from the recorded deformation by means of conversion equations. A 3D "cloud" of points representing the skin surface of the test person is then created from the image information.

The body scanner approach has many advantages over traditional sizing surveys. Scanners are more precise and consistent than even trained human beings are, requiring only 810 seconds per stance to capture the needed data. Also, the entire point cloud becomes available almost instantly, allowing it to be mined and analyzed along with thousands of other clouds, yielding information about the changing shape of our bodies as well as about changes in body measurements. Because the scanning system captures shapes rather than measurements, the most important element of the body scanner is not in itself the method that is used to obtain the point cloud, but the software that extracts the body measurements.

Since the turn of the century, body scanners have been universally used for large international anthropometric surveys. In the CAESAR study (Blackwell et al., 2002) carried out in the United States and Holland, the manual and scanning methods of measurement were subjected to an intensive comparison. In 2002, an extensive study began in Great Britain (SizeUK). In 2003 in France an anthropometric survey with body scanners (IFTH) was started and large survey projects have been in preparation in Sweden and in China since 2004.

The use of 3D body scanners poses a special challenge for the carrying-out of the study or survey. On the one hand, this technology enables for the first time the acquisition of reproducible and accurate skin surface and posture data. At the same time, however, comparability with manual studies and surveys must be ensured. This is why combined measurement methods have since become established. Geometrical markers are attached to relevant skeletal feature points (Robinette, Daanen, & Zehner, 2004). During the scan, both sets of information, the surface data of the subject and the marker positions, are detected by special software solutions, and the body measurements are calculated in a fully automated process (see Figure 8.7).

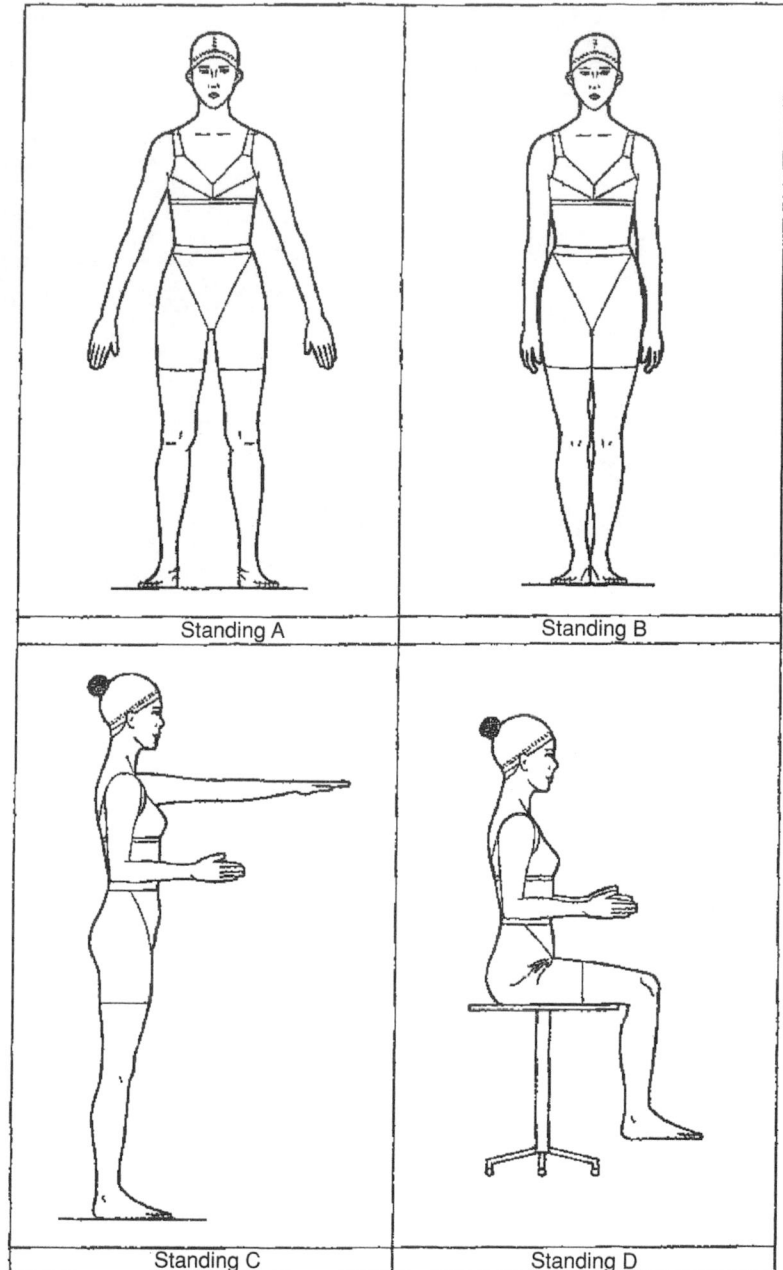

FIG. 8.8.   Defined measuring postures in ISO 20685.

Integrated solutions like this —i.e. body scanner, hardware and matching software—are today offered by a few specialized companies (e.g. Cyberware, Human Solutions and TC2).

ISO 20685 (Draft)—3D Scanning Methodologies for Internationally Compatible Anthropometric Databases

In order to meet the demands that the strong growth of this technology has created, the ISO/TC 159 Technical Committee, Subcommittee SC3, is working on ISO Standard 20685. At present, this is still at the draft stage thanks to the topicality of the technology it is relatively extensive. ISO 20685 goes into details of the technical characteristics of the hardware and software, the clothing of the test persons, and the methods for reducing errors in 3D scanning. Suitable postures for carrying out a study in body scanners are given special attention, because the posture definitions of the ISO 7250 cannot be included in all systems for reasons of space. Here the ISO 20685 suggests four postures, three standing and one sitting (see Figure 8.8) Furthermore, the ISO 20685 also discusses the different body measurement definitions in the ISO 7250 and the ISO 8559. Because the ISO 7250 is primarily defined for ergonomic applications, the ISO 8559 focuses on garment design users. A further important element in the ISO 20685 is the functional properties of the software, by means of which the analysis of the scan data takes place. This also includes the interaction between 3D measurement and combined measurement with landmarks, plus features of manual and automated measurement and the selection, segmentation, and visualization of body parts and landmarks.

THE STATISTICAL ANALYSIS OF THE DATA

Fundamentals of Anthropometric Statistics

Taking only the average of the test data, received data will not suffice. The discipline of statistics provides different ways of representing human measurements, especially in their variability: the range between the highest and the lowest observed measure, the so-called percentile (see following), and the standard deviation. As it makes sense to use the percentile only for smaller and known populations, its application of the percentile is the norm in scientific anthropometry. The idea here is as follows: The persons surveyed are put into a sequence of small to large, going by the required measurement. The X-th percentile is defined by the measurements of the one subject who separates the $X\%$ lower part from the remainder (see Figure 8.9). In anthropometric tables it is usually the 5th, the 50th, or the 95th that is taken (also the 1st and the 99th on occasion). Taking as an example the measurement of the 5th percentile, this means that 5% of the sample is lower than this. The 50th percentile divides the population of the sample into half smaller and half larger. This is not necessarily the mean value, however. This will only be the case if the distribution of the values is symmetric. Nevertheless, experience

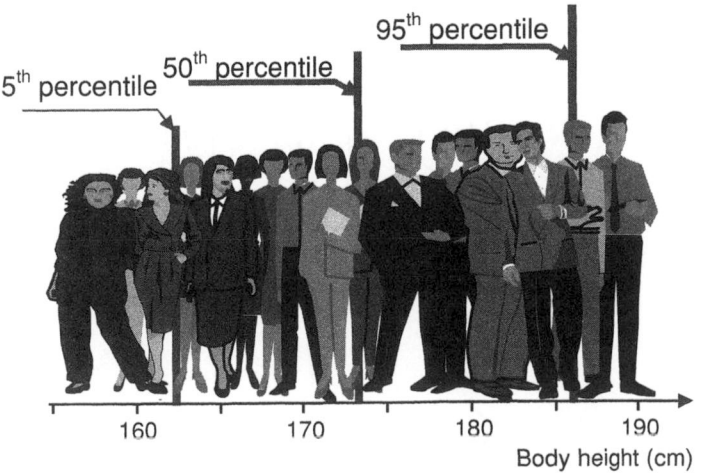

FIG. 8.9. The percentile idea, illustrated by the example of the body height.

has shown that anthropometric distance measurements have a good chance of being normally distributed and therefore symmetric.

The percentile of a body measurement is derived from the distribution of a measurement. The pertinent distribution is determined from all the measurement values of a size. It is from this that the total frequency distribution is derived through integration. The associated percentile value can be taken from the total frequency distribution. At the ordinate (y-axis), the desired percentile value is selected (e.g., 5th percentile) and projected on to the cumulated frequency curve, and the associated measurement value is read from the abscissa (x-axis).

When the number of subjects in the sample is large enough and the sample's composition is representative of the required population, these values are good estimations of the true values of the population. The necessary number of subjects can be estimated by the rules of scientific statistics. Formula 1 gives the necessary number N of subjects in a sample, when the estimated value s of the standard deviation is known (e.g., from a preexperiment). Under the assumption of the standard Normal distributors, the value $z_{a/2'}$ can be determined by this tolerated error probability. n_P is the amount of the preexperiment sample.

$$N = \frac{4 \cdot z_{\alpha/2}^2 \cdot s^2 \cdot n_P}{\Delta_{crit}^2 \cdot (n_P - 1)}$$

For example, in a preexperiment of $n_P = 40$ subjects for body height, a standard deviation of $s = 73.4$ mm had been observed. If the average of the body height of a certain population (e.g., population of the people in the United States) should be investigated, a representative composition (according to human race, social situation, profession, age, and gender) of the people of the United States must be found. In the case this value should be indicated with an accuracy of $\Delta_{crit} = 5$ mm and an error probability of $\alpha = 1\%$ (from statistic table $z_{0.5} = 2.576$)—according to Formula 1 at least 5,885 subjects are necessary. If the demand can be reduced (e.g., acuity only 20 mm, error probability $\alpha \leq 10\%$), the necessary number of subjects would only be 150. In ISO15535 (Annex A) simplified formulars for the calculation of the number of subjects needed on a sample are printed.

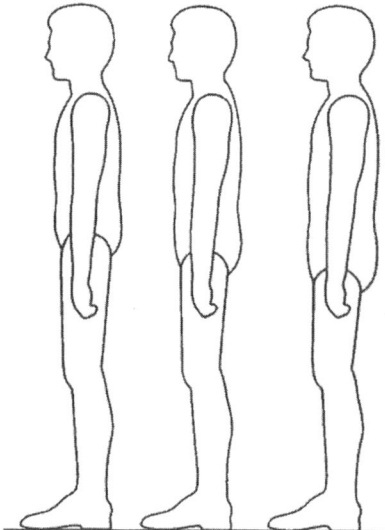

FIG. 8.10.   Different proportion with same body height.

The Combinatorial Analysis of Body Measurements

In all international anthropometric studies and applications, it has become standard to use body height as a criterion. The input of a 5th, 50th, and 95th percentile type always emanates solely from body height. During application, the error is very often made that this percentile value is also transferred to other individual or composite measurements. Here is an example of this: For vehicle interior design nowadays, two-dimensional templates are still popular. If the 95th percentile type is now taken for the design of the roof height of a car (as mentioned earlier, the input relates only to body height), then a wrong result will be produced. For the design of the roof height, the sitting length of the torso (trunk length) is decisive, but this size has relatively little to do with body height (it correlates poorly). This effect is shown in Figure 8.10. Three persons have the same body height, but different relation of leg length and torso height.

More modern anthropometric analyses take the correlation information on body measurements into account. The correlation coefficient specifies whether two measurements have a relatively minimally varying, proportional correlation. The correlation coefficient can take on a range of values from 0 (no correlation) to 1 (direct correlation).

Juergens (1992) shows that coefficients between 0 and 0.7 are (seen from an ergonomic standpoint) so nonrepresentative that the one measurement cannot be derived from the other. In Table 8.4 the correlations between a few other important measurements are itemized in an exemplary manner. You can see that body height with crotch height still correlates well (0.84), but when taken with sitting measurements only a weaker correlation exists (0.74 with sitting height). With shoulder breadth (0.37) and the circumference (0.13, 0.15), a significant correlation is no longer apparent. In contrast, one can see that the sitting dimensions (sitting height and elbow height) show a significantly higher correlation (0.53) than when taken against the body height. A similar effect can be observed in the case of circumferences, whereas correlation with body height hardly exists, in this case a higher intercorrelation can indeed be seen (0.46).

For the practical application, it can be assumed that body height is only a sensible indicator for leg lengths and accessibility statements (Juergens 1992). For measurements that occur in sitting positions, sitting height is a better control value than body height. In the case of circumferences, the same thing occurs—they display a correlation with one another and not with body height. This means that for the design of complex workplaces (e.g., vehicle interiors) we must revert to detailed task-specific tables or to a suitable correlation matrix. However, because this is very complex, other means like computer-supported *manikins* (computer-supported human models) can be used.

TABLE 8.4
Correlation Matrix of Some Body Dimensions (Juergen, 1992)

	Body Height	Crotch Height	Sitting Height	Ellbow Height, Sitting	Shoulder Breadth	Chest Circumference	Waist Circumference
Body Height (4.1.2)	L	0.84	0.74	0.25	0.37	0.13	0.15
Crotch Height (4.1.7)	M	—	0.46	0.02	0.25	0.04	0.00
Sitting Height (4.2.1)			—	0.53	0.35	0.13	0.14
Elbow Height Sitting (4.2.5)				—	0.05	0.06	0.09
Shoulder Breadth (4.2.9)					—	0.21	0.30
Chest Circumference (4.4.9)						—	0.46
Waist Circumference (4.4.10)							—

THE USE OF DATA

It is the objective of anthropometric data procurement to make available high-quality data for the development of ergonomic machines and products. Just as we have seen in the previous chapter, it is a relatively complex matter to select the right test persons and representative body measurements. With these results, easier problems like, for example, the height of a door frame or the distance to an operating lever must be solved quickly and accurately. The development of more complex task problems is another matter entirely. For instance, multifaceted questions must be answered during the development of a vehicle interior—where is the seating position, how is vision looking out the windows, can controls be reached and activated, how is pedal travel, are there collisions with door trims, how clearly can the mirror be seen, and so on. There is no guarantee that the answers to such problems can be given from tables—for such complex matters, more powerful tools are needed.

An effect observed by Kirchner et al. (1990) must be included here as well—the fact that different users produce different results from the same task. On the one hand, this occurs thanks to the different test approaches used in different tables (social layers, age structures, and posture definitions). On the other hand, the transformation of the anthropometric data in the application is interpreted differently by designers and engineers, with the result that results are difficult to reproduce.

Hence, in the case of the application, as in the case of data procurement, standardized methods and working procedures must be adhered to. Juergens (1992) differentiates here between the *direct application* and the *indirect application* of anthropometric data.

Direct Application of Anthropometric Data

In direct application, problem questions are answered in which the following are discussed:

- Whether the body envelope (external contours) of the human being suffices.
- Whether one single measurement for the design of the product to be developed is relevant

The first question comes into play for all proximal (pertaining to close contact with the body) products. Clothing, shoes, or safety equipment can be derived directly from the tables. Nevertheless, the size of access openings for body parts has to be directly dimensioned. In the case of the second question, the measurements are directly related to the technical object to be designed. A simple example of this is the standardized door height of buildings. To enable the majority of the population to go through the door without stooping, a corresponding measurement (e.g., 99th percentile) must be selected. Another example of this is the design of an office table. To enable everyone to reach the papers, telephone, computer, and so forth, on the table, its depth must be applied to the 5th percentile female (subject to functionality demands).

Very often, additional add-ons have to be integrated into the design. A simple example of this kind of application would be the design of the roof height of the highest point in a vehicle cabin: The relevant measurement to take into account in this case would be the maximum body height of the population. In order to get the measurement of the technical object, a safe distance of 10% to 25% must be added, depending on the type of application the vehicle cabin will be used for.

In all these examples, only a limit value for the design of the products is used. With more complex designing tasks, a situation could arise where interior and exterior measurements have to be considered at the same time. For instance, the design of a vehicle cabin must be

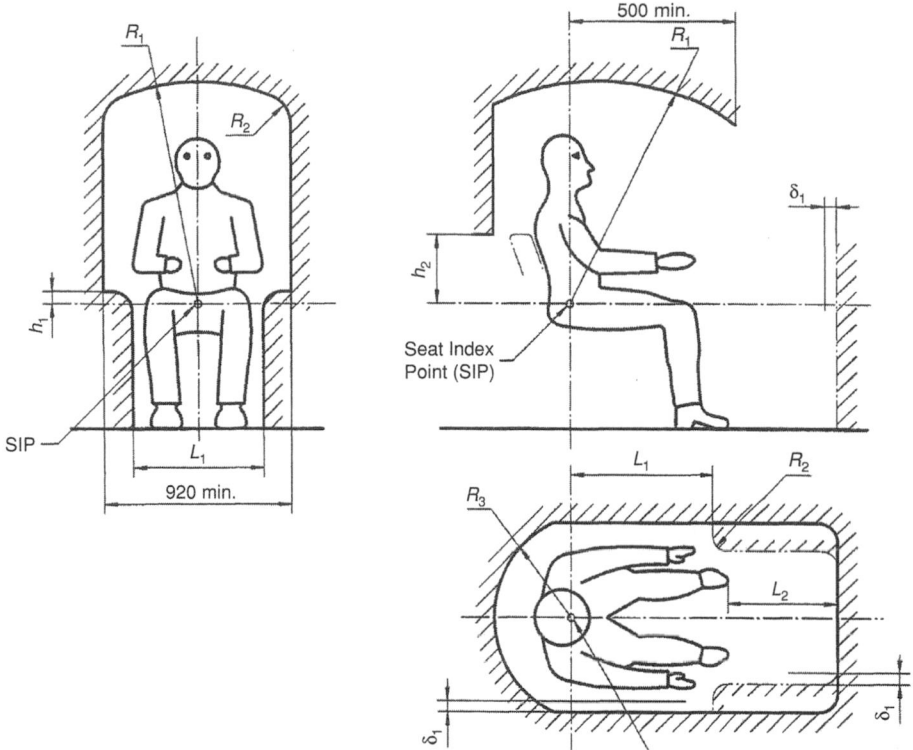

FIG. 8.11. ISO 3411—Recommendations for interior space envelopes.

tackled in such a way that a tall man has enough room inside the cabin, and short persons must also be able to operate the controls. Industrial products such as this should be designed in such a manner that at least 90% of the population can use them without any restrictions or hindrances. The values of the 5th percentile and the values of the 95th percentile are normally taken as limit for this design. As anthropometric tables indicate the measurements for males and females separately, the demand for 90% usability can be fulfilled by taking the values of the 5th percentile for the female and the 95th percentile for the male.

ISO 3411—Earth-Moving Machinery, Human Physical Dimensions of Operators and Minimum Operator Space Envelope

ISO 3411 was developed to simplify the application of anthropometric data for the design of earth-moving machines. What has been written here in the last few paragraphs can be effectively reconstructed here.

In the first step, the statistical physique models are defined. Here, one differentiates between a small operator (5th percentile), a medium operator (50th percentile), and a large operator (95th percentile). In addition, the data are supplemented by the addition of a large operator with heavy clothing (arctic). The most important body measurements are given for each of these four types. In the second step, the relationship to the object to be constructed is established in the ISO 3411:1995. Recommendations for clearances, distances, and widths are given (see Figure 8.11).

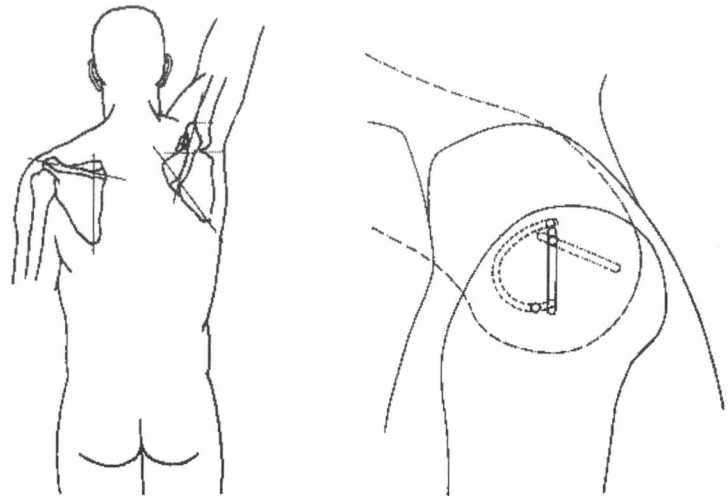

FIG. 8.12.   Mechanical simulation of shoulder joint.

Indirect Application, Templates, and Three-Dimensional Models

Although available tables can suffice for the estimation of the body measurements needed for simple workplace designs, sensible and meaningful design aids must be available for high-value and safety-relevant workplaces. One option that has proven its worth in the last few decades is the use of two-dimensional contour templates of percentile types (5th, 50th, and 95th) of body height. These are available in different sizes in either fixed or movable template versions. Two-dimensional templates were developed to make the anthropometric data of the human being available to the designer (in direct association with the movement potentiality of the future user of the product design).

The development of a body contour template is a complex procedure, because it not only must portray the anthropometric data correctly but also must take body shape into account. Therefore a pure derivation from anthropometric tables is not possible; there are additional body shape analyses to be carried out with the aid of image-providing processes. Furthermore, in the case of movable templates, the correct anthropometric body part lengths must be created for every posture position used. In such templates the very complex movements of human joints (e.g., combined rotation and displacement in the knee joint) must be mechanically simulated. Yet another demand made on movable templates is the correct portrayal of the physiologically correct mobility areas. Here linked (coupled) movements of several body elements must be simulated by means of intelligent mechanics. In Figure 8.12 this is represented by an example of the shoulder joint (Helbig & Juergens, 1977). The forward movement of the arm forces a displacement of the shoulder joint in an upward direction via the coupling with the scapula. If this movement is continued upward, the clavicle (collarbone) will become increasingly involved in the movement.

In Figure 8.13 the "Kieler Puppe" is illustrated. This simulates the complex biomechanics of the most important human joints thanks to intelligent mechanical replacement designs. The back construction of three parts with coupled joints attracts special attention. It allows a very good simulation of the physiological movement of the spinal column (Helbig, & Juergens, 1977).

In the last 40 years, a multitude of templates have been developed for various purposes. A few of these templates have found their way into various national and international standards:

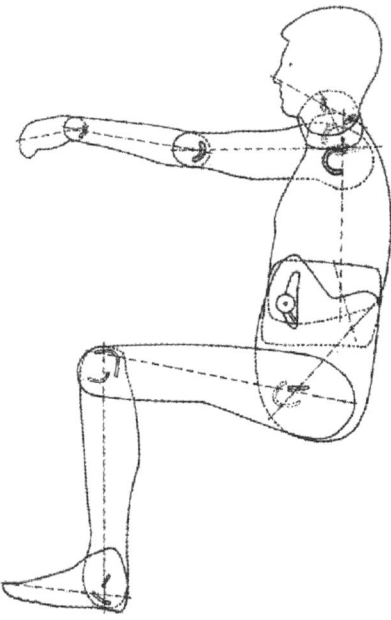

FIG. 8.13.   Template "Kieler Puppe."

The *SAE template* is a 2D template especially developed for use in the automotive industry. The data, design, and application are on record in the standard SAE J826a (Society of Automotive Engineers). This template has become extremely important in the automotive industry, because it is used not only for ergonomic design but also for the authorization of the vehicles themselves. The template consists only of the torso of the 50th percentile American and a 2D leg simulation of the 95th percentile American. In Figure 8.14 the SAE template (J826a) and the related SAE-Standards (J1100, J941, J1052, J1517, J1516) are shown.

The *DIN 33408 body contour template for seats* was modeled on the SAE template. It is derived from the developments of the Kieler Puppe (Helbig & Juergens, 1977) and encompasses altogether six templates in three perspectives: front, side, and top view. A total of five body types are supported. For the male form, the 5th, 50th, and 95th percentiles are used, whereas for the female form, the 1st to the 5th and the 95th percentile are available.

There also exist various templates, which offer a simplified representation of the human being in the form of body contour lines (e.g., DIN 33416 or Bosch template). These have never achieved the same breadth of distribution as the templates described earlier. It must, however, be said that the body contour templates are still extremely widespread in actual practice. This is due to the comparably economical purchase price and the manageability of the templates themselves.

The informational value of the templates is, however, very limited, because neither the physique proportions with regard to future product user population (age, race, living space, secular growth...) can be sufficiently taken into account, nor can task-dependent postures or movements be simulated or evaluated from a comfort point of view.

Digital Human Models (DHM) for Ergonomic Design

The limitations of templates with regard to physiological movement simulation, anthropometric scalability, and the correct computation of anthropometric data correlations have in the last few

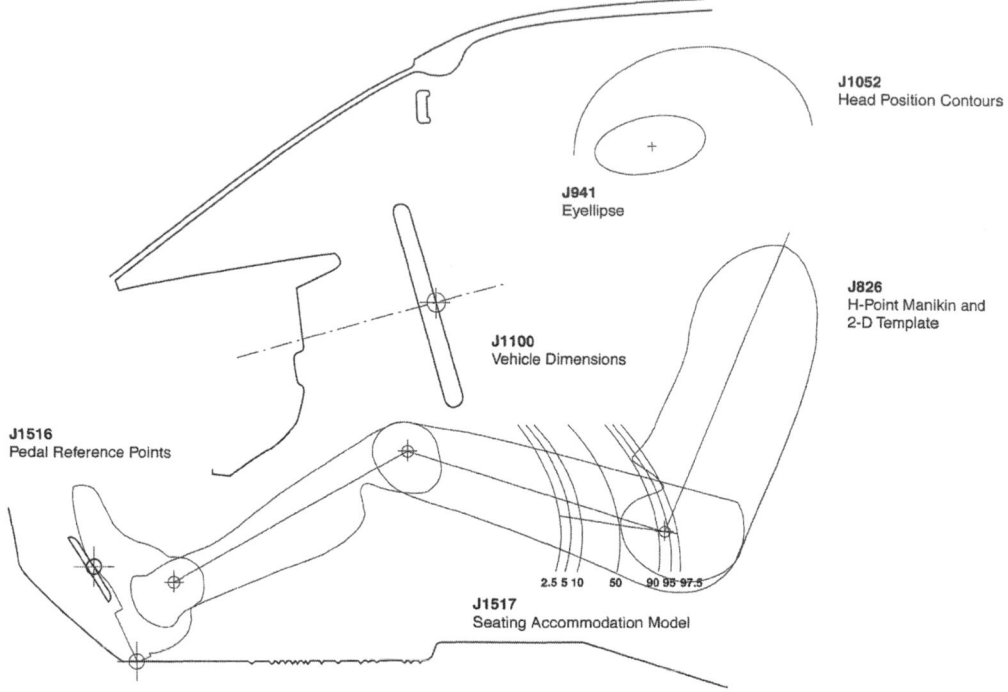

FIG. 8.14.   SAE Template.

years been more or less eliminated by modern computer technology. Since the 1980s, modern 3D CAD technologies have become established—and this in turn leads to a need for 3D aids for design of the human being. Although more than 150 ergonomic manikins are known (Hickey, Pierrynowski & Rothwell, 1985; Porter, Case, Freer, & Bonney, 1993; Seidl & Speyer, 1997) to have been developed, only a few systems have been able to establish themselves successfully on a global industrial basis in the last few years. Thanks to the increasing impact of the DHM, more can now be ascertained about their potential capabilities.

Historical Development of the DHM

The development of computer models of human beings originated with the NASA. In 1967, William A. Fetter, a Boeing employee, developed the optical computer model, First Man. Based on the anthropometric dimensions of a scalable 50th percentile man, it was used to check reach and accessibility in aircraft cockpits. Building on this development, there followed (between 1969 and 1977) the Second Man, Third Man, Fourth Man, Woman, BoeMan, and CARmodels. Developments in the United States were further advanced with the Bubbleman, Tempus, and Jack models from Badler and the Chrysler Cyberman (Figure 8.15). In the middle 1980s, the manikin Safework was developed in Canada, at the Ecole Politechnique in Montreal.

Various manikin approaches were also created and used in Europe. In 1984 in the Laboratoire d' Anthropologie Appliquee et d'Ecole Humaine in Paris, the manikin Ergoman was created, and in Germany in the middle 1980s, the computer model Franky (within the framework of a BMFT program), the Heiner System (at the Technical University in Darmstadt), and the manikin Anybody (and of course its successor ANTHROPOS) were realized. Anybody was modeled on Oszkar, the development of a Hungarian management consultancy. In England, Sammie

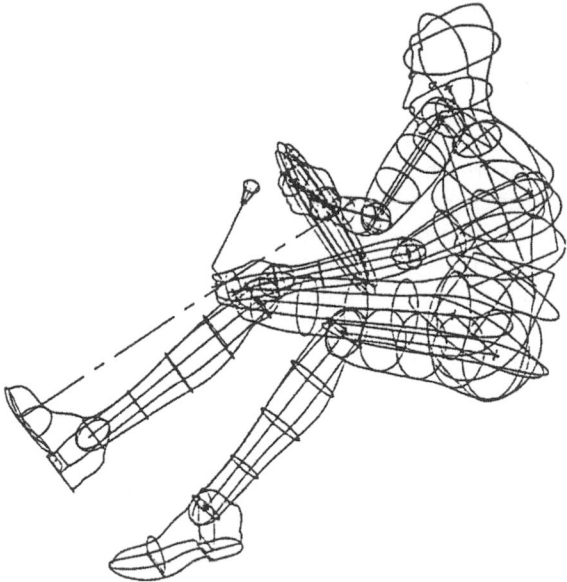

FIG. 8.15.   Digital Man Model Cyberman (General Motors).

was created—the first manikin on the market, for which the license was made industrially available for third parties.

In the middle of the 1980s, functional differentiation of the manikin came into being for the first time. While Sammie focused on the automotive industry, Ergomas and Champ DHM (later marketed as AnySim) were created for factory planning.

One development with a special rating is the RAMSIS manikin. It was developed for the ergonomic interior design of vehicles between 1987 and 1994 under contract to, and in cooperation with, the entire German automotive industry. This industrial project is arguably the most extensive ever for the development of a manikin.

The Five Basic Elements of a Manikin

In principle, manikins constitute the foundation for a multifaceted usability in a wide range of different applications. Here, however, the prerequisite is that the orientation of the model is tailored to the specific application area. The morphology and biomechanics of the human body are made available in the quality required for the specific problem. There are five basic elements that have to be taken into account in manikin development.

The *design of the manikin* takes into account the kinematics of the model, that is, the number and mobility of the joints and associated body elements as well as the accurate detail of the outer surface simulation that corresponds to the human skin. Here there are (discussed in part in the chapter on 2D templates) three significant characteristics for the design of the manikin. The *number of joints* defines the kinematic flexibility of the manikin. The *mathematical–biomechanical simulation of joints* influences the motion paths of all body elements that are coupled with a joint. The selection of *the limit angle ranges* of each joint ensures that even in the case of extreme postures the behavior of the DHM will remain correct. Besides the physiological simulations of the skeleton, the *realistic outer surface simulation* of the human skin surface constitutes the second focal point in a manikin design. The surface of the body of the DHM must not only be able to be correctly altered in an anthropometric and movement-dependent manner, it must also fulfill beauty requirements, because such systems are always used to present the analysis results.

The coupling of the DHM with suitable *anthropometric databases* is the basis of being able to develop efficient products with the DHM. Conventional tables are integrated into most DHMs and most systems have interfaces for their own data input. However, only very few manikins support further modern approaches like (the already discussed) calculation of anthropometric data via correlation coefficients (RAMSIS, Safework). The user can take into account the correlation of individual body data to one another and consequently compute physique models with selectable or relevant critical sizes, which represent the population much more realistically. Enhanced control parameters like age group, regional differentiation, or somatotypes are also selectable. Secular growth simulators; that is, forecasting on the expected mean changes in size (growth in length) of the population groups have also become available in the meantime (RAMSIS).

The third focal characteristic is *biomechanical posture and movement simulation*. In a sitting posture, for instance, a hip joint angle deviation of only 3 degrees leads to a knee upper edge that will deviate more than 2 cm from the real body posture. In many models, postures are set using the "artistic feeling" of the user. Thanks to inverse kinematics, more modern systems now have solution processes available. These processes, leaning on robotics and simulating postures and movements by means of angle or energy minimization, take into account the real movement behavior of the human being inadequately at best. Only a small number of systems have realistic calculation and modeling capabilities. The Ergomas/Ergoman system, for example, has a movement simulation that is based on real observations. RAMSIS has a posture simulator based on high-dimensional distribution functions, which draw upon roughly 60,000 3D data values. Latest studies (Chaffin, 2004) do indeed simulate movement sequences by taking into account the biomechanical characteristics of the body (maximum forces, mass moved, and distributions); but the necessary scientific data substructures are not yet fully available at the time of this writing.

The fourth focal characteristic of a DHM is its *ability to analyze* with regard to the product to be developed. The user wants to obtain information about the ability of his design to match human requirements and perhaps some indications as to quality-increasing measures he could take. Whereas all ergonomic analysis and test procedures were integrated into DHM at the beginning of its development (e.g., OWAS, NIOSH, MtM), modern DHMs specialize more and more in their own field of application. Although DHMs need a fast ergonomic appraisal with the aid of maximum force procedures (e.g., NIOSH), comfort, mirror view, and belt run all have a more major role to play in the designing of systems for, for example, the automotive industry. Nevertheless, a few basic and analysis functionalities from the development of the DHM have meanwhile succeeded in becoming established. In vision simulation, the view of the CAD design is reproduced using the manikin's eyes. In the simulation of spaces within reach (grasp spaces), mathematical 3D hull (enveloping) surfaces for extremities are calculated. The forecasting of maximum operating forces is also one of DHM's domains, because the dependency of the maximum forces on the posture can be correctly taken into account.

The last focal characteristic of the DHM is *seamless integration into the design process*. A DHM can only be sensibly put to use if it is integrated into the design environment—in other words, if it is in the "virtual world" of the product. Today there are two different solution approaches available. The *Standalone-System* is a computer program, which loads product data via standardized CAD interfaces. The advantage of this solution is flexibility, because it can work together with all available systems. One disadvantage to perhaps note is that the user must learn the ropes of a second computer program. With *CAD Integration*, the manikin, with all functionalities, is integrated into the CAD system of the user. He doesn't have to leave his

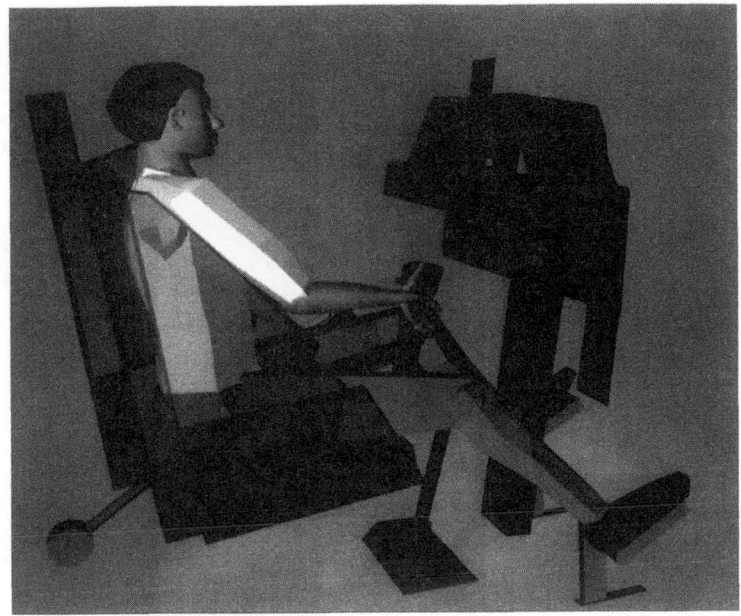

FIG. 8.16.   The DHM Jack.

familiar surroundings and he won't have to take the time to export and import product data. The disadvantage of this solution is the high price involved.

Systems Available in Industry

In the last few years, three ergonomic manikins have become established in professional applications, controlling more than 95% of the market.

The main field of application of the manikin JACK (Figure 8.16) is animation and visualization in vehicle design and architecture. This model consists of 39 body elements; visualization takes place via area segment imagery with textures. The anthropometric database is based on the Human Solutions Library; on NASA data; and on in-house studies, surveys, and tests. Posture and movement simulation enables interactive movement of the entire manikin in realtime. Animation is based on robotics methods, and realistic movement simulation is not stored. Numerous modules are available in the form of analysis tools for factory planning and vehicle development. The system is available either as a stand-alone version or integrated in Unigraphics. Jack is used by John Deere, British Aerospace, Caterpillar, Volvo, and General Dynamics as by well as various universities.

The manikin SAFEWORK was conceived for workplace design in factory planning and product design. The anthropometric database is based on U.S. Army data and encompasses an anthropometric body-type generator that can create statistical test samples. The posture and movement simulation enables the movement of short chains in the model's arms, legs, and torso by means of inverse kinematics. Vision simulation, fixed accessibility areas, a joint-dependent comfort evaluation, maximum force calculation, and a center of gravity analysis are available as analysis tools. SAFEWORK is available solely in the form of an integration unit with CATIA by DassaultSystems. SAFEWORK is used by Chrysler, Boeing, and various universities and academies.

FIG. 8.17.   The DHM RAMSIS.

The manikin RAMSIS (Figure 8.17) was developed for vehicle design. The anthropometric database encompasses a physique typology that takes body measurement correlation, desired percentile models, an international database with more than 10 global regions, and a secular growth model into account. Posture and movement simulation is carried out by task-related animation, enabling the forecasting of the most probable posture. More than 80 functions for the analysis of vehicles and vehicle interiors are available (vision and mirror simulation, seat simulation, accessibility limits, comfort, belt analyses, etc.). The system is available as a stand-alone version for UNIX and Windows, a CAD integration unit with CATIA, and a programming library for independent applications. RAMSIS is used by more than 75% of all car manufacturers (Audi, BMW, Mercedes-Benz, General Motors, Ford, Porsche Volkswagen, Seat, Honda, Mazda, etc.) by aircraft manufacturers (Airbus), and by manufacturers of heavy machinery (e.g., Bomag).

SUMMARY

Anthropometrics is the scientific measurement and statistic analysis of data about human physical characteristics and the application (engineering anthropometry) of these data in the design and evaluation of systems, equipment, manufactured products, and human-made environments and facilities. In the last 100 years or so, knowledge and information about anthropometrics have been collected and made available in numerous scientific publications and various ISO standards.

The carrying-out of anthropometric surveys and studies on a representative population form the basis for all anthropometric tables. Here age, gender, race, and social differentiation must be taken into account during the planning of an anthropometric study of future users. This is taken into consideration in the ISO 15537.

Product-specific body sizes must also be selected during the preliminary stages of an anthropometric study. The most important of these are summarized in the ISO 7250:1996. A significant factor in the quality of study data, and the amount of effort involved in the study, is the use of the correct measuring technology (manual or body scanner). The use of innovative measuring technology by means of body scanners is discussed in the draft of the ISO 20685:2004.

The statistical analysis converts the results of the measurement process into usable data sets. The statistical combination of the data into corresponding percentile values (5th, 50th, and 95th percentiles) has become established as being representative and application-friendly. In the ISO 3411:1995, the application of anthropometric data in the design of vehicle cabins is treated in exemplary fashion.

Besides the application of tables, product development with the aid of 2D templates is also widespread. Here the biomechanics of the human being can also be taken into account during product development. Digital human models (DHM) represent the cutting edge of today's technology. They enable the efficient use of anthropometric data in product design (thanks to biomechanical simulations of the human being), direct use of 3D data, the consideration of modern statistical correlation methods, and product-related functionalities.

REFERENCES

Backwin, H., McLaughlin, S. D. (1964). *Increase in stature—Is the end in sight? Lancet.*

Blackwell, S., Robinette, K., Daanen, H., Boehmer, M., Fleming, S., Kelly, S., et al. (2002). *Civilian American and European surface anthropometry resource (CAESAR)*, Final report, Volume 2.

Bradtmiller, B., Gordon, C. C., Kouchi, M. K., Juergens, H. W., & Lee, Y. (2004). *Traditional anthropometry.* In N. J., Delleman, C. M., Haslegrave, & D. B., Chaffin (Eds.), *Working postures and movements* (pp 18–29). CRC Press LLC.

Bräuer, G. (1988). *Osteometrie.* In R., Martin, & R., Knussmann (Eds.), *Anthropologie: Handbuch der vergleichenden Biologie des Menschen.* Band 1. Fischer Verlag.

Chaffin, D. B. (2004). *Digital human models for ergonomic design and engineering.* In N. J., Delleman, C. M., Haslegrave, D. B., Chaffin (Eds.), *Working postures and movements* (pp. 426–430). CRC Press LLC.

DIN 33402 (1978). *Koerpermasse des Menschen.* Parts 1–3.

Helbig, K., & Juergens, H. W. (1977). *Entwicklung einer praxisgerechten Körperumrissschablone des sitzenden Menschen.* Forschungsbericht Nr. 187 der Bundesanstalt fuer Arbeitsschutz und Unfallforschung Dortmund.

Helbig, K., Juergens, H. W., & Reelfs, H. (1987). *Augen-Kopf-Körper-Interaktion in der Vertikalebene am Beispiel des Mensch-Maschine-Systems.* Ergonomische Studien Nr. 4. Bundesamt fuer Wehrtechnik und Beschaffung.

Hickey, D., Pierrynowski, M. R., & Rothwell, P. (1985). *Man-modeling CAD programs for workspace evaluations.* Lecture Manuscript, University of Toronto.

ISO 14738:2002. *Safety of machinery—Anthropometric requirements for the design of workstations at machinery.*

ISO 15534-1:2000. *Ergonomic design for the safety of machinery—Part 1: Principles for determining the dimensions required for openings for whole-body access into machinery.*

ISO 15534-2:2000. *Ergonomic design for the safety of machinery—Part 2: Principles for determining the dimensions required for access openings.*

ISO 15534-3:2000. *Ergonomic design for the safety of machinery—Part 3: Anthropometric data.*

ISO 15535:2003. *General requirements for establishing anthropometric databases.*

ISO 15537:2004. *Principles for selecting and using test persons for testing anthropometric aspects of industrial products and designs.*

ISO 20685:2004 (Draft). *3D scanning methodologies for internationally compatible anthropometric databases.*

ISO 3411:1995. *Earth-moving machinery—Human physical dimensions of operators and minimum operator space envelope.*

ISO 7250:1996. *Basic human body measurements for technological design.*

ISO 8559:1989. *Garment construction and anthropometric surveys—Body dimensions.*

Juergens, H. W. (1992). *Anwendung anthropometrischer Daten.* Direkte und indirekte Anwendung. In H., Schmidtke (Ed.), *Handbuch der Ergonomie* (A—3.3.1). Hanser Verlag.

Juergens, H. W., Aune, I. A., & Pieper, U. (1989). *Internationaler anthropometrischer Datenatlas*. Schriftenreihe der Bundesanstalt fuer Abreitsschutz. Fb 587.

Juergens, H. W., Habicht-Benthin, B., & Lengsfeld, W. (1971). *Körpermasse 20jähriger Männer als Grundlage für die Gestaltung von Arbeitsgerät, Ausruestung und Arbeitsplatz*. Bundesministerium für Verteidigung, Bonn.

Juergens, H. W., & Matzdorff, I. (1988). *Spezielle industrieanthropologische Methoden*. In R. Martin, & R. Knussmann (Eds.), (1988). *Anthropologie: Handbuch der vergleichenden Biologie des Menschen*. Band 1. Fischer Verlag.

Karoly, L. (1971). *Anthropometrie: Grundlagen der anthropologischen Methoden*. Fischer Verlag.

Kirchner, A., Kirchner, J.-H., Kliem, M., & Mueller, J. M. (1990). *Räumlich-ergonomische Gestaltung*. Schriftenreihe der Bundesanstalt fuer Arbeitsschutz. Verlag für neue Wissenschaft.

Lewin, T., & Juergens, H. W. (1969). *Üeber die Vergleichbarkeit von anthropometrischen Massen*. Zeitschrift Morph. Anthrop. 61.

Martin, R. (1914). *Lehrbuch der Anthropologie*. Jena: Verlag Gustav Fischer.

Martin, R., & Knussmann, R. (1988). *Anthropologie. Handbuch der vergleichenden Biologie des Menschen*. Fischer Verlag.

Porter, J. M., Case, K., Freer, M. T., & Bonney, M. C. (1993). *Computer-aided ergonomics design of automobiles*. In B. Peacock, W. Karwowski (Eds.), Automotive Ergonomics (pp. 43–78). Taylor & Francis.

Robinette, K. M., Daanen, H. A. M., & Zehner, G. F. (2004). Three dimensional anthropometry. In N. J. Delleman, C. M. Haslegrave, D. B. Chaffin (Eds.), *Working postures and movements* (pp. 29–49). CRC Press LLC.

Seidl, A. (1997). *Computer man models in ergonomic design*. In H. Schmidtke (Eds.). Handbuch der Ergonomie (Part A—3.3.3). Bundesamt für Wehrtechnik und Beschaffung. Hanser Verlag.

Seidl, A. (2004). *The RAMSIS and ANTHROPOS Human Simulation Tools*. In N. J. Delleman, C. M. Haslegrave, D. B. Chaffin (Eds.), Working Postures and Movements (pp. 445–453). CRC Press LLC.

Seidl, A., & Speyer, H. (1992). *RAMSIS: 3D-Menschmodell und integriertes Konzept zur Erhebung und konstruktiven Nutzung von Ergonomiedaten*. In VDI-Bericht 948. Das Mensch-Maschine-System im Verkehr (pp. 297–309). VDI-Verlag Düesseldorf.

9

Evaluation of Static Working Postures

Nico J. Delleman
TNO Human Factors, Paris Descartes University

Jan Dul
Erasmus University Rotterdam

INTRODUCTION

Pain, fatigue, and disorders of the musculoskeletal system may result from sustained inadequate working postures and repetitive movements that may be caused by poor work situations. Musculoskeletal pain and fatigue may themselves influence posture control, which can increase the risk of errors and may result in reduced quality of work or production and in hazardous situations. Good ergonomic design and proper organization of work are basic requirements to avoid these adverse effects. ISO 11226 (2000) contains an approach to determine the acceptability of static working postures. The standard has been written by Working Group 2 "Evaluation of working postures" of Subcommittee 3 "Anthropometry and Biomechanics" of Technical Committee 159 "Ergonomics" of the International Organization for Standardization (ISO).

ISO 11226—SCOPE

ISO 11226 (ISO, 2000) establishes ergonomic recommendations for different work tasks. It provides information to those involved in design, or redesign, of work, jobs, and products who are familiar with the basic concepts of ergonomics, in general, and working postures, in particular. The standard specifies recommended limits for static working postures without any or only with minimal external force exertion, while taking into account body angles and time aspects. It is designed to provide guidance on the assessment of several task variables, allowing the health risks for the working population to be evaluated. The standard applies to the adult working population. The recommendations will give reasonable protection for nearly all healthy adults. The recommendations concerning health risks and protection are mainly based on experimental studies regarding the musculoskeletal load, discomfort/pain, and endurance/fatigue related to static working postures.

RECOMMENDATIONS

The standard starts with general recommendations. It is stated that work tasks and operations should provide sufficient physical and mental variation. This means a complete job, with sufficient variation of tasks (for instance, an adequate number of organizing tasks, an appropriate mix of short, medium, and long task cycles, and a balanced distribution of easy and difficult tasks), sufficient autonomy, opportunities for contact, information, and learning. Furthermore, the full range of workers possibly involved with the tasks and operations should be considered, in particular, body dimensions. With respect to working postures, the work should offer sufficient variation between and within sitting, standing, and walking. Awkward postures, such as kneeling and crouching, should be avoided, whenever possible. It is stressed that measures meant to induce variations of posture should not lead to monotonous repetitive work.

EVALUATION PROCEDURE

The main part of the standard consists of specific recommendations to evaluate static working postures. The evaluation procedure considers various body segments and joints independently by one or two steps. The first step considers only the body angles (recommendations are mainly based upon risks for overloading passive body structures, such as ligaments, cartilage, and intervertebral disks). An evaluation may lead to the result *acceptable*, *go to step 2*, or *not recommended*. An evaluation result acceptable means that a working posture is acceptable only if variations of posture are also present (refer above). Furthermore, it is stated that every effort should be made to obtain a working posture closer to the neutral posture, if this is not already the case. An evaluation result go to step 2 means that the duration of the working posture will also need to be considered (recommendations are based on endurance data). Extreme positions of joints should be evaluated as *not recommended* (Figures 9.1 and 9.2).

FIG. 9.1. Typical working posture during sewing machine operation.

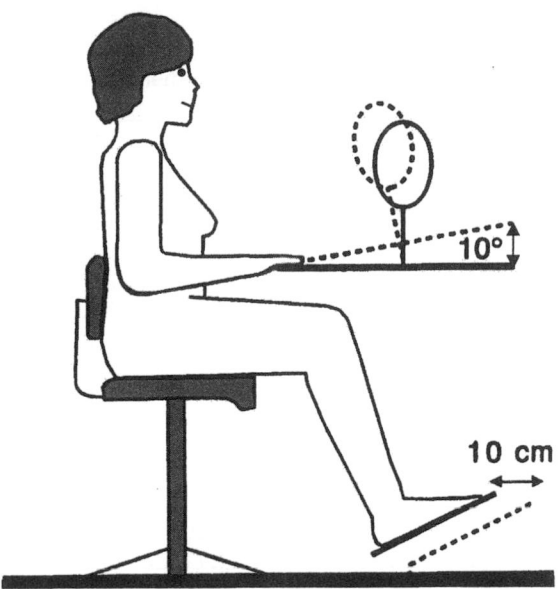

FIG. 9.2.   The four adjustments of the sewing machine workstation tested (2 table slopes × 2 pedal positions).

RELATIONSHIP TO OTHER STANDARDS

Evaluation procedures and recommendations concerning lifting and carrying, pushing and pulling, and handling of low loads at high frequency can be found in ISO 11228-1 (ISO 2003), ISO/DIS 11228-2 (ISO 2005a), and ISO/DIS 11228-3 (ISO 2005b), respectively.

EXAMPLE OF APPLICATION—SEWING MACHINE OPERATION

The working posture of sewing machine operators is characterized by a forward inclined head and trunk and a flexed neck (Figure 9.1). In an effort to improve working conditions through definition of recommendations for adjustment of the workstation, several operators worked for 45 minutes at each of four workstation adjustments, that is, 2 pedal positions × 2 table slopes (Figure 9.2). One pedal position was the average of the pedal positions at the operators' own industrial workstations, whereas the other was 10 cm farther away from the front edge of the table. The latter condition was introduced in order to allow the operator to sit closer to the table and to create a more upright trunk posture. One table slope was 0° (flat table), which is in accordance with the operators' own industrial workstations. The other table slope was 10° (inclined toward the operator), which was introduced in order to create a more upright head posture. In the following, the focus is placed on the neck flexion. All neck flexions measured were within the range of 0° to 25°, which is acceptable according to ISO 11226 (Figure 9.3). However, the average ratings on neck posture provide data for establishing more detailed evaluation criteria (Figure 9.3). A neck flexion of about 15° turned out to be better than smaller neck flexions. These and other new scientific data are described by Delleman, Haslegrave, and Chaffin (2004).

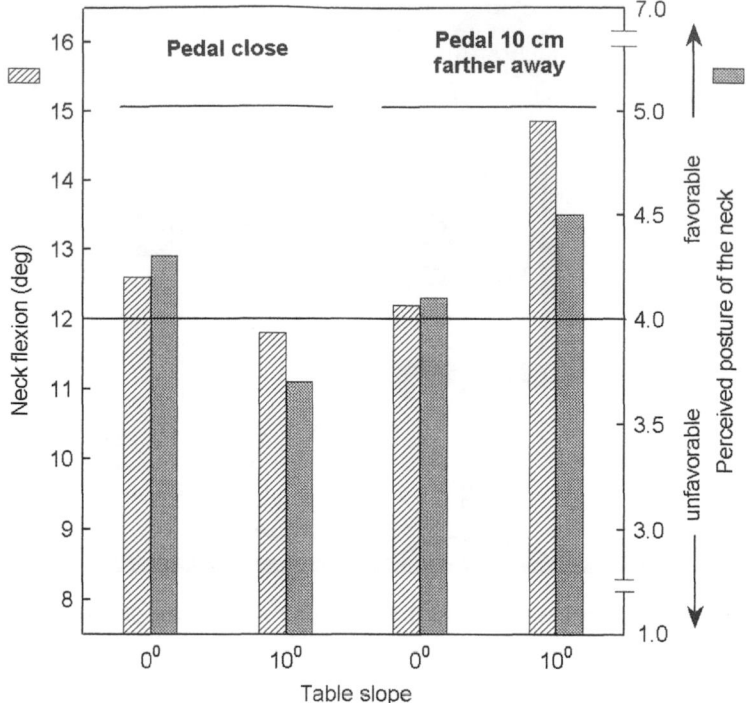

FIG. 9.3.   The four adjustments of the sewing machine workstation tested versus neck flexion and perceived posture of the neck (a score of 4 is exactly between favorable and unfavorable).

REFERENCES

Delleman, N. J., Haslegrave, C. M., & Chaffin, D. B. (Eds.). (2004). *Working postures and movements—Tools for evaluation and engineering.* Boca Raton, FL: CRC Press.

ISO (2000). *ISO 11226 Ergonomics—Evaluation of static working postures.* Geneva, Switzerland: International Organization for Standardization.

ISO (2003). *ISO 11228-1 Ergonomics—Manual handling—Part 1: Lifting and carrying.* Geneva, Switzerland: International Organization for Standardization.

ISO (2005a). *ISO/DIS 11228-2 Ergonomics—Manual handling—Part 2: Pushing and pulling.* Standard under preparation. Geneva, Switzerland: International Organization for Standardization.

ISO (2005b). *ISO/DIS 11228-3 Ergonomics—Manual handling—Part 3: Handling of low loads at high frequency.* Standard under preparation. Geneva, Switzerland: International Organization for Standardization.

10

Evaluation of Working Postures and Movements in Relation to Machinery

Nico J. Delleman
TNO Human Factors, Paris Descartes University

Jan Dul
Erasmus University Rotterdam

INTRODUCTION

About one third of the workers in the European Union (EU) are involved in painful or tiring positions for more than half of their time at work, and close to 50% of the workers are exposed to repetitive hand or arm movements (Paoli & Merllié, 2001). Pain and fatigue may lead to musculoskeletal diseases, reduced productivity, and deteriorated posture and movement control. The latter can increase the risk of errors and may result in reduced quality and hazardous situations.

EN 1005-4—SCOPE AND STATUS

The European Standard 1005-4 (CEN, 2005a) presents guidance when designing machinery or its components parts in assessing and controlling health risks due to machine-related postures and movements, that is, during assembly, installation, operation, adjustment, maintenance, cleaning, repair, transport, and dismantlement. The standard specifies requirements for postures and movements without any or with only minimal external force exertion. The requirements are intended to reduce the risks for nearly all healthy adults, but could also have a positive effect on the quality, efficiency, and profitability of machine-related actions.

The standard has been written by Working Group 4 "Biomechanics" of Technical Committee 122 "Ergonomics" of the European Committee for Standardization (CEN), under a mandate given to CEN by the European Commission and the European Free Trade Association (EFTA). The standard applies to all machines traded within the EU. The technique and principles could, however, be applied elsewhere and for other products as well.

RISK ASSESSMENT

The standard adopts a stepwise risk-assessment approach for assessing postures and movements as part of the machinery design process and provides guidance during the various design

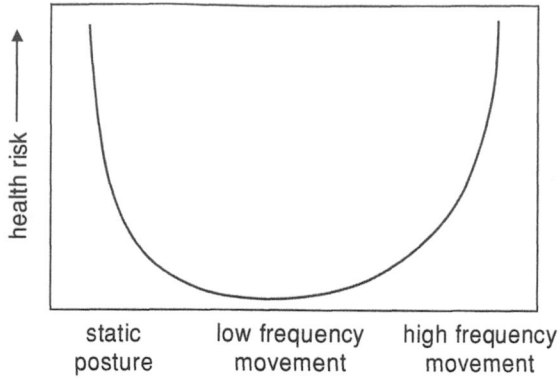

FIG. 10.1.   Flow chart illustrating the risk-assessment approach.

stages (as shown in Figure 10.1). The approach is based on the U-shaped model presented in Figure 10.2, which proposes that health risks increase when the task approaches either end of the curve, that is, if there is a static posture (with little or no movement) or if movement frequencies are high. A distinction is made between:

- Evaluation without users—when there is no full-size model prototype of the machinery or its parts currently available.
- Evaluation with users—when a full-size model prototype of the machinery or its parts is available.

Postures and movements are evaluated according to the following scheme:

- Acceptable—The health risk is considered low or negligible for nearly all healthy adults. No action is needed.
- Conditionally acceptable—There exists an increased health risk for the whole or part of the user population. The risk shall be analyzed together with contributing risk factors, followed as soon as possible by a reduction of the risks (i.e., redesign); or if that is not possible, other suitable measures shall be taken, for example, the provision of user guidelines to ensure that the use of the machine is acceptable.
- Not acceptable—The health risk cannot be accepted for any part of the user population. Redesign to improve the working posture is mandatory.

For postures or movements observed that are assigned the outcome conditionally acceptable, a second step of the evaluation procedure is introduced. Acceptability may depend on the nature and duration of the posture and period of recovery, on the presence or absence of body support, or on the movement frequency. Concerning static postures, the risk assessment is a simplified version of the procedure described in ISO 11226 (2000). Frequency-related risk assessment of movements is based on Kilbom (1994).

RELATIONSHIP TO OTHER STANDARDS

In the early phases of design (Figure 10.2), the standard refers to EN 614-1 (CEN, 1995) and EN-ISO 14738 (CEN/ISO, 2002). Concerning manual materials handling, force exertion, and

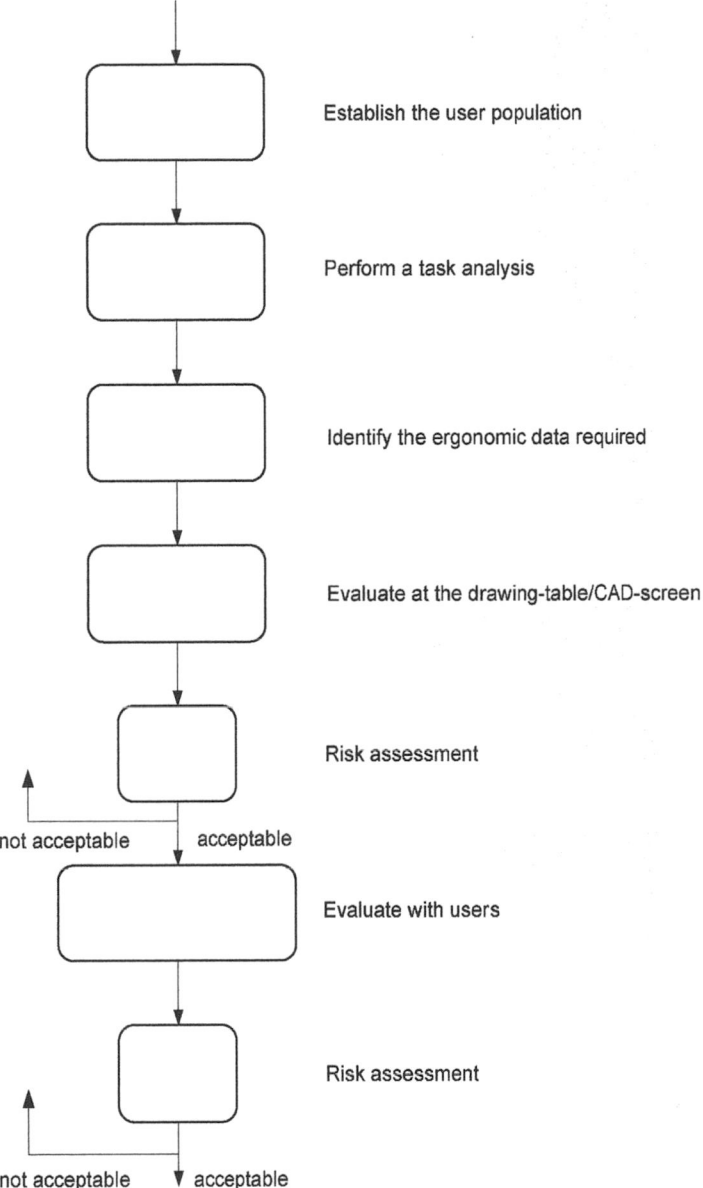

FIG. 10.2.   Diagrammatic model of the health risks associated with postures and movements.

repetitive work in relation to machinery, the designer is referred to the Parts 2, 3, and 5 of EN 1005 (CEN, 2003, 2002, 2005b).

EXAMPLE OF APPLICATION—PRESS OPERATION

In the metal industry, presses are used for forming objects, as well as for cutting superfluous material. Usually an operator has to reach forward in order to place an object into the press, as well as to take it out after the press action (Figure 10.3). In an effort to improve working

FIG. 10.3.   Forward reaching during press operation.

conditions through redesign of presses, several operators processed lightweight objects at four reach distances (70, 80, 90, and 100 cm). Each distance was tested for a period of 25 minutes. Movement frequencies of the trunk and the upper arm were around 10 per minute. Table 10.1 provides the range of forward inclination of the trunk, and the evaluation results by applying the standard as well as the average operators' ratings. It is concluded that the evaluation results according to the standard are supported by the operator's ratings.

Table 10.2 contains the range of forward elevation of the upper arms, and the evaluation results by applying the standard as well as the average operators' rating. According to the standard upper arm elevations into the 20° to 60° zone are acceptable if frequencies are below 10 per minute, and not acceptable if above. In the example, the movement frequency is around 10 per minute, and elevations are in the upper part of the zone (reach distances 80–100 cm). In such a case, it is wise not to draw conclusions immediately. Here, the operators' ratings after 25-minute testing support a conclusion that reach distances up to 90 cm are acceptable. Nevertheless, longer testing is recommended, as the standard states: "Working periods of long duration and high movement frequencies are known to increase health risks due to machine-related working postures and movements. Current knowledge only allows in part for

TABLE 10.1
Forward Inclination of the Trunk, and Evaluation Results According to the Standard and the
Operators for Various Reach Distances

Reach Distance	70 cm	80 cm	90 cm	100 cm
Trunk Inclination Forwards	0–10°	10–20°	20–30°	30–40°
Evaluation Result According to Standard	Acceptable	Acceptable	Not acceptable	Not acceptable
Evaluation Result According to Operators	Favorable	Favorable	Unfavorable	Unfavorable

TABLE 10.2
Forward Elevation of the Upper Arms, and Evaluation Results According to the Standard and the
Operators for Various Reach Distances

Reach Distance	70 cm	80 cm	90 cm	100 cm
Upper Arm Elevation Forwards	30–40°	40–50°	40–50°	40–50°
Evaluation Result According to Standard	Acceptable–Not acceptable	Acceptable–Not acceptable	Acceptable–Not acceptable	Acceptable–Not acceptable
Evaluation Result According to Operators	Favorable	Favorable	Favorable	Favorable–Unfavorable

quantitative evaluation of these risk factors." In addition the standard states: "Particularly if the machine may be used under the conditions mentioned above, it is strongly recommended that maximum improvement of the working posture be achieved, even if the working posture or movement is already assigned the outcome 'acceptable'." For this, it is recommended to do a detailed analysis using the newest scientific data concerning working postures and movements (Delleman, Haslegrave, & Chaffin, 2004).

REFERENCES

CEN. (1995). *EN 614-1 Safety of machinery—Ergonomic design principles—Part 1: Terminology and general principles.* Brussels, Belgium: European Committee for Standardization.

CEN. (2002). *EN 1005-3 Safety of Machinery—Human physical performance. Part 3: Recommended force limits for machinery operation.* Brussels, Belgium: European Committee for Standardization.

CEN. (2003). *EN 1005-2 Safety of machinery—Human physical performance—Part 2: Manual handling of machinery and component parts of machinery.* Brussels, Belgium: European Committee for Standardization.

CEN. (2005a). *EN 1005-4 Safety of Machinery—Human physical performance. Part 4: Evaluation of working postures and movements in relation to machinery.* Brussels, Belgium: European Committee for Standardization.

CEN. (2005b). *prEN 1005-5 Safety of Machinery—Human physical performance. Part 5: Risk assessment for repetitive handling at high frequency.* Standard under preparation. Brussels, Belgium: European Committee for Standardization.

CEN/ISO. (2002). *EN-ISO 14738 Safety of Machinery—Anthropometric requirements for the design of workplaces at machinery.* Brussels, Belgium: European Committee for Standardization or Geneva, Switzerland: International Organization for Standardization.

Delleman, N. J., Haslegrave, C. M., & Chaffin, D. B. (Eds.). (2004). *Working postures and movements—Tools for evaluation and engineering*. Boca Raton, FL: CRC Press.

ISO. (2000). *ISO 11226 Ergonomics—Evaluation of static working postures*. Geneva, Switzerland: International Organization for Standardization.

Kilbom, Å. (1994). Repetitive work of the upper extremity: Part I—Guidelines for the practitioner, and Part II—The scientific basis (knowledge base) for the guide. *International Journal of Industrial Ergonomics, 14*, 51–57, 59–86.

Paoli, P., & Merllié, D. (2001). *Third European survey on working conditions 2000*. Luxembourg: Office for Official Publications of the European Communities.

11

Standards on Physical Work: Demands in the Construction Industry

Henk F. van der Molen
Arbouw Foundation

Nico J. Delleman
TNO Human Factors, Paris Descartes University

INTRODUCTION

Building and construction is one of mankind's oldest activities. Looking at the pace of innovation in other sectors of industry, the construction industry should be characterized as conservative. Work still imposes physical demands, and work organization and working methods remain traditional. In the Netherlands, however, the past decades have shown a growing emphasis on working conditions. In the construction industry, *physical work demands* are the most important cause of absenteeism and disability. More than half of the cases of sick leave among Dutch construction site workers are the result of musculoskeletal complaints and disorders, mostly related to the lower back region (Arbouw, 1990–2002).

According to Hoonakker, Schreurs, Van der Molen, and Kummer (1992), in a group of about 1,000 Dutch construction workers, the following hazards were evident: repetitive work (61%), handling heavy materials (59%), poor working posture (52%), and high-force exertion (49%). In addition, occupational profiles based on the results of periodic medical examinations show a high percentage of complaints related to the musculoskeletal system. To reduce or eliminate health risks relating to the back, neck, and limbs, it is necessary to know which tasks particularly overload construction workers. Consequently, guidelines and standards on physical workloads are mandatory. In 1997, Koningsveld and van der Molen (1997) described the history and future of ergonomics in building and construction at the first international ergonomics symposium, as part of the 13th triennial congress of the IEA in Tampere. It was concluded that radical changes in working methods, work organization, and working conditions are a prerequisite for the future of companies and the well-being of their workers. Ergonomists can help in the improvement process by means of applying ergonomic standards, particularly with respect to physical work demands.

The section Guidelines on Physical Work Demands describes guidelines for assessing all aspects of the physical work demands of building tasks in the construction industry. The assessment of these demands on the basis of an adequate task analysis is a particularly important step in reducing physical work demands. It is then possible to reduce physical work demands using technical, organizational, or individual measures. The following section (Applications of the Guidelines in Standards for Professions) describes the process of applying these instructions in specific standards for construction industry professions—the Arbouw documents (A-documents). The final section contains Dutch legislation relating to lifting on construction sites, based on the A-documents under Application of the Guidelines in Standards for Professions.

GUIDELINES ON PHYSICAL WORK DEMANDS

Background

Most guidelines on physical work demands focus on the lifting of materials. The collective labor agreements for the construction industry in the Netherlands generally specify a maximum weight of 25 kg. In countries such as Australia, the United Kingdom and the United States, it is common practice not to consider the object weight as the only risk factor in quantitative guidelines. Therefore, the so-called Arbouw guidelines follow this multifactor approach by also taking into consideration, for instance, handling frequencies, task duration, or whatever is known to most affect physical load in particular work activities. Quantitative guidelines concerning lifting and carrying, pushing and pulling, static postures, and repetitive work have been developed on request of the parties involved in negotiating the labor agreements. The first edition of the Arbouw guidelines was based on a literature review and discussions among experts on physical workload in 1992 and 1993. Subsequently, 10 health and safety professionals evaluated the guidelines by actually working with them. In 2001, the guidelines were slightly revised in order to include the latest information from standards, reviews, and so forth. The health limits in the revised version of the Arbouw guidelines are based on Waters, Putz-Anderson, Garg, and Fine (1993), Mital, Nicholson, and Ayoub (1993), Kilbom (1994a, 1994b), ISO 11226 (2000), NF X 35-106 (1985), and prEN 1005-4 (2002).

Evaluation Scheme

All construction guidelines are based on the following evaluation scheme:

Green zone: basic, that is, nonincreased health, risk for $\geq$90% of men (P90 men).

Yellow zone: increased health risk: Action may be planned in stages; immediate action is to be preferred. For the guidelines on external force exertion, such as those for lifting and carrying, pushing and pulling, and repetitive arm work, this zone denotes that between 25% and 90% of men (P25 men–P90 men) are able to exert a certain force. Static postures and repetitive movements associated with an increased health risk are included in this zone in the case of a task duration of between 1 and 4 hours.

Red zone: strongly increased health risk: Immediate action is necessary. For the guidelines on external force exertion (refer to previous) the zone denotes that $\leq$25% of men (P25 men) are able to exert a certain force. Static postures and repetitive movements associated with an increased health risk are included in this zone in the case of a task duration of >4 hours.

The boundary between the green zone and the yellow zone is called the *action limit* (AL), whereas the boundary between the yellow zone and the red zone is called the *maximum Arbouw limit* (MAL). The AL–MAL concept has been used earlier for manual lifting (i.e., the National Institute for Occupational Safety and Health [NIOSH], 1981), whereas the green-yellow-red concept has also been described before, for example, for repetitive work (e.g., Hedén et al., 1993). The Arbouw guidelines use both concepts as one, with their own definitions, and in a consistent way for all kinds of physical work demands.

Criteria for AL and MAL

AL represents a health limit; that is, it should be the ultimate goal for action programs to get working conditions into the green zone. MAL was introduced for setting priorities within these programs. Strongly increased health risks (red) are to be tackled first, followed by those conditions associated with an increased health risk (yellow). Nonetheless, it should realized that an essential element of priority setting is also to check whether possible actions are reasonably practicable in economic, organizational, and technical terms.

The MALs for external force exertion (lifting and carrying, whole-body pushing and pulling, and repetitive arm work) are set to a level at which a majority of the workers (75%) are not even able to exert that particular force. Employers do not desire such a situation in particular as it excludes far too many workers from doing the jobs. So there is a strong motivation for setting in motion a process of change, in addition to the obvious reasons for reducing health risks. Currently most Dutch construction workers are male. So the guidelines on maximum forces are based on male population data. It would be possible to formulate guidelines for females as well.

For static postures and repetitive movements, task duration was used to set AL and MAL. Primarily tasks lasting more than 1 hour a day are included in the risk-assessment procedure, based on the results of a review by Kilbom (1994a, 1994b). For repetitive work, primary attention is on movement frequencies ≥ 2 and ≥ 10 per minute, respectively. Positions of body segments and joints are evaluated based on the European Committee for Standardization (CEN) and the International Organization for Standardization (ISO) standards. Those postures and movements observed that are associated with an increased health risk are classified in the yellow zone if the task lasts between 1 and 4 hours, and classified in the red zone if it lasts more than 4 hours. As a matter of course, durations of tasks loading the same body region are to be added together. It is recommended that each evaluation should be interpreted in relation to the level of complaints for the associated body region for the group of workers involved. Although the 4-hour limit, and to a lesser extent also the 1-hour limit, is arbitrary, the authors are of the opinion that the current procedure at least takes the risk factor task duration into consideration to some extent. If application of the guidelines does not disclose an increased health risk, while an increased level of physical complaints for the group of workers involved exists, further analysis is considered necessary.

Guidelines for Five areas of Physical Work Demands (Lifting, Pushing and Pulling, Carrying, Static Load, and Repetitive Work)

Guideline on Lifting (Mainly Based on NIOSH, 1991; Mital et al., 1993; and NF X 35-106, 1985)

The following conditions are *red:*

- Weight more than 25 kg when in a standing position
- Weight more than 10 kg when in a sitting, squatting, or kneeling position
- Weight more than 17 kg when lifting one-handed

- Horizontal location (H) more than 63 cm
- Asymmetry angle (A) more than 135°
- Vertical location (V) more than 175 cm or less than 0 cm
- Frequency (F) more than 15 lifts/minute

The first two tables here distinguish between symmetric and asymmetric lifting, where a lifting index ≤1 is *green*, a lifting index 1 to 3 is *yellow*, and a lifting index ≥3 is *red*. In the case of limited headroom the lifting index found in one of these tables has to be multiplied by a multiplier taken from the third table. An index greater than 1 means an increased health risk, concerning musculoskeletal complaints and disorders in particular. A greater index means a higher risk. The fourth table presents the guidelines for lifting in a sitting, squatting, or kneeling position, or for one-handed lifting.

Lifting index for *symmetric* lifting (asymmetry angle [A] 0°), vertical location (V) 175 cm, vertical travel distance (D) 175 cm, work and lifting duration 8 hours and no recovery time. The figures in parentheses are for lifting under optimal symmetric conditions, that is, vertical location (V) 75 cm, vertical travel distance (D) 0 cm, work and lifting duration 1 hour, and recovery time 7 hours. A lifting index ≤1 is *green*, a lifting index 1 to 3 is *yellow*, and a lifting index ≥3 is *red*.

		Frequency		
Weight	Horizontal Location (H)	5–9 Lifts/Minute	1–5 Lifts/Minute	<1 Lift/Minute
0–5 kg	≤25 cm	2.8 (0.5)	1.2 (0.3)	0.5 (0.3) Green
	25–40 cm	4.2 (0.7)	1.9 (0.5)	0.9 (0.4)
	40–63 cm	7.1 (1.7)	2.9 (0.8)	1.4 (0.6)
				Yellow
5–10 kg	≤25 cm	5.9 (0.9)	2.3 (0.6)	1.1 (0.5)
	25–40 cm	8.3 (1.5)	3.7 (1.0)	1.7 (0.8)
	40–63 cm	14.2 (2.3)	5.9 (1.5)	2.7 (1.3)
10–15 kg	≤25 cm	8.3 (1.4)	3.5 (0.9)	1.6 (0.8)
	25–40 cm	12.5 (2.2)	5.6 (1.4)	2.6 (1.2)
	40–63 cm	21.4 (1.9)	8.8 (2.3)	4.1 (1.9)
				Red
15–20 kg	≤25 cm	11.1 (1.9)	4.7 (1.2)	2.2 (1.0)
	25–40 cm	16.6 (2.9)	7.4 (1.9)	3.5 (1.6)
	40–63 cm	28.5 (4.7)	11.7 (3.0)	5.4 (2.6)
20–25 kg	≤25 cm	13.8 (2.3)	5.8 (1.5)	2.7 (1.3)
	25–40 cm	20.8 (3.7)	9.3 (2.4)	4.3 (2.0)
	40–63 cm	35.7 (5.8)	14.7 (3.8)	6.8 (3.2)

Lifting index for *asymmetric* lifting (asymmetric angle [A] 135°), vertical location (V) 175 cm, vertical travel distance (D) 175 cm, work and lifting duration 8 hours, and no recovery time. The figures in parentheses are for lifting under optimal, asymmetric conditions, that is, vertical location (V) 75 cm, vertical travel distance (D) 0 cm, work and lifting duration 1 hour, and recovery time 7 hours. A lifting index ≤1 is *green*, a lifting index 1 to 3 is *yellow*, and a lifting index ≥3 is *red*. (See next page.)

Weight	Horizontal Location (H)	Frequency (F)		
		5–9 Lifts/Minute	1–5 Lifts/Minute	<1 Lift/Minute
0–5 kg	≤25 cm	4.6 (0.8)	2.0 (0.5) Yellow	0.9 (0.5) Green
	25–40 cm	7.1 (1.3)	3.3 (0.8)	1.5 (1.7)
	40–63 cm	12.5 (2.0)	5.0 (1.3)	2.4 (1.1) Yellow
5–10 kg	≤25 cm	9.1 (1.6)	4.0 (1.1)	1.9 (0.9)
	25–40 cm	14.2 (2.6)	6.7 (1.7)	3.0 (1.4)
	40–63 cm	25.0 (4.0)	10.0 (2.6)	4.8 (2.3)
10–15 kg	≤25 cm	13.6 (2.5)	6.0 (1.6)	2.8 (1.4) Yellow
	25–40 cm	21.4 (3.9)	10.0 (2.5)	4.6 (2.1)
	40–63 cm	37.5 (6.0)	15.0 (4.0)	7.1 (3.4) Red
15–20 kg	≤25 cm	18.1 (3.3)	8.0 (2.1)	3.8 (1.8)
	25–40 cm	29.1 (5.1)	13.3 (3.4)	6.1 (2.9)
	40–63 cm	50.0 (8.0)	20.0 (5.3)	9.5 (4.5)
20–25 kg	≤25 cm	22.7 (4.1)	10.0 (2.7)	4.7 (2.3)
	25–40 cm	35.7 (6.4)	16.6 (4.2)	7.6 (3.6)
	40–63 cm	62.5 (10.0)	25.0 (6.6)	11.9 (5.7)

Limited Headroom Multiplier

	Multiplier
Fully Upright	1.00
95% Upright	1.67
90% Upright	2.50
85% Upright	2.60
80% Upright	2.78

Guidelines for Lifting in a Sitting, Squatting, or Kneeling Position or for Lifting One-Handed Infrequently.

	Weight	
	Green/Yellow Limit	Yellow/Red Limit
Sitting, Squatting, or Kneeling	4.5 kg	10.0 kg
One-Handed	7.5 kg	17.0 kg

For frequent lifting (≥2/minute) and longer durations (>1 hour/day), the guidelines on repetitive work apply.

Guideline on Pushing and Pulling (Mainly Based on Mital et al., 1993, and NF X 35-106, 1985)

Pushing/Pulling With the Whole Body While Walking. The following tables distinguish among pushing to set an object in motion (initial force exertion), pulling to set an object in motion (initial force exertion), and pushing or pulling to keep an object in motion (sustained force exertion). The guidelines are valid for two-handed pushing and pulling for a whole working day (8 hours), and an optimal height of the hands during force exertion (95–130 cm). A detailed analysis is necessary if slipping is likely to occur (or actually occurs) or in the case of high movement speed, awkward posture, asymmetric force exertion (one-handed and course changes), or a bad view of the surroundings (surface, obstacles etc.). All cells of the tables contain a pushing and pulling index, besides the green, yellow, or red evaluation result. An index greater than 1 means an increased health risk, concerning musculoskeletal complaints and disorders in particular. A greater index means a higher risk. Pushing to set an object in motion (initial force exertion). * = usually not possible because slipping is likely to occur.

	Frequency			
Force	2.5×/Minute- 1×/Minute	1×/Minute- 1×/5-Minutes	1×/5 Minutes- 1×/8 Hours	≤1×/8 Hours
0–25 kgf	1.0 Green	1.0 Green	0.9 Green	0.8 Green
25–30 kgf*	*1.2 Yellow*	*1.2 Yellow*	*1.1 Yellow*	1.0 Green
30–45 kgf*	*1.8 Yellow*	*1.7 Yellow*	*1.6 Yellow*	*1.5 Yellow*
45–50 kgf*	2.0 Red	*1.9 Yellow*	*1.8 Yellow*	*1.7 Yellow*
50–65 kgf*	2.6 Red	2.5 Red	2.3 Red	*2.2 Yellow*
>65 kgf*	>2.6 Red	>2.5 Red	>2.3 Red	>2.2 Red

Pulling to set an object in motion (initial force exertion). * = usually not possible because slipping is likely to occur.

	Frequency			
Force	2.5×/Minute- 1×/Minute	1×/Minute- 1×/5 Minutes	1×/5 Minutes- 1×/8 Hours	≤1×/8 Hours
0–20 kgf	1.0 Green	1.0 Green	1.0 Green	1.0 Green
20–40 kgf*	*2.0 Yellow*	*2.0 Yellow*	*2.0 Yellow*	*2.0 Yellow*
40–45 kgf*	2.3 Red	2.3 Red	*2.3 Yellow*	2.3 Yellow
45–50 kgf*	2.5 Red	2.5 Red	2.5 Red	2.5 Yellow
>50 kgf*	>2.5 Red	>2.5 Red	>2.5 Red	>2.5 Red

Pushing or pulling to keep an object in motion (sustained force exertion). * = usually not possible because slipping is likely to occur. — = combination of frequency and distance is not realistic. (See next page.)

		Frequency			
Force	Distance (m)	2.5/Minute-1/Minute	1/Minute-1/5 Minutes	1/5 Minutes-1/8 Hours	≤1/8 Hours
0–5 kgf	2–15	0.8 Green	0.4 Green	0.4 Green	0.3 Green
	15–30	—	0.8 Green	0.4 Green	0.3 Green
	30–60	—	—	0.6 Green	0.4 Green
5–10 kgf	2–15	1.5 Yellow	0.9 Green	0.7 Green	0.6 Green
	15–30	—	1.5 Yellow	0.8 Green	0.6 Green
	30–60	—	—	1.2 Yellow	0.9 Green
10–15 kgf	2–15	2.3 Red	1.3 Yellow	1.1 Yellow	0.9 Green
	15–30	—	2.3 Red	1.3 Yellow	0.9 Green
	30–60	—	—	1.8 Yellow	1.3 Yellow
15–20 kgf	2–15	3.1 Red	1.7 Red	1.5 Yellow	1.2 Yellow
	15–30	—	3.1 Red	1.7 Yellow	1.2 Yellow
	30–60	—	—	2.3 Red	1.7 Yellow
20–25 kgf*	2–15	3.8 Red	2.2 Red	1.9 Yellow	1.5 Yellow
	15–30	—	3.8 Red	2.1 Yellow	1.5 Yellow
	30–60	—	—	2.9 Red	2.2 Red
25–30 kgf*	2–15	4.6 Red	2.6 Red	2.2 Red	1.8 Yellow
	15–30	—	4.6 Red	2.5 Red	1.8 Yellow
	30–60	—	—	3.5 Red	2.6 Red
>30 kgf*	2–15	>4.6 Red	>2.6 Red	>2.2 Red	>1.8 Yellow
	15–30	—	>4.6 Red	>2.5 Red	>1.8 Red
	30–60	—	—	>3.5 Red	>2.6 Red

Pushing and Pulling With the Whole Body While Staying on the Spot. The guidelines for setting an object in motion (refer to the first two tables under Guidelines for Five Areas on Pushing and Pulling) also apply to pushing/pulling with the whole body while staying on the spot.

Pushing and Pulling With the Upper Limbs. The following guidelines are valid if the postures of the trunk and the upper limbs are evaluated as green (refer to the guideline on repetitive work), if the hands do not reach further forwards than three fourths of the maximum reach distance (trunk upright), and if the hands are between pelvis height and shoulder height.

Guidelines for infrequent pushing and pulling with the upper limbs. A and B refer to the various types of force exertion distinguished in the guideline on repetitive work. (See page 216.)

	Force	
	Green/Yellow Limit	Yellow/Red Limit
A	7.5 kgf	17.0 kgf
B	19.0 kgf	43.0 kgf

For frequent pushing/pulling ($\geq$2/minute) and longer durations (>1 hour/day) the guidelines on repetitive work apply.

Guideline on carrying (Mainly Based on Mital et al., 1993)

The following condition is *red*:

- Weight more than 25 kg
- Weight more than 10.5 kg for infrequent one-handed carrying

The following guideline is valid for two-handed carrying for a whole working day (8 hours), and the hands at an optimal height (knuckle height and arms hanging down). All cells of the tables contain a carrying-index, besides the evaluation result green, yellow, or red. An index greater than 1 means an increased health risk, concerning musculoskeletal complaints and disorders in particular. A greater index means a higher risk.

Weight	Distance (m)	Frequency			
		3/Minute-1/Minute	1/Minute-1/5 Minutes	1/5 Minutes-1/8 Hour	$\leq$1/8 Hour
0–10 kg	$\leq$2.0	0.5 Green	0.5 Green	0.4 Green	0.4 Green
	2-8.5	0.8 Green	0.6 Green	0.5 Green	0.4 Green
10–15 kg	$\leq$2.0	0.8 Green	0.7 Green	0.7 Green	0.6 Green
	2–8.5	*1.2 Yellow*	0.9 Green	0.8 Green	0.6 Green
15–20 kg	$\leq$2.0	*1.1 Yellow*	1.0 Green	0.9 Green	0.8 Green
	2–8.5	*1.5 Yellow*	*1.2 Yellow*	1.0 Green	0.8 Green
20–25 kg	$\leq$2.0	*1.3 Yellow*	*1.2 Yellow*	*1.1 Yellow*	1.0 Green
	2–8.5	1.9 Red	*1.5 Yellow*	*1.3 Yellow*	1.0 Green
>25 kg	$\leq$2.0	>1.3 Red	>1.2 Red	>1.1 Red	>1.0 Red
	2–8.5	>1.9 Red	>1.5 Red	>1.3 Red	>1.0 Red

Guideline for Infrequent One-Handed Carrying (Standing or Walking Up to 90 m).

	Weight	
	Green/Yellow Limit	Yellow/Red Limit
Infrequent One-Handed Carrying (Standing or Walking)	6 kg	10.5 kg

Guideline on Static Postures (Mainly Based on ISO 11226, 2000)

It is recommended to use the following guidelines to begin with on tasks lasting longer than 1 hour (continuous or a total of distinct periods) per working day (N.B. shorter task durations cannot be considered safe in all cases). In the case that two or more tasks load the same body region (through static postures or repetitive work), the durations of these tasks are to be taken together. It is recommended to interpret the results of guideline application in relation to the complaints and disorders of the particular body region found for the group of employees involved. An evaluation result yellow and red becomes yellow in the case of a task duration of between 1 and 4 hours, whereas it becomes red in the case of a task duration >4 hours.

The guidelines on lower back, shoulder, and shoulder girdle, as well as on neck and upper back are based on a description of the actual posture (observed and measured) with respect to a reference posture, that is, a sitting or standing posture with a nonrotated upright trunk, a nonkyphotic lumbar spine posture, and the arms hanging freely, while looking straight ahead along the horizontal. Guidelines on sitting also include raised sitting.

Low Back

Trunk Inclination

<0°	*0–20°*	*20–60°*	*>60°*
Yellow/Red**	Green	Green/Yellow/Red*	Yellow/Red

- Asymmetric trunk posture (axial rotation or lateral flexion): *yellow/red*
- For sitting: kyphotic lumbar spine posture (valid for trunk inclination between 0° and 20°): *yellow/red*
- * = with full trunk support: *green*; without full trunk support: consult an expert for evaluating holding time – recovery time regimes (see ISO 11226, 2000); if that is possible, the evaluation result is *yellow* in the case of a task duration of between 1 and 4 hours, whereas it is *red* in the case of a task duration >4 hours (expert guess of the authors); if two or more tasks load the same body region, the durations of these tasks are to be taken.
- ** = with full trunk support: *green*

Shoulder and Shoulder Girdle

Upper Arm Elevation

0–20°	*20°–60°*	*>60°*
Green	Green/Yellow/Red*	Yellow/Red

- Upper arm retroflexion (i.e., elbow behind the trunk when viewed from the side of the trunk), upper-arm adduction (i.e., elbow not visible when viewed from behind the trunk), or extreme upper-arm external rotation: *yellow/red*
- Raised shoulder: *yellow/red*
- * = with full arm support: *green*; without full arm support: consult an expert for evaluating holding time − recovery time regimes (see ISO 11226, 2000); if that is not possible, the evalulation result is *yellow* in the case of a task duration of between 1 and 4 hours, whereas it is *red* in the case of a task duration >4 hours (expert guess of the authors); if two or more tasks load the same body region, the durations of these tasks are to be taken together.

Neck and Upper Back

Green
- Head inclination 0°–25° and neck flexion (i.e., head inclination minus trunk inclination) 0°–25°.

Green/Yellow/Red
- Head inclination 25°–85°
- With full trunk support, consult an expert for evaluating holding time—recovery time regimes (see ISO 11226, 2000); if that is not possible, the evaluation result is *yellow* in the case of a task duration of between 1 and 4 hours, whereas it is *red* in the case of a task duration >4 hours (expert guess of the authors); if two or more tasks load the same body region, the durations of these tasks are to be taken together
- Without full trunk support, the holding time for trunk inclination is critical, and should be evaluated.

Yellow/Red
- Head inclination <0°
- Changing to *green* in the case of full head support
- Head inclination >85°
- Neck flexion (head inclination minus trunk inclination) <0° (i.e., neck extension) or >25°
- Asymmetric neck posture, that is, axial rotation or lateral flexion of the head versus the upper part of the trunk

Other Joints.   The following conditions are *Yellow/Red*:

- Extreme joint positions
- For a kneeling position: no adequate knee protection
- For standing (except when using a buttock rest): flexed knee
- For sitting: knee angle >135°(180° = the upper leg in line with the lower leg), unless the trunk is inclined backwards and fully supported (as in car driving)
- For sitting: knee angle <90°(180° = the upper leg in line with the lower leg)

Pedal Operation.   The following conditions are *yellow/red*:

- Standing
- Sitting, leg-actuated
- Sitting, ankle-actuated, force exertion >5.5 kgf
- Sitting, ankle-actuated, pedal operating already by merely supporting foot weight

General.   The following conditions are *yellow*:

- Standing continuously ≥1 hour/day
- Standing for a total of distinct periods ≥4 hours/day

Guideline on Repetitive Work (Mainly Based on NF X 35-106, 1985, and prEN 1005-4, 2002)

It is recommended to use the following guidelines to begin with on tasks lasting longer than 1 hour (continuous or a total of distinct periods) per working day (N.B. shorter task durations cannot be considered safe in all cases). If two or more tasks load the same body region (through static postures or repetitive work), the durations of these tasks are to be taken together. It is recommended to interpret the results of guideline application in relation to the complaints and disorders of the particular body region found for the group of employees involved. An evaluation result of yellow and red becomes yellow in the case of a task duration of between 1 and 4 hours, whereas it becomes red in the case of a task duration >4 hours.

Low Back

Trunk Inclination

<0°	0°–20°	>20°
Yellow/Red 1	Green	Yellow/Red 1

- Asymmetric trunk posture (axial rotation and/or lateral flexion): *yellow/red*

  1. An increased health risk is present if ≥ 2/minute an evaluation result yellow and red is found. (N.B. Lower frequencies cannot be considered safe in all cases; for example, with longer task duration.)

Shoulder and Shoulder Girdle

Upper Arm Elevation

0°–20°	20°–60°	>60°
Green	Yellow/Red 1	Yellow/Red 2

- Upper-arm retroflexion (i.e., elbow behind the trunk when viewed from the side of the trunk), upper-arm adduction (i.e., elbow not visible when viewed from behind the trunk), or extreme upper-arm external rotation: *yellow/red* 2
- Raised shoulder: *yellow/red* 2 neck flexion (i.e., head inclination minus trunk inclination) 0°–25°.

  1. An increased health risk is present if ≥10/minute an evaluation result yellow and red is found. (N.B. lower frequencies cannot be considered safe in all cases, for example, with longer task duration.)
  2. An increased health risk is present if ≥2/minute an evaluation result yellow and red is found. (N.B. lower frequencies cannot be considered safe in all cases, for example, with longer task duration.)

Neck and Upper Back

Green
- Neck flexion (i.e., head inclination minus trunk inclination) 0°–25°.

Yellow/red:
- Neck flexion (head inclination minus trunk inclination) <0° (i.e., neck extension) or >25°(1)

- Asymmetric neck posture, that is, lateral flexion or extreme axial rotation of the head versus the upper part of the trunk (1).

1. An increased health risk is present if $\geq$2/minute an evaluation result yellow and red is found. (N.B. lower frequencies cannot be considered safe in all cases.)

Other Joints.   The following conditions are *yellow and red*:

- Extreme joint positions (1)
- For a kneeling position: no adequate knee protection
- For standing (except when using a buttock rest): flexed knee (1)
- For sitting: knee angle >135° (1) (180° = the upper leg in line with the lower leg), unless the trunk is inclined backwards and fully supported
- For sitting: knee angle <90° (1) (180° = the upper leg in line with the lower leg)

1. An increased health risk is present if $\geq$2/minute an evaluation result yellow/red is found. (N.B. lower frequencies cannot be considered safe in all cases.)

Pedal Operation.   The following conditions are *yellow and red*:

- Standing
- Sitting, leg-actuated

Force Exertion

A. General force limits, that is, one-handed or two-handed, sitting or standing, almost all directions

	Force	
A Frequency	*Green and Yellow Limit*	*Yellow and Red Limit*
$\geq$2 and <3 /minute	3.0 kgf	6.5 kgf
$\geq$3 and <4 /minute	2.0 kgf	4.0 kgf
$\geq$4 and <5 /minute	1.5 kgf	3.0 kgf
$\geq$5/minute	1.0 kgf	2.5 kgf

B. Specific force limits: fore and aft directions; upward and downward directions; force limits are also valid for combined fore and aft and upward and downward directions. For one-handed pinching and pedal operation, see NF X 35-106 (1985).

	Force	
B Frequency	*Green/Yellow Limit*	*Yellow/Red Limit*
$\geq$2 and <3 /minute	8.0 kgf	18.0 kgf
$\geq$3 and <4 /minute	5.0 kgf	11.5 kgf
$\geq$4 and <5 /minute	3.0 kgf	7.0 kgf
$\geq$5 /minute	2.0 kgf	4.5 kgf

APPLICATION OF THE GUIDELINES IN STANDARDS FOR PROFESSIONS

The Arbouw guidelines are intended for health and safety professionals. The guidelines show the user the most effective way to arrive at the green zone, or in the short term, the yellow zone. In other words, they provide guidance as to whether actions are best directed toward the workplace (fixtures, transport, machines, and tools and objects), the work organization, and the workers. The results of analyses regarding a certain profession or job are discussed with employers and employees in order to arrive at a decision regarding actions that are reasonably easy to implement. Subsequently, a clear document is written for employers, employees, commissioners of work, architects or manufacturers of equipment and tools. Fifteen A-documents (state-of-the-art documents) are available: lifting, paving, scaffolding, installing window panes, steel bending, finishing inclined roofs, finishing terrace roofs, carpentry window frames, carpentry roofs, joining bricks, bricklaying, cabling and piping, tile setting, laying natural stones, and ergonomics of crane cabins.

The recommendations in A-documents include only those solutions that are attainable in practice (http://www.arbouw.nl). Furthermore, the recommendations include improvements that have the greatest effect in the long term. Using the A-documents, employers and workers will be able to observe government regulations satisfactorily. NIOSH is now having some of these documents translated into English.

DUTCH LEGISLATION TAILORED ON CONSTRUCTION WORK

In the Dutch Working Conditions Act, the government lays down general rules regarding physical work demands. The main theme is that employers should prevent or at least limit the risks to the health and safety of workers from physically demanding work. The employer should devote special attention to these risks when drawing up a risk inventory and evaluation. The Working Conditions Act also gives specific rules for temporary and mobile work sites. According to these rules, a safety and health plan must be drawn up for work sites of any size; this plan must also include information about risks and measures to reduce the physical work demands.

The government developed specific policies to reduce physical work demands imposed by lifting and by manual work during construction work. In a policy guideline, the government announced specific legislation aimed at introducing standards for construction work, based mainly on the A-documents (see Application of the Guidelines in Standards for Professions). The full text of this legislation is given in the following.

Policy Guidelines Amendment to Working Conditions Legislation

Amendment of policy guidelines decree with respect to working conditions legislation, in connection with the adoption of a policy guideline relating to lifting on worksites

Article 1

The policy guidelines with respect to working conditions legislation[1] are amended as follows:

---

[1] Supplement to the Netherlands Government Gazette, (2001, p. 239), last amended by decree of April 24, 2002 (Netherlands Government Gazette 84).

Following policy guideline 5.2-2 Physical Demands in Day Nurseries, a policy guideline is inserted that reads as follows:

Policy Guideline 5.3 Lifting on Worksites

Basis: Article 5.3, paragraph 1, of the Working Conditions Decree.

The conditions of Article 5.3, paragraph 1, of the Working Conditions Decree, regarding lifting on work sites as referred to in Article 1.1, paragraph 2, under a, of the Working Conditions Decree, are met if the following guidelines are observed.

General

1. Manual lifting must be avoided or limited as far as reasonably possible.
2. The maximum allowable weight lifted manually by one person is 25 kg.
3. If there is sufficient space, the maximum allowable weight lifted manually by two persons jointly is 50 kg.
4. Until 1 January 2007, points 2 and 3 will not apply to work in the fitting and insulation sector, the furniture industry, the fitting-out sector, and the fitting of stairways in the carpentry sector, on condition that a covenant is concluded with these sectors by 1 January 2003 (31 December 2003 for the fitting and insulation sector). This covenant must contain an agreement that before 1 January 2007, measures will be introduced by which the physical work demands made on employees will be structurally reduced to a level comparable to the standards referred to in points 2 and 3.
5. Until 1 July 2003, points 2 and 3 will not apply to the fitting of exterior sun blinds, on condition that before 1 January 2003, a sector standardization will be developed by which the physical work demands made on employees will be structurally reduced to a level comparable to the standards referred to in points 2 and 3.
6. Until 1 July 2004, points 2 and 3 will not apply to the fitting, renovation and maintenance of lifts, on condition that before 1 January 2004, a sector standardization will be developed by which the physical work demands made on employees will be structurally reduced to a level comparable to the standards referred to in points 2 and 3.

Specific

7. Roof rolls heavier than 25 kg must be transported mechanically. In situations where that is technically or organizationally impossible, roof rolls that are not heavier than 35 kg can be transported manually—contrary to point 2—to a maximum of 5 rolls per person per day.
8. Pavers heavier than 4 kg must not be processed manually.
9. Bricklaying and gluing may only take place without mechanical lifting aids if
   * The free work space is at least 0.6 m;
   * The elements are lighter than 14 kg.;
   * Elements weighing 4 to 14 kg must be lifted using both hands and processed to a maximum of 1.5 m above and not below standing level.
   * Elements lighter than 4 kg must be processed to a maximum of 1.7 m and from at least 0.2 m above standing level.
10. With the exception of
    * Interior bricklaying or gluing directly under or from an upper floor or
    * Bricklaying or gluing at ground level.

11. Concrete reinforcing steel and tools for processing heavier than
    - 17 kg must not be lifted with one hand;
    - 20 kg must not be lifted from less than 50 cm above the ground.
12. Scaffold elements heavier than 23 kg must not be lifted and transported manually by one person.

Article II.   This decree will come into force on 1 January 2003.

This decree will be published in the Netherlands Government Gazette together with the explanation.

Explanation

Sphere of Application.   This policy guideline relates to lifting on work sites and refers to sites as described in the Working Conditions Decree. In Article 1.1, paragraph 2, under "a", a worksite is described as a temporary or mobile site where civil engineering works or building works are carried out, a nonexhaustive description of which is included in Appendix I to the Council Directive 92/57/EEC of June 24, 1992, relating to the minimum safety conditions regarding health and safety at temporary and mobile work sites (PbEG L 245).

In the a forementioned appendix, the following civil engineering works and building works are listed: 1. Excavation work; 2. Groundwork; 3. Construction; 4. Assembly and disassembly of prefabricated elements; 5. Fitting-out or equipping; 6. Refurbishment; 7. Renovation; 8. Repair; 9. Dismantling; 10. Demolition; 11. Upkeep; 12. Maintenance, painting and cleaning work; 13. Redemption.

By way of explanation, it should be noted that window cleaning is not included in "12. Maintenance, painting and cleaning work." Consequently, the policy guideline does not apply to the activities of window cleaners.

Objective, Starting Points.   The policy guideline objective is to avoid or limit the dangers of physical work demands on the health and safety of employees as far as reasonably possible. In this context, physical work demands are understood to mean the work posture, movements, or exertions, consisting of lifting, carrying, or moving, or supporting one or more loads in any other way (Article 1.1, paragraph 4, under "a", of the Working Conditions Decree). For the sake of brevity, the policy guidelines refer to lifting, where lifting, carrying, or moving or supporting one or more loads is meant. In this policy guideline, the A-documents of the Arbouw Foundation have been used as a starting point. These A-documents contain limit values and recommendations to reduce the physical work demands in the construction industry. In 1994, the "Lifting" A-document was published, which includes limit values and recommendations for the manual lifting of loads. In addition, specific A-documents have now been published for various professions and tasks in the construction industry. The following A-documents include passages relating to physical demands: Steel Bending, Paving, Fitting Window Panes, Finishing Inclined Roofs, Cabling and Piping, Carpentry Roofs, Carpentry Window Frames, Finishing Terrace Roofs, Scaffolding, Lifting, Tile Setting. The A-documents series is still being extended. It is therefore possible that when new A-documents are published, supplementary conditions will be added to this policy guideline. In principle, the lifting assessment system in the specific A-documents is identical to that in the "Lifting" A-document. Several striking deviations are incorporated in this policy guideline. The conditions of this policy guideline are met if the lifting and ergonomic recommendations described for the work in question in the A-documents are observed.

General.   The "Lifting" A-document argues that in ideal situations manual lifting must never take place. As a result, the policy guideline indicates that lifting must be avoided or limited as far as reasonably possible. Consequently, the starting point is that employers must prevent employees being exposed to health and safety hazards by physical work demands.

Nevertheless, in certain cases, which are provided for in Point 1 of the policy guideline, it is not possible to avoid this entirely. After all, there may be other pressing interests for which the employer is also responsible, which may be damaged excessively as a result. In that case, the employer should weigh up the interests involved and take a decision according to reasonableness. When weighing up the options, the technical, operational, and economic feasibility of the proposed measures should be considered in relation to the possible social consequences for employees (illness and disability). If heavy or large loads have to be lifted manually, the safety of the employee is paramount. This means that the following important factors must always be observed:

- Loads must be lifted with two hands.
- The load must be easy to grasp.
- The floor must be flat, stable, and not slippery.

In any case, the maximum liftable manual load for one person is 25 kg. The "Lifting" A-document contains even more requirements to obtain the optimal lifting situation. These requirements are derived from the NIOSH formula. In short, this boils down to the following: The maximum permissible weight may be lifted when the result of the NIOSH formula is optimal. The lower the result, and therefore the greater the deviation from the optimal situation, the lower the maximum manually lifted weight becomes. When circumstances are no longer optimal, the maximum liftable manual load (i.e., 25 kg.) will, in principle, be too heavy for manual lifting. The A-document therefore recommends applying the NIOSH formula when assessing the situation and using the result of this calculation as a guideline for the maximum liftable weight. Any loads exceeding the weight permitted by the formula pose a threat to health. In the construction industry, however, situations occur in which the limit value according to the NIOSH formula is exceeded, whereas there are no reasonable alternatives available. As regards these situations, the social partners have agreed that for the time being, values will be used that are feasible. For this purpose, the Maximum Arbouw Limit (MAL) is used in the "Lifting" A-document. This MAL value may increase to a maximum of three times the NIOSH limit. However, the upper limit of 25 kg is always maintained. For the time being, only the upper limit of 25 kg has been included in the policy guideline. However, that does not alter the fact that manual lifting should be avoided or limited as far as reasonably possible, whereas at least the MAL, and preferably the NIOSH limit, should be the goal.

In addition to lifting by one person, there are activities where it is usual, or at least possible, to carry out lifting activities with two persons. Examples include activities such as fitting frames, fitting window panes, or fitting (plastic) catchpits, cesspits, and street furniture (see the relevant A-documents). In these situations, a maximum jointly lifted weight of 50 kg is permitted. Here, too, the following important factors must always be observed:

- Loads must be lifted with two hands.
- The load must be easy to grasp.
- There must be satisfactory hand-load contact.
- The floor must be flat, stable, and not slippery.

Moreover, the surroundings must offer both persons sufficient space to choose an optimal position with respect to the load. In addition, the lifting task must be practically symmetric, enabling an even distribution of the load.

The recommendations for a further restriction of the lift weight if conditions are not optimal, as indicated previously for one person, also apply to lifting by two persons.

With regard to activities in the fitting and insulation sector, the furniture industry, and the fitting-out sector, as well as the fitting of wooden stairways in the carpentry sector, installing, renovating, and maintaining lifts and fitting exterior sun blinds, the provisions of points 2 and 3 of the policy guideline do not (yet) apply. However, the provisions of point 1 continue to apply, so in those cases, too, manual lifting must be avoided or limited as far as reasonably possible. This refers to activities where in accordance with the current state of science and technology, certain physical heavy work cannot yet be avoided.

As a result, this exception is subject to the condition that the fitting and insulation sector, the furniture industry, the fitting-out sector, and the carpentry sector will have reached agreements, within the framework of a covenant, to implement measures by January 1, 2007, to reduce the physical work demands on employees to the level included in this policy guideline. For the time being, the exception is valid until January 1, 2007, when the situation will be reconsidered.

The exception for lifts and exterior sun blinds is subject to the condition that by January 1, 2004, and 1 January 2003, respectively, these sectors will draw up standardizations to avoid and limit the physical demands involved in these activities. Agreements have been reached with these sectors to form a tripartite working group for that purpose. Subsequently, these sectors will be given 6 months extra for announcing and implementing these standardizations within the sector, before the policy guideline comes into force.

As regards the exception for the fitting of stairways, it has been assumed that in line with the "Stairway Fitting Work Instructions" of the Netherlands Carpenters' Association, a maximum of 12 flights of stairs are fitted daily by a two-person team (or a maximum of six flights of stairs per person), and that the working method referred to as "dompen" is used. "Dompen" is a method by which the fitter does not need to lift more than half the weight of the stairway. The exception refers to the fitting of stairways and does not refer to carrying parts of the stairway to the place within the dwelling where it is to be fitted, that is, unloading it from a lorry or container. The exception is included, because here, too, the current state of science and technology means that these activities cannot yet be avoided. It is intended to reach agreement with the sector to develop the necessary research activities in order to reduce the physical demands associated with fitting stairways. When the time comes, the question of whether this research can be used as a basis for more specific conditions relating to the fitting of stairways can be included in the policy guideline.

As regards the exception for exterior sun blinds, it has been assumed that standard-sized sun blinds are fitted by a team comprising at least two persons. The exception refers to hanging and fitting sun blinds in brackets fitted earlier for that purpose. It is also been assumed that sun blinds intended for windows, and so forth, on the ground floor are fitted with the aid of ladders, with employees standing on separate ladders. A tripartite working group is mapping out the specific bottlenecks associated with these activities. This group will propose solutions before January 1, 2003, to structurally reduce the physical work demands on employees in this sector to a level comparable to the standards mentioned in points 2 and 3 of this policy guideline.

As a rule, the installation of lifts takes place at a late stage of construction, when a building is more or less completed. That means that there are sometimes only limited possibilities to move, lift, or position heavy materials mechanically. This applies even more during maintenance and renovation. This exception is intended for actions that cannot yet be carried out in accordance with points 2 and 3 of this policy guideline. A tripartite working group is mapping out the specific bottlenecks associated with these activities. This group will propose solutions before January 1, 2004, to structurally reduce the physical work demands on employees in this sector to a level comparable to the standards mentioned in points 2 and 3 of this policy guideline.

Specific. Practice has shown that a limit of 25 kg for rolls of bituminous roof covering (roof rolls) would impose unacceptable limits on the length of the roll or the thickness of the material. The "Terrace Roofs" A-document indicates that a weight of 35 kg is an acceptable compromise when the work involves only occasional manual lifting and carrying of roof rolls. That has been included in this policy guideline.

The limit values for lifting materials in the other specific parts of this policy guideline have been copied from the "Paving," "Steel Bending," and "Scaffolding" A-documents. As regards the manual processing of pavers, a maximum permissible weight of 4 kg applies. This relates to the high frequency with which the pavers are exposed, which results in the weight making demands on the employee at an earlier stage.

The manual processing of glue blocks for bricklaying and building bricks of 14 kg or more is not permitted. This limit is also mentioned in collective labor agreements. For such elements, the use of technical lifting aids is therefore always necessary. Elements of 4 to 14 kg may only be processed manually if they are lifted with two hands and are processed at a maximum of 1.5 m above and not below standing level. Elements lighter than 4 kg may only be processed manually at a maximum of 1.7 m and from 0.2 m above standing level. Consequently, elements heavier than 4 kg may not be processed manually with one hand.

When laying the first and final layers of bricks internally between upper floors, it is not always possible to adhere to the aforementioned bricklaying heights. This is also the case when laying the first layers of bricks externally from ground level. This policy guideline contains an exception for such cases. Free work space is understood to mean the available depth for bricklayer and materials when laying is taking place from the scaffold or the minimum depth of the bricklaying console when laying is taking place from a console.

REFERENCES

Arbouw. (1990–2002). Surveys sick leave data in the Dutch construction industry.

Broersen, J. P. L., Bloemhoff, A., Van Duivenbooden, J. C., Weel, A. N. H. & Van Dijk, F. J. H. (1992). *Figures of the atlas of health and work perception in the construction industry* (in Dutch). Amsterdam: Arbouw.

Hedén, K., Andersen, V., Kemmlert, K., Samdahl-Høiden, L., Seppänen, H., & Wickström, G. (1993). Model for assessment of repetitive, monotonous work—RMW. In W. S. Marras, W. Karwowski, J. L. Smith, & L. Pacholski (Eds.), *The ergonomics of manual work* (pp. 315–317). Taylor & Francis: London Washington DC.

Hoonakker, P. L. T., Schreurs, P. J. G., Van der Molen, H. F., & Kummer, R. (1992). *Conclusions and evaluation of the research projects concerning the psychosocial workload of six professions in the construction industry* (in Dutch). Amsterdam: Arbouw.

ISO/DIS 11226. (2000). *Ergonomics—Evaluation of static working postures.* Geneva, Switzerland: ISO.

Kilbom, Å. (1994). Repetitive work of the upper extremity: Part I—Guidelines for the practitioner. *International Journal of Industrial Ergonomics, 14,* 51–57.

Kilbom, Å. (1994). Repetitive work of the upper extremity: Part II—The scientific basis (knowledge base) for the guide. *International Journal of Industrial Ergonomics, 14,* 59–86.

Koningsveld E. A. P., & van der Molen, H.F. (1997). History and future of ergonomics in building construction. In *Abstracts from the first international symposium on ergonomics in building and construction.* Part of the 14th IEA congress. Tampere, Finland.

Ministry of Social Affairs and Employment. (2002). *Amendment of policy guidelines decree with respect to working conditions legislation, in connection with the adoption of a policy guideline relating to lifting on work site.* The Hague, Netherlands.

Mital, A., Nicholson, A. S., & Ayoub, M. M. (1993). *A guide to manual materials handling*. London/Washington, DC: Taylor & Francis.

NF X 35-106. (1985). *Ergonomie—Limites d'efforts recommandées pour le travail et la manutention au poste de travail*. Paris: AFNOR.

NIOSH. (1981). *Work practices guide for manual lifting* (NIOSH Technical Report No. 81-122). Cincinnati, OH: U.S. Department of Health and Human Services, National Institute for Occupational Safety and Health.

prEN 1005-4. (2002). *Safety of machinery—Human physical performance—Part 4: Evaluation of working postures and movements in relation to machinery*. Brussels, Belgium: CEN.

Waters, T. R., Putz-Anderson, V., Garg, A., & Fine, L. J. (1993). Revised NIOSH equation for the design and evaluation of manual lifting tasks. *Ergonomics, 36*, 749–776.

12

Safety of Machinery—Human Physical Performance: Manual Handling of Machinery and Component Parts of Machinery

Karlheinz G. Schaub
Darmstadt University of Technology

INTRODUCTION

Manual Materials Handling (MMH) During Machine Operation

In industrialized as well as in developing countries the machinery sector is an important part of the engineering industry and may be one of the industrial mainstays of the economy. However, social costs arise from accidents or sick leave caused directly by hazardous machinery operations. Manual handling of machinery or components parts of machinery can lead to a high risk of injury to the musculoskeletal system if the loads to be handled are too heavy, or handled at high frequencies for long durations or in awkward postures. Manually applied effort is often required by operators working with machines for their intended purpose. Risks exist if the design of the machinery is not in accordance with ergonomic design principles.

When designing and constructing machinery where manual handling is required, the designer of machinery should ensure the placing on the market of safe and ergonomically designed machines. An ergonomic risk assessment in an early design stage of machinery and appropriate measures to reduce the risk by redesign if necessary may help to reach this aim. It also eliminates the need for cost and time intensive machine alterations later on at shop floor level, when health and safety inspectors discover hazardous situations during the intended machine operations.

Ergonomic Evaluation of Manual Materials Handling Tasks

In a worldwide overview many regulations and evaluation tools on manual materials handling exist. Though most of the regulations on manual materials handling are edited by national institutions of occupational health and safety, the regulations and associated evaluation tools differ from one country to the other, as it concerns their complexity, their level of protection and their way of presentation. In order to provide a general global common approach ISO

11228 (Ergonomics—Manual Handling, Parts 1 and 2) was created. Like national or regional regulations (e.g., EU-Directive [europa.eu.int] on manual materials handling (90/269/EEC) international standards address to employers and employees (Meyer et al., 1998).

However, for designers and manufacturers of machinery that involve manual materials handling activities no international regulations or guidelines exist at the moment. In the European Union the Machinery Directive bridges that gap.

The EU-Machinery Directive

Whereas existing national health and safety provisions providing protection against the risks caused by machinery must be approximated in the EU in order to ensure free movement on the market of machinery without lowering existing justified levels of protection in the Member States the provisions of the EU-Machinery Directive (98/37/EC) concerning the design and construction of machinery, essential for a safer working environment, shall be accompanied by specific provisions concerning the prevention of certain risks to which workers can be exposed at work, as well as by provisions based on the organization of safety of workers in the working environment.

The EU-Machinery Directive addresses to designers and manufacturers of machinery solely (not to employers and employees) and demands an EC declaration of conformity—represented by the CE-mark on a machine—to be carried out. An easy way to declare conformity is to pinpoint that machines had been designed in accordance with harmonized CEN standards that had been mandated by the European Commission (see earlier chapters on ergonomically relevant EU-Directives). EN 1005 ("Safety of Machinery—Human Physical Performance") highlights ergonomic requirements concerning the operator's physical effort during machine operation.

Part 2 of this standard deals with "Manual handling of machinery and component parts of machinery." The design criteria given in this standard can be used by the designer when making risk assessments.

EN 1005-2 provides relevant data for the aspects of working posture, load, frequency, and duration.

EN 1005-2 is of relevance for all designers and manufacturers of machinery that involve manual materials handling in any phase of a machine's "life cycle" (construction, transport and commissioning, use and decommissioning) for the "intended user population" in its "intended use" including a "foreseeable misuse."

EN 1005-2 helps to ensure compliance with the EU-Machinery Directive and reduces social costs arising from hazardous machinery operations (Ringelberg & Schaub, 1997; Schaub et al., 1996; Schaub, 1997; Schaub & Landau, 1998).

MANUAL HANDLING OF MACHINERY AND COMPONENT PARTS OF MACHINERY

Subsidiary Standards

Guidance to the work on EN 1005-2 was given by several type A standards mandated under the Machinery Directive:

- EN 292-1 "Safety of Machinery—Basic Concepts, General Principles for Design—Part 1: Basic Terminology, Methodology"

- EN 292-2 "Safety of Machinery—Basic Concepts, General Principles for Design—Part 2: Technical Principles and Specifications" give some rough information on general ergonomic aspects.
- EN 614-1, Safety of Machinery—Ergonomic Design Principles—Part 1: Terminology and General Principles
- EN 1050, Safety of Machinery—Risk Assessment
- EN 1005-1, Safety of Machinery—Human Physical Performance—Terms and Definitions

For a better understanding and proper application of EN 1005-2, it is highly recommended to read the previously mentioned standards first.

Structure of the Standard

EN 1005-2 consists of four main chapters and five annexes. Most of them are required and pre-defined due to drafting rules for CEN standards (e.g., chaps. 0–3 and Annex ZA: Introduction, Scope, Normative References, Definitions and Relationship between this European Standard and the EU Directive for Machinery. Chapter 4 represents the core of the standard: "Recommendations for the Design of Machinery and Component Parts Where Objects are Lifted, Lowered and Carried." See also ENV26385. Annexes A to D (population characteristics and system design, recommended thermal comfort requirements, risk assessment worksheets and bibliography) offer supplementary information.

A foreword introduces the standard. It presents information on the document status and the Technical Committee which has prepared the standard (CEN/TC 122 "Ergonomics"). This standard was mandated by the European Commission and the European Free Trade Association and supports essential requirements of EU Directive(s).

Reference is given to the other parts of EN 1005 "Safety of Machinery—Human Physical Performance":

- Part 1: Terms and Definitions
- Part 2: Manual Handling of Machinery and Component Parts of Machinery
- Part 3: Recommended Force Limits for Machinery Operation
- Part 4: Evaluation of Working Postures in Relation to Machinery
- Part 5: Risk Assessment for Repetitive Handling at High Frequency

Part 5 is still under preparation. This part (Part 2) and Part 4 are currently submitted to CEN enquiry. For definitions in that standard, refer to EN 1005-1.

Introduction and scope

The introduction describes general aspects of manual materials handling and the associated risk of injury for the musculoskeletal system. Parts of the text are taken over from EN 614-1 the corresponding type A standard. The standard requires machinery designers to adopt a three stage approach to:

- Avoid manual handling activities wherever possible
- Utilize technical aids
- Further reduce the inherent level of risk by optimizing handling activities

TABLE 12.1

Population Percentages in Relation to Measurement Criteria and the Object Mass

Options	Psychophysical data indicating tolerability capacity	Measurements of forces indicating limits	Measurements on the maximum metabolic ability limits
10 kg	99 % (F + M)	99 % (F + M)	99 % (F + M)
	99 % F	99 % F	99 % F
	99.9 % M	99.9 % M	99.9 % M
20 kg	95 % (F + M)	95 % (F + M)	95 % (F + M)
	90 % F	90 % F	80 to 85 % F
	99.9 % M	99.9 % M	99 % M
25 kg	85 % (F + M)	85 % (F + M)	85 % (F + M)
	75 % F	75 to 75 % F	70 % F
	99.9 % M	99.9 % M	99 % M

F: Female   M: Male

The standard applies to the manual handling of objects of 3 kg or more (and for carrying of less than 2 m). Lower weights are dealt within EN 1005-5. As addressed to designers and manufacturers, there is little focus onto the carrying of objects. The designers should eliminate the need to carry objects during machinery operation. Whenever necessary, carrying should be limited to a movement distance of one or two steps (less than 2 m). As mandated within the Machinery Directive, the standard provides information on ergonomic design and risk assessments concerning lifting, lowering, and carrying during the whole "life cycle" of a machine, that is, construction, transport and commissioning, use and decommissioning, disposal and dismantling.

The standard does not cover the holding and pushing or pulling of objects. Hand-held machines and manual materials handling while seated are excluded as well.

Recommendations for the Design of Machinery and Component Parts

In order to minimize the health risks emerging from manual handling of machinery or component parts, the designer should establish whether hazards exist. If hazards exist, a risk assessment should be carried out (chap. 4.3 of the Standard). In general, hazards should be removed by excluding the need for manual handling activities. Where this is not possible, technical aids for the handling of machinery and component parts should be provided, and machinery and component parts should be (re)designed in accordance to ergonomic design principles. When machinery, component parts, or technical aids introduce awkward postures, reference should be made to EN 1005-4. When pushing and pulling is introduced, EN 1005-3 should be considered.

Population Characteristics

Table 12.1 applies to the general working population. This information is in accordance with measurements of maximum energetic capacity, subjective estimation of tolerability limits, and objective measurements of physical capabilities.

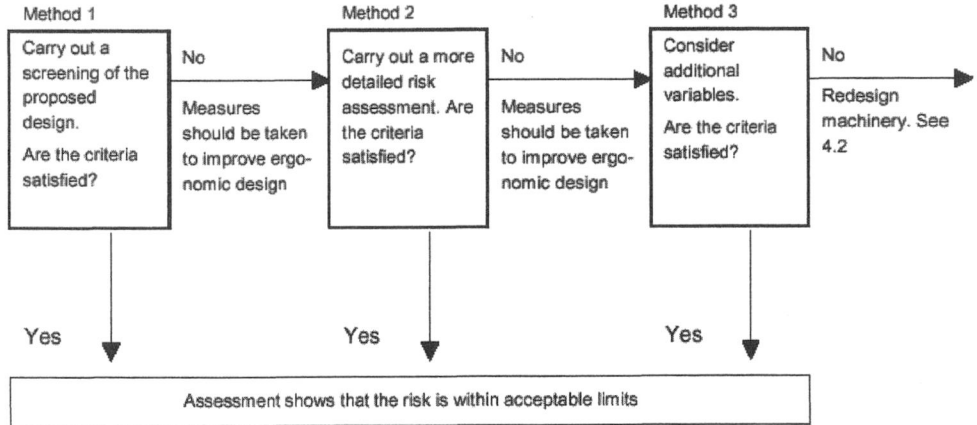

FIG. 12.1.   Flowchart identifying the stepwise approach to assessment.

System Design

For the system design, the following interrelated aspects from EN 1005-2 Annex A should be considered:

- **Constrained postures.** Thought should be given to the design and redesign of machines which lead to constrained working postures and monotonous work. In these circumstances discomfort and fatigue increases rapidly and muscular efficiency falls. In addition machinery should be designed to minimize static postures as far as possible.
- **Acceleration and movement accuracy.** Acceleration places higher force requirements and strain on the body. Movement accuracy increases the time needed for manual handling and increases muscle effort. Accuracy of precise positioning should be supplied by the design.
- **Discomfort, fatigue, and stress on the operator.** Research and experience in industry has shown that preventing discomfort, fatigue, and stress during physically demanding work reduces ill-health and increases output. It is important to consider three factors:
  a. Physiological effort required
  b. The amount of work in constrained postures
  c. The large variation in individual susceptibility to fatigue

Risk Assessments

General Information

The risk assessment model presented, involves three methods. These methods have the same basis, but differ in their complexity of appliance. The most efficient approach is to begin the risk assessment by applying Method 1 (the simplest procedure) and use Methods 2 or 3 only if the assumptions or operational situations identified in method 1 are not met (see Figure 12.1). It method 3 fails, chapter 4.2 of the standard offers recommendations for the redesign of machinery. Each method requires three steps to be carried out as described in the chapter "Worksheets."

TABLE 12.2
Hazards Taken Into Account in the Risk-Assessment Model of EN 1005-2

Hazards[a]	
Posture of the Trunk	Posture of the Arms
Frequency of Operation	Work Duration
One handed Operations[b]	Manual Handling by Two Persons[b]
Coupling Conditions (Worker ⇔ Load)	Additional physically demanding tasks[b]

[a]Assumptions for the working conditions are moderate ambient thermal environment, unrestricted standing posture, good coupling between feet and floor, smooth lifting, objects to be lifted are not very cold, hot, or contaminated.
[b]Only available in risk-assessment method 3.

The risk assessment takes respect of the parameters shown in Table 12.2.

The risk-assessment model follows widely the revised NIOSH (1991) approach (NIOSH, 1994; Andersson, 1999; Garg, 1989; Waters et al., 1999); supplemented by additional parameters from other international sources (Directorate of the Danish Labour, Inspection, Service for Machinary, 1986, FIOSH 1994, 1997) NF X 35-10.

Contrary to other national or international standards, EN 1005-2 (like all harmonized CEN standards realized for support of the Machinery Directive) addresses the designer and manufacturer of machinery only. Therefore, it does not focus on a "general working population" (or 90% of it), but aims toward the "intended user population" (see Table 12.1). On one hand, the target population of these standards could include elderly people and children as well as domestic applications of machinery; on the other hand, the target population could be a highly selected collective of young or medium-age male workers on, for example, an oil-drilling platform in the arctic sea.

Therefore a variable reference mass is introduced into the risk-assessment model. This might conflict on one hand with the struggle for equal chances of male and female workers when applying for jobs, which require considerable physical effort, as the gender is, next to age, one of the primary influences on physical capabilities. On the other hand, highly protective ergonomic regulations could have negative effects on the foundation and development of small and medium-size companies. The reference mass plays a similar role as the mass constant in the NIOSH approach. This reference mass is regarded to be the maximum weight that should be manipulated manually under "ideal" conditions. With reference to the hazards described in Table 12.3, this reference mass is lowered by means of multipliers.

Risk-Assessment Methods

Method 1 is a quick screening method. First, the user has to check whether the operational assumptions made (e.g., two-handed operation only, handling by one person only, and smooth lifting) in this risk assessment correspond with the work situation to be analyzed. If yes, he or she may choose one of three working examples characterized by a critical mass, a critical vertical mass displacement, or a critical frequency. If one of the working examples fits the work situation to be analyzed, the risk assessment was carried out successfully. If not all of the operational assumptions made fit and none of the working examples correspond to the work situation to be analyzed, the user should continue with method 2. Method 2 is an easy to handle method, as opposed to screening Method 1, which indicates risks.

TABLE 12.3
Reference Mass (M_{ref}) Taking Into Consideration the Intended User Population 6.

Field of Applicaiton	M_{ref} [kg]	Percentage of:			Population Group	
		F and M	F_{emales}	M_{ales}		
Domestic Use[a]	5	Data no available			Children and the Elderly	Total population
	10	99	99	99	General domestic population	
Professional Use (General)[b]	15	95	90	99	General working population	General working population
	25	85	70	90	including the young and elderly Adult working popuation	
Professional Use (Exceptional)[c]	30	Data not available			Special working population	special working population
	35					
	40					

[a]When designing a machine for domestic use, 10 kg should be used as a general reference mass in the risk assessment. If children and elderly are included in the intended user population, the reference mass should be lowered to 5 kg.

[b]When designing a machine for professional use, a reference mass of 25 kg should not be exceeded in general.

[c]Although every effort should be made to avoid manual handling activities or reduce the risks to the lowest possible level, there may be exceptional circumstances where the reference mass might exceed 25 kg (e.g., where technological developments or interventions are not sufficiently advanced). Under these special conditions, other measures have to be taken to control the risk according to EN 614 (e.g., technical aids, instructions, or special training for the intended operator group).

In comparison with Method 1, some additional risk factors can be taken into account in Method 2. The user has to calculate a risk index (RI) as a quotient of the actual mass to be handled and recommended mass limit. The recommended mass limit is composed from a reference mass and several multipliers (e.g., horizontal multiplier, vertical multiplier, and distance multiplier), which may be selected from precalculated tables. If the risk index is ≤ 0.85, the risk may be regarded as tolerable (green). A risk index between 0.85 and 1.0 indicates that a significant risk exists (yellow). It is recommended to apply Method 3 in order to identify how the risk may be reduced, to redesign the machinery, or to ensure that the risk is tolerable. A RI of ≥ 1.0 (red) means that redesign is necessary. The design can be improved by changing the situations that lead to low multipliers.

Method 3 is an extended assessment method, which assesses risk in a thorough way and is supplemented by additional risk factors not presented in Methods 1 and 2 (e.g., one-handed lifting; lifting by two persons; and additional physical workload next to lifting, lowering, and carrying). The calculation procedure used in this method is similar to that used in Method 2.

Worksheets

The risk-assessment procedures are described in detail in chapter 4.3 of the Standard. Worksheets for the three methods of the risk-assessment model are contained in Annex C and may be copied from there for application purposes (see Tables 12.4–12.8). The risk-assessment model consists of three methods increasing in the level of complexity. The first method is a quick

TABLE 12.4

Worksheet 1: Risk-Assessment for Method 1

Risk Assessment: Method 1—Screening by Means of Critical Values

EN 1005 Safety of Machinery—Human Physical Performance—Part 2: Manual Handling of Machinery and Component Parts of Machinery

This method provides a quick screening procedure to identify whether the handling operation represent a risk to the operator(s). Step 2 requires one of three critical operational situations (cases 1 to 3) to be selected. The limiting condition is that all assumptions for handling operation are fulfilled.

Step 1: Consider the reference mass.

Identify the intended user population and select the reference mass (M_{ref}) according to the intended user population (Table C.1)

Step 2: Carry out the risk-assessment.

Please tick the following criteria for the handling operation, if met:

❑ Two-handed operation only
❑ Unrestricted standing posture and movements
❑ Handling by one person only
❑ Smooth lifting
❑ Good coupling between the hands and the objects handled
❑ Good coupling between the feet and floor
❑ manual handling activities, other than lifting, are minimal
❑ The object to be lifted are not cold, hot or contaminated
❑ Moderate ambient thermal environment

If one or more of these criteria are not met, refer to Method 2.

If all criteria are met, then select one of the following critical variables, These apply to a work shift of 8 hours or less

Case 1   *Critical mass*

❑ The load the handled does not exceed 70% of the reference mass selected from Table C.1.
❑ Vertical displacement of the load is ≤ 25 cm and between hip and shoulder height.
❑ The trunk is uprigh and not rotated.
❑ The load is kept close to the body.
❑ The frequency of lifts is equal to or less than 0.00333 Hz (1 lift every 5 min).

Case 2   *Critical vertical mass displacement*

❑ The load handled does not exceed 60% of the reference mass selected from Table C.1.
❑ Vertical displacement of the load is not above shoulder height or below knee height.
❑ The trunk is upright and not rotated.
❑ The load is kept close to the body.
❑ The frequency of lifts is ≤ 0.00333 Hz (1 lift every 5 min).

Case 3   *Critical frequency*

❑ The load handled does not exceed 30% of the reference mass selected from Table C.1
❑ Vertical displacement of the load is ≤ 25 cm and between hip and shoulder height.
❑ The frequency of lifts is equal or less than 0.08 Hz (5 lifts every min)
❑ The trunk is upright and not rotated.
❑ The load is kept close to the body.
Or
❑ The load handled does not exceed 50% of the reference mass selected from Table C.1.
❑ Vertical displacement of the load is ≤ 25 cm and between hip and shoulder height.
❑ The frequency of lifts is ≤ 0.04 btz (2.5 lifts every min).
❑ The trunk is upright and not rotated.
❑ The load is kept close to the body.

Step 3:   Select the action required.

If the design fits one of the operational situations (cases 1 to 3) described above, the risk assessment has been carried out successfully.

If none of the operational situations are satisfied, or any of the criteria specified in step 2 are not met, either

⇒ Consider modifying or redesign the machinery or
⇒ Use a more detailed risk-assessment procedure to identify critical risk factors (Methods 2).

TABLE 12.5
Worksheet 2a: Risk Assessment Worksheet for Method 2: Part a

Risk Assessment: Method 2—Estimation by tables
EN 1005 Safety of Machinery—Human Physical Performance—Part 2: Manual Handling of Machinery and Component Parts of Machinery

---

Step 1: Consider the reference mass.
Identify the intended user population and select the reference mass (M_{ref}) according to the intended user population (Table C.1)

Step 2: Carry out the risk assessment.
Please indicate (tick), whether the handling operation meets the following criteria:
- ❑ Two-handed operation only
- ❑ Unrestricted standing posture and movements
- ❑ Handling by one person only
- ❑ Smooth lifting
- ❑ Good coupling between the feet and floor
- ❑ Manual handling activities, other than lifting, are minimal.
- ❑ The objects to be handled are not cold, hot, or contaminated.
- ❑ Moderate ambient thermal environment

If one or more of these criteria are not met, refer to Method 3.
If all criteria are met, then determine the level of risk by:
1. calculating the recommended mass limit (R_{ML2}) using the multipliers provided in the Table C.1
2. calculating the risk index (RI) as follows:

$$\text{risk index } (R_I) = \frac{\text{actual mass}}{R_{ML}} = -\frac{[\text{kg}]}{[\text{kg}]}$$

Step 3: Select the action required.
- $R_I \leq 0.85$ the risk may be regarded as tolerable.
- $0.85 < R_1 < 1.0 \Rightarrow$ significant risk exists. It is recommended that:
  - $\Rightarrow$ Method 3 is applied in order to identify how thr risk may be reduced or
  - $\Rightarrow$ The machinery be either redesigned or
  - $\Rightarrow$ Ensure, that the risk is tolerable.
- $R_I \geq 1.0$ Redesign is necessary. The design can be improved by changing the situations that lead to low multipliers.

---

screening procedure to assess the task. Method 2 has to be applied if the screening procedure indicates risks. This method tasks account of additional risk factors. It is advisable to begin the risk assessment by applying Method 1 (the simplest procedure) and use Method 2 or 3 only if the assumptions or operational situations identified in Method 1 are not met.

Each method requires three steps to be carried out:

Step 1:   Consider the reference mass (see Table 12.3).
Step 2:   Assess the risk factors according to the worksheet.
Step 3:   Identify the action required:
- No action is necessary, if risk level is tolerable.
- Redesign, if the risk level is not tolerable, or check that the risk is tolerable.
- Use a more complex risk-assessment method.

TABLE 12.6

Worksheet 2b: Risk Assessment Worksheet for Method 2: Part b

Risk Assessment: Method 2—Estimation by Tables

*EN 1005 Safety of Machinery—Human Physical Performance—Part 2: Manual Handling of Machinery
and Component Parts of Machinery*

Table C.2—Calculation of the recommended mass limit (R_{ML2})

Reference mass (M_{ref}) R_{ML2} =

| Reference Mass [kg] (see Table C.1) | | | | | | | M_{ref} |

Vertical multiplier (V_M)   ×

Vertical Location (cm)	0	25	50	75	100	130	>175	V_M
Factor	0.78	0.85	0.93	1.00	0.93	0.84	0.00	

Distance Multiplier (D_M)   ×

Vertical Displacement (cm)	25	30	40	50	70	100	>175	D_M
Factor	1.00	0.97	0.93	0.91	0.88	0.87	0.00	

Horizontal Multiplier (H_M)   ×

Horizontal Location (cm)	25	30	40	50	55	60	>63	H_M
Factor	1.00	0.83	0.63	0.50	0.45	0.42	0.00	

Asymmetric Multiplier (A_M)   ×

Angle of Asymmetry (°)	0	30	60	90	120	135	>135	A_M
Factor	1.00	0.90	0.81	0.71	0.62	0.57	0.00	

Coupling Multiplier (C_M)   ×

Quality of Grip	Good	Fair	Poor	C_M
Description	Load length ≤ 40 cm; Load height ≤ 30 cm; good handles or hand-held cutouts. Easy to handle loose parts and objects with wrap-around grasp and without excessive wrist deviation.	Load length ≤ 40 cm, load height ≤ 30 cm; and poor handles or Hand-held cutouts or 90° finger flexion. Easy to handle loose parts and objects with 90° finger flexion and without excessive wrist deviation.	Load length > 40 cm or; load height > 30 cm; or difficult to handle parts or sagging objects or asymmetric center of mass or unstable contents or hard to grasp object or use of gloves.	
Factor	1.00	0.95	0.90	

Frequency multiplier (F_M) dependent from work duration (d)   ×

		Frequency							F_M
	Hz	0.0033	0.0166	0.0666	0.1000	0.1500	0.2000	>0.2500	
	[lifts/min]	0.2	1	4	6	99	12	>15	
Work	d ≤ 1 h	1.00	0.94	0.84	0.75	0.52	0.37	0.00	
Duration (d)	1 h < d ≤ 2 h	0.95	0.88	0.72	0.50	0.30	0.00	0.00	
	2 h < d 8 h	0.85	0.75	0.45	0.27	0.00	0.00	0.00	

=

| $R_{ML2} = M_{ref} \times V_m \times D_M \times H_M \times A_M \times C_M \times F_M$ | = | | [kg] |

TABLE 12.7

Worksheet 3a: Risk Assessment Worksheet for Method 3: Part a

Risk Assessment: Method 3—Calculation by Formula

EN 1005 Safety of Machinery—Human Physical Performance—Part 2: Manual Handling of Machinery and Component Parts of Machinery

---

Step 1: Consider the reference mass.

Identify the intended user population and select the reference mass (M_{ref}) according to the intended user population (Table C.1)

Step 2: Carry out the risk assessments

Please indicate (tick), whether the handling operation meets the following criteria:

❑ Unrestricted standing posture and movements
❑ Smooth lifting
❑ Good coupling between the feet and floor;
❑ The objects to be handled are not cold, hot, or contaminated.
❑ Moderate ambient thermal environment

If one or more of the criteria are not met, consider ways of meeting each of the criteria. Refer to chapter 4 of this Standard.

If all criteria are met, calculate the recommended mass limit (R_{ML}).

Case 1   If the recommended mass limit (R_{ML2}) is already known (calculated during Method 2), then calculate the recommended mass limit (R_{ML}) as follows:

$R_{ML} = R_{ML2} \times O_M \times P_M \times A_1$ [kg],

where

O_M one-handed operation	if true $O_M = 0.6$	otherwise $O_M = 1.0$
P_M two-person operation	if true $P_M = 0.85$	otherwise $P_M = 1.00$
AT additional physically demanding tasks	if true $A_T = 0.8$	otherwise $A_T = 1.0$

Case 2 If the recommended mass limit (R_M) has *not* been calculated, then calculate the recommended mass limit (R_{ML}) as follows:

$R_{ML} = M_{ref} \times V_M \times D_M \times H_M \times A_M \times C_M \times F_M \times O_M \times P_M \times A_T$

The following definitions apply:

$V_M = 1 - 0.003|V - 75|$	if $V < 0$ cm, $V_M = 0.78$	if $V > 175$ cm. $V_M = 0$
$D_M = 0.82 + 4.5/D$	if $D < 25$ cm, $D_M = 1$	if $D > 175$ cm, $D_M = 0$
$A_M = 1 - (0.0032A)$		if $A > 135°$, $A_M = 0$
$H_M = 25/H$	if $H < 25$ cm, $H_M = 1$	if $H > 63$ cm, $H_M = 0$

M	the reference mass from Table C.1 in kg
V	vertical location of the load, in cm
D	vertical displacement of the load, in cm
H	horizontal location of the load, in cm
A	angle of asymmetry, in degree
C_M	coupling multiplier from Table C.1
F_M	frequency multiplier from Table C.2

O_M one-handed operation	if true $C_M = 0.6$	otherwise $O_M = 1.0$
P_M two-person operation	if true $P_M = 0.85$	otherwise $P_M = 1.0$
A_T additional physically demanding tasks	if true $A_T = 0.8$	otherwise $A_T = 1.0$

Calculate the risk index (R_1) as follows:

$$\text{Risk index } (R_I) = \frac{\text{actual mass}}{R_{ML}} = -\frac{\text{[kg]}}{\text{[kg]}}$$

Step 3: Select the action required.

- $R_I \leq 0.85$ the risk may be regarded as tolerable.
- $0.85 < R_, < 1.0$ Significant risk exists, it is recommended to:
  ⇒ redesign the machinery or
  ⇒ ensure, that the risk is tolerable.
- $R_I \geq 1.0$ Redesign is necessary. The design can be improved by changing the situations that lead to low multipliers.

TABLE 12.8
Worksheet 3b: Risk Assessment Worksheet for Method 3: Part b

Risk Assessment: Method 3—Calculation by Formula

EN 1005 Safety of Machinery—Human Physical Performance—Part 2:
Manual Handling of Machinery and Component Parts of Machinery

Table C.3—Coupling Multiplier (CM)

Quality of Grip	Good	Fair	Poor
Description	Load length ≤40 cm; load height ≤30 cm; good handles or hand-hold cut-outs. Easy to handle loose parts and objects with wrap-around grasp and without excessive wrist deviation.	Load length ≤40 cm; load height ≤30 cm; **and** poor handles or Hand-hold cut-outs **or** 90° finger flexion. Easy to handle loose parts and objects with 90° finger flexion and without excessive wrist deviation.	Load length >40 cm **or**; load height >30 cm; **or** difficult to handle parts or sagging Objects or asymmetric center of mass or unstable contents **or** hard-to-grasp object **or** use of gloves.
Factor	1.00	0.95	0.90

Table C.4—Frequency Multiplier (FM)

		\multicolumn{6}{c}{Work Duration d}					

Frequency		2 h < d ≤ 8 h		1 h < d ≤ 2 h		d ≤ 1 h	
[Hz]	[Lifts/minute]	V^a ≤ 75 cm	V ≥ 75 cm	V ≥ 75 cm	V ≥ 75 cm	V < 75 cm	V ≥ 75 cm
≤0.00333	≥0.2	0.85	0.85	0.95	9.95	1.00	1.00
0.00833	0.5	0.81	0.81	0.92	0.92	0.97	0.97
0.01666	1	0.75	0.75	0.88	0.88	0.94	0.94
0.03333	2	0.65	0.65	0.84	0.84	0.91	0.91
0.05000	3	0.55	0.55	0.79	0.79	0.88	0.88
0.06666	4	0.45	0.45	0.72	0.72	0.84	0.84
0.08333	5	0.35	0.35	0.60	0.60	0.80	0.80
0.10000	6	0.27	0.27	0.50	0.50	0.75	0.75
0.11666	7	0.22	0.22	0.42	0.42	0.70	0.70
0.13333	8	0.18	0.18	0.35	0.35	0.60	0.60
0.15000	9	0.00	0.15	0.30	0.30	0.52	0.52
0.16666	10	0.00	0.13	0.26	0.26	0.45	0.45
0.18333	11	0.00	0.00	0.00	0.23	0.41	0.41
0.20000	12	0.00	0.00	0.00	0.21	0.37	0.37
0.21666	13	0.00	0.00	0.00	0.00	0.00	0.34
0.23333	14	0.00	0.00	0.00	0.00	0.00	0.31
0.25000	15	0.00	0.00	0.00	0.00	0.00	0.28
>0.2500	>15	0.00	0.00	0.00	0.00	0.00	0.00

^{a} V is the vertical location.

ACKNOWLEDGMENTS

Dedicated to CEN/TC 122/WG 4, who developed this standard; especially to my colleges from the writing group.

REFERENCES

89/391/EEC. (1989). *Council Directive of 12 June 1989 on the introduction of measures to encourage improvements in the safety and health of workers at work.*

90/269/EEC. (1990). *Council Directive of 29 May 1990 on the minimum health and safety requirements for the manual handling of loads where there is a risk particularly of back injury to workers (fourth individual Directive within the meaning of Article 16 (1) of Directive 89/391/EEC).*

Andersson, Gunnar B. J. (1999). Point of view: Evaluation of the revised NIOSH lifting equation, a cross-sectional epidemiologic study. *Spine 24*(4), 395.

Directorate of the Danish National Labour Inspection Service for Machinery. (1986). Heavy lifts "backaches" compendium 5), Copenhagen.

EN 292-1. *Safety of machinery—Basic concepts, general principles for design—Part 1: Basic terminology, methodology.*

EN 292-2. (1991). *Safety of machinery—Basic concepts, general principles for design—Part 2: Technical principles and specifications.*

EN 614-1. *Safety of machinery—Ergonomic design principles—Part 1: Terminology and general principles.*

EN 614-2. *Safety of machinery—Ergonomic design principles—Part 2: Interactions between the design of machinery and work tasks.*

EN 1005-1. *Safety of machinery—Human physical performance—Part 1: Terms and definitions.*

EN 1005-2. *Safety of machinery—Human physical performance—Part 2: Manual handling of machinery and component parts of machinery.*

EN 1005-3. *Safety of machinery—Human physical performance—Part 3: Recommended force limits for machinery operation.*

EN 1005-4. *Safety of machinery—Human physical performance—Part 4: Evaluation of working postures in relation to machinery.*

EN 1005-5. *Safety of machinery—Human physical performance—Part 5: Risk assessment for repetitive handling at high frequency.*

EN 1050. *Safety of machinery—Risk assessment.*

EN ISO 7730. *Moderate thermal environments—Determination of the PMV and PPD indices and specification of the conditions for thermal comfort.*

Federal German Institution of Occupational Safety and Health (FIOSH). Guideline on Safety and Health Protection During Manual Handling. (1994/1997). Special edition 9 and 43 of the Series of the Federal German Institution of Occupational Safety and Medicine. Berlin.

Garg, A. (1989). An evaluation of the NIOSH Guidelines for Manual Lifting, with special reference to horizontal distance. *American Industrial Hygiene Assocation. J. 50*(3), 157–164.

ISO 11228-1. *Ergonomics—Manual handling—Part 1: Lifting and carrying.* ISO 11228-2, Ergonomics—Manual handling—Part 2: Pushing and pulling.

Meyer, J.-P., Colombini, D., Heden, K., Ringelberg, A., Viikari-Juntura, E., Boocock, M., Lobato, J. R., Schauh, Kh. (1998). European directive (90/269) for the prevention of risks in manual handling tasks. In Health Service Section International Social Security Association (Ed.), 2nd Internationales Symposium, *Low back pain in the health care profession—risk and prevention.* Hamburg, Germany, 10–11 September 1998.

National Institute of Occupational Safety and Health. (1991, May). *Scientific support documentation for the revised 1991 Lifting Equation.* (Technical contract reports). Cincinnati, OH: National Institute for Occupational Safety and Health. U.S. Department of Commerce, National Technical Information Service, Springfield, VA.

National Institute of Occupational Safety and Health (NIOSH). (1994, January). *Applications manual for the revised NIOSH Lifting Equation.* U.S. Department of Health and Human Services, Public Health Service, Centres for Disease Control and Prevention, National Institute for Occupational Safety and Health, Cincinnati, OH. 45226.

NF X 35-10. Acceptable limits of manual load carrying for one person.

Ringelberg, J. A., & Schaub, Kh. (1997). Background of prEN 1005, Part 2. Manual handling. In IEA '97 (Ed.), *From experience to innovation* (Vol. 3, p. 570). The 13th Triennial Congress of the International Ergonomics Association, Tampere, Finnland, 29 June–4 July 1997.

Schaub, Kh., Boocock, M., Grevé, R., Kapitaniak, B., & Ringelberg, A. (1996). *The implemenation of risk assessment models for musculoskeletal disorders in CEN standards.* In N. Fallentin, & G. Sjøgaard (Eds.), *Proceedings of the symposium Risk Assessment for Musculoskeletal Disorders* (pp. 73–74). Nordic Satellite Symposium under the auspices of ICOH '96, Copenhagen, Denmark 13–14 September 1996.

Schaub, Kh. (1997). Manual handling of machinery and component parts of machinery. In IEA '97 (Ed.), *From experience to innovation* (Vol. 3, pp. 574–577). The 13th Triennial Congress of the International Ergonomics Association, Tampere, Finnland, 29 June–4 July 1997.

Schaub, Kh., & Landau, K. (1998). The EU machinery directive as a source for a new ergonomic tool box for preventive health care and ergonomic workplace and product design. In P. A. Scott, R. S. Bridger, & J. Charteris, (Eds.), *Global Ergonomics, Proceedings of the Ergonomics Conference* (p. 219–224). Cape Town, South Africa, 9–11 September.

Waters, T. R., Baron, S. L., Piacitelli, L. A., Andersen, V. P., Skov, T. Haring-Sweeney, M. Wall, D. K., & Fine, L. J. (1999, February). Evaluation of the revised NIOSH lifting equation, *Spine 24*(4), 386–394.

http://europa.eu.int/comm/enterprise/newapproach/standardization/harmstds/reflist/machines.html

IV

Design and Evaluation of Manual Material Handling Tasks

13

Repetitive Actions and Movements of the Upper Limbs

Enrico Occhipinti
Daniela Colombini
Research Unit, Ergonomics of Posture and Movement (EPM)

INTRODUCTION

Working tasks that require manual repetitive actions at high frequency may cause the risk of fatigue, discomfort, and musculoskeletal disorders. A proper risk assessment and management should seek to minimize these health effects by taking into account a variety of risk factors including, in relation to the duration of exposure, the frequency of actions, the use of force, the postures and movements of the body segments, the lack of recovery periods, and other additional factors (Colombini et al., 2001).

To this regard, two parallel Standards are in preparation by CEN and ISO:

- PrEN 1005-5: Safety of Machinery—Human Physical Performance—Part 5: Risk Assessment for Repetitive Handling at High Frequency (CEN, 2004).
- ISO CD 11228-3: Manual Handling—Part 3: Handling of Low Loads at High Frequency (ISO, 2004)

Though the two mentioned drafts are devoted to different targets, they are conceptually similar and can be presented in the same context.

They are at an advanced stage of development from a technical point of view; at the moment writing, the CEN draft underwent a first positive official enquiry, and a second draft has been prepared taking into account the different official comments by CEN members. This second draft is expected now to undergo a second enquiry to be approved as a "harmonized" standard. The ISO draft is now ready to undergo a first official enquiry.

SCOPE OF THE STANDARDS

PrEN 1005-5: Risk Assessment for Repetitive Handling at High Frequency

This European (draft) Standard presents guidance to the designer of machinery or its component parts in controlling health risks due to machine-related repetitive handling at high frequency.

The Standard has been prepared to be a harmonized standard as defined by the European Union "Machinery Directive" and associated European Free Trade Association (EFTA) regulations. It applies only to designers of new machinery and assembly lines for professional use operated by the healthy adult working population. The machinery designer has to specify reference data for action frequency of the upper limbs during machinery operation: The Standard presents a risk-assessment method and gives guidance to the designer on how to reduce health risks for the operator.

ISO CD 11228-3: Handling of Low Loads at High Frequency

This International (draft) Standard establishes ergonomics recommendations for repetitive work tasks involving the handling of low loads at high frequency. The Standard will provide information for all those involved in the design or redesign of work, jobs, and products. It is designed to provide guidance on several task variables, allowing the health risks for the working population to be evaluated. It applies to the adult working population; the recommendations will give reasonable protection for nearly all healthy adults.

MAIN DEFINITIONS

Work task. An activity or activities required to achieve an intended outcome of the work system (e.g., stitching of cloth and the loading or unloading of pallets).

Repetitive task. Task characterized by repeated cycles.

Cycle. A sequence of technical actions that are repeated always the same way.

Cycle time. The time elapsing from the moment when one operator begins a work cycle to the moment that the same work cycle is repeated (in seconds).

Technical action (mechanical). Elementary manual actions required to complete the operations within the cycle, such as holding, turning, pushing, and cutting.

Repetitiveness. Quality of task when a person is continuously repeating the same cycle, technical actions, and movements in a significant part of a normal workday.

Frequency. The number of technical actions per minute.

Force. The physical effort of the operator required to execute the operations (related to the machinery).

Posture and movements. The positions and movements of body segment(s) or joint(s) required to execute the operations related to the machinery.

Recovery time. The period of rest following a period of activity in which restoration of a muscle can occur.

Additional risk factors. Other factors for which there is evidence of a causal or aggravating relationship with work-related musculoskeletal disorders of the upper limbs (e.g., vibration, local pressure, and cold).

CONTENTS OF THE STANDARDS

General Recommendations

Manual repetitive tasks, if unavoidable, should be designed in a way so that activities demanding high frequency can be performed adequately with respect to the force required, the posture of the limbs and the foreseeable presence of recovery periods. In addition tasks and related machines should be designed to allow for variations in movements. Additional factors (like vibration, cold, etc.) have to be considered.

Data from recent epidemiological studies on workers exposed to repetitive movements of upper limbs allow those involved in the design or redesign of workplaces, task and jobs to forecast, from exposure indexes, the occurrence of the consequent upper-limb work-related musculoskeletal disorders (UL-WMSDs); (Colombini et al., 2002; Occhipinti & Colombini, 2004). The adequate situation occurs when the exposure index corresponds to a forecast of occurrence of WMSDs as observed in a working population not exposed to occupational risks for the upper limbs (Colombini, Grieco, & Occhipinti, 1998; Hagberg et al., 1995; NIOSH, 1997).

Risk Assessment

When manual repetitive tasks are unavoidable, then a risk-assessment approach should be adopted. This should follow a four-step approach:

1. Hazard identification
2. Risk estimation by simple methods
3. Risk evaluation by detailed methods (if necessary)
4. Risk reduction

The international literature reports the "frequency of upper limbs action" as connected to other risk factors like force (the more the force, the lower the frequency), posture (the more the joint excursion, the longer the time necessary to carry out an action), and recovery periods (if well distributed during the shift, they increase the recovery of muscles) (Colombini et al., 2001). The technical action is identified as the specific characteristic variable relevant to repetitive movements of the upper extremities. The technical action is factored by its relative frequency during a given unit of time.

The hazard identification and simple risk estimation procedures are largely based on different experiences and proposals of the literature (Colombini, Occhipinti, & Grieco, 2002; Keyserling, Stetson, Silverstein, & Brower, 1993; Silverstein, Fine, & Armstrong, 1987; Schneider, 1995); the detailed risk evaluation procedures are substantially based on the OCRA Index method proposed by the Authors (Occhipinti, 1998; Colombini et al., 2002).

Due to the different scopes and targets, the two mentioned Standards have slight differences when presenting specific procedures for risk assessment: Those aspects will be separately and synthetically detailed in the following paragraphs.

PrEN 1005-5

Hazard Identification

The first stage of the risk assessment is to identify whether hazards exist which may expose individuals to a risk of injury. If such hazards are present, then a more detailed risk assessment is necessary.

In *PrEN 1005-5*, the "no-hazard" option (for the designer) is present when machinery and the related task imply: No cycles or a cyclic task in which perceptual of cognitive activities are clearly prevalent. For all the machinery and task combinations in which cyclic manual activities are foreseen, risk estimation shall be applied. To this end, the designer shall identify and count the technical actions (for each upper limb) needed to carry out the task (NTC); define the foreseeable duration of the cycle time (FCT); consider the foreseeable duration of work and frequency of recovery periods (generally duration of 240–480 minutes of a task during one shift with at least two usual breaks of 10 minutes are to be considered); consider the possibility of rotation on different tasks, when designing a machinery in the context of an assembly line.

Risk Estimation by Simple Methods (Method 1)

The presence of acceptable characteristics for all of the considered risk factors is verified. When the characteristics described are fully and simultaneously present, it is possible to affirm that exposure to repetitive movements is acceptable. Where one or more of the listed characteristics for the different risk factors are not satisfied, the designer shall use a more detailed evaluation. The acceptable characteristics of the risk factors are listed in Table 13.1. It is to be underlined that the final acceptable frequency of action per minute was set to 40, given that the designer should consider a reference organizational scenario (task duration of 240–480 minutes with at least two usual breaks of 10 minutes plus meal break during the shift) and not

TABLE 13.1
List of Acceptable Characteristics of the Risk Factors

Absence of force, or use of force at the same conditions exposed in EN 1005-3

    Absence of awkward postures and movements considering the same conditions exposed in prEN 1005-4 as summarized below:

- The upper-arm postures and movements are in the range between 0° and 20°.
- The articular movements of the elbow and wrist do not exceed 50% of the maximum articular range.
- The kinds of grasp are "power grip," or "pinch" lasting not more than 1/3 of the cycle time."

Low repetitiveness. This occurs when:
- The cycle time is more than 30 seconds.
- The same kinds of action are not repeated for more than 50% of the cycle time.

Absence of additional factors (physical and mechanical factors). This occurs when:
- The task should not include hand/arm vibration, shock (such as hammering), localized compression on anatomical structures due to tools, exposure to cold, use of inadequate gloves for grasping, etc.

Frequency of upper-limb actions (for each arm) is less than 40 actions/min.
- In order to compute the frequency of actions/min, use the following formula:
  $FF = NTC \times 60/FCT$, where:

      FF is the foreseeable frequency of actions per minute.
      FCT is foreseeable duration of the cycle time in seconds.
      NTC is the number of technical actions (for each upper limb) needed to carry out the task.

---

the "best" scenario (almost one break of 10 minutes every hour of repetitive work) that should lead to an higher acceptable frequency of actions per minute.

Risk Evaluation by Detailed Method (Method 2)

If the acceptable conditions underlined in the previous step are not satisfied, the designer shall describe more analytically each risk factor that interferes with the frequency of actions. Because different risk factors can be present in different combinations and degrees, it is possible to expect many levels of risk.

The level of risk is assessed with reference to the OCRA method (Colombini et al., 2002). The OCRA Index, when assessing a single repetitive task in a shift (monotask job), is given by the ratio between the foreseeable frequency (**FF**) of technical actions needed to carry out the task, and the reference frequency (**RF**) of technical actions, for each upper limb. This is a particular procedure for monotask jobs. For multitask jobs, one can refer to a specific annex (see also OCRA Index in ISO draft).

In this context, **OCRA Index = FF/RF**.

The foreseeable frequency (number per minute) of technical actions needed to carry out the task (**FF**) is given by the formula already reported in Table 13.1.

The following formula calculates the reference frequency (numbers per minute) of technical actions (**RF**) on a work cycle base:

$$RF = CF \times Po_M \times Rc_M \times Ad_M \times Fo_M \times (Rc_M \times Du_M),$$

where:

CF = "constant of frequency" of technical actions per minute = 30,
Po_M; Re_M; Ad_M; Fo_M = multipliers for the risk factors postures, repetitiveness, additional, force,
Rc_M = multiplier for the risk factor "lack of recovery period,"
Du_M = multiplier for the overall duration of repetitive task(s) during a shift.

When designing a machinery-related task, evaluate reference frequency of the technical actions within a work cycle that is representative of the task under examination. The analyses shall include the main risk factors that the designer can influence with the consequent choice of a specific multiplier for each risk factor. These multipliers will decrease from 1 to 0 as the risk level increases. The risk factors and the corresponding multiplier, influenced by the designer, are:

- Awkward or uncomfortable postures or movements (posture multiplier) (Po_M)
- High repetition of the same movements (repetitiveness multiplier) (Re_M)
- Presence of additional factors (additional multiplier) (Ad_M)
- Frequent or high-force exertions (force multiplier, Fo_M).

The other factors considered in the formula (**$Rc_M \times Du_M$**) are generally out of the direct influence of the designer, and consequently they will be considered in this context as a constant, reflecting a common condition of repetitive task duration of 240 to 480 minutes/shift with two breaks of 10 minutes plus the lunch break. If other "daily repetitive task duration" or "breaks or recovery periods" scenarios are foreseen (less duration; more recovery periods) reference action frequency can be higher: Special tables are provided to this aim in an annex.

In practice, to determine the reference frequency (per minute) of technical actions (**RF**), proceed as follows:

- Start from CF (30 actions/minute).
- CF (the frequency constant) has to be weighted (by the respective multipliers) considering the presence and degree of the following risk factors: force (**Fo$_M$**), posture (**Po$_M$**), repetitiveness (**Re$_M$**), and additional factors (**Ad$_M$**).
- Apply the constant that considers the multiplier for repetitive task duration (**Du$_M$**) and the multiplier for recovery periods (**Rc$_M$**).
- The value obtained represents the reference frequency (per minute) of technical actions (**RF**) for the examined task in the common condition of at least two breaks of 10 minutes (plus the lunch break) in a shift of maximal 480 minutes.

Posture Multiplier (Po$_M$).   If the conditions described in Method 1 for posture are present, the multiplier factor is 1. If those conditions are not present, use the indications in Table 13.2 for obtaining the specific multiplier.

At the end of the analysis of awkward postures, choose the lowest multiplier Po$_M$ (that corresponds to the worst condition) between the posture and the movements of elbow, wrist, and hand (type of grip).

The designer, at this step, shall consider also shoulder postures and movements. To this end, the designer shall check that:

- The conditions in ISO EN 14738 and prEN 1005-4 are satisfied
- The arms are not held or moved at about shoulder level for more than 10% of cycle time (Punnett, Fine, Keyserling, & Chaffin, 2000)

If one of those two conditions occurs, a risk of shoulder disorders exists and should be accurately considered. However, at this moment there are no available data for identifying a

TABLE 13.2
Multiplier for Awkward Postures (Po$_M$)

Awkward Posture	Portion of the Cycle Time			
	Less Than 1/3 From 1% to 24%	*1/3 From 25% to 50%*	*2/3 From 51% to 80%*	*3/3 More Than 80%*
Elbow Supination ($\geq 60°$) Wrist Extension ($\geq 45°$) or Flexion ($\geq 45°$) Hand Pinch or Hook Grip or Palmar Grip (Wide Span)	1	0.7	0.6	0.5
Elbow Pronation ($\geq 60°$) or Flexion/Extension ($\geq 60°$) Wrist Radioulnar Deviation ($\geq 20°$) Hand Power Grip With Narrow Span (≤ 2 cm)	1	1	0.7	0.6

specific Po$_M$ for shoulders: Consequently Po$_M$ for shoulders cannot be included in the OCRA computation procedure.

Repetitiveness Multiplier (Re$_M$).   When the task requires the performance of the same technical actions of the upper limbs for at least 50% of the cycle time or when the cycle time is shorter than 15 seconds, the corresponding multiplier factor (Re$_M$) is 0.7. Otherwise Re$_M$ is equal to 1.

Additional Multiplier (Ad$_M$).   The main additional factors are (non-exhaustive list) use of vibrating tools, gestures implying countershock (such as hammering), requirement for absolute accuracy, localized compression of anatomical structures, exposure to cold, use of gloves interfering with handling ability, high pace completely determined by the machinery. If additional factors are absent for most of the task duration, the multiplier factor equals 1. Otherwise the additional factor multiplier Ad$_M$ equals:

- 1 if one or more additional factors are present for less than 25% of the cycle time
- 0.95 if one or more additional factors are present for 1/3 (from 25%–50%) of the cycle time
- 0.90 if one or more additional factors are present for 2/3 (from 51%–80%) of the cycle time
- 0.80 if one or more additional factors are present for 3/3 (more than 80%) of the cycle time

Force Multiplier (Fo$_M$).   If the criteria described in Method 1 are satisfied, the multiplier is 1. If these conditions are not met, use Table 13.3 to determine the force multiplier (Fo$_M$) that applies to the average level of force, as a function of time.

The force level (upper row) is given as a percentage of the Maximal Isometric Force (F_b) as determined in EN 1005-3 (step A). As an alternative, a value derived from the application of the CR-10 Borg scale can be used (second row; Borg, 1998). Use a $Fo_M = 0.01$ when the technical actions require "peaks" above 50% of F_b or a score of 5 (or more) in CR-10 Borg scale for almost 10% of the cycle time. The values in the Table 13.3 can be interpolated if intermediate results are obtained.

Predetermined Value (Constant) for the Repetitive Task Duration Multiplier (Du$_M$) and the Multiplier for Recovery Periods (Rc$_M$).   Because the multipliers (Du$_M$ and Rc$_M$),

TABLE 13.3
Multiplier Relative to the Different Use of Force (Fo$_M$)

Force Level in % of F_b	5	10	20	30	40	≥ 50
CR-10 Borg Score	0,5 Very, very weak	1 Very weak	2 Weak	3 Moderate	4 Somewhat strong	≥ 5 Strong/very strong
Force Multiplier (Fo$_M$)	**1**	**0.85**	**0.65**	**0.35**	**0.2**	**0.01**

considered in the formula, are generally out of the direct influence of the designer, they are here considered as unique constant, reflecting a common condition as:

$Du_M = 1$ (multiplier for overall repetitive task duration of 240–480 minutes)
$Rc_M = 0.6$ (for a foreseeable presence of two breaks of 10 minutes and a lunch break in a repetitive task duration of 240–480 minutes per shift). Therefore: $(Rc_M \times Du_M) = 0.6$.

Final Evaluation by Method 2 and Criteria for Risk Reduction.   For jobs with a single repetitive task, the OCRA Index is obtained by comparing, for each upper limb, the foreseeable frequency (FF) of technical actions needed to carry out the repetitive task and the reference frequency (RF) of technical actions, as previously calculated. Table 13.4 supplies the relevant values of the OCRA Index to assess the risk in relation to the 3-zone rating system (green, yellow and red) and to decide for consequent actions to be taken.

The criteria of Table 13.4 were defined in relation to the available literature regarding both the occurrence of UL-WMSDs in working populations not exposed to repetitive movements of the upper limbs and the association between OCRA Index and the prevalence of persons affected (PA) by (one or more) UL-WMSDs. Details about the procedure used for identifying the critical values of OCRA Index are given in a specific annex of the draft. In synthesis, on the basis of recent studies (Occhipinti & Colombini, 2004), the association between the OCRA Index (independent variable) and the prevalence of persons affected (PA) by one or more UL-WMSDs (dependent variable) can be summarized by the following simple regression linear equation:

$$PA = 2.39 \ (\pm \ 0.14) \times OCRA.$$

On the other side, by using the PA variable in a reference not exposed population, reference limits were established starting from the 95th percentile (PA = 4.8%) as the "driver value" for the so-called green limit and from twice the 50th percentile (PA = 7.4%) as the "driver value" for the so-called red limit. Those "driver" values of PA expected in a reference working population (not exposed) have been compared with the regression equation at the level corresponding to the 5th percentile: In such a way, by adopting a prudential criterion of assessment of not acceptable (yellow) or at risk (red) results, it was possible to find the OCRA values corresponding, respectively, to the green and red limits and discriminating green, yellow, and red areas as reported in Table 13.4.

In practice:

- The green limit means that, just above that level, in the exposed working population are forecasted, almost in 95% of cases, PA values higher than the 95th percentile (PA = 4.8%) expected in the reference (not exposed) population.

TABLE 13.4
Classification of OCRA Index Results for Evaluation Purposes

OCRA Risk Index	Zone	Risk Evaluation
≤2.2	Green	Acceptable
2.3 to 3.5	Yellow	Conditionally acceptable
>3.5	Red	Not acceptable

- The red limit means that, just above that level, in the exposed working population are forecasted, almost in 95% of cases, PA values higher than twice the 50th percentile expected in the reference (not exposed) population.

Annexes

The prEN 1005-5 draft is completed by the following annexes:

- Annex A (informative): Identification of technical action
- Annex B (informative): Posture and types of movements
- Annex C (informative): Force
- Annex D (informative): Association between the OCRA Index and the occurrence of UL-WMSDs: criteria for the classification of results and forecast models
- Annex E (informative): Influence of recovery period and work time duration in determining the overall number of reference technical actions within a shift (RTA) and, consequently, the OCRA Index
- Annex F (informative): An application example of risk reduction in a monotask analysis
- Annex G (informative): Definition and quantification of additional risk factors
- Annex H (informative): Risk assessment by Method 2 when designing multitask jobs

ISO CD 11228-3

Hazard Identification

When determining whether hazard is present, attention should be given to the following factors: Repetition, Posture and Movement, Force, Duration and Insufficient Recovery, Additional Risk Factors (Object Characteristics, Vibration, Impact Forces, Environment, Work Organization, and Psychosocial Factors). For each factor, a brief statement explains when it is to be considered as a hazard.

Simple Risk Assessments

This method is useful for performing a simple assessment of monotask jobs. The risk-assessment procedure uses a specific checklist and evaluation model given in an annex.

There are four parts in the estimation procedure: (Part A) preliminary information describing the job task, (Part B) hazard identification and risk estimation checklist, (Part C) overall evaluation of the risk, and (Part D) remedial action to be taken.

The checklist adopts a six-step approach taking into account the four primary physical risk factors (repetition, high force, awkward posture and movements, and insufficient recovery) as well as any other additional risk factors that may be present. Initial consideration is given to the prevalence of work-related health complaints or work changes that may have been implemented by the operator or supervisor.

As a result of the overall classification of risk (Part C), the following action should be taken:

Green Zone: No action is required.
Yellow Zone: The risk shall be further estimated, analyzed together with contributing risk factors, and followed as soon as possible by redesign. Where redesign is not possible, other measures to control the risk shall be taken.
Red Zone: The work could be harmful. It is advisable to evaluate more accurately the task by Method 2. Action to lower the risk (e.g., re-design, work organization, and worker instruction and training) is necessary.

Risk Evaluation by Detailed Method

If the risk assessed by Method 1 is in the yellow or red zone, it is recommended to perform a more detailed risk assessment also for a better choice and follow-up of the remedial measures to be taken. If the job is composed of two or more repetitive tasks (multitask job), it is recommended to use the present method.

The risk evaluation is performed using the same procedures (OCRA method) presented for the prEN 1005-5 draft: The only relevant difference regarding the fact that here, as in the original OCRA method, the actual number of technical actions (ATA) carried out during the work shift and the number of reference technical actions (RTA; for each upper limb) are directly computed taking into account multipliers for "daily duration of repetitive work" and "recovery periods." In practice, the OCRA Index is given by the fomula:

$$\textbf{OCRA Index} = \frac{\text{number of technical actions actually carried out in the shift} \, (\textbf{ATA})}{\text{number of reference technical actions in the shift} \, (\textbf{RTA})}.$$

The overall actual number of technical actions carried out within the shift (ATA) can be calculated multiplying F_j for the net duration (D_J in minutes) of each repetitive task/s analyzed and summing the results of each repetitive task.

$$\textbf{ATA} = \Sigma(\textbf{F}_j \times \textbf{D}_j),$$

where:

D_j is the net duration (in minutes) of the task j;
F_j is the frequency of actions per minute of task j.

The following general formula calculates the overall number of *reference* technical actions within a shift: rta

$$\textbf{RTA} = \sum_{J=1}^{n} [\textbf{CF} \times (\textbf{Fo}_{Mj} \times \textbf{Po}_{Mj} \times \textbf{Re}_{Mj} \times \textbf{Ad}_{Mj}) \times \textbf{D}_j] \times (\textbf{Rc}_M \times \textbf{Du}_M),$$

where:

n = number of repetitive task/s performed during the shift.
j = generic repetitive task.
CF = "constant of frequency" of technical actions per minute = 30.
Fo_{Mj}; Po_{Mj}; Re_{Mj}; Ad_{Mj} = multipliers for the risk factors: force, postures, repetitiveness, additional in each j repetitive task.
Dj = net duration (in minutes) of the repetitive task j.
Rc_M = multiplier for the risk factor "lack of recovery period."
Du_M = multiplier according to the overall duration of all repetitive tasks during a shift.

All the multiplier factors are identical to those given in prEN 1005-5, with the exception of the Rc_M and Du_M multipliers that are determined by the criteria given in the following and detailed in Table 13.5 and Table 13.6.

TABLE 13.5
Elements for the Determination of the Recovery Period Multiplier (Rc_M)

Number of Hours Without Adequate Recovery	0	1	2	3	4	5	6	7	8
Multiplier Rc_M	1	0.90	0.80	0.70	0.60	0.45	0.25	0.10	0

TABLE 13.6
Elements for the Determination of the Duration Multiplier (Du_M)

Total Time (in Minutes) Devoted to Repetitive Tasks During Shift	<120	120–239	240–480	>480
Duration Multiplier Du_M	2	1.5	1	0.5

Recovery Period Multiplier (Rc_M)

A recovery period is a period during which one or more muscle–tendon groups are basically at rest.

The following can be considered as recovery periods:

1. Breaks (official or non official) including the lunch break
2. Visual control tasks
3. Periods within the cycle that leave muscle groups totally at rest consecutively for at least 10 seconds almost every few minutes

For repetitive tasks, the reference condition is represented by the presence for each hour of repetitive task, of work breaks of at least 8 to 10 minutes consecutively or, for working periods lasting less than 1 hour, in a ratio of 5:1 between work time and recovery time.

In relation to these reference criteria, it is possible to consider how many hours, during the work shift, do not have an adequate recovery period. It requires the observation, one by one, of the single hours that make up a working shift: For each hour, a check must be made if there are repetitive tasks and if there are adequate recovery periods. On the basis of the presence or absence of adequate recovery periods within every hour of repetitive work, the number of hours with "no recovery" is counted. Consequently it is possible to determine the Rc_M multiplier according to Table 13.5.

Overall Duration of Manual Repetitive Tasks and Duration Multiplier (Du_M).

Within a working shift, the overall duration of manual repetitive tasks is important to determine the overall risk for upper limbs. When repetitive manual tasks last for a relevant part (4 hours or more) of the shift, the Du_M is equal to 1. In some contexts, however, there may be differences with respect to this more "typical" scenario (e.g., regularly working overtime, part-time work, and repetitive manual tasks for only a part of a shift); the multiplier (Du_M) considers these changes with respect to usual exposure conditions. Table 13.6 gives the values of Du_M in relation with the overall duration of manual repetitive tasks.

Risk Index Calculation and Risk Evaluation.   The OCRA Index is obtained by comparing, for each upper limb, the ATA carried out during the work shift and the RTA.

The risk classification criteria (green, yellow, and red) are identical to those given in prEN 1005-5 and reported in Table 13.4.

Annexes

The ISO CD 11228-3 draft is completed by the following annexes:

- Annex A (informative): Method 1. Simple risk assessment
- Annex B (informative): Details on the OCRA method as used in this Standard (similar to most of annexes in prEN 1005-5)
- Annex C (informative): Risk reduction
- Annex D (informative): Scientific evidence

REFERENCES

Borg, G. A. V. (1998). *Borg's Perceived Exertion and Pain Scales*, Human Kinetic Europe.

CEN. (2004). PrEN 1005-5: Safety of machinery—Human physical performance—Part 5: Risk assessment for repetitive handling at high frequency. *Doc CEN/TC 122/WG4/Version 04/2004.*

Colombini, D., Grieco, A., & Occhipinti, E. (Eds). (1998). *Occupational musculo-skeletal disorders of the upper limbs due to mechanical overload. Ergonomics 41*(9) (Special Issue).

Colombini, D., Occhipinti, E., Delleman, N., Fallentin, N., Kilbom, A., & Grieco, A. (2001). Exposure assessment of upper limb repetitive movements: A consensus document. In W. Karwowski (Ed.), *International Encyclopaedia of Ergonomics and Human Factors* (pp. 52–66). London and New York: Taylor and Francis.

Colombini, D., Occhipinti, E., & Grieco A. (2002). *Risk assessment and management of repetitive movements and exertions of upper limbs: Job analysis, Ocra Risk Index, prevention strategies and design principles* (Vol. 2). Elsevier Ergonomics book series. Amsterdam: Elsevier.

Hagberg, M., Silverstein, B., Wells, R., Smith, M. S., Hendrick, H. W., Carayon, P., et al. (1995). *Work-related musculoskeletal disorders. A reference book for prevention.* I. Kuorinka & L. Forcier (Eds.), London/Philadelphia: Taylor and Francis.

Keyserling, W. M., Stetson, D. S., Silverstein, B., & Brower, M. L. (1993). *A checklist for evaluating ergonomic risk factors associated with upper extremity cumula tive trauma disorders. Ergonomics 36*, 807–831.

ISO CD 11228-3. (2004). Manual handling. Part 3: Handling of low loads at high frequency. *ISO/TC 159/SC3/ Version 07/2004.*

NIOSH, Center for Diseases Control and Prevention. (1997). *Musculoskeletal disorders and workplace factors. A critical review of epidemiologic evidence for WMSDs of the neck, upper extremity and low back* (second printing). Cincinnati: U.S. Department of Health and Human Services.

Occhipinti, E. (1998). OCRA, a concise index for the assessment of exposure to repetitive movements of the upper limbs. *Ergonomics, 41*(9), 1290–1331.

Occhipinti, E., & Colombini, D. (2004). *Metodo OCRA: Aggiornamento dei valori di riferimento e dei modelli di previsione dell'occorrenza di UL-WMSDs nelle popolazioni lavorative esposte a movimenti e sforzi ripetuti degli arti superiori.* La Medicina del Lavoro, *95*(4), 305–319. Milano (JT).

Punnett, L., Fine, L. J., Keyserling, W. M., & Chaffin, D. B. (2000). Shoulder disorders and postural stress in automobile assembly work. *Scandinavian Journal of Work Environ Health, 26*(4), 283–291.

Schneider, S. (1995). OSHA's draft standard for prevention of work-related musculoskeletal disorders. *Appl. Occup. Environ. Journal, 10*(8), 665–674.

Silverstein, B. A., Fine, L. J., Armstrong, T. J. (1987). Occupational factors and carpal tunnel syndrome. *American Journal of Industrial Medicine, 11*, 343–358.

14

Ergonomics of Manual Handling— Part 1: Lifting and Carrying

Karlheinz G. Schaub
Darmstadt University of Technology

INTRODUCTION

Ergonomic and Economic Aspects of Manual Materials Handling

Disorders of the musculoskeletal system are common worldwide. They play an important role in occupational health and are one of the most frequent disorders. In a world characterized by the globalization of national economies, it is desirable to specify recommended limits for manual materials handling activities on an internationally accepted basis. This helps to improve and maintain healthy and safe working conditions which enhances quality and productivity and reduces sick leave and personal suffering from work-related muscular disorders. Harmonized international regulations also help to reduce advantages in economic competition based on poor and hazardous ergonomic design.

Due to the progress in ergonomic knowledge, a comprehensive risk assessment should be used to derive recommended limits for manual materials handling, and attention should be paid to the mass of objects in combination with working postures, frequency, and duration of manual handling which persons may be reasonably expected to exert when carrying out activities associated to manual handling. An ergonomic approach has a significant impact on reducing the risks of lifting and carrying. Of particular relevance is a good design of the work, especially the tasks and the workplace, which may include the use of appropriate aids.

Relation to Other Standards, Regulations, and Guidelines

The manual materials handling is one of the classical topics in the ergonomics of physical work-load. Many national, regional (e.g., European http://europe.eu.int/eur-lex/), and international (e.g., ILO) regulations and guidelines exist in that field.

In the 1970s, the first multifactorial methods for the calculation of load limits for the manual handling of materials were created in Germany (Schaub & Landau, 1997).

In the United States, the NIOSH work practices guide and the corresponding lifting equation was published in 1981 and revised in 1991. Currently, the NIOSH lifting-equation is being used in many other countries in order to derive limitations for manual materials handling tasks.

In Europe, manual materials handling is regulated by the "Council Directive 90/269/EEC of 29 May 1990 on the minimum health and safety requirements for the manual handling of loads where there is a risk particularly of back injury to workers." As this directive offers a good approach for a risk assessment of manual-handling activities, it served as a basis for the development of ISO 11228-1 and is described in detail later.

Next to this directive, manual materials handling is addressed in Europe by the European Union (EU) Machinery Directive (98/37/EC) that addresses designers and manufacturers of machinery, namely, EN 1005, a set of harmonized CEN standards that supports this directive.

LIFTING AND CARRYING

Part 1 of ISO 11778 provides a step-by-step approach to estimating the health risks of manual lifting and carrying; at each step, recommended limits are proposed. In addition, practical guidance for manual handling is given in the annexes.

The risk-assessment model presented estimates the risk associated with a manual material handling task. It also takes into consideration the hazards (unfavorable conditions) related to the manual lifting and the time spent with manual handling activities. Unfavorable conditions could be high masses to be manipulated or awkward postures required during the lifting process, such as twisted or bent trunks or far reach. This Standard provides information on both repetitive and nonrepetitive lifting.

Subsidiary Standards

Guidance to the work on ISO 11228-1 was given by several ISO and CEN Standards mentioned in the following and listed as normative references:

- ISO 7730, Moderate Thermal Environments—Determination of the PMV (predicted mean vote) and PPD (predicted percentage of dissatisfied) Indices and Specification of the Conditions for Thermal Comfort
- ISO 11226, Ergonomics—Evaluation of Working Postures
- ISO 14121, Safety of Machinery—Principles of Risk Assessment
- ISO/IEC Guide 51, Safety Aspects—Guidelines for Their Inclusion in Standards
- EN 614-1, Safety of Machinery—Ergonomic Design Principles—Part 1: Terminology and General Principles
- EN 614-2, Safety of Machinery—Ergonomic Design Principles—Part 2: Interactions Between the Design of Machinery and Work Tasks
- EN 1005-2, Safety of Machinery—Human Physical Performance—Part 2: Manual Handling of Machinery and Component Parts of Machinery
- EN-ISO 7250, Basic List of Definitions of Human Body Measurements for Technical Design

Structure

ISO11228-1 consists—next to the Foreword and Introduction—of four main chapters and four annexes (including bibliography). Most of them are required and predefined in drafting rules. Chapter 3 offers definitions that apply within this Standard.

Chapter 4, "Recommendations," represents the core of the Standard and contains the following subheadings:

- Ergonomic Approach
- Risk Assessment
- Risk Estimation and Risk Evaluation
- Manual Lifting
- Cumulative Mass of Manual Lifting and Carrying
- Risk Reduction
- Additional Considerations

Annexes A to C offer detailed supplementary information as the main part is kept relatively short.

ISO 11228 "Ergonomics—Manual handling" consists of the following parts:

- Part 1: Lifting and Carrying
- Part 2: Pushing and Pulling
- Part 3: Handling of Low Loads at High Frequency

Introduction and Scope

The introduction describes general aspects of manual materials handling and the associated risk of injury for the musculoskeletal system. An ergonomic approach that includes guidelines for a good design of work is implemented in this standard and aims to have a substantial impact on reducing the risk of lifting and carrying. Recommended limits in the standard consider the intensity, the frequency, and the duration of tasks and are derived from four major research approaches: the epidemiological, the biomechanical, the physiological, and the psychophysical. The Standard provides information for designers, employers, employees, and other persons engaged in work, job, and product design.

The Standard applies to the manual handling of objects of 3 kg or more in vocational and nonoccupational activities. Lower weights are dealt with in Part 3 of this Standard. The Standards provide information for designers, employers, employees, and others involved in work, job, and product design. The Standard applies to an 8-hour working day and to moderate walking speed on horizontal level surfaces.

The Standard does not cover the holding and pushing or pulling of objects. One-handed lifting and lifting by two or more people as well as hand-held machines and manual materials handling while seated are excluded as well.

Ergonomic Approach

In order to minimize the health risks emerging from manual handling, manual handling should be avoided whenever possible. Where this is not possible, a risk assessment should ensure that health risks are on an acceptable low level. Guidance for the ergonomic design of manual materials handling tasks is given in Annex A of this standard. Other considerations for the ergonomic design of manual materials handling tasks are mentioned in chapter 4, "The EU-Manual Handling Directive."

Risk Assessment, Risk Estimation, and Risk Evaluation

In general, a risk assessment consists of five stages: peril recognition, hazard identification, risk estimation, risk evaluation, and risk reduction. For further information, reference is made to ISO 14121, EN 1050, and ISO/IEC Guide 51. (For information about hazard identification, refer to Annex A of the Standard.) The risk assessment presented in the following takes into consideration the mass, the grip, and the position of the object, as well as the frequency and duration of the handling task.

The risk assessment consists of a step-by-step approach for the evaluation of manual lifting and carrying. When recommended limits are exceeded, the tasks analyzed should not be carried out manually, but adopted in a way that the risk-assessment model will be satisfied (see Figure 14.1).

Providing information and training for employees should not be used as the only measure to ensure a safe way of manual handling.

Risk estimation and evaluation is presented as a step-by-step model, consisting of five steps. When the questions asked in each step have to be answered with "no," adaptation of the manual handling task is necessary. (For appropriate actions refer to Annex A of the Standard.)

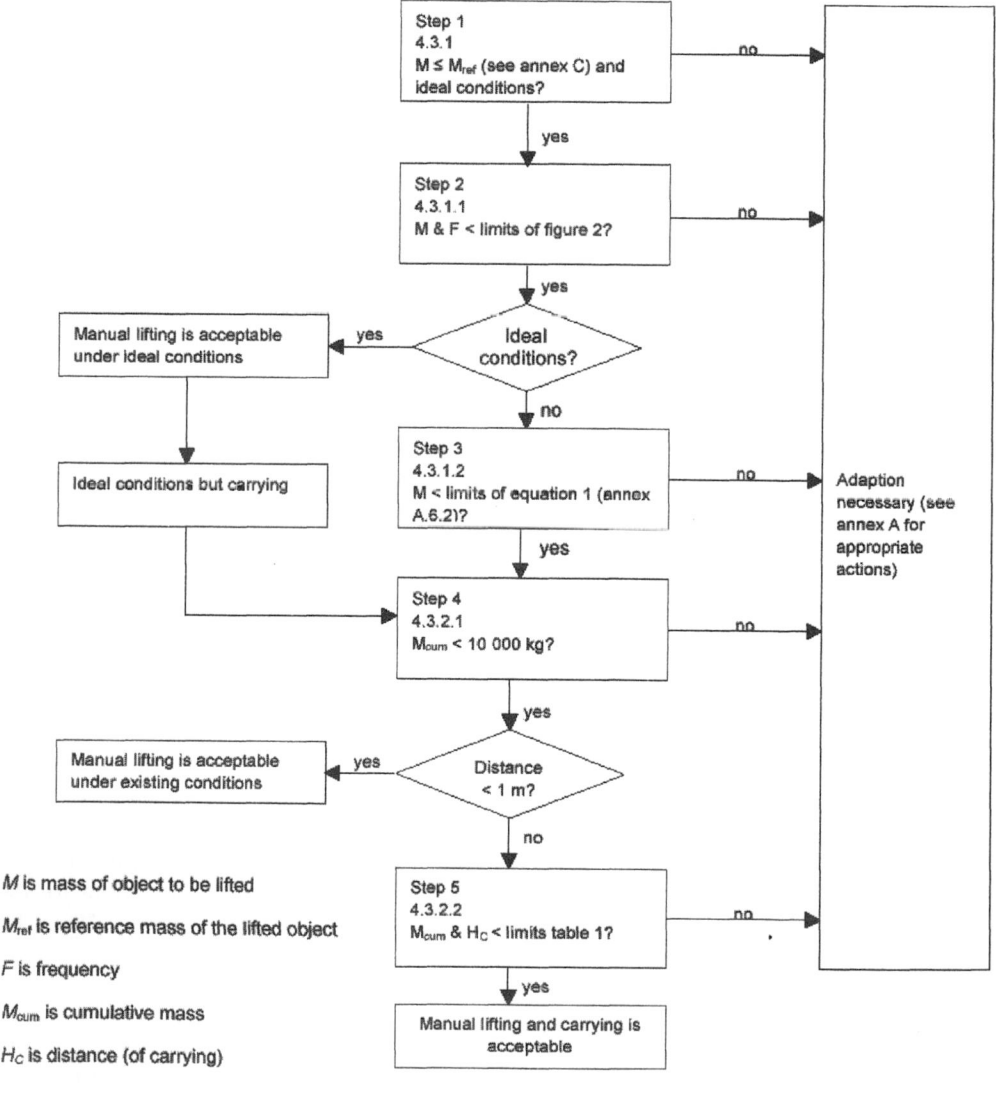

FIG. 14.1.   Step model for the risk-assessment procedure.

TABLE 14.1
"Ideal" Condition for Manual Handling

Moderate ambient thermal environment
Two-handed operation only
Unrestricted standing posture
Handling by one person only
Smooth lifting
Good coupling between the hands and the objects handled
Good coupling between the feet and the floor
Manual handling activities, other than lifting, are minimal.
The objects to be lifted are not cold, hot, or contaminated.
Vertical displacement of the load is ≤ 0.25 m and does not occur below or above the knuckle
Shoulder height
Trunk upright and not rotated
Load kept close to the body

TABLE 14.2
Reference Mass (M_{ref}) Taking Into Consideration Different Populations

Field of Applicaiton	M_{ref} (Kg)	Percentage of			Population Group	
		F and M	Females	Males		
Domestic Use[a]	5	Data not available			Children and the elderly	Total population
	10	99	99	99	General domestic population	
Professional Use (General)[b]	15	95	90	99	General working population including the young and old	General working population
	25	85	70	90	Adult working population	
Professional Use (Exceptional)[c]	30	Data not available			Special working population	Special working population
	35					
	40					

[a]When designing a machine for domestic use, 10 kg should be used as a general reference mass in the risk assessment. If children and elderly are included in the intended user population, the reference mass should be lowered to 5 kg.
[b]When designing a machine for professional use, a reference mass of 25 kg should not be exceeded in general.
[c]Although every effort should be made to avoid manual handling activities or reduce the risks to the lowest possible level, there may be exceptional circumstances where the reference mass might exceed 25 kg (e.g., where technological developments or interventions are not sufficiently advanced). Under these special conditions, other measures have to be taken to control the risk according to EN 614 (e.g., technical aids, instructions, and special training for the intended operator group).

Steps 1 to 3 apply to the limits while lifting; Steps 4 and 5 consider the carrying and are described in detail in Figure 14.1.

Step 1

As an initial screening for *nonrepetitive lifting* tasks under ideal conditions (see Table 14.1), the mass of the object to be manipulated should be considered in relation to the reference mass as described in Table 14.2. This table is taken over from EN 1005-2. If the mass to be manipulated exceeds the reference mass, adaptation is necessary.

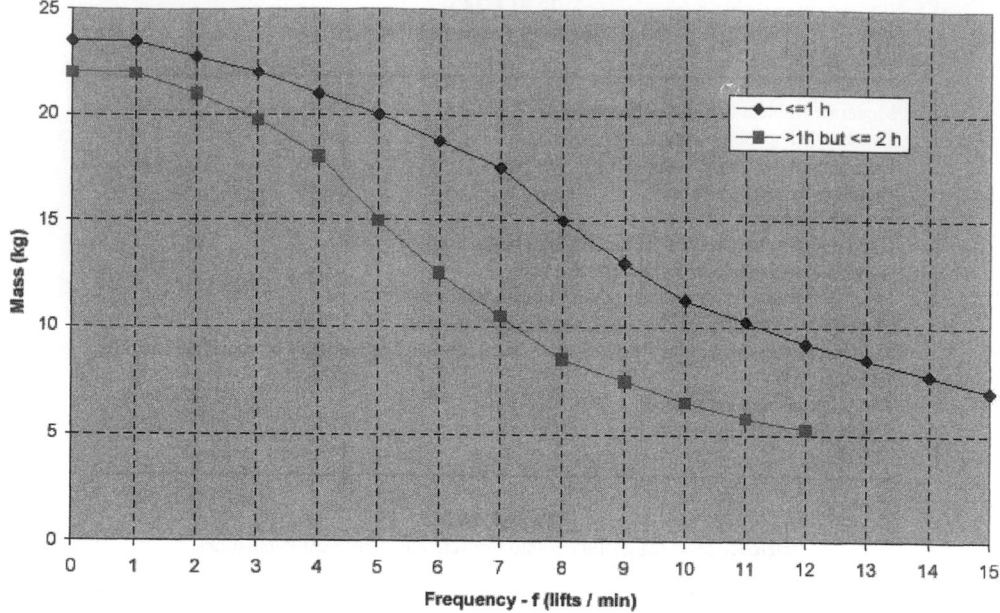

FIG. 14.2.   Maximum frequency for manual lifting related to mass of the object under ideal conditions for two different lifting durations, in correspondence with Table 3.

To estimate the influence of an unfavorable posture, use the risk-assessment model equation in step 3 with a frequency multiplier of "1." The horizontal multiplier will indicate the severity of a possible far reach. Vertical, distance, and asymmetry multipliers will show the negative influence of a twisted or bent trunk.

Step 2

In any case, but especially for *repetitive lifting tasks*, the mass of the object must be considered in combination with the lifting frequency as described in Figure 14.2. This figure offers limitations for mass—frequency combinations and lifting durations of less than 1 hour per day or 1 to 2 hours per day, respectively. The absolute maximal frequency for lifting is 15 lifts per minute. In this case, the total duration of liftings shall not exceed 1 hour per day, and the object mass shall not exceed 7 kg.

For repetitive manual lifting under ideal conditions, lifting durations of ≤ 2 hours per day, the risk assessment for lifting has been finished successfully. Otherwise step 3 must carried out. In case of the absence of carrying, the risk assessment may stop here (Table 14.3).

Step 3

The risk assessment to be carried out in this step is presented in Table 14.4 (see also ISO 11228-1, Annex A.6.2). The risk assessment described follows widely the NIOSH '91 approach, except that the NIOSH "mass constant" is replaced by the variable "reference mass" described in Table 14.2. It also considers other methods (Schaub & Landau, 1997).

The recommended limits are derived from a risk assessment model with the following assumptions listed in Table 14.4.

TABLE 14.3

Frequency Multiplier (F_M) of Equation (1)[a]

| | Continuous, Repetitive Lifting Task duration | | | | | |
| | ≤ 1 hour | | 1 but < 2 hours | | > 2 but < 8 hours | |
Frequency (lifts/min)	V < 0.75 m	V ≥ 0.75 m	V < 0.75 m	V ≥ 0.75 m	V < 0.75 m	V ≥ 0.75 m
≥ 0.2	1.00	1.00	.95	.95	.85	.85
0.5	.97	.97	.92	.92	.81	.81
1	.94	.94	.88	.88	.75	.75
2	.91	.91	.84	.84	.65	.65
3	.88	.88	.79	.79	.55	.55
4	.84	.84	.72	.72	.45	.45
5	.80	.80	.60	.60	.35	.35
6	.75	.75	.50	.50	.27	.27
7	.70	.70	.42	.42	.22	.22
8	.60	.60	.35	.35	.18	.18
9	.52	.52	.30	.30	.00	.15
10	.45	.45	.26	.26	.00	.13
11	.41	.41	.00	.23	.00	.00
12	.37	.37	.00	.21	.00	.00
13	.00	.34	.00	.00	.00	.00
14	.00	.31	.00	.00	.00	.00
15	.00	.28	.00	.00	.00	.00
> 15	.00	.00	.00	.00	.00	.00

[a]If $V < 0.75$ m, $F_M = 0.00$.

TABLE 14.4

Assumptions Made in the Risk-Assessment Model

Are only valid for two-handed, smooth lifting with no sudden acceleration effects (i.e., jerking)

Cannot be used for tasks where the worker is partly supported (e.g., one foot not on the floor)

Width of the object 0.75 m or less for populations with smaller statues (body height)

Are only valid for unrestricted lifting postures

Are only valid when good coupling exists (i.e., hand holds are secure, and shoe/floor slip potential is low)

Are only valid under favorable conditions

The primary task variables include the following data (see Figure 14.3):

- Object mass (M) in kg
- Horizontal distance (H) in meters (m), measured from the midpoint of the line joining the ankles
- To the midpoint at which the hands grasp the object while in the lifting position
- Vertical location (V) in meters (m), determined by measuring the distance from the floor to the point at which the hands grasp the object

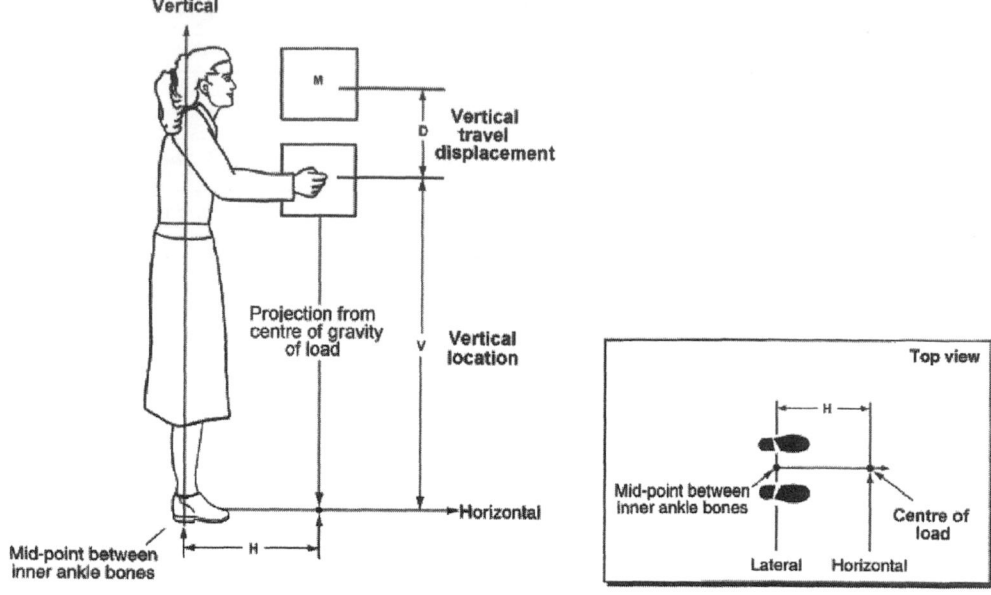

FIG. 14.3.   Task variables.

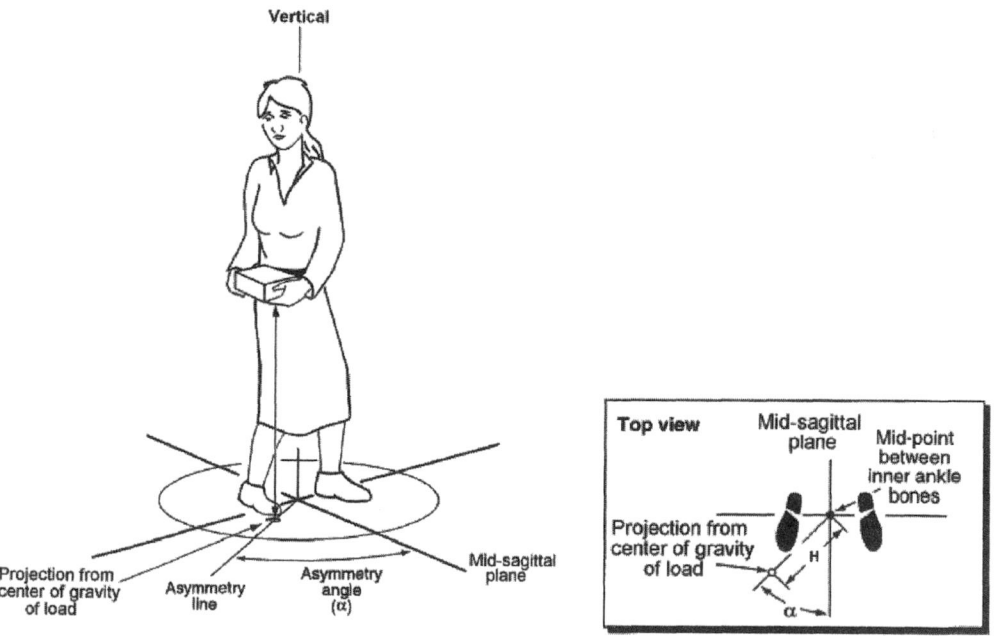

FIG. 14.4.   Angle of asymmetry.

- Vertical travel displacement (D) in meters (m), from origin to destination of lift
- Frequency of lifting (F), average number of lifts per minute (min)
- Duration of manual lifting in hours (h)
- Angle of asymmetry (A) in degrees (in Figure 14.4 as α)
- Quality of gripping (C)

The limit for the mass of the object is derived using the following equation:

$$M \leq M_{ref}{}^1 \times HM \times VM \times DM \times AM \times FM \times CM, \qquad (14.1)$$

where H_M is the horizontal distance multiplier, derived from Equation (2); V_M is the vertical location multiplier, derived from Equation (3); D_M is the vertical displacement multiplier, derived from Equation (4); A_M is the asymmetry multiplier, derived from Equation (5); F_M is the frequency multiplier (see Table 14.3); C_M is the coupling multiplier for quality of gripping see Table A.3.

The multipliers for Equation 1 are obtained from Equations 2 to 5 and Table 14.3 to Table 14.6. If such a multiplier exceeds a value of 1, its value should be taken as 1.

$$
\begin{array}{llll}
H_M = 0.25/H & \text{If } H \leq 0.25, \text{ then} & H_M = 1 & \\
& \text{If } H > 0.63, \text{ then} & H_M = 0 & (14.2) \\
V_M = 1 - 0.3 \times |0.75 - V| & \text{If } V > 1.75, \text{ then} & V_M = 0 & \\
& \text{If } V < 0, \text{ then} & V_M = 0 & (14.3) \\
D_M = 0.82 + 0.045/D & \text{If } D > 1.75, \text{ then} & D_M = 0 & \\
& \text{If } D < 0.25, \text{ then} & D_M = 1 & (14.4) \\
A_M = 1 - 0.0032 \times A & \text{If } A > 135°, \text{ then} & A_M = 0 & (14.5)
\end{array}
$$

The equation has to be calculated for both the start and the end point of each task. End-point calculations may only be of importance if there is a definite precision placement involved. If the item is thrown into place without undue stress on the body in the extended position, then calculating end-point value is not necessary.

The appropriate frequency multiplier, F_M, is determined by first considering the continuous duration of the repetitive lifting task and then considering the duration of the rest period that immediately follows the repetitive lifting task.

The categories of continuous, repetitive lifting tasks, their durations, and the required duration of the rest period that is to immediately follow the lifting task are provided in Table 14.5.

It is critical to note that the combination of the work period and the rest period must be jointly considered to be a work–rest cycle, wherein the rest period provides sufficient opportunity for the worker to recover following a continuous period of lifting-related work. Accordingly, if two successive work periods are separated by a rest period of inadequate duration, then the worker cannot adequately recover, and the entire period—the two work periods plus the rest period—must be treated as if it were a single, continuous work period.

The determination of F_M is then accomplished via entry into Table 14.3. The use of Table A.2 requires three components of information: the duration of the continuous, repetitive lifting task; the vertical location (V) of the hands on the object-to-be-lifted at the beginning of the lift; and the rate of lifting (number of lifts per minute).

The quality of gripping is defined as:

Good: if the object can be grasped by wrapping the hand comfortably around the handles or hand-hold cutouts of the object without significant deviations from the neutral wrist posture or the object itself without causing excessive wrist deviations or awkward postures

---

[1]Taking into account the field of application and the intended user population it will be necessary to select an alternative appropriate reference mass as shown in Table 14.2.

TABLE 14.5
Continuous Lifting Tasks and Their Required Rest Periods

Categories	Definitions	Required Resting Period
Short Duration	≤ 1 hour	≥ 120% of duration of the continuous, repetitive lifting task
Medium Duration	> 1 hour but < 2 hours	≥ 30% of duration of the continuous, repetitive lifting task
Long Duration	≥ 2 hours but ≤ 8 hours	No amount is specified; normal morning, afternoon, and lunch breaks are presumed

Note: For respective frequency coefficients, see Table 14.3.

TABLE 14.6
Coupling Multiplier for the Quality of Gripping (C_M)

Quality of Gripping	Height < 0.75 m	Height ≥ 0.75 m
Good	1.00	1.00
Fair	0.95	1.00
Poor	0.90	0.90

Fair: if the object has handles or cutouts that do not fulfill the criteria of good quality of gripping or if the object itself can be grasped with a grip in which the hand can be flexed about 90°
Poor: if the criteria of good or fair quality of gripping are not fulfilled

Step 4

Under ideal conditions, the cumulative mass (product of mass and frequency of carrying, which may have been limited in steps 1 and 2), manipulated per shift should not exceed 10.000 kg.

For carrying, the reference mass should never exceed 25 kg, and the frequency is limited to a maximum of 15 times per minute.

If the carrying distance is less than 1 m, manual handling is acceptable under the existing conditions. For longer distances (i.e., 20 m), it is substantially lower (6.000 kg / 8 hours). For nonideal conditions and carrying distances greater then 1 m, step 5 has to be applied.

This step considers data from NF × 35-106, N × 35-109, Grieco et al 1997, Bongwald et al 1995; Gay et al 1978, Genaidy & Ashfour 1987,

Step 5

Maximum cumulative masses for several distances and frequencies are contained in Table 14.7. If the cumulative masses described in Table 14.1 are not exceeded, the entire risk assessment has been carried out successfully. Otherwise adaptation is necessary.

TABLE 14.7
Recommended Limits for Cumulative Mass Related to Carrying Distance
(for General Working Population)

Distance (m)	Maximum Cumulative Mass			
	kg/min	*kg/hr*	*kg/8 hr*	*Example of M*f*
20	15	750	6 000	5 kg × 3/min 15 kg × 1/min 25 kg × 0.5/min
10	30	1 500	10 000	5 kg × 6/min 15 kg × 2/min 25 kg × 1/min
4	60	3 000	10 000	5 kg × 12/min 15 kg × 4/min 25 kg × 1/min
2	75	4 500	10 000	5 kg × 15/min 15 kg × 5/min 25 kg × 1/min
1	120	7 200	10 000	5 kg × 15/min 15 kg × 8/min 25 kg × 1/min

Note1. In the calculation of the cumulative mass, a reference mass of 15 kg is used for general working population and frequency of carrying—15 times/min.

Note2. The total cumulative mass of lifting and manual carrying should never exceed 10,000 kg/day, whichever is the daily duration of work.

Note3. Twenty-three kilograms is included in the 25 kg.

Table 14.7 provides the limits in kilograms per minute which should protect against the excess of local load, in kilograms per hour, which should protect against excess of general load and in kilograms per 8 hours, which limits the long-term risk. The limits are not the simple multiplications, because the risks for short term, medium term, and long term are qualitatively different. The last column of the table takes the examples of different combinations of mass and frequency. These examples show that the limits in kilograms per minute cannot be always applied because of the limits of maximal mass and frequency (5 kg × 15/min = 75 kg/min even for a distance of 1 m, and 25 kg cannot be lifted more than 1/min – see Figure 14.2).

In the practical application of the Standard, first the limits of maximal mass and frequency have the priority; when those limits are respected, the limits to carrying have to be applied.

Under unfavorable environmental conditions, or when lifting from and to low levels, for example, below knee height or when the arms are lifted above the shoulders, the recommended limits for cumulative mass for carrying in Table 14.7 should be substantially reduced (at least by one third).

Risk Reduction

Risk reduction can be achieved by reducing the need for manual handling activities or by minimizing or excluding hazards resulting from the task, the object, the workplace, the work organization, or the environmental conditions. Examples are given in Annexes A.2 to A.5 of the Standard.

Annexes

Although the main part of the standard is relatively compact and describes for the risk assessment mainly the five-step approach, Annex A offers detailed background information on the following:

- Avoidance of manual handling (A1)
- Design of task, workplace, and work organization (A2)
- Design of the object (A3)
- Design of the handling of live objects (A4)
- Design of the work environment (A5)
  (for "ideal conditions for manual materials," see Table 14.1)
- Assessment method for recommended limits for mass, frequency, and object position for repetitive and nonrepetitive lifting tasks (A6). Risk assessments are based on the assumptions listed in Table 14.4.
- Individual considerations (A7)
- Information and training (A8)

Annex B offers two examples of an assessment and ergonomic approach of manual handling of (live) objects. Annex C offers the table for the reference mass (see Table 14.2).

WORKSHEETS

Unfortunately neither the main part nor the annexes contain any worksheets that may be copied for application purposes.

THE EU-MANUAL HANDLING DIRECTIVE

This Directive, which is the fourth individual Directive within the meaning of Article 16(1) of Directive 89/391/EEC, lays down minimum health and safety requirements for the manual handling of loads where there is a risk, particularly of back injury to workers. It obliges the employer to take appropriate organizational measures, or use the appropriate means, in particular mechanical equipment, in order to avoid the need for the manual handling of loads by workers. Where the need for the manual handling of loads by workers cannot be avoided, the employer shall take the appropriate organizational measures, use the appropriate means, or provide workers with such means in order to reduce the risk involved in the manual handling of such loads, having regard to Annex I of this directive. Wherever the need for manual handling of loads by workers cannot be avoided, the employer shall organize workstations in such a way as to make such handling as safe and healthy as possible and:

- Assess, in advance if possible, the health and safety conditions of the type of work involved and, in particular, examine the characteristics of loads, taking account of Annex I
- Take care to avoid or reduce the risk particularly of back injury to workers, by taking appropriate measures, considering in particular the characteristics of the working environment and the requirements of the activity, taking account of Annex I

Workers or their representatives shall be informed of all measures to be implemented, pursuant to this Directive, with regard to the protection of safety and of health.

Employers must ensure that workers or their representatives receive general indications and, where possible, precise information on:

- The weight of a load
- The center of gravity of the heaviest side when a package is eccentrically loaded

Employers must ensure that workers receive in addition proper training and information on how to handle loads correctly and the risks they might be open to, particularly if these tasks are not performed correctly, having regard to Annexes I and II.

Annex I describes characteristics of the load.

The manual handling of a load may present a risk particularly of back injury if it is:

- Too heavy or too large
- Unwieldy or difficult to grasp
- Unstable or has contents likely to shift
- Positioned in a manner requiring it to be held or manipulated at a distance from the trunk, or with a bending or twisting of the trunk
- Likely, because of its contours or consistency, to result in injury to workers, particularly in the event of a collision

A physical effort may present a risk particularly of back injury if it is:

- Too strenuous
- Only achieved by a twisting movement of the trunk
- Likely to result in a sudden movement of the load
- Made with the body in an unstable posture

The characteristics of the work environment may increase a risk particularly of back injury if:

- There is not enough room, in particular vertically, to carry out the activity
- The floor is uneven, thus presenting tripping hazards, or is slippery in relation to the worker's footwear
- The place of work or the working environment prevents the handling of loads at a safe height or with good posture by the worker
- There are variations in the level of the floor or the working surface, requiring the load to be manipulated on different levels
- The floor or foot rest is unstable
- The temperature, humidity, or ventilation is unsuitable

The activity may present a risk particularly of back injury if it entails one or more of the following requirements:

- Overfrequent or overprolonged physical effort involving in particular the spine
- An insufficient bodily rest or recovery period
- Excessive lifting, lowering, or carrying distances
- A rate of work imposed by a process which cannot be altered by the worker

Annex II describes individual risk factors. The worker may be at risk if he or she:

* Is physically unsuited to carry out the task in question
* Is wearing unsuitable clothing, footwear or other personal effects
* Does not have adequate or appropriate knowledge or training

The complete text of this directive may be downloaded in several languages from http://europa.eu.int/eur-lex/.

ACKNOWLEDGMENTS

This work is dedicated to the members of ISO/TC 159/SC 3/WG 4, who developed this standard, especially to my colleagues from the writing group.

REFERENCES

Andersson, G. B. J. (1999, February). Point of View: Evaluation of the revised NIOSH lifting equation: A cross-sectional epidemiologic study. *Spine, 24*(4), 395.

Applications Manual for the Revised NIOSH Lifting Equation. NIOSH, 1999. Cincinnati, OH: CDC.

90/269/EEC. (1990). Council Directive of 29 May 1990 on the minimum health and safety requirements for the manual handling of loads where there is a risk particularly of back injury to workers (fourth individual Directive within the meaning of Article 16 (1) of Directive 89/391/EEC).

EN 614-1. Safety of machinery—Ergonomic design principles—Part 1: Terminology and general principles.

EN 614-2. Safety of Machinery—Ergonomic design principles—Part 2: Interactions between the design of machinery and work tasks.

EN 1005-1. Safety of machinery—Human physical performance—Part 1: Terms and definitions.

EN 1005-2. Safety of machinery—Human physical performance—Part 2: Manual handling of machinery and component parts of machinery.

EN 1050. *Safety of machinery—Risk assessment.*

EN-ISO 7250, Basic list of definitions of human body measurements to technical design.

Fritsch, W., Enderlein, G., Aurich, I., & Kurschwitz, S. (1975). Einflusse beruflicher Faktoren auf die gynäkologische Mobilität und Tauglichkeit. *Zeit Sohxift fur die grante Hggien, 21*, 825.

Garg, A. (1989). An evaluation of the NIOSH guidelines for manual lifting, with special reference to horizontal distance. *Am. Ind. Hyg. Assoc. J., 50*(3), 157–164.

Garg, A., Chaffin, D., & Herrin, G. D. (1978). Prediction of metabolic rates for manual materials handling jobs. *American Industrial Hygiene Association Journal, 39*(8), 661–674.

Genaidy, A. M., & Ashfour, S. S. (1987). Review and evaluation of physiological cost prediction models for manual materials handling. *Human Factors, 29*(4), 465–476.

Grieco A., Occhipinti E., Colombini D., & Molteni G. (1997). Manual handling of loads: The point of view of experts involved in the application of EC Directive 90/269. *Ergonomics, 40*(10), 1035–1056.

Hettinger, T. (1981). *Heben und Tragen von Lasten. Gutachten über Gewichtsgrenzen für Männer, Frauen und Jugendliche.* Der Bundesminister für Arbeit und Sozialordung. Bonn.

Hettinger, T., Müller, B. H., & Gebhardt, H. (1989). *Ermittlung des Arbeitsenergieumsatzes bei dynamisch muskulärer Arbeit.* Bundesanstalt für Arbeitsschutz (Hrsg.), Fa 22, Wirtschaftsverlag NW, Bremerhaven.

ISO 7730. *Moderate thermal environments—Determination of the PMV and PPD indices and specification of the conditions for thermal comfort.*

ISO 11226. *Ergonomics—Evaluation of working postures.*

ISO 14121. *Safety of machinery—Principles of risk assessment.*

ISO 1228 Ergonomics part 1: Lifting and carrying part 2: pushing and pulling part 3: Handling of law loads at high frequency.

ISO/IEC Guide 51. *Safety aspects—Guidelines for their inclusion in standards.*

Jäger, M., & Luttmann, A. (1991). Compressive strength of lumbar spine elements related to age, gender and other influencing factors. In P. A. Anderson, D. J. Hobart, & J. V. Danoff (Eds.), *Electromyographical kinesiology* (pp. 291–294). Amsterdam: Elsevier Science Publishers B.V. (Biomedical Division).

Mital, A., Nicholson, A. S., & Ayoub, M. M. (1997). *A guide to manual materials handling* (2nd ed.). Taylor & Francis.

Monroe Keyserling, W. (1989). Analysis of manual lifting tasks: A qualitative alternative to the NIOSH Work Practices Guide. *Am. Ind. Hyg. Assoc. J., 50*(3), 165–173.

Schaub, Kh., & Landau, K. (1997). Computer-aided tool for ergonomic workplace design and preventive health care. In *Human Factors and Ergonomics in Manufacturing, 7*(4), 269–304.

Snook, S. H. (1978). The design of manual handling tasks. *Ergonomics, 21*, 963–985.

Snook, S. H., & Ciriello, V. M. (1991). The design of manual handling tasks: Revised tables of maximum acceptable weights and forces. *Ergonomics, 34*(9), 1197–1213.

Snook, S. H., Irvine, C. H., & Bass, S. F. Maximum weights and work loads acceptable to male, industrial workers. A study of lifting, lowering, pushing, pulling, carrying and walking tasks.

Waters, T. R., Putz-Anderson, V., Garg, A., Fine, L. J. (1993). Revised NIOSH equation for the design and evaluation of manual lifting tasks. *Ergonomics, 36*(7), 749–776.

Waters, T. R., Baron, S. L., Piacitelli, L. A., Andersen, V. P., Skov, T. Haring-Sweeney, M. et al. (1999, February). Evaluation of the revised NIOSH lifting equation. *Spine, 24*(4), 386–394.

http://europe.eu.int/eur-lex/

15

Ergonomics of Manual Handling—Part 2: Pushing and Pulling

Karlheinz G. Schaub
Darmstadt University of Technology

Peter Schaefer
München University of Technology

INTRODUCTION

Relation to Other Standards, Guidelines, and Regulations

Probably due to the fact that pushing and pulling is physically less demanding than lifting and carrying, there is a lack of risk-assessment methods on a quantitative basis by means of a formula. It is the aim of ISO 11228-2 to bridge that gap. Several approaches exist for the evaluation of pushing and pulling tasks in a worldwide overview.

In Europe, pushing and pulling as two types of manual materials handling are regulated by the Council Directive 90/269/EEC "on the minimum health and safety requirements for the manual handling of loads where there is a risk particularly of back injury to workers." Next to this directive, manual materials handling is addressed by the European Union (EU) machinery directive (98/37/EC), namely, by EN 1005, a set of harmonized CEN standards that supports this directive. The next chapters offer additional information about the dual European system of health and safety at work and the manual handling directive.

The Dual Concept of Health and Safety in Europe

On their way to building up a political union, the EU Member States decided to add social components to their various branches of the community that were predominantly of economic nature for many years. It was intended that these social components should also include measurements for preventive health care and aspects of ergonomic workplace and product design. The adoption in 1986 of the Single European Act gave new impetus to the occupational health and safety measures taken by the community. This was the first time health and safety at work had been directly included in the EEC Treaty of 1957 and was done through the new Articles 100a and 118a. Article 100a requires harmonization of national legislation. The objective is to remove all barriers to trade in the single market and allow free movement of goods and people across borders. In principle, Article 100a does not permit Member States to set higher requirements for their products than those laid down by the directives.

Of most significant importance to the level of protection in the EU Member States is that directives adopted under Article 118a lay down minimum requirements concerning health and safety at work. According to this principle, the Member states may raise their level of protection, if desired. They also must raise their level of protection, if it is lower than the minimum requirements set by the directives. Beyond this, they have the obligation to maintain and introduce more stringent protective measures than required by the directives.

Articles 100a and 118a contribute to the improvement of the Member States' working environments as well as equal or better protection of their workers. Directives under Article 100a are intended to ensure the placing on the market of safe products, and under Article 118a they are intended to ensure the healthy and safe use of the products at the workplace (EC 1993). Articles 100a and 118a had been reformulated as Articles 95 and 137 in the Maastricht treaty.

The Framework Directive on health and safety at work (89/391/EEC) including the corresponding Individual Directives (e.g., Visual Display Units [VDU] work, manual materials handling, and personal protective equipment) as well as the Machinery Directive (98/37/EC; formerly 89/392/EEC) serve for the implementation of the Articles 95 and 137 of the Maastricht treaty. Whereas the Framework Directive addresses to employers and employees, the Machinery Directive focuses onto the designer and manufacturer of machinery only.

The Manual Handling Directive

This Directive, which is the fourth individual Directive within the meaning of Article 16(1) of Directive 89/391/EEC, lays down minimum health and safety requirements for the manual handling of loads where there is a risk, particularly of back injury to workers. It obliges the employer to take appropriate organizational measures, or use the appropriate means, in particular mechanical equipment, in order to avoid the need for the manual handling of loads by workers. Where the need for the manual handling of loads by workers cannot be avoided, the employer shall take the appropriate organizational measures, use the appropriate means, or provide workers with such means in order to reduce the risk involved in the manual handling of such loads, having regard to Annex I of this directive.

Wherever the need for manual handling of loads by workers cannot be avoided, the employer shall organize workstations in such a way as to make such handling as safe and healthy as possible and:

- Assess, in advance if possible, the health and safety conditions of the type of work involved, and, in particular, examine the characteristics of loads, taking account of Annex I
- Take care to avoid or reduce the risk, particularly of back injury, to workers, by taking appropriate measures, considering in particular the characteristics of the working environment and the requirements of the activity, taking account of Annex I

Workers or their representatives shall be informed of all measures to be implemented, pursuant to this Directive, with regard to the protection of safety and of health.

Employers must ensure that workers or their representatives receive general indications and, where possible, precise information on the weight of a load and the center of gravity of the heaviest side when a package is eccentrically loaded.

Employers must ensure that workers receive in addition proper training and information on how to handle loads correctly and the risks they might be open to, particularly if these tasks are not performed correctly, having regard to Annexes I and II.

Annex I describes characteristics of the load. The manual handling of a load may present a risk particularly of back injury if it is:

- Too heavy or too large
- Unwieldy or difficult to grasp
- Unstable or has contents likely to shift
- Positioned in a manner requiring it to be held or manipulated at a distance from the trunk, or with a bending or twisting of the trunk
- Likely, because of its contours or consistency, to result in injury to workers, particularly in the event of a collision

A physical effort may present a risk particularly of back injury if it is:

- Too strenuous
- Only achieved by a twisting movement of the trunk
- Likely to result in a sudden movement of the load
- Made with the body in an unstable posture

The characteristics of the work environment may increase a risk particularly of back injury if:

- There is not enough room, in particular vertically, to carry out the activity
- The floor is uneven, thus presenting tripping hazards, or is slippery in relation to the worker's footwear
- The place of work or the working environment prevents the handling of loads at a safe height or with good posture by the worker,
- There are variations in the level of the floor or the working surface, requiring the load to be manipulated on different levels,
- The floor or foot rest is unstable
- The temperature, humidity, or ventilation is unsuitable

The activity may present a risk particularly of back injury if it entails one or more of the following requirements:

- Over frequent or overprolonged physical effort involving, in particular, the spine
- An insufficient bodily rest or recovery period
- Excessive lifting, lowering, or carrying distances
- A rate of work imposed by a process that cannot be altered by the worker

Annex II describes individual risk factors. The worker may be at risk if he or she:

- Is physically unsuited to carry out the task in question
- Is wearing unsuitable clothing, footwear, or other personal effects
- Does not have adequate or appropriate knowledge or training

The complete text of this directive may be downloaded in several languages from http://europa.eu.int/eur-lex/.

GENERAL APPROACHES

This standard is under development and has reached the status of a draft. When the working group the international standard has finished the draft, it will be sent out as a committee draft Standard (ISO/CD 11228-2) for voting.

Structure of the Standard

The Standard follows the predefined typical structure of ISO Standards. After the foreword, introduction, scope, and normative references, chapter 3 offers definitions required for that Standard. Chapter 4 as the main part of the Standard offers recommendations for the avoidance of hazardous manual handling tasks and the execution of a risk-assessment composed of two risk-assessment methods. Method 1—an easily applicable checklist—is described in detail in Annex A. Method 2—a more sophisticated quantitative approach—is described in detail in this chapter. Annex E offers an example for Method 2.

Annex A describes in detail Method 1 of the risk assessment. Method 1 is an easy-to-handle checklist, that allows the rating of a task for high or low risks, to verbally describe the problems arising from the task, and to formulate possible remedial actions to lower the existing risk.

The method presented takes into consideration the following:

- The task
- The loads or objects to be moved
- The working environment
- Other factors
- Management and organizational issues

Annex C offers biomechanical considerations of pushing and pulling. Annex D deals with measures for risk reduction and covers the following:

- Avoidance of repetitive handling
- Design of the work: task, workplace, and work organization
- Design of object, tool, or material handled
- Design of the working environment
- The worker capabilities

Annex E finally offers examples for the application of the more sophisticated risk-assessment Method 2.

Basic Philosophy

A major part of the Standard aims to offer a comprehensive risk assessment for pushing and pulling tasks. Depending on the sources reviewed, the cardiovascular system might be the bottleneck for pushing and pulling tasks, as well as the muscular or the skeletal system (Schaefer et al. 2000). Especially for high loads, the initial forces required might exceed or come close to the maximal force limits, whereas the sustained forces required are on a very low level only, if the environmental (moderate ambient thermal environment, good coupling in between the feet and the floor, smooth floor without obstacles, etc.) and task conditions (no sudden accelerations, nor obstacles, no curves with narrow radius, etc.) are appropriate. It was believed that, contrary to lifting and carrying, pushing and pulling would not be of biomechanical relevance (Mital, Ayoub, & Nicholson 1993). Recent investigations of Jäger (2001), however, showed that, especially for pulling tasks, high lumbar spine loads might result from pulling tasks.

When pushing and pulling objects, the posture adapted and the location of force exertion have a substantial impact on maximum voluntary contraction (MVC) (Schaub et al. 1997). The posture adopted and the location of force exertion depend on the position of the handles and the stature of the operator. So the biomechanical load situation for a given task may vary among the operators dependent on their stature. In order to allow an overall individual independent risk-assessment, it was decided to realize a risk-assessment model that considers muscular as well as skeletal bottlenecks, and that takes into consideration the effect of the stature onto maximal force limits and the postures adapted.

Data Sources

Relevant tables from DIN 33411-5, a comprehensive national German Standard (18 tables with maximal action forces for the 5th, 10th, 15th, 50th, and 95th force percentiles of a grand total of 188 different types of force exertions), were used to describe the influence of the relative working height (gripping height as percentage of stature) onto the maximal force capabilities (see Figures 15.1 and 15.2). For a conservative assessment, the values for the 5th force percentiles were chosen.

The procedure of EN 1005-3 derives acceptable load limits out of maximal static force limits by means of several multipliers that consider the velocity, frequency and action time of force exertions, the task duration and a "risk" multiplier.

Snook's tables will be used to set up multipliers for the influence of frequency and duration for maximum limits of pushing and pulling tasks.

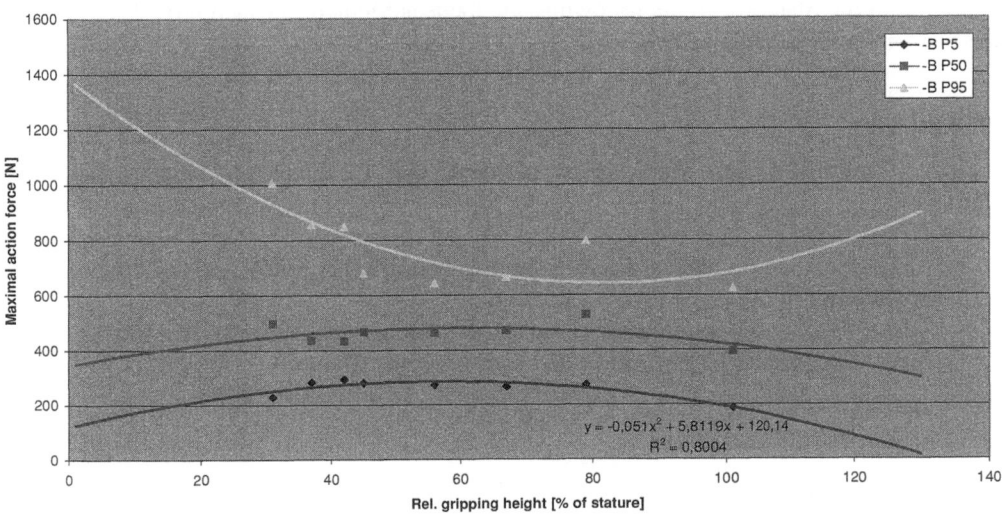

Sagittal Push (-B)
Force atlas - tables 21 (& 24) gripping distance 500mm and
DIN 33411-5 - tables 9 (& 12)

$$y = -0.051x^2 + 5.8119x + 120.14$$
$$R^2 = 0.8004$$

FIG. 15.1. Maximal static push action forces in relation to gripping height and force percentiles.

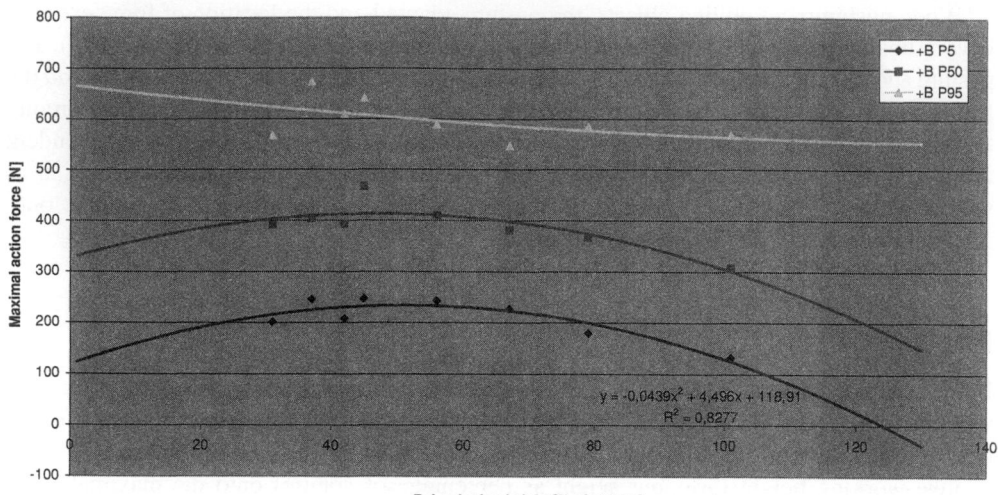

Sagittal Pull (+B)
Force atlas - tables 21 (& 24) gripping distance 500mm and
DIN 33411-5 - tables 9 (& 12)

FIG. 15.2.   Maximal static pull action forces in relation to gripping height and force percentiles.

RISK-ASSESSMENT MODEL

Approaches

ISO 11228-2 provides two different ways of risk assessment. The first way (Method 1) is a checklist-based procedure that yields very rough and rather pessimistic estimates. Results of Method 1 may be understood in such a way that loads accommodated in the green zone are surely green. But any accommodation in red zones may turn out to be green or yellow as well. That's why Method 1 will not be presented here. The checklists are easy to apply and may be found in the Standard. In the following, the focus is Method 2, which provides a detailed assessment and evaluation of risk.

Innovations

Method 2 introduces two characteristic elements that are absolutely new in this field. These are:

- User-oriented assessments—limits reflecting demographic profiles of any optional user population
- skeletal load limts—in addition to muscular aspects, skeletal limits considering spinal aspects in particular (e.g., compressive load of lumbar spine)

User-Oriented Assessment

Industrial production generally seeks to meet the demands of a wide and heterogeneous clientele as precisely as possible. Hence, as user profiles change, the design of industrial products is changing as well. To ensure good interaction between man and machine, technical aids generally should reflect major characteristics of envisaged user populations—for example, demographic profiles as described by age, gender, and stature distributions. That kind of target group orientation not only adresses classical anthropometrics but also influences manipulation forces and thus may directly affect health protection. The general problem is how to protect different user populations by the same safety level—in other words, how to make sure that user groups all over the world are really taking comparable risks.

That's why ISO 11228-2 introduced force limits reflecting demographic profiles in particular. These characteristics are given by distributions of age, gender, and stature.

Such an evaluation procedure is absolutely new in the field of load rating. Traditional approaches either refer to pure male or pure female populations or to a fixed mix of both (e.g., NIOSH, 1981; Siemens, 1969). It is an outstanding characteristic of ISO 11228-2 that its special load limits react directly to the slightest changes in demographic profiles.

Skeletal Load Limts

This risk estimation approach adopts a multidisciplinary approach giving suitable consideration to biomechanical, physiological, and psychophysical parameters. The biomechanical approach considers force exertions in relation to both individual strength capabilities and the risk of injury. Such injuries not only include muscular diseases but also cover skeletal diseases, that is, why this risk estimation approach additionally considers lumbar spine compression in relation to lumbar spine strength for different age populations.

How the Model Works

Method 2 is a quite complex procedure and provides a breakdown of risk assessment into four major parts (see Figure 15.3):

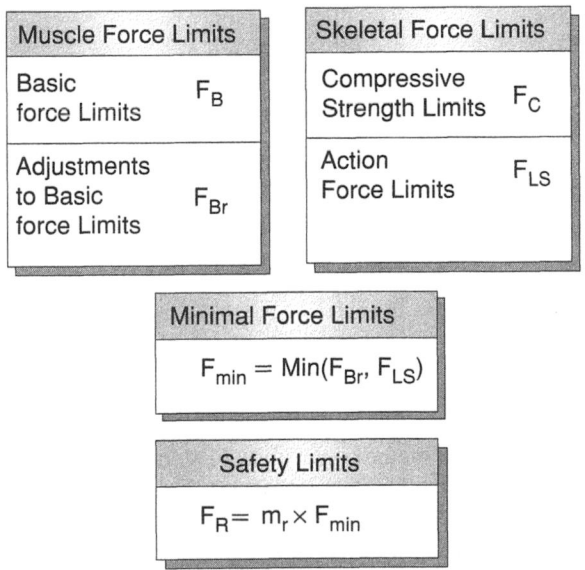

FIG. 15.3.   Risk-assessment procedure (Method 2) in ISO 11228-2.

Part 1: Muscle-Based Force Limits F_{Br}

Part 1 adopts a two-step approach:

Step 1:   Basic force limits (F_B). Basic force limits (F_B) are adjustable strength limits of the intended user population taking into account the effects of age, gender, and stature.

Step 2:   Adjusted force limits (F_{Br}). Adjusted force limits (F_{Br}) reflect characteristics of the work task, such as velocity, frequency, and duration. The effects of these characteristics are accounted for by a set of mutipliers (m_v, m_f, and m_d):

$$F_{Br} = m_v \times m_f \times m_d \times F_B.$$

Part 2: Skeletal-Based Force Limits (F_{LS})

Part 2 provides load limits (F_{LS}) that are based on the limited compressive strength of the lumbar spine. The procedure integrates the following two steps:

Step 1:   Compressive strength limit (F_C). In a first step, compressive strength limits are determined taking into account the effects of age and gender of the forseeable user population.

Step 2:   Action force limits (F_{LS}). Step 2 introduces a relation between compressive strength limits (F_C) determined by step 1 and action forces as observed at the workplace.

This way the action force limit (F_{LS}) makes sure that compressive strength limits (F_C) of the lumbar spine generally are not exceeded at work.

Part 3: Minimal Limits (F_{min})

Part 3 selects the minimum between muscle-based force limits (F_{Br}) and skeletal-based force limits (F_{LS}):

$$F_{min} = Min (F_{Br}, F_{LS}).$$

Part 4: Safety Limits (F_R)

Safety limits (F_R) are based on capacity limits (F_{min}) and suitable risk multipliers (m_r), such that

$$F_R = m_r \times F_{min}.$$

Muscle Force Limits

Basic Force Limits

Part 1 of the risk-assessment method specifies a calculation procedure providing basic force limits that change with changing characteristics of optional user populations. Required characteristics are given by distributions of age, gender, and stature.

The technical procedure is a step by step approach as described in the following.

Objectives.   The procedure calculates force limits when pushing or pulling, at selected absolute handle heights, and regarding specified target populations.

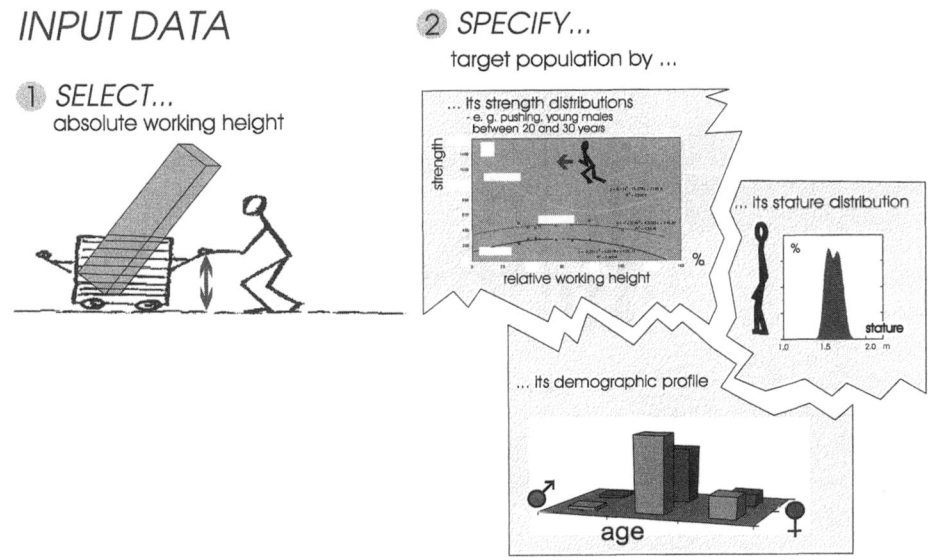

FIG. 15.4. Input data of risk-assessment model.

Input Data. To get started, the procedure first requires a set of input data (see Figure 15.4). In detail these are the following:

- Selected absolute handle height above ground—for example, 1.2 m
- Characteristics of the envisaged user population, that is, stature distribution and distributions of age and gender
- Strength distributions from tables or experimentally found of young females (20 years $\leq$ age < 30 years), when pushing or pulling, and at selected relative handle heights

Predicting Statures. In a first step, the procedure predicts statures (Figure 15.5) resulting from various working situations. Generally there is a relation between these statures on the one side and relative working heights (h_{rel}) on the other side when working on given absolute levels (h_{abs}):

$$S = \frac{h_{abs}}{h_{rel}} \, m \qquad \text{with: S: stature.}$$

Practically, the procedure works as follows (see Fig. 15.5):

- Select absolute working height—for example, 1.2 m above ground
- Select a set of relative working heights covering the range man is able to work. Intervals between selected heights should be reasonably spaced.
- Predict statures for each relative working height selected previously, for example, when working 1.2 m above ground and relative working height is 120%, then stature is 1.71 m.

Finally, there is a set of statures allowing work at one and the same absolute working height while selecting various relative working heights.

PREDICTING STATURES

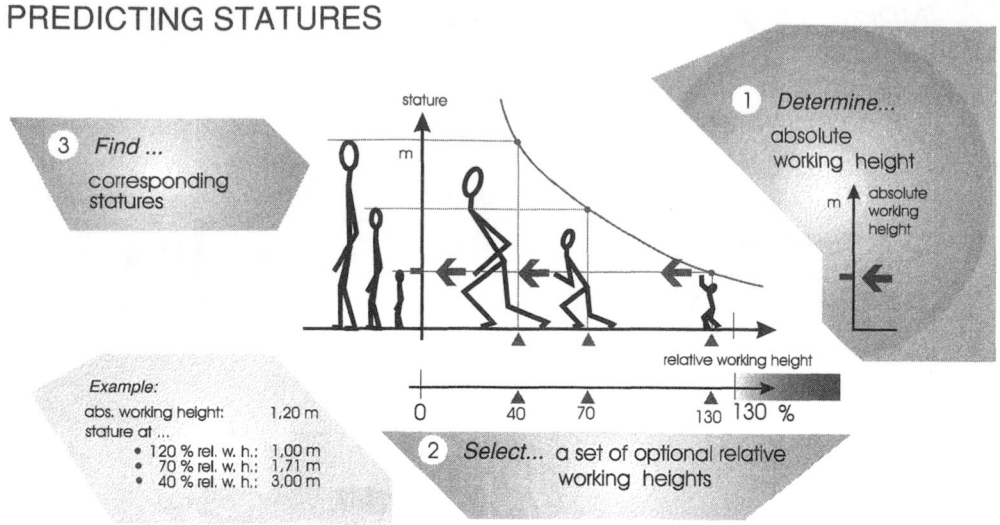

FIG. 15.5.   Statures required when working at selected relative heights.

FINDING WEIGHTING MULTIPLIERS

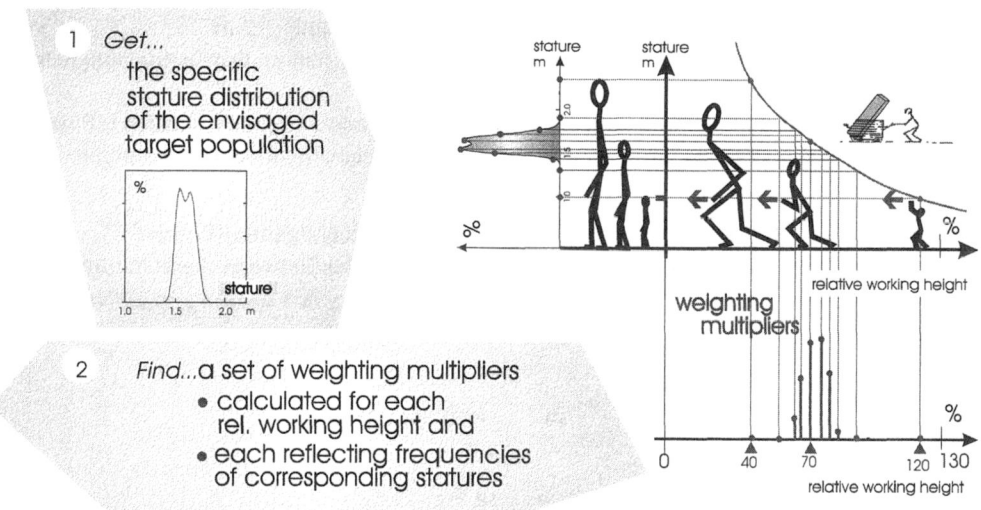

FIG. 15.6.   Multipliers weighting relative working heights.

Weighting Multipliers.   The chance stature (S) will be observed certainly depends on the specific stature distribution to be found in an envisaged target population. On the other side, stature (S) is defined when working at given relative height (h_{rel}) and on a given absolute level (h_{abs}). That's why the aforementioned stature probabilities describe incidence rates of those corresponding working situations in the same way. Such probabilities may define weighting multipliers to be determined for each relative height in particular (Figure 15.6).

INTRODUCING STRENGTH DISTRIBUTIONS

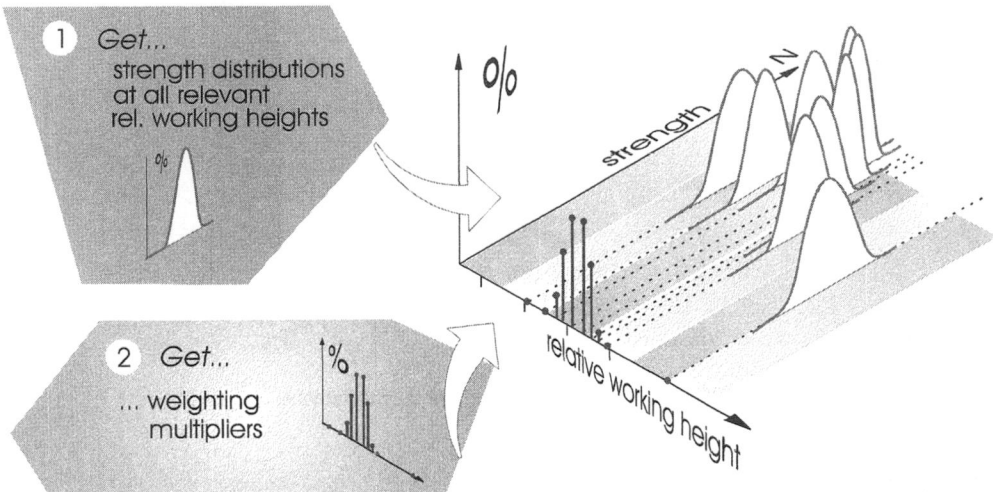

FIG. 15.7.   Distributions describing strength at selected relative working heights.

The way these weighting multipliers will be found is explained in Figure 15.6. The major steps are:

- Get a specific stature distribution of the envisaged target population
- Select a set of relative working heights
- Predict corresponding statures
- Determine probabilities that these statures will be found in the target population
- Assign multipliers to stature probabilities

Introduction of Strength Distributions.   Strength distributions may be given by percentile functions (e.g., 15th, 50th, and 85th percentiles) depending on relative working heights (see Fig. 15.1 and Fig. 15.2). This is the experimental base of a set of distributions to be determined at selected relative working heights. In the end, each relative working height accommodates a specified strength distribution and an appropriate weighting multiplier (see Figure 15.7).

Weighting Strength Distributions.   A next step adjusts reference distributions at each relative working height by aforementioned weighting multipliers—each multiplier modifies his or her distribution at his or her relative working height (see Figure 15.8). Such modifications reflect the probabilities of these relative working heights to be found among the envisaged user population. The procedure is in particular:

- Get strength distributions at each relative working height
- Get according weighting multipliers
- Multiply each distribution by its "own" multiplier

Demographic Fitting.   Up to this point, results strictly represent reference groups only—for example, females between 20 and 30 years of age. That's why another step adjusts these reference distributions to real distributions that may be observed within real user populations.

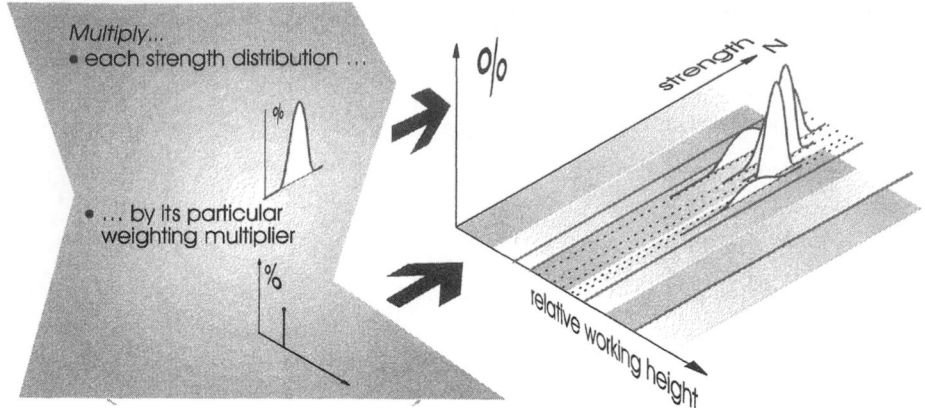

FIG. 15.8.   Weighting of reference distributions.

Demographic fitting

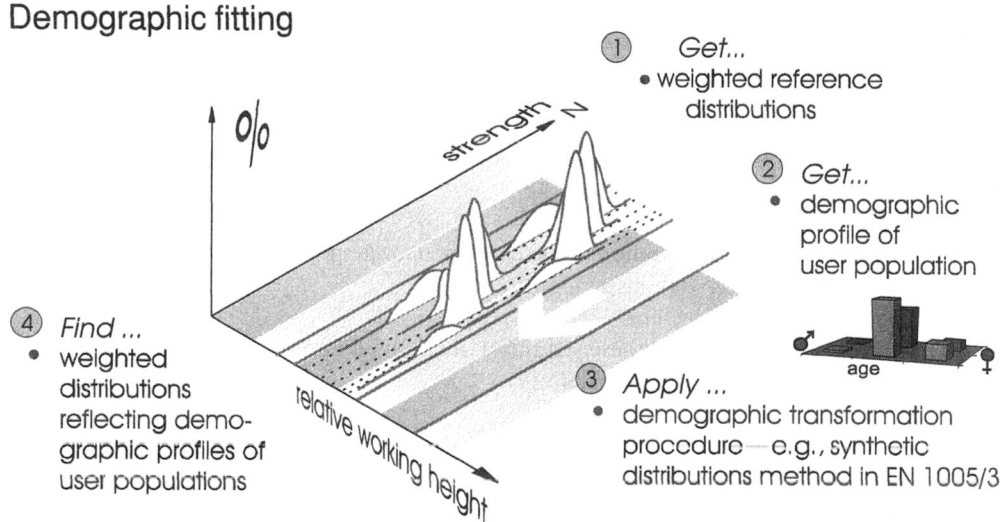

FIG. 15.9.   Adjusting distributions to demographic profiles.

This work is done by the "Synthetic Distributions Method," a special procedure that was standardized by EN 1005-3 and was introduced into ISO 11228-2 as well. A general overview is given in Figure 15.9. The procedure simply calculates modified distributions reflecting the effects of age and gender within optional user populations. These modified distributions are no longer approximations to normal.

Combined Distribution Functions.   Physical strength of any user population as a total may be described by combined distributions shaped by demographic profiles and suitable sets of relative working heights reflecting stature distributions in particular. Those distributions may be found by a simple sum integrating weighted distributions of all relative working heights.

Finally a second integration of these combined distributions with increasing strength yields combined distribution functions (see Figure 15.10).

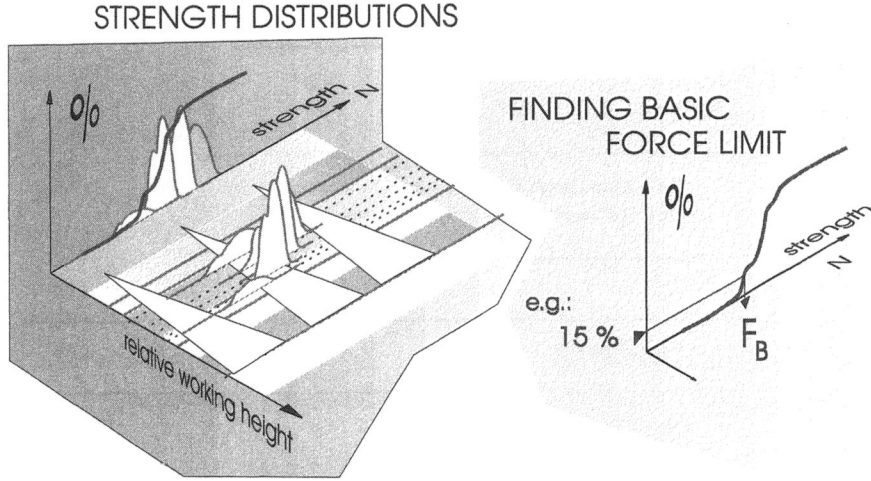

FIG. 15.10.   Finding combined strength distributions and basic force limits.

To define load limits in a reproducible way, percentile approaches were applied. Such an approach starts by combined strength distribution functions as found within well-defined target populations at selected activities. Such an approach allows force limits to be found easily in the following way (see Figure 15.10):

- Specify particular percentages of the user population that should be able to do the activities under consideration (here: 85%)
- Determine the percentage of people not capable—here: $110 - 85 = 15\%$
- Read the corresponding force limit (F_B) directly out of appropriate distribution functions

Activity Limits

Basic force limits (F_B) are estimate short-term capacity limits. Usually that kind of limit is far from any real activities of daily living. Hence, these short-term limits should be adjusted to real human life. In ISO 11228-2, this is done by a set of multipliers accounting for the effects of velocity, frequency, and duration (see Figure 15.11). This approach adopts the multiplier system of EN 1005/3 in European Standardization. Possibly this particular multiplier approach in ISO 11228-2 will be replaced by another procedure based on more recent findings.

Velocity Multiplier (m$_v$).   Human force-generating capacity is reduced with increasing speed of concentric movements. This is realized by velocity multiplier (m_V) as determined by Table 15.1.

Frequency multiplier (m$_f$).   Frequently repeated actions increasingly may lead to fatigue effects resulting in reversible reductions of human strength. Principally these effects depend on action time and cycle frequency—where each cycle duration may be broken down into its action time and its individual break. A frequency multiplier (m_f) was defined to describe fatigue effects. Details are found in Table 15.2. An overview is given in Figure 15.11.

TABLE 15.1
Velocity Multiplier (m_v) Depending on Moving Speed

m_v	1.0	0.8
Movement	No	Yes

"No" Action implies no or a very slow movement
"Yes" Action implies an evident movement

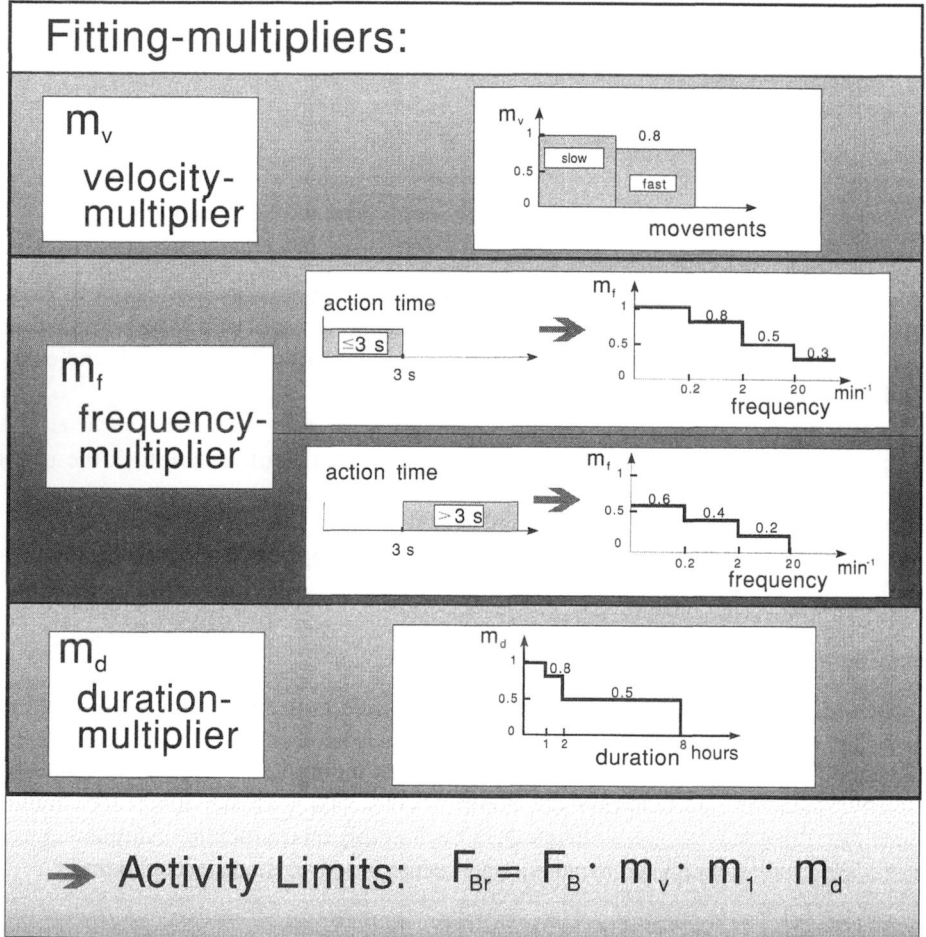

FIG. 15.11. A set of multipliers designed to adjust basic force limits (F_B) to real activities of daily living.

Duration multiplier (m_d).   Human force generating capacity is further reduced by the overall time spent on the job. Actions that are loading muscle structures may add up causing local fatigue. To regard those local effects it's not only overall action time that should be considered—in particular durations of similar actions seem to be important. That's why "similar

TABLE 15.2

Frequency Multiplier (m_f) Defined by Action Time and Cycle Frequency

Action Time (min)	< Frequency of Actions (min^{-1}) >			
	< 0.2	0.2–2	2–20	> 20
≤ 0.05	1.0	0.8	0.5	0.3
> 0.05	0.6	0.4	0.2	Not applicable

TABLE 15.3

Duration Multiplier m_d Depending on Cumulated Duration of Similar Actions

Duration	Duration (Working Time) in Similar Actions (hours)		
	≤ 1 hour	1–2 hours	2–8 hours
m_d	1.0	0.8	0.5

actions" are defined in the Standard by actions that are very similar in character (i.e., pushing and pressing).

Those duration effects are described by multiplier m_d. Here "duration" always refers to total duration of similar activities including interruptions.

Activity Limits (F_{Br}). Limits adjusted to real activities of daily living may be found by simple reduction of basic force limits F_B. In particular, F_B is reduced by the following set of multipliers accounting for the effects of velocity, frequency, and duration:

$$F_{Br} = F_B \times m_v \times m_f \times m_d,$$

where F_B is the basic force limit, m_v is the velocity multiplier (Table 15.1), m_f is the frequency multiplier (Table 15.2), and m_d is the duration multiplier (Table 15.3).

Skeletal Force Limits

In addition to muscular approaches (muscle force limits F_B and F_{Br}) ISO 11228-2 adopts spinal aspects as well. However, the procedure described here is a tentative approach and probably will be replaced by a more sophisticated method. Basically, this preliminary approach has two major steps:

Step 1:   estimates compressive force limits of lumbar spine
Step 1:   Finds action force limits

Compressive Force Limits

Human spinal strength apparently depends on effects of age and gender (see Figure 15.12; Jäger, 2001). That's why spinal limits should reflect these effects; that is, when demographic profiles change, spinal limits should change as well. To this end, this spinal approach basically adopts the same demographic philosophy that was realized in the muscle force approach (see Fig. 15.9).

The major steps are as follows:

- Start out with compressive strength as an experimental base (see Figure 15.12).
- Find regressions describing effects of age in both genders.
- Introduce age classes.
- Calculate distribution parameters (percentiles or average and standard deviation) of compressive strength in each age class.
- Generate log-distributions of compressive strength in all age classes.
- Check demographic profile of target population using age-intervals in Figure 15.12.
- Find weighting multipliers accounting for the demographic weight of each age class.
- Multiply each age class distribution by its weighting multiplier.

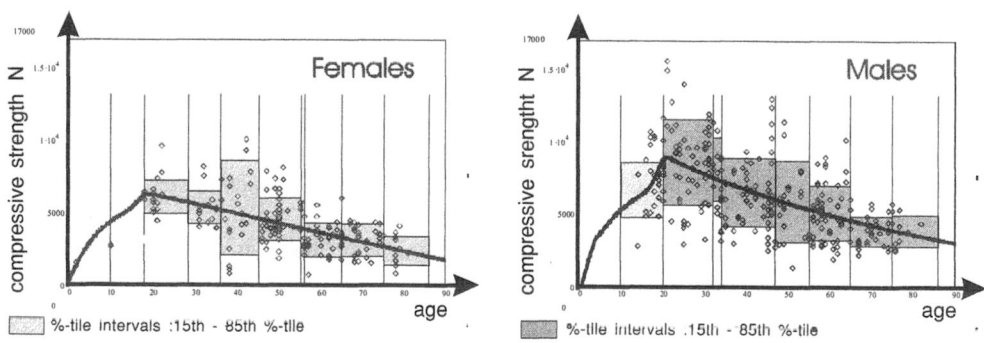

FIG. 15.12.   Compressive strength of lumbar spine (Jäger, 2001; Rosenberg, 2003).

Compressive force limits of lumbar spine

RATIO males/females %	ALL AGES 15 - 64 y kN	"ACTIVE" SENIORS 50 - 64 y kN
0 : 100	3.0	2.4
25 : 75	3.2	2.5
59 : 41	3.6	2.9
75 : 25	3.8	3.0
100 : 0	4.2	3.4

Target population:
EU 12 (1993)—working population of the 12 EU countries in 1993

FIG. 15.13.   Precalculated LBS force limits varying with selected user populations.

- Sum up all weighted age class distributions to get total spinal strength distributions of both genders.
- Integrate total strength distributions with increasing strength to get total Strength distribution functions of males and females.
- Determine 15th percentile to find compressive strength limits of lumbar spine.

Certainly those lumbar spine limits change with changing user populations. In Figure 15.13 a variety of precalculated LBS (lumbar spine) limits may be found estimating a set of preselected situations. Its characteristics are two different age groups—general EU-working population and "active" EU-seniors and a selection of specified ratios between males and females.

In fact these limits are changing in a rather "natural" way with changing profiles of target populations (see Figure 15.13).

Action Force Limits

In the following step, limitations of externally applied forces are determined in such a way that compressive force limits of lumbar spine are not exceeded. The procedure to determine those action force limits has not yet been designed in detail. That's why the following method provides a very rough approach that will be improved on in the near future.

At the moment the following procedure applies:

- *Determine* . . . average stature ($\bar{h}$) of target population suggesting anthropometry of an average user;

  e.g.,

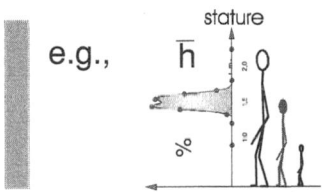

- *Select* . . . absolute working height h_w;

  e.g.,

- *Predict* . . . most probable working posture of an average user;

  e.g.,

- *Find* . . . shoulder joint angle (SJA);
- *Find* . . . force angle (FA);

  i.e.,

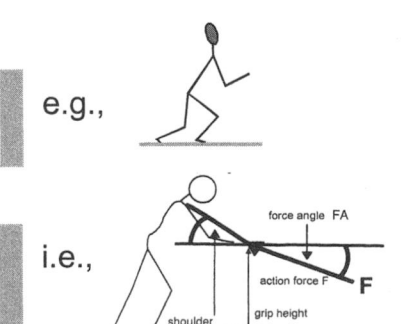

- *Check* ... appropriate charts—if available; for example, see Figure 15.14.

- *Determine* ... action force limit (F_{LS}) by the chart selected.

e.g.,

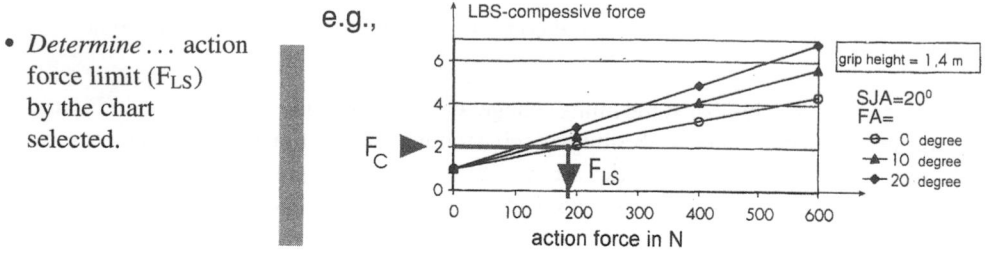

Minimal Force Limits

On this stage the evaluation procedure provides two capacity limits in particular made to assess the activities under consideration—these are muscle force limits (see Fig. 15.9) and skeletal load limits (see Fig. 15.12). That's why the procedure has to decide now which limit to adopt for further processing. Such a decision will be made on a safety basis. To make sure that no less than a predefined majoritiy is protected either way, the lower one of both limits should be selected.

That's why the procedure determines the minimum of muscular force limit F_{Br} and skeletal load limit F_{LS}:

$$F_{min} = Min(F_{Br}, F_{LS}).$$

PUSHING   PULLING

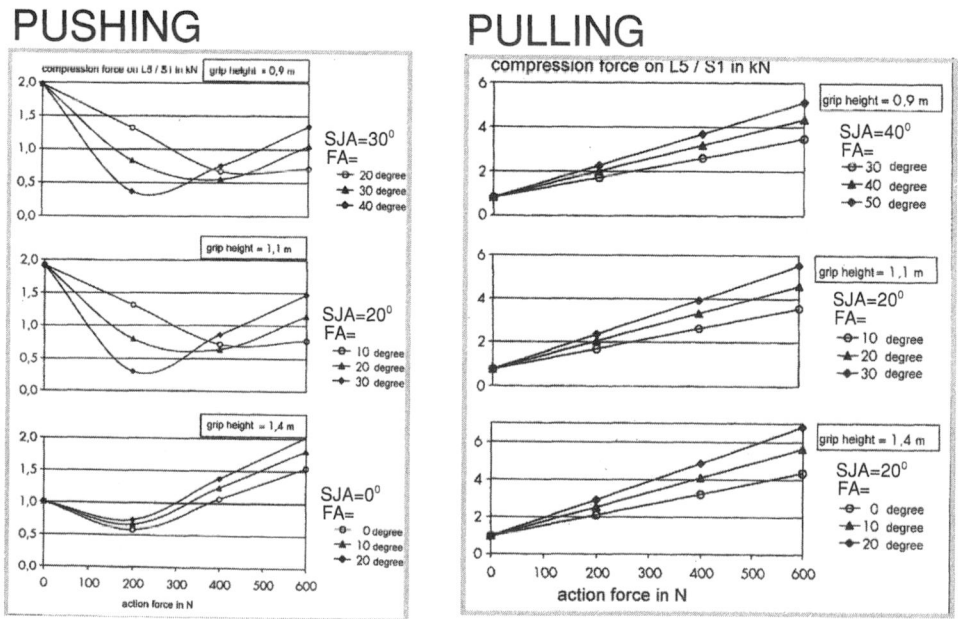

FIG. 15.14.   Compressive strength of lumbar spine depending on action forces at selected pushing or pulling activities (Jäger, 2001).

This way, resultant capacity limits F_{min} make sure that a specified safety level is not undercut in the evaluation procedure.

Safety Limits

Both muscle force limits (F_{Br}) and skeletal load limits (F_{LS}) are capacity limits allowing wide majorities to do specified activities of daily living. On the other hand, that kind of force limit does not necessarily guarantee safe jobs. Certainly health risks may be found below maximal force levels as well. Hence, above-capacity limits F_{min} should be transformed into safety limits. This is done by two multipliers m_r^{red} and m_r^{green} defining green and red safety limits F_R:

$$F_R^{red} = m_r^{red} \cdot F_{min} \quad \text{with } m_r^{red} = 0,7$$
$$F_R^{green} = m_r^{green} \cdot F_{min} \quad m_r^{green} = 0,5$$

Risk Assessment

Above safety limits F_R^{red} and F_R^{green} are drawing up a three-zone model providing three categories to assess easily any real load (L):

red zone:	load $L \geq F_R^{red}$
yellow zone:	$F_R^{green} < loadL < F_R^{red}$
green zone:	load $L \leq F_R^{green}$.

Interpretations depend on the zones accomodating load L, in particular:

Green zone:	No action is required.
Yellow zone:	Risk estimation shall be continued including other relevant risk factors and followed as soon as possible by redesign. Where redesign is not possible, other measures to control the risk shall be taken.
Red zone:	Action to lower the risk (e.g., redesign, work organization, worker instruction, and training) is necessary.

Example

To demonstrate the way ISO 11228-2 works, please see Figure 15.15. This example calculates basic force limits (F_B) when pushing. For demonstration purposes, limits were made for American and Japanese test populations widely differing in stature distributions. Further strength distributions of both nations were assumed to be identical. In each of both nationalities, limits were determined for a variety of predefined subpopulations specified in particular by a mix of age and gender.

Age Mixes

In this example, there are three age mixes arranging the general working population in three different ways:

- Junior—population concentrates in age group 1 (age < 20 years)
- All ages—equal distributions in all three age groups
- Senior—population concentrates in age group 3 (50 < age ≤ 65 years)

EXAMPLE

Scenario
- activity: pushing
- population: American vs. Japanese
- absolute operating height: 1,5 m

Assumptions
- identical strength distributions in both nationalities
- load limits: 15th %-tile

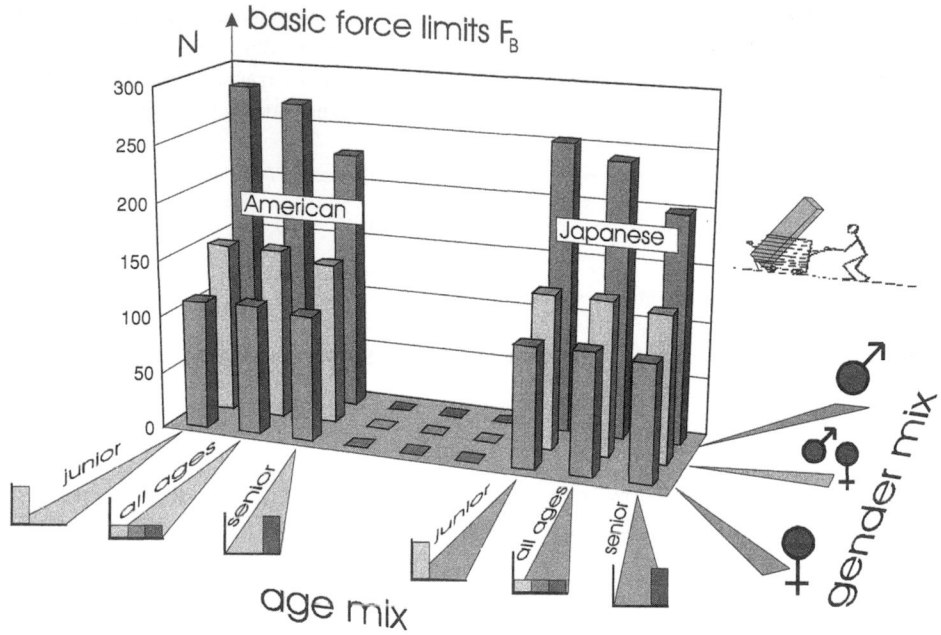

FIG. 15.15. Adaptive load limits of ISO 11228-2 varying within and between two different user populations.

Gender Mixes

Each of aforementioned age mixes, realizing three different ratios between males and females, is realized:

- Mix 1:   males/females =   0 : 100%
- Mix 2:   males/females =   50 : 50%
- Mix 3:   males/females = 100 : 0%

A combination of these mixes in age and gender further yields a 3×3 array as shown in Figure 15.15.

Finally basic force limits F_B are calculated

- at the place of each element in the array
- regarding both the American and the Japanese population.

Interpretations.   Results in Figure 15.15 quantitatively demonstrate the wide adaptivity of that innovative ISO procedure. Its limits do not only reflect given demographic profiles—in both nations limits are decreasing with increasing age and increasing female representation. Its limits additionally depend on "nationalities" as an effect of stature distributions in particular.

PERSPECTIVES

The risk-assessment procedure in ISO 11228-2 generally demonstrates the use of what might be understood as "target group ergonomics." That kind of target group orientation in the field of ergonomics provides very flexible limits adjustable to optional demographic profiles. This makes sure that each particular user group may be protected on a specified safety level—there is little overprotection and little underprotection. Global assessments usually do not work so precisely. Above all, it must be realized that any overprotection or underprotection may produce considerable costs—apart from all kinds of individual misfits. In this more economical sense, ISO 11228-2 seems to be very cost-effective. That is why it may be an interesting perspective to introduce that kind of target group ergonomics into other fields of ergonomics as well.

REFERENCES

DIN 33411-5. Human physical strength—maximal static action forces—Values.

90/269/EEC. (1990). Council Directive of 29 May 1990 on the minimum health and safety requirements for the manual handling of loads where there is a risk particularly of back injury to workers (fourth individual Directive within the meaning of Article 16 (1) of Directive 89/391/EEC).

EN 292-1. Safety of machinery—Basic concepts, general principles for design—Part 1: Basic terminology, methodology.

EN 292-2:1991. Safety of machinery—Basic concepts, general principles for design—Part 2: Technical principles and specifications.

EN 614-1. Safety of machinery—Ergonomic design principles—Part 1: Terminology and general principles.

EN 614-2. Safety of machinery—Ergonomic design principles—Part 2: Interactions between the design of machinery and work tasks.

EN 1005-1. Safety of machinery—Human physical performance—Part 1: Terms and definitions.

EN 1005-2. Safety of machinery—Human physical performance—Part 2: Manual handling of machinery and component parts of machinery.

EN 1005-3. Safety of machinery—Human physical performance—Part 3: Recommended force limits for machinery operation.

EN 1005-4. Safety of machinery—Human physical performance—Part 4: Evaluation of working postures in relation to machinery.

EN 1005-5. Safety of machinery—Human physical performance—Part 5: Risk assessment for repetitive handling at high frequency.

EN 1050. Safety of machinery—Risk assessment.

ISO 7730. Moderate thermal environments—Determination of the PMV and PPD indices and specification of the conditions for thermal comfort.

ISO 11226. Ergonomics–Evaluation of working postures.

ISO 11228-1. Ergonomics—manual handling—Part 1: Lifting and carrying.

ISO 14121. Safety of machinery—Principles of risk assessment.

ISO/IEC Guide 51. Safety aspects—Guidelines for their inclusion in standards.

Jäger M. (2001). Belastung und Belastbarkeit der Lendenwirbelsäule im Berufsalltag, Fortschr.-Ber. VDI Reihe 17 Nr. 208, VDI Verlag Düsseldorf, ISBN 3-18-320817–2.

Mital, A., Nicholson A. S., & Ayoub M. M. (1997). A guide to manual materials handling. 2nd edition. London: Taylor & Francis.

NIOSH. (1981). Work practices guide for manual lifting. U.S. Department of Health and Human Services (Technical Report, Publication No. 81-122). Cincinnati, OH:

Rosenberg, S. (2003). Human lumbar spine—structural stabilities and load limits, under publication. Inst. of Ergonomics, TU München, Boltzmannstr. 15, 85747 Garching, Germany.

Schaefer, P., Kapitaniak, B., & Schaub, Kh. (2000). Custom made load limits shaped by age, gender and stature distributions—A new ISO-approach. In Proceedings of the IEA 2000/HFES 2000 Congress, 4-349–4-351.

Schaub, Kh., Berg, K., & Wakula, J. (1997). Postural and workplace related influences on maximal force capacities. In IEA '97 (Ed.), From experience to innovation (Vol. 4, pp. 219–221). Tampere, Finland: The 13th Triennial Congress of the International Ergonomics Association.

Siemens. (1969). Lastentransport von Hand, Mitteilung aus dem Labor für angewandte Arbeitswissenschaften, Nr. 6.

16

Recommended Force Limits for Machinery Operation: A New Approach Reflecting User Group Characteristics

Peter Schaefer
München University of Technology

Karlheinz G. Schaub
Darmstadt University of Technology

INTRODUCTION

Industrial production generally seeks to meet the demands of a wide and heterogene clientele as precisely as possible. Hence, as user profiles change, the design of industrial products is changing as well. To ensure good interaction between man and machine, technical aids generally should reflect major characteristics of envisaged user populations (e.g., demographic profiles as described by age and gender). That kind of target group orientation not only adresses classical anthropometrics but also influences manipulation forces and thus may directly affect health protection.

The general problem is how to protect different user populations by the same safety level— in other words, how to make sure that all user groups are really taking comparable risks. The European Union (EU) is defining safe products by CEN standardization. Some of these CEN safety limits largely depend on the profile of the intended target population—so do force limits in particular. That's why EN 1005/3 introduced force limits reflecting demographic profiles in particular. These characteristics are given by distributions of age and gender.

Such an evaluation procedure is absolutely new in the field of load rating. Traditional approaches either refer to pure male or pure female populations or to a fixed mix of both (e.g., NIOSH, 1981; Siemens, 1969). It is an outstanding characteristic of EN 1005/3 that its special load limits react directly to the slightest changes in demographic profiles.

BASIC ELEMENTS

This particular load evaluation procedure in EN 1005/3 combines two basic elements.

Percentile Approach

To define load limits in a reproducible way, percentile approaches were applied. Such an approach simply starts by isometric strength distributions as found within well-defined target populations and selected activities. Results may be readily displayed by strength distribution functions. Such an approach allows force limits to be found easily in the following ways:

- Specify particular percentages of the user population supposed to be capable of doing the activities under consideration (here: 85%)
- Read the corresponding force limit directly out of appropriate distribution functions

Synthetic Disributions

To realize target group sensibility of force limits, a way was found to synthesize strength distributions of arbitrary target populations as described by age and gender. The aforementioned percentile approach finally yields force limits adjusted to optional user populations.

ELEMENTARY STEPS

Percentile approach and the idea to synthesize strength distributions are two major elements on the way to adjustable force limits. These elements are part of a three-step evaluation procedure.

Step 1: Basic Force Limits

In a first step, basic force limits are calculated. To this end, a combined strength distribution function of the intended user population is synthesized. Force limits may be found by predefined percentiles of these distributions. In EN 1005/3, the 15th percentile was selected. Such a choice makes sure that an 85% majority of the envisaged user population is able to do its job.

Step 2: Activity Limits

The aforementioned force limits describe human strength sustained for a few seconds only. That's why such short-term capacity limits must be adjusted to real user activities. In EN 1005/3, this is done by a set of multipliers reflecting real user situations.

Step 3: Safety Limits

Generally, capacity limits do not necessarily ensure a safe job. To realize safety limits, a special risk multiplier was introduced transforming the capability limits into safety limits (Figure 16.1).

BASIC FORCE LIMITS F_B

Basic force limits may be directly calculated when starting by human strength data. They reflect some basic short-term force-generating capacity. To this purpose, EN 1005/3 provides two alternative ways to calculate F_B.

Procedure 1

Procedure 1 is a quick way to determine force limits. This procedure assumes equal representations of males and females and may be applied

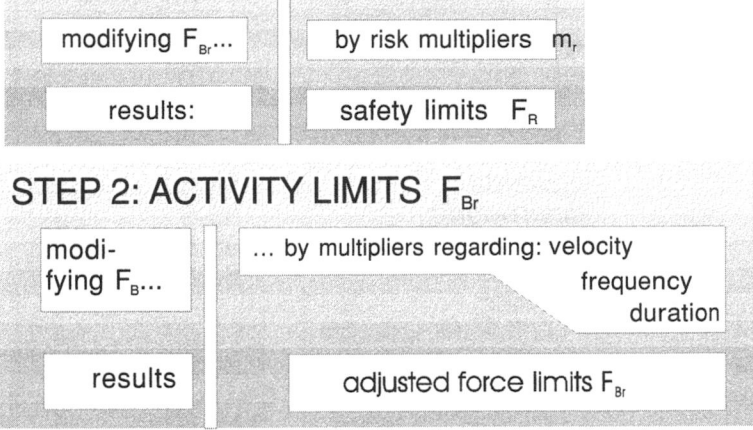

STEP 3: SAFETY LIMITS F_R

modifying F_{Br}...	by risk multipliers m_r
results:	safety limits F_R

STEP 2: ACTIVITY LIMITS F_{Br}

modi-fying F_B...	... by multipliers regarding: velocity frequency duration
results	adjusted force limits F_{Br}

STEP 1: BASIC FORCE LIMITS F_B

target population...	... is not known	... is reasonably well known
calculation procedure	Procedure 1	Procedure 2
demographic profile	total working popu-lation (m/f: 50/50 %)	as specified by the designer
results	basic force limit F_B	

FIG. 16.1. A three-step model to put up force limits.

- If the envisaged user profile is similiar to the general European population
- If demographic details simply are not available

Procedure 2

On the other hand, procedure 2 applies if age and gender distributions of an envisaged user population are reasonably well known. In this case it's up to procedure 2 to provide load limits that reflect demographic profiles in particular.

Procedure 1

Input Parameters

In a first step, manipulations of the human operator should be analyzed to pinpoint most hazardous activities. Such an analysis yields distribution parameters (average and standard deviation) of human strength that may be determined either by tables or by measurements. These strength data ideally should represent the adult European working population. Practically, it is recommended to start with distribution parameters of female reference groups only. These female distribution parameters allow reasonably good predictions of force limits F_B including both genders. Such a procedure is integrated in EN 1005/3 (see Figure 16.2).

Activity		$\bar{F}$ [N]	σ [N]
	hand work (one hand) power grip	278.0	62.2
arm work (sitting posture, one arm)			
- upwards		58.0	18.4
- downwards		88.6	33.2
- outwards		65.5	26.2
- inwards		85.6	24.6
- pushing			
- with trunk support		312.0	84.8
- without trk. support		78.0	42.7
- pulling			
- with trunk support		246.0	45.7
- without trk. support		67.9	33.5
whole body work (standing posture)			
- pushing		233.7	81.0
- pulling		164.6	44.9
pedal work (sitting posture, with trunk support)			
- ankle action		293.4	104.7
- leg action		542.5	156.2

FIG. 16.2. A selection of strength distribution parameters (reference group: adult female population).

Its basic characteristics and inputs are:

- *Target population:* adult (e.g., European) working population
- *Reference group:* adult female population

- *Distribution parameters:* average force $\bar{F}$ and standard deviation σ of reference group

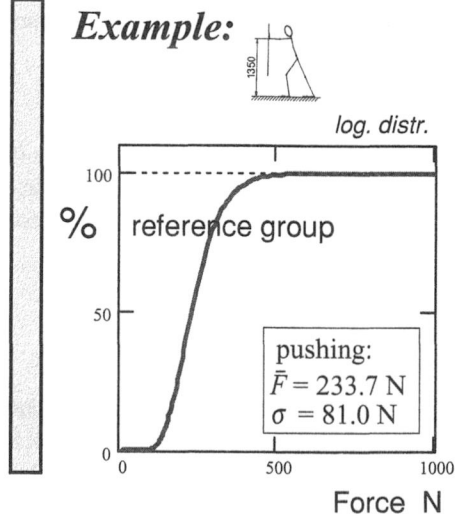

Example:

log. distr.

pushing:
$\bar{F} = 233.7$ N
$\sigma = 81.0$ N

Approximation. If data of female reference groups are not available, distribution parameters of young females (between 20 and 30 years) may be used approximatively (see procedure 2).

Calculations

Force Distribution.   Average and standard deviation define strength distribution functions DF(x). Such an approximation to normal is a good and easy way to determine force limits in most practical applications.

Logarithmic Transformation.   Force limits are more reliable when shifting over to logarithmic normal distributions:

$$\bar{F}_{\ln} = \ln \bar{F} \qquad \sigma_{\ln} = \ln \frac{\bar{F} + \sigma}{\bar{F}}$$

> **Example**
> $$\bar{F}_{\ln} = \ln 233.7 = 5.45$$
> $$\sigma_{\ln} = \ln \frac{233.7 + 81}{233.7}$$
> $$= 0.30$$

Strength Percentiles.   Starting by aforementioned distribution parameters, $\bar{F}_{\ln}$ and $\sigma_{\ln}$, logarithmic force percentiles ($\bar{F}_{\ln \%}$) may be calculated:

> **Example**
> $$F_{\ln 15\%} = 5.45 - 0.5244 \cdot 0.30$$
> $$= 5.30$$
> $$F_{\ln 1\%} = 5.45 - 2.0537 \cdot 0.30$$
> $$= 4.84$$

$$F_{\ln \%} = \bar{F}_{\ln} + z_\% \cdot \sigma_{\ln}.$$

Referring to the 15th or 1st percentile of the target group, $z_\%$ amounts to

$$z_{15\%} = -0.5244$$
$$z_{1\%} = -2.0537$$

with $z_{15\%} = z_{15\%}^{\text{gen. popul.}} = z_{30\%}^{\text{females}} = -0.5244$ and $z_{1\%} = z_{1\%}^{\text{gen. popul.}} = z_{2\%}^{\text{females}} = -2.0537$.

A simple transformation back to linear finally yields appropriate percentiles $F_\%$:

$$F_\% = e^{F_{\ln \%}} \text{N}$$

> **Example**
> $$F_{15\%} = e^{5.3} = 200 \,\text{N}$$
> $$F_{1\%} = e^{4.84} = 127 \,\text{N}$$

Results

Both percentiles $F_{15\%}$ and $F_{1\%}$ are defining basic force limits F_B:

$$F_B = \begin{cases} F_{15\%} & \text{for professional use} \\ F_{1\%} & \text{for domestic use} \end{cases}$$

> **Example**
> $$F_B = 200 \, N$$
> *for professional use*

These force limits allow specified activities up to 85% or 99% of the adult (e.g., European) working population without exceeding their physical capacity.

Procedure 2

Input Parameters

Forces.   To be started, procedure 2 requires distribution parameters of a defined female reference group (also see Figure16.3).
The details are:
reference group:

- females
- $20 \leq$ age ≤ 30 years

distribution parameters:

- average force $\bar{F}$ and
- standard deviation σ of reference group

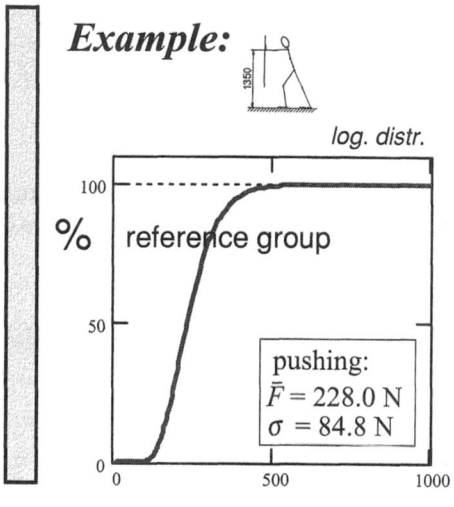

Example:

log. distr.

reference group

pushing:
$\bar{F} = 228.0$ N
$\sigma = 84.8$ N

Force N

Certainly, reference forces change with changing individual strategies and activities.

Approximation.   If no data of young female reference groups are available, distribution parameters of procedure 1—adult female population—may be used, approximately.

User Demography.   Next, the demographic profile of the envisaged user population is specified. Results may be arranged in a pattern of age and gender:

females : $n_{f1}\%$: age < 20 years
$n_{f2}\%$: $20 \leq$ age ≤ 50 years
$n_{f3}\%$: $50 <$ age ≤ 65 years

males : $n_{m1}\%$: age < 20 years
$n_{m2}\%$: $20 \leq$ age ≤ 50 years
$n_{m3}\%$: $50 <$ age ≤ 65 years

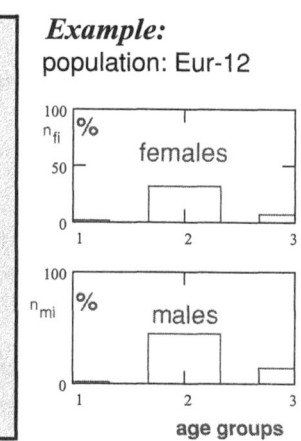

Example:
population: Eur-12

females

males

age groups

Activity		$\bar{F}$ [N]	σ [N]
	hand work (one hand) power grip	270.0	54.1
in ◄ ► out push (head figure) ▲ ▼ pull	arm work (sitting posture, one arm)		
	- upwards	56.0	18.4
	- downwards	86.0	33.2
up (torso figure) ▲ ▼ down	- outwards	63.5	26.2
	- inwards	83.4	24.6
	- pushing		
	- with trunk support	303.0	81.0
	- without trk. support	75.5	42.7
	- pulling		
	- witht runk support	242.0	44.9
	- without trk. support	65,7	33.5
(standing figure)	whole body work (standing posture)		
	- pushing	228.0	84.8
	- pulling	161.0	45.7
(pedal figures)	pedal work (sitting posture, with trunk support)		
	- ankle action	282.0	96.5
	- leg action	528.5	157.6

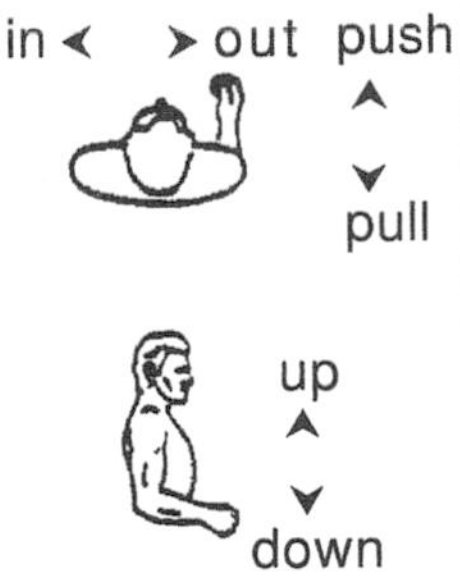

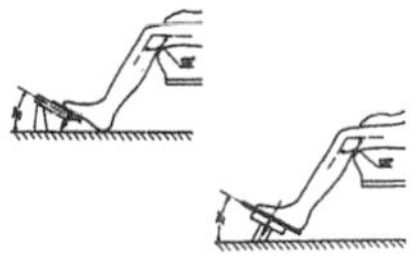

FIG. 16.3. A selection of strength distribution parameters (reference group: females between 20 and 30 years of age.)

with

n_{fi}, n_{mi} percentages of subgroups as found within any user population.

n_{f1}= 1.6 %	n_{m1}= 2.0 %
n_{f2}= 31.6 %	n_{m2}= 43.8 %
n_{f3}= 7.6 %	n_{m3}= 13.4 %

Procedure

Procedure 2 calculates force limits adjusted to user populations as previously specified. The procedure works as follows.

Generation of Subgroup Distributions.   Force averages and standard deviations of all other subgroups are simply calculated by aforementioned reference parameters $(\bar{F}, \sigma)$, and some appropriate multipliers (a_{xx}, s_{xx}) expressing relations between age and gender (Schaefer, et al., 1997; Rühmann & Schmidtke, 1992):

Females
Average forces: $\bar{F}_{fi} = \bar{F} * a_{fi}$
SD: $\sigma_{fi} = \sigma * s_{fi}$

Males
Average forces: $\bar{F}_{mi} = \bar{F} * a_{mi}$
SD: $\sigma_{mi} = \sigma * s_{mi}$

with

Averages a_{xx}				Standard Deviations s_{xx}			
Age Groups	1	2	3	Age Groups	1	2	3
Females a_{fi}	0.96	1.00	0.93	Females s_{fi}	1.03	1.00	0.96
Males a_{mi}	1.95	2.16	1.70	Males s_{mi}	1.57	1.65	1.81

and

$i = 1 \ldots 3$:   Age groups

a_{xx}, s_{xx}:   Subgroup multipliers

$\bar{F}, \sigma$:   Average force and standard deviation of reference group

Example

Age Groups	1	2	3
$\bar{F}_{fi}$	172.8	180.0	167.4
σ_{fi}	61.8	60.0	57.6
$\bar{F}_{mi}$	351.0	388.8	306.0
σ_{mi}	94.2	99.0	108

Logarithmic Distributions.   When approaching lower force levels above approximation to normal yields increasingly poor results at lower percentiles (e.g., 1%). In this case logarithmic distributions are more realistic. An easy transformation is calculating a new set of logarithmic distribution parameters:

$$females: \quad \bar{F}_{fi}^{L} = \ln(\bar{F}_{fi}) \qquad \sigma_{fi}^{L} = \ln \frac{\bar{F}_{fi} + \sigma_{fi}}{\bar{F}_{fi}}$$

$$males: \quad \bar{F}_{mi}^{L} = \ln(\bar{F}_{mi}) \qquad \sigma_{mi}^{L} = \ln \frac{\bar{F}_{mi} + \sigma_{mi}}{\bar{F}_{mi}}$$

Generating Subgroup Distribution, Functions.

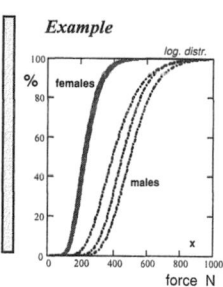

Example

females:

$$DF_{fi}(x) = \frac{1}{\sigma_{fi}^{L} \sqrt{2\pi}} \int_{-\infty}^{\ln x} e^{-z_{fi}^2/2} dz$$

with

$$z_{fi} = \frac{\ln x - \bar{F}_{fi}^{L}}{\sigma_{fi}^{L}}$$

x: forces

males:

$$DF_{mi}(x) = \frac{1}{\sigma_{mi}^{L} \sqrt{2\pi}} \int_{-\infty}^{\ln x} e^{-z_{mi}^2/2} dz$$

with

$$z_{mi} = \frac{\ln x - \bar{F}_{mi}^{L}}{\sigma_{mi}^{L}}$$

Weighting and Combining Subgroup Distributions.

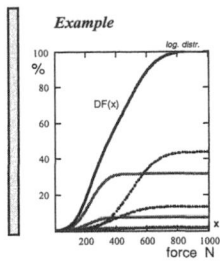

Example

$$DF(x) = \sum_i \left(n_{fi} DF_{fi}(x) + n_{mi} DF_{mi}(x) \right) / 100$$

Percentiles. DF(x) is the combined distribution function of all subgroups depending on force x. So force limits may be found by calculating the 15th or the 1st percentile of DF(x):

$$DF(x) = \begin{cases} 0.15 & \text{for professional use} \\ 0.01 & \text{for domestic use} \end{cases}$$

$$\Rightarrow \text{force } x$$

Results. Above percentile approach yields basic force limits F_B:

$$F_B = x[\text{N}]$$

Example

$$F_B = 200.2 \quad \text{N}$$

These limits allow specified activities up to 85% or 99% of any optional user population without exceeding their physical capacity.

ACTIVITY LIMITS

Basic force limits F_B are describing short-term capacity limits. Usually that kind of limit is far from any activities of daily living. Hence, these short-term limits should be adjusted to real human life. In EN 1005/3, this is done by a set of multipliers accounting for effects of velocity, frequency, and duration (see Figure 6.4).

Velocity Multiplier m_V

Human force generating capacity is reduced with increasing speed of concentric movements. This is covered by velocity multiplier m_V as determined by Table 16.1.

TABLE 16.1

Velocity Multiplier m_v Depending on Moving Speed

m_v	1.0	0.8
Movement	No	Yes

"No"	Action implies no or a very slow movement.
"Yes"	Action implies an evident movement.

TABLE 16.2
Frequency Multiplier m_f Defined by Action Time and Cycle Frequency

Action Time (min)	Frequency of Actions(min^{-1})			
	< 0.2	0.2–2	2–20	> 20
≤ 0.05	1.0	0.8	0.5	0.3
> 0.05	0.6	0.4	0.2	Not applicable

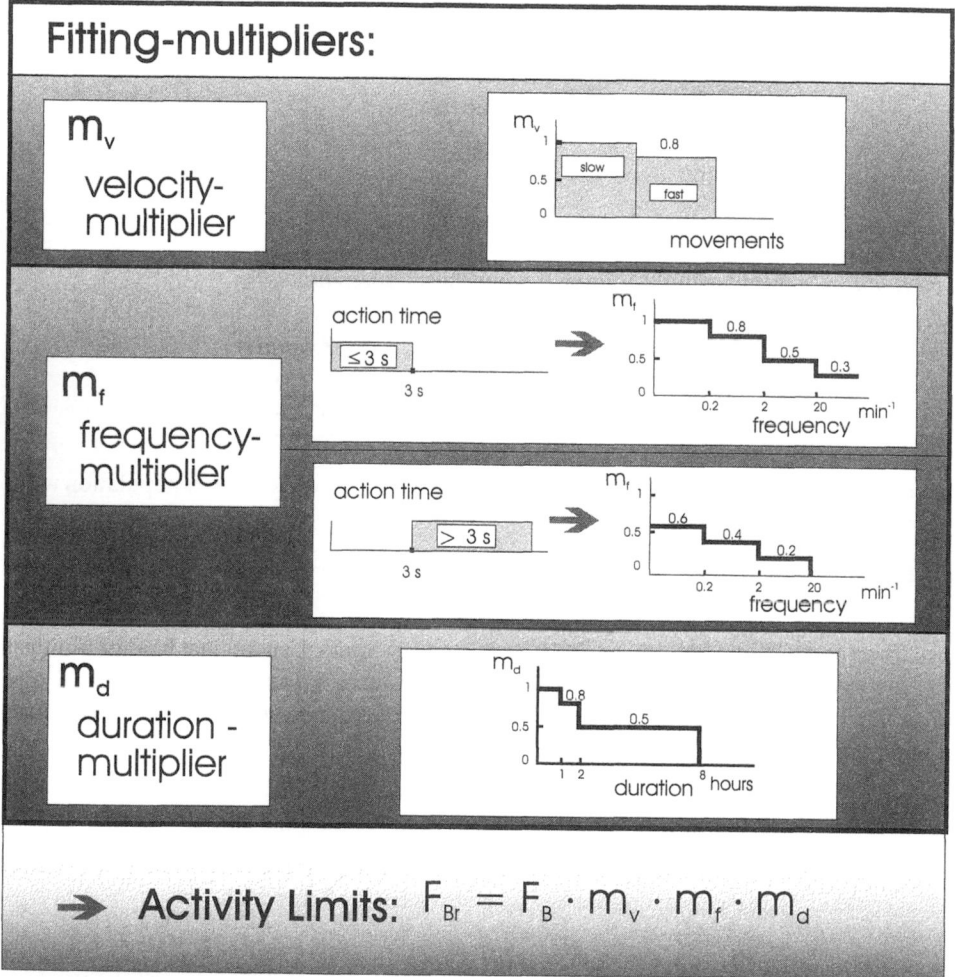

FIG. 16.4.   A Set of multipliers designed to adjust basic force limits F_B to activities of daily living.

Frequency Multiplier m_f

Frequently repeated actions increasingly may lead to fatigue effects that usually make human force generating capacity decrease. Principally, these effects depend on action time and cycle frequency—where each cycle duration is made up by its action time and its individual break.

TABLE 16.3
Duration Multiplier m_d Depending on the Cumulated
Duration of Similar Actions

Duration	duration (Working Time) in Similiar Actions (hr)		
	≤1 hr	1–2 hr	2–8 hr
m_d	1.0	0.8	0.5

A frequency multiplier m_f was defined to describe fatigue effects. Details may be found in Table 16.2. An overview is given in Figure 16.4.

Duration Multiplier m_d

Human force generating capacity is further reduced by the overall time spent on the job. Actions loading muscle structures may add up causing local fatigue. To regard those local effects, it is not only overall action time that should be considered—but also, in particular, durations of similar actions. That is why "similar" actions are defined in the Standard by actions that are very similar in character (i.e., pushing and pressing).

Those duration effects are described by multiplier m_d (see Table 16.3). Here "duration" always refers to total duration of similar activities including interruptions.

Activity Limits F_{Br}

Limits adjusted to activities of daily living may be found by simple reduction of basic force limits F_B. In particular, F_B is reduced by the previous set of multipliers assessing effects of velocity, frequency, and duration:

$$F_{Br} = F_B \times m_v \times m_f \times m_d,$$

where F_B is basic force limit, m_v is velocity multiplier, m_f is frequency multiplier, and m_d is duration multiplier.

SAFETY LIMITS

Activity limits are capacity limits allowing wide majorities specified activities of daily living. That kind of force limit does not necessarily guarantee safety on the job. Certainly health risks may be found below maximal force levels as well. Hence, above capacity limits should be reduced to safety limits. This is done by two different multipliers, m_r^{red} and m_r^{green}, defining green and red safety limits F_R:

$$F_R^{red} = m_r^{red} \cdot F_{Br} \qquad \text{with} \quad m_r^{red} = 0.7$$
$$F_R^{green} = m_r^{green} \cdot F_{Br} \qquad m_r^{green} = 0.5$$

TABLE 16.4
Risk Multiplier m_r Defining Risk
Zones

Risk Zone	m_r
Recommended	≤ 0.5
Not Recommended	0.5–0.7
To Be Avoided	>0.7

RISK ASSESSMENT

Above safety limits F_R^{red} and F_R^{green} define a three zone model providing categories for assessing easily any real load L:

$$\text{Red zone:} \qquad \text{load L} \geq F_R^{red}$$

$$\text{Yellow zone:} \qquad F_R^{green} < \text{load L} < F_R^{red}$$

$$\text{Green zone:} \qquad \text{load L} \leq F_R^{green}$$

Interpretations depend on the zones accommodating load L in particular: (see also Table 16.4)

Green zone: Recommended zone—the risk of disease or injury is negligible. No intervention is needed.

Yellow Zone: Zone is not recommended—risks of disease or injury cannot be neglected. Additional risk estimation should be done considering other risk factors. Risk analysis may yield acceptable risks even in the yellow zone. If, on the other hand, analysis is reconfirmed a risk re-design or other measures may be needed.

Red zone: Zone to be avoided—risks of disease or injury are obvious and cannot be accepted. Further activities to lower risks are necessary.

EXAMPLE

To demonstrate the way EN 1005/3 works, an example is given in Figure 16.5. This example presents selected activity limits F_{Br} when pushing as described by the input data in Figure 16.5. These limits are allocated to a set of predefined target populations. Each of these populations is specified in particular by a mix of age and gender.

Age Mixes

Mix 1: young people—all people accommodated by age group 1
Mix 2: all ages—equal distributions in all three age groups
Mix 3: seniors only—all people accommodated by age group 3

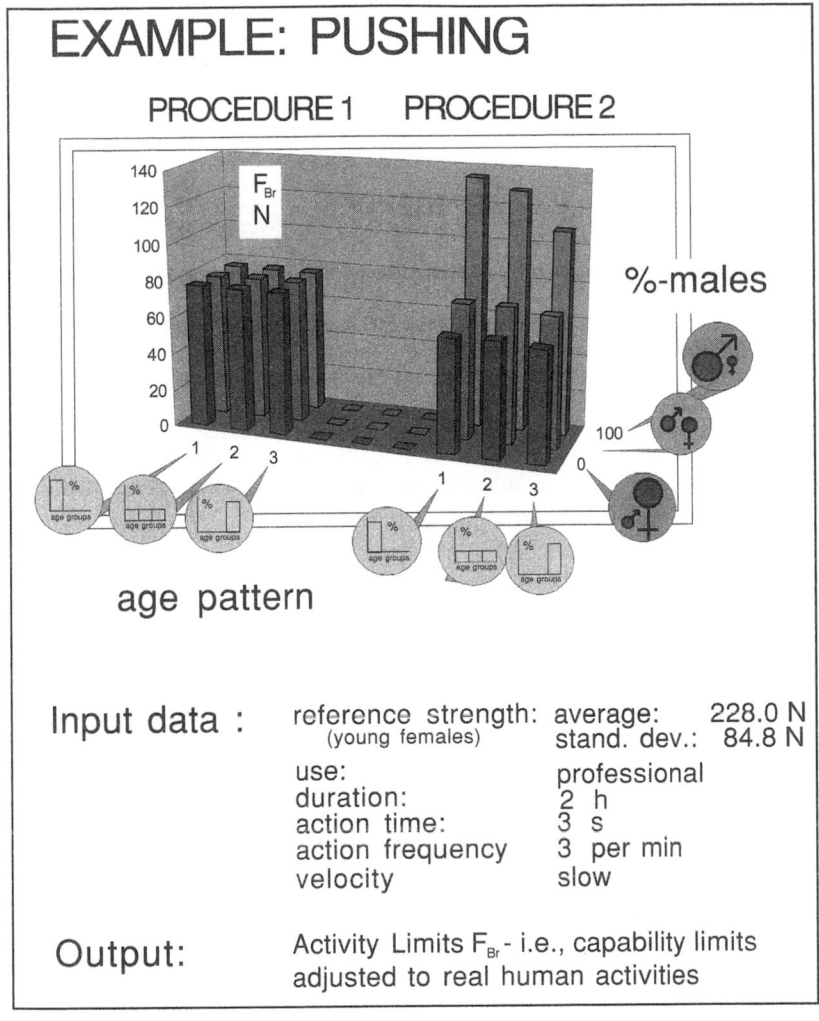

FIG. 16.5. Activity limits calculated with selected demographic profiles.

Gender Mixes

Each of the previous age mixes realizes three different ratios between males and females:

Mix 1: males and females = 0% : 100%
Mix 2: males and females = 50% : 50%
Mix 3: males and females = 100% : 0%

A combination of the previous mixes in age and gender further yields a 3 × 3 array as shown in Figure 16.5. Finally activity limits are calculated at the place of each element in the array and by procedure 1 and procedure 2.

Interpretation

Results in Figure 16.5 quantitatively demonstrate the way force limits depend on given demographic profiles. Obviously limits decrease not only with increasing age but also when changing from male to female populations.

Such a "natural behavior" of force limits is provided by procedure 2 only. It is obvious that procedure 1 is absolutely resistant to any changes in demographic profiles. Hence, these results demonstrate that procedure 2 is able to reflect demographic profiles quantitatively, whereas procedure 1 continues old load rating traditions.

REFERENCES

EN 1005-3. Safety of machinery—Human physical performance—Part 3: Recommended force limits for machinery operation.

NIOSH. (1981). *Work Practices Guide for Manual Lifting*. Cincinnati, OH: U.S. Department of Health and Human Services (Technical Report, Publication No. 81-122).

Rühmann H., & Schmidtke H. (1992). Körperkräfte des Menschen, Otto Schmidt KG Köln, Dokumentation Arbeitswissenschaft, Bd. 31, ISBN 3-504-65637-9.

Schaefer, P., Rudolph, H., & Schwarz, W. (1997). Variable force limits for optional target populations—A new approach realized in CEN—*Standardization, Proceedings of the 13th IEA Conference, Tampere, 4*, 533–535.

Siemens. (1969). Lastentransport von Hand, Mitteilung aus dem Labor für angewandte Arbeitswissenschaften, Nr. 6.

17

Guidelines for the Prevention of Work-Related Musculoskeletal Disorders: The Italian Experience

Enrico Occhipinti
Daniela Colombini
Antonio Grieco
Research Unit, Ergonomics of Posture and Movement (EPM)

INTRODUCTION

Introduction in Italy, in 1994, of general and more ergonomically oriented norms, derived from a series of EU Directives in OSH at the workplace (European Directives 391/89, 90/269, 270/90), called the attention of several stakeholders (employers, trade unions, OH&S professionals, and public authorities) to work-related musculoskeletal disorder (WMSD) prevention issues.

In particular, acknowledgment of European Directive 90/269 (1990) concerning manual handling of loads resulted in the adoption of spinal risk assessment and management procedures affecting over 4 million workers engaged in physically heavy tasks (about 20% of the laborforce).

On the other hand, the increasing reports of occupational upper-limb WMSDs (by now in second place among most frequently reported work-related diseases) induced the national and regional authorities, in the absence of a specific regulation on the subject, to issue guidelines for assessing and managing the risk from upper-limb repetitive movements potentially involving about 6 million workers (30% of general workforce), mainly from the manufacturing industry.

On this basis, the major points of the following Italian guidelines are summarized and discussed:

- The guidelines for the application of European Directive 90/269 on manual handling of loads, prepared by the Authors and officially adopted by the Conferenza dei Presidenti delle Regioni Italiane (1996) and by the National Institute for Safety and Prevention at Work (ISPESL)
- The guidelines (and a related application program) for prevention of upper-limb WMSDs connected to repetitive movements and exertions, prepared by the Authors and officially adopted by the Lombardy Regional Government (2004). In this view, it is worth underlining that so far no national governmental guidelines have been issued and, of the Italian

regions, Lombardy is the most densely populated (9 million inhabitants) and industrialized (more than 4 million workers).

GUIDELINE CONTENT

Manual Handling of Loads (Directive EC 90/269)

Exposure Assessment

Risk assessment, that is the individual or collective probability of contracting dorsolumbar spinal disorders due to manual load handling, is one of the pillars of preventive measures required by new European regulations and intervention methodologies in the field of ergonomics.

It is worth recalling that an extremely simplified interpretation of the assessment concept has become standard practice in workplaces and in other applications according to which, for example, load handling may be assessed solely on the basis of the load weight (as stated in old national regulations and standards).

This was widespread practice in Italy: In fact, a weight limit (30 kg) was introduced in the legislation to incorporate the EC Directive in Italian law, hence, the simplification that all objects weighing less than 30 kg may be handled in "safety"!

With such a scenario, when defining the appropriate tools for risk (or, better, exposure) assessment, the Authors had to redefine the requirements for assessment validity and applicability, not only leaving aside the rigors of a sophisticated scientific approach as a prerogative of research elites but also opposing the oversimplification demanded by operators in the field (Grieco, Occhipinti, Colombini, & Molteni, 1997)

Assessment of Lifting Tasks. As regards assessment of manual lifting, the models proposed by the guideline are based on the revised NIOSH equation (Waters, Putz Anderson, Garg, & Fine, 1993). Obviously, this occurs via major adaptations and changes in the original model as, on the other side, was suggested by a prEN draft standard (prEN 1005-2; CEN, 2002).

For large-scale application of the model, it seemed (and actually was) useful to propose graphic and procedural simplifications as reported in Table 17.1, which shows a sheet for collection and processing of all the data required for calculating the lifting index. Note that on the proposed sheet, reference is made to a weight whose details are to be identified according to local situations and the degree of protection that is to be assured to adult working population in accordance with the Directive.

The sheet, and so the underlying operating procedure, has been widely appreciated in Italy, and it is similar to the one reported in the final version of prEN 1005-2 (CEN, 2002). It is important to underline that this document, on account of the impossibility of achieving a univocally established load constant (maximum recommended weight under ideal lifting conditions), proposes a range of possible constants indicating, as required, the relevant "target" population as well as the degree of its presumed protection.

It should be further stressed that adopting the NIOSH model to make assessments in the field of manual load lifting tasks did, however, pose some problems that can be schematically described in the following.

1. The American authors themselves (Waters et al., 1993) emphasize that the procedure is not applicable in some situations: Such caution is quite understandable from a strictly scientific viewpoint, but in some cases caution may be overcome by making assumptions based on empirical data. When, for example, the load is lifted with one arm only, the prEN 1005-2 proposes introducing a further multiplier of 0.6. If lifting is carried out by two or more operators, it was proposed to consider, as the weight actually lifted, the weight of the object divided by the number of operators, and for the recommended weight to introduce a further

TABLE 17.1
Datasheet for the Evaluation of a Lifting Task (Adapted from Italian Guidelines)

LOAD CONSTANT (KG)

USER POPULATION	MALE	FEMALE		LC

VERTICAL MULTIPLIER (VM)

HEIGHT (cm)	0	25	50	75	100	130	150	>175
FACTOR	0.78	0.85	0.93	1.00	0.93	0.84	0.78	0.00

VM

DISPLACEMENT MULTIPLIER (DM)
VERTICAL LOAD DISPLACEMENT DISTANCE BETWEEN ORIGIN AND DESTINATION OF LIFTING

DISTANCE (cm)	25	30	40	50	70	100	170	>175
FACTOR	1.00	0.97	0.93	0.91	0.88	0.87	0.86	0.00

DM

HORIZONTAL MULTIPLIER (HM)

DISTANCE (cm)	25	30	40	50	56	60	>63
FACTOR	1.00	0.83	0.63	0.50	0.45	0.42	0.00

HM

HORIZONTAL DISTANCE BETWEEN HANDS AND MIDPOINT OF ANKLES

ASYMMETRIC MULTIPLIER (AM)

ANGULAR DISPLACEMENT OF LOAD (IN DEGREES)

ANGLE	0°	30°	60°	90°	120°	135°	>135°
FACTOR	1.00	0.90	0.81	0.71	0.62	0.57	0.00

AM

COUPLING MULTIPLIER (CM)

COUPLING MULTIPLIER	QUALITY	GOOD	POOR
	FACTOR	1.00	0.90

CM

FREQUENCY MULTIPLIER (FM)- *ACTION FREQUENCY (N. OF ACTIONS/MIN.)* **VERSUS DURATION**

FREQUENCY MULTIPLIER	FREQUENCY	0.20	1	4	6	9	12	>15
	CONTINUOUS < 1 HOUR	1.00	0.94	0.84	0.75	0.52	0.37	0.00
	CONTINUOUS 1-2 HOURS	0.95	0.88	0.72	0.50	0.30	0.21	0.00
	CONTINUOUS 2-8 HOURS	0.85	0.75	0.45	0.27	0.15	0.00	0.00

FM

ACTUAL LOAD (in kg)   RWL = LCxVMxDMxHMxAMx CMxFM   kg

multiplier of 0.85. Such adjustments tend to reply to widespread applicative problems that would remain unsolved. In this sense, further proposals aimed at favoring an everincreasing practical applicability of the method are no doubt to be encouraged.

2. In many working situations, the same group of workers has to carry out different lifting tasks often in the same workshift. The different lifting tasks may be irregular in a given period of time in the workshift (e.g., in a warehouse with picking activities) or according to established

time sequences (e.g., when an operator works every 1–2 hours on an assembly line, first loads the line, then unloads the finished products, and then packs them). In such cases, the analytical procedure for each task is not suitable to summarize the overall exposure of the individual worker (or of the group of) to lifting. Therefore these cases require an analytical procedure for multiple tasks which is obviously more complex (Waters, Putz Anderson, & Garg, 1994).

3. The NIOSH assessment procedure is not well suited to application in various working sectors (typically nonindustrial sectors), sometimes on account of the characteristic of the lifted load, the great variability of lifting tasks, their frequent association with other manual handling tasks (trolley pulling or pushing) , and finally the presence of other risk factors for the lumbar spine (e.g., whole-body vibrations). Agriculture, transport and delivery of goods, and assistance to individuals who are not self-sufficient (at home or in the hospital) are typical examples. In these situations, though the NIOSH lifting index is useful, validated procedures for integrated exposure assessment are not yet available, hence, the need for further research and proposals of specific simplified exposure assessment procedures aimed also at managing risk factors.

Assessment of Other Manual Handling Activities (Pulling, Pushing, and Carrying). No equally consolidated procedures, based also on multidisciplinary approaches, like the NIOSH procedure for lifting, are available in the literature for the assessment of exposure to manual load handling such as pulling, pushing, or carrying. Considering this aspect, it was decided to use the data derived from the specific application of psychophysical methods summarized by Snook and Ciriello (1991). This was mainly due to three reasons:

1. Such data were also expressed with reference to the percentiles of potentially satisfied (even if not necessarily protected) population. In particular, where it was possible to select data on "satisfaction" of 90% of population, we were able to provide reference values that had a "cover" value comparable to that resulting from the application of lifting index.
2. Data from psychophysical studies were also used to develop the NIOSH formula (NIOSH, 1981; Waters et al., 1993) to assess lifting tasks: In particular, they were also used to evaluate the degree of "protection" or better "satisfaction" associated with use of the recommended weight limit.
3. The data from psychophysical studies were expressed by Snook and Ciriello (1991) with reference not only to the two genders but also to structural variables (height of pushing and carrying areas and distance) as well as to organizational variables (frequency and duration of tasks) which produced well-defined application methods according to the different working situations.

Manual Handling Index and Its Consequences. We have seen that it is always possible to calculate, albeit with a variety of assessment procedures according to the kind of analyzed manual handling activity, a synthetical exposure index (manual handling index = MHI), as follows:

$$MHI = \frac{\text{actually handled weight (force)}}{\text{recommended weight (force) as a function of major situation variables}}.$$

Such a synthetical manual handling index, even if determined by semiquantitative assessment procedures, may become an effective tool not only, and not so much, for defining the exposure level of one worker (or group of workers) involved in manual handling, but also for defining the consequent preventive measures in accordance both with European regulations and, more generally, with correct prevention strategies.

To reach the latter goal, it is convenient to classify MHI results at least according to a model having more than two levels. This is because the level of approximation (both intrinsic and in conditions of application) of the suggested methods and procedures calls for a certain amount of caution, in particular as regards borderline results around the value of 1. This three-zone model (or traffic light model), appeared to be useful in this sense: accordingly, the MHI results could be classified as follows:

- MHI up to 0.75 = *Green zone*: There is not a particular exposure for the working population and therefore no collective preventive actions are required
- 0.76 = MHI = 1.25 = *Yellow zone*: This is the borderline zone where exposure is limited but may exist for some of the population. Prudent measures are to be taken especially in training and health surveillance of operators. Wherever possible, it is suggested to limit exposure so as to return to the green zone.
- MHI higher than 1.26 = *Red zone*: Exposure exists and is significantly present. The higher the MHI value, the higher the exposure for increasing numbers of the population. MHI values may determine priority of prevention measures that must in any case be taken to minimize exposure toward the yellow zone. Training and active health surveillance of operators must be undertaken in any case.

Health Surveillance Strategies.   This paper is aimed at ergonomists; therefore health surveillance problems are discussed only considering the general aspects that may be of interest to our readers.

The guideline provides reference rules synthetically reported:

- Active health surveillance should regard all thoracolumbar spinal diseases,
- Active health surveillance of spinal work-related musculoskeletal diseases (WMSDs) can be performed in different steps:
    (a) The first step envisages, for all exposed subjects, administration of questionnaires or anamnestic interviews according to models that are already available in the literature.
    (b) The second step envisages a clinical examination of the spine only for subjects classified as positive ones in the previous anamnestic survey. This examination can be made by the occupational physician in the company medical department using a standardized set of specific clinical tests and maneuvers reported in the literature.
    (c) The third step applies to those subjects, identified in the two previous steps, requiring more specialistic tests (neurological, orthopedical, etc.) or instrumental tests (image diagnostics, Electromyography, etc.) in order to complete the individual diagnostic procedure.
- The frequency of health surveillance (first + second step) checks may be established according to relative exposure indices as well as health results obtained in the latest "round" of examinations. Generally speaking, because health surveillance is concerned with slowly evolving chronic degenerative diseases, 3- to 5-year checks are adequate in most cases.
- One of the goals of health surveillance, from a collective viewpoint, is to check whether in a given working population, exposed to a specific risk, the occurrence of spinal WMSDs is other than expected. In order to make such comparisons, adequate reference data on the whole working population are needed. The guideline report data on the prevalence of positive cases (defined according to established criteria) for cervical, thoracic, and lumbar-sacral spine in a group of workers with low or zero, present or past, exposure to occupational risk factors for the spine (manual materials handling, fixed postures, and whole body vibrations). Data are subdivided by gender and 10-year age classes.

• Another goal of specific health surveillance at individual level is the earliest possible identification of subjects affected by spinal disorders for whom it would not be advisable to allow exposure levels that were defined as permissible for healthy subjects. The guideline gives detailed criteria to manage those cases.

Risk Management and Workplace (Re)design.   The guideline gives full details regarding criteria and examples of task and workplace (re)design for reducing the need or almost the risks connected to manual handling activities. Because the issues presented are very common in the international literature and in other specific guidelines, they are skipped in this presentation.

Prevention of WMSDs Connected to Upper limbs' Repetitive Exertions and Movements

Program for Implementation.   These guidelines come jointly with a 3-year experimental plan (Regione Lombardia, 2004) involving its application in approximately 2,000 manufacturing industries of Lombardy, identified on the basis of kind of production (mainly in mechanical, electromechanical and electronic, textile, clothing, food and meat, and plastics and rubber processing) and number of employed workers (over 50).

The plan as well as the guidelines were agreed on between Public Authority (with functions of Labour Inspectorate), Employers' Associations and Trade Unions. The plan defines the general goal of risk assessment and management actions application in identified companies as well as a series of actions to be carried out by the different protagonists involved. They can be summarized as follows:

• Definition of regional guidelines agreed on between Prevention Regional System and social actors
• Start and finalizing of an education and training programme of all public operators (laborinspectors) and operators from OSH services of concerned companies
• Assistance provided by laborinspectors in applying guidelines
• Preliminary risk assessment and possible consequent actions in accordance to guidelines carried out by enterprises
• Monitoring of the state of progress of the project
• Implementing of the recording regional system of reported WMSDs
• Implementing of a regional data Web site on risks and injuries caused by upper-limb repetitive movements and, more specifically, on preventive solutions adopted to the benefit of all potential users
• Critical check of the outcome of the experimental project and revision of guidelines in view of a generalized application in all manufacturing sectors.

Last, the plan defines process, output, and outcome indicators to check trend and results with time.

Guidelines.   The guidelines, taking into account the general indications in European Directive 391/89, state that each employer shall also consider the risk associated with upper-limb repetitive movements when generally assessing work-related risks. If such a risk is present, a specific program is to be started to reduce the risk.

Therefore guidelines, after providing the (epidemiologic, legal, technical) state of the art on this subject, provide indications on:

• Risk identification
• Risk estimate and assessment

TABLE 17.2
General Flowchart for the Application of the Guidelines

```
┌──────────────────┐                          ┌──────────────────┐
│ THE UNIT IS IN   │       ┌──────┐           │ IT IS ANYWAY     │
│ THE FIELD OF     │──────▶│  NO  │──────────▶│ NECESSARY TO     │
│ APPLICATION      │       └──────┘           │ PROCEED WITH     │
└──────────────────┘                          │ EVALUATION       │
         │                                     └──────────────────┘
    ┌────────┐              ┌──────┐          ┌──────┐
    │  YES   │              │ YES  │          │  NO  │
    └────────┘              └──────┘          └──────┘
         │                                          │
         ▼                                          ▼
┌──────────────────┐   ┌──────┐      ┌──────────────────┐
│ SIGNALS OF A     │──▶│  NO  │─────▶│ NO               │
│ POSSIBLE RISK    │   └──────┘      │ INTERVENTION     │
└──────────────────┘                 └──────────────────┘
         │
    ┌────────┐
    │  YES   │
    └────────┘
         │
         ▼
┌──────────────────┐   ┌──────────────────┐
│ RISK IDENTIFICATION │▶│ NOT              │
│ AND ESTIMATION   │   │ SIGNIFICATIVE    │
└──────────────────┘   └──────────────────┘
         │
         ▼
┌──────────────────┐
│ RISK IS          │
│ PRESENT          │
└──────────────────┘
```

- Health surveillance
- Medical–legal and insurance consequences
- (re)Design of tasks, workplaces, and working facilities in view of risk reduction

The general process as indicated in the guidelines is summarized in Table 17.2. A preliminary assessment of possible risk develops along three successive steps:

- Identification of "problematic jobs"
- Risk assessment
- Analytical risk assessment (in selected cases)

Risk Assessment

As to *identification of "problematic jobs,"* whose exposure assessment shall be carried out in the concerned working sectors, the following criteria hold valid:

- The worker/s has/have a nearly daily exposure to one or more indicators of possible exposure reported in Table 17.3
- There are reported cases (one or more also taking into account the number of workers involved) of diagnosed WMSDs of upper limbs.

As to *exposure estimate,* all workplaces and processing already identified as "problematic" are to be first analyzed through simplified assessment tools. With this purpose, use can be made of appropriate investigation tools, available in the literature mostly as checklists that have to be filled in by specially trained staff. An OCRA checklist is enclosed as well as the related instructions for use and interpretation of results (Colombini, Occhipinti, & Grieco, 2002).

As to exposure, the final score of the OCRA checklist can be interpreted according to the classification scheme (based on the so-called trafic light model) reported in Table 17.4. As to *risk analytical assessment,* it may be necessary in some specific situations. There is not a precise rule fixing when a task or a workplace needs a more detailed investigation: As a consequence, this decision is up to discretion and individual fortuitous requirements.

TABLE 17.3

Signals of a Possible Exposure to Repetitive Movements and Exertions of the Upper Limbs
("Problem Job" When One or More Signals Are Present)

1. Repetitivness. Task(s) organized in cycles lasting up to 30 seconds or requiring the same upper-limb movement (or brief group of movements) every few seconds, for at least 2 hours in the shift.
2. Use of force. Task(s) requiring the repetitive use of force (at least once every 5 minutes), for at least 2 hours in the shift. To this, consider the following criteria: handling of object weighing more than 2.7 kg; the handling, between thumb and forefinger, of objects weighing over 900 g; the use of tools requiring the application of quite maximal force.
3. Bad postures. Task(s) requiring the repetitive presence of extreme postures or movements of the upper limbs, such as, uplifted arms, deviated wrist, or rapid movements, for at least 1 hour continuously or 2 hours in the shift.
4. Repeated impacts. Task(s) requiring the use of the hand like a tool for more than 10 times in a hour, for at least 2 hours in the shift

TABLE 17.4

Classification of OCRA Checklist Results Into Four Areas for
Risk Exposure Level Assessment

Checklist Score	Ocra Index	Risk Classification
Up to a 7.5	2.2	Green, yellow/green = no risk
7.0–11.0	2.2–3.5	Yellow = low risk
11.1–22.5	3.6–9	Medium red = medium risk
≥ 22.6	≥ 9.1	High red = high risk

Nevertheless the decision orientation criteria are reported in the following:

- More detailed investigation can be excluded when the results and data from risk assessment are sufficiently sound, coherent with the other contextual information and, in particular, more able to address in sufficient detail the consequent actions with respect to different risk determinants.
- Risk-detailed investigation should be carried out in all the cases when the risk estimate results are uncertain or do not correspond to other contextual information (e.g., WMSD occurrence), or when more data are required to define the consequent preventive actions, or when it is necessary to establish more precisely a connection between risk and damage in acknowledging a upper-limb (UL)-WMSD as an occupational disease.

The preferential tool for investigating in detail the risk is the so-called OCRA Index method (Occupational Repetitive Action; Colombini et al., 2002). A special enclosure includes some considerations concerning the OCRA index use as a probabilistic prediction tool of induced health effects (UL–WMSD) and for risk classification.

Health Surveillance Strategies. The guidelines provide detailed indications and tools for implementing and managing active health surveillance and developing all the medical–legal and insurance fulfillments resulting from identification of fully diagnosed UL-WMSDs cases. This handbook being addressed to ergonomists and health surveillance strategies being similar to those summarized for spine disorders, details on this aspect are not reported.

Task and Workplace (Re)design. When both exposure assessment and the study of UL-WMSDs have revealed a significant risk associated with repetitive or strenuous movements of the upper limbs, the need arises to implement specific measures aimed at re-designing tasks, procedures, workplaces, and equipments. These measures are often urgent and complex and are generally based on three types of co-ordinated and virtually simultaneous actions being carried out: structural modifications, organizational changes, and personnel training, as reported in Table 17.5. Although the structural measures are almost universally accepted and widely recommended, actions involving organizational changes do not always meet with unanimous consent, nor does the scientific literature provide concrete examples. The guidelines provide

TABLE 17.5
General Description of Different Kinds of Preventive Actions

Structural Modifications
  The use of ergonomic tools
  An optimal arrangement of the work station, furnishings, and layout
Improve aspects related to the excessive use of force, awkward posture, and localized compressions

Organizational Modifications
  An ergonomically designed job (pace, breaks, and alternating tasks)
Improve aspects related to:
  Movements performed frequently and repetitively for prolonged periods
  Absence or inadequacy of recovery periods.

Training
  Suggestions concerning breaks
  Appropriate information on specific risks and injuries
  Concrete methods for performing tasks and utilizing proper techniques
Are additional to the other interventions

TABLE 17.6
Brief Recommendations for Reducing the
Frequency of Technical Actions (but not
Productivity)

Avoid Useless Actions:
  Added arbitrarily by the worker
  Due to manufacturing flaws
  Due to obsolete technologies
Distribute Actions Between Both Limbs
Reduce the Repetition of Identical Actions
  By processing preassembled pieces
  By introducing semiautomatic steps
  By replacing manual tasks with hi-tech solutions
Reduce Auxiliary Actions
  By creating intersections between the
  conveyor belt and the work bench

---

criteria and some concrete examples for re-designing jobs and preventing disorders caused by repetitive movements of the upper limbs. Reference is made to the three areas mentioned previously, and specific indications are given for each area, based on the abundant literature already available on structural modifications. A section is also devoted to the subject of possible organizational changes, already investigated and applied in some field experiments and whose criteria, regarding the reduction of pace (without reducing productivity), are synthetically reported in Table 17.6. Last, guidelines are supplied for training programmes designed to support the previous two classes of actions (i.e., structural and organizational) and devoted to workers.

REFERENCES

CEN. (2002). PrEN 1005–2: Safety of machinery—Human physical performance—Part 2: Manual handling of machinery and components parts of machinery.

Colombini, D., Occhipinti, E., & Grieco, A. (2002). *Risk assessment and management of repetitive movements and exertions of upper limbs: Job analysis, Ocra risk index, prevention strategies and design principles* (Vol. 2). Elsevier Ergonomics Book Series Amsterdam (NL).

Conferenza dei Presidenti delle Regioni Italiane. (1996). *Linee guida per l'applicazione del D. Lgs.* 626/94. (pp. 359–418). Edizioni Regione Emilia Romagna. Ravenne (IT).

Council Directive N. 90/269. (1990). Minimum health and safety requirements for the manual handling of loads where there is a risk particularly of back injury to workers. *Official Journal of the European Communities*, N.L 156/9, 21.6.90.

Grieco, A., Occhipinti, E., Colombini, D., & Molteni G. (1997). Manual handling of loads: The point of view of experts involved in the application of EC Directive 90/269. *Ergonomics, 40*(10), 1035–1056.

NIOSH. (1981). *Work Practices Guide for manual lifting.* Cincinnati, OH: U.S. Department of Health and Human Services (Technical Report N. 81-122).

Regione Lombardia. Official Bullettin (BURL). Linee guida regionali per le prevenzione delle patologie muscoloscheletniche connesse con movimenti e sforzi ripetuti degli arti superiori. Decreto dirigenziale No. 18140. Supplemento Straordinario del 16 mazzo 2004. Milano (IT).

Snook, S. H., & Ciriello, V.M. (1991). The design of manual handling tasks: revised tables of maximum acceptable weights and forces. *Ergonomics, 36*(9), 1197–1213.

Waters, T., Putz Anderson, V., Garg A., & Fine L. (1993). Revised NIOSH equation for the design and evaluation of manual lifting tasks. *Ergonomics, 36*(7), 749–776.

Waters T., Putz Anderson V., & Garg A. (1994). *Application manual for the revised NIOSH Lifting Equation.* Cincinnati, OH: U.S. Department of Health and Human Services.

18

Assessment of Manual Material Handling Based on Key Indicators: German Guidelines

Ulf Steinberg
Gustav Caffier
Falk Liebers
Federal Institute for Occupational Safety and Health

INTRODUCTION

This chapter presents a practice-based method of describing and evaluating the working conditions which prevail in the manual handling of loads. This simple method is geared to the recognition and removal of bottlenecks. Because it only covers the major activity indicators, it is called the key indicator method. This method was developed and tested from 1996 to 2001 in connection with the implementation of the European Union (EU) framework and individual directives on occupational safety and health in German national law. It consists of two independent, but formally adaptable parts for lifting, holding, and carrying, and for pulling and pushing. The method was drawn up in the Federal Institute for Occupational Safety and Health in close collaboration with the Committee of the Laender for Occupational Safety and Health (Länderausschuss für Arbeitsschutz und Sicherheitstechnik—LASI) with the involvement of numerous companies, scientists, accident insurance bodies, and trade unions. The method can be used only to assess risks with the aim of preventing work-related health risks from the manual handling of loads. In the 6 years since its first publication, this method has enjoyed a wide acceptance among possible users and a correspondingly broad application in Germany.

REQUIREMENTS FROM EUROPEAN OCCUPATIONAL SAFETY AND HEALTH LAW

German occupational safety and health law, on the basis of the common statutory instruments of the EU, requires an assessment of the working conditions and documentation of the results of this assessment. In particular are the following Council Directives,

- **Council Directive 89/391** of 12 June 1989 on the introduction of measures to encourage improvements in the safety and health of workers at work

317

- **Council Directive 90/269/EEC** of 29 May 1990 on the minimum health and safety requirements for the manual handling of loads where there is a risk particularly of back injury to workers (fourth individual Directive within the meaning of Article 16 (1) of Directive 89/391/EEC),

which have been incorporated in national law. These laws essentially contain protective goals while avoiding differentiated limits value and methodological regulations. They give the employer the responsibility for safe working conditions. For their part, the employees are obliged to conduct themselves in a safe manner.

Because in Germany about 90% of all companies are small- and medium-sized enterprises with a small workforce, a complete safety service from trained specialists is not possible. Many tasks are performed by the employers themselves or by persons within the company specially assigned for the purpose. For this reason, appropriate support for the competence of the corporate practitioners, taking due account of available resources, was a major criterion for the development of the key indicator method described in this chapter.

The application of this method is not mandatory in law. However, there is an application recommendation of the *Committee of the Laender for Occupational Safety and Health*, the body representing the authorities of the German Federal states responsible for supervising government occupational safety and health law.

ANALYSIS OF DEFICIENCIES

Risk estimates for physical load situations are, in methodological terms, very demanding, and they always involve compromises. Among the three cornerstones of practicability, error of judgment, and damage model there is an explosive area of conflict. Rationally arrived at results are particularly important in everyday judgments, for which few resources in terms of funds and time are normally estimated. As far as possible, simple algorithms are preferred, which supply a concrete result with few measurements and computation specifications. Complex relations which have a reciprocal effect and are not easy to measure are simplified. Unfortunately the concrete figures calculated exhibit a deceptive precision which is not realistic. And unfortunately the methodological directions do not indicate the nature and the scope of the possible error. The resulting, uncritical judgments can have far-reaching consequences.

Risk-assessment methods for physical load situations therefore have to fulfill the following indicators in practical terms:

- Value-neutral description of the most important activity indicators
- Reliable coverage of these indicators with the lowest possible effort
- Revelation and rough quantification of relevant risks
- Indication of design bottlenecks
- Comprehensibility and retraceability of the judgment by the user
- Low effort for documentation
- Calculability of assessment errors

CRITICAL CONSIDERATION OF AVAILABLE ASSESSMENT PROCEDURES

Starting with the experience gained from many years of practical ergonomic work and scientific, methodologically critical studies, a large number of methods available worldwide were tested in a research project with the aim of making a specific application recommendation.

The following methods were evaluated, covering as they do the period 1959 to 1996:

- Transport formula. Spitzer, Hettinger, and Kaminiski (1982)
- Schultetus-Burandt method (also known as Siemens-Burandt), determination of admissible limit values for forces and torques. (Siemens AG, 1981)
- Formulae for calculating the energy conversion during physical work. Garg, Chaffin, and Herrin (1978)
- Work Practice Guide for Manual Lifting, calculation of control and allowability limits for load weights. NIOSH (1981)
- Manual load displacement in standing position – determination of recommended limits values. BOSCH (1982)
- Regulations and code of practice – manual handling No. 8. Occupational Health and Safety Act. Melbourne OSHA (1988)
- Determination of maximum muscle strain when lifting and carrying. Heben und Tragen. VDI/Daimler-Benz AG (1987)
- Further development of the Siemens-Burandt method in the REFA chemistry experts' committee REFA (1987)
- Simplified procedure on the basis of the NIOSH approach of 1981. Pangert and Hartmann (1989)
- Der Dortmunder (computer-aided tools for the biochemical analysis of the strain on the spine when loads are manipulated) Jäger, Luttmann, Laurig (1992)
- National Standard for Manual Handling. NOHSC (1990)
- Manual handling – manual handling operations regulations. HSE (1992)
- Revised NIOSH equation for the design and evaluation of manual lifting tasks," Calculation of a recommended limit load weight. NIOSH (1993)
- ISO-CD 11228 Ergonomics, Manual handling, lifting and carrying, calculation of recommended limit load weights. ISO TC 159 (1994)
- ErgonLift, PC-aided calculation of biomechanical and energy-related strain figures. Laurig, Schiffmann (1995)
- Formula for the three-dimensional calculation of compression forces. McGill, Normann, and Cholewicki (1996)
- prEN 1005 Safety of machinery – Human physical performance, Part 2: Manual handling of machinery and component parts of machinery, calculation of recommended limit load weights when handling machines. CEN TC 122 (1993)

The titles of the methods already make clear that they were developed for different objectives. Initially the establishment of practicability criteria had priority in connection with the humane design of work. Over the past few years, the risk estimate has become the center of attention. Against the background of the national economic relevance of persistently high sickness frequency rates with respect to the spine and the preventive notion of the EU Directives in matters of employee protection, there arose the need to conduct a prospective assessment of possible health risks.

An analysis of the studies published in the technical literature on the numerous methods revealed that these methods only tentatively satisfy the requirements that arise in practice. The principal problems that occurred time and again were that the methodological models are not comprehensible for the practical user, that methods are too often not practicable because of restrictions on their application, and that they are too laborious and that possible application errors are not defined. Apart form the high and, in practice, hardly achievable effort required, this gives rise to critical application situations. The users, who are normally well practiced at their workplace, do not have a clear view of the mostly complicated overall system and apply it purely schematically, if at all. This can be the cause of errors of judgment with grave consequences for personnel or the economic situation. The rejection of these methods by many of those involved is correspondingly great.

DEVELOPMENT OF THE KEY INDICATOR METHOD

Model Approach

If we examine the methods indicated in chapter 3, it can be seen that checklists and analytically biomechanical or energy-related approaches dominate in the assessment of the risk from the manual handling of loads. With the biomechanical approach, the load on the lumbar spine is determined (without taking account of other joints or increasing muscular fatigue). In the energy-related approach, the aim is to minimize the strain on the heart and the circulatory system, ignoring biomechanical aspects.

With the development of the key indicator method, an attempt was made to facilitate a holistic approach and, at the same time, to implement the abstract model approach for the users in concrete circumstances. The initial notions were concerned with the fact that chronic damage to the muscular and skeletal system does not progress unnoticed, but emerges over a long period in the form of a variety of complaints. Every form of constrained posture (identical muscular tension and joint positions over an extended period) leads to pain and should as far as possible be avoided by changing the posture. This discomfort becomes more evident in extreme positions of the joints. Greater physical forces, regardless of whether they involve forces of posture or action, lead to a strain on muscles, tendons, tendon roots, and ligaments. These strains are first perceived as fatigue and then, if they have to be applied over an extended period, as pain. From this complex unit of physical forces, postures, and positions of the joints, as well as duration and frequency, there arise specific feelings of stress, which result in avoidance reactions if a tolerance limit is passed. If, owing to restrictions, there is no possibility of avoidance, reactions of overload can be expected.

Starting with these circumstances, which can be understood from personal experience, the key indicator method was developed on the basis of biomechanical and energy-related knowledge and experience. With this method, the probability of damage to the muscular and skeletal system from the manual handling of loads is evaluated. The basis for the evaluation is the acting dose. The damaging potential of load weight, posture, or working conditions thus depends on duration and frequency. The model approach of the key indicator method is based on an integrative assessment of these key indicators of the manual manipulation of loads. This approach only takes account of chronic damage and not accident-like events. The indicators of a case of damage include clinically relevant functional disturbances or pain. The direct consequences may be complaints and illnesses; the indirect consequences are work incapacity, occupational disablement, early retirement, or fluctuation.

The target variable of the method is the assessment of the risk from the manual handling of loads in the form of a risk score. This is determined by allocating a rating point to the individual key indicators according to how marked they are and then linking them in a simple computation (see Figure 18.1).

Test of Method

To test the draft key indicator method, two different approaches were selected (revised NIOSH equation for design and evaluation of manual lifting tasks; checklist with 24 items according to the annex of the Council Directive 90/269/EEC on the minimum health and safety requirements for the manual handling of loads where there is a risk particularly of back injury to workers).

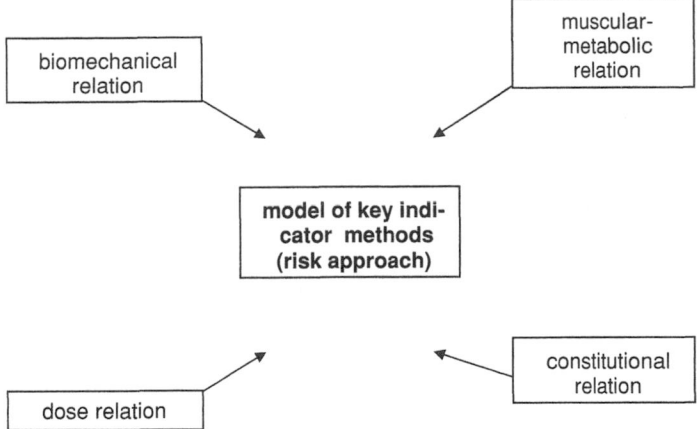

FIG. 18.1. Model of the key indicator method.

They were then subjected simultaneously to a critical application test in 1995 and 1996. In 51 companies from various sectors throughout Germany, 168 workplaces were assessed in parallel using the three methods. A total of 112 individuals were involved, with different training and interests. The test encompassed the analytical-statistical part and, in particular, an open discussion among all those involved on the spot.

The validity test accounted for much of the overall test. It was only possible indirectly to check whether the key indicator method is capable of depicting a largely realistic health risk. Theoretically, the frequency, of diseases of the muscular and skeletal system should also increase as the rating points rise, in accordance with the stress–strain formula. This linear effect was not measurable under practical conditions in Germany. Deployment of employees more in accordance with their abilities, early change of workplace, and reduction of evidently high strain rarely lead to specific disorders. Instead the "healthy worker effect" prevails. Where there are high physical strains, one encounters mainly well-trained employees with a good constitution. The test of validity was conducted taking data on the circumstances relating to stress–strain disorders derived from studies, expert opinions on the assessment results, and comparisons with backed-up assessments using other methods and plausibility criteria.

The result of this project was a clear decision in favor of the key indicator method. The experience using the "revised NIOSH [1993] equation for the design and evaluation of manual lifting tasks" essentially accord with those of Dempsey (1999, 2002). By far, the largest number of activities assessed are mixed forms involving lifting, holding, carrying, and displacement of loads on the same level. Because of the limiting application criteria of this method, they cannot be assessed. The experience with application of the checklist from the annex of the Council Directive 90/269/EEC was negative. Without any quantitative differentiation of the indicators, the evaluation was exclusively a matter of the user's own discretion. It was therefore just as impossible to conduct a risk evaluation as it was to ascertain the specific needs with regard to action.

The development of the method was published in detail in Steinberg et al. (1998, 2000) and was discussed at numerous expert congresses. The results of the application test led to a revision of the draft key indicator method, and this was published in 1996 as a discussion proposal (LASI, 1996; Steinberg and Windberg, 1997). From 1996 to 2001 broad-based application tests were conducted. The experience with these led to a more detailed statement of the matter, which was brought to a conclusion for the time being with the LASI, 2001, published version.

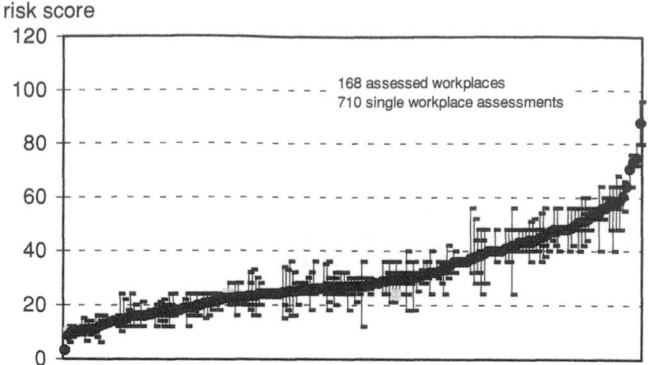

risk score

FIG. 18.2.   Mean values (filled circle) and range widths (a reference line per workplace, single values as stroke) for the point totals determined using the key indicator method, 168 workplaces, and 710 single assessments by different users (workplaces sorted in ascending order according to the level of the mean point totals).

Initially a method was considered which takes equal account of lifting, holding, and carrying, and of pulling and pushing. In the course of the application test, it became clear, however, that this is not possible. The differences in the nature of the load and possible stresses are too great. In connection with the publication of the final version of the key indicator method in 2001, this was clearly limited to lifting, holding, and carrying. For pulling and pushing, a formally adaptable draft was published in 2002 (LASI, 2002) for public discussion and general application trials.

Error Examination

The reliability examinations conducted when the method was being tested revealed relatively small differences in assessment for the key indicator method. If an assessment is conducted of individual persons who have precise knowledge of the working sequences being assessed, and if the application regulations are observed, incorrect judgments are very seldom made. When the users were compared, the lowest error proneness was found among charge-hands, foremen, and engineers.

Even so, because of the rough gradations in the load description and with correct recording, errors in the range of ±10% always had to be included in the calculation (see Figure 18.2). Assuming incorrect values, for example, incorrect frequency, much higher error rates are possible through to totally meaningless results. The focus is therefore on the correct recording of indicators and a critical plausibility test. It is possible to increase the reliability of assessment by integrating it in a methods inventory (chap. 6).

PRESENTATION OF THE KEY INDICATOR METHOD

Description of the Activities

The key indicator method comprises the two parts "lifting, holding, carrying loads" and "pulling, pushing loads." Both parts of the method contain descriptions of the nature and markedness of the relevant activity indicators, the so-called key indicators. This is used to define indicators that exert a major influence on the assumed effect complexes. The selection

TABLE 18.1
Overview of Key Indicators

Lifting, Holding, Carrying	Pulling, Pushing
Duration, Frequency	Duration, frequency
Load weight	Mass to be moved and transport vehicle
	Positioning accuracy and speed of motion
Posture	Posture
Working conditions	Working conditions

of the indicators is therefore geared initially to their influence on the cause-and-effect relationship. It is also important, however, that these indicators can be reliably recorded under practical conditions. The use of measured values would basically be desirable, but it is practically not financially feasible because of the great expense it involves. This applies in particular for the measurement of physical forces, postures and working conditions. Table 18.1 gives an overview of the key indicators.

The general practical understanding of the terms has clear priority over scientific precision. This is why the term "force" is not used for pulling and pushing. The anticipated forces are roughly circumscribed instead via the indicators of weight, means of transport, positioning accuracy, and working conditions.

In the tables of the forms, the markedness of the key indicators is quantified roughly by classification into value ranges. The spectrum of value ranges largely reflects the practical conditions completely. When these value ranges are exceeded, it is possible to extrapolate as appropriate. Within the roughly incremented scales, it is possible to interpolate as appropriate. The rough classification of the markedness of the individual indicators contained in the key indicator methods reflects two aspects. On the one hand, exact and quantitative measurements of the markedness of the indicators (e.g., posture or duration) within 1 working day are normally only possible with great effort and time input. Qualified estimates of the markedness of the indicators on the basis of a sound knowledge of the working conditions are more effective. On the other hand, it is not monotonous activities which are encountered (lifting of a defined load over a certain period with a certain frequency), but mixed activities (lifting, carrying, and holding of different loads with varying frequency). Even with exact measurements via working sequence studies, relatively rough classifications are needed in such cases for an evaluation of the manual handling task.

Evaluation

The evaluation is conducted both for lifting, holding, and carrying and for pulling and pushing in identical form as a risk rating. The probability of damage to the muscular and skeletal system is indicated—not its severity or localization. The evaluation is conducted separately from the description by multiplying the time rating by the sum of key indicator rating points. The number of points represents a measure of the prevailing risk. Between the ranges there are fluid transitions.

The evaluation model takes account of four factors: (a) biomechanics, (b) muscular-metabolic effort, (c) acting dose, and (d) variations in constitution. Although all four factors do not stand in isolation, they reflect the specific effects in each case.

The factor of *biomechanics* takes into account, in particular, the mechanical load on bones and joints from the postural and action forces applied. The forces to be transferred in the

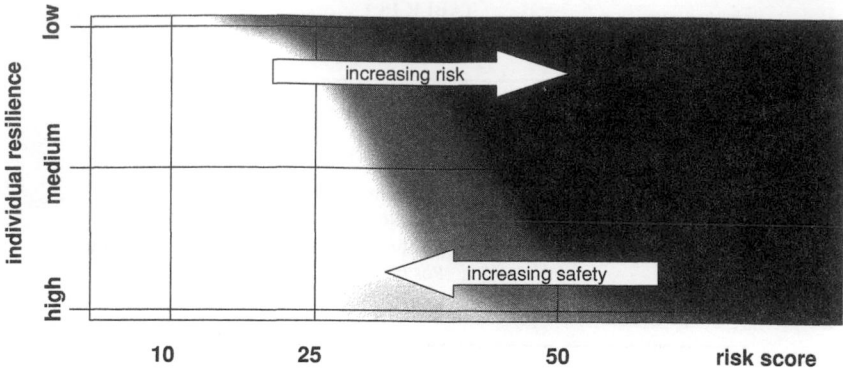

FIG. 18.3. Consideration of the differing individual resilience in the evaluation model of the key indicator methods.

skeletal system are a measure of the internal strain and possible overstrain on individual structural elements. Possible long-term consequences are degenerative changes, and short-term overstrain may lead to fractures and cracks.

The biomechanical components are taken into consideration through the load weight, positioning accuracy, speed of motion, and posture.

The *muscular-metabolic* component relates to the activity of the muscles. Direct hazards are only possible with major cases of overload (sprains and torn muscle fiber). Otherwise, the muscles react under load situations with reversible fatigue. Increasing fatigue is, however, the cause of loss of strength and deteriorating coordination. The progress of work and hence also the biomechanical loads thus worsen. The great significance of this component is that it is perceived by the employee and that it can be measured directly (in individual cases with appropriate effort). The physiological effort under working conditions can be measured by the heartbeat and oxygen intake.

The muscular-metabolic component is taken into consideration mainly via the duration and frequency and load weight, positioning accuracy, speed of motion, and postures.

The *dose relation* is obtained by considering the duration of the action of the biomechanical load (dose = internal force $*$ time [Nh]) or the muscular-metabolic effort (energy = performance $*$ time [Wh]).

The dose is taken into consideration mainly in terms of the duration.

Whereas the three components mentioned relate to activity, the *constitutional* prerequisite is considered in relation to individuals. The relationship of work strain and physical resilience must be noted. Muscular strength, endurance, physical type, and skill vary considerably. Healthy employees with sturdy bone structure and well-trained muscles are less at risk under the same load situations. The connection between individual resilience and the load situation evaluation is shown in Figure 18.3.

The evaluation is based formally on the calculated connection of the scaled activity description. Concrete figures are always calculated. They must be seen in the overall context. Two aspects are of particular importance: the fluid transitions between the ranges and the possible assessment errors (Figure 18.4).

The basic design objective should be to adhere to the range up to 25 points. This means that health damage from the manual handling of loads will largely be excluded. But it is neither physiologically necessary nor economically appropriate to assume this limit as an absolute rule. The limit is a protection in particular for employees with low resilience. For

Risk range	Risk score	Description
1	< 10	Low load situation: health risk from physical overload unlikely to appear.
2	10 to < 25	Increased load situation: Physical overload is possible for less resilient persons. For this group re-design of workplace is appropriate.
3	25 to < 50	Highly increased load situation. Physical overload also possible for persons with normal resilience. Re-design of the workplace is recommended.
4	50	High load situation. Physical overload is likely to appear. Workplace re-design is essential.

FIG. 18.4.   Evaluation table as per forms.

employees able to take physically greater burdens values around 35 are probably still acceptable (Figure 18.3). The design of work should therefore always take account of the efficiency present in the workforce structure. These ideas are based to a large extent on the assessment model of the *Work Practice Guide for Manual Lifting* (NIOSH, 1981).

Evaluations of the test results indicate in this context an occasional, but serious problem: The assessment results signalize a possible risk which is not accepted as such or which cannot be eliminated with a reasonable amount of effort. Reactions are rejection of the assessment method or willful changes to the rating scales.

Nearly always, these problems arise from an excessively narrow way of looking at the matter and insufficient consideration of the assessment model. The Risk Range 2 is defined as *increased load situation, physical overload is possible for less resilient persons. For this group re-design of workplace is appropriate.* Risk Range 3 is defined as *highly increased load situation, physical overload also possible for persons with normal resilience. Re-design of the workplace is recommended.*

When terms were being drawn up, close attention was paid to these situations. The term *load situation* relates to work demands and hence depends on the individual. The term *strain* in this paper is the value-free reaction of an individual organism to a load as it acts. *Physical overload*, however, describes the potential negative consequences of a load situation. *Resilience* describes the ability to deal with a load situation present without any adverse stress. *Normal resilience* takes account of the very wide range of individually different, physical efficiency. Muscle mass, endurance, body mass, state of health, and motivation are, alongside sex and age, some of the variables that have to be taken into account. *Possible overload* means that damage will not always necessarily occur, but only if there is an incorrect relationship between load situation and resilience. These terms must be interpreted in the correct context.

For practical purposes, this means that, in the range of 25 to 50 points, precisely this incorrect relationship must be noted. Because neither the employer nor the company doctor is hardly in a position to make an exact estimate of resilience, an "early warning system" is of great importance. By making a sensitive record of perceived stress and health complaints on the part of employees, the load situation survey can be supplemented in terms of biomonitoring. Support by company doctors is effectively possible in the form of selected orthopedic examinations. The primary design objective then is not adherence to the 25-point range, but an acceptable

326 STEINBERG, CAFFIER, AND LIEBERS

physical strain that reliably avoids damage. If the employees do not experience excessive stress, if there are no health complaints, and if there is no evidence of a higher sickness rate, there is no need for action. The prerequisite is that this must actually be verified. From 50 points, however, there is always a need for action.

Focusing on a favorable total point rating should not conceal the actual goal—the design, that is, the avoidance of bottlenecks. The focus of attention should therefore be on reducing high individual rating points.

It should be said here by way of warning that the simple key indicator methods primarily serve to delimit the problem. If more extensive design measures have been judged to be necessary, reasonable preparations should then always be made for investments.

Formal Structure

Both key indicator methods are designed in their original form as one-page worksheets. For reasons of practicality the complete overview is important. The worksheets can be filled in directly and filed away as documentation. The reverse contains important instructions for use. A description of activities and an evaluation are drawn up in two separate stages. The description drawn up in the first stage is value-free. This means that the description remains valid without restriction at a later time when changes may have been made to statutory regulations and assessment procedures.

The forms and the related instructions for use are attached as annexes.

Limits of Application

The key indicator methods are intended only for the identification of bottlenecks and needs for action.

Evaluations of the whole work or a whole day are thus not possible. For this purpose, special analyses based on precise work sequence studies are needed in order to take account of potential synergy and compensation effects. Such studies should always remain the task of ergonomically qualified persons. The possible consequences of incorrect judgments cannot be estimated by nonspecialists in ergonomics.

Use for questions with legal implications is also not permissible.

In the long term, it is also conceivable that the load situation data collected may provide the basis for more extensive data analyses. Spot-check analyses in Germany already show that the overall situation for the manual handling of loads is determined more by unfavorable postures and high time fractions than by great load weights. The efforts made over the past few decades to reduce the great load weights are already bearing fruit.

INCREASE OF ASSESSMENT RELIABILITY BY INTEGRATION IN A METHODS INVENTORY

Assessment errors can basically never be discounted. The assessment methods are incomplete and are subject to limitations in their application. This also applies to the key indicator methods, even though their errors are comparatively slight. The user left to his or her own resources would not in practice like to be confused by methodological inadequacies or to have to put in an unreasonably great effort.

One simple way out of this problem is to combine a number of methods that complement or check one another methodologically. For the onsite work of ergonomists and company doctors in Germany, a four-part methods inventory (Caffier & Steinberg, 2000) has been developed. It involves an average time input of less than 2 hours per workplace or employee in order to record the following:

- The objective physical strain *(key indicator method)*
- The perceived stress *(questionnaire with 47 items according to Slesina [1987])*,
- The existing health complaints *(Nordic questionnaire according to Kuorinka et al. [1987])* and
- Orthopedic findings *(orthopaedic multistage diagnosis according to Grifka, Peters, and Bär [2001])*.

The methods can also be used singly. As with the key indicators method, the value for the corporate practitioner is not in the precise calculation and comparative evaluation, but in the highlighting of relationships and the influencing factors to be considered.

DISCUSSION AND EXPERIENCE WITH APPLICATION

After 7 years experience with application, the following can be said:

- The method is currently used very widely. It has been included in numerous methodological recommendations, including in German-speaking countries outside Germany itself. It is seen by many practitioners as a source of support for their professional expertise.
- The assessment results are nearly always plausible and enable one to conclude the need for action in a short time. Interpretation of the results is occasionally conducted too schematically, however, on the basis of the classification into one of the four risk ranges and neglects the fluid transitions between the ranges.
- The proportion of relevant misapplications and misjudgments is small. The most frequent errors are assessments based on inadequate knowledge of the activity, computation errors, application to complicated work sequences, uncritical applications, and failure to give heed to the instructions. The form is frequently copied and applied without methodological explanation.
- A random compilation of the assessment results gives initial, Germany-wide overviews of the nature and scope of the physical strain involved in the manual handling of loads (Figure 18.5).
- The fears prevalent in the mid-1990s that multiple burdens would arise for companies from the obligation to assess have not become a reality. In the majority of cases, the key indicator method is acknowledged to be an efficient tool.
- In several cases, the method has been applied far beyond its use limits for complex work designs and for appraisal purposes in legal disputes. The method is unsuitable for this purpose.
- In some cases, the method has been modified autonomously by users. Partly for frivolous reasons—the simple appearance is associated with the possibility of simple adaptation—and also with the intention of influencing the results of the assessment. The developers of the method have not agreed and do not agree to such modifications. All experience and criticisms are collected, discussed, and decided on a consensus basis.

The prime aim in drawing up the key indicator method is to provide a suitable tool for assessing the working conditions and for justifying necessary preventive actions. A further intention of

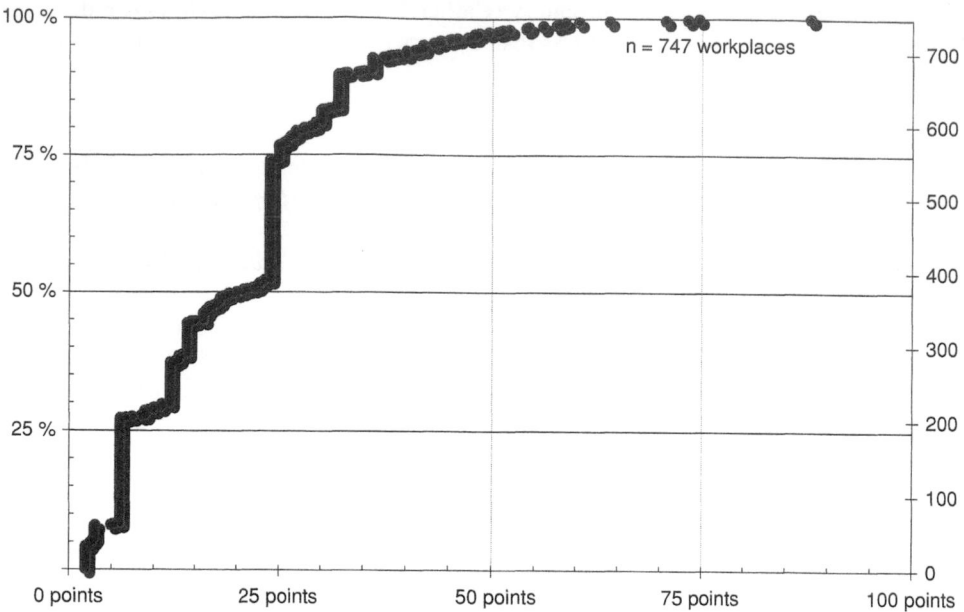

FIG. 18.5.   Ordered distribution of risk scores in the total stock evaluated. *[Distribution of risk scores for 747 workplaces assessed using the key indicator method "Lifting, carrying, and holding loads" (in ascending order, ranges up to 25 points and up to 50 points marked)]*

the method developers was to reduce the existing shortfall of knowledge regarding the nature and extent of work-related load situations to the muscular and skeletal system and the cause-and-effect relationship in the emergence of disorders of the muscular and skeletal system. The data routinely collected were intended to be useful when drawing up a standardized description of the load situation. At the present time, the assessment results for 747 activities involving the manual handling of loads are available in the form of a random sample (Figure 18.5). Because the distribution of the rating points for the first random sample from 1997 of 168 activities has not basically changed (see Figure 18.2), it can be taken as approximately representative for the load situation in Germany. The distribution is generally accepted as reflection of the real situation. It documents the fact that the critical limit of 50 points is exceeded in less than 10% of the workplaces where loads are handled manually.

INTERNET PRESENTATION

Updates of the methods are accessible at the Internet address http://www.baua.de. These pages also show the link with the whole methods inventory. Interactive calculation programs, partly combined with database applications, supplement the package.

APPENDIX 1: WORKSHEET FOR THE KEY INDICATOR METHOD FOR LIFTING, HOLDING, AND CARRYING

ASSESSMENT OF MANUAL HANDLING TASKS BASED ON KEY INDICATORS Version 2001

Where there are a number of individual activities with considerable physical strains, they must be estimated separately

Workplace/Activity: _____

1st step: Determination of time rating points (Select only one column i)

Lifting or displacement operations (< 5 s)		Holding (> 5 s)		Carrying (> 5 m)	
Number on working day	Time rating points	Total duration on working day	Time rating points	Overall length on working day	Time rating points
< 10	1	< 5 min	1	< 300 m	1
10 to < 40	2	5 to 15 min	2	300 m to < 1km	2
40 to < 200	4	15 min to < 1 hr	4	1 km to < 4 km	4
200 to < 500	6	1 hrs to < 2 hrs	6	4 to < 8 km	6
500 to < 1000	8	2 hrs to < 4 hrs	8	8 to < 16 km	8
≥ 1000	10	≥ 4 hrs	10	≥ 16 km	10

Examples: • laying bricks, • placing workpieces into a machine, • taking boxes out of a container and putting them onto a conveyor belt

Examples: • holding and guiding a cast iron slag while working on a wheel stand, • operating a hand grinding machine, • operating a wrench etc.

Examples: • furniture removal, • clothing/or scaffolding parts to a building site

1) "Effective load" means in this context the real action force which is necessary for moving load. This action force does not correspond to the load mass in each case. When lifting a carton, only 50 % of the load mass will have an effect on worker and when using a cart only 10 %.

2nd step: Determination of rating points of load, posture and working conditions

Effective load[1] for men	Load rating point	Effective load[1] for women	Load rating point
< 10 kg	1	< 5 kg	1
10 to < 20 kg	2	5 to <10 kg	2
20 to < 30 kg	4	10 to <15 kg	4
30 to < 40 kg	7	15 to < 25 kg	7
≥ 40 kg	25	≥ 25 kg	25

Typical posture, position of load[2]	Posture, position of load	Posture rating point
	+ Upper body upright, not twisted + When lifting, holding, carrying and lowering the load is close to body	1
	+ Slightly bending forward or twisting the trunk + When lifting, holding, carrying and lowering load is near to medium to body	2
	+ Low bending or far bending forward + Slightly bending forward with simultaneous twisting of trunk + Load far from the body or above shoulder height	4
	+ Bending far forward with simultaneous twisting of trunk + Load far from body + Restricted stability of posture when standing + Crouching or kneeling	8

2) To determine the posture rating points the typical posture during manual handling must be used. For example when there are different postures with load a mean value must be used – not occasional extreme values.

Working conditions	Working conditions rating point
Good ergonomic conditions, e.g. sufficient space, no physical obstacles within the workspace, even's level and solid flooring, sufficient lighting, good gripping conditions	0
Space for movement restricted and unfavourable ergonomic conditions (e.g. 1: space for movement restricted by too low high or working area less than 1,5 m² or 2: posture stability impaired by uneven floor or soft ground)	1
Strongly restricted space of movement and/or instability of centre of gravity of load (e.g. transfer of patients)	2

3rd step: Evaluation

The rating points relevant to this activity are to be entered and calculated in the diagram:

Load rating points	
+ Posture rating points	
+ Working conditions rating points	
= Total	

Total × Time rating points = Risk score

On the basis of the rating calculated and the table below it is possible to make a rough evaluation. [3] Regardless of this provisions of the Maternity Leave Act apply.

Risk range	Risk score	Description
1	< 10	Low load situation, physical overload unlikely to appear.
2	10 bis < 25	Increased load situation, physical overload is possible for less resilient persons[4]. For that group redesign of workplace is helpful.
3	25 bis < 50	Highly increased load situation, physical overload also possible for normal persons. Redesign of the workplace is recommended.
4	× 50	High load situation, physical overload is likely to appear. Workplace redesign is necessary[5].

Basically it must be assumed that as the number of point rating rises, so the risk of overloading the musculo-skeletal system increases. The boundaries between the risk ranges are fluid because of the individual working techniques and performance conditions. The classification may therefore only be regarded as an orientation aid. More exact analyses require specialist ergonomic knowledge.

4) Less resilient persons in this context are persons older than 40 or younger than 21 years, newcomers to the job or people suffering from illness.

5) Design requirements can be determinated with reference to the number of point in the table. By reducing the weight, improving the execution conditions or shortening the strain time, elevation stress can be avoided.

Check of the workplace necessary for other reasons: ☐

Reasons: _____

Date of assessment: _____   Assessed by: _____

Ed. by «Federal Institute for Occupational Safety and Health and • Committee of the Länder for Occupational Safety and Health (•Bundesanstalt für Arbeitsschutz und Arbeitsmedizin • BAuA und •Länderausschuss für Arbeitsschutz und Sicherheitstechnik · LASI) 2001

APPENDIX 2: INSTRUCTIONS FOR USE WHEN ASSESSING WORKING CONDITIONS USING THE KEY INDICATOR METHOD FOR ACTIVITIES INVOLVING LIFTING, HOLDING, AND CARRYING

> **Caution!**
>
> This procedure only serves as an orienting assessment of the working conditions for lifting and carrying loads. Nevertheless, good knowledge of the manual handling task being assessed is absolutely essential when determining the time rating, load rating, posture rating, and the rating of the working conditions. If this knowledge is not present, no assessment may be made. Rough estimates or suppositions lead to incorrect results.

The assessment is basically conducted for manual handling tasks and must be related to 1 working day. If load weights or postures change within an individual activity average values must be formed. If *a number of manual handling tasks* with substantially different load manipulations arise within the overall activity, they must be *estimated* and documented *separately*.

Three steps are necessary in the assessment: First: determination of the time rating points; second, determination of the rating points for the key indicators, and third, evaluation.

In the determination of the rating points it is basically permitted to form intermediate steps (interpolation). A frequency of 40 produces the time rating point 3, for example. The only exception is the effective load of $\geq$ 40 kg for a man and $\geq$ 25 kg for a woman. These loads uncompromisingly yield a load rating of 25.

First Step: Determination of the Time Rating Points

The time rating points are determined with reference to the table separately for three possible forms of load handling:

- For manual handling tasks characterized by the *regular repetition of short lifting, lowering, or displacement operations*, the number of operations is a determinant for the time rating points.
- For manual handling tasks characterized by the *holding* of loads, the total duration of the holding is taken: *Total duration = number of holding operations × duration of a single holding operation.*
- For manual handling tasks characterized by the *carrying* of a load, the total distance covered with the load is taken. An average speed when walking of 4km/hr, 1 m/sec is assumed.

Second Step: Determination of the Rating Points of Load, Posture, and Working Conditions

Load Weight

- The load rating points are determined with reference to the table separately for *men and women*.
- If, in the course of the manual handling task being assessed, different loads are handled, an *average value* may be formed where the greatest single load for men does not exceed

40 kg and for women 25 kg. For comparison purposes, peak load values can also be used. Then, however, the reduced frequency of these peaks must be taken as a basis, and in no account the total frequency.

- In the case of *lifting, holding, carrying, and setting-down activities*, the effective load must be taken. The effective load mass here is the weight force which the employee actually has to cancel out. The load is therefore not always equal to the weight of the object. When a box is tilted, only about 50% of the weight of the box acts.
- When loads are being *pushed and pulled* a separate assessment is necessary.

Posture

The rating points of posture are determined with reference to the pictograms in the table. The *characteristic postures during the handling of loads* must be used for the individual activity. If different postures are adopted as work progresses, an average value can be formed from the posture rating points for the manual handling task being assessed.

Working Conditions

To determine rating points of the working conditions, the working conditions that predominate most of the time must be used. Occasional discomfort which has no safety significance will not be taken into account. Safety-relevant indicators must be documented in the text box *"Check of the workplace for other reasons."*

Third Step: Evaluation

Each task is evaluated on the basis of an *activity-related risk score* (calculation by addition of the rating points of the key indicators and multiplication with the time rating points).

- The *basis for evaluation* comprises biomechanical mechanisms of action combined with dose models. Account is taken here of the fact that the internal strain on the lower spine depends to a crucial extent on the extent to which the trunk is leaning forward and on the load weight and that it increases with increasing load duration or frequency, side bending, or twisting.
- *Summarized evaluations* are difficult with a number of manual handling tasks because they go beyond the informative scope of this orientation analysis. They normally require more extensive work analysis procedures to obtain a risk assessment.
- *Design needs that can be concluded*. From this risk estimate there is immediate evidence of design needs and approaches. Basically, the causes of high rating points must be eliminated. Specifically, these are organizational regulations in the case of high time rating points, reduction of the load weight, or the use of lifting aids in the case of high load rating points or the improvement of ergonomic conditions in the case of high posture rating points.

APPENDIX 3: WORKSHEET FOR THE KEY INDICATOR METHOD FOR PULLING, AND PUSHING

Assessment of pulling and pushing based on key indicators Version Sept. 2002

The overall activity must be broken down into individual activities. Each individual activity involving major physical strain must be assessed separately.

Workplace/Activity:

1st step: Determination of time rating points (select only one column)

Pulling and pushing over short distances or frequent stopping (single distance up to 5 metres)		Pulling and pushing over longer distances (single distance more than 5 metres)	
Number on working day	Time rating points	Total distance on working day	Time rating points
< 10	1	< 300 m	1
10 to < 40	2	300 m to < 1km	2
40 to < 200	4	1 km to < 4km	4
200 to < 500	6	4 to < 8km	6
500 to < 1000	8	8 to < 16km	8
≥ 1000	10	≥ 16 km	10
Examples: operation of manipulators, setting up machines, distribution of meals in a hospital		Examples: garbage collection, furniture transport in buildings on rollers, unloading and transporting of containers	

2nd step: Determination of rating points of mass, positioning accuracy, speed, posture and working conditions

Mass to be moved (load weight)	Without load is rolled	Barrow	Carriage, roller, trolley without fixed rollers (only castor-style rollers)	Rail cars, hand carts, roller tables, carriages with fixed rollers	Manipulator, rope balancer
rolling					
< 50 kg	0.5	0.5	0.5	0.5	0.5
50 to < 100 kg	1	1	1	1	1
100 to < 200 kg	1.5	2	2	1.5	2
200 to < 300 kg	2	4	3	2	4
300 to < 400 kg	3		4	3	
400 to < 600 kg	4		5	4	
600 to < 1000 kg	5			5	
≥ 1000 kg					
sliding					
< 10 kg	1				
10 to < 25 kg	2				
25 to < 50 kg	4				
> 50 kg					

Gray areas:
Critical because acceleration or movement of industrial truckload depends very much on skill and physical strength.

White area 1 without number:
Basically to be avoided because the necessary action forces can easily exceed the maximum physical forces.

Positioning accuracy

Low
- no specification of travelling distance
- load can roll to a stop or runs against a stop

High
- load must be accurately positioned and stopped
- travelling distance must be adhered to exactly
- frequent changes in direction

Note: The average walking speed is approx. 1 m/s

Speed of motion

slow (< 0.8 m/s)	fast (0.8 bis 1.3 m/s)
1	2
2	4

Posture[1]

Trunk upright, not twisted	1
Trunk slightly bending forward or slightly twisted (one-sided pulling)	2
Body inclined low in direction of motion Squatting, kneeling, bending	4
Combination of bending and twisting	8

[1] The typical posture must be used. The greater trunk inclination possible when starting up, braking or shunting can be ignored if it only occurs occasionally.

Working conditions

Good: → floor or other surfaces level, firm, smooth, dry → no incline → no obstacles in workspace → rollers or wheels run easily, no evident wear in the wheel bearings	0
Restricted: → floor soiled, a little uneven, soft → slight incline up to 2° → obstacles in workspace which have to be bypassed → rollers or wheels soiled, no longer run easily, bearings worn	2
Difficult: → unpaved or roughly paved roadway, potholes ,severe soiling → inclines of 2 to 5° → industrial trucks have to be torn loose when starting up → rollers or wheels soiled, bearings run sluggishly	4
Complicated: → steps, stairs → inclines >5° → combinations of indicators from 'restricted' to 'difficult'	8

Indicators not mentioned in the table may be added as appropriate.

3rd step: Evaluation

The rating points relevant to this activity are to be entered and calculated in the diagram.

Mass/action aid task		
+ Positioning accuracy/ speed of motion		
+ Posture rating point		
+ Working conditions rating point		
= Total		

Total	×	Time rating point	=	Risk score
	×	1.3	=	

for women employees:

On the basis of the rating point calculated and the table below it is possible to make a rough evaluation.

Risk range [2]	Risk score	Description
1	< 10	Low load situation, physical overload unlikely to appear.
2	10 to < 25	Increased load situation, physical overload is possible for less resilient persons[3]. For that group redesign of workplace is helpful.
3	25 to < 50	Highly increased load situation, physical overload also possible for normally resilient persons. Redesign of workplace is recommended.
4	≥ 50	High load situation, physical overload is likely to appear. Workplace redesign is necessary.

Published by: Federal Institute for Occupational Safety and Health and Committee of the Länder for Occupational Safety and Health...

332

APPENDIX 4: INSTRUCTIONS FOR USE WHEN ASSESSING WORKING CONDITIONS USING THE KEY INDICATOR METHOD FOR ACTIVITIES INVOLVING PULLING, AND PUSHING

Caution!

This procedure serves for an orienting assessment of working conditions with the pulling and pushing of loads. Nevertheless, good knowledge of the manual handling task being assessed is absolutely essential when determining the time rating and the rating points for mass, positioning accuracy, speed, posture, and working conditions. If this knowledge is not present no assessment may be made. Rough estimates or suppositions lead to incorrect results.

The assessment is basically conducted for individual activities and relates to 1 working day. If load weights or postures change within an individual activity, average values must be formed. If *a number of manual handling tasks* with substantially different load manipulations arise within the overall activity, they must be *estimated* and documented *separately.*

Three steps are necessary in the assessment: First, determination of the time rating points; second, determination of the rating points for the key indicators; and third, evaluation.

In the determination of the rating points, it is basically permitted to form intermediate steps (interpolation). A frequency of 40 produces the time rating point 3, for example.

First Step: Determination of Time Rating Points

The time rating points are determined, with reference to the table, separately for pulling and pushing over short distances with frequent stopping and pulling and pushing over longer distances.

- For pulling and pushing over short distances with frequent stopping the frequency is taken as the basis.
- For pulling and pushing over longer distances the total distance is taken as the basis.

The limit value for the individual distance of 5 metres should be regarded as a rough aid. In cases of doubt a decision should be taken according to which criterion arises more frequently: start-up and braking or extended pulling.

Second Step: Determination of Rating Points of Mass, Positioning Accuracy, Posture, and Working Conditions

Mass to Be Moved

The determination is conducted with reference to the table, taking account of the mass to be moved (weight of means of transport plus load) and the nature of the means of transport (industrial truck, aid). Very often drawbar-less trolleys with rollers are used. A distinction is drawn here between (steerable) steering rollers and (nonsteerable) fixed rollers.

If different loads are handled in the course of the individual activity to be assessed, an *average value* may be formed. For comparison purposes, peak load values may also be used. Then the lower frequency of these peaks must be taken as a basis and on no account the overall frequency.

Positioning Accuracy and Speed of Motion

The determination is conducted with reference to the table. The speed "fast" is equivalent to normal walking. If in special cases there are clearly faster speeds, the table can be extended as appropriate and a 4 or 8 can be given. Interpolations are permissible.

Posture

The posture rating points are determined with reference to the pictograms in the table. The *characteristic postures during the handling of loads* must be used for the individual activity. If different postures are adopted, an average value may be formed from the posture rating points for the manual handling task being assessed.

Working Conditions

To determine the rating points of the working conditions, the working conditions that predominate most of the time must be used. Occasional discomfort which has not safety significance will not be taken into account.

Third Step: Evaluation

Each task is evaluated with reference to an *activity-related risk score* (calculation by addition of rating points for the key indicators and multiplication by the time rating points). If women perform this task, the rating points are multiplied by a factor of 1.3. This takes account of the fact that women have on average about two thirds of the capacity of men.

- *The basis for the evaluation* is the probability of health damage. The nature and level of damage is not defined more closely. Account is taken of biomechanical and physiological action mechanisms combined with dose models. It is taken that the internal strain on the muscular and skeletal system depends to a crucial extent on the physical forces to be applied. These physical forces are determined by the weight of the object to be moved, the acceleration values, and the floor surface resistances. Unfavorable postures and increasing load duration or frequency increase the internal load.
  The instructions in the grey box on page 2 of the form must be adhered to.
- *Summarized evaluations* are *difficult* in the case of a number of manual handling tasks because they go beyond the informative scope of this orientation analysis. They normally require more extensive procedures of risk assessment.
- *Design needs that can be concluded.* From this risk estimate there is immediate evidence of design needs and approaches. Basically the causes of high rating points must be eliminated. Specially these are:
  - For time rating points, organizational regulations
  - For high mass rating points, reduction of load weight or use of suitable industrial trucks
  - For high rating points for speed of motion and positioning accuracy, use of wheel guides and stop buffers or reduction in workload
  - For high posture rating points, improvement in workplace design
  - The working conditions should always be "good."

REFERENCES

BOSCH. (1982). Lastenumsetzung von Hand im Stehen—Ermittlung empfohlener Grenzwerte. Stuttgart: Robert Bosch GmbH.

Caffier, G., & Steinberg, U. (2000). Praxisgerechtes Methodeninventar zur Umsetzung der Lastenhandhabungsverordnung. *Arbeitsschutz aktuell. 2,* 69–74.

CEN TC 122. (1993). Ergonomics: Draft prEN 1005-2 Safety of machinery—Human physical performance. Part 2: Manual handling of machinery and component parts of machinery.

Dempsey, P. G. (1999). Utilizing criteria for exposure and compliance assessment of multiple task manual material handling jobs. *International Journal of Industrial Ergonomics 24,* 405–416.

Dempsey, P. G. (2002). Usability of the revised NIOSH lifting equation. *Ergonomics 45,* 817–828.

Garg A., Chaffin, D. B., & Herrin, G. D. (1978). Prediction of metabolic rates for manual material handling jobs. *Am. Ind. Hyg. Assoc. 39*, 661–674.

Grifka, J., Peters, T., & Bär, H.-F. (2001). Mehrstufendiagnostik von Muskel- Skelett-Erkrankungen in der arbeitsmedizinischen Praxis. Bremerhaven: Wirtschaftsverl. NW 2001 (Publication series of the Federal Institute for Occupational Safety and Health: Offprint, p. 62).

Handhaben von Lasten. (1987). REFA-Fachausschusse Chemie. 2nd ed. Darmstadt: REFA-Bundesverband.

Heben und Tragen. (1987). Ermittlung maximaler Muskelbelastung nach VDI. Daimler Benz AG.

HSE. (1992). Manual handling—Manual handling operations regulations 1992, Guidance on regulations L23. Sheffield: HSE Information Centre.

ISO TC 159/SC 4/WG 4: ISO CD 11228 (1994). Ergonomics—Manual handling. Part 1: Lifting and carrying. Draft.

Jäger, M., Luttmann, A., Laurig, W. (1992). Ein computergestütztes Werkzeug zur biomechanischen Analyse der Belastung der Wirbelsäule bei Lastenmanipulationen: "Der Dortmunder." *Medizinisch-orthopädische Technik 112*, 305–309.

Kuorinka, I., Jonsson, B., Kilbom, A., Vinterberg, H., Biering-Sorensen, F., Andersson, G. et al. (1987). Standardisied Nordic questionaires for the analysis of musculoskeletal symptoms. *Appl. Ergonom. 18*, 223–237.

LASI. (1996). *Handlungsanleitung zur Beurteilung der Arbeitsbedingungen beim Heben und Tragen von Lasten.* (1st ed., LASI publication 9). Länderausschuss für Arbeitsschutz und Sicherheitstechnik (Eds).

LASI. (2001). *Handlungsanleitung zur Beurteilung der Arbeitsbedingungen beim Heben und Tragen von Lasten.* (4th revised ed., LASI publication 9). Länderausschuss für Arbeitsschutz und Sicherheitstechnik (Eds.).

LASI. (2002). *Handlungsanleitung zur Beurteilung der Arbeitsbedingungen beim Ziehen und Schieben von Lasten.* (LASI publication LV29). Länderausschuss für Arbeitsschutz und Sicherheitstechnik (Eds.).

Laurig, W., & Schiffmann, M. (1995). ErgonLIFT: Rechnerunterstützte Methodik zur Gefährdungsbewertung und Prävention beim manuellen Handhaben von Lasten. Bielefeld: Schmidt.

McGill, S. M., Normann, R. W., & Cholewicki, J. (1996). A simple polynomial that predicts low-back compression during complex 3-D tasks. *Ergonomics 39*, 1107–1118.

NIOSH. (1981). *Work practice guide for manual lifting.* Washington, DC: U.S. Gov. Print Office 1981 (DHHS [NIOSH] publication, 81–122).

NOHSC: 1001. (1990). National Standard for Manual Handling and National Code of Practice for Manual Handling [NOHSC: 2005 (1990)]. (Eds.), National Occupational Health and Safety Commission. Canberra: Australian Government Publishing Service.

OHSA. (1988). Regulations and code of practice—manual handling Nr. 8. Occupational Health and Safety Act 1985. Melbourne.

Pangert, R., & Hartmann, H. (1989). Ein einfaches Verfahren zur Bestimmung der Belastung der Lendenwirbelsäule am Arbeitsplatz. Zentralblatt f. Arbeitsmedizin, Arbeitsschutz and Prophylaxe, 191–194.

Siemens AG (1981). *Ermitteln zulässiger Grenzwerte für Kräfte und Drehmomente*, Arbeitsblatt. Munich: Siemens AG 1981.

Slesina, W. (1987). Arbeitsbedingte Erkrankungen und Arbeitsanalyse – Arbeitsanalyse unter dem Gesichtspunkt der Gesundheitsvorsorge. Stuttgart: Enke.

Spitzer, H., Hettinger, T., & Kaminski, G. (1982). *Tafeln für den Kalorienumsatz bei körperlicher Arbeit* (6th ed.). Berlin: Beuth.

Steinberg, U., Behrendt, S., Bradl, I., Caffier, G., Gebhardt, Hj., Liebers, F.; et al. (2000). Erprobung und Evaluierung des Leitfadens Sicherheit und Gesundheitsschutz bei der manuellen Handhabung von Lasten. Bremerhaven: Wirtschaftsverl. NW 2000. (Publication series of the Federal Institute for Occupational Safety and Health: Research, Fb 897)

Steinberg, U., Caffier, G., Mohr, D., Liebers, F., & Behrendt, S. (1998). Modellhafte Erprobung des Leitfadens Sicherheit und Gesundheitsschutz bei der manuellen Handhabung von Lasten. Bremerhaven: Wirtschaftsverl. NW 1998. (Publication series of the Federal Institute for Occupational Safety and Health: Research, Fb 804)

Steinberg, U., & Windberg, H.-J. (1997). Leitfaden Sicherheit und Gesundheitsschutz bei der manuellen Handhabung von Lasten. Bremerhaven: Wirtschaftsverl. NW 1997 (Publication series of the Federal Institute for Occupational Safety and Health: Offprint, 43)

Waters, T. R., Putz-Anderson, V., Garg, A., & Fine, L. F. (1993). Revised NIOSH equation for the design and evaluation of manual lifting tasks. *Ergonomics 36*, 749–776.

V

HUMAN–COMPUTER INTERACTION

19

Standards, Guidelines, and Style Guides for Human–Computer Interaction

Tom Stewart
System Concepts, Ltd.

INTRODUCTION

Standards, guidelines, and style guides generally exist to improve the consistency of the user interface and to improve the quality of interface components. They help specifiers to procure systems and system components which can be used effectively, efficiently, safely, and comfortably. They also help restrict the unnecessary variety of interface hardware, software, and technology and ensure that the benefits of any variations are fully justified against the costs of incompatibility, loss of efficiency, and increased training time for users. Even standards that are still under development can have an impact on hardware and software development. The major suppliers play an active part in generating the standards, and increasingly they are incorporating the guidance on good practice into products before the standards themselves are published.

In this chapter, we describe a number of human–computer interaction (HCI) standards which have been developed internationally and explain how they can be used in conjunction with other guidelines and style guides to improve the user experience.

Although designing usable systems requires far more than simply applying standards, guidelines, and style guides, nonetheless, they can make a significant contribution by promoting consistency, good practice, common understanding, and an appropriate prioritization of user interface issues.

Consistency

Anyone who uses computers knows only too well the problems of inconsistency between applications and often even within the same application. Inconsistency, even at the simplest level, can cause problems. Just three examples:

- Press the <escape> key in one place and you are safely returned to your previous menu choice. In another place, you are unceremoniously "dropped" to the operating system, the friendly messages disappear, and you lose all your data.

- On the Web, inconsistency is rampant. Even something as straightforward as a hypertext link may be denoted by underlining on one site, by a mouseover on a second site, and by nothing at all on a third site.
- Different and confusing keyboard layouts sit side by side in many offices.

Standards, guidelines, and style guides play an important part in helping address these issues by collating and communicating agreed best practices for user interfaces and for the processes by which they are designed and evaluated. They can provide a consistent reference across design teams or across time to help avoid such unpleasant experiences. Indeed, in other fields, consistency, for example, between components that should interconnect, is the prime motivation for standards. It is certainly a worthwhile target for user interface standards.

Good Practice

In many fields, standards provide definitive statements of good practice. In user interface design, there are many conflicting viewpoints about good practice. Standards, especially International Standards, can provide independent and authoritative guidance. International Standards are developed slowly, by consensus, using extensive consultation and development processes. This has its disadvantages in such a fast moving field as user interface design, and some have criticized any attempts at standardization as premature. However, there are areas where a great deal is known which can be made accessible to designers through appropriate standards, and there are approaches to user interface standardization, based on human characteristics, which are relatively independent of specific technologies.

The practical discipline of having to achieve consensus helps moderate some of the wilder claims of user interface enthusiasts and helps ensure that the resulting standards do represent good practice. The slow development process also means that standards can seldom represent the leading edge of design. Nonetheless, properly written, they should not inhibit helpful creativity.

Common Understanding

Standards themselves do not guarantee good design, but they do provide a means for different parties to share a common understanding when specifying interface quality in design, procurement, and use.

- *For users*, standards allow them to set appropriate procurement requirements and to evaluate competing suppliers' offerings.
- *For suppliers*, standards allow them to check their products during design and manufacture and provide a basis for making claims about the quality of their products.
- *For regulators*, standards allow them to assess quality and provide a basis for testing products.

Appropriate Prioritization of User Interface Issues

One of the most significant benefits of standardization is that it places user interface issues squarely on the agenda. Standards are serious business, and whereas many organizations pay little regard to research findings, few organizations can afford to ignore standards. Indeed in Europe, and increasingly in other parts of the world, compliance with relevant standards is a mandatory requirement in major contracts.

A Note on Terminology

HCI standards, guidelines, and style guides represent three approaches to improving the usability of systems. They are not mutually exclusive categories. For example, many HCI standards simply provide agreed guidelines rather than specify requirements. Some style guides are implemented in such a way that they have become standards from which designs cannot vary.

However, generally (and indeed in this chapter), we use the terms as follows:

- Guidelines—recommendations of good practice which rely on the credibility of their authors for their authority
- Standards—formal documents published by standards making bodies which are developed through some form of consensus and formal voting process
- Style guides—set of recommendations from software providers or agreed within development organizations to increase consistency of design and to promote good practice within a design process of some kind.

HCI STANDARDS

Sources of HCI Standards

Many standards bodies have been in existence for some time and are organized according to traditional views of technology and trade. Software is used as part of systems that involve a range of technologies. The purpose of this section is to introduce one of the key standards organizations that is working on standards relevant to HCI and describe its main activities briefly.

In most people's minds, one of the most basic and fundamental objectives of standardization is to minimize unnecessary variations. Ideally, for any product category, there is one standard which should be satisfied, and products which meet that standard give their owners or users some reassurance about quality or about what standards makers refer to as "interoperability." Thus, yachtsmen in Europe who buy a lifejacket which meets EN 396 might reasonably expect it to keep them afloat if they have the misfortune to fall overboard in the Florida Keys. Similarly, an office manager in the United States who orders A4 paper for a photocopier might reasonably expect paper which meets that standard (ISO 216:1975) to fit, even though it is not the typical size used locally.

Which brings us to a rather important point. It is often difficult to achieve a single agreed-on standard, and a common solution is to have more than one standard. An obvious example concerns paper size where there are the ISO A series (A0, A1, etc), the ISO B series (B0, B1 etc) as well as U.S. sizes (legal, letter, etc). Although this solves the standards makers' problems in agreeing on a single standard, it is an endless source of frustration for users of the standard—as anyone who has forgotten to check the paper source in an e-mailed document can testify.

However, there is another reason why there are more standards than one might imagine, especially when it comes to user interface design issues. The reason is that computer technology forms the basis of many different industries, and standards can have an important impact on market success.

But it is not just at the international level that there appears to be some duplication. In the United Kingdom (UK), the British Standards Institution mirror committee to SC4 published an early version of the first six parts of ISO 9241 as a British Standard BS 7179: 1990. The prime reason for this was to provide early guidance for employers of users of visual displays, who wanted to use standards to help them select equipment which met the requirements in the Schedule to the Health and Safety (Display Screen Equipment) Regulations 1992. These regulations are the UK implementation of a European Community Directive on the minimum safety and health requirements for work with display screen equipment (90/270/EEC). Of course, as a spin-off, the British Standards Institution was able to generate revenue from selling these standards several years before the various parts of ISO 9241 became available as British Standards.

A similar process has taken place in the United States with the Human Factors and Ergonomics Society (HFES) developing HFS 100 on Visual Display Terminal Ergonomics as an ANSI-authorized Standards Developing Organisation. More recently, there are two HFES standards development committees working on HFES 100 (a new version of HFS 100) and on HFES 200 which addresses user interface issues. It includes sections on accessibility, voice, and telephony applications; color and presentation; and slightly rewritten parts of the software parts of ISO 9241.

International ergonomics standards in HCI are being developed by the International Organization for Standardization (ISO). The work of ISO is important for two reasons. First, the major manufacturers are international and therefore the best and most effective solutions need to be international. Second, the European Standardization Organization (CEN) has opted for a strategy of adopting ISO standards wherever appropriate as part of the creation of the single market. CEN standards replace national standards in the European Union and European Free Trade Area member states.

The International Organization for Standardization (ISO) comprises national standards bodies from member states (see www.iso.ch for more information). Its work is conducted by technical and subcommittees which meet every year or so and are attended by formal delegations from participating members of that committee. In practice, the technical work takes place in Working Groups of experts, nominated by national standards committees but expected to act as independent experts. The standards are developed over a period of several years and in the early stages, the published documents may change dramatically from version to version until consensus is reached (usually within a Working Group of experts). As the standard becomes more mature (from the Committee Draft Stage onwards), formal voting takes place (usually within the parent subcommittee), and the draft documents provide a good indication of what the final standard is likely to look like. Table 19.1 shows the main stages.

The International Organization for Standardization (ISO)

In the late 1970s, the kind of concern about the ergonomics of visual display terminals (also called visual display units) which stimulated German standards (see Appendix 1), became more widespread, especially in Europe.

The prime concern at that time concerned the possibility that prolonged use (especially of displays with poor image quality) might cause deterioration in users eyesight. (Note: Since then, several studies have shown that aging causes the main effect on eyesight, and because display screen work can be visually demanding, many people only discover this deterioration when they experience discomfort from intensive display screen use. This can incorrectly lead them to attribute their need for glasses to their use of display screens.)

When a new International Standards work item to address this concern was proposed, the Information Technology committee decided that this was a suitable topic for the recently formed ergonomics committee ISO/TC 159. The work item was allocated to the subcommittee ISO/TC 159/SC4 Signals and Controls, and an inaugural meeting was held at BSI in Manchester in 1983. The meeting was well attended with delegates from many countries, and a few key decisions were made.

TABLE 19.1

The main stages of international standards development

WI	Work Item—an approved and recognized topic for a working group to be addressing which should lead to one or more published standards
WD	Working Draft—a partial or complete first draft of the text of the proposed standard
CD	Committee Draft—a document circulated for comment and approval within the committee working on it and the national mirror committees. Voting and approval are required for the document to reach the next stage.
DIS	Draft International Standard—a draft standard which is circulated widely for public comment via national standards bodies. Voting and approval are required for the draft to reach the final stage.
FDIS	Final Draft International Standard—the final draft is circulated for formal voting for adoption as an International Standard.
IS	International Standard. The final published standard

Note: Documents may be reissued as further CDs and DISs

TABLE 19.2

The Working Groups of ISO/TC159/SC4

WG1	Fundamentals of Controls and Signalling Methods
WG2	Visual Display requirements
WG3	Control, workplace, and environmental requirements
WG4	Task requirements (disbanded)
WG5	Software ergonomics and human–computer dialog
WG6	Human-centered design processes for interactive systems
WG8	Ergonomics design of control centers

At that time, there was a proliferation of office-based systems, and SC4 decided to focus on office tasks (word processing, spreadsheet, etc.) rather than try to include computer-aided design or process control applications. It also decided that we would need a multipart standard to cover the wide range of ergonomics issues which it believed needed to be addressed in order to improve the ergonomics of display screen work. A number of working groups was established to carry out the technical work of the subcommittee. Table 19.2 lists the current working groups of ISO/TC159/SC4.

Little did any of those present realize that it would be nearly 7 years before the first parts of ISO 9241 would be published and that it would take to the end of the century to publish all 17 parts. Table 19.3 shows the published parts of ISO 9241, and Table 19.4, the other published standards which were part of the ISO/TC159/SC4 work programme.

HCI STANDARDS UNDER DEVELOPMENT

Although the ISO 9241 standards have represented a major part of the output of ISO/TC159/SC4, a number of other standards is under development at this time (December 2004) and these are listed in Table 19.5.

TABLE 19.3
ISO 9241 Standards Published by ISO/TC159/SC4 Ergonomics of Human–System
Interaction Including Amendments and Revisions

ISO 9241

ISO 9241-1:1997 Ergonomic Requirements for Office Work With Visual Display
   terminals (VDTs)—Part 1: General Introduction ISO 9241-1:1997/Amd 1:2001
ISO 9241-2:1992 Ergonomic Requirements for Office Work With Visual Display
   terminals (VDTs)—Part 2: Guidance on Task Requirements
ISO 9241-3:1992 Ergonomic Requirements for Office Work With Visual Display
   terminals (VDTs)—Part 3: Visual Display Requirements ISO 9241-3:1992/Amd 1:2000
ISO 9241-4:1998 Ergonomic Requirements for Office Work With Visual Display
   terminals (VDTs)—Part 4: Keyboard Requirements ISO 9241-4:1998/Cor 1:2000
ISO 9241-5:1998 Ergonomic Requirements for Office Work With Visual Display
   terminals (VDTs)—Part 5: Workstation Layout and Postural Requirements
ISO 9241-6:1999 Ergonomic Requirements for Office Work With Visual Display
   terminals (VDTs)—Part 6: Guidance on the Work Environment
ISO 9241-7:1998 Ergonomic Requirements for Office Work With Visual Display
   terminals (VDTs)—Part 7: Requirements for Display With Reflections
ISO 9241-8:1997 Ergonomic Requirements for Office Work With Visual Display
   terminals (VDTs)—Part 8: Requirements for Displayed Colors
ISO 9241-9:2000 Ergonomic Requirements for Office Work With Visual Display
   terminals (VDTs)—Part 9: Requirements for Non-Keyboard Input Devices
ISO 9241-10:1996 Ergonomic Requirements for Office Work With Visual Display
   terminals (VDTs)—Part 10: Dialog Principles
ISO 9241-11:1998 Ergonomic Requirements for Office Work With Visual Display
   terminals (VDTs)—Part 11: Guidance on Usability
ISO 9241-12:1998 Ergonomic Requirements for Office Work With Visual Display
   terminals (VDTs)—Part 12: Presentation of Information
ISO 9241-13:1998 Ergonomic Requirements for Office Work With Visual Display
   terminals (VDTs)—Part 13: User Guidance
ISO 9241-14:1997 Ergonomic Requirements for Office Work With Visual Display
   terminals (VDTs)—Part 14: Menu Dialog
ISO 9241-15:1997 Ergonomic Requirements for Office Work With Visual Display
   terminals (VDTs)—Part 15: Command Dialogs
ISO 9241-16:1999 Ergonomic Requirements for Office Work With Visual Display
   terminals (VDTs)—Part 16: Direct Manipulation Dialog
ISO 9241-17:1998 Ergonomic Requirements for Office Work With Visual Display
   Terminals (VDTs)—Part 17: Form Filling Dialogs

How to Use ISO 9241 Standards

Although it was not made explicit at the time, SC4 had an underlying set of assumptions about
HCI design activities and how the standards would support these. These activities included:

- Analyzing and defining system requirements
- Designing user–system dialogs and interface navigation
- Designing or selecting displays
- Designing or selecting keyboards and other input devices
- Designing workplaces for display screen users
- Supporting and training users
- Designing jobs and tasks

TABLE 19.4

Other Standards Published by ISO/TC159/SC4 Ergonomics of Human–System Interaction

Other Standards

ISO 11064-1:2000 Ergonomic Design of Control Centers—Part 1: Principles for the
  Design of Control Centers
ISO 11064-2:2000 Ergonomic Design of Control Centers—Part 2: Principles for the
  Arrangement of Control Suites
ISO 11064-3:1999 Ergonomic Design of Control Centers—Part 3: Control Room
  layout ISO 11064-3:1999/Cor 1:2002
11064-4:2004 Ergonomic Design of Control Centers—Part 4: Layout and Dimensions
  of Workstations
ISO 13406-1:1999 Ergonomic Requirements for Work With Visual Displays Based on Flat
  Panels—Part 1: Introduction
ISO 13406-2:2001 Ergonomic Requirements for Work With Visual Displays Based on Flat
  Panels—Part 2: Ergonomic Requirements for Flat Panel Displays
ISO 13407:1999 Human-Centered Design Processes for Interactive Systems
ISO 14915-1:2002 Software Ergonomics for Multimedia User Interfaces—Part 1:
  Design Principles and Framework
ISO 14915-2:2003 Software Ergonomics for Multimedia User Interfaces—Part 2:
  Multimedia navigation and Control
ISO 14915-3:2002 Software Ergonomics for Multimedia User Interfaces—Part 3:
  Media Selection and Combination
ISO/TS 16071:2003 Ergonomics of Human–System Interaction—Guidance on
  Accessibility for Human–Computer Interfaces
ISO/TR 16982:2002 Ergonomics of Human–System Interaction—Usability Methods
  Supporting Human-Centered design
ISO/PAS 18152:2003 Ergonomics of Human-System Interaction—Specification for the
  Process Assessment of Human–System Issues
ISO/TR 18529:2000 Ergonomics—Ergonomics of Human–System Interaction—
  Human-Centered Lifecycle Process Descriptions (available in English only)

TABLE 19.5

Other Main Standards Under Development by ISO/TC159/SC4 Ergonomics of
Human–System Interaction

Standard	*Status*
11064-5—Ergonomic Design of Control Centers—Part 5: Human–System Interfaces	Delayed
11064-6—Ergonomic Design of Control Centers—Part 6: Environmental Requirements	FDIS
11064-7—Ergonomic Design of Control Centers—Part 7: Principles for the Evaluation of Control Centres	DIS
1503 (rev)—Ergonomics Requirements for Design on Spatial Orientation and Directions of Movements	WI agreed
16071 Ergonomics of Human–System Interaction—Guidance on Software Accessibility	CD due in 12/04
23973—Software Ergonomics for World Wide Web User Interfaces	DIS in preparation

TABLE 19.6
How Parts of ISO 9241 Were Intended to Be Used in HCI Design Activities

HCI Activity	Relevant Part of ISO 9241
Analyzing and Defining System Requirements	ISO 9241-11:1998 Guidance on usability
Designing User–System Dialogs and Interface Navigation	ISO 9241 10:1996 Dialog principles
	ISO 9241-14:1997 Menu dialogs
	ISO 9241-15:1998 Command dialogs
	ISO 9241-16:1999 Direct manipulation dialogs
	ISO 9241-17:1998 Form-filling dialogs
Designing or Selecting Displays	ISO 9241-3:1992 Display requirements
	ISO 9241-7:1998 Requirements for displays with reflections
	ISO 9241-8:1997 Requirements for displayed colors
	ISO 9241-12:1998 Presentation of information
Designing or Selecting Keyboards and Other Input Devices	ISO 9241-4:1998 Keyboard requirements
	ISO 9241-9: 2000 Requirements for non-keyboard input devices
Designing Workplaces for Display Screen Users	ISO 9241-5:1998 Workstation layout and postural requirements
	ISO 9241-6:1998 Guidance on the work environment
Supporting and Training Users	ISO 9241-13:1998 User guidance
Designing Jobs and Tasks	ISO 9241-2:1992 Guidance on task requirements

Table 19.6 shows how it was anticipated that the standards would be used to support these activities.

Revisions to 9241

Following the completion of the seventeen part ISO 9241, work is underway on a major revision and restructuring to incorporate other relevant standards and make the ISO 9241 series more usable. This article describes the new structure, the principles agreed for the revision process and gives the current status of the new parts (as of December 2004).

Although computer technology has changed dramatically over the period the original ISO 9241 has been under development (more than 20 years), many of the ergonomics issues remain similar. For example, when Part 14 (menu dialogs) was originally planned, menus were usually displayed on character-based screens and menu choices were typically made by selecting numbered choices by keystroke. However, by the time the standard was finished, menus were a common part of graphical interfaces and items were selected by pointing devices. Nonetheless, the guidance on menu structures, how many options should be presented, and so on, remained applicable because it related to how people make choices and interpret information and these have not changed much in the time.

But, changes in the technology and the way we use it have made it difficult for the 9241 standard to keep up to date. The development of flat panel displays meant that new display standards had to be developed. The rapid pace of change meant that we could not always wait for the technology to stabilize before developing standards, so we developed some design process standards. And of course, ISO itself requires standards to be reviewed after 5 years.

TABLE 19.7

The Structure and the Current Status of the Parts (as of December 2004)

New ISO 9241 Ergonomics of Human–system Interaction

Part	Title	Status and Notes
1	Introduction	Reserved number—could be a TR to allow frequent updates
2	Job Design	Reserved number for revision and extension of old Part 2
11	Hardware and Software Usability	Reserved number for revision and extension of old Part 11 to include hardware explicitly
20	Accessibility and Human–System Interaction	Approved WI—technical work to start at meeting in Tokyo in December
21–99	Reserved Numbers	No plans to allocate at present
100	Software Ergonomics	Reserved number for series of software ergonomics standards
110	Dialog Principles	DIS (voting ends 01/05, revision of old Part 10)
200	Human System Interaction Processes	Reserved number for revision and extension of ISO 13407 and other process standards
300	Displays and display-related hardware	Reserved number for series of display ergonomics standards (will become Introduction)
301	Introduction	CD approved, DIS in preparation but will be renumbered 300
302	Terms and Definitions	CD approved, DIS in preparation
303	Ergonomic Requirements	CD approved, DIS in preparation
304	User Performance Test Methods	CD approved, DIS in preparation
305	Optical Laboratory Test Methods	CD approved, DIS in preparation
306	Field Assessment Methods	CD approved, DIS in preparation
307	Analysis and Compliance Test Methods	CD approved, DIS in preparation
400	Physical Input Devices—Ergonomics Principles	CD approved, DIS in preparation
410	Design Criteria for Products	CD expected 12/04
420	Ergonomic Selection Procedures	CD expected 12/04
500	Workplace Ergonomics	Reserved number for revision and extension of old Part 5
600	Environment Ergonomics	Reserved number for revision and extension of old Part 6
700	Special Application Domains	Reserved number for series of ergonomics standards for specific application domains, for example, process control

As a result, ISO/TC159/SC4 has been working to develop a new set of standards which build on the strengths of the previous work but which are also easier to use by standards users.

The agreed title for the new ISO 9241 is "The Ergonomics of Human System Interaction." This title was selected to demonstrate the broadening of the scope from office tasks and to align the standard with the overall title and scope of SC4.

We also wanted to build on the "branding" of ISO 9241 which has become recognized as a benchmark, particularly in Europe.

Table 19.7 shows the structure and the current status of the parts.

Strengths and Limitations of HCI Standards

It is important to be aware of the strengths and limitations of standards. They cannot be understood (nor therefore used effectively) in isolation from the context in which they were developed. It is important to realize that:

- Standards are developed over an extended period of time.
- It Is easy to misunderstand the scope and purpose of a particular standard.
- Standards making involves politics as well as science.
- The language of standards can be obscure.

But,

- Structure and formality can be a help as well as a hindrance.
- The benefits do not just come from the standards themselves.
- Being international makes it all worthwhile.

Standards Are Developed Over an Extended Period of Time

One of the reasons why the process is slow is that there is an extensive consultation period at each stage of development with time being allowed for national member bodies to circulate the documents to mirror committees and then to collate their comments. Another reason is that Working Group members can spend a great deal of time working on drafts and reaching consensus only to find that the national mirror committees reject their work when it comes to the official vote. It is particularly frustrating for project editors to receive extensive comments (which must be answered) from countries who do not send experts to participate in the work. Of course, the fact that the work is usually voluntary means that it is difficult to get people to agree to work quickly.

However, there are some benefits which come directly from the slow pace of the process. One benefit is that when the technology is moving quicker than the standards makers can react, it does make it clear that certain types of standards may be premature. For example, ISO 9241-14:1997 Menu Dialogues was originally proposed when character-based menu driven systems were a popular style of dialog design. Its development was delayed considerably for all manner of reasons. But these delays meant that the final standard was relevant to pull down and pop-up menus which had not even been considered when the standard was first proposed.

Another benefit is that during the development process, those who may be affected have the opportunity to prepare for the standard. Thus by the time ISO 9241-3:1992 Display requirements was published, many manufacturers were able to claim that they already produced monitors which met the standard. They had not been in that position when the standard was first proposed (although some argued that they would have been improving the design of their displays anyway). Certainly the standards provided a clear target for both demanding consumers and quality manufacturers.

It Is Easy to Misunderstand the Scope and Purpose of a Particular Standard

HCI standards have been criticized for being too generous to manufacturers in some areas and too restrictive in other areas. The "overgenerous" criticism misses the point that most standards are setting minimum requirements and in ergonomics standards makers must be very cautious about setting such levels. However, there certainly are areas where being too restrictive is a problem. Examples include:

- **ISO 9241-3:1992 Ergonomics Requirements for Work with VDTs: Display Requirements.** This standard has been successful in setting a minimum standard for display screens which has helped purchasers and manufacturers. However, it is biased toward Cathode Ray Tube (CRT) display technology. An alternative method of compliance based on a performance test (which is technology independent) has now been published which should help redress the balance.
- **ISO 9241-9:1999 Ergonomics Requirements for Work with VDTs: Non-keyboard input devices.** This standard has suffered because technological developments were faster than either ergonomics research or standards making. Although there has been an urgent need for a standard to help users to be confident in the ergonomic claims made for new designs of mice and other input devices, the lack of reliable data forced the standards makers to slow down or run the risk of prohibiting newer, even better solutions.

Standards Making Involves Politics as Well as Science

Although ergonomics standards are generally concerned with such mundane topics as keyboard design or menu structures, they nonetheless generate considerable emotion among standards makers. Sometimes this is because the resulting standard could have a major impact on product sales or legal liabilities. Other times the reason for the passion is less clear. Nonetheless, the strong feelings have resulted in painful experiences in the process of standardization. These have included:

- **Undue influence of major players.** Large multinational companies can try to exert undue influence by dominating national committees. Although draft standards are usually publicly available from national standards bodies, they are not widely publicised. This means that it is relatively easy for well-informed large companies to provide sufficient experts at the national level to ensure that they can virtually dictate the final vote and comments from a country.
- **"Horse trading" and bargaining to achieve agreement.** End user's requirements can be compromised as part of "horse trading" between conflicting viewpoints. In the interests of reaching agreement, delegates may resort to making political tradeoffs largely independent of the technical merits of the issue.

The Language of Standards Can Be Obscure

In ISO, the formal rules and procedures for operating seem to encourage an elitist atmosphere with standards written for standards enthusiasts. ISO has recognized this and is attempting to make the process more customer focused but such changes take time. These procedures and rules reinforce elitist tendencies and sometimes resulted in standards which leave much to be desired in terms of brevity, clarity, and usability. There are three contributory factors:

- **The use of stilted language and boring formats.** The unfriendliness of the language is illustrated by the fact that although the organization is known by the acronym ISO, its full English title is the International Organization for Standardization. The language and style are governed by a set of Directives and these encourage a wordy and impersonal style.
- **Problems with translation and the use of "Near English."** There are three official languages in ISO—English, French, and Russian. In practice, much of the work is conducted in English, often by nonnative speakers. The result of this is that the English used in standards is often not quite correct—it is "near English." The words are usually correct, but

the combination often makes the exact meaning unclear. These problems are exacerbated when the text is translated.

- **Confusions between requirements and recommendations.** In ISO standards, there are usually some parts which specify what has to be done to conform to the Standard. These are indicated by the use of the word "shall." However, in ergonomics standards, we often want to make recommendations as well. These are indicated by the use of the word "should." Such subtleties are often lost on readers of standards, especially those in different countries. For example, in the Nordic countries, they follow recommendations (shoulds) as well as requirements (shalls), so the distinction is diminished. In the United States, they tend to ignore the "shoulds" and only act on the "shalls."

Structure and Formality Can Be a Help as Well as a Hindrance

One of the benefits of standards is that they do represent a rather simplified and structured view of the world. There is also a degree (sometime excessive) of discipline in what a standard can contain and how certain topics can be addressed. Manufacturers (and ergonomists) frequently make wildly different claims about what represents good ergonomics. This is a major weakness for our customers who may conclude that all claims are equally valid and there is no sound basis for any of it. Standards force a consensus and therefore have real authority in the minds of our customers. Achieving consensus requires compromises, but then so does life.

The formality of the standards means that they are suitable for inclusion in formal procurement processes and for demonstrating best practice. In the UK at least, parts of ISO 9241 may be used by suppliers to convince their customers that visual display screen equipment and its accessories meet good ergonomic practice. Of course, they can also be "abused" in this way with overeager salesmen misrepresenting the legal status of standards, but that is hardly the fault of the standards makers.

The Benefits Do Not Just Come From the Standards Themselves

There are several ways in which ergonomics standardization activities can add value to user interface design apart from the standards themselves which are the end results of the process.

In 1997, the U.S. National Institute of Standards and Technology (NIST) initiated a project (Industry USability Reporting [IUSR]) to increase the visibility of software usability. They were helped in this endeavor by prominent suppliers of software and representatives from large consumer organizations. One of the key goals was to develop a common usability reporting format (Common Industry Format [CIF]). This is currently being processed as an ANSI standard through NCITS. CIF has been developed to be consistent with ISO 9241 and ISO 13407 and is viewed by the IUSR team as "an implementation of that ISO work." This activity in itself should have a major impact on software usability (http://www.nist.gov/iusr).

In the hardware arena, many people are aware of the TCO 99 sticker which appears on computer monitors and understand that it is an indication of ergonomic and environmental quality. What they may not know is that TCO is the Swedish Confederation of White Collar Trades Unions and that ISO 9241 was used as a major inspiration for its original specification. They publish information in English, and details are available on their Web site at http://www.tco.se/eng/index.htm

Being International Makes It All Worthwhile

Although there are national and regional differences in populations, the world is becoming a single market with the major suppliers taking a global perspective. Variations in national standards and requirements not only increase costs and complexity but also tend

TABLE 19.8
Members of ISO/TC159/SC4 Ergonomics of Human-System Interaction

"P" Members	Austria	Belgium	Canada	Czech Republic
	China	Denmark	Finland	France
	Germany	Ireland	Italy	Japan
	Korea	Netherlands	Norway	Poland
	Slovakia	Spain	Sweden	Thailand
	United Kingdom	United States of America		
"O" members	Australia	Hungary	Mexico	Romania
	Tanzania			

to compromise individual choice. Making standards international is one way of ensuring that they have impact and can help improve the ergonomics quality of products for everyone. That has to be a worthwhile objective. Table 19.8 shows the member countries of ISO/TC159/SC4.

GUIDELINES

Background

Whereas European interest in the early 1980s seemed to focus on computer hardware ergonomics, in the United States there was a growing interest in user interface software.

Whereas computer hardware seemed like a possible target for formal standardisation, the highly contextual nature of best practice made formal standards at best premature and at worst dangerous traps for perpetuating obsolete practices in this rapidly developing technology.

Rather than attempt to develop formal standards, several human computer interaction groups and individual researchers started to assemble collections of guidelines, tips, and hints into books and compendiums. We describe three of the most influential in this section.

Smith & Mosier's Guidelines for Designing User Interface Software

In 1986 Sidney Smith and Jane Mosier published *Guidelines For Designing User Interface Software* for the U.S. Air Force. With 944 guidelines, this document remains the largest collection of publicly available user interface guidelines in existence. These guidelines draw extensively from four sources: Brown et al., (1983), Engel and Granda (1975), MIL-STD-1472C (revised; 1983), and Pew and Rollins (1975).

Their report provides user interface guidelines in six categories:

1. Data entry
2. Data display
3. Sequence control
4. User guidance
5. Data transmission
6. Data protection

One example of a guideline from this document is:

1.3 DATA ENTRY: Text

1.3/10 + Upper and Lower Case Equivalent in Search

Unless otherwise specified by a user, treat upper and lower case letters as equivalent in searching text

Example: "STRING", "String", and "string" should all be recognized/accepted by the computer when searching for that word.

Comment: In searching for words, users will generally be indifferent to any distinction between upper and lower case. The computer should not compel a distinction that users do not care about and may find difficult to make. In situations when case actually is important, allow users to specify case as a selectable option in string search.

Comment: It may also be useful for the computer to ignore such other features as bolding, underlining, parentheses and quotes when searching text.
See also: 1.0/27 3.0/12

Although focused on character user interfaces (CUIs), many of the guidelines are still relevant to today's user interfaces—especially web sites.

Shneiderman's User Interface Guidelines

Ben Shneiderman published a landmark text in 1987, *Designing the User Interface*. This book contains many tables of guidelines and includes detailed explanations and background research to justify each guideline. He presented "Eight golden rules of dialog design," which he explained as representing underlying principles of design that were applicable to most interactive systems. These principles were:

- **Strive for consistency**—in particular use consistent sequences of actions and use the same terminology wherever appropriate. He claimed that this was the most frequently violated yet "the easiest one to repair and avoid." Certainly it ought to be easy to ensure consistency, but as we discuss later with respect to style guides, it can prove difficult to ensure that unhelpful inconsistencies do not creep into designs, especially where distributed teams are involved.
- **Enable frequent users to use shortcuts**—abbreviations, special keys, and macros can all be appreciated by frequent knowledgeable users. One of the traps of Windows, Icons, Mouse, and Pop-up menus (WIMP) interfaces is that they are very easy to demonstrate to senior management and superficially appear simple to use. We have seen several examples where experienced users have become extremely frustrated by mouse-intensive systems—in some cases to the extent of suffering work-related upper-limb disorders.
- **Offer informative feedback**—visual feedback can be particularly effective. Shneiderman (1987) goes on to discuss the value of direct manipulation in this context.
- **Design dialogs to yield closure**—organizing actions into groups with a beginning, middle, and end plays to a basic psychological desire for closure and provides a sense of satisfaction when tasks are completed.
- **Offer simple error handling**—where possible design systems so that users cannot make serious errors and ensure that error messages really help the users.

- **Permit easy reversal of actions**—it may not always be possible, but there is nothing more reassuring for users than knowing that they can undo an action which has unintended and sometimes dramatic consequences.
- **Support internal locus of control**—another basic psychological desire is for control, and the more systems are able to provide users with control, the more satisfying they will be, especially for experienced users. Some modern office software packages break this rule, and many of us experience extreme frustration when a "clever" piece of software insists on reformatting what we have carefully laid out on a page.
- **Reduce short-term memory load**—although the "seven plus or minus two chunks" may be an oversimplified description of our memory limitations, many interfaces demand extraordinary feats of memory to perform simple tasks.

Nielsen's Usability Heuristics

In 1990, Rolf Molich and Jakob Nielsen (1990) carried out a factor analysis of 249 usability problems to derive a set of "heuristics" or rules of thumb that would account for all of the problems found. Nielsen (1994a) further revised these heuristics, resulting in the 10 guidelines listed in the following.

1. **Visibility of system status**—The system should always keep users informed about what is going on, through appropriate feedback within reasonable time.
2. **Match between system and the real world**—The system should speak the user's language, with words, phrases, and concepts familiar to the user, rather than with system-oriented terms. Follow real-world conventions, making information appear in a natural and logical order.
3. **User control and freedom**—Users often choose system functions by mistake and will need a clearly marked "emergency exit" to leave the unwanted state without having to go through an extended dialog. Support undo and redo.
4. **Consistency and standards**—Users should not have to wonder whether different words, situations, or actions mean the same thing. Follow platform conventions.
5. **Error prevention**—Even better than good error messages is a careful design that prevents a problem from occurring in the first place.
6. **Recognition rather than recall**—Make objects, actions, and options visible. The user should not have to remember information from one part of the dialog to another. Instructions for use of the system should be visible or easily retrievable whenever appropriate.
7. **Flexibility and efficiency of use**—Accelerators—unseen by the novice user—may often speed up the interaction for the expert user, such that the system can cater to both inexperienced and experienced users. Allow users to tailor frequent actions.
8. **Aesthetic and minimalist design**—Dialogs should not contain information that is irrelevant or rarely needed. Every extra unit of information in a dialog competes with the relevant units of information and diminishes their relative visibility.
9. **Help users recognize, diagnose, and recover from errors**—Error messages should be expressed in plain language (no codes), precisely indicate the problem, and constructively suggest a solution.
10. **Help and documentation**—Even though it is better if the system can be used without documentation, it may be necessary to provide help and documentation. Any such information should be easy to search, focused on the user's task, list concrete steps to be carried out, and not be too large.

Nielsen (1994b) describes a method for structuring a guidelines-based user interface review. Nielsen's method uses Molich and Nielsen's (1990) user interface heuristics, and indeed Nielsen terms this method an "heuristic evaluation." The method, however, can be applied using any set of HCI Guidelines. For further information on evaluation techniques, see chapter 57. "Inspection-Based Evaluations" by Gilbert Cockton, Darryn Lavery, and Alan Woolrych.

STYLE GUIDES

Although few organizations have the history, the infrastructure, or the stamina to impose and police rigid user interface standards, style guides can be developed to reduce the unnecessary variation caused by dispersed design teams and extended system development timescales.

The main stages in the process involve the following.

Choosing the Right Guidelines

We have already pointed out that there are many good sources of guidelines, including a number of proprietary style guides provided by major vendors. There is no right answer as to which guidelines to select. It depends on the specific system under development and the style of interface.

Tailoring the Guidelines Into Specific Design Rules for Your Application

For instance, a guideline which states that displays should be consistently formatted might be translated into design rules that specify where various display features should appear, such as the display title, prompts and other user guidance, error messages, command entries, and so forth. For maximum effectiveness, guideline tailoring must take place early in the design process before any actual design of user interface software. In order to tailor guidelines, designers must have a thorough understanding of task requirements and user characteristics. Thus task analysis is a necessary prerequisite of guidelines tailoring.

The process of developing, reviewing, and agreeing style guides can be a positive process for enhancing organizational communication, especially across traditional organizational barriers.

Implementing the Style Guide

Many style guides offer little more than general recommendations on good practice. The problem with these is that they take considerable interpretation and may therefore still result in different parts of a system behaving differently. The mere presence of a style guide does not ensure consistency. Designers have to choose to conform or be disciplined to conform for the benefits to be achieved.

One of the best ways of encouraging them to follow a guide is to support it with a code library and provide lots of good examples for designers to follow. Interactive demonstrations can be particularly valuable.

Policing and Maintaining the Style Guide

In practice, it is better to motivate and encourage designers to follow good examples in style guides than to rely on postdesign monitoring and policing operations. As we have pointed out earlier, international standards are slowly being developed which address many different aspects of user interface design—both hardware and software. These standards can be used to provide support for in-house measures. In our experience, senior managers are more likely to take style guide and user interface issues seriously if they know that there are public standards that support them. We have incorporated checklists based on parts of ISO 9241 in some style guides that we have developed. Appendix 2 shows a checklist based on the seven key principles in ISO 9241-10: 1996 *Dialogue Principles*. Each principle has been rephrased as a question (in bold) with specific questions below. The "correct" answer is usually "yes," although, there may be occasions where the recommendation is not applicable or not possible; for example, it may not be possible; to *"undo"* a *"commit"* action. Appendix 3 shows a similar checklist for user guidance based on ISO 9241-13: 1998 *User Guidance*.

The checklists can be used by designers when reviewing their own work as well as by those involved in signing-off the design.

CONCLUSION

One of the recurring themes in this chapter is what might be called Stewart's law of usability standards—the easier it is to formulate the usability standard or guideline, the more difficult it is to apply in practice. By this, I mean that a simple guideline like "allow the user to control the pace and sequence of the interaction" has a great deal of backing as a general guideline. In practice, of course, there are many situations where the rule fails, where the answer is "it depends," and where the context makes it more appropriate for the system to control some part of the interaction. The designer wishing to follow such a simple guideline needs to interpret the guideline in the context of the system, and this requires insight and thought. The alternative approach, where the guideline is preceded by statements defining the context, for example, "if the user is x and the task is y and the environment is z, then do abc" become so tedious and confusing that they are quickly ignored. In the ISO 9241 series described earlier, there has been an attempt to define a middle course which involves giving specific practical examples as an aid to design. In ISO 13407 *Human Centered Design Processes for Interactive Systems*, the standard is concerned with the process itself.

ISO 13407-1999 *Human-Centered Design Processes for Interactive Systems* Provides

Guidance for project managers to help them follow a human-centered design process. By undertaking the activities and following the principles described in the Standard, managers can be confident that the resulting systems will be usable and work well for their users.

The Standard describes four principles of human-centered design:

- Active involvement of users (or those who speak for them)
- Appropriate allocation of function (making sure human skill is used properly)
- Iteration of design solutions (allowing time for iteration in project planning)
- Multidisciplinary design (but beware large design teams)

and four key human-centered design activities:

- Understand and specify the context of use (make it explicit—avoid assuming it is obvious)
- Specify user and organizational requirements (note there will be a variety of different viewpoints and individual perspectives)
- Produce design solutions (note plural, multiple designs encourage creativity)
- Evaluate designs against requirements (involves real user testing, not just convincing demonstrations)

In order to claim conformance, the Standard requires that the procedures used, the information collected, and the use made of results are specified (a checklist is provided as an annex to help). This approach to conformance has been used in a number of parts of ISO 9241, because so many ergonomics recommendations are context specific. Thus, there is often only one "shall" in these standards which generally prescribes what kind of evidence is required to convince another party that the relevant recommendations in the Standard have been identified and followed.

We believe this is one way of ensuring that usability standards remain relevant when technology changes and also offer practical help to designers and developers.

APPENDIX 1: HISTORY—DEUSCHES INSTITUT FÜR NORMUNG (DIN)

Thirty years ago, the German National Standards organization started to publish a series of standards which shook the computer world. These standards, DIN 66-234, were published in a number of parts and collectively addressed the ergonomics problems of Visual Display Terminals and their workplaces.

If we ask why there was widespread concern, especially from the computer manufacturers, many of whom happened to be based in the United States, then we receive two answers. One answer, which was popular at the time, was that the Standards were based on too little and too recent research. A particular issue that received such criticism was the requirement that the thickness of the keyboard should be restricted to 30 mm. A number of manufacturers reported studies disputing the importance of keyboard thickness, arguing with the proposed dimension and demonstrating that users showed preferences for quite different arrangements.

Of course, it should not be overlooked that the 30-mm keyboard thickness was a very difficult target to reach at that time. Most key mechanisms themselves required greater depth, and major manufacturers had substantial investment in tooling keyboards to quite different thicknesses.

The second answer is that the very idea of ergonomics requirements affecting sales directly was completely alien to many of the suppliers. Certainly, the large manufacturers employed ergonomists, human factors engineers, and psychologists in their research and development departments. Certainly there was a growing recognition that the human aspects of computer technology were important. But at that time, price performance was the main objective, and it came as a major culture shock for the computer industry that ergonomics standards could have such a major impact on whether a product would sell.

Note that it was not the DIN standard itself but its integration into workplace regulations (ZH 618 Safety Regulations for Display Workplaces in the Office Sector, published by the Central Association of Trade Cooperative Associations) that gave the ergonomics requirements such "teeth." Failure to comply with these regulations leaves an employer uninsured against industrial compensation claims.

DIN 66 234 also contained a number of parts that dealt exclusively with software issues. For example, Part 3 dealt with the grouping and formatting of data; Part 5, with the coding of information; and Part 8, with the principles of dialog design. Although these were more in the form of recommendations, they too were heavily criticized, particularly for their broad scope and their inhibitory effect on interface design.

Indeed, a major criticism of most early standards in this field was that they were based on product design features such as height of characters on the screen. Such standards were specific to current technology, for example, cathode ray tubes (CRT), and did not readily apply to other technologies. They may therefore inhibit innovation and force designers to stick to old solutions.

APPENDIX 2: TASK DESIGN CHECKLIST BASED ON
ISO 9241-10:1996 DIALOG PRINCIPLES

	Yes	No
Is the dialog suitable for the user's task and skill level?	☐	☐
Does the sequence match the logic of the task?	☐	☐
Are there any unnecessary steps that could be avoided?	☐	☐
Is the terminology familiar to the user?	☐	☐
Does the user have the information they need for the task?	☐	☐
Is extra information available if required? (*keep dialog concise*)	☐	☐
Does the dialog help users perform recurrent tasks?	☐	☐
Does the dialog make it clear what the user should do next?	☐	☐
Does the dialog provide feedback for all user actions?	☐	☐
Are users warned about (and asked to confirm) critical actions?	☐	☐
Are all messages constructive and consistent?	☐	☐
Does the dialog provide feedback on response times?	☐	☐
Can the user control the pace and sequence of the interaction?	☐	☐
Can the user choose how to restart an interrupted dialog?	☐	☐
Does the dialog cope with different levels of experience?	☐	☐
Can users control the amount of data displayed at a time?	☐	☐
Is the dialog consistent?	☐	☐
Are the appearance and behavior of dialog objects consistent with other parts of the dialog?	☐	☐
Are similar tasks performed in the same way?	☐	☐
Is the dialog forgiving?	☐	☐
Does the dialog provide "undo" (and warn when not available)?	☐	☐
Does the dialog prevent invalid input?	☐	☐
Are error messages helpful?	☐	☐
Can the dialog be customized to suit the user?	☐	☐
Does the dialog offer users different ways of working?	☐	☐
Does the dialog provide helpful defaults?	☐	☐
Can users choose different levels of explanation?	☐	☐
Can users choose different data representations? (e.g., show files as icons or lists)	☐	☐
Does the dialog support learning?	☐	☐
Does system feedback help the user learn? (e.g., menu items that indicate shortcut key combinations)	☐	☐
Is context-sensitive help provided? (where possible)	☐	☐

APPENDIX 3: CHECKLIST BASED ON ISO 9241-13:1998 USER GUIDANCE

	Yes	No
General		
Can the user guidance be readily distinguished from other information?	☐	☐
Do system-initiated messages disappear when no longer applicable?	☐	☐
Do user-initiated messages remain until the user dismisses them?	☐	☐
Are messages specific and helpful?	☐	☐
Can the user continue while the guidance is displayed?	☐	☐
Are important messages distinctive?	☐	☐
Can users control the level of guidance they receive?	☐	☐
Wording		
Do messages describe results before actions? (e.g., to clear screen, press *del*)	☐	☐
Are most messages worded positively? (i.e., what to do not what to avoid)	☐	☐
Are messages worded in a consistent grammatical style?	☐	☐
Is the guidance written in a short, simple sentences?	☐	☐
Are messages written in the active voice?	☐	☐
Is the wording user oriented?	☐	☐
Prompts—To Indicate That the System Is Waiting for Input		
Do prompts indicate the type of input required?	☐	☐
Is online help available to explain prompts? (if required)	☐	☐
Are prompts displayed in consistent positions?	☐	☐
Does the cursor appear automatically at the first prompted field?	☐	☐
Feedback—To Indicate That the System Has Received Input		
Does the system always provide some form of feedback for all user actions?	☐	☐
Is normal feedback unobtrusive and nondistracting?	☐	☐
Are the type and level of feedback suitable for the skills of the users?	☐	☐
Does the system provide clear feedback on system state? (e.g., waiting for input)	☐	☐
Are selected items always highlighted?	☐	☐
If the action requested is not immediate, is there feedback that requests have been accepted (and confirmation when they are complete), for example, remote printing?	☐	☐
Does the system show progress indicators when appropriate?	☐	☐
Is system response feedback appropriate? (not too fast or too slow)	☐	☐
Status Information—to Indicate What the System Is Currently Doing		
Is appropriate status information available at all times?	☐	☐
Is status information always displayed in a consistent location?	☐	☐
Does the system always indicate when user input is not possible?	☐	☐
Are system modes clearly indicated?	☐	☐
Error Prevention and Validation		
Are the function keys consistent across the system?	☐	☐
Does the system anticipate problems and warn the user appropriately?	☐	☐
Does the system warn the user when data might be lost by a user action?	☐	☐
Is "undo" provided, where appropriate?	☐	☐
Can users modify or cancel input prior to input?	☐	☐
Can users edit wrong input (rather than have to reenter the complete field)?	☐	☐

Does the field level validation:
Immediately indicate that there is an error? ☐ ☐
Position the cursor at the beginning of the first incorrect field? ☐ ☐
Indicate all fields in error (including cross-field errors)? ☐ ☐

Error Messages
Can users get more help easily if required? ☐ ☐
Do error messages indicate what is wrong and what should be done? ☐ ☐
Can the user tell when an error message has reoccurred? ☐ ☐
Do error messages disappear when the error has been corrected? ☐ ☐
Can users remove error messages prior to correction if they wish? ☐ ☐
Do error messages appear in a consistent location? ☐ ☐
Can error messages be moved if they obscure part of the screen? ☐ ☐
Do error messages appear as soon as the wrong input has been entered? ☐ ☐
Can users turn off confirmation screens? ☐ ☐
Can users adjust the volume of warning tones or messages? ☐ ☐

Online Help
Is the online help context sensitive? ☐ ☐
Is the system-initiated online help unobtrusive? ☐ ☐
Can users turn off system-initiated online help? ☐ ☐
Can the users initiate online help by a simple consistent action? ☐ ☐
Does the system accept synonyms and close spelling matches when the ☐ ☐
    user searches for the appropriate area of online help?
Is the online help clear, understandable, and specific to the users' tasks? ☐ ☐
Does the online help provide both descriptive and procedural help? ☐ ☐
Can the user easily go between the task screens and the online help? ☐ ☐
Can the users configure online help to suit their preferences? ☐ ☐
Can the user easily return to the task? ☐ ☐
Are there suitable features to help users find appropriate help? ☐ ☐
Are there quick methods for:
    Going directly to another screen ☐ ☐
    Browsing ☐ ☐
    Exploring linkages between topics ☐ ☐
    Returning to the previous help page ☐ ☐
    Returning to a home location ☐ ☐
    Accessing a history of previously consulted topics ☐ ☐
If the online help system is large, are any of the following provided to aid
search:
    String search of a list of topics? ☐ ☐
    Keyword search of online help text? ☐ ☐
    Hierarchical structure of online help text? ☐ ☐
    Map of online help topics? ☐ ☐
If the online help system has a hierarchical structure, is there:
    An overall indication of the structure? ☐ ☐
    Easy access to any level in the hierarchy? ☐ ☐
    An obvious and consistent method of accessing more detail? ☐ ☐
    A quick means of accessing the main (parent) topic? ☐ ☐
    Are topics self-contained? that is, not dependent on reading previous sections ☐ ☐
If the information is scrollable, does the topic remain clear? ☐ ☐

Does the context-sensitive help provide information on:
    The current dialog step?
    The current task?   ☐   ☐
    The current applications?   ☐   ☐
    The task information presented on the screen?   ☐   ☐
Does the online help explain objects, what they do, and how to use them?   ☐   ☐
Does the online help indicate when it is not available on an object?   ☐   ☐

ACKNOWLEDGMENT

Some material in this chapter was also used in the Human Computer Interaction Handbook edited by Julie Jacko and Andrew Sears.

REFERENCES

Brown, C. M., Brown, D. B., Burkleo, H. V., Mangelsdorf, J. E., Olsen, R. A., & Perkins, R. D. (1983, June 15). *Human factors engineering standards for information processing systems (LMSC-D877141)*. Sunnyvale, CA: Lockheed Missiles and Space Company.

Cockton, G., Lavery, D. & Woolrych, A. (2003), Inspection-Based Evaluations. In J. A. Jacho, & A. Sears (Eds.), *The Human-Computer Intraction Handbook*. Mahwah, NJ: Lawrence Erlbaum Associates.

Engel, S. E., & Granda, R. E. (1975, December). *Guidelines for man/display interfaces* (Technical Report TR 00.2720). Poughkeepsie, NY: IBM.

MIL-STD-1472C, Revised. (1983, September 1). *Military standard: Human engineering design criteria for military systems, equipment and facilities*. Washington, DC: Department of Defense.

Molich, R., & Nielsen, J. (1990, March). Improving a human-computer dialogue. *Communications of the ACM 33(3)*, 338–348.

Nielsen, J. (1994a). Enhancing the explanatory power of usability heuristics. *Proc. ACM CHI'94 Conf.* (Boston, MA, April 24–28), 152–158.

Nielsen, J. (1994b). Heuristic evaluation. In J. Nielsen, and R. L. Mack (Eds.), *Usability inspection methods*. New York: Wiley.

Pew, R. W., & Rollins, A. M. (1975). *Dialog specification procedures* (Report 3129, revised). Cambridge, MA: Bolt Beranek and Newman.

Shneiderman, B. (1987). *Designing the user interface—Strategies for effective human-computer interaction*. Reading, MA: Addison-Wesley.

20

Human Factors Engineering of Computer Workstations: HFES Draft Standard for Trial Use

Thomas J. Albin
Auburn Engineers, Inc.

INTRODUCTION

BSR/HFES 100 Human Factors Engineering of Computer Workstations (HFES 100) is a specification of the recommended human factors and ergonomic principles related to the design of the computer workstation. HFES 100 is primarily intended for fixed, office-type computer workstations for individuals who are moderate to intensive computer users. These design specifications are intended to facilitate the performance and comfort of computer workers.

HFES 100 addresses the interaction between the computer user and the hardware components of the operator–machine system. It addresses input devices, output devices, and furniture, as well as the integration of these components into a human user–computer hardware system.

Specifications are given for individual components. As an example, the specifications for furniture components describe a range of adjustments necessary to accommodate users who vary in size and who use multiple working postures.

HFES 100 provides guidance as to how these individual workstation components are to be integrated into a whole, or system that is ergonomically sound. It is possible to utilize ergonomically well-designed computer workstation components that individually conform to the specifications of HFES 100, but to combine them inappropriately. As an extreme example, a desk designed for standing work only might mistakenly be installed in an area where all other components were intended for sitting work only. The chapter on Installed Systems is intended to prevent such mismatches.

Given that the individual components and the system are able to accommodate a variety of users and postures, HFES 100 also provides guidance as to how to accommodate an individual user. As an example, it provides guidance as to how a chair that is designed to accommodate a variety of sizes of users and a variety of postures may be adjusted to fit an individual.

Finally, it provides guidance as to desirable conditions in the environment around the computer workstation.

THE NEED FOR A DESIGN STANDARD
FOR COMPUTER WORKSTATIONS

The computer is a basic and pervasive tool in all aspects of life. According to the U.S. Census Bureau, three fourths of all households in the United States had computers in 1997. Further, according to the same source, about 71% of all children used a computer at school. Finally, U.S. Census data indicate that half of all employed adults used a computer in their workplace (Newberger, 1999).

Given that half of all workers in the United States use computers, designing computer workstations so as to facilitate computer worker performance offers the prospect of significant benefits. Research shows a clear association between the performance of computer workers and the ergonomics of the workstation. As an example, Dainoff (1990) found that performance increased by approximately 20 percent when the workstation components were adjusted to accommodate the individual user.

The U.S. Census Bureau (U.S. Census Bureau, 2001) estimates that there were 105,000,000 individuals employed in all businesses in the United States in 1997 and that the total payroll for these individuals was approximately $3 trillion (U.S. Census Bureau, 2000). Because one-half of all employed individuals used computers in the workplace that year, the payroll for computer workers in 1997 was approximately $1.5 trillion. Although one would not expect such dramatic facilitation of performance in every case as was demonstrated by Dainoff (1990), the scale is so large that even small changes are significant. The performance benefit realized from good ergonomic practices in computer workstations dictates that it is in the employers' interest to provide computer workplaces that are ergonomically well designed.

Many employers do not have ergonomics or human factors professionals available to assist them in providing ergonomically sound computer workplaces. HFES 100 is a concise resource document for employers to use in providing well-designed computer workplaces, when selecting components, when integrating the components into a system, and when fitting a workstation to an individual user.

HFES is divided into four major chapters: "Installed Systems," "Input Devices," "Visual Displays," and "Furniture."

INSTALLED SYSTEMS

The "Installed Systems" chapter describes how to put together all the workstation components into a system that matches the capabilities of the intended user. It is intended primarily for the System Installer, that is, the individual or individuals responsible for selecting the components and installing them as an integrated system. As such, it deals primarily with issues such as hardware components, noise, thermal comfort, and lighting.

HFES 100 (Human Factors and Ergonomics Society, 2002) lists the following four key issues that the System Installer must address:

- Ensure that all components individually comply with the appropriate specifications of HFES 100
- Ensure that the selected components are compatible with one another when installed
- Ensure that the workstation properly fits the intended user
- Ensure that the users are informed about the proper use and adjustment of the workstation components

Some individuals may fall outside the design specifications for the individual workstation components given in HFES 100. As an example, the furniture specifications are written to accommodate individuals ranging between 5th percentile females and 95th percentile males. The "Installed Systems" chapter offers some guidance as to how to accommodate these individuals.

INPUT DEVICES

The "Input Devices" chapter addresses issues such as physical size, operation force, and handedness in the design of the following input devices:

- Keyboards
- Mouse and puck devices
- Trackballs
- Joysticks
- Styli and light pens
- Tablets and overlays
- Touch-sensitive panels

VISUAL DISPLAYS

The Visual Displays chapter covers monochrome and color CRT and flat-panel displays. It is not intended to apply to displays on items such as photocopy machines and telephones. It addresses the human factors issues related to the design of displays such as viewing characteristics, contrast, and legibility.

FURNITURE

The "Furniture" chapter describes specifications for workstation components such as chairs and desks. There are productivity and comfort benefits associated with changing posture while performing computer work, and computer workers have been observed to frequently change position while working. Consequently, the furniture designer needs to anticipate a range of movement and change in working posture.

HFES 100 describes the various postures that a furniture designer should consider. There are four Reference Postures: reclined sitting, upright sitting, declined sitting, and standing. The computer worker will frequently change the position of their body parts while working in any of the Reference Postures. The expected ranges of movement of body parts are described as User Postures.

ANTHROPOMETRY

The anthropometric data utilized in the Furniture and Installed Systems chapters were developed from the database of U.S. Army personnel (Gordon et al., 1989). A description of the anthropometric methods and models used is provided as an appendix to the "Furniture" chapter.

REFERENCES

Dainoff, M. (1990). Ergonomic improvements in VDT workstations: Health and performance effects. In S. Sauter, M. Dainoff, & M. Smith (Eds.), *Promoting health and productivity in the computerized office* (pp. 49–67). Philadelphia: Taylor and Francis.

Gordon, C. C., Churchill, T., Clauser, T. E., Bradtmiller, B., McConville, J. T., Tebbets, I. et al. (1989). *1988 Anthropometric survey of U.S. Army personnel: Summary statistics interim report* (Tech. Report NATICK/TR-89-027).

Human Factors and Ergonomics Society. (2002). Board of Standards Review/Human Factors and Ergonomics Society 100—Human factors engineering of computer workstations—Draft Standard for Trial Use. Human Factors and Ergonomics Society, Santa Monica, CA.

Newberger, E. C. (1999). *Computer use in the united states. Population characteristics.* U.S. Department of Commerce, Economics and Statistics Administration. U.S. Census Bureau.

U.S. Census Bureau. (2000). 1997 *Economic census: Comparative statistics for the United States 1987 SIC basis.* U.S. Census Bureau. http://landview.census.gov.epcd/ec97sic/E97SUS.HTM

U.S. Census Bureau. (2001). *Statistics about business size (including small business) from the U.S. census bureau.* U.S. Census Bureau. http://www.census.gov/epcd/www/smallbus.html

21

Practical Universal Design Guidelines: A New Proposal

Kazuhiko Yamazaki
IBM Japan Ltd.

Toshiki Yamaoka
Wakayama University

Akira Okada
Osaka City University

Sohsuke Saitoh
Human Factor Co., Ltd.

Masatoshi Nomura
NEC Corporation

Koji Yanagida
SANYO Design Center Co., Ltd.

Sadao Horino
Kanagawa University

INTRODUCTION

Universal design (UD) practices are gaining importance in products design. Because there have been no established universal design methods, designers often tend to use their own experiences and intuition in extracting problems and creating new universal designs. In view of these situations, the Ergo-Design Technical Group of the Japan Ergonomics Society (EDTG/JES) organized a working group for proposing integrated UD methods that could be used by designers. This working group developed new practical UD guidelines together with the members of EDTG/JES and collaborators including ergonomics experts in universities and relevant industrial designers working in the private sectors. The working group utilized various opportunities for discussing the UD needs such as the annual conferences of the Society and annual conferences of Kantoh District (Tokyo area) chapter of the Society, as well as regular and special seminars organized by the Ergo-Design Technical Group. The working group thus examined proposals of the members for practical UD guidelines. This chapter describes the new guidelines including the basic design principles and the detailed methods for accomplishing new designs meeting the requirements of the proposed guidelines.

These guidelines propose ready-to-use procedures for implementing a logical and systematic design approach for universal design. They provide practical guidance for designers and others who are looking for practical design tools.

The following summarizes the basic policies for the development of practical UD guidelines, and gives the contents of the guidelines. This includes a concrete guidance about how to apply the guidelines. In developing the guidelines, the highest priority has been placed on developing methods of extracting user requirements which are considered critical for developing a universal design. The working group was organized for a specified period in the EDTG/JES. The working group consisted of the following members: Chair of the WG: Toshiki Yamaoka (Wakayama University). Members of the WG: Akira Okada (Graduate School of Osaka City University), Sosuk Saito (Human Factor Co., Ltd.), Masatoshi Nomura (NEC Corporation), Hiroharu Yanagida (Sanyo Electric Co., Ltd.), and Kazuhiko Yamazaki (IBM Japan, Ltd.) and Chair of EDTG/JES: Sadao Horino (Kanagawa University).

The guidelines consist of two parts: Part 1: Practical Guide and Part 2: Reference Information. These two parts cover the following topics.

Part 1: Practical Guide

1. Definition of UD and the basic approach
2. UD design processes
3. UD user segments table
4. UD matrix
5. Methods of extracting user requirements
6. Methods of the construction and design of the UD concept
7. Methods of UD evaluation

Part 2: Reference Information

1. Information for the UD user segments table and UD data
2. UD matrix samples
3. Application of the UD matrix
4. UD action checklist
5. Case studies on the use of the Practical Universal Design Guidelines
6. Related information

The Practical Guidelines are intended for direct use by designers and have been prepared in accordance with the following basic policies: (a) Design can be implemented without particular preliminary knowledge of UD. (b) The Practical UD Guidelines are produced for actual UD users (business owners, and company employees), and not merely for the promotion of the UD concept alone. (c) The Practical UD Guidelines are based on a pragmatic approach. (d) The Practical UD Guidelines are based on human-centered design methods. (e) The Practical UD Guidelines are based on logical design methods. (f) Following the Practical UD Guidelines should allow customization. (g) The guidelines comply with the IEC/ISO Guide 71.

OUTLINE OF THE PRACTICAL UD GUIDELINES

The following introduces the outline of the Practical UD Guidelines.

The Definition of Universal Design and the Background for Emerging UD Needs

The term *universal design* is defined as follows: to design products, environments and information with the aim of providing all users in a fair and equitable manner with satisfactory services responding to their varied needs. The need for UD has emerged due to the following background situations: (a) Human-centered design has become increasingly important, whereas products tend to incorporate more functions and advanced technologies. (b) Companies are highly appreciated when they respond to special needs and market needs put forward by the aging society, meet various growing ISO-related standards and legal provisions of laws and regulations, and fulfill their own social functions and responsibilities. (c) There are fundamental changes in the social environment, such as an aging population, widening diversity in the use of products, and enhanced public expectations for efficient welfare spending.

Clearly, these situations have extensive implications for the relevant design processes of products produced for a wide range of users.

Methods of UD Implementation by Companies

To promote UD, it is necessary to disseminate the idea that "UD activities reflect the corporate philosophy of a company, represent its business operations, and signify its presence" so that the significance of UD must be recognized throughout each company. UD can bring about direct and indirect effects on a company in meeting the varied needs of the users of their products.

Direct effects include:

- Improvement of product features
- Enhancement of the corporate image
- Reduction of total costs
- Development of new businesses and products

Indirect Effects Include:

- Market expansion
- Dissemination of social responsibilities of corporations and other organizations.

Effective promotion of UD requires the establishment of steps (procedures) to achieve successful UD implementation and the planning of aggressive activities (action) in the final stage. A company can start with whatever it can undertake without following established procedures, but it is desirable to set up suitable internal organizations (structure) and infiltrate (awareness raising) the UD concept throughout the company members including the executive manager and employees at all levels in order to solidify unity within the company. (a) Procedures: Establishment of the study of successful cases, the whole design processes, design methods, and the check system for design review. (b) Action: Promotion of the "top-down" design approach, active proposals for new design ideas, dissemination of the approach to organizations and people outside the company. (c) Structure: Establishment of the project, creation of a company-wide council, selection of a department in charge of UD promotion. (d) Awareness rising: Production of manuals, holding of lectures and seminars for employees and dissemination of information concerning UD applications conducted by other companies and institutions.

UNIVERSAL DESIGN PROCESS

Universal design means the design of products and environments usable by all people to the greatest extent possible, without the need for adaptation or specialized designs. "All people" include a whole range of people different in age, culture, location, and disability, but universal design does not mean that all people can use all kinds of products or services. Some severely disabled individuals will still need specific modifications. The goal of universal design is not only to eliminate physical barriers but also to eliminate mental barriers considering economic factors such as availability, manufacturing costs, and price. In reality, it is not easy to adopt

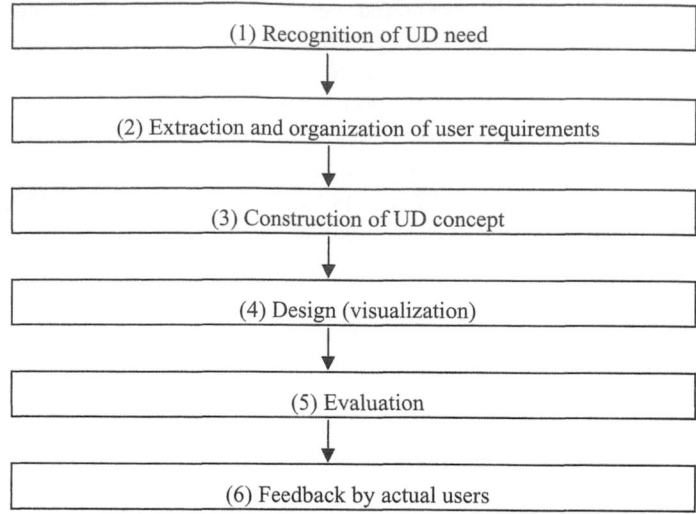

FIG. 21.1.   Universal design development process.

a universal design approach to products design because universal design means covering a wide range of users. It is very difficult for product designers to find answers to questions such as: "What should be the design requirements?" and "How should ideas and designs be evaluated?" The situation becomes very complicated, making it hard for product designers to adopt universal design principles.

The proposed Practical Universal Design Guidelines utilize the user segments table to define user groups, and an UD matrix for determining relevant design requirements, creating a new design concept, and evaluating the ideas. The user segments table consists of all user segments from the universal design viewpoint. The UD matrix consists of user groups, user tasks, and all requirements as specified by user groups and user tasks.

The process of the proposed UD guidelines involves defining user groups by using the user segments table, defining user tasks by using basic user tasks, and then formulating the UD matrix for the defined user groups and user tasks. The next steps are forming a design concept by using the UD matrix with priorities for each identified requirement, creating a detailed design based on the design concept and evaluating the ideas by using the UD matrix. The final step is the evaluation of the design proposal by real users.

Development Process

Before the actual design process starts, it is important to share the understanding of the importance and merits of universal design practices with all the related people involved in the products development. Without such a shared understanding of universal design practices, the design process will not succeed smoothly.

As shown in Figure 21.1, the typical development process is as follows:

- Recognition of the need for UD (awareness raising and information sharing)
- Extraction and identification of user requirements
- Construction of the UD concept
- Design (visualization)
- Evaluation
- Feedback from actual users

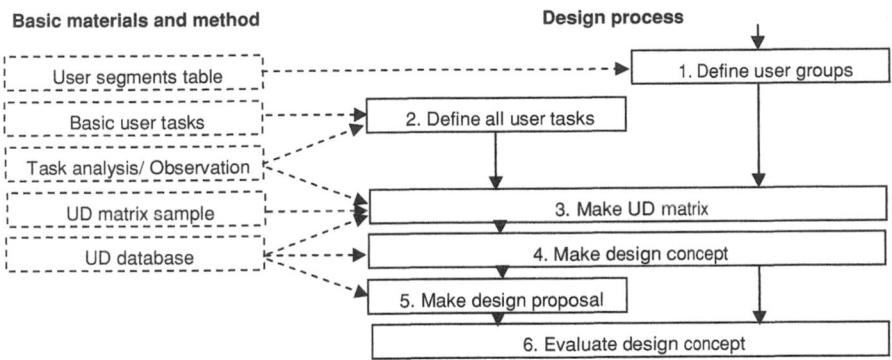

FIG. 21.2. Design process.

Design Process

As shown in Figure 21.2, the process of the proposed Practical UD Guidelines is to define user groups by applying the user segments table, defining user tasks by examining basic user tasks and then composing an UD matrix for the defined user groups and the defined user tasks. The next steps are forming a design concept by using a UD matrix with priorities for each of the requirements, creating a detailed design based on the design concept, and evaluating the ideas by using the UD matrix. The last step corresponds to user evaluation of the final proposed design.

Thus, the detailed design process in this sequence is as follows:

- Define target user groups by referring to the user segments table
- Define all user tasks by referring to basic user tasks
- Prepare the UD matrix by referring to the sample UD matrix
- Establishing UD concept and design method
- Make the design proposal
- Evaluate the design proposal

Define target user groups by referring to the user segments table The user segments table is prepared in order to help designers to find out necessary types of users of the product to be considered and to put these types in some target user groups in the UD matrix. The UD user segments constitute of a matrix of rows and columns. "Function 1, 2," of columns and "User type 1, 2," of rows in the left table correspond to "Factors to be considered" and "User type" in the user segments table, respectively.

- Clarify the user attributes and user functions to be adapted to the users, by referring to the user segments table.
- Define several target user groups for each product.

Define All User Tasks by Referring to Basic User Tasks

Basic user tasks are typical ones and they include preparation, start, acquisition of information, cognition, judgment, understanding, operation, completion of work, and maintenance. The basic user tasks are prepared on the UD matrix.

- Define typical user scenario for targeted products.
- Define all user tasks based on the basic user tasks

It is possible to define some user tasks that aren't derived from basic user tasks. When adding more details or more reality, it is better to use the task analysis method or the direct observation method. It is also better to define the user tasks after selecting one of the user scenarios from among several user scenarios.

Prepare the UD Matrix by Referring to the Sample UD Matrix

UD matrix is the table, which the X-axis (rows) is used for user groups, the Y-axis (columns) is for each user task, and each cell represents user requirements.

* The sample UD matrix on Practical Universal Design Guidelines ensures how to create a new UD matrix
* Determine the user scenario and the user environment.
* Select the target users, based on Step 1, in the target user section of the UD matrix template.
* Enter the user tasks from Step 2 in the user tasks section of the UD matrix template
* Determine the requirements for each cell of the UD matrix. It is useful to use an UD database to define all of the requirements.

Establishing UD Concept and Design Method

There are two ways to structure and express an UD concept. Based on extracted user requirements, an UD concept can be structured by a bottom-up system or by a top-down system (by the planning party). An UD concept is ranked into three levels ((A): shall be realized, (B): should be realized, (C)) should be realized if possible), and this ranking is utilized in the subsequent design and evaluation stages. User requirement items determined by using an UD matrix or simplified user requirement extraction method are changed into visual forms through brainstorming while taking cost and technical aspects into consideration.

Make the Design Proposal

Create ideas to meet each requirement, and then consolidate all the ideas into one system design. This step should use the UD database to make the ideas.

Evaluate the Design Proposal

Evaluate each ideas and the design proposals based on the design concept by designers. And the design proposal should be evaluated by the team and by real users.

Compact Design Process

The compact design process is a simplified way for a designer to design a product from the universal design viewpoints. Without making an UD matrix, the designer may utilize an existing UD matrix with necessary modifications and also utilize an action checklist to evaluate the design ideas.

As shown in Figure 21.3, the compact design process starts with looking at existing UD matrix samples to understand basic requirements. For this purpose, the designer needs to select an existing UD matrix from the samples of UD matrix on the Practical Universal Design Guidelines. Alternatively, the designer may utilize the same kind of UD matrix which was created for a previous product by the designer. The rest of the process is almost the same as the standard design process. Instead of evaluation by a team, the designer will be able to use an action checklist contained in the Practical Universal Design Guidelines.

Basic material and method   **Design process**

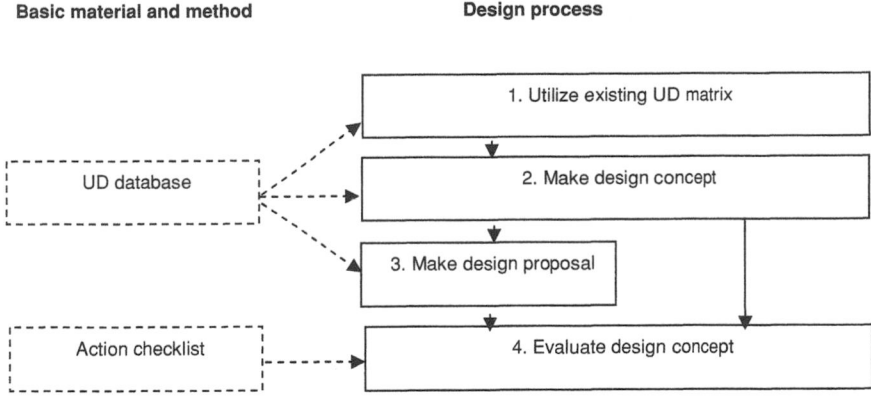

FIG. 21.3. Compact design process.

METHODS OF EXTRACTING USER REQUIREMENT

Target users are narrowed down by using a user segments table or its simplified version, and relevant user requirements are extracted based on the defined UD matrix. Other simplified methods of extracting user requirements under examination include: (a) the use of an action checklist, (b) extraction of user requirements based on user participation (workshop style), (c) making a simplified UD matrix, (d) simplified task analysis, and (e) direct observation. However, when a UD matrix is used, there is no need for using a simplified user requirement extraction method.

UD User Segments Table

A user segments table (user segments table [Table 21.1], user segments material) is produced for the purpose of extracting users at UD and users requiring special considerations and also organizing user types (groups) to be considered in the product design stage. There are two types of UD user segments tables: (A) user segments table and (B) user segments material.

UD Matrix

An UD matrix (Table 21.2) is used to effectively extract UD requirements according to different situations. User groups are indicated along the horizontal axis (rows), and individual

TABLE 21.1
Structure of the User Segments Table

Factors (Seeing (Eyesight)—), Hearing—to Consider		
	↓	
User type (the aged, disabled person, etc.) ⟶	Example: User with impairment in seeing	Example: User of wheelchair

TABLE 21.2
Structure of a UD Matrix

Three Aspects of a Product	UD Principles	Flow of the Basic User Tasks	User Tasks	User Groups (Users with Impairment in Seeing, Wheelchair Users, etc.)
Usable	(1) Easy to acquire information	(1) Preparation (2) Start	Task (1)	
			Task (2)	
	(2) Easy to understand	(3) Acquire information	Task (3)	
	(3) Reduce load on mind and body	(4) Cognition, Judgment, Understanding	Task (4)	⟶ Requirement
	(4) Safety	(5) Operation	—	
	(5) Maintenance	(6) Completion of work	—	
		(7) Maintenance		
Useful	(1) Proper price	(specification)———————		
	(2) Ecology	(specification)———————		
	(3) Function	(specification)———————		
	(4) Performance	(specification)———————		
Desirable	(1) (Beauty)	(specification)———————		
	(2) Pleasant to use	(specification)———————		
	(3) want to own	(specification)———————		

tasks relating to the three facets of a product and operations are classified along the vertical axis (columns). Then, individual UD-related requirements for each particular user group are indicated in intersecting cells. UD-related requirements vary depending on products and target users. In addition, classification based on product aspects and tasks results in organized determination of requirements with minimal omissions. It also allows easier prioritization. As a result, the following benefits can be realized. (a) Overall UD image can be understood visually. (b) Prioritization of requirements can be realized for an easy establishment of the design concept. (c) Solutions to problems are more easily discovered. (d) The use of required databases can be facilitated; this makes it possible to minimize undesired omissions and facilitate the checking process. (e) Individual applications to different products are made possible even for designers without previous experiences. (f) Examination of new applications and functions in a new product development is made possible.

USER SEGMENTS TABLE

Approach to the user segments table

The user segments table is prepared in order to help designers identify the various types of user of a product being designed. These user types selected will be put into relevant target user groups on a UD matrix (Nomura et al., 2002). As shown in Table 21.3, this user segments table

TABLE 21.3
User Segments Table

Cognitive		Others	Demographic				Culture			Environment (Non-user)
J Knowledge/ Decision	*K* Intellect	*L* Other abilities	*M* Age	*N* Gender	*O* Economy	*P* Qualification	*Q* Language	*R* Custom	*S* Nationality/ Religion	*T* Those around user
Aged		Aged								
	User with a cognitive impairment									Heart pacemaker
Visitor from different culture area		Allergy		Male/Female user of product for women/men			Foreigner	Foreigner	Foreigner	
User Under Urgent Situations										
Infant or child		Infant or Child	Minor							Infant or Child
Novice/ inexperienced user					Financially poor person	Person with no driver's license	User who cannot understand Japanese language	Foreigner (exports)	Foreigner (exports)	Nonsmoker Neighbor staff for maintenance

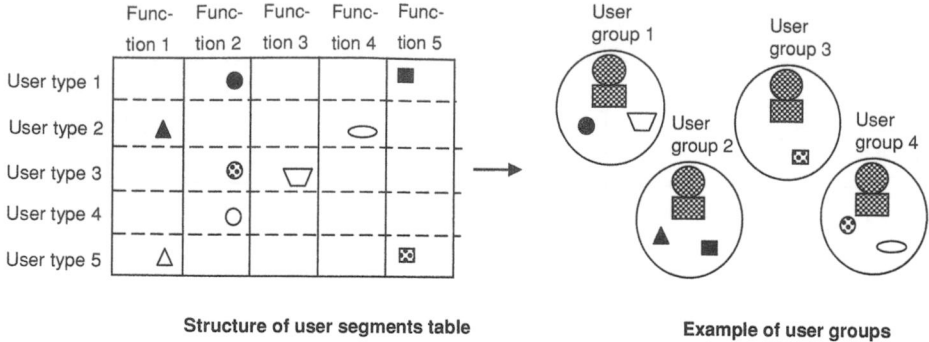

Structure of user segments table   Example of user groups

FIG. 21.4.   Selection of target user groups from the user segments table.

constitutes a matrix of rows and columns as shown in the table. The columns consist of the items related to human physical and mental functions (visual, acoustic, motor, cognitive, etc.) and demographic (gender, economy, etc.), cultural (language, custom, etc.) and environmental factors, all of which designers may need to take into account for a given product. The rows list the basic types of users, for example, people with disability, people with temporary disability, children and so on. For the purpose of helping a better understanding of the table, some cells in the matrix list typical or general user examples. And, additional notes and related human characteristics data for some user examples are listed in extra tables and figures as the appendixes in the Practical Universal Design Guidelines.

"Function 1, 2," of columns and "User type 1, 2," of rows in the left table correspond to "Factors to consider" and "User type" in the user segments table, respectively. And several figures in the cells represent user examples.

Identifying user groups

As shown in the Figure 21.4 designers select target users that should be taken into account in designing the product after referring to user examples listed in the cells or in the respective columns and rows in this table. And these users are classified into user groups according to their needs they have that must be accommodated in the product design. At that point user groups are applied in the UD matrix. In making it simpler to select target users, a simplified table of user segments is also proposed in the Practical Universal Design Guidelines. This simplified table consists of ready-made target user groups that are general and basic. This table will be presented in the final version of the practical guideline.

UD MATRIX

What Is a UD Matrix?

A UD matrix is a matrix to be formulated by the designer in considering all the relevant requirements for the target users concerned. It helps designers easily pinpoint the requirements for UD that may vary in different situations. The X-axis (rows) is used for user groups; the Y-axis (columns), for the three aspects of products and for each user task. Thus, each cell contains individual requirements for UD.

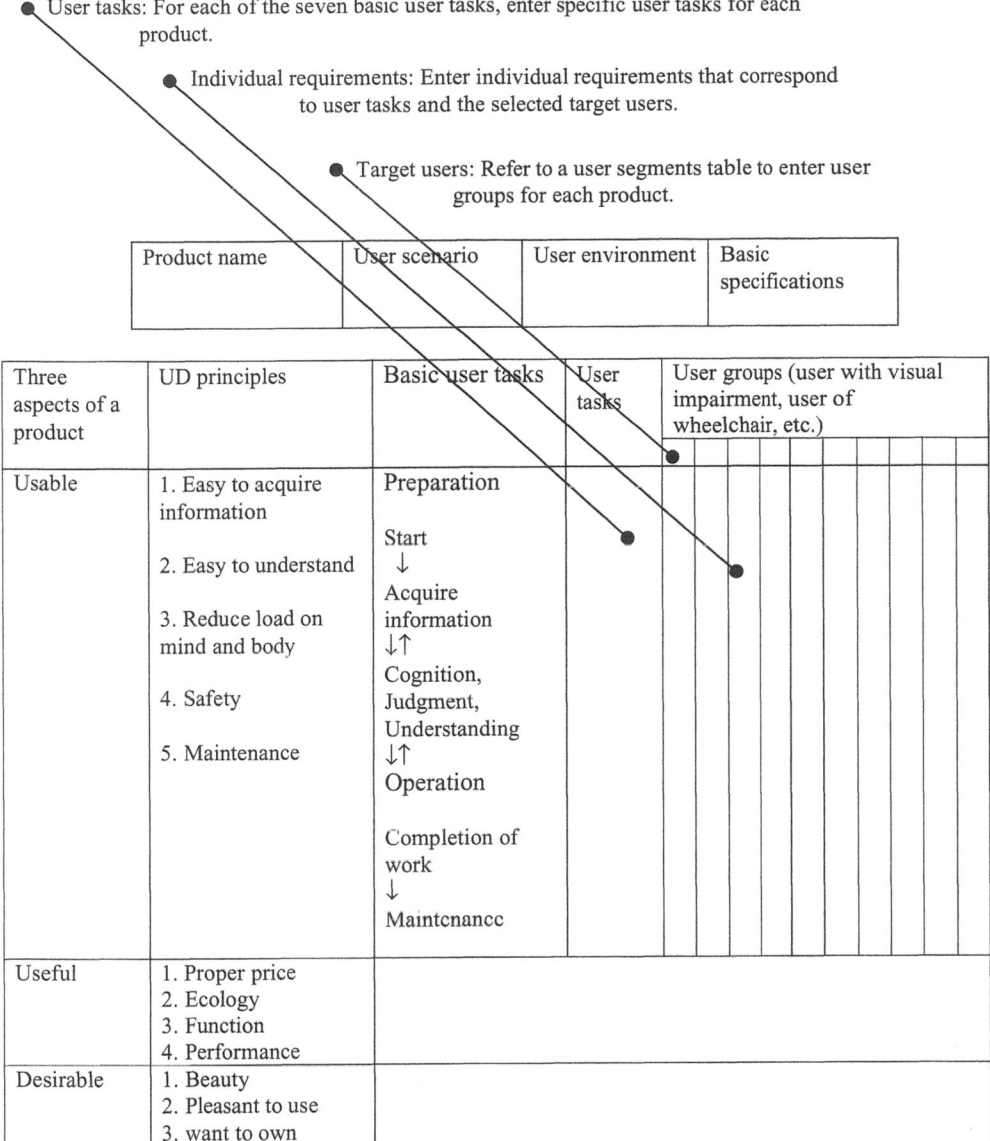

● User tasks: For each of the seven basic user tasks, enter specific user tasks for each product.

● Individual requirements: Enter individual requirements that correspond to user tasks and the selected target users.

● Target users: Refer to a user segments table to enter user groups for each product.

Product name	User scenario	User environment	Basic specifications

Three aspects of a product	UD principles	Basic user tasks	User tasks	User groups (user with visual impairment, user of wheelchair, etc.)									
Usable	1. Easy to acquire information 2. Easy to understand 3. Reduce load on mind and body 4. Safety 5. Maintenance	Preparation Start ↓ Acquire information ↓↑ Cognition, Judgment, Understanding ↓↑ Operation Completion of work ↓ Maintenance											
Useful	1. Proper price 2. Ecology 3. Function 4. Performance												
Desirable	1. Beauty 2. Pleasant to use 3. want to own												

FIG. 21.5.   How to develop an UD matrix?

How To Create a UD Matrix?

As shown in Figure 21.5, in creating a UD matrix for a particular product to be designed, it is advisable to follow the procedures below on a ready-made format:

- Enter the product classification, the user environment, the user scenario as well as basic specifications foreseen.
- For each of the seven basic user tasks, enter user tasks for each product in the column on the right.

- Refer to the user segments table so as to identify user groups for each product, and enter them in the columns across the top.
- Enter individual requirements in each of the matrix cells. Refer to entries on the left and above to enter minimum general requirements.
- Use a database, if necessary, when identifying individual requirements.

Three Aspects of a Product and UD Principles

The fundamental requirement in developing and supplying products is ensuring fairness. It is therefore necessary to consider three aspects: usable, useful, and desirable. The UD principles that correspond to these three aspects should be taken into account. The examples of related items which are of interest to the designer when examining individual requirements are shown in the following sections. The designer must follow these principles for realizing UD. These principles can be incorporated into a checklist in the form of an action checklist to be used by designers.

Usability

- A product should be designed so as to make it easy to acquire information.
- Information about how to operate should be presented so as to be identified easily and, where appropriate, through more than one sensory channel.
- Brightness, size, colors, and contrast of markings should be appropriate and conducive to understanding.
- Layout, clues, emphasis, and mapping of information presented should be appropriate.
- Information for users should be presented in an easily recognizable position.
- Information for users should be presented at an appropriate timing.

Easy to Understand

- Operating points, markings, and moving parts should be easily recognizable.
- Expressions and language should be easy to understand.
- Sufficient information for understanding and judgment should be provided.
- Way of using the designed product should be such that it is familiar to users or felt as natural by users.
- Appropriate feedback to users about the results of operation done should be provided during or immediately after the operation.

Reduce Physical and Mental Load

- Operating steps should be reduced to the minimum through appropriate omissions or automation.
- All operations should be done with one hand or one finger.
- Enough space should be provided for access regardless of the users' posture or body shape.
- Controls should be within easy reach and easily visible.
- The shape and size of controls and mechanical stress in handling them should be designed so that the product can be operated with adequate force.
- Proper hitches and antislip devices should be located where the operation could be slippery.
- The product should support the part of the body or the load handled if necessary.

Safety

- Sufficient time allowance should be provided and due considerations should be given to the operation in order to prevent erroneous operation.
- Safety measures should be provided to prevent potential operational errors and avoid escalation to a system malfunction or a higher degree danger.
- Markings and alarm sounds against danger should be presented in such a way that users do not fail to understand its meaning.
- Users should not be required to take any action that might involve any serious or immediate danger.

Maintenance

- Due considerations should be given to maintenance actions that might be taken by users.
- Appropriate workspace, working posture, work time, and repairability should be secured for maintenance by specialists.

Usefulness

Proper Price

- The product must allow its purposeful use with its standard model alone or by adding additional simple options with no extra costs incurred.

Ecology

- The product should be well constructed in terms of durability and being recyclable.

Availability of all Necessary and Adequate Functions

- All relevant mode control mechanisms and relevant adjustment functions should be available for different user groups.
- Multipurpose or easy to additionally operate functions should be available to meet diverse access needs.
- The number of functions should not be too many to make operations complicated.

The Product Performance Should Be Provided to the Extent Required and Should Be Sufficient Enough

- The product should allow its use without any concerns over malfunctioning.
- Sufficient levels of performance should be available to meet various access needs.
- Basic performance is not obsolete in comparison with other standard products.

Attractiveness

Beauty

- Consideration to UD should enhance beauty.

Pleasantness in Use

- Consideration to weakened physical function should not necessarily mean any negative image.
- Consideration to UD should enhance joy of use.

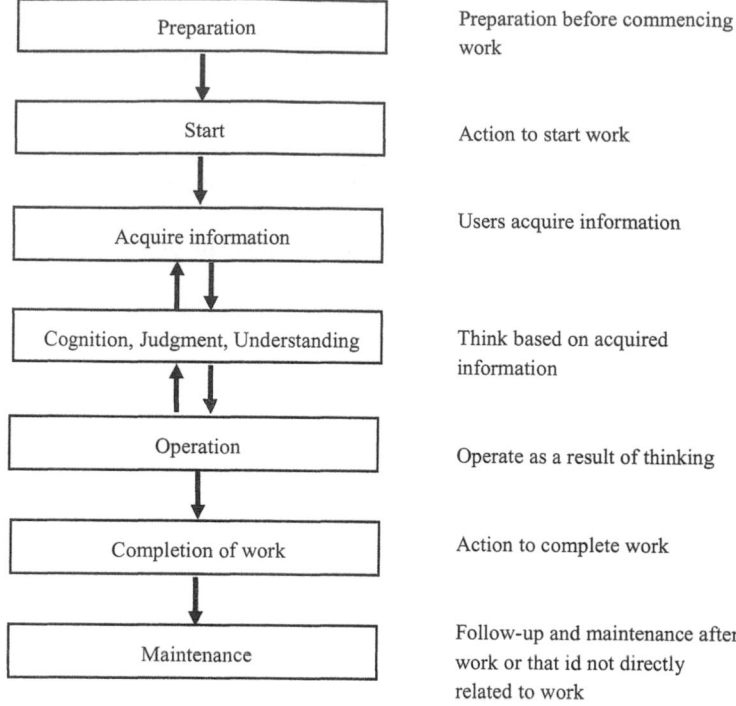

Preparation	Preparation before commencing work
Start	Action to start work
Acquire information	Users acquire information
Cognition, Judgment, Understanding	Think based on acquired information
Operation	Operate as a result of thinking
Completion of work	Action to complete work
Maintenance	Follow-up and maintenance after work or that id not directly related to work

FIG. 21.6.   Basic user tasks.

Wanting to Own

- The product should be attractive not only visually but also aurally and tactually.
- Attachment to the product should be enhanced by showing considerations to users' characteristics.

Basic User Tasks

Regarding "usability" among three aspects of a product, UD requirements for usability vary in individual user tasks. Although concrete user tasks vary in each product, the basic user tasks have something in common. Their real order may differ depending on the tasks performed, with some tasks omitted as appropriate. The theoretical order for basic user tasks is shown in Figure 21.6.

Benefits of an UD Matrix

There are significant benefits derived from applying a UD matrix. This may be confirmed in the stage of developing a UD matrix. To facilitate this stage, UD matrix samples have been arranged as examples for some products (an automatic teller machine, a cellular phone, a laptop personal computer, washing machine, etc.). Consequently, the following benefits are expected when a UD matrix is applied.
   An UD matrix will:

- Provide a whole picture of UD at a glance,
- Help designers form a concept as they prioritize requirements with reference to the matrix,

- Help find clues to solving problem,
- Make it possible to use a relevant database,
- Make it easy to check requirements, since there are few omissions,
- Apply to individual examples of each product, and
- Help examine new uses and functions for new products.

DESIGN EVALUATION METHOD

The purpose of design evaluation is to verify whether a design plan is compatible with the UD concept. In a general evaluation method, omission of design examination in connection with the concept is checked, and a UD action checklist can be used to confirm the appropriateness of the overall UD process, the product and the ease of its use. In the final stage, with the participation of target users (healthy users and physically challenged users), design evaluation of the developed system is conducted to extract any previously undiscovered points by using a checklist and relevant tools. New items extracted through the evaluation process are examined, and the design is reworked for improvement. To achieve a higher level of UD performance, it is necessary to establish this development cycle.

REFERENCES

Stephanidis, C. (2001). *User interfaces for all.* UK.: Lawrence Erlbaum Associates.

Yamazaki K. (2000). *Universal design approach to portable computers* (in Japanese). (IPSJ Tech. Rep. Vol. 2000. No. 81.33-38).

Yamaoka, T., Yamazaki, K., Okada, A., Saitoh, S., Nomura, M., & Yanagida, K. (2002). A proposal for universal design practical guidelines (1) Framework for universal design practical guidelines. *International Conference for Universal Design 2002*, 606–613.

Okada, A., Saitoh, S., Nomura, M., Yanagida, K., Yamaoka, T., & Yamazaki, K. (2002). A proposal for universal design practical guidelines (2) User segments table for universal design practical guidelines. *International Conference for Universal Design 2002*, 614–617.

Nomura, M., Yanagida, K., Yamaoka, T., Yamazaki, K., Okada, A., & Saito, S. (2002). A proposal for universal design practical guidelines (3): Using an UD matrix to pinpoint the requirements of UD, *International Conference for Universal Design 2002*, 718–722.

Yamazaki, K., Okada, A., Saitoh, S., Nomura, M., Yanagida, K., & Yamaoka, T. (2002). A proposal for universal design practical guidelines (4) Design process for Universal design practical guidelines. *International Conference for Universal Design 2002*, 666–670.

22

Virtual Environment Usage Protocols

Kay M. Stanney
University of Central Florida

David A. Graeber
The Boeing Company

Robert S. Kennedy
RSK Assessments, Inc.

INTRODUCTION

Technological advances over the past decade have laid the foundation for ubiquitous computing. One such advancement that could potentially address people's information and general needs is virtual environment (VE) technology (Stanney, 2002). Virtual environments allow users to be immersed into three-dimensional (3D) digital worlds, surrounding them with tangible objects to be manipulated and venues to be traversed, which they can experience from an egocentric perspective. Through the concrete and familiar, users can enact known perceptual and cognitive skills to interact with a virtual world; there is no need to learn the contrived conventions of more traditional graphical user interfaces. Virtual environments also extend the realm of computer interaction, from the purely visual to multimodal communication that more closely parallels human–human exchanges. Virtual environment users not only see visual representations but can also reach out and grab objects, "feel" their size, rotate them in any given axis, hear their movement, and even smell associated aromas. Such experiences do not have to be in solitude, as VE users can take along artificial autonomous agents or collaborate with other users who also have representations within the virtual world. Taken together, this multisensory experience should afford natural and intuitive interaction.

The previous paragraph describes an ideal, but unfortunately, not the current state of the art. In today's virtual environments, users are immersed in an experience with suboptimal visual resolution, inadequate spatialization of sound, encumbering interactive devices, and misregistration of tracking information (Durlach & Mavor, 1995; Stanney, Salvendy et al., 1998). These technological shortcomings engender adverse physiological effects and pose usability concerns that require additional study to identify how best to design and use VE technology. Substantial research is needed before the functionality, reliability, and appropriateness of VE technology are understood to the point that this technology can be used by a wide variety of people. This research should determine how to design and utilize VE interfaces that will enable the broadest possible spectrum of citizenry to interact easily and effectively with these systems (National Research Council, 1997).

The fundamental question is: Can an understanding of human physiological responses to VE technology be developed such that design guidelines and usage protocols can be devised that render VE interfaces usable by the broad spectrum of people who may wish to utilize this technology? Efforts by many researchers (Cobb Nichols, Ramsey, & Wilson, 1999; DiZio & Lackner, 2002; DiZio & Lackner, 1997; Harm, 2002; Howarth & Finch, 1999; Lawson, Graeber, Mead, & Muth, 2002; Stanney, Kennedy, & Kingdon, 2002; Stanney, Kingdon, Nahmens, & Kennedy, 2003; Viirre & Bush, 2002; Wann & Mon-Williams, 2002; Welch, 2002; Wilson, 1997; Wilson, Nichols, & Haldane, 1997; Wilson, Nichols, & Ramsey, 1995) have made substantial gains toward this goal. Heeding the warning of Biocca (1992), who indicated that cybersickness (McCauley & Sharkey, 1992), a form of motion sickness in VE systems, could be a snake lingering in the underbrush and threatening the widespread diffusion of VE technology, researchers have set out to overcome a number of challenging objectives. This research has focused on developing tools to measure the adverse effects of VE exposure (Kennedy & Stanney, 1996; Stanney, Kennedy, Drexler, & Harm, 1999), examining psychometrics of cybersickness (Graeber, 2001a, 2001b; Kennedy, Lane et al., 2001; Kennedy, Stanney, & Dunlap, 2000; Kingdon, Stanney, & Kennedy, 2001; Stanney & Kennedy, 1997a, 1997b; Stanney, Kingdon, Nahmens, & Kennedy, 2003; Stanney, Lanham, Kennedy, & Breaux, 1999), developing usage protocols (Stanney, Kennedy, & Kingdon, 2002), investigating system related issues that influence cybersickness (Stanney & Hash, 1998; Stanney, Kingdon, Nahmens, & Kennedy, 2003; Stanney, Salvendy et al., 1998), examining the efficacy of readaptation mechanisms for recalibrating those exposed to VE systems (Champney, Stanney, & Kennedy, under revision), as well as examining the influences of cybersickness on human performance (Stanney, Kingdon, Graeber, & Kennedy, 2002; Stanney, Kingdon, & Kennedy, 2001), among other related pursuits.

Although tremendous progress has been and continues to be made toward these objectives, during the course of these pursuits, a nettlesome problem has reared that could substantially limit the accessibility of VE technology. The problem is that a substantial number of people cannot withstand prolonged exposure to a VE and thus quickly withdraw from these systems. This exodus is predictably due to the adverse effects associated with VE exposure, which may be as minor as a headache or as severe as vomiting or intense vertigo (Cobb et al., 1999; Howarth & Finch, 1999; Regan & Price, 1994; Stanney, Kennedy et al., 1999; Stanney, Kingdon, Nahmens, & Kennedy, 2003; Stanney, Salvendy et al., 1998; Wilson, 1997; Wilson et al., 1997). It is on rare occasion that VE exposure causes an emetic response (about 1.5%); while many of those exposed experience some level of nausea, disorientation, or oculomotor problems (Lawson et al., 2002; Stanney, Kingdon et al., 1998, 2003; Stanney et al., 1998). Other problems, such as sleepiness and visual flashbacks also occur (Lawson et al., 2002). Approximately 80% to 95% of those exposed to a VE report some level of adverse symptomatology (Stanney, Salvendy et al., 1998). These disturbances lead a substantial proportion of those exposed to VE systems to prematurely cease their interaction, thus impeding general accessibility to this technology. In fact, dropout rates as high as 50% have been found in exposures of 1 hour in length (Stanney, Kingdon, Nahmens, & Kennedy, 2003). In general, dropout rates ranging from 5% to 50% have been found in a number of VE studies (Cobb et al., 1999; DiZio & Lackner, 1997; Howarth & Finch, 1999; Regan & Price, 1994; Singer, Ehrlich, & Allen, 1998; Stanney, Lanham et al., 1999; Wilson et al., 1995, 1997). Based on this tendency to have dropouts, the U.S. Army Research Institute (Knerr et al., 1998) has suggested that VE exposures should be limited to 15 minutes, a time period that may be too short for many training, educational, or analysis-based applications. An alternative is for those who utilize this technology to screen out individuals who are susceptible to adverse effects, leaving it available only to those who are nonsusceptible. Such an approach was considered by Chevron, who used a VE CAD tool to build 3D earth models for such purposes as seismic imaging, structural mapping, and reservoir characterization (Kowalik, personal communication, August 20, 2001) and will likely be adopted by others. Yet, such accessibility limitations should not be tolerated. All children should be able to enjoy the excitement of an immersive VE learning experience (Moshell & Hughes, 2002). All employees should be able to leverage their VE design tools (Davies, 2002). All medical, military, and other such personnel should be able to utilize VE training systems (Knerr, Breaux, Goldberg, & Thurman, 2002; Satava & Jones, 2002). By developing a solid understanding of physiological responses to VE technology, limited accessibility to VE technology could potentially be circumvented. From this understanding user-centered design principles and usage protocols for VE systems could be identified, which if applied successfully should render VE systems suitable for a broad spectrum of people. (Note: VE exposure should likely be avoided by certain populations, including those susceptible to photic seizures and migraines, those displaying co-morbid features of various psychotic, bipolar, paranoid, substance abuse, claustrophobic, or other disorders where reality testing and identity problems are evident, and those with preexisting binocular anomalies [Rizzo, Buckwalter, & van der Zaag, 2002; Stanney, Kennedy, & Kingdon, 2002; Viirre & Bush, 2002; Wann & Mon-Williams, 2002].)

In 1997, the Life Sciences Division at NASA Headquarters, Washington, DC, funded a study, one of the primary objectives of which was to characterize the current state of knowledge regarding aftereffects in virtual environments (Stanney, Salvendy et al., 1998). Two of the most critical research issues identified by this effort were to (a) establish VE design guidelines that minimize adverse effects and (b) establish means of determining user susceptibility to cybersickness and aftereffects. This chapter seeks to address both of these issues so that this knowledge can be used to support more effective human–VE interaction. The realization of VE usage protocols and design guidelines are essential because recent advances in VE applications are quite impressive and have the potential to advance many fields from education to medicine. In fact, a recent review of VE applications by Stone (2002) provides clear evidence that VE technology has advanced sufficiently to support a wide range of applications in such areas as engineering, micro- and nanotechnology, data visualization, ergonomics and human factors design, manufacturing, military, medicine, retail, and education. Development of an understanding of human physiological responses to VE technology can ensure such applications are designed such that adverse effects to their users are minimized.

DEFINITION OF TERMS

Adaptation: Decline in the amplitude of a sensory response in the presence of a constant prolonged stimulus.

Aftereffect: Any effect of virtual environment exposure that is observed after a participant has returned to the physical world.

Cybersickness: Sensations of nausea, oculomotor disturbances, disorientation, and other adverse effects associated with virtual environment exposure.

Habituation: Gradual decline in sensory response to repeated stimulus exposure.

Virtual environment (VE): A three-dimensional dataset describing an environment based on real-world or abstract objects and data.

Full Name of the Standard

Virtual Environment Usage Protocols

OBJECTIVE AND SCOPE

The current chapter focuses on contributing to the advancement of VE technology by promoting a better understanding of human physiological limitations that impede widespread use of this technology and identifying techniques that can render VE systems more usable, useful, and accessible. This is done through investigation of factors known to affect the level of adverse effects associated with VE exposure, including system design parameters (i.e., degrees of freedom [DOF] of user control, and scene complexity), usage variables (i.e., exposure duration, and intersession interval), and individual characteristics (i.e., susceptibility and gender). The major objectives being addressed via this chapter and their expected significance are as follows:

- Increase exposure durations and overall proportion of exposure duration completed via theorized design guidelines and usage protocols (i.e., exposure management techniques) that depress the stimulus intensity of a VE system and drive acclimation.
- Decrease dropout rates via theorized design guidelines and usage protocols that afford repeated VE exposures with systematically determined intersession intervals.
- Create equal opportunity for VE system use despite motion sickness susceptibility.

DISCUSSION OF FACTORS EFFECTING VIRTUAL ENVIRONMENT USAGE

Table 22.1 provides a synopsis of the extent and severity of adverse effects associated with VE exposure.

Based on the studies summarized in Table 22.1, the most conservative predictions would be that at least 5% of all users and potentially up to half of those exposed will not be able to tolerate prolonged use of current VEs and that a substantial proportion would experience some level of adverse effects. Further, females, younger (<23 years) and older (>40 years) individuals, and those highly susceptible to motion sickness may be particularly bothered by VE exposure. In addition to the problems listed in Table 22.1, the repercussions associated with such adverse effects may also include unequal opportunities for VE accessibility among the moderate to highly motion-sickness-susceptible population (Graeber, 2001a; 2001b), decreased user acceptance and use of VE systems (Biocca, 1992), decreased human performance (Kolasinski, 1995; Lawson et al., 2002; Stanney, Kingdon, Graeber, & Kennedy, 2002), and acquisition of improper behaviors (e.g., reduction of *rotational* movements [i.e., roll, pitch, and yaw] to quell side effects; Kennedy, Hettinger, & Lilienthal, 1990).

To summarize the findings in Table 22.1, VE systems can be hampered by intense malaise, high attrition rates, limited exposure durations, possibility for rejection of the system by the intended user population, conceivable decrements in human performance, and unequal opportunities for use. Although the exact causes of these problems remain elusive, they are thought to be a result of system design (e.g., scene content and user control strategies) and technological deficiencies (e.g., lag, distortions, and limited sensorial cues), as well as individual susceptibility (see review of factors by Kolasinski, 1995, Stanney, 1995, and Stanney, Salvendy et al., 1998). The most widely accepted theory is that mismatches (due to system design issues or technological deficiencies) between the sensorial stimulation provided by a VE or simulator

TABLE 22.1

Brief List of Problems Associated With Exposure to VE Systems

- 80% to 95% of individuals interacting with a head mounted display (HMD) VE system report some level of side effects, with 5% to 50% experiencing symptoms severe enough to end participation, approximately 50% of those dropouts occur in the first 20 min and nearly 75% by 30 min (Cobb et al., 1999; DiZio & Lackner, 1997; Howarth & Finch, 1999; Regan & Price, 1994; Singer, Ehrlich, & Allen, 1998; Stanney, Kennedy, & Kingdon, 2002; Stanney, Lanham et al., 1999; Stanney, Kingdon, Nahmens, & Kennedy, 2003; Wilson et al., 1995, 1997).

- Virtual environment exposure can cause people to vomit (about 1.5%), and approximately three-fourths of those exposed tend to experience some level of nausea, disorientation, and oculomotor problems (Cobb et al., 1999; DiZio & Lackner, 1997; Howarth & Finch, 1999; Regan & Price, 1994; Singer, Ehrlich, & Allen, 1998; Lawson et al., 2002; Stanney, Kingdon, Nahmens, & Kennedy, 2003; Stanney, Salvendy et al., 1998; Wilson et al., 1995; Wilson et al., 1997).

- With prolonged (>45-min) VE exposure oculomotor problems may become more pronounced, whereas nausea and disorientation tend to level off (Stanney, Kingdon, Nahmens, & Kennedy, 2003).

- Subjective report of sickness post VE exposure is 2.5 to 3 times greater than sickness reported by pilots training in motion-based and non-motion-based U.S. Navy fixed wing and helicopter simulators (Stanney & Kennedy, 1997a; Stanney, Salvendy et al., 1998).

- Before the age of 2, children appear to be immune to motion sickness, after which time susceptibility increases until about the age of 12, at which point it declines again (Money, 1970). Those over 25 are thought to be about half as susceptible as they were at 18 years of age (Mirabile, 1990). However,
  - older individuals (>40 years) exposed to VE systems can be expected to experience substantially more nausea, as well as greater oculomotor disturbances and disorientation as compared to younger individuals (Stanney, Kingdon, Nahmens, & Kennedy, 2003).
  - younger individuals (<23 years) exposed to VE systems may experience more oculomotor disturbances than older populations (Stanney, Kingdon, Nahmens, & Kennedy, 2003).

- Females exposed to VE systems can be expected to be more susceptible to motion sickness than males and to experience higher levels of oculomotor and disorientation symptoms as compared to males (Graeber, 2001a; Kennedy, Lanham, Massey, Drexler, & Lilienthal, 1995; Stanney, Kingdon, Nahmens, & Kennedy, 2003). In general, females tend to adapt more slowly to nauseogenic stimulation (McFarland, 1953; Mirabile, 1990; Reason & Brand, 1975).

- Individuals susceptible to motion sickness can be expected to experience more than twice the level of adverse effects to VE exposure as compared to nonsusceptible individuals (Stanney, Kingdon, Nahmens, & Kennedy, 2003).

- Individuals exposed to VE systems can be expected to experience lowered arousal (e.g., drowsiness and fatigue) upon postexposure (Lawson et al., 2002; Stanney, Kingdon, Nahmens, & Kennedy, 2003).

- Flashbacks (i.e., visual illusion of movement or false sensations of movement, when *not* in the VE) can be expected to occur quite regularly to those exposed to a VE system (Lawson et al., 2002; Stanney, Kingdon, Nahmens, & Kennedy, 2003).

- Prolonged disorientation (e.g., dizziness and vertigo) can be expected after VE exposure, with symptoms potentially lasting more than 24 hour (Stanney & Kennedy, 1998; Stanney, Kingdon, Nahmens, & Kennedy, 2003).

and that expected due to real world experiences (and thus established neural pathways) are thought to be the primary cause of motion sickness (a.k.a Sensory Conflict Theory; Reason, 1970, 1978; Reason & Brand, 1975). From the individual susceptibility perspective, age, gender, prior experience, individual factors (e.g., unstable binocular vision, individual variations in interpupillary distance (IPD), and susceptibility to photic seizures and migraines), drug and alcohol consumption, health status, and ability to adapt to novel sensory environments are all thought to contribute to the extent of symptoms experienced (Kolasinski, 1995; McFarland, 1953; Mirabile, 1990; Reason & Brand, 1975; Stanney, Kennedy, & Kingdon, 2002; Stanney, Salvendy et al., 1998). Although considerable research into the exact causes is requisite and has been ongoing for decades in its various forms (seasickness, motion sickness, simulator sickness, space sickness, and cybersickness; Chinn & Smith, 1953; Crampton, 1990; Kennedy & Fowlkes, 1992; McCauley & Sharkey, 1992; McNally & Stuart, 1942; Reason, 1970, 1978;

Reason & Brand, 1975; Sjoberg, 1929; Stanney, Salvendy et al., 1998; Tyler & Bard, 1949; Wendt, 1968), there are currently few if any efforts focusing on the development of means of reducing adverse effects by "acclimating" users to the VE stimulus. Yet, stimulus–response studies, such as the classical conditioning studies of Pavlov (1928), offer a potential paradigm through which to realize such acclimation. There is, however, a lack of understanding about the factors that drive VE stimulus intensity and acclimation (i.e., stimulus depression) versus sensitization (i.e., heightened response) to such a stimulus, such that this knowledge can be used to identify VE usage protocols that minimize adverse effects. In fact, current usage of VE technology generally treats users as if they are immune to motion sickness or possess low motion sickness susceptibility and are capable of rapid acclimation to novel sensory environments. This is not the case, as evidenced by the intensity and extent of side effects listed in Table 22.1. Additional research is required because, although the beginnings of VE design guidelines are under development (Stanney, Kennedy, & Kingdon, 2002) and expectations of use have been identified (Stanney, Kingdon, Nahmens, & Kennedy, 2003), means of prolonging exposure via acclimation or other such conditioning strategies are still required to ensure VE systems are accessible by a broad spectrum of users.

To leverage and compare conditioning strategies, a means of characterizing VE stimulus intensity is required. Research focused on developing an understanding of system design, technological, and individual drivers of adverse effects associated with VE exposure can be used to make predictions on VE stimulus intensity (see Table 22.2). One can see from Table 22.2 that there are a diverse number of factors influencing VE stimulus intensity. By developing an understanding of these factors, means of acclimating users to a VE stimulus could potentially be identified.

The factors reviewed in Table 22.2 can be used to characterize the intensity of a VE stimulus. The technological factors (e.g., system consistency, lag, update rate, mismatched IPDs, and unimodal and intersensorial distortions) are system specific and may be resolved to some extent as technology progresses, whereas the system design (e.g., DOF of user control, scene complexity), usage (e.g., exposure duration, intersession interval), and individual factors (e.g., susceptibility, gender) are not dependent on technological progress and thus will likely have an enduring influence on human–VE interaction. Among these factors, exposure durations of 15 minutes or more, very short (<2 days) or extended (>5 days) intersession intervals, high DOF of user control, and complex visual scenes are all known to lead to a more intense VE stimulus. In terms of user control, Held (1965) and Reason (1978) provide evidence to suggest that motion sickness can be overcome if users have control over their movements and receive an appropriate sensory response (e.g., visual, vestibular, or proprioceptive) to their actions. Essentially, in response to a neural mismatch (i.e., sensory conflict), users create a new "neural store" with which incoming sensory information is compared, eventually resulting in a new neural match (i.e., the reafference copy resulting from effector stimulation [i.e., stimuli resulting from one's own muscular activity] overwrites the efference copy [i.e., motor impulses]). However, Stanney and Hash (1998) found that high DOF of motion, although being superior to passive motion in minimizing cybersickness, may not be the best solution to the sickness problem. Under such conditions, VE users may not be able to efficiently update the neural store with the abundant amount of sensory information resulting from their unrestricted movements. Thus, by allowing users *streamlined* control (i.e., only those DOF necessary for supporting an activity) within a VE, the neural store may be updated quickly in response to streamlined reafference.

Individual susceptibility also plays a critical role. Stanney, Kingdon, Nahmens, and Kennedy (2003) found that susceptible individuals can be expected to experience more than twice the level of adverse effects to VE exposure as compared to nonsusceptible individuals. McFarland (1953) found females to be five times and children nine times more motion sickness susceptible than adult males. Thus, to gauge the intensity of a VE stimulus, individual susceptibility

TABLE 22.2
Factors Influencing VE Stimulus Intensity

- Adverse effects associated with VE exposure are positively correlated with exposure duration (Kennedy, Stanney, & Dunlap, 2000). Lanham (2000) has shown that sickness increases linearly at a rate of 23% per 15 minutes. Dropouts occur in as little as 15 min of exposure (Cobb et al., 1999; DiZio & Lackner, 1997; Howarth & Finch, 1999; Regan & Price, 1994; Singer et al., 1998; Stanney, Kennedy, & Kingdon, 2002; Stanney, Lanham et al., 1999; Stanney, Kingdon, Nahmens, & Kennedy, 2003; Wilson et al., 1995; Wilson et al., 1997).
- Intersession intervals of 2 to 5 days are effective in mitigating adverse effects, while intervals less than two or greater than five days are ineffective in reducing symptomatology (Kennedy, Lane et al., 1993; Watson, 1998).
- Repeated exposure intervals within a session less than 2 hr apart appear to heighten adverse effects upon reentry (Graeber, 2001a).
- As the amount of user movement control in terms of DOF and head tracking increases so too does the level of nausea experienced (Stanney & Hash, 1998; Stanney, Kingdon, Nahmens, & Kennedy, 2003).
  - Complete user movement control (6 DOF) can be expected to lead to 2.5 times more dropouts than streamlined control (3 DOF).
  - Movements in rotational axes (e.g., roll, pitch, and yaw) may be more provocative than linear movement (Money & Myles, 1975).
- The rate of visual flow (i.e., visual scene complexity) may influence the incidence, and more so the severity of motion sickness experienced by an individual (Kennedy & Fowlkes, 1992; McCauley & Sharkey, 1992).
  - Complex visual scenes may be more nauseogenic than simple scenes with complex scenes possibly resulting in 1.5 times more emetic responses; however, scene complexity does not appear to affect dropout rates (Dichgans & Brandt, 1978; Kennedy, Dunlap, Berbaum, & Hettinger, 1996; Stanney, Kingdon, Nahmens, & Kennedy, 2003).
  - Such affects may be exacerbated by a large field-of-view (FOV) (Kennedy & Fowlkes, 1992), high spatial frequency content (Dichgans & Brandt, 1978), and visual simulation of action motion (i.e., vection [Kennedy, Dunlap et al., 1996]).
- Various technological factors thought to influence how provocative a VE is include: system consistency (Uliano, Kennedy, & Lambert, 1986); lag (So & Griffin, 1995); update rate (So & Griffin, 1995) mismatched IPDs (Mon-Williams, Rushton, & Wann, 1995); and unimodal and intersensorial distortions (both temporal and spatial [Welch, 1978]).
- Individual factors thought to contribute to an individual's motion sickness susceptibility include age, gender, prior experience, visual predispositions (e.g., unstable binocular vision, individual variations in IPD, and susceptibility to photic seizures and migraines), drug and alcohol consumption, health status, and ability to adapt to novel sensory environments (Kolasinski, 1995; McFarland, 1953; Mirabile, 1990; Reason & Brand, 1975; Stanney, Kennedy, & Kingdon, 2002; Stanney, Salvendy et al., 1998).
- Individuals who have experienced an emetic response associated with carnival rides can be expected to experience more than twice the level of adverse effects to VE exposure as compared to those who do not experience such emesis (Stanney, Kingdon, Nahmens, & Kennedy, 2003).
- Individuals with higher preexposure drowsiness will be more likely to experience drowsiness upon post-VE exposure, and those exposed for 60 min or longer can be expected to experience more than twice the level of drowsiness as compared to those exposed for a shorter duration (Lawson et al., 2002; Stanney, Kingdon, Nahmens, & Kennedy, 2003).
- As drowsiness increases one can expect a greater severity of flashbacks (Lawson et al., 2002; Stanney, Kingdon, Nahmens, & Kennedy, 2003).
- Body mass index (BMI) does not tend to be related to VE sickness symptoms; however, those with higher BMIs may be less prone to experience an emetic response (Stanney, Kingdon, Nahmens, & Kennedy, 2003).

---

must be considered. Kennedy and Graybiel (1965) have developed a Motion History Questionnaire (MHQ) that could prove useful in assessing individual susceptibility. From its inception, the MHQ's purpose was to ascertain levels of motion sickness susceptibility to a variety of provocative motion challenges using an individual's history of motion experiences. As a result of the MHQ's utility, an array of scoring keys have been developed (Kennedy & Fowlkes, 1992) for various motion challenges (e.g., high- and low-frequency vertical motion, Coriolis stimulation, simulators, etc.), and more recently for circular vection (Graeber, 2001b) and virtual environments (Kennedy, Lane et al., 2001). The latter study demonstrated that the MHQ has promise as a predictive tool for VE sickness but indicated that it needs to be further refined for these purposes. A recent analysis (Kingdon, 2001) indicated that the predictability of this tool could be potentially enhanced by coupling questionnaire items with a psychophysical test battery to assess postural stability, dexterity, and ability of the human visual system to perform smooth pursuit tracking (i.e., gaze nystagmus), similar to the roadside battery test used to assess sobriety (Tharp, Burns, & Moskowitz, 1981). Such enhanced predictability of individual susceptibility could assist in more fully characterizing the influences of VE stimulus intensity.

The summary in Table 22.2 suggests that the sickness inducing characteristics of a VE stimulus could potentially be reduced by shortening exposure duration, maintaining an intersession interval of 2 to 5 days, reducing DOF of user movement control to only those necessary to support an activity, and simplifying visual scenes. Table 22.2 further suggests that the exact strategy that is most effective may depend on individual susceptibility. If these tactics are coupled with conditioning approaches, reductions in adverse effects and associated dropout rates can be expected.

Prior research has implemented various conditioning techniques to reduce adverse effects in, and hasten acclimation to, a variety of altered sensory environments. Conditioning regimens, including adaptation (one prolonged exposure to the full intensity of a stimulus), habituation (repeated exposures to the full intensity of a stimulus), cognitive therapies (use of education and biofeedback), dual adaptation (repeated exposures until abatement of side effects and negative aftereffects), incremental adaptation (an incrementing of stimulus intensity within one exposure), and incremental habituation (an incrementing of stimulus intensity across multiple exposures) have been applied to underwater, zero gravity, and acrobatic flight maneuver environments, artificial optical distortions (e.g., prism lenses and optokinetic drums), Pensacola Slow Rotation Room (SRR), and simulators with varying degrees of success (see review by Graeber, 2001a). Although such conditioning regimens have been investigated, they have primarily been curtailed to adaptation and habituation regimens, despite the proven performance of other conditioning regimens to mitigate motion sickness in various sensory environments and lackluster success of adaptation regimens for reducing cybersickness in VE systems. In particular, incremental conditioning approaches have been very successful in minimizing side effects in an array of environments including the SRR (Cramer, Graybiel, & Oosterveld, 1978; Graybiel, Deane, & Colehour, 1969), combat flight training (Bagshaw & Stott, 1985), and artificial optical distortions (Hu, Stern, & Koch, 1991), as well as being recommended as an effective method for mitigating simulator sickness (Kennedy et al., 1987) and cybersickness (Graeber, 2001a; Welch, 2002). Incremental conditioning protocols have been shown to be effective in not only reducing adverse effects but also increasing the rate (Lackner & Lobovitz, 1978) and extent (Ebenholtz & Mayer, 1968) of acclimation obtained in comparison to adaptation and habituation protocols.

The Dual Process Theory (DPT) of neural plasticity (Groves & Thompson, 1970; Prescott, 1998; Prescott & Chase, 1999) provides a theoretical means for understanding how particular conditioning regimens to various motion sickness inducing environments may affect the magnitude of a sensory conflict (Graeber, 2001a). In the context of a VE system, the DPT would state that there are two opponent processes undertaken upon VE stimulus onset (see Figure 22.1). The two processes are depression (i.e., lessening or acclimation) and sensitization (i.e., heightening) of sickness outcome, which are carried out in parallel through different tracks. The DPT theory suggests that the depression process occurs along the stimulus–response (S–R) pathways, whereas the sensitization process is undertaken through the "state" system (i.e., the central nervous system [CNS]). During VE stimulus processing, a confluence occurs where the processes converge to yield the observed response (e.g., net outcome, a.k.a. sickness severity). The observed response is suggested to be a function of the integration of these two paths and their characteristics based on VE stimulus strength, number of VE stimulus exposures, and degree of neural plasticity in an individual.

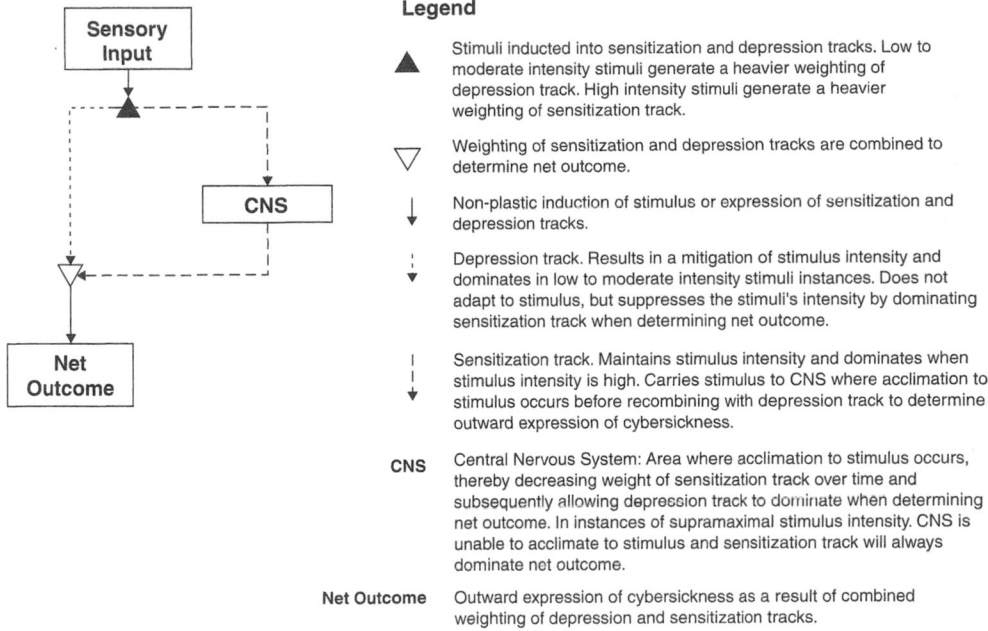

FIG. 22.1.   DPT points of induction and expression of sensitization and depression.

In essence, based on DPT theory, VE stimulus intensity should set the relationship between the opponent processes, which should then determine the proclivity of acclimation in the CNS and the degree to which each opponent process contributes to the net outcome. Concurrently, number of exposures should co-determine the rate of acclimation in the CNS and subsequent depression of the sensitization tracks' weighting. As VE stimulus intensity increases the sensitization opponent process should play a more dominant role in determining the net outcome, whereas the effect of the depression opponent process should wane. The aspect of DPT of concern to VE usage protocols is primarily the sensitization opponent process and how it may be shaped, and subsequently the depression curve as well, through behavioral modification. It is herein suggested that the sensitization opponent process' effect on net outcome determines the degree of side effects experienced (i.e., net outcome is equivalent to degree of cybersickness). In other words, as VE stimulus intensity increases, the influence of the sensitization opponent process increases thereby exerting greater control over the net outcome (i.e., heightened cybersickness). This is similar to the idea of long-term potentiation (Carlson, 2001), which is a long-term increase in the excitability of a neuron due to a particular input, particularly if exposure to the input is repeated with a brief interstimulus interval.

By using an incremental approach, in which a VE stimulus is incrementally increased in intensity within one exposure (incremental adaptation) or across multiple exposures (incremental habituation), it is suggested that the VE stimulus intensity may be low enough that the depression track would dominate the net outcome and afford more rapid acclimation to the stimulus in the CNS, thereby mitigating outward expression of cybersickness in the net outcome. If a VE user is able to complete an adaptive process at each increment of stimulus intensity (i.e., depressing sensitization to preexposure levels or below), then each stepwise increase in VE stimulus intensity, assuming the increase is within the bounds of moderate to rapid CNS acclimation, should keep the magnitude and duration of sensitization to a minimum. In theory, this should afford attainment of exposure to VE stimulus intensities that would normally result in supramaximal sensitization (i.e., a VE stimulus intensity of sufficient strength to inhibit acclimation in the CNS, yielding a continued state of heightened sensitization that does not depress and thus predictably leads to cessation of exposure) without experiencing supramaximal sensitization. Further research is needed to verify this supposition.

Therefore, according to DPT, using an incremental approach should result in both minimal sensitization that dissipates relatively quickly and near maximal depression that will gradually dominate the net behavioral output over time (see Figure 22.1). In other words, by incrementing the intensity of the VE stimulus, depression is facilitated rather than sensitization, which allows one to capitalize on humans' inherent neural plasticity and maximize the rate and extent of acclimation to a VE stimulus.

Graeber (2001a) investigated the efficacy of an incremental approach to minimize motion sickness associated with exposure to a vection drum. The results suggest that by employing an incremental approach it may be possible to substantially reduce adverse effects and dropout rates. Graeber (2001a) found dropouts were reduced to approximately 20% among high motion sickness susceptibles, compared to a dropout rate of 50% for high susceptibles employing nonincremented protocols. Although this work demonstrated the potential efficacy of such conditioning strategies in a vection drum, further study is needed to determine their suitability for VE systems. More specifically, conditioning strategies for incrementally increasing VE stimulus intensity by manipulating both system design and usage variables need to be identified.

VIRTUAL ENVIRONMENT USAGE PROTOCOL

Based on the previous review of VE related usage factors, it is suggested that the optimal conditions to engender acclimation to VE exposure can be identified and characterized by understanding the affects of system design (e.g., DOF of user initiated control and visual scene complexity), usage (e.g., exposure duration and intersession interval), and individual factors (e.g., susceptibility and gender) on VE stimulus intensity. With this understanding, conditioning strategies conducive to facilitating prolonged VE exposure can be developed. Thus, a systematic VE usage protocol can be developed that minimizes risks to users. A comprehensive VE usage protocol should consider each of the following steps (Stanney, Kennedy, & Kingdon, 2002).

1. Following the guidelines in Table 22.3, design VE stimulus to minimize adverse effects.
2. Following the guidelines in Table 22.4, quantify VE stimulus intensity of target system and compare to quartiles in Table 22.5. Table 22.5 provides quartiles of sickness symptoms based on 29 VE studies (Cobb et al., 1999; Kennedy, 2001; Kennedy, Jones, Stanney, Ritter, & Drexler, 1996; Stanney, 2001; Stanney & Hash, 1998) that evaluated sickness via the Simulator Sickness Questionnaire (SSQ; Kennedy, Lane, Berbaum, & Lilienthal, 1993). If a given VE system is of high intensity (say the 50th or higher percentile, with a total SSQ score of 20 or higher), significant dropouts can be expected.
3. Following the guidelines in Table 22.6, identify individual capacity of target user population to resist adverse effects of VE exposure.
4. Following the guidelines in Table 22.6, set exposure duration and intersession interval.
5. Warnings. Provide warnings for those with severe susceptibility to motion sickness, seizures, migraines, cold, flu, or other ailments.
6. Educate users as to the potential risks of VE exposure. Inform users of the insidious effects they may experience during exposure, including nausea, malaise, disorientation, headache, dizziness, vertigo, eyestrain, drowsiness, fatigue, pallor, sweating, increased salivation, and vomiting. Depending on VE content, potential adverse psychological effects may also need to be considered.

TABLE 22.3

Addressing System Factors That Influence the Strength of a VE Stimulus

---

- Ensure any system lags and latencies are stable; variable lags and latencies can be debilitating.
- Avoid high levels of user movement in terms of DOF of user control (i.e., neural store can be updated quickly in response to streamlined reafference).
  - Reduce DOF to those necessary for the activity being supported (e.g., don't provide pitch unless needed)
- Minimize display and phase lags (i.e., end to end tracking latency between head motion and resulting update of the display).
- Optimize frame rates.
- Provide adjustable IPD.
- When large FOVs are used, determine if it drives high levels of vection (i.e., perceived self-motion); if it does consider reducing the FOV.
- If high levels of vection are found and they lead to high levels of sickness, then reduce the spatial frequency content of visual scenes.
- Provide multimodal feedback that minimizes sensory conflicts (i.e., provide visual, auditory, haptic and kinesthetic feedback appropriate for situation being simulated).

TABLE 22.4

Steps to Quantifying VE Stimulus Intensity

---

1. Get an initial estimate. Talk with target users (not developers) of the system and determine the level of adverse effects they experience.
2. Observe. Watch users during and after exposure and note comments and behaviors.
3. Try the system yourself, particularly if you are susceptible to motion sickness, obtain a first-hand assessment of the adverse effects.
4. Measure the drop out rate. If most people can stay in for an hour without symptoms, then the system is likely benign; if most people drop out within 10 min, then the system is probably in need of redesign.
5. Measure. Use simple rating scales to assess sickness and visual, proprioceptive, and postural measures to assess aftereffects.
6. Compare. Use Table 22.5 to determine how the system under evaluation compares to other VE systems.
7. Report. Summarize the severity of the problem, specify required interventions (e.g., warnings and instructions), and set expectations for use (e.g., target exposure duration and inter session intervals).
8. Expect dropouts. With a high-intensity VE stimulus, dropout rates can be high.

TABLE 22.5

Virtual Environment Sickness Quartiles

Quartile	SSQ Score
25th	15.5
50th	20.1
75th	27.9
95th	33.3
99th	53.1

TABLE 22.6

Factors Affecting Individual Capacity to Resist Adverse Effects of VE Exposure

---

- Limit initial exposures. For medium to strong VE stimuli, limit initial exposures to a short duration (e.g., 10 min or less).
- Adaptation. Set inter-session exposure intervals 2 to 5 days apart to enhance individual adaptability.
- Adaptation and Habituation. Use an incremental approach, in which a VE stimulus is incrementally increased in intensity within one exposure (incremental adaptation) or across multiple exposures (incremental habituation).
  - Determine whether users are able to complete an adaptive process at each increment of stimulus intensity (i.e., depressing sensitization to preexposure levels or below); if not lower stimulus intensity.
- Repeat exposure: Avoid repeated exposure intervals occurring less than 2 hr apart if adverse effects are experienced in an exposure.
- Age. Expect little motion sickness for those under age 2; expect greatest susceptibility to motion sickness between the ages of 2 and 12; expect motion sickness to decline after 12, with those over 25 being about half as susceptible as they were at 18 years of age.
- Gender. Expect females to be more susceptible than males (perhaps as great as three times more susceptible).
- Anthropometrics. Consider setting VE stimulus intensity in proportion to body weight and stature.
- Individual susceptibility. Expect individuals to differ greatly in motion sickness susceptibility and use the MHQ or another instrument to gauge the susceptibility of the target user population.
  - Couple MHQ questionnaire items with the roadside battery test used to assess sobriety (Tharp et al., 1981) or other psychophysical test batteries (e.g., postural stability, dexterity, and gaze nystagmus) to enhance its predictability.
- Drug and alcohol consumption. Limit VE exposure to those individuals who are free from drug or alcohol consumption.
- Rest. Encourage individuals to be well rested before commencing VE exposure. Inform individuals that if they become drowsy they may experience a greater frequency and severity of flashbacks.
- Predisposition: Inform individuals who have experienced an emetic response associated with carnival rides that they may become ill during VE exposure.
- Ailments. Discourage those with cold, flu or other ailments (e.g., headache, diplopia, blurred vision, sore eyes, or eyestrain) from participating in VE exposure; encourage those susceptible to photic seizures and migraines, as well as individuals with preexisting binocular anomalies to avoid exposure.
- Fitness and Overall Health: Many factors (e.g., alcohol, flu, hangover, sleep loss, high stress, etc.) contribute to overall health and fitness and thus can be expected to exacerbate sickness in a VE; whereas each individual factor may minimally influence adverse effects if they are concatenated, then you can expect that there will be sickness that, rather than being attributable to the VE, may be due to a combination of conditions in the individual.
- Clinical user groups. Obtain informed sensitivity to the vulnerabilities of these user groups (e.g., unique psychological, cognitive, and functional characteristics). Encourage those displaying co-morbid features of various psychotic, bipolar, paranoid, substance abuse, claustrophobic, or other disorders where reality testing and identity problems are evident to avoid exposure.

---

7. Inform users as to the potential adverse *aftereffects* of VE exposure. Inform users that they may experience disturbed visual functioning, visual flashbacks, as well as unstable locomotor and postural control for prolonged periods following exposure. Relating these experiences to excessive alcohol consumption may prove instructional.

8. Inform users that if they start to feel ill they should terminate their VE interaction because extended exposure is known to exacerbate adverse effects (Kennedy, Stanney, & Dunlap, 2000).

9. Prepare users. Donning an HMD is a jarring experience (Pierce, Pausch, Sturgill, & Christiansen, 1999). Depending on the complexity of the virtual world, it can take 30 to 60 seconds to adjust to the new space (Brooks, 1988). Prepare users for this transition by informing them that there will be an adjustment period.

10. Adjust environmental conditions. Provide adequate air flow and comfortable thermal conditions (Konz, 1997). Sweating often precedes an emetic response, thus proper air flow can enhance user comfort. In addition, extraneous noise should be eliminated, as it can exacerbate ill-effects.

11. Adjust equipment to minimize fatigue. Fatigue can exacerbate the adverse effects of VE exposure. To minimize fatigue, ensure all equipment is comfortable and properly adjusted for fit. HMDs should fit snugly and be evenly weighted about a user's head, stay in place when unsupported, and avoid uneven loading to neck and shoulder muscles. Many HMDs have adjustable head straps, IPDs, and viewing distance between the system's eyepieces and user's eyes. Ensure users optimize these adjustments to obtain proper fit. Tethers should not obstruct movements of users. DataGloves and other effectors should not induce excessive static loads via prolonged unnatural positioning of the arms or other extremities.

12. Avoid provocative movements. For strong VE stimuli, warn users to avoid movements requiring high rates of linear or rotational acceleration and extraordinary maneuvers (e.g., flying backward) during initial interaction (McCauley & Sharkey, 1992).

13. Monitor users. Throughout VE exposure, an attendant should be available at all times to monitor users' behavior and ensure their well-being. An attendant may also have to assist users if they become stuck or lost within the virtual world, as often happens.

14. Look for red flags. Indicators of impending trouble include excessive sweating, verbal frustration, lack of movement within the environment for a significant amount of time, and less overall movement (e.g., restricting head movement). In addition, before they become ill users will tend to become very quiet and will likely totally stop talking. Users demonstrating any of these behaviors should be observed closely, as they may experience an emetic response. Extra care should be taken with these individuals post-exposure. Note: it is beneficial to have a bag or garbage can located near users in the event of an abrupt emetic response.

15. Termination. Set criteria for terminating exposure. Exposure should be terminated immediately if users verbally complain of symptoms and acknowledge they are no longer able to continue. Also, to avoid an emetic response, if telltale signs are observed (i.e., sweating and increased salivation), exposure should be terminated.

16. Postexposure care. Some individuals may feel ill or be unsteady upon postexposure. These individuals may need assistance when initially standing up after exposure.

17. Debriefing. After exposure the well-being of users should be assessed. Measurements of their hand-eye coordination and postural stability should be taken. Similar to field sobriety tests (Tharp et al., 1981), these can include measures of balance (e.g., standing on one foot, walking an imaginary line, and leaning backward with eyes closed), coordination (e.g., alternate hand clapping and finger to nose touch while the eyes are closed), and eye nystagmus (e.g., follow a light pen with the eyes without moving the head).

18. Releasing. Set criteria for releasing users. Specify the amount of time after exposure that users must remain on premises before driving or participating in other such high-risk activities. In our lab a 2-to-1 ratio is used; post-exposure users must remain in the laboratory twice the amount of exposure time to allow for recovery. Do not allow individuals who fail debriefing tests or are experiencing adverse aftereffects to conduct high-risk activities until they have recovered (e.g., have someone drive them home).

19. Follow-up. Call users the next day or have them call to report any prolonged adverse effects.

CONCLUSIONS

The risks associated with VE exposure are real and include ill effects during exposure, as well as the potential for prolonged aftereffects. To minimize these risks, VE system developers should quantify and minimize VE stimulus intensity, identify the capacity of the target user population to resist the adverse effects of VE exposure, and then follow a systematic usage protocol. This protocol should focus on warning, educating, and preparing users, setting appropriate environmental and equipment conditions, limiting initial exposure duration and user movements, monitoring users and looking for red flags, and setting criteria for terminating exposure, debriefing, and release. Adopting such a protocol may minimize the risk factors associated with VE exposure.

ACKNOWLEDGMENTS

This material is based on work supported in part by the Office of Naval Research (ONR) under Grant N000149810642, the National Science Foundation (NSF) under Grants DMI9561266 and IRI-9624968, and the National Aeronautics and Space Administration (NASA) under Grants NAS9-19482 and NAS9-19453. Any opinions, findings, and conclusions or recommendations expressed in this material are those of the authors and do not necessarily reflect the views or the endorsement of the ONR, NSF, or NASA.

REFERENCES

Bagshaw, M., & Stott, J. R. (1985). The desensitization of chronically motion sick aircrew in the Royal Air Force. *Aviation, Space, and Environmental Medicine, 56*(12), 1144–1151.

Biocca, F. (1992). Will simulation sickness slow down the diffusion of virtual environment technology? *Presence: Teleoperators and Virtual Environments, 1*(3), 334–343.

Brooks, Jr., F. P. (1988). Grasping reality through illusion: Interactive graphics serving science. *ACM SIGCHI Conference Proceedings*, 1–11.

Carlson, N. R. (Ed.). (2001). *Physiology of Behavior* (7th ed.). Boston: Allyn & Bacon.

Champney, R., Stanney, K. M., & Kennedy, R. S. (under revision). *Virtual environment readaptation mechanisms.* University of Central Florida.

Chinn, H. I., & Smith, P. K. (1953). Motion sickness. *Pharmacological Review, 7,* 33–82.

Cobb, S. V. G., Nichols, S., Ramsey, A. D., & Wilson, J. R. (1999). Virtual reality-Induced symptoms and effects (VRISE). *Presence: Teleoperators and Virtual Environments, 8*(2), 169–186.

Cramer, D. B., Graybiel, A., & Oosterveld, W. J. (1978). Successful transfer of adaptation acquired in a slow rotation room to motion environments in navy flight training. *Acta Otolayrngol, 85,* 74–84.

Crampton, G. H. (Ed.). (1990). *Motion and space sickness.* Boca Raton, FL: CRC Press.

Davies, R. C. (2002). Applications of systems design using virtual environments. In K. M. Stanney (Ed.), *Handbook of virtual environments: Design, implementation, and applications* (pp. 791–806). Mahwah: NJ: Lawrence Erlbaum Associates.

Dichgans, J., & Brandt, T. (1978). Visual-vestibular interaction: Effects on self-motion perception and postural control. In R. Held, H. W. Leibowitz, & H. L. Teuber (Eds.), *Handbook of sensory physiology, Vol. VIII: Perception* (pp. 756–804). Heidelberg: Springer-Verlag.

DiZio, P., & Lackner, J. R. (1997). Circumventing side effects of immersive virtual environments. In M. Smith, G. Salvendy, & R. Koubek (Eds.), *Design of computing systems: Social and ergonomic considerations* (pp. 893–896). Amsterdam, Netherlands: Elsevier Science Publishers, San Francisco, CA (August 24–29).

Durlach, B. N. I., & Mavor, A. S. (1995). *Virtual reality: Scientific and technological challenges.* Washington, DC: National Academy Press.

Ebenholtz, S. M., & Mayer, D. (1968). Rate of adaptation under constant and varied optical tilt. *Perceptual and Motor Skills, 26,* 507–509.

Graeber, D. A. (2001a). *Use of incremental adaptation and habituation regimens for mitigating optokinetic side effects.* Unpublished doctoral dissertation, University of Central Florida.

Graeber, D. A. (2001b). *Application of the Kennedy and Graybiel Motion History Questionnaire to predict optokinetic induced motion sickness: Creating a scoring key for circular vection* (Tech. Report No. TR-2001-03). Orlando, FL: Naval Air Warfare Center, Training Systems Division.

Graybiel, A., Deane, F., & Colehour, J. (1969). Prevention of overt motion sickness by incremental exposure to otherwise highly stressful coriolis accelerations. *Aerospace Medicine, 40,* 142–148.

Groves, P. M., & Thompson, R. F. (1970). Habituation: A dual process theory. *Psychological Review, 77*(5), 419–450.

Harm, D. L. (2002). Motion sickness neurophysiology, physiological correlates, and treatment. In K. M. Stanney (Ed.), *Handbook of virtual environments: Design, implementation, and applications* (pp. 791–806). Mahwah, NJ: Lawrence Erlbaum Associates.

Held, R. (1965). Plasticity in sensory-motor systems. *Scientific American, 72*, 84–94.

Howarth, P. A., & Finch, M. (1999). The nauseogenicity of two methods of navigating within a virtual environment. *Applied Ergonomics, 30*, 39–45.

Hu, S., Stern, R. M., & Koch, K. L. (1991). Effects of pre-exposure to a rotating optokinetic drum on adaptation to motion sickness. *Aviation, Space, and Environmental Medicine, 62*, 53–56.

Kennedy, R. S. (2001). *Unpublished research data.* RSL Assessments, Inc., Orlando, FL.

Kennedy, R. S., Berbaum, K. S., Lilienthal, M. G., Dunlap, W. P., Mulligan, B. E., & Funaro, J. F. (1987). Guidelines for alleviation of simulator sickness symptomatology. *Aviation, Space, and Environmental Medicine, 60*, 10–16.

Kennedy, R. S., Dunlap, W. P., Berbaum, K. S., & Hettinger, L. J. (1996). Developing automated methods to quantify the visual stimulus for cybersickness. *Proceedings of the Human Factors and Ergonomics Society 40th Annual Meeting* (pp. 1126–1130). Santa Monica, CA: Human Factors & Ergonomics Society.

Kennedy, R. S., & Fowlkes, J. E. (1992). Simulator sickness is polygenic and polysymptomatic: Implications for research. *International Journal of Aviation Psychology, 2*(1), 23–38.

Kennedy, R. S., & Graybiel, A. (1965). *The Dial test: A standardized procedure for the experimental production of canal sickness symptomatology in a rotating environment* (Rep. No. 113, NSAM 930). Pensacola, FL: Naval School of Aerospace Medicine.

Kennedy, R. S., Hettinger, L. J., & Lilienthal, M. G. (1990). Simulator sickness. In G. H. Crampton (Ed.), *Motion and space sickness* (pp. 247–262). Boca Raton, FL: CRC Press.

Kennedy, R. S., Jones, M. B., Stanney, K. M., Ritter, A. D., & Drexler, J. M. (1996, June). *Human factors safety testing for virtual environment mission-operation training* (Final Report, Contract No. NAS9-19482). Houston, TX: NASA Johnson Space Center.

Kennedy, R. S., Lane, N. E., Berbaum, K. S., & Lilienthal, M. G. (1993). Simulator sickness questionnaire: An enhanced method for quantifying simulator sickness. *International Journal of Aviation Psychology, 3*(3), 203–220.

Kennedy, R. S., Lane, N. E., Grizzard, M. C., Stanney, K. M., Kingdon, K., Lanham, S., & Harm, D. L. (2001, September 5–7). Use of a motion history questionnaire to predict simulator sickness. *Driving Simulation Conference 2001.* Sophia-Antipolis (Nice), France.

Kennedy, R. S., & Stanney, K. M. (1996). Postural instability induced by virtual reality exposure: Development of a certification protocol. *International Journal of Human-Computer Interaction, 8*(1), 25–47.

Kennedy, R. S., Stanney, K. M., & Dunlap, W. P. (2000). Duration and exposure to virtual environments: Sickness curves during and across sessions. *Presence: Teleoperators and Virtual Environments, 9*(5), 466–475.

Kingdon, K. (2001). *Effects of low stereo acuity on performance, presence, and sickness within a virtual environment.* Unpublished Master's Thesis, University of Central Florida, Orlando, FL.

Kingdon, K., Stanney, K. M., & Kennedy, R. S. (2001, October 8–12). Extreme responses to virtual environment exposure. *The 45th Annual Human Factors and Ergonomics Society Meeting* (pp. 1906–1910). Minneapolis/St. Paul MN.

Knerr, B. W., Breaux, R., Goldberg, S. L., & Thurman, R. A. (2002). National defense. In K. M. Stanney (Ed.), *Handbook of Virtual Environments: Design, implementation, and applications* (pp. 857–872). Mahwah, NJ: Lawrence Erlbaum Associates.

Knerr, B. W., Lampton, D. R., Singer, M. J., Witmer, B. G., Goldberg, S. L., Parsons, K. J. et al. (1998). *Virtual Environments for dismounted soldier training and performance: Results, recommendations, and issues.* (ARI Tech. Rep. No. 1089). Alexandria, VA: U.S. Army Research Institute for the Behavioral and Social Sciences.

Kolasinski, E. M. (1995). *Simulator sickness in virtual environments* (ARI Tech. Rep. 1027). Alexandria, VA: U.S. Army Research Institute for the Behavioral and Social Sciences.

Konz, S. (1997). Toxicology and thermal discomfort. In G. Salvendy (Ed.), Handbook of human factors and ergonomics (2nd ed., pp. 891–908). New York: Wiley.

Lackner, J. R., & Lobovitz, D. (1978). Incremental exposure facilitates adaptation to sensory rearrangement. *Aviation, space and environmental medicine, 49*, 362–264.

Lawson, B. D., Graeber, D. A., Mead, A. M., & Muth, E. R. (2002). Signs and symptoms of human syndromes associated with synthetic experiences. In K. M. Stanney (Ed.), *Handbook of virtual environments: Design, implementation, and applications* (pp. 791–806). Mahwah, NJ: Lawrence Erlbaum Associates.

McFarland, R. A. (1953). *Human factors in air transportation: Occupational health & safety.* New York: McGraw-Hill.

McCauley, M. E., & Sharkey, T. J. (1992). Cybersickness: Perception of selfmotion in virtual environments. *Presence: Teleoperators and virtual environments, 1*(3), 311–318.

McNally, W. J., & Stuart, E. A. (1942). Physiology of the labyrinth reviewed in relation to seasickness and other forms of motion sickness. *War Medicine, 2*, 683–771.

Mirabile, C. S. (1990). Motion sickness susceptibility and behavior. In G. H. Crampton (Ed.), *Motion and space sickness* (pp. 391–410). Boca Raton, FL: CRC Press.

Mon-Williams, M., Rushton, S., & Wann, J. P. (1995). Binocular vision in stereoscopic virtual-reality systems. *Society for Information Display International Symposium Digest of Technical Papers, 25,* 361–363.

Money, K. E. (1970). Motion sickness. *Psychological Reviews, 50*(1), 1–39.

Money, K. E., & Myles, W. S. (1975). Motion sickness and other vestibulo-gastric illnesses. In R. F. Naunton (Ed.), *The vestibular system* (pp. 371–377). New York: Academic Press.

Moshell, J. R., & Hughes, C. E. (2002). Virtual environments as a tool for academic learning. In K. M. Stanney (Ed.), *Handbook of virtual environments: Design, implementation, and applications* (pp. 791–806). Mahwah, NJ: Lawrence Erlbaum Associates.

National Research Council. (1997). *More than screen deep: Toward every-citizen interfaces to the nation's information infrastructure.* Washington, DC: National Academy Press.

Pavlov, I. P. (1928). *Lectures on conditioned reflexes* W. H. Gantt (Trans.), New York: International.

Pierce, J. S., Pausch, R., Sturgill, C. B., & Christiansen, K. D. (1999). Designing a successful HMD-based experience. *Presence: Teleoperators and virtual environments, 8*(4), 469–473.

Prescott, S. A. (1998). Interactions between depression and facilitation within neural networks: Updating the dual-process theory of plasticity. *Learning & Memory, 5*(6), 446–466.

Prescott, S. A., & Chase, R. (1999). Sites of plasticity in the neural circuit mediating tentacle withdrawal in the snail Helix aspersa: Implications for behavioral change and learning kinetics. *Learning & Memory, 6,* 363–380.

Reason, J. T. (1970). Motion sickness: A special case of sensory rearrangement. *Advancement in Science, 26,* 386–393.

Reason, J. T. (1978). Motion sickness adaptation: A neural mismatch model. *Journal of the Royal Society of Medicine, 71,* 819–829.

Reason, J. T., & Brand, J. J. (1975). *Motion sickness.* New York: Academic Press.

Regan, E. C., & Price, K. R. (1994). The frequency of occurrence and severity of side-effects of immersive virtual reality. *Aviation, Space, and Environmental Medicine, 65,* 527–530.

Rizzo, A. A., Buckwalter, G., & van der Zaag, C. (2002). Virtual environment applications in clinical neuropsychology. In K. M. Stanney (Ed.), *Handbook of virtual environments: Design, implementation, and applications* (pp. 1027–1064). Mahwah, NJ: Lawrence Erlbaum Associates.

Satava, R. M., & Jones, S. B. (2002). Medical applications of virtual environments. In K. M. Stanney (Ed.), *Handbook of virtual environments: Design, implementation, and applications* (pp. 721–730). Mahwah, NJ: Lawrence Erlbaum Associates.

Singer, M. J., Ehrlich, E. A., & Allen, R. C. (1998). *Effect of a body model on performance in a virtual environment search task* (ARI Tech. Rep. 1087). Alexandria, VA: U.S. Army Research Institute for the Behavioral and Social Sciences.

Sjoberg, A. A. (1929). Experimental studies of the eliciting mechanism of sea sickness. *Acta Oto-laryngolica, 13,* 343–347.

So, R. H., & Griffin, M. J. (1995). Effects of lags on human operator transfer functions with head-coupled systems. *Aviation, Space, and Environmental Medicine, 66,* 550–556.

Stanney, K. M. (1995). Realizing the full potential of virtual reality: Human factors issues that could stand in the way. *Virtual Reality Annual International Symposium '95* (pp. 28–34). Los Alamitos, CA: IEEE Computer Society Press.

Stanney, K. M. (2001). *Unpublished research data.* University of Central Florida, Orlando, FL.

Stanney, K. M. (Ed.). (2002). *Handbook of virtual environments: Design, implementation, and applications.* Mahwah, NJ: Lawrence Erlbaum Associates.

Stanney, K. M., & Hash, P. (1998). Locus of user-initiated control in virtual environments: Influences on cybersickness. *Presence: Teleoperators and Virtual Environments, 7*(5), 447–459.

Stanney, K. M., & Kennedy, R. S. (1997a, September 22–26). Cybersickness is not simulator sickness. *Proceedings of the 41st Annual Human Factors and Ergonomics Society Meeting* (pp. 1138–1142). Albuquerque, NM.

Stanney, K. M., & Kennedy, R. S. (1997b). The psychometrics of cybersickness. *Communications of the ACM, 40*(8), 67–68.

Stanney, K. M., & Kennedy, R. S. (1998, October 5–9). Aftereffects from virtual environment exposure: How long do they last? *Proceedings of the 42nd Annual Human Factors and Ergonomics Society Meeting* (pp. 1476–1480). Chicago, IL.

Stanney, K. M., Kennedy, R. S., Drexler, J. M., & Harm, D. L. (1999). Motion sickness and proprioceptive aftereffects following virtual environment exposure. *Applied Ergonomics, 30,* 27–38.

Stanney, K. M., Kennedy, R. S., & Kingdon, K. (2002). Virtual environments usage protocols. In K. M. Stanney (Ed.), *Handbook of virtual environments: Design, implementation, and applications* (pp. 721–730). Mahwah, NJ: Lawrence Erlbaum Associates.

Stanney, K. M., Kingdon, K., Graeber, D., & Kennedy, R. S. (2002). Human performance in immersive virtual environments: Effects of duration, user control, and scene complexity. *Human Performance, 15*(4), 339–366.

Stanney, K. M., Kingdon, K., & Kennedy, R. S. (2001). Human performance in virtual environments: Examining user control techniques. In M. J. Smith, G. Salvendy, D. Harris, & R. J. Koubek (Eds.), *Usability evaluation and interface design: Cognitive engineering, intelligent agents and virtual reality* (Vol. 1 of the *Proceedings of HCI International* 2001, pp. 1051–1055). Mahwah, NJ: Lawrence Erlbaum Associates.

Stanney, K. M., Kingdon, K., Nahmens, I., & Kennedy, R. S. (2003). What to expect from immersive virtual environment exposure: Influences of gender, body mass index, and past experience. *Human Factors, 45*(3), 504–522.

Stanney, K. M., Lanham, S., Kennedy, R. S., & Breaux, R. B. (1999, September 27–October 1). Virtual environment exposure drop-out thresholds. *The 43rd Annual Human Factors and Ergonomics Society Meeting* (pp. 1223–1227). Houston, TX.

Stanney, K. M., Salvendy, G., Deisigner, J., DiZio, P., Ellis, S., Ellison, E. et al. (1998). Aftereffects and sense of presence in virtual environments: Formulation of a research and development agenda. Report sponsored by the Life Sciences Division at NASA Headquarters. *International Journal of Human-Computer Interaction, 10*(2), 135–187.

Stone, R. J. (2002). Applications of virtual environments: An overview. In K. M. Stanney (Ed.), *Handbook of virtual environments: Design, implementation, and applications* (pp. 827–856). Mahwah, NJ: Lawrence Erlbaum Associates.

Tharp, V., Burns, M., & Moskowitz, H. (1981). *Development and field test of psychophysical tests for DWI arrest* (DOT Final Rep. ODT HS 805 864). Washington, DC.

Tyler, D. B., & Bard, P. (1949). Motion sickness. *Physiological Review, 29*, 311–369.

Uliano, K. C., Kennedy, R. S., & Lambert, E. Y. (1986). Asynchronous visual delays and the development of simulator sickness. *Proceedings of the Human Factors Society 30th Annual Meeting* (pp. 422–426). Dayton, OH: Human Factors Society.

Viirre, E., & Bush, D. (2002). Direct effects of virtual environments on users. In K. M. Stanney (Ed.), *Handbook of virtual environments: Design, implementation, and applications* (pp. 581–588). Mahwah, NJ: Lawrence Erlbaum Associates.

Wann, J. P., & Mon-Williams, M. (2002). Measurement of visual aftereffects following virtual environment exposure. In K. M. Stanney (Ed.), *Handbook of virtual environments: Design, implementation, and applications* (pp. 731–749). Mahwah, NJ: Lawrence Erlbaum Associates.

Welch, R. B. (2002). Adapting to virtual environments. In K. M. Stanney (Ed.), *Handbook of virtual environments: Design, implementation, and applications.* Mahwah, NJ: Lawrence Erlbaum Associates.

Wendt, G. R. (1968). *Experiences with research on motion sickness* (NASA Special Publication No. SP-187). Pensacola, FL: Fourth Symposium on the Role of Vestibular Organs in Space Exploration.

Wilson, J. R. (1997). Virtual environments and ergonomics: Needs and opportunities. *Ergonomics, 40*(10), 1057–1077.

Wilson, J. R., Nichols, S., & Haldane, C. (1997, August 24–29). Presence and side effects: Complementary or contradictory? In M. Smith, G. Salvendy, & R. Koubek (Eds.), *Design of computing systems: Social and ergonomic considerations* (pp. 889–892). Amsterdam, Netherlands: Elsevier Science Publishers, San Francisco, CA.

Wilson, J. R., Nichols, S. C., & Ramsey, A. D. (1995). Virtual reality health and safety: Facts, speculation and myths. *VR News, 4*, 20–24.

23

Location and Arrangement of Displays and Control Actuators

Robert W. Proctor
Purdue University

Kim-Phuong L. Vu
California State University, Long Beach

INTRODUCTION

Performance of an operator is highly dependent on the location and arrangement of displays and controls. A well-designed interface takes into account the positioning and grouping of displays and controls, and the relationships between displays and their associated controls. The display arrangement should allow the operator to detect and identify critical visually displayed information with minimal effort. This information needs to be in a form that can be associated easily with the appropriate control that affects the system function captured by the display. The control should be located where it is readily accessible and can be operated comfortably. To accomplish these goals, the display and control arrangements need to be designed according to principles of perception, response selection, motor control, and biomechanics.

DEFINITION OF RELEVANT TERMS

a. Accessibility index—A quantitative measure of control layout that takes into account the ranked frequency of use of the controls and their distances from the operator.
b. Acuity—The ability to perceive and distinguish details in a visual image. Acuity is highest in central vision and decreases dramatically in peripheral vision.
c. Fitts' law—A law used to estimate movement time (MT) as function of the target width (W) and distance (D). $MT = a + b \log_2(2D/W)$, where a and b are constants.
d. Functional groupings—Using grouping principles to organize displays or controls so that those relating to common functions are grouped together.
e. Grouping principles—Organizational principles determining which parts of a display or control arrangement are perceived as belonging together and which are perceived as separate objects.
f. Head-up displays—Head-up displays superimpose a virtual image of the information display on the central area in which the outside world is viewed. The purpose is to allow

the operator to be able to check displays without having to look away from the outer environment.

g. Line of sight—A hypothetical line from the eyes to the point on which they are fixating. The farther a display is angled from the line of sight, the more difficult it is to identify.

h. Link analysis—An analysis that can be applied to display panel design to determine the locations in which components of a display panel should be placed. In a link analysis, displays with high frequency of use or importance are placed in central locations, and displays that are scanned in sequence are placed in adjacent locations.

i. Mental model—A dynamic representation of an event or scenario that reflects the person's understanding of the situation. Mental models direct the comprehension of new information, reasoning, solution of problems, and decisions made under uncertainty.

j. Population stereotypes—Expected relationships between movements of controls and their effects on a display or system.

k. Reach envelope—The distance from the operator within which controls should be located in order for a certain percentage of the population (typically 95%) to be able to reach them easily.

l. Saccadic eye movement—A rapid, ballistic eye movement made to a predetermined location, either intentionally or reflexively, to the onset of a stimulus or change in some property.

m. Spatial compatibility—Performance is better when the spatial relations in a display are mapped in a corresponding manner to the spatial relations of a response panel or control.

n. Stimulus-response compatibility—Differences in the speed and accuracy with which responses to stimuli can be selected for different stimulus–response arrangements and mappings. This is sometimes called display–control compatibility.

o. Useful field of view (UFOV)—UFOV is a measure of visual attention that specifies the range of the visual field in which people can select an object from multiple stimuli. UFOV allows assessment of visual processing speed, divided attention, and selective attention. It decreases with age and is a useful predictor of driving accidents.

p. Visual angle—The angle at the eye subtended by the image of an object. The visual angle is computed as $\tan^{-1}$(size/distance).

q. Visual field—The area in which visual stimuli can be detected. Acuity varies across the visual field, being highest in foveal vision.

FULL NAME OF THE STANDARDS (GUIDELINES)

a. DoD Military Standards—1472D Human Engineering Design Criteria for Military Systems Equipment and Facilities, Section 5.

b. Ergonomic Requirements for the Design of Signals and Control Actuators—Part 1: Human Interactions with Displays and Control Actuators (ISO/CD 9355-1)

c. Ergonomic Principles for the Design of Signals and Control Actuators—Part 2: Displays (ISO/CD 9355-2)

d. Man–Systems Integration Standards, NASA-STD-3000, Volume 1, Section 9.

e. Safety of Machinery; Ergonomics Requirements for the Design of Displays and Control Actuators; Part 1: General Principles for Human Interactions with Displays and Control Actuators (CEN EN 894-1; PREN 894-1; BSI BS EN 894-1)

f. Safety of Machinery—Ergonomics Requirements for the Design of Displays and Control Actuators—Part 2: Displays (CEN EN 894-2; PREN 894-2; BSI BS EN 894-2)

g. Safety of Machinery—Ergonomics Requirements for the Design of Displays and Control Actuators—Part 3: Control Actuators (CEN EN 894-2; PREN 894-3; BSI BS EN 894-3)

Location and Arrangement of Displays

When designing display panels, placement of the components is crucial because of the relation of acuity to location in the visual field. A visual display must be in the visual field if it is to have any chance of being detected and responded to. For young adults, the field of view is approximately 180°, but it decreases to about 140° in older adults. Within the visual field, acuity varies drastically. It is highest in central, foveal vision (approximately 1° of visual angle) and decreases sharply as stimulus location moves further into the visual periphery. Surrounding the fovea is a somewhat larger region of 5°, called the parafovea, in which acuity is still high but not as high as for foveal images.

The central 30° of the field is focal vision, for which images can be brought into foveal vision through saccadic eye movements. The remainder of the visual field is called ambient vision. It plays a role in guiding movement and maintaining spatial orientation, but head movements are required for images in this region to be brought into foveal vision. Studies have shown that the useful field of view is reduced considerably when multiple stimuli are present and need to be processed.

When the head and eyes are oriented directly ahead, the line of sight is horizontal. However, this posture is uncomfortable because it requires the neck and eye muscles to be tense. It is more comfortable to relax the muscles for both the head and the eyes. The head is in the most comfortable position when it is inclined slightly forward, making the normal line of sight relative to the head 10° to 15° below horizontal. When the eyes are also relaxed, the normal line of sight is 25° to 30° below horizontal.

Because time and effort are required to move the head and eyes, it is important to minimize the amount of movement that is necessary to process the information from a display arrangement. A link analysis can be used to help determine where the individual components of a display panel should be placed. As applied to display design, components are arranged according to their importance, frequency, and sequence of use, with adjacent components being those between which the strongest sequential scanning links exist.

The distance of displays from operators is also a factor to consider. Both accommodation and vergence angle vary as a function of distance from the operator. It is important to minimize the need for re-accommodation and change in vergence angle when switching among display components or between displays and the external environment. One benefit of a head-up display, for which the displays are located on the windshield, is that it reduces the differences in accommodation and vergence when switching between the outside world and the display panel, relative to a head-down display. However, this benefit has its associated costs such as creating clutter, being difficult to see against the environmental background, and reducing visibility of objects in the outer environment.

Location and Arrangement of Controls

Because controls often need to be seen for the correct one to be selected, the visual concerns that apply to displays also apply to control panels. However, anthropometric and biomechanical factors with regard to placement of controls are more important. If a control is placed where it cannot be reached and operated easily, this may present problems for the operator. When designing for reach, the 5th percentile reach values are often used because they accommodate all but the very smallest users. A distinction is also often made between the reach envelope of males and females, because males are of larger physical stature on average. The immediate reach envelope refers to the region in which controls can be reached without bending, and the maximum reach envelope refers to the region in which controls can be reached with bending. The exact values for the reach envelope vary as a function of the task, the height at which the control is placed, and whether the operator is standing or sitting. For example, at table level, Sengupta and Das (2000) estimated the maximum straight ahead reach when seated to be 58.3 and 64.5 cm for females and males, respectively. These values varied as a function of height from tabletop and degrees of angle horizontally from straight ahead. The reach envelope for standing operators was approximately 4 cm longer for males and 6 cm for females.

In addition to ensuring that an operator can reach a control, it is important that the operator not confuse controls. If an operator must select one among several controls, the controls should be separated and distinct from one another to minimize the possibility that an incorrect control will be actuated. As with displays, failure to place controls that are used in sequence together may result in excessive movements of the operator and cause fatigue.

Display–Control Relations

Most standards focus on the placement and arrangement of displays and controls, but less on their relation. However, the process of determining what action to take in response to displayed information is a major component of response time. It has been known since at least the time of Fitts and Seeger's (1953) classic study on stimulus–response compatibility that the time for response selection is minimal when spatial compatibility is high. Performance is best when displays and their corresponding controls are configured in similar arrangements and each display is mapped to the spatially corresponding control. If display–control arrangements are not highly compatible, the operator may be delayed in responding, or may even take the wrong action.

OBJECTIVE AND SCOPE OF THE CHAPTER

Displays, controls, and their relations have been studied since the earliest days of Human Factors and Ergonomics as a distinct field. Consequently, much is known about factors that influence performance with different display–control configurations, and numerous guidelines have been developed that reflect this knowledge. In addition, there has been considerable research on stimulus–response compatibility and related effects in recent years from which recommendations can be made but that has not yet been incorporated into specific standards and guidelines.

The objective of this chapter is to provide a thorough description of existing standards and guidelines regarding location and arrangements of displays and controls. In addition, we discuss recommendations that can be made on the basis of current knowledge. In all cases, we discuss the findings that underlie the standards, guidelines, and recommendations.

DISCUSSION OF SPECIFIC STANDARDS AND GUIDELINES

Arrangement and Location of Displays

Display Position

Displays should be positioned according to the following guidelines:

- *Identification:* The display needs to be positioned where it can be detected and identified readily.
- *Visibility:* The operator should be able to see all the essential displays from his or her normal workplace position.
  ○ Primary instruments should be positioned within 30° of the line of sight to avoid excessive movements of the eyes, head, or body.

- ° Visual warning signals should also be positioned within 30° of the line of sight to allow the operator to see the signal readily even if it is not used often.
- ° Auditory warning signals do not have to be positioned within the normal line of vision since they can be detected independent of where the operator is looking. Auditory warning signals can be used to direct where an operator should look.
- ° Secondary instruments should be positioned within 60° of the line of sight. This will allow the operator to read the displays by moving the eyes, without having to change the position of the head.
- ° Infrequently used instruments do not need to lie within the normal line of vision.
- *Derivatives of Change:* When different derivatives of change need to be displayed, displays should be positioned from left to right in order of increasing derivatives.
- *Viewing Distance:* The viewing distance should be at least 40 cm for many office tasks, with 50 cm being preferable. Closer distances may cause strain on the eyes.
- *Maximum Line of Sight Angle:* The line of sight should be no more than 60° above or below the horizontal. For a display to be legible, the size and distance need to be taken into account, as well as the location in the visual field relative to the line of sight.
- *Glare and Reflection:* The display should be position in a location that minimizes glare and reflection from surrounding objects.

For references see ISO 9355-2; BS EN 29241-3; Ivergård, 1989; Engineering Data Compendium (Volume 3, no. 9.355)

Display Layout

- *Related Information:* Displays of highly related information sources should be located close to each other. There are many ways in which different sources of information can be highly related. For example, displays can provide correlated information or information pertinent to the same system function.
- *Importance:* Displays that are the most important should be placed in foveal vision. Important displays should compose a "central zone" on the display panel. This zone should ideally be within 30° of visual angle.
- *Frequency:* Frequently used displays should be in the primary visual field and adjacent to each other.
- *Sequence:* Displays that are typically viewed sequentially should be placed close together, in the order in which they are scanned. People have a bias to scan along the horizontal dimension more than the vertical dimension.
- *Consistency:* When using different display panels or screens for similar or related tasks, the same elements should be positioned in the same locations. This will improve performance by allowing the operator to go to the desired locations automatically, thus minimizing visual search time and memory load.
- *Perceptual Grouping:* The Gestalt perceptual organizational principles should be used to group-related displays together and to separate them from other groupings. Commonly used principles are:
- ° Similarity: Display elements intended to be perceived as grouped together should be similar in appearance.
- ° Spatial proximity: Elements intended to be perceived a group should be placed close together.
- ° Common fate: Elements intended to be perceived as a group should move in the same direction and speed.
- ° Connectedness: Elements intended to be perceived as a group should be connected by lines. This is sometimes called use of sensor lines.

° Common region: Elements intended to be perceived as a group should be within a common boundary. This is often used for over-the-counter medications, where a single dose is surrounded by a perforation boundary.

- *Display integration:* For settings in which information about complex systems at different levels of abstraction needs to be displayed, integrating across different views is a challenge. For tasks such as fault diagnosis that require problem solving, presenting the information integrated in a single display yields better performance than presenting the information for the separate views in distinct windows.
- *Clutter Avoidance:* There should be adequate spacing between all display components on a display panel. Avoid excessive placements of display elements within a small area. Clutter is of particular concern for head-up displays, where each new display element or symbol that is added increases the difficulty of perceiving the outer environment.

For references see ISO 9355-1 and 2; Andre and Wickens, 1990; Burns, 2000; Donk, 1994; Proctor & Van Zandt, 1994; Wickens, Gordon, & Liu, 1999.)

Arrangement and Location of Controls

Position of Controls

- *Identification:* The control needs to be positioned where it can be detected and identified readily.
- *Visibility:* The operator should be able to see all the essential controls from the normal workplace position.
- *Accessibility:* Controls should be located in a position that allows easy and comfortable access by the operator. For example, the slope of a keyboard should be positioned between 0° and 25°, and the operator should be able to make additional adjustments. The keyboard should also be placed in a stable position that is independent of the display.
- *Expectancy:* Controls should be positioned where people would expect them to be located. Population stereotypes can help determine the expectancy for the targeted users.

(For references see ISO 9355-1 and 3; BS EN 29241-3; Ivergård, 1989; Engineering Data Compendium [Volume 3, no. 12.301].)

Control Layout

- *Importance:* Controls that are most important (i.e., primary controls) should be located close to the operator.
- *Frequency:* The most frequently used controls should be located close to the operator, inside the immediate reach envelope where they can be reached without bending. Infrequently used controls can be placed at further distances.
- *Sequence of Use:* Controls operated in sequence should be placed close together and preserve the sequential relation.
- *Accessibility:* All controls should be arranged within the reach envelope to allow easy access. Avoid placing controls where operators must make awkward movements.
- *Tactile landmarks:* Controls should be located near tactile landmarks. This is especially needed for operators with visual disabilities.
- *Grouping:* Related controls should be grouped together, and unrelated controls should be separated in order for the operator to easily identify groups of controls. Examples of grouping include:
  ° Group controls with similar functions together on the control panel.
  ° Make related controls equal in size.

- *Movement Space:* The layout of the controls should not cause the operator to experience any discomfort when operating a control or moving between controls.
- *Movement Time:* Movement time between controls should be minimized. Movement time is an increasing function of movement distance and a decreasing function of target width. It can be estimated using Fitts' Law.
- *Corresponding Control Labels:* Labels on control panels should be larger than usually recommended so that both younger and older adults will be able to see them. If guidelines for control labels were based on young adults, the characters will need to be a minimum of 20% larger than recommended to accommodate older adults. Labels also should have good contrast, with a 10:1 contrast ratio of labels with their background recommended.
- *Symmetry:* Design the panel as symmetrically as possible.
- *Expectancy:* The control should be located in a position that the operator expects to avoid excessive search for the control. Population stereotypes can help determine the targeted users' expectancies.
- *Consistency:* When more than one control panel is used or operators have to switch between different control panels, the specific controls should be located consistently across panels.
- *Clutter avoidance:* Adequate distances should be maintained between controls.

(For references see BS 7179:Part4:1990; ISO 9355-1 and 3; Ivergård, 1989, p. 103; Smith, 1996, p. 276.)

Display–Control Relations

Although many guidelines and standards are established for location and arrangements of displays and controls independently, the display–control layout should also optimize the interrelationship between display–control pairs.

- *Stimulus–Response Compatibility:* Stimulus–response compatibility refers to the natural tendency to respond faster and more accurately with some mappings of stimuli to responses than with others. When designing display–control layouts, it is important to map displays to their associated controls in a compatible manner. Maintaining compatibility is particularly important when workload or stress is high, because performance with incompatible mappings will deteriorate more than that with compatible mappings. However, maintaining compatibility is not as easy as it sounds, because compatibility varies as a function of tasks and their goals. The specific compatibility relations follow:
  - *Relative Location:* It is important to maintain the relative location of displays and their associated controls. Compatibility effects occur with respect to relative locations, not just absolute locations.
  - *Frames of Reference:* Compatibility effects occur with respect to many frames of reference. These frames of reference include: body midline, direction of attention, and location relative to environmental cues and objects. When multiple displays or controls are organized into groups, locations can be coded at both the global and local levels.
  - *Parallel Versus Orthogonal:* Parallel mappings (e.g., displays in left–right locations mapped to controls in left–right locations) yield better performance than orthogonal mappings (e.g., displays in top–bottom locations mapped to controls in left–right locations). Designers should make an effort to align displays and controls along the same axis.
  - *Orthogonal Mappings:* When parallel mappings are not possible, and displays are mapped orthogonally to their associated controls, the mapping of "top" with "right" and "bottom" with "left" often yields better performance than the mapping of "top"

with "left" and "bottom" with "right." However, the preferred mapping is affected by location of the controls relative to the display and to other controls.

- ○ *Pure Versus Mixed Mappings:* Performance is better when the same mapping is used for all display–control pairs than when different mappings are used. The most ideal layout is one in which all display–control mappings are compatible. When mappings are mixed, performance often suffers more for the compatible relation than for the incompatible one.
- ○ *Rules:* When displays cannot be mapped compatibly to their spatially corresponding controls, a mapping that produces a systematic relation (e.g., opposite) is better than a random mapping. For example, performance is much better with a mirror-opposite relation between displays and controls than with a random one. This is because the individual associations between displays and controls do not have to be remembered.
- ○ *Prevalence Effects:* When the display–control configuration can be coded along two dimensions at once, a prevalence effect may occur in which coding with respect to one dimension dominates the other. The dominant dimension is likely to be the one made salient by both the display and control environments.
- ○ *Simon Effects:* Spatial correspondence effects occur when location is irrelevant to the task. For example, when stimuli and responses vary in left–right locations and the relevant stimulus attribute is nonspatial (e.g., identity: S or H), responding is faster when the location of the stimulus and response corresponds spatially than when it does not.
- ○ *Intentions and Task Goals:* Compatibility effects are not automatic consequences of physical relations. Rather, they depend on the operators' intentions and the task's goals.
- ○ *Dimensional Overlap:* Compatibility effects occur not only for spatial stimuli and responses but for any situation in which the stimulus dimension overlaps with the response dimension. This overlap may involve conceptual similarity (when the display and control sets can be categorized along the same dimension), perceptual similarity (when the display and control sets are physically similar), and structural similarity (e.g., when the display and control sets maintain an ordered relationship).
- • *Proximity:* Displays should be close to their associated controls.
- • *Movement:* The direction of movement for a control should be consistent with the direction of movement for both the feedback indicator and the system movement. The following are movement principles based on population stereotypes:
  - ○ *Clockwise-to-Right-or-Up:* Clockwise control movement is expected to move a pointer toward the right for horizontal displays and upward for vertical displays.
  - ○ *Warrick's Principle:* The pointer is expected to move in the same direction as the side of the control that is nearest to the display.
  - ○ *Clockwise-to-Increase:* Clockwise control is expected to cause an increase in the reading of the display.
  - ○ *Clockwise-Away:* Clockwise control is expected to cause the pointer to move away from the control.
- • *Practice:* Performance with an incompatible display–control configuration improves with practice, but typically is worst than it would be with the same amount of practice if the configuration were compatible.
- • *Learning and Transfer:* If an operator has previous experience with a related display–control configuration, the previously learned relationship may transfer to the current situation. This transfer may facilitate performance if the display–control relation conforms to the operator's expectations, but interfere with performance if it does not.

- *Display–Control Relationships:* When the display and its associated control are on the same panel, the control should be placed under the display if possible. If this arrangement is not possible, the control should be placed to the right of the display; because most people are right-handed, the display will not be obstructed when the operator reaches for the control. When multiple displays need to be monitored while adjusting a single control, the control should be placed under the display in the middle in a manner that will allow the control to be operated without obscuring the displays.
- *Redundancy:* The display–control configuration should contain redundant coding of information for important relations.
- *Controllability:* The display–control configuration should help guide the operator through the tasks and allow the operator direct control over the system.
- *Concurrent Use:* Displays that are monitored simultaneously while manipulating a control should be placed where the operator can easily see them.
- *Mental Model:* A designer should be able to predict which display–control mappings would be most compatible based on knowledge of the operators' mental representation of the task and system.

(For references see ISO 9355-1 and 3; BS EN 29241-3; Ivergård, 1989; Engineering Data Compendium [Volume 3]; Andre & Wickens, 1990; Cho & Proctor, 2003; Hommel & Prinz, 1997; Proctor, 2001; Proctor & Reeve, 1990; Vu & Proctor, 2003.)

EXAMPLES OF APPLICATION

- *Multifunction Displays:* In a multifunction display, operators proceed through a hierarchy of information presented on a computer screen by pressing or clicking buttons. With such displays, how to map the hierarchy to the buttons for successive screens is a significant human factors problem. Several of the general design principles mentioned previously, such as frequency and sequence of use, are applicable. Moreover, search time can be reduced dramatically by minimizing the average distance between buttons for successive screens, and maximizing repeated selections of the same buttons. Optimization methods can be applied to select the best mapping relative to a defined cost function.

(For a reference see Francis, 2000.)

- *Aircraft Display–Control Layout:* A British Midland Airways Boeing 737-400 aircraft crashed on January 8, 1989, when the functioning right engine was shut down instead of the failing left engine. The investigators noted that one factor contributing to this error was the layout of the engine displays relative to the controls. The engine displays were grouped into left and right rectangular blocks arranged on the cockpit instrument panel. The left grouping contained the primary instruments for both engines, and the right grouping contained the secondary instruments for both engines. Within each grouping, the dials for the left engine were in the left column, and those for the right engine were in the right column. The throttle controls for the left and right engines were aligned with the primary and secondary groupings, respectively. Thus, left–right display–control compatibility was maintained relative to the columns within each group, but not relative to the left and right groups themselves. With this configuration, spatial compatibility was maintained within the local instrument group, but not for the global grouping. Because the global blocks were aligned with the controls, it was probably more important to maintain global compatibility in this case than to maintain local compatibility.

(For a reference see Learmount & Norris, 1990.)

- *Stove-Top Designs:* A common arrangement of the four burners on a stove-top is rectangular. There are numerous ways in which the controls can be arranged and mapped to the burners. Not too surprisingly, when the control arrangement is isomorphic to the burner arrangement, a compatible mapping of the burners to controls produces fast and accurate performance, because the relative location of the display configuration matches that of the control configuration. However, performance with this arrangement suffers dramatically with an incompatible burner-control mapping. The cost associated with this incompatibility can be reduced by connecting each burner to its control with sensor lines. More commonly, the controls are arranged linearly along the horizontal dimension, and the control configuration is thus not isomorphic with the display arrangement. In this case, there is no natural mapping of burners to controls. This ambiguity can be reduced and performance improved by staggering the burners so that they are in four distinct locations on the horizontal dimension. This produces burner–control correspondence along the horizontal dimension based on relative location. Alternatively, performance can be improved by using sensor lines to connect the burners in the rectangular arrangement to their controls in the horizontal arrangement. This example illustrates that there are many alternative ways to address potential problems of incompatibility.

(For a reference see Chapanis & Yoblick, 2001.)

REFERENCES

Andre, A., D., & Wickens, C. D. (1990). *Display-control compatibility in the cockpit: Guidelines for display layout analysis.* Aviation Research Laboratory Technical Report ARL-90-12/NASA-A^{3}I-90-1. Champaign, IL: University of Illinois.

Boff, K. R., & Lincoln, J. E. (1988). *Engineering data compendium: Human perception and performance.* Wright-Pattern AFB, OH: Harry G. Armstrong Aerospace Medical Research Laboratory.

Burns, C. M. (2000). Putting it all together: Improving display integration in ecological displays. *Human Factors, 42,* 226–241.

Chapanis, A., & Yoblick, D. A. (2001). Another test of sensor lines on control panels. *Ergonomics, 44,* 1302–1311.

Cho, Y. S., & Proctor, R. W. (2003). Stimulus and response representations underlying orthogonal stimulus-response compatibility effects. *Psychonomic Bulletin & Review, 10,* 45–73.

Donk, M. (1994). Human monitoring behavior in a multiple-instrument setting: Independent sampling, sequential sampling or arrangement-dependent sampling. *Acta Psychologica, 86,* 31–55.

Fitts, P. M., & Seeger, C. M. (1953). S-R compatibility: Spatial characteristics of stimulus and response codes. *Journal of Experimental Psychology, 46,* 199–210.

Francis, G. (2000). Designing multifunction displays: An optimization approach. *International Journal of Cognitive Ergonomics, 4,* 107–124.

Hommel, B., & Prinz, W. (Eds.). (1997). *Theoretical issues in stimulus-response compatibility.* Amsterdam: North-Holland.

Ivergård, T. (1989). *Handbook of control room design and ergonomics.* London: Taylor & Francis.

Learmount, D., & Norris, G. (1990). Lessons to be learned. *Flight International, 31 October–6 November,* 24–26.

Proctor, R. W. (2001). Stimulus-response compatibility. In W. Karwowski (Ed.), *International encyclopedia of ergonomics and human factors* (Vol. 1, pp. 486–489). London: Taylor and Francis.

Proctor, R. W., & Reeve, T. G. (Eds.). (1990). *Stimulus-response compatibility: An integrated perspective.* Amsterdam: North-Holland.

Proctor, R. W., & Van Zandt, T. (1994). *Human factors in simple and complex systems.* Boston: Allyn & Bacon.

Sengupta, A. K., & Das, B. (2000). Maximum reach envelope for the seated and standing male and female for industrial workstation design. *Ergonomics, 43,* 1390–1404.

Smith, W. J. (1996). *ISO and ANSI: Ergonomic standards for computer products.* Upper Saddle River, NJ: Prentice-Hall.

Vu, K.-P. L, & Proctor, R. W. (2003). Naïve and experienced judgments of stimulus-response compatibility: Implications for interface design. *Ergonomics, 46*, 169–187.

Wickens, C. D., Gordon, S. E., & Liu, Y. (1998). *An introduction to human factors engineering.* New York: Longman.

LIST OF OTHER RELEVANT STANDARDS AND GUIDELINES
WITH FULL REFERENCE

a. Aircrew Station Control Panels (NATO STANAG 3869 ED 1 AMD 4).

b. Construction and Industrial Equipment, Instrument Face Design and Location For (SAE J209).

c. Control Panel, Aircraft, General Requirements for (MIL-C-81774A-1).

d. Design Objectives for CRT Displays for Part 23 Aircraft (SAE ARP 4067).

e. Deutsches Institut für Normung: 666234: VDT Workstations, Part 3: Grouping and formatting of data.

f. Display, Head-Up, General Specification for (MIL-D-81641 NOTICE 2).

g. Earth-Moving Machinery—Instrumentation and Operator's Controls—Part 0: General Introduction and Listing (AS2956.0)

h. Earth-Moving Machinery—Instrumentation and Operator's Controls—Part 2: Operating Instrumentation (ISO 6011, AS 2956.2).

i. Earth-Moving Machinery—Instrumentation and Operator's Controls—Part 5: Zones of Comfort and Reach for Controls (ISO 6682, AS 2956.5).

j. Ergonomics Aspects of Indicating Devices; Types, Observation Tasks, Suitability (DIN 33413 PT1)

k. Flight Deck Panels, Controls, and Displays, Part 8: Flight Deck Head-up Displays (ANSI/SAE ARP 4102/8).

l. Human Engineering (MIL-STD-1472F).

m. Human Engineering Considerations in the Application of Color to Electronic Aircraft Displays (ANSI/SAE ARP 4032).

n. Instruments and Controls in Motor Truck Cabs, Location and Operation of (SAE J680).

o. Location and Direction of Motion of Operator's Controls for Agricultural Tractors and Self-Propelled Agricultural Machines (AS 1246).

p. Man-Machine Interface (MMI)—Actuating Principles (IEC 447, CENELEC EN 60447).

q. Man-Systems Integration Standards (NASA-STD-3000 REV B VOL II).

r. Operator Controls and Displays on Motorcycles, Recommended Practice (SAE J107).

s. Photometric Guidelines for Instrument Panel Displays that Accommodate Older Drivers, Information Report (SAE J2217).

t. Remote Control and Display Unit Design (ARINC 561-11 SEC 10.0).

u. Visual Display Requirements (BS EN 29241-3).

VI

Management of Occupational Safety and Health

24

The Benefits of Occupational Health and Safety Standards

Denis A. Coelho
João Carlos de Oliveira Matias
University of Beira Interior

INTRODUCTION

In this chapter, occupational health and safety is approached from the viewpoints of both human factors and ergonomics and standardization, including management systems standards. The field of Occupational Health and Safety (OHS) is summarily characterized, and a historical perspective of its evolution is given, in relationship to Human Factors and Ergonomics (HFE). Although several components of what makes up the field of OHS today preexisted as specialized fields before the foundation of the discipline of HFE (following World War II), and some of them were included in the genesis of HFE, OHS and HFE have developed as two distinct, although somewhat overlapping, areas of activity. Links and commonalities can be found today between the two. Environmental conditions (such as noise, climate, or lighting) is an example of an area that is dealt with within the discipline of HFE, and is also a central concern of OHS. Although inadequate postures and movements at work and work psychosocial factors impinge on OHS, HFE is equipped with knowledge and methods to perform the design of work and to deal with these work design factors. Regarding legislation, regulations, and standardization, OHS has gained from a head start in comparison to HFE (in the United States, the Occupational Safety and Health Act of 1970; at an international level, the ILO R164 recommendation of 1981; in the EU, the directive 89/391 on safety and health at work of 1989), which supports the inclusion of this discussion on OHS standards within the realm of HFE standards, given the intersecting interests of the two disciplines. Regarding economic considerations, some HFE interventions have been the object of cost–benefit analyses reported in the literature. In what concerns OHS, the literature has focused on the cost of occupational accidents, injuries, illnesses, and fatalities. Estimates of costs and benefits of complying to (or adopting) OHS standards are also available, in the form of literature dealing with the mandatory U.S. OSHA standards.

Central to the implementation of OHS is the assessment of hazards and risks and their resolution or reduction, besides the consideration of cost–benefit analyses, which may provide an important input in decision making regarding alternative risk control strategies and measures.

Despite the differing legal statuses of specific OHS standards (whether legal or voluntary), statistical overviews of reported occupational fatalities, injuries, and illnesses in the United States and the EU support the consideration of OHS as a paramount issue in organizations. This chapter focuses on OHS legislation, regulations, and voluntary standards in the EU and the United States, as well as on management systems. Great emphasis is given to OHSAS 18001 as a specification for an OHS management system. It is seen as an effective tool for guiding the design of a continuously improved OHS management system in an organization. Lessons learned from quality management systems reinforce the interest in a management system in the OHS area of an organization. Emphasis is also given to the advantages of integrating management systems within an organization, (quality, environmental, and OHS).

DEFINITION OF RELEVANT TERMS (PLUS ABBREVIATIONS)

BSI-OHSAS 18001:1999—Occupational Health and Safety Assessment Series: Occupational health and safety management systems—Specification, Incorporating Amendment No. 1, British Standards Institution, 13 December 2002
Hazard—(BSI-OHSAS 18001:1999) source or situation with a potential for harm in terms of human injury or ill health, damage to property, damage to the environment, or a combination of these.
HFE—Human Factors and Ergonomics
ILO—International Labour Organization (an agency of the United Nations)
Occupational Health and Safety—(BSI-OHSAS 18001:1999) conditions and factors that affect the well-being of employees, temporary workers, contractor personnel, visitors, and any other person in the workplace.
OHS—Occupational Health and Safety
Organization—(BSI-OHSAS 18001:1999) company, operation, firm, enterprise, institution or association, or part thereof, whether incorporated or not, public or private, that has its own functions and administration.
OSHA—Occupational Safety and Health Administration (Department of Labor—United States of America)
Risk—(BSI-OHSAS 18001:1999) combination of the likelihood and consequence(s) of a specified hazardous event occurring.
SMEs—small and medium enterprises
Standard—Procedural normative document intended for wide dissemination and implementation that may have a voluntary nature of implementation in organizations (this is the typical use of the word in Europe), but may also have a compulsory nature of implementation (such is the case for mandatory OSHA standards in the United States)

OCCUPATIONAL HEALTH AND SAFETY

OHS is a field of activity in organizations where several professional specialties collaborate. Safety engineering, occupational medicine, and ergonomics are principal specialties involved in the field of OHS. Assuring health and safety in the workplace is typically a joint effort of a number of professional activities. The purpose of OHS may be seen as assuring the good health and safety of people at work, thus preventing fatalities, injuries, and illnesses brought about through exposure to work-related health and safety hazards. Assuring OHS is accomplished by creating a controlled work environment. In the words of the EU Commission's (2002) strategy on health and safety at work for the period from 2002 to 2006, this entails developing an approach that is both global and preventive, geared to promoting well-being at work, and going beyond the mere prevention of specific risks.

Evolutional Perspective of Occupational Health and Safety

Work hazards to safety and health have been acknowledged throughout history. Loss of hearing by workers processing stone, bronze, and iron was already recorded 2,000 years ago in the Roman Empire (Tytyk, 2004), but it was only during modern times, subsequently to the processes of industrialization in Europe and the United States, that occupational health and safety concerns began to be dealt with systematically. Although ergonomists and human factor specialists only came onto the industrial scene after World War II, industrialization was to bring the advent of a set of professions whose activities would foster improvements to the health and safety conditions of workers. The principles of work organization, introduced by Frederic W. Taylor (1856–1915) in American Industry, promoted the division of industrial labor into monotonous tasks with repetitive movements, alienating the operators from their work (Saha, 1998). At that time, factory owners often viewed workers as unreliable, not efficient enough, and requiring supervision and force; whereas children, along with women and men, labored long hours in unsafe, unhygienic conditions (Tytyk). A number of new industrial professions developed concurrently, as a response to the need for productivity improvements, and thus, profit gains. The professions of work hygienist, illumination technician, and occupational physician were some of the novel specialties that appeared in the course of industrialization. The activity of these professionals promoted improvements to the otherwise appalling work safety and health conditions of factory workers.

At an international level, consciousness of the importance of preventing occupational injuries and illnesses led to action starting in the early 20th century. Since 1919, the International Labour Conference has been issuing, for ratification by member countries, a set of conventions and recommendations concerning occupational safety and health and the working environment. Starting with a fragmented, patchy, coverage of specific hazards in selected sectors of activity (e.g., white lead in painting—1921; marking of weight for packages transported by vessels—1929; or safety provisions in building—1937), the scope of the instruments issued gradually enlarged. The occupational safety and health recommendation of 1981 (ILO R164: 1981) encourages a systematic approach to prevention, extended to all branches of economic activity and all categories of workers. Currently, the International Labour Organization (ILO, the first specialized agency of the United Nations, established in 1946) collaborates with the International Ergonomics Association (IEA) in developing the forthcoming ILO instrument on ergonomics and the prevention of musculoskeletal disorders. The two institutions are also working together in developing the IEA/ILO publications "Checkpoints on ergonomics" on a cross-section of industry application areas (Caple, 2004). This ongoing collaboration provides evidence of widespread recognition of the fact that effective application of ergonomics in work design, while promoting a balance between worker characteristics and task demands, provides worker safety and physical and mental well-being.

Relationship Between Human Factors and Ergonomics and Occupational Health and Safety

As a scientific discipline and a professional specialty, HFE is somewhat younger than other disciplines involved in OHS. This notwithstanding, HFE professionals are especially requested to deal with specific OHS problems of the working environment. In an article about changes in OHS education (Dijk, 1995), coming from an occupational medicine background, mentions safety experts, occupational physicians, occupational nurses, occupational hygienists, work and organization specialists and occupational physiotherapists as the professionals collaborating, at that time in the Netherlands, in the activities of providing OHS services to organizations. Ergonomists and work psychologists were expected to shortly join the aforementioned professionals in applying their selves to OHS problems, particularly those dealing with work-related disorders like musculoskeletal disorders and mental breakdown, burnout, and the delayed consequences of psychic trauma as an effect of aggression or depressing events at work. Assertively, in a review of occupational risk management, Zimolong (1997) acknowledged that workplace design that does not take ergonomic principles into account is likely to lead to an increase in errors and accidents.

In a report about OHS in Spain, Sesé et al. (2002) consider a number of OHS aspects of the workplace that are used to characterize the working environment: physical exposure (noise, vibration, and thermal stress); postures and movements (lifting and handling heavy objects, repetitive movements, and pain or tiring postures); exposure to chemical substances, carcinogens, neurotoxic chemicals, and infectious biological agents. They also considered psychosocial factors (e.g., work rate, violence at work, or monotonous work), as well as accidents and occupational diseases (produced by chemical agents, skin diseases produced by substances and chemical agents, produced by inhalation of substances, infectious and parasitic occupational diseases, produced by physical agents and systemic diseases). A number of these categories are HFE themes and objects of study, with an array of ergonomic methods of intervention and design applicable in these areas. These include, from the OHS categories listed, the physical environment, postures and movements, and psychosocial factors. Interestingly, standards concerned with the physical environment have generally not been produced under the heading of ergonomics. This has often led to a technical or engineering approach, where standards have concentrated on the physical aspects of the environment with little detail concerning human responses or ergonomics methods (Parsons, 1995). In this regard, it is important to emphasize that HFE is different from most bodies of knowledge that have supported the discipline. While the primary purpose of anthropology, cognitive science, psychology and sociology is to understand and model human behavior, the main purpose of ergonomics is design (Helander, 1997). Still, research in ergonomics can be both applied and basic, because it is often also necessary to engage in basic research to seek support for design activities, and in doing so, ergonomics has created a sizable body of knowledge, one that is constantly being enlarged. The involvement of HFE in assuring OHS at the workplace, whether at the design stage or in postdesign interventions, is hence appropriate and desirable given the relevance of the knowledge and methods of HFE to many important dimensions of OHS.

Within organizations, opportunities for collaboration between OHS and HFE may be many. Given the closeness of the two disciplines and their overlapping activities, the links between them can serve as the basis on which to build synergistic operational programs within organizations. In a study discussing ergonomic design principles and programs in terms of a loss management viewpoint, Amell, Kumar, and Rosser (2001) show how an Occupational Injury and Illness surveillance program may be used both in determining whether ergonomic intervention is required, and as a means of evaluating the intervention. Notwithstanding a preventive approach to ergonomics and to OHS, the approach Amell et al. present is admittedly suited to workplaces and work activity where ergonomic principles were not employed in the design stage.

Economics of Interventions

The economic worth of HFE and OHS interventions has been the object of analysis. Beevis and Slade (1970) identified a dozen cases where ergonomics benefits had been expressed in financial terms, and had resulted in cost savings. The authors stated that the examples available at the time were scarce, and that few of those published included economic data. In their opinion, this reflected the difficulty of converting the usual ergonomic criteria of human performance into costs, for example, when assigning a monetary value to a reduction in the incidence of mistakes or accidents. Publishing on the same theme 33 years later, Beevis (2003) verifies that assigning costs and benefits to ergonomics applications remains very arduous. Nevertheless, ergonomists should be able to discuss potential benefits of ergonomics applications with clients or project managers. To that end, Beevis (2003) points out that, when ergonomics is considered as part of a risk-reduction strategy, it may be possible to estimate the reduction in the risks of costs, or to estimate the reduction of the risk of loss of anticipated benefits. In the process–safety field, cost–benefit analysis of risk-reduction systems is done by examining the reduction in expected annual loss from the implementation of those systems and comparing this with the annual cost of implementing such systems to determine whether the cost is justified (Antes, Miri, & Flamberg, 2001). This approach to cost–benefit analysis is quantitative and as such does not take into account intangible factors, such as the public and personnel perception of alternative risk-reduction systems (Antes et al., 2001). Moreover, the result of any cost–benefit analysis is only as accurate as the assumptions made to determine the costs of implementation and the anticipated benefits. In establishing the economic case for HFE or OHS interventions, estimating the economic value of the anticipated benefits is bound to be more troublesome than estimating the cost of implementing the applications or recommendations.

In assessing costs and benefits of human factors, Rouse and Boff (1997) consider the variation in aspects such as human performance, accidents, or absenteeism as determinants of the likelihood of benefits resulting from improvements in health and safety at the workplace. Translating those determinants into economic terms, in the context of a specific organization, could be a daunting task, especially if overall processes of the organization have not been characterized and work tasks are not standardized. A literature search on the theme of costs and benefits in respect to OHS measures returned a few studies which focused, however, on the cost of occupational accidents, injuries, illnesses, and fatalities (Dembe, 2001; Leigh, Cone, & Harrisson, 2001; Rikhardsson & Impgaard, 2004; Weil, 2001). In these studies, the analysis of those costs (to society at large, to companies, or to individuals) is used in establishing a case for implementation of preventive measures to safeguard health and safety at the workplace. The existence of occupational fatalities, injuries, and illnesses also raises moral questions, especially when the former are viewed as an element of cost–benefit analyses. Another study (Seong & Mendeloff, 2004) deals with the methods used to estimate the benefits of U.S. OSHA (mandatory) safety standards. The controversial OSHA Ergonomic Program Standard of the United States (it was several years in the making and generated controversy at almost every step along the way, being overturned by the U.S. Congress in 2001), is the theme of an article (Biddle & Roberts, 2004) dealing with the cost effectiveness of that program. In these two last studies, the starting point is the comparison of the costs incurred by employers at large in implementing the standards (at nationwide level in the United States) to cost savings and prevention of injuries and fatalities brought about by the implementation. The values presented in both studies are based on OSHA estimates and the methods used to arrive at those estimates are also analyzed. In preparation for the promulgation of a new mandatory occupational safety and health standard, OSHA seeks information on the costs and benefits of various potential preventive strategies intended to face the particular hazards involved in the standard (U.S.-DOL, Fact Sheet No. OSHA 92-14). Although establishing the cost effectiveness of ergonomic and OHS mandatory standards is not a straightforward task, in the U.S. endeavors are carried out to that end by taking nation-wide statistical values into account to produce cost and benefit estimates for OSHA standards. These estimates corroborate the cost effectiveness at a macroeconomic level of the regulations intended for compulsory implementation. Nevertheless, in the United States, OSHA's estimates of the benefits of new mandatory standards are the object of debate because alternative methods and sources of compiled data can be used to establish those projections.

Large-scale losses in organizations, such as those arising from major fires or explosions, or otherwise involving loss of life, make the accidents immediately behind those losses very visible. The losses accruing from these accidents may, accordingly, be very thoroughly costed. Less well understood, however, is the nature and extent of loss from accidents of a more routine nature. Most accidents in industry typically injure but do not kill people, while damaging and interrupting production processes (EU Commission, 2004). Costs of this type of accident can often be extended to sick pay, increased insurance premiums, or maintenance budgets. Few organizations have the mechanisms to identify these extended costs separately and fewer still actually identify and examine the costs of accidents systematically (Rikhardsson & Impgaard, 2004). Many employers mistakenly believe that they are covered by insurance for most of the costs arising from accidents, but often uninsured costs far exceed insured costs (EU Commission, 2004). Additionally, there are also intangible costs due, for example, to loss of business image, customer satisfaction, employee morale, or goodwill (Antes, Miri, & Flamberg, 2001). These intangibles are hard, if not impossible, to quantify in financial terms. For these reasons, accident costs are usually incomplete. It may thus be difficult to demonstrate the financial benefits of accident reduction measures, in some cases.

Risks to Occupational Health and Safety

A central goal in the implementation of OHS is the resolution or reduction of risks, which should follow a thorough identification of workplace hazards and the assessment of their degree of risk. Besides being hazards at the workplace, some events and situations may also represent threats to property integrity or have a potential to cause environmental damage. Hence, safety engineering is another discipline with intersecting activities in OHS, besides HFE. The hazards with which OHS is primarily concerned are those that can directly affect the well-being (health- and safetywise) of people at work. A hazardous event or situation can have negative consequences to people at work, including illnesses (whether acute or chronic) and accidents causing physical injury (nonfatal or fatal), and may also have psychological impacts on people (e.g. anxiety, stress, or depression). An array of hazards can be present in the workplace. A provisional categorization of hazards to health and safety at work is shown in Table 24.1. Risks may accrue

TABLE 24.1
Provisional Categorization of Workplace Hazards to Health and Safety

Type of Hazard	Hazards Falling in this Category
Dangerous Substances and Agents	Exposure to chemicals, carcinogens, toxins and infectious and parasitic biological agents
Physical Exposures	Noise, vibration, radiation and inadequate illumination and thermal environment
	Hazards associated to other (physical) ergonomic factors, such as working postures, workload, or materials handling
Integrity Threats	Fire, explosion, and reactivity hazards
	Threats to physical integrity in the form of hazards that can result in cuts, falls, projections, crushing, asphyxiating, or electrocution
Psychosocial Factors	Harassment and violence at work
	Inadequate job design in terms of work rate, job demands and content, repetitive work, mental workload, or interacting technology

differently from exposure to single hazards than from the interaction of hazards (e.g., interaction among dangerous substances and agents, or between these and physical exposures, as well as interactions between inadequately set factors in the working environment, whether of a physical or psychosocial nature). Moreover, many of the hazards are not directly perceivable and their negative consequences have to be inferred from knowledge, personal experience or reliance on documentation (Zimolong, 1997). Preventive action is hence necessary, and organizations should handle both obvious and nonobvious hazards systematically and in a continuous manner. In planning for ongoing activities of hazard identification, risk assessment, and risk resolution or reduction (control), organizations should establish and maintain procedures pertaining to activities and facilities, which include, according to the requirements set out in the OHSAS 18001 specification (BSI-OHSAS 18001:1999):

- Routine and nonroutine activities
- Activities of all personnel having access to the workplace (including contractors and visitors)
- Facilities at the workplace, whether provided by the organization or others

Procedures provide a description that can be used as an external memory in guiding activities according to principles safeguarding OHS, and as such are an instrument for the management of risk. In general, measures taken within organizations for the management of risk should follow the principle of elimination of hazards where practicable, followed in turn by risk reduction (either by reducing the likelihood of occurrence or the potential severity of injury or damage), with the adoption of personal protective equipment (PPE) only as a last resort (BSI-OHSAS 18002:2002). The magnitude of risks can be assessed using the definition of risk presented by Rodrigues and Guedes (2003). It estimates risk by considering risk as the combination of the probability of a hazardous event or situation occurring and its severity in terms of seriousness of consequences. Defined in this way, risk is directly proportional to both probability and severity. The greater the probability of occurrence of a hazardous event or situation, or the greater the severity of the consequences, the greater the risk. Likewise, the smaller the probability or the severity, the smaller the risk. Once an organization is in a position to appreciate all significant OHS hazards and to estimate their probability and severity, risks can then be classified, permitting identification of those that must be eliminated as well as those that should be controlled by specific measures within the organization. According to Antes et al. (2001), organizations have many different methods for analyzing hazards and risks, with the most common being what-if analysis, checklist reviews, hazard and operability (HAZOP) procedures, and failure modes and effects analysis (FMEA). Once risks are identified and categorized, the organization must decide which preventive measures or mitigation strategies to implement. Cost–benefit analysis of alternative risk control measures can help organizations to select the most effective means of risk resolution or reduction.

Risk control measures in OHS entail changes to the workplace and the work environment. These changes typically have involved setting up signs, disseminating guidance materials, providing incentives, setting performance goals, establishing training programs, intervening in work design and performing ergonomic (re)design of workplaces and work environments. Control of new and emerging OHS risks of a psychosocial nature, linked to social change, may necessitate intervention in other spheres, namely, human relations, both within organizations and in a broader context. Emerging disorders such as stress, depression, anxiety, violence at work, harassment, and intimidation are becoming increasingly prominent in Europe in education, health, and social services (EU Commission, 2002). These disorders cannot be linked solely to exposure to a single particular hazard, but are likely to result from the interaction of a

TABLE 24.2

Overview of Occupational Accident and Injury and Health Problems and Illness Statistics
for the EU and the United States

EU (15 Countries, 2000)[a]	USA (2002)[b]
Total number of employees covered in data	
142.2 million	137.7 million
Total accident-related deaths at work/Fatal work injury count	
5,237	5,525
Occupational accidents leading to more than three days absence from work: 4.8 million	Work-related injuries and illnesses that resulted in days away from work: 1.44 million
Total number of accidents, including those which did not involve absence from work: 7.6 million	Total nonfatal work-related injuries: 4.4 million
Total work-related health problems: 7.7 million	Total nonfatal work-related illnesses: 0.29 million

[a] *Source:* Eurostat, European Social Statistics (EU-15 countries: Austria, Belgium, Denmark, Finland, France, Germany, Greece, Ireland, Italy, Luxembourg, Netherlands, Portugal, Spain, Sweden, and United Kingdom).

[b] *Source:* U.S. Department of Labor, Bureau of Labor Statistics (the figures shown for work-related injuries and illnesses only concern private industry).

wide set of factors, including the degree of acceptance of human diversity within organizations, hierarchical relations, work design, working time arrangements, and commuter-related fatigue.

The Need for Occupational Health and Safety

OHS has progressed significantly in recent decades. The U.S. Occupational Safety and Health Administration stated that it has, since its creation in 1970, driven the work-related fatality rate down by 62% and promoted a reduction in overall injury and illness rates of 42% (OSHA, 2003). An estimated current annual cost of 170 billion U.S. dollars for occupational injuries and illnesses (OSHA, 2003) is irrefutable evidence that occupational accidents and work-related health problems are expensive in modern society. According to Dupré (2001), Eurostat estimated that every year, occupational accidents in the EU resulted in 150 million lost working days, and that a further 350 million days were lost because of work-related health problems (these figures are from 1998 and 1999 and a 15 country EU). In economic terms, this means a projected 2.6% to 3.8% of the collective EU gross national production (GNP) is lost every year (EU OSHA, 2001). Statistical data on occupational accidents and illnesses in the United States (for the year 2002) and the EU (for the year 2000), shown in Table 24.2, provides insight into the dimension of the problem.

More than 5,000 people lose their lives at work every year in the EU, and in the United States, fatalities directly related to work have a similar magnitude. In addition, OSHA (2003) estimates that perhaps as many as 50,000 workers die in the United States each year from illnesses in which workplace exposures are a contributing factor. In the EU, statistics indicate that every year more than 75,000 people are so severely disabled that they can no longer work (Saari, 2001). Injury insurance systems are designed to protect those injured and their dependents, but the costs of work-related accidents and illnesses extend not only to those directly affected (financially and emotionally) but also to employers and organizations (lost productivity and negative impact on the public's perception) and to countries, given their significant impact on the economy. Ultimately, these costs fall on citizens at large, as taxpayers and consumers.

Countries, organizations, and employees have a myriad of tools at hand for promoting health and safety at the workplace. Some of these are laws and regulations, which are mandatory and enforced by government agencies. Methods and standards that can guide organizations in complying to legislation, or in attaining performance levels in OHS above minimum legal requirements, including OHS management systems geared to continuous improvement, are also available. The sections of this chapter that follow focus on existing standards (both mandatory and voluntary) in the area of health and safety in the workplace in the United States and the European Union. The OHSAS 18001 specification, supporting the implementation of a system for the management of OHS within organizations, is also discussed.

OCCUPATIONAL HEALTH AND SAFETY LEGISLATION, REGULATIONS AND VOLUNTARY STANDARDS

Normative documents in OHS may be divided into three main types. Those that are legislative or regulatory are mandatory; there are standards that are intended for voluntary implementation by organizations, and internal procedures developed within organizations. In this section, legislation and voluntary standards will be dealt with in more detail than internal procedures. Still, the latter warrant some attention as possible precursors to standards, especially those developed within organizations of a significant size and which encourage prevention as the suitable approach to OHS. Benchmarking of best in class organizations may be at the origin of the diffusion of some internal procedures to other organizations, depending on the results evidenced by their original application and their usefulness. Internal OHS procedures may assume great importance in those cases where they become adopted as voluntary standards, although in a somewhat modified version. This version is usually prepared by a specialized technical committee within a standards publishing organization (i.e., European standards in ergonomics are produced by working groups of CEN TC 122 "Ergonomics"). The creation of voluntary standards may also take place in reaction to specific legislation, providing guidance in interpreting and implementing it. Mandatory standards (legislative and regulatory) in the EU and in the United States, standards for voluntary implementation and OHSAS 18001 constitute the focus of this section.

Legislation and Regulations

The principle underlying the development of legislation and regulations in OHS is that all workers are entitled to work in healthy, safe and hygienic conditions. Mandatory standards and legislation are developed to guarantee these rights, which assist each and every worker. The range of the success and benefits of implementation of mandatory standards and legislation may depend however on the thoroughness of inspection activities. Inspection can work as a motivation for the implementation of the minimum requirements set forth in legislation and regulations, but to that end it ought to be effective not only at the initial licensing stage of businesses, but also in continuity. Inspection can be decisive in reaching the objectives underlying the development of laws and regulations in OHS. If employers expect that they will not be inspected, while not having rooted a culture of health and safety at work in their organization, they may be tempted to neglect OHS regulatory or legislative requirements. Health and safety at the workplace are also deeply rooted in organizations' culture. In this regard and in the EU, it is recognized that inspectors have a crucial role to play as agents of change to promote better compliance, first through education, persuasion, and encouragement and through increase in enforcement activities, where necessary (EU Commission, 2004).

Employers must apply all the minimum mandatory requirements to satisfy their obligations of assuring basic health and safety conditions to their employees. Employers have the duty to evaluate and control risks, in terms of not only safety but also regarding workers, health. Therefore, it owes to them to adopt measures, to give instructions, to inform the workers or their representatives and to organize health and safety activities at work. Employees have the right to communicate to labor inspectors situations or conditions where minimum OHS requirements have not been met. Besides their own rights, legal and regulatory provisions on OHS enunciate that employees also have duties, which include executing the instructions and recommendations of their employer, in addition to demonstrating a collaborative attitude in assuring OHS conditions.

Legislation and regulatory mandatory standards in the occupational health and safety domain deal with a wide range of areas. The aspects covered include the organization of prevention services, the organization of work, the protection of more vulnerable worker groups, the prevention of exposure to specific hazards, the prevention of industrial accidents, fire protection, and industry-specific preventive measures. In what follows, OHS legislation and regulation in the EU and the United States are presented seperately. For the EU, the focus will be drawn on the EU directives on OHS, which are disseminated and adopted in the internal law of the EU member states. For the United States, the role of OSHA in developing mandatory standards promulgated by the US Congress will be presented.

Legislation on Safety and Health at Work in the European Union

Prevention is the guiding principle for occupational health and safety legislation in the EU. According to a communication from the EU Commission (2004), the goal of instilling a culture of prevention rests on the double foundation that the minimum requirements provide a level playing field for businesses operating within the large European domestic market and provide a high degree of protection to workers, avoiding pain and suffering and minimizing the income foregone for enterprises as a result of preventing occupational accidents and diseases. Hence, in order to avoid accidents and occupational diseases, EU wide minimum requirements for health and safety protection at the workplace have been adopted. In the member states of the EU, OHS legislation comes from two different sources. One is the legislation developed by each member state in its internal law. The other is the obligatory adoption (transposition) of EU directives to the internal law of every member state. Because the legislative systems covering safety and health at the workplace differed widely among the individual member states, and needed improvement, the Council of the European Union adopted, by means of directives, minimum requirements for encouraging improvements to guarantee the protection of the health and safety of workers (e.g., directives 89/391/EEC and 94/33/EC). The adoption of these directives was not intended to permit any reduction in levels of the protection already achieved in individual member states prior to the transposition of the directives. They aimed, rather, at harmonizing national provisions on the subject, which often included technical specifications and self-regulatory standards, resulting in different levels of safety and health protection and permitting competition at the expense of safety and health (directive 89/391/EEC). Hence, EU countries are committed to encouraging improvements in OHS conditions and to harmonizing OHS conditions across member states.

Directive 89/391/EEC, on the introduction of measures to encourage improvements in OHS, is the European OHS framework directive. As such, it serves as a basis for other (individual) directives, which together with the framework directive form a coherent whole. Directive 89/391/EEC lays down the principles for the introduction of measures to encourage improvements in the safety and health of workers and provides a framework for specific workplace environments, developed in individual directives. The OHS framework directive and the individual directives that spring from it are listed in Table 24.3. Besides the framework directive and its individual directives, there are still other complementary directives in the OHS domain, of which examples are given in Table 24.4.

The transposition of framework directive 89/391/EEC, and of the individual directives on OHS to the internal law of EU member states, has led to the harmonization of the legislative framework in OHS across the EU. The shift of paradigm represented by the EU health and safety legislation (which envisaged moving beyond a technology-driven approach to accident, injury, and illness prevention to a policy much more focused on both the personal behavior and organizational structures) is recognized as having had a major impact in the national OHS systems (EU Commission, 2004). The impact has been greater in those countries that had either less developed legislation in the field or legislation based on corrective principles rather than on a preventive approach to assuring OHS (EU Commission). For the Nordic EU member states (Denmark, Finland, and Sweden), transposition did not require major adjustments because they had already rules in place that were in line with the OHS EU directives. In Austria, Belgium, France, Germany, the Netherlands, and the United Kingdom, the framework directive served to complete or refine existing national legislation. In the EU-15 countries of the South (Greece, Italy, Portugal, and Spain) as well as in Ireland and Luxembourg, the framework directive had considerable legal consequences, because prior to its adoption these countries possessed antiquated or inadequate legislation on the subject. Enlargement in the year 2004 has brought in new member states to what is now a EU comprising 25 countries. In many of the 10 new member states the prevention culture has yet to be rooted (EU Commission). Transposition of the EU directives on OHS to their internal law will constitute a decisive step toward instilling a generalized preventive approach to OHS. Member states that already benefited from the EU legislation to modernize their occupational health and safety rules emphasize the following innovative aspects of the framework directive (EU Commission):

- The broad scope of application, including the public sector
- The principle of objective responsibility of the employer
- The requirement that a risk assessment shall be drawn up and documented
- The obligation to establish a prevention plan based on the results of the risk assessment
- The recourse to prevention and protective services
- Workers' rights to information, consultation, participation, and training

Some member states also report on difficulties in enforcing the legislation adopted as a result of transposing the EU directives. Although framework directive 89/391 recognizes the need to avoid imposing administrative, financial and legal constraints, which would hold back the creation and development of SMEs, problems in implementing this legislation in SMEs in practice have been reported (EU Commission, 2004). These problems are ascribed to specific administrative obligations, formalities, and financial burdens, as well as the time required to develop appropriate measures, which gave rise to a certain negative reaction from SMEs, especially in Belgium, Denmark, Germany, the Netherlands, Sweden, and the United Kingdom.

TABLE 24.3

EU Framework Directive on Occupational Health and Safety and Its Individual Directives

Content	Directive
Framework: Introduction of measures to encourage improvements in the safety and health of workers at work	89/391/EEC
Workplaces: Minimum health and safety requirements for the workplace	89/654/EEC
Use of work equipment: Minimum safety and health requirements for the use of work equipment by workers at work	89/655/EEC[a]
Use of personal protective equipment: Minimum health and safety requirements for the use by workers of personal protective equipment at the workplace	89/656/EEC[b]
Manual handling: Minimum health and safety requirements for the manual handling of loads where there is a risk particularly of back injury to workers	90/269/EEC
Work with screen display equipment: Minimum safety and health requirements for work with screen display equipment	90/270/EEC
Carcinogens: Protection of workers from the risks related to exposure to carcinogens at work	90/394/EEC[c]
Biological agents: Protection of workers from risks related to exposure to biological agents	2000/54/EC
Safety signs: Minimum requirements for the provision of safety and health signs at work	92/58/EEC
Pregnant workers: Introduction of measures to encourage improvements in the safety and health at work of pregnant workers and workers who have recently given birth or are breastfeeding	92/85/EEC
Mineral-extracting industries (drilling): Minimum requirements for improving the safety and health protection of workers in the mineral-extracting industries through drilling	92/91/EEC
Mineral-extracting industries: Minimum requirements for improving safety and health protection of workers in surface and underground mineral-extracting industries	92/104/EEC
Fishing vessels: Minimum safety and health requirements for work on board fishing vessels	93/103/EC
Chemical agents: Protection of the health and safety of workers from the risks related to chemical agents at work	98/24/EC
Explosive atmospheres: Minimum requirements for improving the safety and health protection of workers potentially at risk from explosive atmospheres	99/92/EC
Physical agents—vibration: Minimum health and safety requirements regarding the exposure of workers to the risks arising from physical agents (vibration)	2002/44/EC
Physical agents—noise: Minimum health and safety requirements regarding the exposure of workers to the risks arising from physical agents (noise)	2003/10/EC
Temporary or mobile construction sites: Implementation of minimum health and safety requirements at temporary mobile or construction sites	92/57/EEC
Transport activities: Minimum safety and health requirements for transport activities and workplaces on means of transport	_[d]

[a] Modified by directives 95/63/EC and 2001/45/EC.
[b] Amended by directives 93/68/EEC, 93/95/EEC, and 96/58/EC.
[c] Modified by directive 97/42/EC and amended by directive 99/38/EC.
[d] This is not a directive but an amended proposal for a council decision.

The overriding objective of framework directive 89/391 and that of its individual directives is achieving a high level of protection of the safety and health of workers. This goal can only be met if all actors involved (employers, workers, workers' representatives, and national enforcement authorities) cooperate to pursue the efforts that are necessary to attain a comprehensive, effective, and correct application of the content of the EU directives on OHS. EU directives

TABLE 24.4

Examples of Other EU Directives in the Domain of OHS (the Directives Shown Here Are Not Individual Directives of Framework Directive 89/391/EEC)

Content	Directive
Temporary workers: Supplements the measures to encourage improvements in the safety and health at work of workers with a fixed-duration employment relationship or a temporary employment relationship	91/383/EEC
Medical treatment on board vessels: Minimum safety and health requirements for improved medical treatment on board vessels	92/29/EEC
Young People: Protection of young people at work	94/33/EC
Electrical equipment for use in potentially explosive atmosphere in mines susceptible to firedamp	98/65/EC

also acknowledge that technical standards constitute an indispensable reference in adopting procedures and measures to meet legal requirements in OHS.

Mandatory Standards on Safety and Health at Work in the United States

In the United States, the Occupational Safety and Health Act of 1970 authorizes the Secretary of Labor, through the Occupational Safety and Health Administration (OSHA), to set mandatory occupational safety and health standards applicable to businesses affecting interstate commerce (US-DOL, Fact Sheet No. OSHA 92-14). In its Web site (http://www. osha.org), OSHA states that its mission is to assure the safety and health of American workers by setting and enforcing standards; providing training, outreach, and education; establishing partnerships; and encouraging continual improvement in workplace safety and health. OSHA regulations are mandatory standards designed to reduce on-the-job injuries and limit workers' risk of developing occupational diseases. According to Fact Sheet No. OSHA 92-14 of the U.S. Department of Labor, the impetus to develop a new standard can come from several sources, including public petitions, the U.S. Congress, information from governmental departments, referral from the Environmental Protection Agency, OSHA's own initiative or requests from OSHA advisory committees. In preparing a new standard, OSHA seeks information to determine the extent of a particular hazard or group of hazards, investigates currently used and potential protective measures, and estimates costs and benefits of various protective strategies. OSHA uses several sources of information, including surveys, meetings with employers and employer groups, and focus group discussion with workers from many plants and industries across the United States. OSHA also consults with the interest groups potentially affected by an envisaged new standard, including industry and labor, which meet to hammer out agreements serving as the basis for a proposed rule.

Individual states in the USA are encouraged to establish and maintain their own job safety and health programs, but only if they are "at least as effective" as federal standards, and subject to federal approval. States are also allowed to develop standards covering areas or issues not regulated by OSHA's (federal) standards, but only if they are "required by compelling local conditions" and will not "unduly burden interstate commerce" (US-DOL, Fact Sheet No. OSHA 92-14). Interestingly, during the first 2 years following its creation, OSHA was authorized to promulgate national consensus standards and other federal standards as OSHA standards. National consensus standards came from voluntary standards developed by such organizations

TABLE 24.5

Subparts of OSHA's General Industry Standards, Excluding Reserved Subparts
(Standards-29 CFR—Part 1910—Occupational Safety and Health Standards)

Subpart	Heading
A	General
B	Adoption and Extension of Established Federal Standards
D	Walking—Working Surfaces
E	Means of Egress
F	Powered Platforms, Manlifts, and Vehicle-Mounted Work Platforms
G	Occupational Health and Environmental Control
H	Hazardous Materials
I	Personal Protective Equipment
J	General Environmental Controls
K	Medical and First Aid
L	Fire Protection
M	Compressed Gas and Compressed Air Equipment
N	Materials Handling and Storage
O	Machinery and Machine Guarding
P	Hand Portable Powered Tools and Other Hand-Held Equipment
Q	Welding, Cutting, and Brazing
R	Special Industries
S	Electrical
T	Commercial Diving Operations
Z	Toxic and Hazardous Substances

TABLE 24.6

Some Industry-Specific OSHA Standards

Title	Standards—29 CFR
Construction: Safety and Health Regulations for Construction	Part 1926
Agriculture: Occupational Safety and Health Standards for Agriculture	Part 1928
Maritime Sectors:	
Occupational Safety and Health Standards for Shipyard Employment	Part 1915
Marine Terminals	Part 1917
Safety and Health Regulations for Longshoring	Part 1918

as the American National Standards Institute and the National Fire Protection Association (US-DOL, Fact Sheet No. OSHA 92-14).

Most OSHA regulations (Standards-29 CFR) cover hazards that may be present in a wide set of industries. These standards are compiled as the OSHA General Industry Standards (Standards-29 CFR–part 1910–Occupational Safety and Health Standards). The nonreserved subparts of these standards are listed in Table 24.5. OSHA has also promulgated other standards in OHS, some of which apply to a single industry, of which examples are shown in Table 24.6. Besides Part 1910 and the industry specific parts shown on Table 24.6, OSHA has produced a number of other regulations, which are relevant to its activity. Examples are shown on Table 24.7. OSHA is not limited to creating and enforcing standards. Because the agency was allegedly burdening employers with rules, inspections, and penalties, it decided to adopt an

TABLE 24.7

Other OSHA Regulations (Standards—29 CFR)

Title	Standards—29 CFR
State Plans for the Development and Enforcement of State Standards	Part 1902
Inspections, Citations, and Proposed Penalties	Part 1903
Recording and Reporting Occupational Injuries and Illness	Part 1904
Rules of Procedure for Promulgating, Modifying or Revoking OSHA Standards Standards	Part 1911
Advisory Committees on Standards	Part 1912
Safety Standards Applicable to Workshops and Rehabilitation Facilities	Part 1924
Safety and Health Standards for Federal Service Contracts	Part 1925
Discrimination against Employees under OSHA Act of 1970	Part 1977
Identification, Classification, and Regulation of Carcinogens	Part 1990

alternative approach that includes providing onsite consultative assistance to employers who want help in establishing OHS conditions, thus offering a choice to employers. According to OSHA (2003), voluntary, cooperative relationships among employers, employees, unions, and OSHA can be a useful alternative to traditional OSHA enforcement and an effective way to reduce worker deaths, injuries, and illnesses. OSHA also provides guidelines and other voluntary standards to assist in implementing regulations and improving OHS conditions beyond minimum regulatory requirements.

Voluntary Standards in Occupational Health and Safety

Voluntary standards play a crucial role in the adoption and implementation of procedures and measures required by applicable legislation and regulations in the OHS domain. Budgetary limitations and consequent resource and staff shortages are at the root of the difficulty the national organisms responsible for the production of laws and rulings have to accompany scientific and technological development. Although adoption of voluntary standards in OHS does not automatically confer immunity from legal obligations, some voluntary standards serve as guidelines for interpretation and consolidation of specific OHS regulations or legislation. This leads to a complementarity between legislation and regulations and voluntary standards in several activities. Voluntary standards are sound technical solutions to safety and health problems without creating additional cost and operations burdens to governments (McCabe, 2004). Legislators and regulators impel voluntary safety and health standards because they may also be introduced more easily and more quickly than laws and regulations. In spite of the voluntary adhesion to this kind of standards, these standards are procedures, rules and measures, which are recognized and accredited by specialized organisms. Voluntary standards are accredited by national and international organisms, such as ANSI (American National Standards Institute), AFNOR (Association Française de Normalisation), BSI (British Standards Institute), DIN (Deutsches Institut für Normung), ISO (International Standards Organization), or CEN (European Committee for Standardization). These organizations, and other organizations of the same kind acting in other countries and regions of the world, develop standards internally and also adopt standards with a national, foreign, or international origin. There are many voluntary standards applicable to OHS. Coverage spans from personal protective equipment to fire and explosion, safety of machinery, evaluation methodologies (vibration, acoustics, and noise), signals and safety signs, limit-values of exposure, and management of OHS within organizations. In what follows, the development and accreditation of voluntary standards in

TABLE 24.8

Examples of ANSI-Accredited Standards in the Domain of OHS, With an Indication of Their
Original Developer

Standard	Developer
ANSI Z87.1-2003—Occupational and Educational Personal Eye and Face Protection Devices	ASSE—American Society of Safety Engineers
ANSI N13.15-1985—Dosimetry Systems, Performance of Personnel Thermoluminescence	HPS—Health Physics Society
NFPA 70E-2004—Standard for Electrical Safety Requirements for Employee Workplaces	NFPA—National Fire Protection Association
UL 2351-2004—Standard for Safety for Spray Nozzles for Fire-Protection Service	UL—Underwriters Laboratories

OHS in the United States and the EU are discussed in separate sections. Examples of voluntary standards in OHS are also given.

Voluntary Standards in Occupational Health and Safety in the United States

In the United States, no single government agency has control over standards. Some bodies of the U.S. government require companies to use certain voluntary standards, but there is no federal government agency that controls the voluntary standards. The standards are created by many different organizations and are not ratified or governed by any government agency. It is the American National Standards Institute (ANSI), a private, nonprofit organization, which administers and coordinates the U.S. voluntary standardization and conformity assessment system. ANSI is not a governmental agency or a regulatory body. According to the information presented in its website (http:// www. ansi.org), ANSI itself does not develop American National Standards, but it provides all interested U.S. parties with a neutral venue to come together and work toward common agreements. ANSI approves the process by which standards become accepted. Accreditation by ANSI signifies that the procedures used by a standards developing body satisfy the essential requirements for American National Standards. ANSI also states in its Web site that it has currently around 270 accredited standards developer groups representing approximately 200 distinct organizations in the private and public U.S. sectors, where these groups work cooperatively to develop voluntary national consensus standards and American National Standards. In the domain of safety, health, and protection, of the workers several developers are active. Examples of ANSI accredited standards within the OHS domain are listed in Table 24.8, which also shows the organization that was at the origin of each of the standards listed.

Voluntary Standards in Occupational Health and Safety in the European Union

There are three European standards organizations that are recognized as competent to adopt harmonized standards at the level of the EU. The European Committee for Standardization (CEN), the European Committee for Electrotechnical Standardization (Cenelec), and the European Telecommunications Standards Institute (ETSI) are the three bodies accrediting standards that meet requirements expressed in EU directives. Some European Standards were developed in support of the policy, adopted in 1985, which stipulated that legislative harmonization in the form of EU directives should be limited to the essential requirements and that writing of the detailed technical specifications necessary for the implementation of directives was entrusted to the European voluntary standards organizations. On behalf of governments, of the European Commission, or of the EFTA Secretariat, the European standards organizations can be requested to develop standards. In the more recent EU directives, such as directive 98/37/EC (safety of machinery), voluntary standards accredited by CEN and Cenelec are explicitly recommended as a valid tool in implementing the requirements set forth in the directives. These voluntary standards are listed for each of the applicable EU Directives in the European Commission's Web site that is dedicated to enterprises (http://europa.eu.int/comm/enterprise/). Compliance with harmonized standards does not, however, translate into immunity from legal obligations. Caution should hence be exercised because CEN and Cenelec harmonized standards may not cover all health and safety requirements of a given directive. Thus, lawful obligations should always be assessed independently of the analysis of the requirements of the voluntary standards.

According to the information posted in its Web site (http://www.cenorm.be), CEN aims to draw up voluntary European standards and to promote corresponding conformity of products and services in areas other than electrical and telecommunications, which are covered by the activity of Cenelec and ETSI. CEN's mission is to promote voluntary technical harmonization of standards in Europe in conjunction with worldwide bodies and its European partners, having as purpose to diminish trade barriers; to promote safety; to allow interoperability of products, systems, and services; and to promote common technical understanding. In this way, EU member states and EFTA countries subsequently adopt the standards adopted by CEN. Table 24.9 presents some European standards on OHS and examples of corresponding national standards resulting from the adoption of the former. The exchange of standards at an international level, with the consequent adoption of voluntary standards developed by organisms other than the European Standards organizations, is evident in some of the European standards. As an example, standard EN ISO 12100-2:2003 (Safety of Machinery—Basic Concepts, General Principles for Design—Part 2: Technical Principles) was adopted as a European standard from an original ISO standard. The same standard is adopted by European countries and transposed

TABLE 24.9

Examples of CEN-Accredited European Standards in the Domain of OHS and Corresponding National Standards Resulting From Their Transposition in France, Germany, and the UK

European Standard	France: AFNOR	Germany: DIN	UK: BSI
EN 54-12:2002—Fire Detection and Fire Alarm Systems—Part 12: Smoke Detectors—Line Detectors Using an Optical Light Beam	NF EN 54-12	DIN EN 54-12	BS EN 54-12:2002
EN 169:2002—Personal Eye-Protection—Filters for Welding and Related Techniques—Transmittance Requirements and Recommended Use	NF EN 169	DIN EN 169	BS EN 169:2002
EN 294:1992—Safety of Machinery—Safety Distance to Prevent Danger Zones Being Reached by the Upper Limbs	NF EN 294	DIN EN 294	BS EN 294:1992
EN 13861:2002—Safety of Machinery—Guidance for the Application of Ergonomics Standards in the Design of Machinery	NF EN 13861	DIN EN 13861	BS EN 13861:2002

to their national standardization systems (e.g., BS EN ISO 12100-2:2003, in the UK). The national standards institutions of European countries also produce their own standards, which do not come from the harmonization process led by CEN, Cenelec or ETSI, nor do they proceed from ISO (e.g., German Standard DIN 58214:1997—Eye-Protectors—Helmets—Terms, Forms and Safety Requirements). National standards institutions also adopt standards, which were not previously drawn up by any of the three European standards organizations, directly from international standards.

Management of Occupational Health and Safety

Traditionally, occupational health and safety management entailed little more than reacting to accidents and work injuries, as part of a reactive culture in workplace health and safety. However, over time, it has been possible to verify improvements in OHS conditions, especially because of legislation and regulations that have been developed in this domain. In the EU, legislation has clearly had a positive influence on OHS conditions. It has contributed to instilling a culture of prevention, despite some flaws, which according to the EU Commission (2004) still hold back the achievement of the full potential of this legislation. Throughout the EU, attitudinal changes concerning the behavior and the awareness of people have to take place at a large scale in order to fully implement in practice the concept of prevention in the management of OHS within organizations. In the United States, OSHA has been developing and enforcing mandatory OHS standards for more than 30 years, promoting during this time, a very significant reduction in work-related fatalities, injuries and illnesses. Despite these achievements, the most recent U.S. statistics still spell out a stark reality, as do the statistics from the EU (Table 24.2). An impetus for further improvements could come from further disseminating a culture of a systematic approach to OHS, where all the people in the organization are included, in a process of continual improvement. Improvements are undoubtedly necessary, especially in moving beyond the traditional reactive approach to a preventive systematic approach to the management of OHS in organizations. Differences between a systematic approach to the management of OHS in organizations and the traditional reactive approach are shown in Table 24.10, after Bottomley (1999). An OHS management system that puts a systematic

TABLE 24.10
Differences Between a Reactive and a Systematic Approach to the Management of OHS in Organizations

A Reactive Approach to the Management of OHS in Organizations Entails That . . .	A Systematic Approach to the Management of OHS in Organizations Entails That . . .
Hazards are dealt with reactively.	Hazards are identified preventively.
Risk controls are dependent on individuals.	Risk controls are described in procedures.
Risk controls are not linked to each other.	Risk controls are linked by a common method.
OHS activity happens but is not planned.	OHS activity is planned.
Controls are reviewed after an incident.	Controls are monitored and reviewed regularly.
Responsibilities are not defined.	Responsibilities are defined for everyone.
Focus is only given to risks on the organization's own "backyard."	Public and supplier risks are managed in a planned way.
There is no company policy on OHS to communicate.	Company policy on OHS is communicated.

Note: Adapted from B. Bottomley (1999).

approach to prevention into practice can be part of the management systems of any organization. Such a management system can provide a set of procedures and tools that make the management of risks to health and safety at work more efficient, and that enable organizations to improve performance by controlling their risks to OHS.

In implementing an OHS program based on a policy of prevention and continuous improvement, a fundamental issue to be dealt with is, according to Zimolong (1997), how to provide the long-term commitment of employees and management. Participative management and a suitable leadership style may be useful to ensure the cooperation of all members, promoting commitment and involvement in OHS activities over time. There are several factors that may motivate organizations to implement an OHS management system in such a way that leads to acquiring a culture of prevention and continuous improvement. OHS regulations and legislation in the United States, and in the EU, encourage prevention and continuous improvement through periodic revision of hazards and resolution or control of risks to workplace safety and health. Organizations may also have learned lessons from implementing quality management and environmental management systems, through ISO 9001 and ISO 14001 certification, respectively. Benefits of ISO 9001 certification have been reported in literature (Bryde & Slocock, 1998; Lipovatz, Steno, & Vaka, 1999; Neergaard, 1999). Important benefits motivated by certification to ISO 9001 include disseminating a systematic approach that promotes consistency in operations and controlled documentation, encourages teamwork, personnel training, and improved motivation and organization. Some problems linked to ISO 9001 certification were also reported, including its high cost, the extra documentation that is necessary, and the resistance to change of some people in the organization (Quazi & Padibjo, 1998).

A review of empirical research, which was conducted in the UK by Wright (1998), identified what motivates managers to proactively manage OHS (using a planned and systematic approach). The review suggested that there are two decisive factors that motivate both SMEs and large organizations to initiate OHS improvements. These factors are the fear of losing credibility and a belief that it is necessary and morally correct to comply with OHS regulations. Other factors identified were the aim to improve staff morale and productivity, the extension of modern concepts such as Total Quality Management to all areas of responsibility, and the integration of OHS into quality and environmental management systems. The need to avoid or reduce the immediate costs associated with ill health and injuries could also be a very strong motivating factor. In the EU, many enterprises excuse the absence of an OHS management system because of its implementation costs (EU Commission, 2004). But this should not be the case because EU legislation (EU framework directive 89/391/EEC) does not call for sophisticated management systems. It simply encourages applying basic management principles in the field of OHS. The size of the organization and its economic sector of activity are not critical in determining the potential use of an OHS management system, although, according to Bottomley (1999), these may determine the manner and style of implementation.

The implementation of management systems based on standards and specifications in organizations at a worldwide level has had an incremental development over the years. Quality management systems based on the ISO 9001 standard are currently very much disseminated. A similar path is being followed in certification to ISO 14001 (environmental management system) by organizations. The OHS management system based on Occupational Health and Safety Assessment Series 18001 is also being implemented in organizations at a worldwide level. In many of these organizations, quality and environmental management systems were already in place or were implemented simultaneously with the implementation of the OHS management system. An OHS management system based on OHSAS 18001 provides a set of procedures and tools that enable an efficient management of OHS, promoting continual improvements in organizational performance through the ongoing processes of hazard identification, risk assessment, and control of risks to OHS.

Occupational Health and Safety Management Systems: Specification and Guidelines for Implementation

Demand for a recognizable OHS management system standard, against which management systems could be assessed and certified, was at the origin of the development of the OHSAS (Occupational Health and Safety Assessment Series) 18001 specifications. OHSAS 18001 was developed collaboratively by a group representing standards institutions and other organizations with concerns on OHS, with an international and national base (Germany, Ireland, Japan, Mexico, Norway, Singapore, South Africa, Taiwan, and the UK). It has been issued in several countries and in several languages, in some cases as a standard and in others as a specification.

The OHSAS 18001 specification (BSI-OHSAS 18001:1999) and its accompanying guidelines for implementation (BSI-OHSAS 18002:2000) entail a preventive and proactive approach to the management of OHS. Emphasis is given to ensuring the identification, evaluation, and control of risks to OHS, in a continuously improved manner. In what follows, the applicability and envisaged benefits of the OSHAS 18001 specification for OHS management systems are illustrated. An overview of the requirements set out in the specification is given and the compatibility of an OHS management system with other management systems in organizations is also discussed.

Applicability and Envisaged Benefits

The OHSAS 18001 specification is deemed applicable to organizations that may wish to establish an OHS management system to eliminate or minimize risk to employees and other interested parties who may be exposed to OHS risks associated with its activities (BSI-OHSAS 18001:1999). This OHSAS specification is applicable to the implementation, maintenance, and continual improvement of an OHS management system. An organization may also implement it to assure itself of its conformance with its stated OHS policy or to demonstrate such conformance to others (BSI-OHSAS 18001:1999). Organizations seeking certification and registration of their OHS management systems by an external organization, or willing to make a self-determination and declaration of conformance with this OHSAS specification may also choose to implement it. Although the requirements in OHSAS 18001 are intended to be incorporated into any OHS management system, the extent of their application will depend on the OHS policy of the organization, the nature of its activities, and the risks and complexity of its operations (BSI-OHSAS 18001:1999).

The specification does not state specific OHS performance criteria, nor does it give detailed specifications for the design of a management system, but the advantages of certification in OHSAS 18001 demonstrate business and social benefits of health and safety improvements. Some case studies have been produced by the British Standards Institute and are posted in its Web site (http://emea.bsi-global.com/OHS/CaseStudies/). The UK Health and Safety Executive has also published in its own Web site (http://www. hse.gov.uk) a series of case studies setting out the business case for good health and safety management. The case studies prepared by these two institutions from the UK cover a variety of industry sectors, including pharmaceutical, construction, manufacturing, petrochemical, aerospace, defense, utilities, food and service companies, as well as organizations in the public sector. The benefits that are reported in the case studies spring from improvements in operational efficiency, productivity, and public image; from reduction in lost working days; from reduction in accidents and medical claims; from recognition by insurers; and from heightened workers' satisfaction and involvement in assuring OHS.

Overview of Requirements

There are 18 requirements set out in the OHSAS 18001 specification. Table 24.11 presents a summary of all the requirements in the OHSAS 18001 specification. The guidelines for the implementation of OHSAS 18001 (BSI-OHSAS 18002:2000) explain the requirements, detailing their intent and the typical inputs, process, and outputs that can be used in implementing each of the requirements. Two of the 18 requirements portrayed in OHSAS 18001 are discussed in this section. There is the requirement for defining an occupational health and safety policy statement. This requirement establishes the overall sense of direction and sets the principles of action for an organization. This policy has to be authorized by top management and should clearly state overall health and safety objectives and a commitment to improving health and safety performance. It should also be consistent with other management policies of the organization (e.g., pertaining to quality or environmental aspects). The OHS policy statement of the organization should be reviewed periodically to ensure that it remains relevant and appropriate to the organization. The policy should be communicated to all employees and to other groups or individuals affected by the OHS performance of the organization. In many countries, OHS legislation and regulations demand consultation and participation of employees in their organization's OHS management systems.

Another requirement set out in the OHSAS 18001 specification concerns establishing and maintaining procedures for the ongoing identification of hazards, the assessment of risks, and the implementation of necessary control measures. The OHSAS 18001 specification and the guidelines for its implementation do not make recommendations on how these activities should be conducted, but establish principles by which the organization can determine whether its processes are suitable and sufficient. The hazard identification, risk assessment, and risk control processes form the core of the whole OHS management system, enabling the organization to continuously identify, evaluate, and control its OHS risks. These processes should be carried out in respect to normal operations within the organization, and also in what concerns abnormal operations. The latter include periodic or occasional operations such as plant cleaning and maintenance, as well as plant startups and shutdowns. Potential emergency conditions should also be covered in hazard identification, risk assessment, and risk control processes. Depending on the competencies that exist within the organization, it may be necessary for the organization to seek external advice or services in support for conducting these processes. Following the performance of the processes of hazard identification, risk assessment, and risk control, any corrective or preventive actions identified as being necessary should be monitored to assure their timely completion. The results of the processes of hazard identification, risk assessment, and risk control should also be considered as input for the establishment of revised or new OHS objectives in the organization.

Compatibility With Other Management Systems

The OHSAS 18001 specification was developed to be compatible with ISO 9001 and ISO 14001 standards. Reviews are to be done whenever new revisions to either ISO 9001 or ISO 14001 are made, to ensure continuing compatibility, in order to facilitate the integration of quality, environmental and OHS management systems in organizations. In the year 2000, a revision to ISO 9001 was done (the earlier version was dated from 1994), which promoted a greater approximation to Total Quality Management philosophy. Emphasized, among other things, was continuous improvement, which is required in a formal manner by the three sets of standards. In this revision of ISO 9001, emphasis was also given to the management of resources, through the comprehensive treatment of elements such as information, communication, infrastructure, and work environment. In what concerns the latter, a change has been made in the 2000 version of the standard for a quality management system toward emphasizing

TABLE 24.11

Overview of Requirements Set Out in the OHSAS 18001 Specification, According to
BSI-OHSAS 18001:1999

OHSAS 18001 Requirement on	*Some Elements of the Requirement*
(4.1) General Requirements	The organization shall establish and maintain an OHS management system (. . .)
(4.2) OHS Policy	There shall be an occupational health and safety policy authorized by the organization's top management (. . .)
(4.3.1) Planning for Hazard Identification, Risk Assessment, and Risk Control	The organization shall establish and maintain procedures for the ongoing identification of hazards, the assessment of risks, and the implementation of necessary control measures. (. . .)
(4.3.2) Legal and Other Requirements	The organization shall establish and maintain a procedure for identifying and accessing the legal and other OHS requirements that are applicable to it. (. . .)
(4.3.3) Objectives	The organization shall establish and maintain documented occupational health and safety objectives, at each relevant function and level within the organization. (. . .)
(4.3.4) OHS Management Programme(s)	The organization shall establish and maintain (an) OHS management programme(s) for achieving its objectives (. . .)
(4.4.1) Structure and Responsibilities	The roles, responsibilities and authorities of personnel who manage, perform and verify activities having an effect on the OHS risks of the organization's activities, facilities and processes, shall be defined (. . .)
(4.4.2) Training, Awareness, and Competence	Personnel shall be competent to perform tasks that may impact on OHS in the workplace. Competence shall be defined in terms of appropriate education, training and/or experience. (. . .)
(4.4.3) Consultation and Communication	The organization shall have procedures for ensuring that pertinent OHS information is communicated to and from employees and other interested parties. (. . .)
(4.4.4) Documentation	The organization shall establish and maintain information, (. . .) that: a) describes the core elements of the management system and their interaction; and b) provides direction to related documentation.
(4.4.5) Document and Data Control	The organization shall establish and maintain procedures for controlling all documents and data required by this OHSAS specification (. . .)
(4.4.6) Operational Control	The organization shall identify those operations and activities that are associated with identified risks where control measures need to be applied (. . .)
(4.4.7) Emergency Preparedness and Response	The organization shall establish and maintain plans and procedures to identify the potential for, and responses to, incidents and emergency situations and for preventing and mitigating the likely illness and injury that may be associated with them. (. . .)
(4.5.1) Performance Measurement and Monitoring	The organization shall establish and maintain procedures to monitor and measure OHS performance on a regular basis (. . .)
(4.5.2) Accidents, Incidents, Nonconformances, and Corrective and Preventive Action	The organization shall establish and maintain procedures for defining responsibility and authority for: a) the handling and investigation of: - accidents; - incidents; - nonconformances; b) taking action to mitigate any consequences arising from accidents, incidents or nonconformances (. . .)
(4.5.3) Records and Records Management	The organization shall establish and maintain procedures for the identification, maintenance and disposition of OHS records, as well as the results of audits and reviews. (. . .)
(4.5.4) Audit	The organization shall establish and maintain an audit programme and procedures for periodic OHS management system audits to be carried out (. . .)
(4.6) Management review	The organization's top management shall (. . .) review the OHS management system, to ensure its continuing suitability, adequacy and effectiveness. (. . .)

[a]Heading numbers shown in parentheses indicate the applicable section of the OHSAS 18001 specification.

Note: Permission to reproduce extracts from OHSAS 18001:1999 was granted by BSI. British Standards can be obtained from BSI Costumer Services, 389 Chiswick High Road, London W4 4AL. Tel: +44 (0)20 8996 9001. e-mail: cservices@bsi-global.com

human resources within the organization, with the introduction of a new concept of working environment. The importance of human resources and their working environment for the quality of products is explicitly emphasized. There are several requirements that are common to the three management systems, such as prevention, conformance, continuous improvement, process management, or leadership. For these reasons, organizations running certified ISO compliant quality and environmental management systems, or on the process of implementing these, can more easily implement a system of management of occupational health and safety based on the OHSAS 18001 specification.

Some organizations may benefit in having integrated management systems. Others may prefer to adopt different systems based on similar management principles. Given the absence of a universally accepted single management system standard or specification that brings together the currently separate standards and specification for quality management, environmental management, and occupational health and safety management, an alternative way to reap the benefits of a simultaneous approach to the management of the three systems could be pursued. It consists of integrating these systems in practice inside the organization (Matias & Coelho, 2002). To be of worth for organizations, the benefits of such integration should be greater than the sum of the partial benefits of the independently managed systems. Disadvantages should also be relatively smaller. There may also be barriers to such integration. Given that an organization has functioned with the systems as separate entities, fear of change may exist and manifest itself as opposition to the integration, which will affect company organization. This situation is clearly more relevant for organizations that have previously attained success with separate management systems. Favorable arguments for the integration of the different management systems spring from economies of scale in integrating information channels and documentation. Information bottlenecks can preexist due to communication problems occurring in disperse management systems. These disperse systems in the organization may, however, share common goals, such as continuous improvements, or prevention of nonconformance and accidents. With a unified management system, the interests of all the organization's stakeholders (employees, customers, shareholders, suppliers, and society) would be more conveniently satisfied (Figure 24.1).

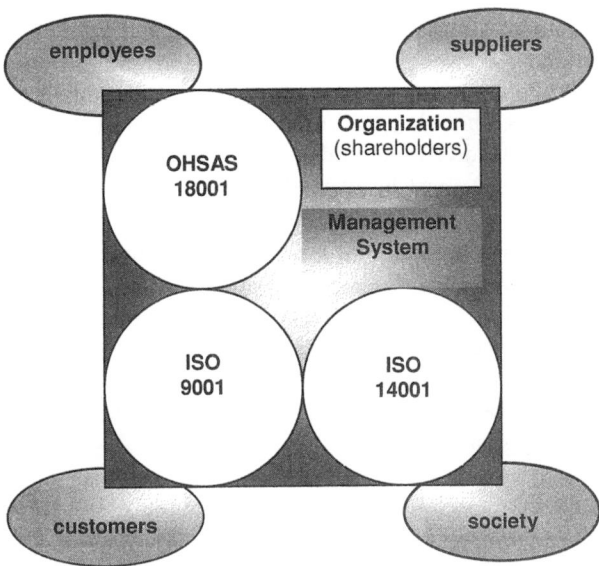

FIG. 24.1. The organization's stakeholders and the integration of certified management systems within the organization.

CONCLUSION

OHS can be accomplished by creating a controlled work environment, where risks to health and safety have been resolved or are under control. The involvement of HFE in assuring OHS at the workplace, whether at the design stage or in postdesign interventions, is appropriate and desirable given the relevance of the knowledge and methods of HFE to many important dimensions of OHS. Within organizations, opportunities for collaboration between OHS and HFE may be many. Effective application of ergonomics in work design, while promoting a balance between worker characteristics and task demands, provides worker safety and physical and mental well-being. In establishing the economic case for HFE or OHS interventions, estimating the economic value of the anticipated benefits is bound to be more troublesome than estimating the cost of implementation. Similarly, although establishing the cost effectiveness of ergonomic and OHS mandatory standards is not a straightforward task, endeavors are carried out to that end in the United States, by taking nationwide statistical values into account to produce cost and benefit estimates for OSHA standards. At the level of organizations, the costs of accidents are usually incompletely assessed, when comparing them with the costs of accident prevention for the purpose of determining the benefits of accident reduction. The costs of work- related accidents and illnesses extend not only to those directly affected but also to organizations and to countries, given their significant impact on the economy. Ultimately, these costs fall on citizens at large, as taxpayers and consumers. Preventive action in OHS is undoubtedly necessary, and organizations should handle both obvious and nonobvious hazards systematically and in a continuous manner.

OHS regulations and legislation in the United States and in the EU encourage prevention and continuous improvement through periodic revision of hazards and resolution or control of risks to workplace safety and health. Through the adoption of EU directives on OHS to their internal law, EU countries are committed to encouraging improvements in OHS conditions and to harmonizing OHS conditions across the EU. In the EU, legislation has clearly had a positive influence on OHS conditions. It has contributed to instilling a culture of prevention, although further disseminating a culture of a preventive and systematic approach to OHS is still needed, moving beyond the traditional reactive approach. Inspectors have a crucial role to play as agents to promote compliance to OHS legislation and regulations, not only through enforcement activities but also through education, persuasion, and encouragement. People also have a very important role in promoting their safety and health at their workplace. If people can agree about something, then they are more likely to do it, than if they are simply told to. A proactive attitude toward OHS should not only be seen as a concern of management, or as a law or regulation that must be obeyed, but it is recommended for everyone in an organization.

Voluntary standards assume an important role in the development and improvement of work conditions, because they complement mandatory standards and legislation, supporting their implementation. EU directives acknowledge that technical standards constitute an indispensable reference in adopting procedures and measures to meet legal requirements in OHS. In the United States, OSHA provides guidelines and other voluntary standards to assist in implementing regulations and improving OHS conditions beyond minimum regulatory requirements set forth in OSHA mandatory standards.

An OHS management system based on OHSAS 18001 provides a set of procedures and tools that enable an efficient management of OHS, promoting continual improvements in organizational performance through the ongoing processes of hazard identification, risk assessment, and control of risks to OHS. The advantages of certification in OHSAS 18001 demonstrate business and social benefits of health and safety improvements. These include improvements in operational efficiency, productivity, and public image; reduction in lost working days;

accidents and medical claims; recognition by insurers; and heightened workers' satisfaction and involvement in assuring OHS.

The acceptance and dissemination of ISO 9001 and ISO 14001 certified management systems, and the benefits they promoted in organizations, opened the path for the implementation and certification of other management systems in organizations. The OHSAS 18001 specification was developed to be compatible with ISO 9001 and ISO 14001 standards, with several requirements that are common to the three management systems, including prevention, continuous improvement, and process management. Organizations running ISO compliant quality and environmental management systems, or on the process of implementing them, can thus more easily implement a system of management of occupational health and safety based on OHSAS 18001. This set of three management systems can be integrated within the organization, with benefits springing from integrating information channels and documentation.

REFERENCES

Amell, T. K., Kumar, S., & Rosser, B. W. J. (2001). Ergonomics, loss management, and occupational injury and illness surveillance. Part 1: Elements of loss management and surveillance. A review. *International Journal of Industrial Ergonomics, 28*, 69–84.

Amended Proposal for a Council Decision Concerning the Minimum Safety and Health Requirements for Transport Activities and Workplaces on Means of Transport—Individual Directive Within the Meaning of Article 16 of Directive 89/391/EEC, COM/93/421FINAL—SYN 420, Official Journal C 294, 30/10/1993, p. 4.

ANSI N13.15-1985—Dosimetry Systems, Performance of Personnel Thermoluminescence, HPS—Health Physics Society, McLean, VA.

ANSI Z87.1-2003—Occupational and Educational Personal Eye and Face Protection Devices, ASSE—American Society of Safety Engineers, Des Plaines, IL.

Antes, M. K., Miri, M. F., & Flamberg, S.A. (2001). Selection and design of cost-effective risk reduction systems. *Process Safety Progress, 20*(3), 197–203.

Beevis, D. (2003). Ergonomics—Costs and benefits revisited. *Applied Ergonomics, 34*, 491–496.

Beevis, D., & Slade, I. M. (1970). Ergonomics—costs and benefits. *Applied Ergonomics, 1* (2), 79–84 (republished in *Applied Ergonomics, 34*(2003), 413–418).

Biddle, J., & Roberts, K. (2004). More evidence on the need for an ergonomic standard. *American Journal of Industrial Medicine,45*, 329–337.

Bottomley, B. (1999). Occupational Health and Safety Management Systems: Strategic Issues Report. National Occupational Health and Safety Commission, Canberra, Australia. Available online at http://www.nohsc.gov.au.

Bryde, D. J., & Slocock B. (1998). Quality Management Systems Certification: A survey. *International Journal of Quality and Reliability Management, 15*(5), 467–480.

BS EN ISO 12100-2:2003—*Safety of machinery—Basic concepts, general principles for design—Part 2: Technical principles.* British Standards Institute, London, UK.

BSI-OHSAS 18001:1999—*Occupational Health and Safety Assessment Series: Occupational health and safety management systems—Specification, Incorporating Amendment No. 1.* British Standards Institution (13 December 2002), London, UK.

BSI-OHSAS 18002:2000—*Occupational Health and Safety Management Systems—Guidelines for the implementation of OHSAS 18001.* British Standards Institution (13 December 2002), London, UK.

Caple, D. C. (2004). IEA International Development. *Proceedings of the Ergonomics Congress of Portuguese Speaking Countries.* Funchal, Portugal, Jul. 26–28, 1994. CD-ROM, APERGO—Portuguese Ergonomics Association, Lisbon, Portugal.

Council of the European Communities Directive 89/391/EEC of 12 June 1989 concerning the introduction of measures to encourage improvements in the safety and health of workers at work. *Official Journal L 183, 29/06/1989*, 1–8.

Council of the European Communities Directive 89/654/EEC of 30 November 1989 concerning the minimum safety and health requirements for the workplace. *Official Journal L 393 of 30/12/1989*, 1–12.

Council of the European Communities Directive 89/655/EEC of 30 November 1989 concerning the minimum safety and health requirements for the use of work equipment by workers at work. *Official Journal L 393, 30/12/1989*, 13–17.

Council of the European Communities Directive 89/656/EEC of 30 November 1989 concerning the minimum health and safety requirements for the use by workers of personal protective equipment at the workplace. *Official Journal L 393, 30/12/1989*, 18–28.

Council of the European Communities Directive 90/269/EEC of 29 May 1990 concerning the minimum health and safety requirements for the manual handling of loads where there is a risk particularly of back injury to workers. *Official Journal L 156, 21/06/1990*, 9–13.

Council of the European Communities Directive 90/270/EEC of 29 May 1990 concerning the minimum safety and health requirements for work with display screen equipment. *Official Journal L 156 of 21/06/1990*, 14–18.

Council of the European Communities Directive 90/394/EEC of 28 June 1990 concerning the protection of workers from the risks related to exposure to carcinogens at work. *Official Journal L 196 of 26/07/1990*, 1–7.

Council of the European Communities Directive 91/383/EEC of 25 June 1991 supplementing the measures to encourage improvements in the safety and health at work of workers with a fixed-duration employment relationship or a temporary employment relationship. *Official Journal L 206, 29/07/1991*, 19–21.

Council of the European Communities Directive 92/29/EEC of 31 March 1992 on the minimum safety and health requirements for improved medical treatment on board vessels. *Official Journal L 113, 30/04/1992*, 19–36.

Council of the European Communities Directive 92/57/EEC of 24 June 1992 concerning the implementation of minimum safety and health requirements at temporary or mobile construction sites. *Official Journal L 245, 26/08/1992*, 6–22.

Council of the European Communities Directive 92/58/EEC of 24 June 1992 Concerning the minimum requirements for the provision of safety and/or health signs at work. *Official Journal L 245 of 26/08/1992*, 23–42.

Council of the European Communities Directive 92/85/EEC of 19 October 1992 on the introduction of measures to encourage improvements in the safety and health at work of pregnant workers and workers who have recently given birth or are breastfeeding. *Official Journal L 348, 28/11/1992*, 1–8.

Council of the European Communities Directive 92/91/EEC of 3 November 1992 concerning the minimum requirements for improving the safety and health protection of workers in the mineral-extracting industries through drilling. *Official Journal L 348 of 28/11/1992*, 9–24.

Council of the European Communities Directive 92/104/EEC of 3 December 1992 concerning the minimum requirements for improving the safety and health protection of workers in surface and underground mineral-extracting industries. *Official Journal L 404 of 31/12/1992*, 10–25.

Council of the European Union Directive 93/68/EEC amending the Council Directive 89/686/ EEC. *Official Journal L 220, of 30/8/1993*, 1.

Council of the European Union Directive 93/95/EEC amending for the second time Council Directive 89/686/EEC. *Official Journal L 276, of 9/11/1993*, 11.

Council of the European Union Directive 93/103/EC of 23 November 1993 concerning the minimum safety and health requirements for work on board fishing vessels. *Official Journal L 307 of 13/12/1993*, 1–17.

Council Directive 94/33/EC of 22 June 1994 concerning the protection of young people at work. *Official Journal L 216, 20/08/1994*, 12–20.

Council of the European Union Directive 95/63/EC of 5 December 1995 amending Directive 89/655/EEC. *Official Journal L 335 of 30/12/1995*, 28–36.

Council Directive 96/58/EC amending for the third time Council Directive 89/686/EEC. *Official Journal L 236, of 18/9/1996*, 44.

Council of the European Union Directive 97/42/EC of 27 June 1997. *Official Journal L 179 of 08/07/1997*, 4–6 (amended in *Official Journal C 123 of 22/04/1998*, 21).

Council of the European Union Directive 98/24/EC of 7 April 1998 on the protection of the health and safety of workers from the risks related to chemical agents at work. *Official Journal L 131, 05/05/1998*, 11–23.

Commission of the European Union Directive 98/65/EC of 3 September 1998 adapting to technical progress Council Directive 82/130/EEC on the approximation of the laws of the Member States concerning electrical equipment for use in potentially explosive atmospheres in mines susceptible to firedamp. *Official Journal L 257, 19/09/1998*, 29–34.

Council of the European Union Directive 99/38/EC of 29 April 1999 amending Directive 90/394/EEC. *Official Journal L 138, 01/06/1999*, 66–69.

Dembe, A. E. (2001). The social consequences of occupational injuries and illnesses. *American Journal of Industrial Medicine, 40*, 403–417.

Dijk, F. J. H. van (1995). From input to outcome: changes in OHS education. *Safety Science, 20*, 165–171.

DIN 58214:1997—*Eye-protectors—Helmets—Terms, forms and safety requirements*. Berlin: Deutsches Institut für Normung (DIN). 1997.

Directive 98/37/EC of The European Parliament and of The Council of 22 June 1998 on the approximation of the laws of the Member States relating to machinery. *Official Journal of the European Communities L 207, 23/07/1998*, 1–46.

Directive 2003/10/EC of the European Parliament and of the Council of 6 February 2003 on the minimum health and safety requirements regarding the exposure of workers to the risks arising from physical agents (noise). *Official Journal L 042, 15/02/2003*, 38–44.

Dupré, D. (2001). Statistics spell it out. *Magazine of the European Agency for Safety and Health at Work, 4,* 5–7. Available online at http://osha.eu.int.

EN 54-12:2002—*Fire detection and fire alarm systems—Part 12: Smoke detec tors—Line detectors using an optical light beam.* Brussels, Belgium: European Committee for Standardization (CEN).

EN 169: 2002—*Personal eye-protection—Filters for welding and related techn iques—Transmittance requirements and recommended use.* Brussels, Belgium: European Committee for Standardization (CEN).

EN 294:1992—*Safety of machinery—Safety distance to prevent danger zones being reached by the upper limbs.* Brussels, Belgium: European Committee for Standardization (CEN).

EN 13861:2002—*Safety of machinery—Guidance for the application of ergonomics standards in the design of machinery.* Brussels, Belgium: European Committee for Standardization (CEN).

EN ISO 12100-2: 2003—*Safety of machinery—Basic concepts, general principles for design—Part 2: Technical principles.* Brussels, Belgium: European Committee for Standardization (CEN).

EU Commission. (2002). Commission of the European Communities COM(2002) 118 final. *Adapting to change in work and society: A new Community strategy on health and safety at work 2002–2006* (Communication).

EU Commission. (2004). Commission of the European Communities COM(2004) 62 final. Communication on the practical implementation of the provisions of the Health and Safety at Work Directives 89/391 (Framework), 89/654 (Workplaces), 89/655 (Work Equipment), 89/656 (Personal Protective Equipment), 90/269 (Manual Handling of Loads) and 90/270 (Display Screen Equipment).

EU OSHA. (2001). Economic Impact of Occupational Safety and Health in the Member States of the European Union. European Agency for Safety and Health at Work.

European Parliament and Council of the European Union Directive 99/92/EC of 16 December 1999 on minimum requirements for improving the safety and health protection of workers potentially at risk from explosive atmospheres. *Official Journal L 23, 28/01/2000,* 57–64.

European Parliament and Council of the European Union Directive 2000/54/EC of 18 September 2000 concerning the protection of workers from risks related to exposure to biological agents at work. *Official Journal L 262, 17/10/2000,* 21–45.

European Parliament and Council Directive 2001/45/EC of the 27 June 2001 amending Council Directive 89/655/EC. *Official journal L 195, 10/07/2001,* 46–49.

European Parliament and Council Directive 2002/44/EC of the of 25 June 2002 concerning the minimum health and safety requirements regarding the exposure of workers to the risks arising from physical agents (vibration), Joint Statement by the European Parliament and the Council. *Official Journal L 177, 06/07/2002,* 13–20.

Helander, M. G. (1997). The human factors profession. In G. Salvendy (Ed.). *Handbook of human factors and ergonomics,* (2nd ed., pp. 3–16). New York: Wiley.

ILO R164:1981—Recommendation R 164 concerning Occupational Safety and Health and the Working Environment. International Labour Organization. Available online at http://www. ilo.org.

ISO 9001:2000—*Quality management systems: Requirements.* International Organization for Standardization (ISO). Geneva, Switzerland.

ISO 14001:1996—*Environmental management systems—Specification with guidance for use.* International Organization for Standardization (ISO), Geneva, Switzerland.

Leigh, J. P., Cone, J. E., & Harrison, R. (2001). Cost of occupational injuries and illnesses in California. *Preventive Medicine, 32,* 393–406.

Lipovatz, D., Steno F., & Vaka, A. (1999). Implementation of ISO 9000 quality systems in Greek enterprises. *International Journal of Quality and Reliability Management, 16*(6), 534–551.

Matias, J. C. O., & Coelho, D. A. (2002). The integration of the standards systems of quality management, environmental management and occupational health and safety management. *International Journal of Production Research, 40*(15), 3857–3866.

McCabe, J. (2004). Voluntary Safety Standards: In the Public Interest? CFA Consumer Assembly 2004. Washington DC, March 11, 2004, Available on-line at http://www.ansi.org.

Neergaard, P. (1999). Quality Management: a survey on accomplished results. *International Journal of Quality and Reliability Management, 16*(3), 277–289.

NFPA 70E-2004—Standard for Electrical Safety Requirements for Employee Workplaces, NFPA—National Fire Protection Association, Quincy, MA.

Occupational Safety and Health Act of 1970. Public Law 91-596. 84 STAT. 1590. 91st Congress, S.2193. December 29, 1970, as amended through January 1, 2004. Available on- line at http://www.osha.gov.

OSHA. (2003). All about OSHA Occupational Health and Safety Administration. U.S. Department of Labor. Information booklet available online at http://www.osha.gov.

Parsons, K. C. (1995). Ergonomics of the physical environment—International ergonomics standards concerning speech communication, danger signals, lighting, vibration and surface temperatures. *Applied Ergonomics, 26*(4), 281–292.

Quazi, H. A., & Padibjo, S. R. (1998). A journey toward total quality management through ISO 9000 certification. *International Journal of Quality and Reliability Management, 15*(5), 489–508.

Rikhardsson, P. M., & Impgaard, M. (2004). Corporate cost of occupational accidents: An activity-based analysis. *Accident Analysis and Prevention, 36*, 173–182.

Rodrigues, C., & Guedes, J. F. (2003). Linhas de orientação para a interpretação da norma OHSAS 18001/NP 4397 [Guidelines for the interpretation of the OHSAS 18001/NP 4397 standard]. APCER—Associação Portuguesa de Certificação. Available online at http://www.apcer.pt.

Rouse, W. B., & Boff, K. R. (1997). Assessing cost/benefits of human factors. In G. Salvendy (Ed.), *Handbook of human factors and ergonomics* (2nd ed., pp. 1617–1633). New York: Wiley.

Saari, J. (2001). Accident prevention today. *Magazine of the European Agency for Safety and Health at Work, 4*, 3–5. Available online at http://osha.eu.int.

Saha, A. (1998). Technological innovation and Western values. *Technology in Society, 20*, 499–520.

Seong, S. K., & Mendeloff, J. (2004). Assessing the accuracy of OSHA's projections of the benefits of new safety standards. *American Journal of Industrial Medicine, 45*, 313–328.

Sesé, A., Palmer, A. L., Cajal, B., Montaño, J. J., Jiménez, R., & Llorens, N. (2002). Occupational safety and health in Spain. *Journal of Safety Research, 33*, 511–525.

Standards-29 CFR—Part 1902—State Plans for the Development and Enforcement of State Standards, Occupational Safety and Health Administration (OSHA), U.S. Department of Labor.

Standards-29 CFR—Part 1903—Inspections, Citations, and Proposed Penalties, Occupational Safety and Health Administration (OSHA), U.S. Department of Labor.

Standards-29 CFR—Part 1904—Recording and Reporting Occupational Injuries and Illness, Occupational Safety and Health Administration (OSHA), U.S. Department of Labor.

Standards-29 CFR—Part 1910—Occupational Safety and Health Standards, Occupational Safety and Health Administration (OSHA), U.S. Department of Labor.

Standards-29 CFR—Part 1911—Rules of Procedure for Promulgating, Modifying or Revoking OSHA Standards, Occupational Safety and Health Administration (OSHA), U.S. Department of Labor.

Standards-29 CFR—Part 1912—Advisory Committees on Standards, Occupational Safety and Health Administration (OSHA), U.S. Department of Labor.

Standards-29 CFR—Part 1915—Occupational Safety and Health Standards for Shipyard Employment, Occupational Safety and Health Administration (OSHA), U.S. Department of Labor.

Standards-29 CFR—Part 1917—Marine Terminals, Occupational Safety and Health Administration (OSHA), U.S. Department of Labor.

Standards-29 CFR—Part 1918—Safety and Health Regulations for Longshoring, Occupational Safety and Health Administration (OSHA), U.S. Department of Labor.

Standards-29 CFR—part 1924—Safety Standards Applicable to Workshops and Rehabilitation Facilities, Occupational Safety and Health Administration (OSHA), U.S. Department of Labor.

Standards-29 CFR—Part 1925—Safety and Health Standards for Federal Service Contracts, Occupational Safety and Health Administration (OSHA), U.S. Department of Labor.

Standards-29 CFR—Part 1926—Safety and Health Regulations for Construction, Occupational Safety and Health Administration (OSHA), U.S. Department of Labor.

Standards-29 CFR—Part 1928—Occupational Safety and Health Standards for Agriculture, Occupational Safety and Health Administration (OSHA), U.S. Department of Labor.

Standards-29 CFR—Part 1977—Discrimination against Employees under OSHA Act of 1970, Occupational Safety and Health Administration (OSHA), U.S. Department of Labor.

Standards-29 CFR—Part 1990—Identification, Classification, and Regulation of Carcinogens, Occupational Safety and Health Administration (OSHA), U.S. Department of Labor.

Tytyk, E. (2004). Evolutionary background for humanizing technology. *Human Factors and Ergonomics in Manufacturing, 14*(3), 307–319.

UL 2351-2004—*Standard for Safety for Spray Nozzles for Fire-Protection Service*, UL—Underwriters Laboratories, Research Triangle Park, NC.

US-DOL, U.S. Department of Labor. Setting occupational safety and health standards. Program Highlights, Fact Sheet No. OSHA 92-14. Available online at http://www.osha.gov.

Weil, D. (2001). Valuing the Economic Consequences of Work Injury and Illness: A Comparison of Methods and Findings. *American Journal of Industrial Medicine, 40*, 418–437.

Wright, M. S. (1998). Factors Motivating Proactive Health and Safety. Prepared by ENTEC UK L^td for the Health and Safety Executive (Contract Research Report 179/1998), UK. Available online at http://www.hse.gov.uk.

Zimolong, B. (1997). Occupational Risk Management. In G. Salvendy (Ed.). *Handbook of Human Factors and Ergonomics* (2nd ed., pp. 989–1020). New York: Wiley.

25

World Trends in Occupational Health and Safety Management Systems

Pranab K. Nag
Anjali Nag
National Institute of Occupational Health

INTRODUCTION

The safety and health of the workplace is a decisive factor in making an organization function effectively. An emphatic worldwide endeavor is under way, applying a management framework, in order to implement cost-effective occupational health and safety (OHS) and assist governments and employer and employe representatives to better target their priorities in prevention of workplace injuries and diseases (Frick, Jensen, Quinlan, & Wilthagen, 2000). Revolving around the families of management standards, such as ISO 9001:2000 and ISO 14001:1996, 2004 the systems approach has drawn attention among the standards organizations, accreditation bodies, and the national regulatory agencies in formalizing, implementing, and evaluating OHS management systems (OHSMS). Several OHSMS frameworks have been proposed, aiming at building a documented approach to sustainable change and OHS improvement in the enterprises (International Occupational Hygiene Association[IOHA], 1998). Therefore, by conforming to an OHSMS, the organization can make visible its business objectives of quality management (Kara, 1996; Nag, 2002). This contribution elucidates the key elements of OHSMS and the issues surrounding the development of the OHSMS models, outlines the key documentation and implementation requirement, and certification of the management system, and briefly examines the global acceptance and challenges of the OHS programs and systems in the context of diverse national initiatives.

WHAT IS AN OHS MANAGEMENT SYSTEM?

An OHSMS is a planned, documented, and verifiable process of managing hazards and risks of health and safety in the workplace, increasing productivity by reducing the direct and indirect costs associated with accidents, and increasing the quality of products and services. What makes it a *system* is the deliberate linking and sequencing of processes that govern the way for a repeatable and identifiable way of managing OHS. What makes it a *management system* is the allocation of accountabilities, responsibilities, and resources by the top management through to all employees to enable decision on OHS matters. An ideal OHSMS must provide a direction to OHS activities, in accordance with the organizational policies, regulatory requirements, industry practices and standards, including negotiated labor agreements. Therefore, the purposes of establishing OHSMS in the organization are to align OHS objectives with business objectives of an organization; integrate OHS programs and systems into the business systems; establish a logical framework on which to establish an OHS program; devise a set of effective policies, targets, programs, and procedures; provide an auditable reference for performance benchmarking, and establish a continual improvement framework.

Bottomley (1999), in his information paper to Australian National Occupational Health and Safety Commission, cited the management elements for an effective system as below:

- *Organization, responsibility, and accountability*—(senior manager involvement, line manager or supervisor duties, specialist personnel, management accountability, performance measurement, and company OHS policy);
- *Consultative arrangements*—(health and safety representatives; issue resolution— employee, employer, and OHS representatives; joint OHS committees and employee participation)
- *Specific program elements*—(health and safety rules and procedures, training program, workplace inspections, incident reporting and investigation, hazard prevention, data collection and analysis or record keeping, OHS promotion and information provision, purchasing and design, emergency procedures, medical and first aid, monitoring and evaluation, and work organization issues)

IOHA (1998) analyzed different OHS standards and models and, in its report to the International Labor Office for an international OHSMS framework, adopted a universal OHSMS assessment instrument (Redinger & Levine 1998), that is comprised of 27 variables:

- *Initiation (OHS inputs)*—(management commitment and resources, regulatory compliance and system conformance, accountability, responsibility and authority, and employee participation)
- *Formulation (OHS process)*—(occupational health and safety policy, goals and objectives, performance measures, system planning and development, baseline evaluation and hazard or risk assessment, and OHSMS manual and procedures)
- *Implementation or operations (OHS process)*—(training system, technical expertise, and personnel qualifications; hazard control system; process design; emergency preparedness and response system; hazardous agent management system; preventive and corrective action system; and procurement and contracting)
- *Evaluation (feedback)*—(communication system, document and record management system, evaluation system, auditing and self-inspection, incident investigation and root cause analysis, and health or medical program and surveillance)
- *Improvement/integration (open system elements)*—(continual improvement, integration, and management review)

Apart from the elements included, their linkages are critical to form a management system. The emphasis remains with the system of work, covering the potential risks that an organization faces. Purchasing, design, and contractor management are the mainstream business operations that fall under operational control. The documentation requirements and OHS management responsibilities are integrated into the management structure. The scheduling of activity, monitoring, and review, including the internal audits and performance measurement, look at the system planning and its outcomes. An organization may opt to have its OHSMS audited against a certification standard by either an internal auditor or a second- or third-party auditor. An independent external audit is conducted by an accredited certifying body to establish whether the OHSMS meets the specific standard. Many OHSMS standards and guidelines are not intended for certification purposes. The outlines of some of the management models are included in Table 25.1.

OHSMS DEVELOPMENT—ENTERPRISE AND NATIONAL MODELS

Responsible Care

The United States Chemical Manufacturers Association designed the health and safety code of management practice—Responsible Care (1992). With the broad outline—program management, identification, and evaluation; prevention and control; and communication and training, the Responsible Care is a multidisciplinary means to enable member enterprises to operate in a manner that protects and promotes the OHS of employees, contractors, and the public, and protects the environment.

The British Responsible Care management system (1998) represents the Chemical Industry Association's version of the International Council of Chemical Association's Responsible Care program. Although the U.S. Responsible Care code has no reference to OHS goals and objectives, regulatory compliance and system conformance, hazardous agent management, and so forth, the British document appears more comprehensive. It includes a guidance and self-assessment document and describes the key requirements of an OHS and environment management system that should be incorporated into the management systems of an enterprise. The Responsible Care does not explicitly address medical program and surveillance; however, the guidance includes the requirements of British HSE—HS (G) 65 (1993), Eco-Management and Audit scheme (EMAS, 1993), ISO 14001 (1996, 2004), and BS 8800 (1996).

Community Eco-Management and Audit Scheme (EMAS)

The Council of the European Communities Regulation (1993) developed the Community Eco-Management and Audit Scheme (EMAS) to provide environmental management guidance to organizations. The EMAS includes 21 articles covering eco-management objectives, environmental statement, auditing and validation, accreditation of environmental verifiers, registration of sites, participation of small and medium-size enterprises and other sectors, information, infringements, competent bodies, and committees. Although the EMAS is not an OHS-related system, its framework provides scope for integrating other national, European, and international standards. EMAS is voluntary and is used in third-party certification audits.

ISO 14001—Environmental Management System

The emergence of the ISO 14000 (1996, revised, 2004), models is a global business initiative to create voluntary agreement on environmental management. The ISO 14001 specifies the requirements, and ISO 14004 provides the guidelines for its implementation. The standard represents a structured approach in setting environmental objectives and targets and demonstrating that they have been achieved. A management commitment to compliance with applicable legislation and regulations is required, along with the commitment to continual improvement. The ISO 14001 requires objective evidence to demonstrate that the system is operating effectively in

TABLE 25.1.a

Brief Outlines of the OHS and Environmental Management Systems

Responsible Care Management System (UK) (1998)	EMAS (1993)	ISO 14001 (1996) (revised, 2004)	AIHA (1996)	US OSHA program	E&P Forum (oil & gas companies) (1998)	BS 8800-ISO 14001 (1996)	AS/NZS 4801 (Australia/New Zealand) (1997)	SafetyMAP (1999)
2.0 Leadership commitment	1. EMAS objectives	4.2 Environmental policy	4.1 Responsibility (policy, responsibility, MR, review, resources)	**VPP (1988)**	1. Leadership commitment	**HS(G)65 approach**	4.1 Commitment and policy (leadership and commitment; initial OHS review, OHS policy)	1. Building and sustaining commitment
3.0 Policy	2. Definitions	4.3 Planning (environmental aspects, legal and other requirements, objectives and targets, management programs)	4.2 OHSMS (procedures, planning, performance measures)	a. Management commitment and planning	2. Policy and strategic objectives	4. OHSMS initial status review		2. Documenting strategy
4.0 Identifying requirements (regulatory and other requirements, risk assessment, defining program)	3. Participation	4.4 Implementation and operation (resources and roles, responsibility, authority, training, awareness and competence, communication, documentation	4.3 Compliance and conformance review (goals and objectives, continuous improvement)	b. Hazard assessment	3. Organization, resources and documentation (structure and responsibilities, MR, resources, competence, contractors, communication, documentation control)	4.1 OHS policy	4.2 Planning (identification of hazards, assessment and control of risk, legal and other requirements, objectives and targets, performance indicators, management plans)	3. Design and contract review
5.0 Planning (HS&E objectives, setting targets for improvement, planning for control, performance criteria, emergency preparedness)	4. Auditing and validation		4.4 OHS Design control (organizational and technical interfaces, design input, output, administrative and engineering control, design changes, verification and validation)	c. Hazard correction and control	4. Evaluation and risk management (identification, evaluation and recording of hazards and effects, objectives	4.2 Organizing (responsibilities, organizational arrangements, documentation)		4. Document control
6.0 Organization (structure and	5. Environmental statement		4.5 Document control	d. Safety and health training		4.3 Planning and implementing (risk assessment, legal and other requirements, management arrangements)	4.3 Implementation (ensuring capability,	5. Purchasing
	6. Accreditation and supervision of environmental verifiers		4.6 Purchasing	e. Employee participation		4.4 Measuring performance		6. Working safety by system
	7. List of accredited environmental verifiers		4.7 OHS communication systems	**Safety and Health Program (USA) (1989)**		4.5 Audit		7. Monitoring standards
	8. Site registration			a. Scope and application		4.6 Periodic status review		8. Reporting and correcting deficiencies
	9. List of registered sites			b. Basic obligation				9. Managing movement and materials
	10. Statement of			c. Management leadership and				10. Collecting and using data

11. Auditing of management systems
12. Developing skills and competencies

support action, risk assessment and control, contingency preparedness and response)
4.4 Measurement and evaluation (inspection, testing and monitoring; audits, corrective and preventive action)
4.5 Review and continual improvement

ISO 14001 approach
4. OHSMS initial status review
4.1 OHS policy
4.2 Planning (risk assessment, legal and other requirements, management arrangements)
4.3 Implementation and Operation (structure and responsibility, training, awareness and competence, communications, OHSMS documentation and operational control, emergency preparedness and response)
4.4 Checking and corrective action (monitoring and measurement, corrective action, records, audit)
4.5 Management review

ard performance criteria, risk reduction measures)
5. Planning (asset integrity, procedures, work instructions, management of change, contingency and emergency planning)
6. Implementation and monitoring (activities, monitoring, records, non-compliance and corrective action, incident reporting and follow-up)
7. Auditing and reviewing

employee participation
d. Hazard assessment and control
e. Information and training
f. Program evaluation
g. Dates
h. Definitions

4.8 Hazard identification and traceability
4.9 Process control
4.10 OHS inspection and evaluation (receiving inspection, evaluation, records)
4.11 Control of inspection, measurement and test equipment
4.12 OHS Inspection and evaluation status
4.13 Control of nonconforming processes or devices (review and disposition)
4.14 Corrective and preventive action
4.15 Handling, storage and packaging of hazardous materials
4.16 Control of OHS records
4.17 Internal audits
4.18 OHS Training
4.19 Operations and maintenance services
4.20 Statistical techniques

and operational control, emergency preparedness and response)
4.5 Checking and corrective action (Monitoring and measurement; evaluation of compliance, nonconformity, corrective and preventive action, records, audit)
4.6 Management review

participation
11. Costs and fees
12. Relationship with national and international standards.
13. Promotion of companies' participation.
14. Inclusion of other sectors
15. Information
16. Infringements
17. Annexes
18. Competent bodies
19. Committee
20. Revisions
21. Entry into force

responsibility, MR, resources, competence, documentation, communications)
7.0 Implementation and control (people, purchasing, contractors, manufacturing, storage, transportation, distribution, management of change)
8.0 Monitoring (measurement, checking and inspection, internal audit, improvements, records)
9.0 Management review

TABLE 25.1.b

Brief Outlines of the OHS and Environmental Management Systems

Spain (UNE 81900)	Norway Management Standard (1996)	The Netherlands (NPR 5001)	Ireland (KS NR 11-97) (1997)	OHSAS 18001 (1999)	ILO-OSH (2001)	Japan (JISHA 1997)
4.1 OHS policy	4.1 External requirements (stakeholder, legal requirements)	4.3 Maintenance and development of OHSMS	4. OHSMS principles	4.2 OHS policy	2. National framework for OSH-MS (national policy and tailored guidelines)	0. Corporate policy and CEO's policy
4.2 OHS management system	4.2 Management responsibility (leadership commitment, policy and objectives, system documentation)	5 First and periodic review of OHS status (initial and periodic management review)	5. Initial status review	4.3 Planning (planning for hazard identification, risk assessment and control, legal and other requirements, objectives, OHS management programs)	3. OSHMS **policy** (OSH policy, worker participation)	1. Establishment's policy and director's Policy
4.3 Responsibilities (management and staff responsibility and resources, management review)	4.3 Management system structure (organization, documentation)	5.3 Risk assessment (initial and periodic risk assessment)	6.1 OHS policy		**Organizing** (responsibility, accountability, competence and training, documentation, communication)	2. OHS management programs
4.4 risk evaluation (record of legal, regulatory and other requirements, hazard identification, risk evaluation and control, maintenance of risk control measures)	4.4 Communication (internal and external)	6 Decision-making and planning (OHS policy, plan of action)	6.2 Planning (legal and other requirements, objectives and targets, planning for OHS)	4.4 Implementation and operation (structure and responsibility, training, awareness and competence, consultation and communications, documentation and data control, operational control, emergency preparedness and response)	**Planning and implementation** (initial review, system planning, development and implementation, OSH objectives, hazard prevention, control measures, management of change, emergency prevention, preparedness and response, procurement, contracting)	3. OHS management manuals
4.5 Prevention planning (OHS management program)	4.5 Planning (human resource management, establishing targets, management program, management of change, loss prevention, contingency and emergency preparedness)	7 Organization and implementation	6.3 Implementation and operation (structure and responsibility, consultation, training, awareness and competence, communication, documentation, operational control, management and Control of contractors, emergency preparedness and response)			4. OHS management organization system
4.6 OHS handbook and documents	4.6 Operational and functional management (marketing, development, procurement, processing)	7.1 Organization (responsibilities, authority and resources, OHS targets, consultation with staff, representatives, qualification, training and information, expert assistance, information management, internal and external information and communication)				5. OHS committees
4.7 Performance control (active control, verification, ensuring control, nonconformity and corrective actions)	4.7 Measuring and evaluation (process performance, evaluation, assessment, audit and review, records)	7.2 Implementation of plan of action	6.4 Checking and corrective action (monitoring and measurement, accidents and incidents of noncompliance with OHSMS—corrective and preventive action, audit, records)	4.5 Checking and corrective action (performance measurement and monitoring, accidents, incidents, nonconformances, corrective and preventive action, records management, audit)	**Evaluation** (performance monitoring and measurement, investigation of work-related injuries, OSH performance, audit, management review)	6. Guidance to subcontractors on the premises
4.8 Safety and health records	4.8 Improvement (nonconformance handling, preventive and corrective action, improvement)	8 Performance measurement (reviewing OHSMS, evaluation of plan of action)	6.5 Management review	4.6 Management review	**Action for improvement** (preventive and corrective action, continual improvement)	7. OHS education
4.9 Evaluation of OHSMS (audits, review of OHSMS)						8. OHS daily activities
						9. Machinery and equipment safety
						10. Chemical safety
						11. Working environment
						12. Work management
						13. Medical surveillance
						14. Promotion of physical and mental health
						15. Comfortable workplaces
						16. Measures for workers in need of special care
						17. Traffic safety of commuters
						18. Analysis of accident causes
						19. Emergencies

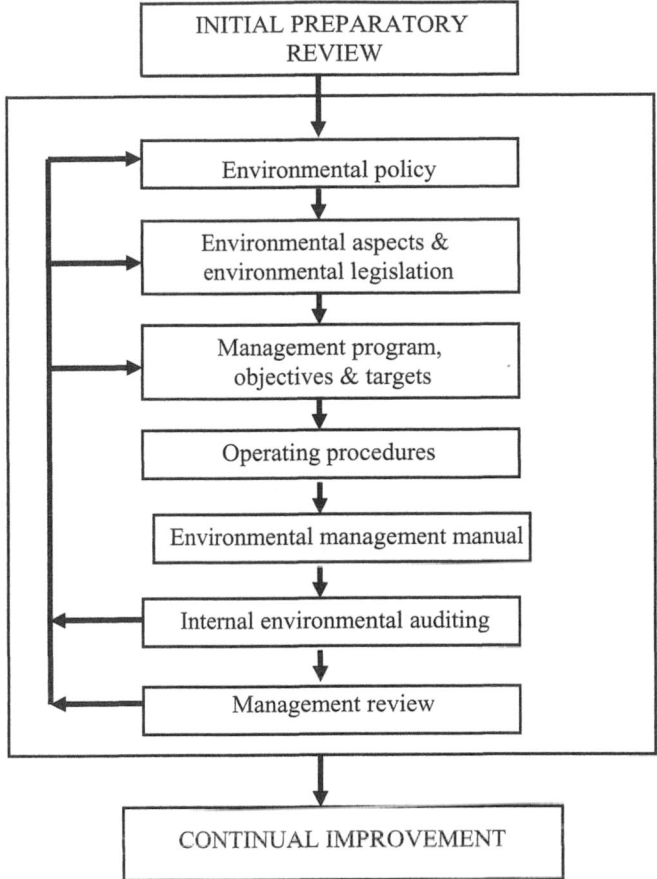

FIG. 25.1. The structure of the ISO 14001 environmental management system.

conformance with the standard. The ISO 14001:2004 revision adds further clarification on legal and other requirements, internal and external communication, calibration, evaluation of compliance, etc. This generic management system (Figure 25.1) can be adopted to other systems and applied to an organization, large or small—independent of whether its *product* is a service, a business enterprise, a public administration, or a government sector. It is suitable for suppliers' declaration of conformity, assessment of conformity by an external stakeholder such as a business client, and for certification of conformity by an independent certification body. The ISO 14001 has been widely used as template for OHSMS development by different standards bodies.

US OSHA and Other OHS Management Programs

The United States, Department of Labor, Occupational Safety and Health Administration (OSHA) has long encouraged states and territories to develop their own safety and health plans under the Occupational Safety and Health Act (1970). There are currently 26 state plans: 23 cover both private and public (state and local government) employment, and 3 states (Connecticut, New Jersey, and New York) cover the public sector only. The State of California (Cal OSHA) introduced the general OHS requirements and programs (1995), whose document was originated in a State Senate bill, SB-198, that required developing and implementing an OHS program in an enterprise with greater than 25 employees.

The OSHA promulgated the Voluntary Protection Programs (VPP) as a means to recognize and approve worksites with exemplary OHS management programs (U.S. Federal Register, 1988). The three levels of VPP are Star, Merit, and Demonstration designed to recognize outstanding achievements by enterprises that have successfully incorporated OHS programs into their total management system. The OSHA Safety and Health Program Management guidelines (U.S. Federal Register, 1989) identify five general elements—management commitment, employee involvement, worksite analysis, hazard prevention and control, safety and health training. The guidelines aim at instituting and maintaining an organizational program that provides systematic policies, procedures, and practices to protect employees from, and allow them to recognize and control general workplace hazards, specific job hazards, and potential hazards.

The OSHA envisages a strategic management plan for the period 2003 to 2008, that by implementing OSHA's goals every employer and employee will recognize that OHS adds value to American businesses, workplaces, and workers' lives. The OSHA's Safety and Health Achievement Recognition Program (SHARP) includes employer's consultation visit, demonstration of exemplary achievements in abating workplace hazards, and development of a safety and health program. The OSHA's Strategic Partnership Program and Alliance Program are the newer members of its cooperative programs. Whereas SHARP and VPP entail one-on-one relationships between OSHA and individual worksites, the voluntary strategic partnerships seek a broader impact by building cooperative relationships among OSHA, employers, employee representatives and others. The Alliance Program is open to all groups, including trade or professional organizations, businesses, labor organizations, educational institutions, and government agencies. The program requirements are that the OSHA and the participating organizations must define, implement, and meet a set of short- and long-term goals for (a) training and education, (b) outreach and communication, and (c) promoting the dialog on workplace safety and health.

The American Industrial Hygiene Association released OHSMS guidance document (AIHA 1996; Redinger & Levine, 1996; ANSI accredited) for designing, implementing, and evaluating OHS management systems. For sector-specific application, the American Textile Manufacturers Institute (1994) introduced a 10-point E3 program that includes corporate environmental policy, management responsibility and accountability, objectives and targets, regulatory compliance, emergency response plan, and environmental audit. NSF International (Ann Arbor, MI), an accredited certifying body, introduced NSF 110 (1995) for the purpose of environmental management, including components of OHS.

E&P Forum (Oil and Gas Companies) and Other Business Charters

The International Exploration and Production Forum (E&P Forum) guidelines (1998) of the international association of oil and gas companies apply to developing and applying health, safety, and environmental management systems in exploration and production operations worldwide. A comprehensive management system addresses core structural issues for programs and procedure development, including a controlled documentation system. With similar applications, the American Petroleum Institute (1993) brought out strategies for today's environmental partnership, focusing on pollution prevention, operating and process safety, community awareness concerning safety, health and environmental issues, emergency readiness and response planning, and proactive government interaction. The International Chamber of Commerce (1991) published the business charter for sustainable development in environmental management, which comprises of 16 corporate principles to help businesses in improving their environmental performance and contributing to sustainable development. For integrated environmental management, the charter includes unique issues like transfer of environmentally sound technology and management methods. These management system guidelines are sufficiently generic to make adaptable to different types of enterprises, systems, and their cultures.

NATIONAL STANDARDS

British HSE Guidance and BS 8800

Following the British Health and Safety Work Act, the Health and Safety Executive (HSE) updated the guidance, HS (G) 65 (1993) for enterprise level OHS management (Figure 25.2). The document outlines a system framework for describing the issues that organizations need to address to manage health and safety effectively. Together with the approved codes of practice on various regulations, which include a requirement for arrangements to manage OHS, the HS (G) 65 approach provides the basis to employers on managing health and safety.

The British Standards Institution released BS 8800 (1996), guidance on OHSMS to improve OHS performance in organizations and on how the management of OHS may be integrated with other aspects of business, and to assist organizations to establish responsible image within the marketplace. Structurally, the BS 8800 standard is based on the general principles of good management, having increased visibility on two approaches. The first approach is based on the HSE guidance—HS (G) 65—and the second approach is based on ISO 14001 standards. The guidance presented in each approach is essentially the same, the only difference being

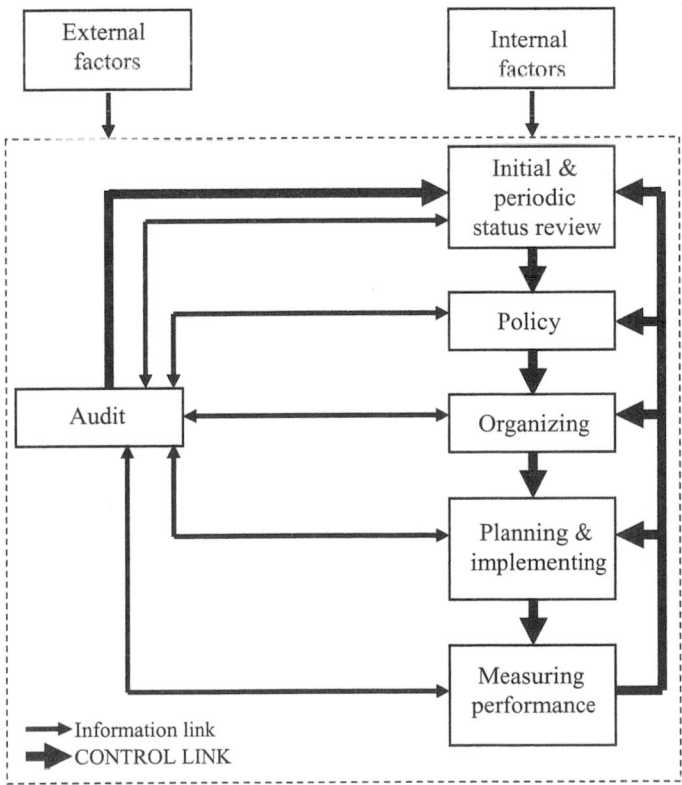

FIG. 25.2.   Element of OHSMS in HS (G) 65 approach.

the order of presentation of system elements. The BS 8800 contains annexes on the practical guidance, including planning and implementing, risk assessment, and measuring performance. The standard states explicitly that it should not be quoted as if it were a specification, and that it should not be used for certification purposes.

Australia and New Zealand

The Standards Australia/Standards New Zealand (1997) jointly issued AS/NZS 4801 and 4804 standards on OHSMS. The underlying structure of the management system is similar to ISO 14001/4. Within the national OHS improvement framework, most states and territories of Australia use the standard to structure OHSMS model. The Joint Accreditation System of Australia and New Zealand (JAS-ANZ) controls certification. The Safety Management Achievement Program (SafetyMAP), developed by the Health and Safety Organization (HSO), Victoria, Australia (1999), is a practical tool for an organization to (a) assess the scope and effectiveness of its OHS policies and operations, (b) plan improvements in the operations, (c) develop benchmarking standards, and (d) gain recognition of the standards achieved. Although it uses audit criteria similar to ISO 9001:2000, the goals and objectives and the performance measures are not explicit in the system. The SafetyMAP is used in an HSO certification program at three levels—initiation, transition, and advanced level of certification.

Spain

Asociacion Española de Normalizacion y Certificacion, UNE 81900 (1996) documented prevention of occupational risks and general rules for implementation of an OHSMS. Three documents included are the auditable specification standard, audit process, and a vocabulary document. In overall analysis, the Spanish certification standard is sufficiently generic to be adaptable to any organization regardless of its size or the nature of its activities.

Norway

Norges Standardisengsforbund 96/402803 (1996) issued management principles that cover multiple issues in a single document, such as:

> Process quality having an influence on the quality of products and services; protection of the environment, including protection against pollution and waste; OHS as well as safety of products and services.

The standard facilitates operation within the framework of statutory requirements and evaluation of an organization's operations against established and recognized international standards, as appropriate.

The Netherlands

Nederlands Normalisatie-Instituut, Guide NPR 5001 (1997) includes the OHSMS functions, namely:

- Guaranteeing effective OHS management through a planned approach to risks prevention and management, implementation of the plans, and integration into the business management

- Encouraging people in the organization to contribute to controlling the risks and the step-by-step OHS management
- Prompting the organization to learn from experience in order to continue improving OHS
- Building relationships outside the organization (clients, Labor Inspectorate)
- Understanding the way in which the organization meets the statutory OHS obligations.

In the Dutch system, every employer is obliged to have a contract with a certified OHS Advisory Service ("Arbodienst"), in building up and maintaining an OHSMS by making the risk assessment, implementation plan, and carrying out a regular evaluation of the OHS situation. The Dutch Labour Inspectorate has a basic approach of enforcing actions directed on the failures in the OHSMS, in combination with enforcing actions directed on failures at workplace level.

Ireland

To make compatible with the management systems of ISO 9000, ISO 14000, and BS 8800 Standards, a Code of Practice for an OHSMS was issued by the National Standards Authority of Ireland (NSAI, 1997). Like the Dutch standard, the Irish code of practice is not intended for certification purposes. It may be implemented in support of compliance with OHS legislation, which is regulated by the Irish Health and Safety Authority under the Safety, Health and Welfare Act (1989).

OHSAS 18001:1999

With the background of national standards and certification schemes on OHSMS, a specification standard, Occupational Health and Safety Assessment Series (OHSAS 18001:1999) has been developed by a consortium of national standards and certification bodies, and specialist consultants. The schematic of the system elements of OHSAS 18001 specifications is illustrated in Figure 25.3. An accompanying publication, OHSAS 18002 (2000), is a guidance document, which corresponds directly to the specifications. The British Standards Institution

FIG. 25.3.   Element of OHSMS in OHSAS 18001 approach.

states that the creation of OHSAS was necessary to reduce the marketplace confusion and meet demands for creating a global framework to apply in any organization that wishes to:

> Establish an OHSMS to eliminate or minimize risk to employees and other interested parties who may be exposed to OHS risks associated with its activities; implement, maintain and continually improve an OHSMS; assure itself of its conformance with its stated OHS policy; demonstrate such conformance to others; seek certification of its OHSMS by an external organization; or make a self-determination and declaration of conformance with this OHSAS specification.

The national standards and proprietary certification schemes, and other participations, which were used to create the OHSAS 18001 specification, are:

- National Standards Authority of Ireland, NSAI SR 320 (1997). Recommendation for an occupational health and safety (OH and S) management system
- British Standards Institution, BS 8800 (1996). Guide to occupational health and safety management systems
- Bureau Veritas Quality International (BVQI SafetyCert). Occupational safety and health management standard
- Det Norske Veritas, DNV Standard for certification of OHSMS (1997).
- Lloyds Register Quality Assurance (LRQA SMS 8800; 1998). Health and safety management systems assessment criteria
- SGS Yarsley International Certification Services, SGS & ISMOL ISA 2000 (1997). Requirements for safety and health management systems
- Asociacion Española de Normalizacion y Certificacion, UNE 81900 (1996). Prevention of occupational risks: General rules for implementation of an occupational safety and health management system
- Standards Australia and Standards New Zealand, AS/NZS 4801/4804 (1997), Occupational health and safety management systems—Specification with guidance for use
- Nederlands Normalisatie-Instituut, Technical Report NPR 5001 (1997). Guide to an occupational health and safety management system
- National Quality Assurance
- SFS Certification
- South African Bureau of Standards (1998)
- International Safety Management Organization Ltd
- Standards and Industry Research Institute of Malaysia, Quality Assurance Services
- Draft BSI PAS 088, Occupational health and safety management systems.

In addition, other reference publications that provide information and guidance include ISO guidelines for auditing quality and environmental systems.

ILO-OSH MS

Recognizing the positive impact of introducing OHSMS at the enterprise level, the International Labor Office brought out a consensus model (ILO-OSH, 2001), according to the principles defined by the ILO's tripartite constituents (governments, employers, and workers). The readers may note that the ILO-OSH model is referred to as the occupational safety and health management system (OSH-MS). In this document, we continue to use the abbreviation as OHSMS. The ILO-OSH model reflects the ILO values that advance the objectives of labor conventions, for example, occupational safety and health (1981), chemicals (1990), occupational health

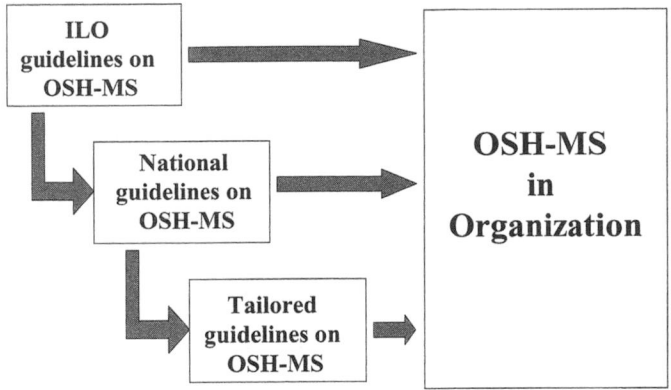

FIG. 25.4.   ILO-OSH national framework for OHS management system.

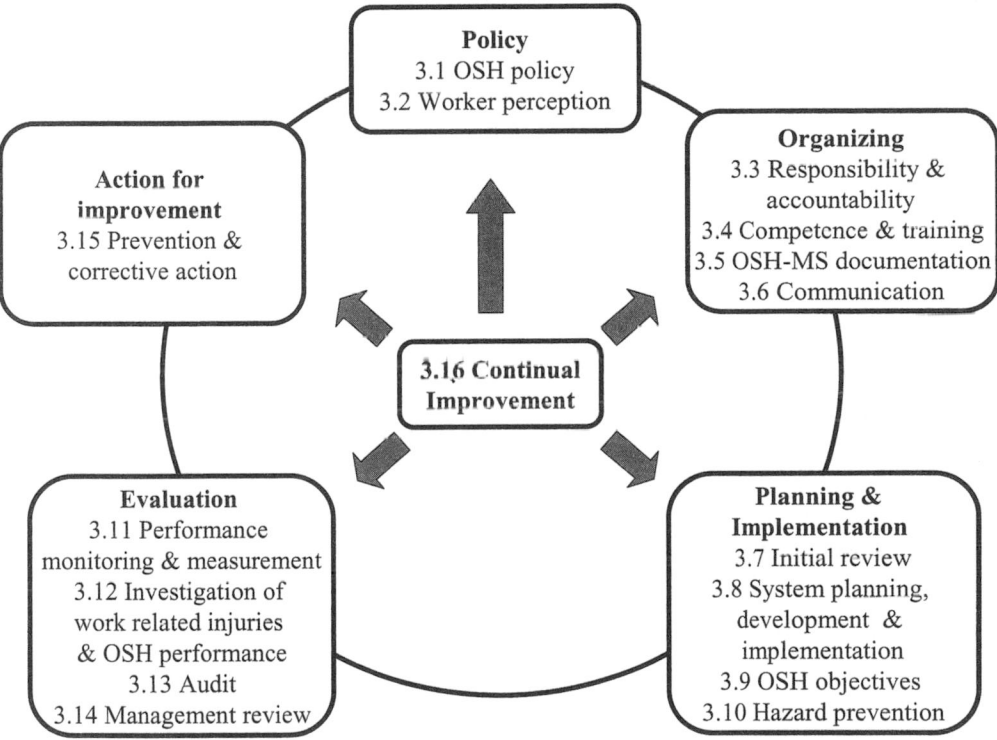

FIG. 25.5.   Elements of ILO-OSH guidelines on OHS management system.

services (1985), prevention of major industrial accidents (1993), asbestos (1986), safety and health in construction and mines (1988, 1995), and others.

Structurally, the ILO-OSH model is distinctively different from the OHSAS 18001, by the unique focus of the ILO model in developing national OHS policy and guidelines, including a tailored enterprise-level management framework (Figure 25.4). For enterprise-level application, the ILO-OSH model is constituted by 16 elements, which are grouped under five sections, namely, policy, organizing, planning and implementation, evaluation, and action for improvement (Figure 25.5). Essentially, the management elements have been ordered in Edwards Deming's PDCA (plan-do-check-act) cycle.

TABLE 25.2
Analysis of Different OHS and Environment Management System Standards

Management Systems	Leadership Commitment	OHS Policy	OHS Strategic Objectives	Accountability, Responsibility, and Authority	Regulatory Compliance and System Conformance	Performance Measures	System Planning for Hazard Identification	Risk Assessment	Hazard Control System	Emergency Preparedness and Response	Training System	Operational Control Procedure	Communication System/Feedback Channels	Document and Records Management	Monitoring and Evaluation	Auditing and Self-Inspection	Incident Investigation, Nonconformances—Root Cause Analysis	Medical Program and Surveillance	Continual Improvement	Integration	Management Review
Responsible Care, UK (1998)	●[a]	●	●	●	●	○[a]	●	●	●	●	●	○	●	●	●	●	●		●	●	●
EMAS (1993)	●	●	●	●	●	●	●	●	○	●	●	●	○	●	●	●	●		●	●	●
ISO 14001 (1996, revision 2004)	●	●	●	●	●	●	●	●	○	●	●	●	●	●	●	●	○		●	●	●
VPP (1988)	●	●	●	●		●	●	●	○	●	●	●	●	●	●	●	●	●	●	●	●
OSHA Safety & Health Program (1989)	●	●	●	●	●	●	●	●	○	●	●	●	●	●	●	●	●	●	●	●	●
AIHA (1996)	●	●	●	●		●	●	●	○	●	●	●	●	●	●	●	●		●	●	●
E&P Forum (oil/gas companies) (1998)	●	●	●	●	●	●	●	●	○	●	●	●	●	●	●	●	●	●	●	●	●
HS (G) 65 (1993) /BS 8800 (1996)	●	●	●	●	●	●	●	●	○	●	●	●	●	●	●	●	●		●	●	●
AS/NZS 4801 (1997)	●	●	●	●	●	○	●	○	○	●	●	●	●	●	●	●	●		●	●	●
SafetyMAP (1999)	○	●	○	●	○	○	●	●	●	●	●	●	●	●	●	○	●		●	●	○
Spain, UNE 81900 (1996)	●	●	●	●	●	●	●	●	○	●	●	●	●	●	●	●	●	●	●	●	●
Norway, Management standard (1996)	●	●	●	●	●	●	●	●	●	●	●	●	●	●	●	●	●		●	●	●
The Netherlands, NPR 5001 (1997)	●	●	●	●	●	●	●	●	○	●	●	●	●	○	●	●	●		●	●	●
Ireland, NSAI (1997) KS-NR 11-97	●	●	●	●	●	●	●	●	○	●	●	●	●	●	●	●	●		●	●	●
OHSAS 18001 (1999)	●	●	●	●	●	●	●	●	●	●	●	●	○	●	●	●	●		●	●	●
ILO-OSH (2001)	●	●	●	●	●	●	●	●	○	●	●	●	○	●	●	●	●	○	●	●	●
Amer. Textile Manuf. Institute (1994)	○	●	●	●	●	●	●	●	●	●	●	●	○	●	●	●	●	●	●		●
ICC Business charter (1991)	●	●	●	●	●						●	●	●						●	○	●
Japan, JISHA (1997)	○	●											○			●	●		●	○	
India, Factories Act (1948/1988)	●	○		●			●	●	●	○	●	○	○		●	●	●	●		○	●

[a] ●, explicitly stated; ○, implied or partly addressed.

Comparative Analysis

Across nations, there are a variety of OHSMS models, with a mix of auditable and nonauditable standards, and guidance documents (IOHA, 1998; Levine & Dyjack, 1997; Nag & Nag 2003a). The national and enterprise level OHS and environmental management models exhibit wide coverage of the system elements, as summarized in Table 25.2. The status of most OHSMS models, either a local system or a system certified by an accredited body, are the enterprise level framework, with due cognizance of the applicable regulatory requirements. The management systems, whether certified or not, are therefore a means of meeting legislative obligations, but they are not substitutes for the OHS laws. The Responsible Care, EMAS, OSHA programs, E&P Forum, and other sector-specific models have wide differences in the structure and content. The ISO 14001 environmental management system has been adopted as a generic template by many Standards bodies to develop OHSMS models; however, the key OHSMS variables that are missing in ISO 14001-based models are employee participation, medical surveillance, and health programs.

The Australia/New Zealand AS/NZS 4801, Norwegian and Spanish standards, SafetyMAP, AIHA OHSMS, EMAS, and the Responsible Care programs are considered as the strong auditable systems. The AIHA model contains 20 primary sections, corresponding to the ISO 9001:1994 version of the quality assurance system. The American Textile Manufacturers Institute's (1994) E3 program and NSF 110 (1995) refer to environmental management; however, the content and the structural orientation of the systems are generally devoid of many key OHS management variables. The sector-specific models, such as E&P Forum (oil and gas companies) and the American Petroleum Institute (1993), address core structural issues for program and procedure development, including a documentation system. Particularly, the E&P Forum is comprehensive in addressing the traditional OHS issues and the management system approaches, such as leadership commitment, resource allocation, management review, continual improvement, and integration with other organizational systems.

The OHSAS 18001 and ILO-OSH (2001) models are competitive in approach. The OHSAS specification has been made compatible with the ISO 9000/ISO 14000 standards in order to align and integrate quality, environment, and OHS management systems in organizations. For the enterprise-level application, the overall contents and elements of the ILO-OSH model largely resemble the elements of BS 8800/OHSAS 18001 frameworks, with specific mention to employee participation. The OHSAS 18001 remains broadly focused on "interested parties" in scope. The ILO model is not intended for certification purposes, whereas the OHSAS auditable standard has widely been adopted for certification purposes. The correspondences of the OHSAS 18001 and ILO-OSH are given in Table 25.3.

Key Features in OHSMS Documentation

An organization has been advised to implement OHSMS for the systematic identification, evaluation, and prevention or control of workplace hazards, including potential hazards. As the size of a worksite or the complexity of an operation increases, the need for written guidance increases to clearly communicate policies and consistently apply guidelines and instructions. Because the basic premises of the OHSMS frameworks (e.g., OHSAS 18001 and ILO-OSH 2001), are built on the voluntary approach, these do not set any rigid technical requirements. An organization may develop flexible strategies within its unique operating environment. The objective is to outline the scope of the management system and to bring an orderly arrangement of interdependent activities that define the processes in an organization. Generally, the OHSMS

TABLE 25.3
Correspondence Between OHSAS 18001 and ILO-OSH (2001) Guidelines

Clause	OHSAS 18001 (1999)	Clause	ILO-OSH (2001)
1	Scope	1.0	Objectives
2	Reference publications		
3	Terms and definitions		
4	OHS management system elements	3.0	OSH management system
4.1	General requirements	3.0	OSH management system
4.2	OH&S policy	3.1	OSH policy
4.3	Planning	3.7	Initial review
		3.8	System planning, development and implementation
4.3.1	Planning for hazard identification, risk assessment and risk control	3.10	Hazard prevention
		3.10.1	Prevention and control measures
		3.10.2	Management of change
		3.10.5	Contracting
4.3.2	Legal and other requirements	3.7.2	Initial review
		3.10.1.2	Hazard prevention
4.3.3	Objectives	3.8	System planning, development and implementation
		3.9	OSH objectives
		3.16	Continual improvement
4.3.4	OH&S management programs	3.8	System planning, development and implementation
4.4	Implementation and operation		
4.4.1	Structure and responsibility	3.3	Responsibility and accountability
		3.8	System planning, development and implementation
4.4.2	Traning, awareness and competence	3.2	Worker participation
		3.4	Competence and training
4.4.3	Consultation and communications	3.2	Worker participation
		3.6	Communication
4.4.4	Documentation	3.5	OSH managaement system documentation
4.4.5	Document and data control	3.5	OSH managaement system documentation
4.4.6	Operational control	3.10.2	Management of change
		3.10.4	Procurement
		3.10.5	Contracting
4.4.7	Emergency preparedness and response	3.10.3	Emergency prevention, preparedness and response
4.5	Checking and corrective action		
4.5.1	Performance measurement and monitoring	3.11	Performance monitoring and measurement
4.5.2	Accidents, incidents, nonconformances, and corrective and preventive action	3.12	Investigation of work related injuries, etc.
		3.15	Preventive and corrective action
4.5.3	Records and records management	3.5	OSH managaement system documentation
4.5.4	Audit	3.13	Audit
4.6	Management review	3.14	Management review

documentation might be followed in consistent with the ISO 9001:2000/ISO 14001 systems, with three levels of documentation in a pyramidal structure (Figure 25.6).

The top level is the OHSMS manual, referring to the policies of OHS management. The second level is made up of the standard operating procedures (SOPs)—task procedures, health and safety procedures, risk assessment and specific instructions to the processes and products. The third level is referred to as records management that essentially means the documented evidence of OHS performance.

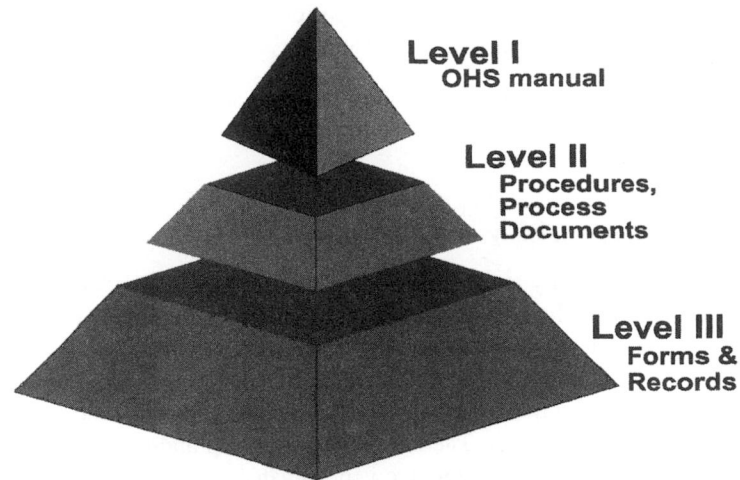

FIG. 25.6.   Pyramidal arrangement of the OHSMS document.

The structural content of the OHSMS documentation depends on the model advocated (Nag, 2002). At the enterprise level, the OHSMS variables that are central to a systems approach are the following:

- Leadership commitment, OHS policy, and strategic objectives
- Accountability, responsibility, and authority
- Regulatory compliance and system conformance
- Performance measures
- System planning for hazard identification and risk assessment
- Hazard control
- Emergency preparedness and response
- Training system
- Operational control procedure
- Communication system/feedback channels
- Document and records management
- Monitoring and evaluation
- Auditing and self-inspection
- Incident investigation, nonconformances—root cause analysis
- Medical program and surveillance
- Continual improvement
- Integration
- Management review

Leadership Commitment, OHS Policy, and Strategic Objectives

The leadership commitment to occupational health and safety may be operationally defined. The OHS policy represents the foundation for developing, implementing, and improving an organization's OHSMS. This documented policy reflects the formal leadership commitment for good OHS practices and management. It sets out an overall sense of direction, goals and objectives, and principles of action to meet OHS performance expectations of an organization, and therefore these must be measurable. For example,

> Reduction in risk levels; introduction of additional features into the OHSMS; the steps taken to improve existing features, or the consistency of their application; elimination or reduction in the frequency of undesired incident(s).

Accountability, Responsibility, and Authority

The responsibility, authority and accountability are the variables, which address the manner an organization defines the roles of personnel who are involved in OHSMS management. Therefore, the establishment of organizational structures is a part of the management commitment. Each OHS functions (e.g., budget, health and safety committee, medical services, emergency preparedness, and training) shall have defined the authority and responsibility, and how each are related with the other. The executive ownership and accountability is a critical feature. A senior management representative is designated, and the designee is held directly accountable for each aspect of the OHS program, for example, anticipation, recognition, evaluation, and control of OHS hazards. The role and responsibilities shall be documented and communicated to all employees and other relevant parties.

Regulatory Compliance and System Conformance

The government regulations as well as the nongovernment standards to which an organization subscribes impose requirements on OHSMS and, therefore, can affect the way an OHSMS is designed, implemented, and operated. Broadly, both regulatory compliance and conformance to other system requirements lay the groundwork for a systematic management approach. It seeks to focus on:

> The broad framework and responsibility for action on OHS, and on the employer in particular; emphasis to consultative arrangements, disclosure of information; and back-up enforcement and prosecution action.

The organization must have procedures for (a) identifying and accessing information, and identifying which activities in the organization are affected by applicable legal and other requirements; and (b) monitoring the implementation of controls consequent to changes in OHS legislation.

Performance Measures

The ability to measure OHS performance over time is a prerequisite in eliminating injuries and illnesses, and verifying continuous improvement. For valid and reliable performance measurements, the indicators, variables, measurement units, and their logical relationships are required to be established. In terms of the indicators to measure, the distinction has been made in the OHS literature between leading and trailing indicators, and therefore the organization needs to identify the key indicators. These should include, but not be limited to, parameters that determine whether:

> OHS policy and objectives are achieved; risk controls have been implemented; lessons learnt from OHSMS failures, including hazardous events (accidents and illnesses); training, communication and consultation programs for employees and interested parties are effective; and information that can be used to review and improve OHSMS is being produced and used.

System Planning for Hazard Identification and Risk Assessment

The system planning addresses the initial OHSMS development, ongoing revision, and modification of the system to achieve the required level of performance and continuous improvement. The organization requires a documented OHS management program (strategies and plans of actions), and primarily, it must take account of all significant OHS hazards in its domain. In OHSAS 18001, the hazard is defined as a source or situation with a potential for harm in terms of injury or ill health, damage to property and workplace environment, or a combination of these. The documented procedure(s) should be established and records maintained for:

- Identification of hazards and determination of the level of the risks associated with the hazards
- Description of, or reference to, the measures to monitor and control the risks, particularly those are not tolerable
- The OHS objectives and actions to reduce identified risks, and follow-up activities to monitor progress in their reduction
- Identification of the competency and training requirements to implement the control measures
- Control measures as part of the operational control

Hazard Control

The hazard control is broadly defined to reduce or eliminate occupational hazards. The methods of the control system are generally recognized as administrative controls, personal protective equipment (PPE), or engineering controls. The hazard control also covers management of hazardous agent, including such things as radioactive materials, noise, heat, cold, lasers, and hazardous wastes.

The OHSAS 18001 emphasizes hazard identification and risk assessment, and it remains less definitive in hazard control. In many of the OHSMS models, approaches to management of hazardous agents are not explicit. The OHSAS 18001 specifies that the management of risk should reflect the principle of the elimination of hazards where practicable, followed by risk reduction (reducing the likelihood of occurrence or potential severity of injury or damage), with the adoption of *PPE as a last resort*. On the other hand, ILO-OSH (2001) mentions the implementation of preventive and protective measures to control hazards and risks, in an order of priority from elimination of the hazard or risk to provision of PPE.

Emergency Preparedness and Response

Emergency preparedness and response refers to the manner in which the organization prepares for and responds to emergencies and accidents. The organization requires the following: to assess potential accident and emergency response needs, develop plan and procedures to cope with them, test its planned responses, and seek to improve the effectiveness of emergency responses. This may include:

- Identification of potential accidents and emergencies
- Identification of the personnel to take charge during the emergency; responsibility, authority, and duties of personnel with specific roles (e.g., fire-fighters, first-aid staff, and toxic spillage specialists);
- Documented emergency plans and procedures, including evacuation procedures
- Identification of hazardous materials, and emergency action required

- Interface with external emergency services including communication with statutory bodies, and the public
- Availability of vital information during the emergency, for example, plant layout drawings, hazardous material data, and contact telephone numbers
- Emergency equipment list (alarm systems, means of escape, fire-fighting equipment, first aid equipment, and its test records)
- Records and review of the practice drills and recommended actions arising from the reviews

Operational Control Procedure

The organization requires the establishment of procedures to control its identified risks (including those introduced by contractors or visitors) and the documentation of instances where a failure to do so could lead to accidents or deviations from the OHS policy and objectives. The operational control procedures should be reviewed periodically, for its suitability and effectiveness. The examples of areas in which risks typically arise are:

- *Purchase or transfer of goods and services and use of external resources* (e.g., purchase or transfer of hazardous materials, documentation for handling of machinery and materials, evaluation of the OHS competence of contractors, and OHS provisions for new plant or equipment).
- *Hazardous tasks* (e.g., identification of hazardous tasks, predetermination of working methods, prequalification of personnel, and permit-to-work systems to control exposures to hazardous tasks).
- *Hazardous materials* (e.g., identification of inventories, storage locations; storage provisions and control of access; and access to MSDS).
- *Maintenance of safe plant and equipment* (e.g., maintenance of plant and equipment; segregation and control of access; testing of OHS related equipment; local exhaust ventilation systems, medical facilities, PPE, guarding; shutdown, fire suppression; materials handling equipment; radiological sources, safeguards, and monitoring devices).

Training System

The training has been an integral component of OHS management, having qualified personnel with direct OHSMS responsibilities as well as external consultants who may provide OHS services to the organization. An effective procedure is required for ensuring the competence of personnel to carry out their designated functions. The documentation may include (a) competency requirements for individual roles, (b) analysis of training needs and programs for employees, and (c) training records and records of evaluation of the effectiveness of training. However, ILO-OSH (2001) makes mention that training should be provided to all participants at no cost and should take place during working hours, if possible.

Communication System and Feedback Channels

An effective communication system with defined feedback channels encourages participation in good OHS practices. The organization's OHS policy and objectives need support from all those affected by its operations, by a process of consultation and communication. A viable communication system should identify how, and to whom, information on the functioning of the OHSMS will be transmitted. For example:

Formal management and employee consultations through OHS committees; employee involvement and initiatives in hazard identification, risk assessment and control; employee OHS representatives with defined roles in communication with management, including involvement in accident and incident investigation, site inspection; OHS briefings for employees and other interested parties; notice boards containing OHS performance data, newsletter, poster, etc.

Document and Records Management

An essence of a management system is a well-structured documentation to ensure that the system is effective and efficiently operated. A well-functioning document and record management system allows effective communication of policies and procedures and demonstrates an organization's ability that it is achieving what it said it would. In other words, the document and record management system provides one of the key indicators of whether the OHSMS is currently in conformance. All documents and data critical to the operation of the OHSMS should be identified and controlled. The written procedures should define the controls for the identification, approval, issue, and removal of OHS documentation, together with the control of OHS data. Typically, it should include:

> Document control procedure, including assigned responsibilities; document master lists or indexes; list of controlled documentation, OHS records, and its location.

The documented procedure to control OHS records should demonstrate that the OHSMS operates effectively, and the processes have been carried out under safe conditions. For example, the procedure should be defined for the identification, maintenance, and disposition of OHS records.

Monitoring and Evaluation

The monitoring and evaluation system of an OHSMS encompasses several areas of concern, such as performance monitoring, auditing and self-inspection, incident investigation, root cause analysis, the availability of medical surveillance and health programs, and management review. The documentation may include:

> Procedure(s) for monitoring and measuring; critical equipment lists, including equipment inspection checklists; workplace conditions standards, inspection schedules and checklists; measuring equipment lists, measurement procedures; calibration records, maintenance activities and results; OHSMS checklists, audit outputs and non-conformance reports; and evidence of the results of implementing the procedure(s).

Auditing and Self-Inspection

The auditing and self-inspection are specific activities whereby an organization can assess the changes in OHS hazards and continuously evaluate the effectiveness of the OHSMS to respond to the changes. It brings essential information to other OHSMS components, including the training, hazard control, and preventive and corrective action systems. An internal audit program allows reviewing its own conformity to the specification it subscribes. The planned audits should be carried out by personnel from within the organization or by external personnel. Although the ILO-OSH (2001) recommends consultation on the selection of auditors, the OHSAS 18001 requires that the personnel undertaking the audits should be in a position to do so impartially and objectively. The documentation may include:

> OHSMS audit plan/program and procedures; OHSMS audit reports, including non-conformance reports, corrective action requests; closed-out non-conformance reports; and evidence of the reporting of the audit results to top management.

Incident Investigation, Nonconformances—Root Cause Analysis

Incident investigation refers to the activities in determining the origin and cause(s) of accidents and incidents. The organization shall have established documented procedures and records maintained for reporting and evaluating and investigating accidents, incidents, and nonconformances. The documentation may include:

Accidents/incidents and non-conformances; non-conformance register and reports; investigation reports, including hazard identification, risk assessment and risk control reports; management review input; evidence of evaluation of the corrective and preventive actions taken.

The root cause analysis is applied in moving up the causal chain in detecting, analyzing, and eliminating potential cause of nonconformities. The lack of procedures for, or documentation of, the use of root cause analysis may be traceable to, for example, nonconformance related to OHS policy, management commitment, and training.

Medical Program and Surveillance

The medical program and surveillance refers to the activities associated with providing occupational health services within the organization, and the development and operation of a health surveillance and promotion program. These are key aspects in an OHSMS, which provide feedback to the hazard control system. Many of the existing OHSMS models evolved from the ISO 14001 templates do not explicitly mention the elements of medical program; however, these remain implied at the core of the system.

Continual Improvement

The central concept of continual improvement can be defined as the process of improving the OHSMS, with the ultimate goal of eliminating or minimizing the potential risks of workplace injury and illness. The ILO-OSH (2001) includes a separate subclause on continual improvement; however, the OHSAS 18001 detailed the processes that should be taken to achieving continual improvement.

Integration

An ideal characteristic of an OHSMS is its flexibility that facilitates integration of the management system within the core business functions of the organization. Also, the relative effectiveness of an OHSMS model depends on its compatibility with other management systems in practice. The OHSMS models, developed on the ISO 14001 template, and ILO-OSH model include system elements, which are connected through feedback channels. This would mean that the issues and aspects of the OHSMS are at the core of the organizational activities—the employer, employees, and other stakeholders are committed to OHS matters. Irrespective of whether an organization decides to adopt OHSAS 18001, ILO-OSH, or other systems, the organization that is certified to process-based models of ISO 9001:2000 or ISO 14001 has already in place the generic framework to integrate quality, environment, health and safety, and other management systems. The goal is to establish cost-effective operational structures for streamlined and synergistic delivery of OHS services, within the scope of the business plan of the organization.

Management Review

Management review is a key attribute of strong management commitment to OHS. This is the means whereby the overall performance of an OHSMS is evaluated. This involves evaluating the ability of OHSMS to meet the overall needs of the organization, its stakeholders, its employees, and regulating agencies. The top management should periodically review whether the operation of the OHSMS remains suitable for achieving the organization's stated OHS policy and objectives, and whether the OHS policy continues to be appropriate. The documentation may include:

Minutes of the review meeting; revisions to the OHS policy and objectives; corrective/improvement actions, with assigned responsibilities and targets for completion; review of corrective action; and areas to emphasis in the planning of future internal OHSMS audits.

OHSAS 18001 CERTIFICATION

Certification is the process by which certification bodies examine compliance with the relevant specification of a standard and, where satisfied, issue certification to that effect. Guidance standards on the other hand are not part of the certification process. The organization that has established, implemented, and maintained OHSMS meeting the specification is eligible to apply for OHSAS 18001 certification. The certification bodies set out principles against which an audit is conducted to verify conformance with the specification of the system. Most OHSMS models include audit criteria to enable organizations to undertake self-assessment prior to independent auditing. The certification offers independent verification and auditing that an organization has become more self-regulatory in promoting OHS and taken reasonable measures to reduce workplace injuries and potential risks. The OHSAS 18001 can be utilized for *self-declaration* where an organization can arrange for auditing of their system outside of the certification infrastructure.

The OHSAS 18001 certification process is schematically shown in (Figure 25.7). In preaudit review of an organization's application, the certification body conducts a document review and an optional site visit to see the scope and planning of the entire OHSMS. In completion of the preaudit review, the organization undertakes emendations in the system, and once the follow-up actions have been executed, the organization resubmits its revised OHS manual and related documentation for OHSMS audit. The objective of the second audit is to confirm that the organization conforms to requirements of the OHSAS 18001. It is a minimum requirement that an organization undertakes one internal audit and management review prior to the OHSMS audit. In case of major nonconformities identified during the audit, the organization requires to go through a re-audit. All nonconformities shall be closed out within the specified time frame, and based on which a recommendation for the award of the certification is made. During the 3-year validity period of the certificate, at least three annual surveillance audits are conducted to ensure that the certified organization continues to comply with the OHSAS 18001 requirements. The organization may make use of combined quality, environment, and OHSMS routine surveillance audits to optimize time and resources. A full reassessment for the renewal of certificate shall be conducted every 3 years, about 3 months before expiry of the certificate. Upon successful completion of reassessment, the organization will be issued a new certificate valid for a further 3 years.

Global Acceptance of the National Initiatives

There has been unanimity that the safety, health, and environmental issues are integral to other priorities such as employment, industrial relations, and enterprise development (Frick et al., 2000; Nag, 2002). At the present time, there exist a lot of regulations as well as auditable and nonauditable OHSMS guidelines, which have been differently advocated in the enterprises. What is emphasized in the models is that these are templates and that individual enterprises may

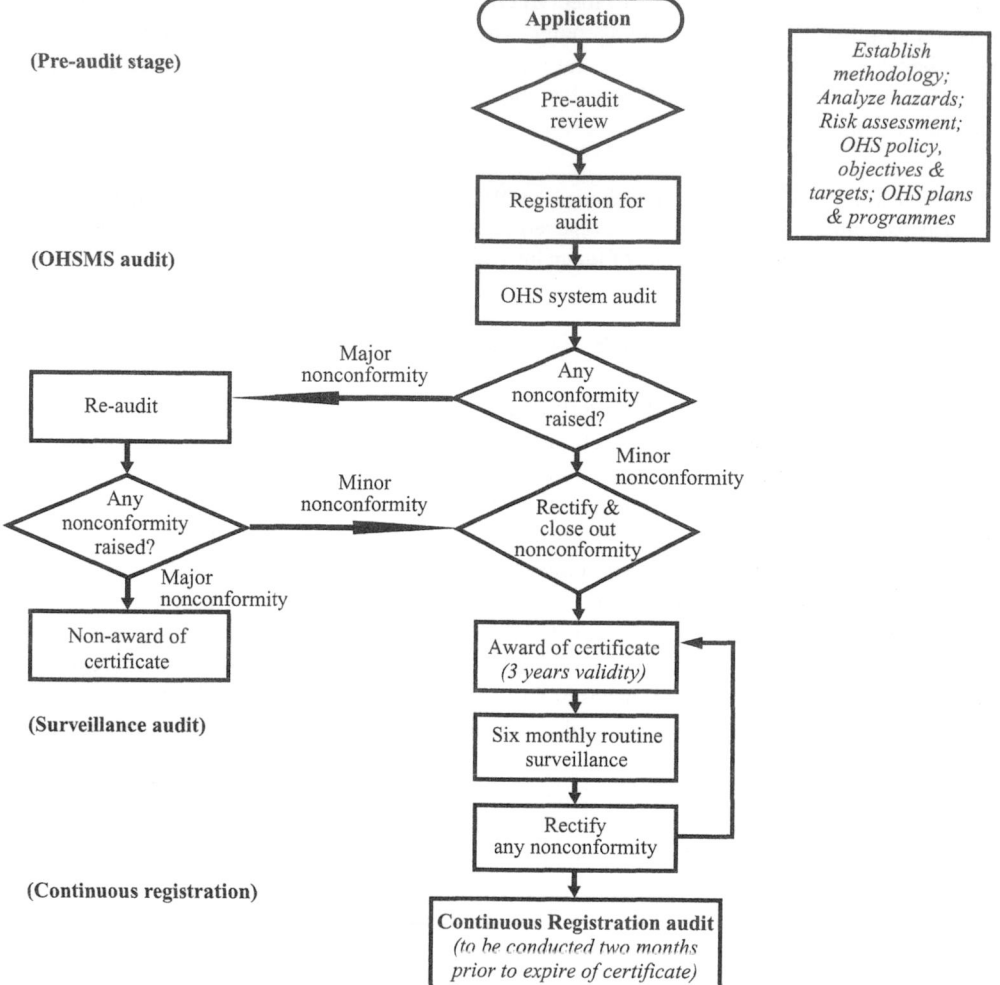

FIG. 25.7. The OHSAS 18001 certification process.

tailor an OHSMS consistent with the national context, business needs, and culture. Limitations exist that there is no consensus on a globally harmonized strategy on OHS management. No one approach uniformly applies to a given country or a sector of employment. The broad outlines of the national and enterprise-level OHSMS approaches are the following:

- Mandatory OHSMS with regulatory and advisory measures
- Voluntary OHSMS standards with national-level certification support
- National OHSMS models promoted through statutory bodies
- Self-regulatory OHS management

Mandatory OHSMS With Regulatory Measures

Several countries favored the mandatory OSHMS in combination with regulatory and advisory measures. In Australia, all states and territories have similar OHS legislation, which places clear duties on employers to provide a safe workplace. The OHS Act 2000 (supplemented by a new OHS Regulation 2001) of the New South Wales, Australia, clarifies that the employers must identify, assess, and eliminate or control risks, and consult with workers on OHS matters.

In Hong Kong, under the Safety Management Regulation (1999), the "general duty provisions" is imposed by laws that place the maintenance of OHS as a continuous legal responsibility on all those who have control over the workplace environment. The proprietors or contractors of selected industrial undertakings are required to develop, implement, and maintain a safety management system, having 14 process elements (e.g., management commitment, training of workers, evaluation of hazards, safety procedures, and programs to protect workers from hazards). The OHS Council as a statutory body is responsible for safety audit schemes, including auditor accreditation.

Indonesia and Singapore are unique in making OHSMS mandatory in specified undertakings. The Indonesia Ministry Regulation (1996) directs to implement OHSMS in the enterprises with 100 or more workers or high workplace risk. The Ministry issues certificates to industries, recommended by an independent audit body. In Singapore, the mandatory implementation of OHSMS applies to shipyards, specified construction worksites, and three classes of factories in the manufacturing sector. No mandatory certification, ministerial audit, and accreditation guidelines have been specified (Kogi & Kawakami, 2002).

The European survey on OHSMS, carried out by senior labor inspectors committee (Konkolewsky, 1999), generally indicated that the Scandinavian countries make obligatory for all enterprises and public institutions, to integrate preventive actions of OHS into the management system of the enterprise (e.g., Internal Control regulations in Norway, 1992). The voluntary certificates may also form part of the documentation relating to systematic OHS and environmental activities in the enterprises. Emphasis has been placed on the Labor Inspectorate to change the inspection methods and strategies, from the traditional onsite detailed inspections to a system-oriented approach. The Norwegian regulation recognizes that higher quality in the working life requires the involvement and active participation of top management.

The Brazilian Ministry of Labor guidelines, NR-9 (1997), established the obligatoriness of implementation of an environmental risks prevention program (PPRA) by employers and institutions that have workers under their responsibilities. The regulation focuses on preserving workers' health and integrity, considering the anticipation, recognition, and evaluation of environmental risks. The Brazilian regulation has its limitation to recognize it as an OHSMS, because it lacks many system elements such as management commitment, OHS goals and objectives, performance measures, programs and procedures, documentation, records management, and auditing.

Voluntary OHSMS Standards With Certification Support

Up to December 2004, some 760,970 management systems conforming to ISO 9001/ISO 14001 standards have been implemented and certified worldwide. China alone represents about 19% of the total number of certification issued up to this date. No official estimate is available of the certification in OHSAS 18001 and other OHSMS models. The certifying bodies that represent OHSAS 18001 have about 80% of the market place certification in ISO 9001/ISO 14001, and this suggests the scope of development of the OHS management systems.

The voluntary OHSMS standards with national-level certification support have been adopted by many countries in Europe (UK, Spain, Norway, Denmark, the Netherlands, Poland, and Ireland) and in the Asia-Pacific regions. For example, the Australia and New Zealand national strategies are complementary to other enterprise-level activities, such as promoting Safety-MAP, certification to AS/NZS 4801 and OHSAS 18001. The Thai Ministry of Labour and Social Welfare and the Ministry of Industry cooperated in setting up an OHSMS standard (TIS 18000:1999), and the Thai Industrial Standards Institute provides general rules on certification. The Polish National Labor Inspection and its worker protection program, PL 9407 (1996), described a collection of the best practices and models to deal with OHS management in small and medium-sized enterprises. The Polish standardization authority established PN-N-18001/18002 standards, defining general requirements of OHSMS and guidelines for risk assessment.

National OHSMS Models Promoted Through Statutory Bodies

Many countries adopted an approach in promoting national OHSMS models through statutory bodies. The Government of China, the State Administration of Work Safety, issued (December 2001) the national OHS management system guidelines based on ILO-OSH (2001). The OHSMS guidance committee of the State Commission of Economy and Trade Trial Standards develops national guidelines, and the accreditation organizations and auditors are certified by the guidance committee.

In Japan, the national 5-year industrial accident prevention plans are effective for long, and with the enactment of the Industrial Safety and Health Act (1972), the Ministry of Health, Labor, and Welfare is responsible for establishing the industrial accident prevention programs (ILO 2003). The Ministerial ordinance (1997) published by the Japan Industrial Safety and Health Association (JISHA) provides guidance to employers for establishing OHSMS with definite corporate policy and responsibility (Kogi & Kawakami 2002). The guidelines apply to enterprises of all scales and of all types of industry. The core concepts addressed in the system are the self-regulatory OHS management by employers, with due regard to the traditional safety and health activities in workplaces, such as 4S activity and foreseeing danger activity, "Hiyari-hatto" activity. There is no official certification of OHSMS, and it only certifies those who are trained in OHSMS.

The Ministry of Labor of the Republic of Korea introduced the Industrial Safety and Health Act (1990, amendment 2000), and under the Act, the Korea Occupational Safety and Health Agency (KOSHA) promotes OHSMS issuing KOSHA 2000 guidance and certification in line with the OHSAS 18001. The KOSHA 2000 program has been developed to fit in with the situations in Korea in assisting the establishment of the voluntary management system at workplaces. The Government has begun a 5-year plan for workplace accidents and disease prevention, emphasizing ensuring OHS regulations and promoting self-management to reflect effective OHS practices (ILO, 2003).

The South African National Occupational Safety Association (NOSA) published the 5-Star OHS management system (1984/1998), which is a compliance-based system to assist enterprises in carrying out financial, legal, and other responsibilities in running their business. The companion audit workbook covers references of OHS goals and objectives, and performance measurement. More emphasis has been given to environmental issues, based on ISO 14001.

Self-Regulatory OHS Management

The self-regulatory OHS management remains the core concept in many countries and sectors. The provisions that prevail in India and Malaysia are supportive of the approach. The Malaysian OHS Act (1994) extends the legal basis in protecting workers to all sectors of economic activity. The Act established the National Council for Occupational Safety and Health as a tripartite forum to develop and review strategies and practical programs. The certification of OHSAS 18001 by a private program is available. A medium-term national program called "Safework Malaysia" is underway that has multiple communication channels linked closely with the national 5-year economic plan.

In India, there are several provisions concerning enabling steps toward improved OHS management. The OHSAS 18001 certification is on the rise in the Indian enterprises. The statutory standards given in the Indian Factories Act (1948, amendment in 1988) and other related legislation, such as the Environment (Protection) Act, have jurisdiction referring to directions and procedures with respect to industrial installations, work environment, and OHS guidelines. The central government promulgated the Act, whereas the state governments promulgate rules on each section of the Act. The rule provisions cover industrial site appraisal, compulsory disclosure, hazardous processes, accident, work-related diseases, OHS surveys, emergency standards, and employee participation in safety management. Three schedules include a list of 20 industries involving hazardous processes, the permissible levels of certain chemicals, and a list of 29 notifiable diseases. The provisions in the Indian Factories Act, however, lack elements of a management system in defining OHS goals and objectives, performance measures, programs and procedures, and continual improvement.

Prospects and Challenges of National Programs

The current trend in OHS management is moving away from the detailed prescriptive approach to a more goal-setting approach. Many national governments formulated national OHS strategy and programs to manage health and safety. As cited in ILO (2003), the Australian Federal ministry of Employment and Workplace relations (May 2002) released the National Occupational Heath and Safety Strategy 2002–12. With the vision—*a safe New Zealand, becoming injury free*, the Government of New Zealand published the Injury Prevention Strategy (2003) that provides a guide for action by a range of government and nongovernment agencies, communities and individuals. The Government of UK launched an initiative, *Revitalizing Health and Safety* (1999), which sets out the objectives for the national health and safety system, aiming at developing new ways to establish and maintain an effective health and safety culture in a changing economy.

The populous nations like China and India have socioeconomic compulsions as regards to delivery of OHS practices to a large workforce in the small enterprises and informal sectors. For example, in the Indian nation, there are about 360 million workforces, and the unregistered small enterprises employ a workforce several times larger than that in the registered enterprises. The background situation in small enterprises is different as compared to large enterprises (Nag & Nag, 2003b), such as greater accident frequency, less knowledge of OHS law obligations, limited financial and human resources, and poor specialization in the performance of the management functions. The vast farming sector will continue to remain an implied challenge for effective OHS transfer. The existing national OHS laws and management models do not exclude small enterprises and informal sectors; however, the absence of long-term links among the constituents (governments, employers, workers, and workplace players) is the primary bottleneck in structuring contextual programs and systems. Conceivably, the small-step progress is indispensable, subject to evolving national OHS policy frameworks and tailoring action-oriented enterprise-level programs. The permeation of OHS to the vulnerable sectors may be possible through local infrastructure extensions, such as block development, primary health services, public network building on good practices, external support and economic incentives, and training at all levels.

REFERENCES

American Industrial Hygiene Association (AIHA). (1996). Occupational Health and Safety Management *System: An AIHA Guidance Document* (No. AIHA OHSMS 96/3/26), pp. 31.
American Petroleum Institute. (1993). Strategies for Today's Environmental Partnership.

American Textile Manufacturers Institute. (1994). E3 program (environmental management). Washington DC.

Asociacion Espanola de Normalizacion y Certificacion—UNE 81900. (1996, December). Prevention of occupational risks: General rules for implementation of an occupational safety and health management system.

Australia, New South Wales, The Occupational Health and Safety (OHS) Act 2000 (supplementary OHS Regulation 2001).

Australia, Victoria, Health and Safety Organization (HSO). (1999). Victoria; Safety Management Achievement Programme (SafetyMAP).

Bottomley, B. (1999). *Occupational health & safety management systems: Information paper.* National Commonwealth of Australia: Occupational Health & Safety Commission.

Brazil. Ministry of Labor guidelines. (1997). NR-9 (PPRA)—Environmental Risk Prevention Programme.

British Health and Safety Executive. (1993). Successful health and safety management: HS(G) 65, HMSO, ISBN 07176 0425 X.

British Standards Institution, BS 8800. (1996). Guide to Occupational Health and Safety Management Systems.

British Standards Institution, BSI-OHSAS 18001:1999. Occupational health and safety management systems—Specifications.

British Standards Institution, BSI-OHSAS 18002:2000. Occupational health and safety management systems—Guidelines for the implementation of OHSAS 18001.

Bureau Veritas Quality International (BVQI SafetyCert)—Occupational safety and health management standard.

China, The State Administration of Work Safety, OHS Management System guidelines, December 2001.

Det Norske Veritas. (1997). DNV Standard for certification of OHSMS.

European Union. (1993). The Council of the European Communities Regulation 1836/93, No. L 168—Community Eco-Management and Audit Scheme (EMAS).

Frick, K., Jensen, P. L., Quinlan, M., & Wilthagen, T. (Eds). (2000). *Systematic occupational health and safety management.* Amsterdam: Elsevier.

Hong Kong. (1999, November): Factories and Industrial Undertakings (Safety Management) Regulation.

India. (1998). Ministry of Environment and Forests, The Environment (Protection) Act.

India. (amendment 1988). Ministry of Labour, The Factories Act. 1948.

Indonesia. (1996). The Ministry Regulation (PER. 05/MEN/1996), Implementation of OHSMS.

International Chamber of Commerce. (1991). The business charter for sustainable development—Principles for environmental management. Paris, France.

International Exploration and Production Forum (E&P Forum). (1998). Report No. 6.36/210—Guidelines for the development and application of health, safety and environmental management systems.

International Labor Office, Geneva, Convention No 115, Radiation protection (1960); 135, Workers' representatives (1971); 136, Benzene (1971); 139, Occupational cancer (1974); 148, Working environment (1977); 155, Occupational safety and health (1981); 161, Occupational health services (1985); 162, asbestos (1986); 167, safety and health in construction (1988); 170, Chemicals (1990); 174, Prevention of major industrial accidents (1993); 176, safety and health in mines (1995)

International Labor Office. (2001). Guidelines on Occupational Safety and Health Management Systems (ILO-OSH), Geneva (ISBN: 92-2-111634-4).

International Labor Office. (2003). Promotional framework for occupational safety and health, Report IV (1). Geneva.

International Occupational Hygiene Association (IOHA). (1998). Occupational Health and Safety Management Systems, Review and Analysis of International, National, and Regional Systems and Proposals for a New International Document (prepared for the International Labour Office, Geneva; contributed by: H. Dalrymple, C. Redinger, D. Dyjack, S. Levine, & Z. Mansdorf).

International Standards Organization, ISO 14001:1996 (revision, Nov 15, 2004). Environmental management systems—Requirement with guidance for use. Geneva.

International Standards Organization, ISO 9001:2000. Quality management systems—Requirements, Geneva.

Ireland. (1997). National Standards Authority of Ireland, NSAI SR 320, Recommendation for an occupational health and safety (OH and S) management system.

Japan. (1997, March). The Ministerial ordinance on OHSMS guidance by the Japan Industrial Safety and Health Association (JISHA).

Kara, S. (April 3, 1996). One size fits all: Unifying ISO management. *Chemical Week*, 27–36.

Kogi, K., Kawakami, T. (2002). Trends in occupational safety and health management systems in Asia and the Pacific. *Asian-Pacific Newsletter—Occupational Health and Safety, 9*(2), 42–47.

Konkolewsky, H. H. (1999). European Survey on OSH-MS. (carried out by "Senior Labour Inspectors Committee, SLIC), European Agency for Safety and Health at work, Bilbao, Spain.

Korea Occupational Safety and Health Agency. (1999, July). KOSHA 2000 Program Certification System.

Korea. (1990) Republic of, Ministry of Labor, Industrial Safety and Health Act, Chapter II—Safety and health management systems (amended 2000).

Levine, S. P., & Dyjack, D. T. (1997). Critical features of an auditable management system for an ISO 9000-compatible occupational health and safety standard. *American Industrial Hygiene Journal, 58,* 291–298.

Lloyds Register Quality Assurance (LRQA SMS 8800)—Health & safety management systems assessment criteria.

Malaysia. (1994). Occupational Safety and Health Act 1994.

Nag, P. K., & Nag, A. (2003a). A national priority on occupational health and safety management system. *ICMR Bulletin,* 33, 15.

Nag, P. K., & Nag, A. (2003b). Head injury—The primary factor to fatalities in Indian industries. In *Ergonomics in the digital age* (IEA: Seoul), Vol. 3, ISBN 89-90838-10-X-98530.

Nag, P. K. (2002). The Management Systems—Quality, Environment, Health and Safety: ISO 9000:2000, ISO 14001, OHSAS 18001. pp. 683 (Mumbai: Quest Publications).

Nederlands Normalisatie-Instituut, NPR 5001. (1997). Dutch Technical Report—Guide to an occupational health and safety management system.

Norges Standardisengsforbund 96/402803. (1996, August). Management principles for enhancing quality of products and services, occupational health & safety, and the environment.

Norway. (1992). Royal Decree, Regulations relating to systematic health, environmental and safety activities in enterprises (Internal Control Regulations).

NSF International, NSF 110. (1995). (Environmental Management System Standard), Ann Arbor, MI.

Poland standardization authority, PN-N-18001. (1999). General requirements for occupational safety and health management systems.

Poland standardization authority, PN-N-18002. (2000). Guidelines for occupational risk assessment.

Poland. (1996, November). Labour Inspectorate, Worker Protection Programme PL 9407, Safety and health management in SME's: Best EU practices regarding safety and health management in small and medium enterprises (SMEs).

Redinger, C. F., & Levine, S. P. (1996). New frontiers in occupational health and safety: A management systems approach and the ISO Model. AIHA Publications, Fairfax, VA.

Redinger, C. F., & Levine, S. P. (1998). Development and evaluation of the Michigan occupational health and safety management system assessment instrument: A universal OHSMS performance measurement tool. *American Industrial Hygiene Journal, 59,* 572–581.

SGS Yarsley International Certification Services, SGS & ISMOL ISA 2000. (1997). Requirements for safety and health management systems.

South Africa, Reg. No. 51/0001/08. HB 0.0050E. The National Occupational Safety Association (NOSA 5 Star) Safety & Health Management System, 1984/98.

Spain. (1996, December). Asociacion Espanola de Normalizacion y Certificacion, UNE 81900, Prevention of occupational risks: General rules for implementation of an occupational safety and health management system.

Standards Australia/Standards New Zealand AS/NZS 4804/4801. (1997). Occupational health and safety management systems—Specification with guidance for use.

Thailand. (1999). The Ministry of Labour, Social Welfare and the Ministry of Industry, The Thai Industrial Standard on Occupational Health and Safety Management System (TIS 18000).

U.K. Chemical Industries Association. (1998). *Responsible care management system* (3rd ed.).

USA. Chemical Manufacturers Association. (1992). Employee Health and Safety Code— Responsible Care: A resource guide for the employee health and safety code of management practice.

USA, Department of Labor and Industrial Relations. (1995). California State; Cal OSHA Title 12, Subtitle 8, Part 2, Chapter 60–2; General Safety and Health Requirements: Safety and Health Programs.

USA, Occupational Safety and Health Administration (OSHA). (1988). Federal Register, 12/4/1988; Voluntary Protection Programs.

USA, Occupational Safety and Health Administration (OSHA). (1989, January 26). Safety and Health Program Management Guidelines, Standard—Federal Register, 29 (CFR), 1926 Subpart C, 54(18): 3094–3916.

26

European Union's Legal Standard on Risk Assessment

Kaj Frick
National Institute for Working Life

RISK ASSESSMENT TO BE UNDERSTOOD
WITHIN ITS WIDER SETTING

The European Union's (EU's) standard on Risk Assessment is formally a simple regulation (in the following, "Risk Assessment" refers to this standard as opposed to all other forms of "risk assessment"). It consists of two paragraphs in the EU Council Directive 89/391/EEC of 12 June 1989 "on the introduction of measures to encourage improvements in the safety and health of workers at work" (usually known as the Framework Directive). Article 6.3(a) states that *"the employer shall ... evaluate the risks to the safety and health of workers."* And article 9.1(a) adds, *"The employer shall be in possession of an assessment of the risks of safety and health at work."*

However, this is not a real standard. It is neither an internationally nor a nationally accepted definition of a procedure. Nor is it a mandatory, legal regulation. All EU directives are addressed to the member states, not to their citizens. The Framework Directive thus had to be transposed into national laws before its requirements came into legal force (at different times in the various member states). The transposition processes depend very much on domestic politics and their results may, and do, vary in important aspects. This is the case also for the Risk Assessment (see, further, Walters, 2002a, on the background, comparison, and analysis of the Directive, its transposal and partly its implementation in seven member states).

That the Framework Directive is not a proper standard does not diminish its importance. It is indeed the framework for regulating workplace occupational health and safety (OHS) within the EU. Its regulation of Risk Assessment is therefore fundamental for the management of ergonomic and other risks at work. However, the brief and principal requirement in the Directive makes it necessary to place Risk Assessment in its wider setting, to understand what risks are to be assessed and how. This is the more important as the requirements of the Framework Directive as a whole are very demanding, including for what risks have to be assessed and how.

To clarify the EU requirement on Risk Assessment, this chapter:

- Discusses the concept of risk assessment and its development
- Describes how the new process-oriented OHSM strategy, notably Directive 89/391/EEC, has evolved from this development and from other changes in OHS policies
- Describes the type and content of the Framework Directive's Risk Assessment, especially which risks are to be assessed and how
- Describes the slow implementation of, and outlines some principles of, how to interpret and apply the Risk Assessment "standard," including in small firms

SCIENTIFIC AND POLITICAL ROLES OF RISK ASSESSMENT

What Is Risk Assessment?

Although the formulations vary much between the many writers (see following), there is a large degree of principal consensus that *risk assessment is a combination of an analysis and an evaluation of risks of the use of a certain process, substance or technology*. It consists of the following steps:

Risk Analysis

1. Define the object that is to be assessed, for example, a substance, a work process, or a production system, and the circumstances and conditions of its use. (Some writers call this "formulate the alternatives," or "system description.")
2. Identify the possible health and safety risks of the object. (According to the EU's official "Guidance on Risk Assessment at Work," this consists of: "Collect information," "Identify hazards," and "Identify those at risk"; European Commission, 1996, p. 13.)
3. Estimate the level of this risk, by systematic qualitative or quantitative means, such as exposure measurements or statistics on the probability of accidents (EU guidance: "Identify patterns of exposure of those at risk").
4. Summarize 2 and 3 into how many and how serious diseases or injuries are likely to be caused by the risk object (EU guidance: "Probability of harm/severity of harm in actual circumstances").

Risk Evaluation

5. The expected risk of harm is compared to the relevant criteria of acceptable risk. At the workplace level this can be, for example, regulations or internal health and safety goals. Societal risk evaluations often try to compare the risk of harm to the social gains by using the substance or the technology.

The identification of the risks and the evaluation if these can be accepted or not ends the risk assessment. For many writers, this is the first half of "health and safety (or risk) management," the second half of which is "risk control" or "risk management." Although it is not expressed in formal risk management terms, this is also the case of the EU Framework Directive's standard on Risk Assessment (see section under Risk Assessment and the New OHS Management Strategy). Such a control starts with an evaluation of alternative methods to prevent, eliminate, or reduce the risks to the acceptable level, often by means of a cost–benefit analysis. After the optimal measures have been chosen, these have to be implemented. Then their effects have to be monitored and, if necessary, new measures have to be taken. Finally, the whole process of risk management should be audited and improved. Risk assessment is thus a technique within a process of assessing and managing the risks, a process which is continuous or at least repeated when any conditions of the assessment are changed.

Risk Assessment as Science

The aforementioned systematic procedure is a modern construction. However, risks are an eternal aspect of human life. We have therefore always had to assess them—usually unconsciously—in order to minimize harm to our health, families, crops, businesses, or whatever is important to us. Our split-second grasp of the traffic—number, distance, and speed of cars, slipperiness, visibility, width of the street, obstacles, and so forth—as we cross a street is a daily example. Since ancient times, thinkers have also sought to draw general conclusions from such assessments. The Roman writer Vitruvius advised against the use of lead pipe for drinking water, as he had noted that exposure to lead fumes was linked to blood diseases (British Medical Association, 1992, p. 1). Ramazzini is perhaps the most famous example. In his work "on the diseases of workers" (1700), he carefully observed and assessed the health risks of some 50 vocations.

But explicit and systematic methods of risk assessment are much more recent. What started as risk analysis around 1970 (Otway, 1985; Renn, 1985, p. 113) has since then evolved into a broad array of instruments. This development is illustrated by Ogden's (2003) example of how the analysis and regulation of asbestos risks have evolved considerably since 1968. We now have methods to assess, analyze, evaluate, estimate, and control or manage the problems of risks, safety, hazards, or exposures in modern production. The professionalisation of this methodology was marked by the formation of a Society for Risk Analysis in the United States in 1980 and in Europe in 1986 (Dwyer, 1991, p. 255).

There are now more than 100 titles on risk assessment in the library of the Swedish National Institute for Working Life alone (http://arbline.arbetslivsinstitutet.se) and close to another 100 if we add the similar concepts of "hazard" and of "evaluation" and "analysis." The most important differences are in which types of risks these works discuss and what level of analysis they emphasize, that is, what balance between risk assessments as advice to societal decisions versus to workplace improvements. Among the type of risks, on the one hand, chemical and, on the other, accident risks dominate. Many are medical–toxicological works on how to assess the risks of diseases and environmental damages of hazardous substances (e.g., Sadhra & Rampal, 1999; Vollmer, Giannoni, Sokull-Klütgen, & Kracher, 1996). Exposures to other work environment risks—such as noise, radiation and heat—are assessed with the same dose-response principles (e.g., Holmér, 2000).

Technically oriented risk assessments of sequences of events, which may turn into accidents, also abound. They include special assessments of machine safety (IEC, 1995; ISO, 1999) and of risks of disasters in chemical factories, nuclear power plants, offshore oil platforms, and other high-risk facilities (Ansell & Wharton, 1992; Cox & Cox, 1996; Stewart & Melchers, 1998; HSE, 1998; Harms-Ringdahl, 2001). These are sometimes developed into mathematical models of probabilistic risk assessment (Kumamato & Henley, 1996; Mosleh & Bari, 1998). Of the assessments of special risks, the ergonomic ones may combine the mechanical–sequential logic of analyzing work systems with the dose–response logic of exposure to different types, durations, and severities of musculoskeletal strain (Colombini, Occhipinti, & Grieco, 2002; HSE, 1994). A similar combination of perspectives, often with the help of checklists, is used in risk assessments of special tasks or industries (e.g., Krüger, Louhevaara, & Nielsen, 1998; Stern, 1980). There are also writers with a broad perspective on business risks in general, who combine, for example, chemical disasters with fraud and fire (e.g., Reason, 1997).

If we instead cut them along the level of analysis, HSE (1999), Roberts-Phelps (1999), and Jeynes (2002) give very hands-on advice on how to assess health and safety risks in workplaces.

Many checklists are even more concrete in structuring what risks to look for and how to assess the results (see, e.g., www.pk-rh.com/en/index.html and www.prevent.se/english/). Others have a general perspective on the varying risks they discuss but also aim to help workplaces to assess (and control) these (e.g., Cox & Cox, 1996; IEC, 1995; Harms-Ringdahl, 2001; Holmér, 2000). Several writers—like Otway and Peltu (1985), Ansell and Wharton (1992), Hansson (1993), Kumamato and Henley (1996), Stewart and Melchers (1998), Vollmer et al. (1996), and Sadhra and Rampal (1999)—mainly focus on the societal principles and politics of risk analysis and assessment.

With such a variety of problems and social levels, it is not surprising that we lack an international standard to define risk assessment. The influential guideline, BS 8800 (1996, p. 4; on OHSM systems) only contains a brief definition of this as being "the overall process of estimating the magnitude of risk and deciding whether or not the risk is tolerable or acceptable." And according to the EU guidance (European Commission, 1996, p. 11), risk assessment is "the process of evaluating the risk to health and safety of workers while at work arising from the circumstances of the occurrence of a hazard at the workplace." Although this guidance describes seven steps of the assessment procedure—and another seven for the management of identified risks—these too are general, as exemplified earlier. (See also the similar definition in the ILO Guidelines on OHS management systems from 2001, which are discussed later).

There are definitions of risk assessment for special purposes (e.g., in ISO, 1999, on machine safety). There is also a general one of risk analysis, that is, "the systematic use of available information to identify hazards and estimate the risk to individuals or populations, property or the environment" (IEC, 1995). Other works give more exhaustive, but also quite varied, descriptions of what risk assessment is, as in Wharton (1992, p. 7), British Medical Association (1992, p. 19), Cox and Cox (1996, p. 32), Rampal and Sadhra (1999, p. 19) and Harms-Ringdahl (2001, p. 43). See also the comparisons by Rakel (1996) and by Boix and Vogel (1999, p. 20). Nevertheless, there is a large principal consensus in these works behind the description of risk assessment given earlier.

Risk Assessment as a Response to Public Concerns

Risk analysis and assessment, as systematic instruments for scientists and other experts, are largely responses to growing public concerns of the risks to health and safety and to the environment of advanced industrial production. Carson's (1962) alarm on the death of nature was a major turning point in the debate. The slow thalidomide disaster of the early 1960s gave another jolt to public trust in how those in power and their experts handled risks. Especially from the 1970s onwards, it has been revealed that the production or use of diverse substances—such as vinyl chloride and asbestos—may be, and often is, very dangerous (Markowitz & Rosner, 2002). Already in 1972, the United Nations conference on the Human Environment in Stockholm therefore "placed risk management issues on the same high level on the international agenda as peace, trade, finance and economic development" (Majone, 1985, p. 41).

Likewise, human development and use of advanced technologies has from the beginning been accompanied by disasters, as mines, buildings, bridges and dams collapsed, ships sank and (later) steam boilers exploded, and trains and airplanes crashed. In industrial production, munitions and explosives factories have regularly exploded (which, e.g., gave rise to the DuPont safety philosophy; Mottel, Long, & Morrison, 1995). Later, the growth of the chemical industry resulted in, for example, the IG Farben explosion in Germany in 1921, which killed 550 people and a series of chemical fires and explosions in Texas in 1947 with 561 dead (Perrow, 1984, pp. 120, 105). Safety in the nuclear power industry has long been contested, and even more so after Three Mile Island was a near disaster in 1979 and Chernobyl became one 1986. With the gradual development of international media, these and other disasters became more and more public knowledge. Seveso in 1976, Bhopal in 1984 (which killed some 10,000 people; Eckerman, 2004), Challenger in 1986 and Piper Alpha in 1988 are some of the most well known. This resulted in a debate on high-risk technologies (Perrow) or even on the modern industrialized world as a risk society (Beck, 1992).

Otway (1985, p. 3) summarizes the development and how the experts reacted:

"As contemporary societies began to produce a dazzling array of consumer goods, an awareness was forming that they were also producing a variety of toxic new chemicals, polluting the environment, and destroying vital natural resources. The public impacts of industrial accidents were also without precedence in size and kind."

"Many scientists and engineers were puzzled by lay group challenges to informed expert opinion. They believed that regulatory decisions would be less controversial if they could be given a firmer 'factual' basis. Their attempts to be more technically rigorous led to the emergence of the new 'science' of *risk analysis*, the use of available data, supplemented by calculation, extrapolation, theory, and expert judgment, to define the risks to people due to their exposure to hazardous materials or operations."

Risk Assessment as Contested Politics

Otway (1985) covers the two main aspects of risk assessment: on the one hand, its scientific, expert nature of a systematic method to identify and evaluate risks and, on the other, the political setting of such assessments. As they are instruments to guide decisions on risks, they are inherently political. The debate on and practice of risk assessment thus also concern issues of power, competence, trust, and legitimacy among the stakeholders of risk. The latter are primarily:

– The risk producers, for example, owners-managers of nuclear or chemical plants
– The risk exposed, for example, workers or consumers
– The politicians, who may regulate between risk-producers and risk exposed but who also may be risk-producers themselves
– The scientists–experts, who advise on how to handle–minimize the risks.

Perrow (1984, pp. 306–328) and Dwyer (1991, pp. 237–269) further analyze the dual functions—the risk—of risk analysis and assessment. On the one hand, these are tools to advice the risk-producers how to control the risks. On the other, they are part of a strategy to limit the reactions and regulations of such technologies. Advanced calculations proving the risks to be acceptable and methods of correct risk communication are means to reassure the public and their political representatives. However, risk analysis and evaluation are difficult, and the experts' methods and outcomes are contested. Attempts to construct "criteria (for risk acceptance) which are usable, transparent and agreed by all parties concerned" (Cox & Cox, 1996, p. 34) are often rejected (e.g., by Perrow, pp. 309–310). And Hansson (1993) criticizes all the assumptions of risk analysis, including that they define their decision problems too narrowly, that they cannot obtain reasonably accurate probability estimates of the resulting risks, and that their evaluations of the alternatives (i.e., risk acceptance criteria) are biased (cf. also the discussions in Otway & Peltu, 1985).

The need to analyze and evaluate risks is not contested, nor is the need for good methods to do so. Risks can be diminished but rarely abolished. They often have to be balanced against each other or against various gains. Systematic methods to guide risk decisions are usually better than intuitive impressions, especially in complex technical systems. The issue is thus not the use as such but the "risk" of poor quality and of practical misuse of risk assessments. A comprehensive discussion of this is outside our scope (see, e.g., Hansson, 1993). However, some clarification is necessary, as the EU's standard on Risk Assessment is also a political decision with varying and debated applications (Bercusson, 1996, chap. 23; Boix & Vogel, 1999; Karageorgiou, Jensen, Walters, & Wilthagen, 2000).

With linking issues, risk assessments are criticized because of their:

1. Reliability: Are their results accurate, for example, that the production and use of this substance or the technology entails the predicted level of risk?
2. Validity: Whose knowledge is relevant in assessing the risks?
3. Sincerity: Are the assessments used in good faith, to minimize risks, or are they means to promote the interests of risk producers (or their opponents)?
4. Responsibility: Are risks assessments (and control) to be voluntary—for risk producers, advised by experts—or are they to be regulated?

1. The *reliability* claims of "systemic safety [including risk assessment] to be able to plan production to reduce risks of accidents to an absolute minimum" (Dwyer, 1991, p. 244) have been questioned on, at least, three grounds:

- Theoretically, the evidence on which to assess the risks is mainly retroactive data, for example, systematic statistics on past incidences. Yet, the assessment aims to predict the likelihood of future risks. It therefore requires a considerable element of fantasy, which makes it more or less inexact. The definition of the risk object is also, by necessity, a subjective selection of factors and variables, whereas all others are disregarded in the assessment (cf. Hansson, 1993).
- Empirically, experts often disagree in their assessments—that is, in what type of risk a certain object entails—even when these are based on the same evidence. One contentious issue is the assessment's level of uncertainty and how to handle this (Hansson, 1993). The input data are also often unreliable (see, e.g., Collinson, 1999, on systematic nonreporting of oil rig incidents).
- Risk predictions do not work if their conditions are not met. Technical changes may be made, outer circumstances may vary and maintenance may deteriorate, all without much notice of their risk effects. How possible changes should be included in the definition of the risk object is an open issue. But disaster inquiries (e.g., Cullen, 1999; Hopkins, 2000) demonstrate that, even in technically advanced organizations, safety claims and safety practices are often not the same.

2. This uncertainty also questions the *validity* of the risk expertise. As other professionals, risk experts try to capture their market by emphasizing the unique importance of their competence (Dwyer, 1991, pp. 255–260). However, with demonstrated shortcomings in their assessments, risk-exposed and other laypersons have—often successfully—asserted that assessments should include their experience and views of the production and its risks. Assessments are also criticized for a poor validity in their risk evaluations, notably a too narrow view of the costs and benefits involved (e.g., Dorman, 1996; Dwyer, p. 238; Hansson, 1993; Perrow, 1984, pp. 308–310; Ruttenberg, 1981a).

3. The *sincerity* of an assessment is in turn related to the validity of practical competence. Not only is the involvement of the risk exposed in the assessment contested, but also there is a tradition of keeping them uninformed of the risks. The French state was not alone in limiting public knowledge of the operation of high-risk technologies (Dwyer, 1991, p. 295; Lagadec, 1981). However, growing media and public interest have made such policies by governments and risk producers less tenable (Dwyer, p. 260). Secrecy can turn into public scandal, with

grave political consequences. For example, fear of another scandal was a major reason behind the recent shift in French policies on OHS risks, due to the "asbestos crisis" (Rivest, 2002, p. 100). Professional methods for "risk communication" have instead been developed to convince the public that the risks assessments of the experts, usually on behalf of the risk producers, are correct. But even special communication methods often have to work against distrust, when it is revealed that the public has often been actively deceived of the nature and severity of the risks to which it has been exposed (Brodeaur, 1974; Markowitz & Rosner, 2002; cf. also the tobacco companies' denial of the addiction risks of nicotine).

4. Finally, the *responsibility* for the risk assessment is also contested. Many risk experts have claimed that proper application of their advanced methods was the best way to control risks and have therefore opposed public regulation (Dwyer, 1991, p. 245; Perrow, 1984, p. 307). "Professionalism with strong ethics is better than laws" (Concha, 1983; in Dwyer, p. 292). However, they—and risk producers that opposed regulation—have largely been unsuccessful. The development of "scientific" risk assessment has generally not been a strategy of "self-regulation," in which the risk producers could avoid having external regulations imposed on their operations. Instead, regulations have in some cases been demonstrated to not only reduce risk levels but also promote technological and organizational innovations within risk-producing industries (Olsen, 1992; Ruttenberg, 1981b; Ashford & Caldart, 1997, chap. 10). See previous section on the discussion of regulation versus voluntary standards under Risk Assessment as a Response to Public Concerns.

Likewise, the development of modern risk assessment is strongly linked to the growth of regulation (Otway & Peltu, 1985; Rampal & Sadhra, 1999, p. 4). The assessment of a widening array of risks have been mandated by more and more regulations in both in the United States and the European Union (EU), as a standard tool to minimize societal harm from various risks (Ashford & Caldart, 1997, chaps. 2 and 4; Otway, 1985; and Walters, 2002a, chap. 2). Within the EU, "the current occupational health and safety legislation . . . depends on a risk assessment approach to managing and controlling hazards" (Jeynes, 2002, p. 13).

RISK ASSESSMENT AND THE NEW OHS MANAGEMENT STRATEGY

Risk Assessments of Products and of Production in EU

The evolvement of Risk Assessment as *the* mandated method to control risks at work is part of the general development of OHS regulations within the EU. The common market was from the beginning also a social project. Occupational health and safety were included in the founding Rome treaty of 1957 (article 118). As other EU policies, those on OHS started slowly in the 1960s, with the Directive 67/540/EEC on dangerous substances, as an early example. An important step was the so-called Seveso Directive (82/501/EEC), on major-accident hazards involving dangerous substances (modernized as 96/82/EC). It requires manufacturers to identify the (major-accident) hazards of their operations, though this was more linked to the environmental then to the OHS debate.

As a part of the EU activism after the Single European Act of 1986, new forms were developed to regulate the *work environment* (a term which intentionally marks a broad definition of the OHS concept; Walters, 2002a, p. 43). This resulted in the parallel and overarching Framework Directive (89/391/EEC) and Machinery Directive (89/392/EEC). Especially during the late 1980s and early 90s, these were followed by several directives on special risks, which were issued in accordance with the former. The one on Material Handling (90/269/EEC) is one of several examples. But other directives on conditions of work were also adopted, for example, on Working Time (93/104/EEC; see further Vogel, 1994; and Walters, 2002a, on the background and principles of the Framework Directive).

The OHS Directives have the dual purposes to avoid risks and to promote the integration of the single market. On the one hand, the Directives are standards in the market sense. Harmonized EU requirements on the prevention of risks of the use of goods are to abolish the varying national regulations, which act as trade barriers. And local risk assessments of production are also to promote the single market by equalizing the terms of competition among the member states. At the same time, the upward harmonization of national regulations is also a social policy to improve occupational health and safety, other working conditions and the environment within the EU.

The concept of Risk Assessment became prominent with the Framework Directive. However, in various formulations, such assessments have been fundamental also in earlier OHS directives (e.g., in 83/477/EEC on protection against asbestos risks). The EU legislation on risk assessment can be divided in either general assessments of specific risks or local assessments of general risks:

• The use of specific products—mainly machinery and chemical substances—is to be assessed as general risk sources. To secure that all products placed on the market are safe, and therefore shall be allowed free trade, the producers are to assess (control, inform about, etc.) the risks of all foreseeable use of a these risk sources. With some changes in the terminology, such product risk assessments have been required since the Directive on dangerous substances in 1967. A large number of amendments and new directives—notably the Machinery Directive in 1989—have gradually made them stricter, both in coverage of risks products and in the required assessments and controls. In the case of substances, the common EU requirements are about classification, labeling, packaging, marketing, and other border-crossing trade aspects. However, workers' exposures to such substances are determined by the local conditions of production. The corresponding regulations of occupational exposure limits (OEL) are predominantly set at national levels, but with growing EU, attempts to coordinate the definition of acceptable risks, that is, the levels of OELs (see 98/24/EC and 2000/39/EC). Machinery, on the other hand, is more to be used as delivered. Thus, sufficient protections against all possible risks—that is, both risk information and the material safeguards, are to be part of the harmonized EU regulations. The specification of these protections is to be determined through European standards, which has resulted in a large standardization process (Walters, 2002a; p. 44).

• To secure safe and sound conditions in individual production units, those responsible are to assess (and control or at least reduce) their general risks. The Seveso Directive launched the principle that manufacturers with potentials for major accidents should assess these risks. The Framework Directive expanded this into a general principle that employers shall assess and manage all OHS risks of their operations (including the use of chemicals, as guided by national OELs). It has since been included in a number of daughter and other directives on working conditions. These types of Risk Assessments may interact with risk assessments of machinery. For example, changes in machinery at a workplace may require a new risk assessment according to the Machinery Directive, while the surrounding safety and health is to be assessed according to the Framework Directive (see further Walters, 2002a, pp. 41–44, on this interaction).

Workplace Risk Assessments Are Tools Within OHS Management

The mandatory workplace Risk Assessments place the responsibility to detect and abate the problems with the employers as risk producers. Rather, they aim to make employers in practice assume the responsibility to prevent work related ill health, which they have had since the beginning of OHS regulation. According to quality control principles, the employers—and their managers—can much more effectively prevent the health risks than can the authorities as representatives of the workers. It is therefore not enough for risk-producing employers to passively await the authorities' assessments and ensuing prescriptive orders, which define the risks and how to abate them. Such a strategy of external control of OHS risks never worked well. With modern high-risk technologies, it has become even more untenable (as discussed earlier).

As mentioned, the evolvement of Risk Assessment was largely a result of growing regulation. The regulation—especially of workplace risks—is itself part of the broader development of OHS management, as a process-oriented prevention strategy. This emphasizes that OHS risks are much more effectively and efficiently controlled at their sources, that is, in the planning and management of the operations causing the risks. To manage the prevention, employers therefore have to set up appropriate quality control procedures, including not only Risk Assessments but also training and a proper distribution of responsibilities and authorities.

Especially since the late 1980s, there has been a development and dissemination of such methods to manage OHS. This has taken the form of active marketing of voluntary OHSM systems (especially in Anglo-Saxon countries), of national and international standards and guidelines for such systems, and of hybrid implementation of OHSM mixing voluntary and mandatory means (see also later on the ILO Guidelines on OHS management systems). Above all, OHSM regulations (rarely constructed as complex systems, as they are also to be applicable also to small firms) have been adopted by the EU and many other OECD countries—with the United States still a notable exception—and by more and more developing nations (see Frick, Jensen, Quinlan, & Wilthagen, 2000, on the background, purposes, and problems of regulated OHSM).

Contested Implementation of OHSM

The spread of OHSM marks a shift from specified, prescriptive risk-regulations to generic, performance-oriented ones (Gunningham & Johnstone, 2000). This is not only a functional issue. As mentioned, earlier, politics and policies of regulation also affect the balance of power over the risk assessments and the resulting controls. Most controversial is probably in how far the increased risk assessment by the employers will be accompanied with a corresponding decrease in the authorities' regulation and enforcement. The producers of equipments already largely certify themselves—according to the Machinery Directive—to comply with the risk regulations on the product market. It is feared that a too lenient interpretation and implementation of regulated OHSM—including of the Framework Directive's required Risk Assessment—may result in the same development on the labor market. Employers may in practice certify themselves to be in compliance with the OHS regulations, with little or no external control of labor inspections (Nichols & Tuckers, 2000; cf. also the debate on voluntary OHSM programs in the United States in Needleman, 2000).

In other words, is the requirement that employers assess and control "their" risks a functional decentralization to increase the effectiveness of OHSM (as was discussed earlier)? Or is the state shedding its responsibility to protect its citizens, that is, the workers, under the (OHS) law? In the latter case, it is argued that letting the risk producers assess themselves is placing an excessive trust in the capability and interests of employers, who operate under the economic pressure of market competition (Nichols & Tucker, 2000; Dorman, 2000; cf. the critique of the machinery standardization process in TUTB, 1996; and in Koukoulaki & Boy, 2003).

The Framework Directive Is the Foremost OHSM Regulation

The balance between success (improved prevention) and sham (deregulation, with little control of prevention) in the practice of the OHSM strategy remains an open question. The answer largely depends on the varying conditions of its implementation, as discussed in Frick et al. (2000). However, Risk Assessment is unavoidably part of this open and contested implementation. This is especially so, as the Framework Directive is the foremost example of these political OHSM regulations. OHS policies all over EU are to be governed according to the principles of this directive.

Within the Directive, the mandatory Risk Assessment is to be the starting point for its intended comprehensive OHSM. Employers are to chart and assess all risks of their production, to be able to take on the responsibility of controlling them. A thorough Risk Assessment is to both enable and motivate them (and their managers) to improve the work environment. The regulations therefore mandate–guide employers in *how* to assess and control the risks and less in *which* these risks are. Information on possible risks, and on suitable prevention measures, is instead to be gathered as a part of the OHS management process.

TYPE AND CONTENT OF THE FRAMEWORK DIRECTIVE'S RISK ASSESSMENT

Not a Normal Ergonomic Standard

The EU Framework Directive is indeed a framework, with "general principles concerning the prevention of occupational risks . . . as well as general guidelines for the implementation of the said principles" (Article 1.2). It is an ambitious modernization of OHS policies in the EU but also a political compromise. Some of its requirements are intentionally ambiguous, and they all have to be adapted to the varying national structures and traditions (see, further, Walters, 2002a).

The following discussion aims to clarify the Risk Assessment intended by the Framework Directive. It is illustrated with some national varieties, but these form no exhaustive analysis of the differences between the EU states (more are presented in the national chapters and in the comparison by Walters, 2002a, 284–288; and in the analysis of the national transposition by Vogel, 1994, 1998). To know what employers have to comply with, legal scholars and OHS practitioners have to study the national regulations and the various guidances on these.

The paragraphs on Risk Assessment earlier are also of principal nature, not guidance on how to do it. This requirement must instead be understood within its setting of the Directive. Formally this is addressed to the member states. These must not only bring into force laws and regulations to comply with its requirements (Article 18.1) but also ensure adequate control and supervision of its workplace implementation (Article 4). The obligation of workers is mainly to "take care as far as possible of his own safety and health and that of other persons affected by his acts or omissions" (Article 13). But above all, the Directive contains principles of how employers in all economic sectors shall secure a good work environment. Some of their major duties are to:

- Ensure the safety and health of workers in every respect related to the work (Article 5.1).
- Provide the organization and means necessary for this (Article 6.1).
- Evaluate and prevent all OHS risks, including organizational ones (Article 6.2-3).
- Implement this through a prevention hierarchy, starting with avoiding the risks and ending with personal protectives and safety training (i.e., safe behavior) as last resorts (Article 6.2).

- Either, themselves, have sufficient competence to comply with the Directive's requirement or enlist such competence (Article 7).
- Organize effective worker involvement in the OHSM, through information, training, consultation, and participation (Articles 10–12). However, this and other OHS measures may in no circumstances involve the workers in any cost. Participation thus must be extensive but free from costs to the workers (Article 6.5).

The Risk Assessment is an integrated aspect of this prescribed OHSM. As such, it differs in many respects from the majority of ergonomic standards in this book. Most of these also focus on their material content, for example, as quantitative anthropometrical or biomechanical measures (e.g., many ISO and CEN standards; cf. also Fallentin et al., 2001). Risk Assessment (and OHSM in general), on the other hand, is not quantitative at all. It deals with the process of OHS improvement, not the content of the risks to be prevented. And although most ergonomic standards are voluntary (at least formally), the Framework Directive and its Risk Assessment is the foremost mandatory OHS regulation within the EU (Walters, 2002a).

The difference is not absolute. The wide variety in standards is discussed in the earlier chapters (and in the comparative evaluation by Fallentin et al., 2001). Some ergonomic standards are mandatory by law, not agreements between experts or the concerned parties (see earlier section on regulatory versus voluntary standards and their different logics). Ergonomic standards may also include process elements, such as programs of assessment and intervention, especially to counter complex problems like repetitive work (see earlier discussion on material versus process—or program—standards). The EU Directive on Manual Handling (90/289/EEC) is an example of a regulated process standard (in fact a miniversion of the Framework Directive). Yet, it is illuminating to contrast its type and content to material and voluntary ergonomic standards. The following table is simplified—and thus slightly exaggerated—but should still illuminate their principal differences:

TABLE 26.1
Differences Between Mandatory Risk Assessment and Voluntary Ergonomic Standards

	Standard	
Aspects of Standard	Mandatory Risk Assessment	Voluntary Ergonomic
Type		
Aim	Prevention process	Ergonomic result
Logic behind	Political–legal process	Experts' analysis
Applicability	All employers	Special risks/voluntary
Compliance through	Inspection–enforcement	Market/certificates
Content		
Risks		
Scope of Risks	All = organization & technology	Technology (usually)
Level of Risks	Ensure safety & health	Acceptable risk
Individualization	Individual adaptation	Human averages
Assessment		
Who Assesses	Social partners	Experts
Knowledge of Risks	Mandatory competence	Ignorance accepted?
Goal of Assessment	Prevention hierarchy	Safe person before safe place?

Only Risks Within the Employment Relation

The differences in type between mandatory and voluntary ergonomic standards and the process regulations have already been discussed, and the means and level of compliance with the Risk Assessment is dealt with later in this chapter. However, before we go into the content of the Risk Assessment, it is important to note both the width and the limits of its application. On the one hand, it applies to all employers, not only to some employers with special risks or to those who choose to use it. The European Court of Justice made this clear in its ruling against Germany (ECJ C-5/00). By exempting employers with 10 or fewer workers from the duty to keep documents containing the results of a Risk Assessment, the German legislation had not properly transposed the Framework Directive.

On the other, the focus on the employment relation is also a growing limitation. As nearly all labor law, the Framework Directive is based on 19th-century industrial relations. To assess and prevent risks at work is a duty for the employers, but only to *their* workers. Yet, in modern production, both sides of the employment relationship wither. Companies produce more and more in networks, by, for example, outsourcing or splitting into separate legal entities (Castells, 1996, chap. 3; Larsson, 2000). On the labor side, casual workers replace long-term employees (Quinlan & Mayhew, 2000). The production—and its work environment—is thus more and more controlled by contract relations under trade law instead of by employment relations under labor law. And contract partners are rarely obliged to assess the OHS risks of the producing workers.

There are some regulatory and voluntary possibilities to assess risks to workers, outside of the employment relation. Producers-employers in the UK have to prevent injuries also to third parties, for example, visitors (HSE, 1997, p. 45; see also Johnstone, 1999, for an interpretation of the British and Australian duty of care to cover contract relations). "Voluntarily," large customers also increasingly use contracts as a market mechanism to spread, that is, require, systematic OHSM from their contractors and suppliers (Walters, 2001, pp. 345–355). As any voluntary standard, risk assessment can thus be applied more freely on the market than by regulation, but then with no assurance that it conforms to the requirements of the Framework Directive.

What Risks to Assess?

Scope of the Risks

These requirements are strict. The risks include everything that may harm the workers. The Framework Directive integrates the Scandinavian broad definition of the work environment (Walters, 2002a, p. 43), which explicitly includes organizational risks. Article 6:2(g) states that *"The employer shall . . . develop a coherent overall prevention policy which covers technology, organization of work, working conditions, social relationships and the influence of factors related to the work environment."* And Article 6.3(a) mandates employers to *"evaluate the risks to the safety and health of workers, inter alia in the choice of work equipment, the chemical substances or preparations used, and the fitting-out of workplaces."*

The European Court of Justice has upheld this wide risk definition. It ruled that working time influenced the health of workers. It was therefore legitimate for the EU to regulate this under Article 118a of the EEC Treaty (Directive 93/104/EEC; ECJ case C-84/94). The court confirmed that every risk that may affect the health and safety of workers is part of the work environment (and shall thus be assessed). In this, it followed the advice of the advocate general, who—among other things—stated that the base, the preventive aim of the Framework Directive is *"far from a view in which the protection of workers is limited to physical or chemical factors."* He also referred to EU's previous acceptance of WHO's wide health concept as a base for its

OHS directives (e.g., in Directive 92/85/EEC on the rights of pregnant women). The court confirmed that *all* risks have to be assessed when it ruled against Italy (in ECJ C-49/00) that the list of examples in Article 6.3 (a)—which hade been translated into Italian Law—was not enough, as *inter alia* had been excluded. Finally, it noted, *"that the occupational risks which are to be evaluated by employers are not fixed once and for all, but are continually changing in relation, particularly, to the progressive development of working conditions and scientific research concerning such risks."*

The inclusion of all risks has important consequences. The Risk Assessment has also to identify and evaluate organizational risks of (for example) social relations at work, management style, gender segregation, organization and distribution of work tasks, pay systems, working time schedules, workers' autonomy in performing tasks, violence and threats, and mental and physical work load. All of these—and more—organizational factors have been documented to affect the health of workers (see, e.g., Johnson & Johansson, 1991; and Marklund, 2001). Likewise, the more traditional technical–chemical risks that are objects of national technical regulations or lists of OELs are (only) examples of such risks. The Risk Assessment may well have to consider other "traditional" risks (as is made clear in, e.g., the German law, Schaapman, 2003, p. 138).

The wide work environment concept is included in the regulations on Risk Assessment in most EU countries or in their general OHS laws, for example, in the Netherlands (Arbeidsomstandighedenwet, 1999) and in France (Rivest, 2002, p. 98). In the Danish Workplace Assessment (Jensen, 2002) and the Swedish Systematic Work Environment Management (Frick, 2002, p. 231), the inclusion of organizational prevention was even a major aim. The scope of risks to assess in the UK is possibly more restrictive (see, e.g., Walters, 2002a; pp. 251–258). The formal transposition (Management of Health and Safety at Work Regulations; HSE, 1992) only orders employers to carry out a suitable and sufficient assessment. The OHS authority is ambiguous in its information on how to interpret this. On the one hand, the general OHSM information includes organizational factors and lists several of these (in HSG-65; HSE, 1997; p. 13). On the other, the guide on Risk Assessment (HSE, 1999, p. 6) informs employers that they need to be able to show that they "dealt with all the obvious significant hazards," which may exclude a duty to assess less obvious but possibly serious risks (see following on the requirement to have sufficient competence to identify all risks).

Risks to Individual Workers

The risks have to be assessed to individual workers, not only to workers in general. Article 6.2 (d) states that the prevention shall be based on a number of general principles, including *"adapting the work to the individual"*; and Article 9.1(a), that the Risk Assessment shall include risks *facing groups of workers exposed to particular risks*. And *"particularly sensitive risk groups must be protected against dangers which specially affect them"* (Article 15). The risks to pregnant women, immigrant workers with poor language, older or inexperienced workers, and any other sensitive-exposed group thus have to be specially assessed. But the work environment must also be compared to workers' individual sensitivities, competences, capabilities, and so forth, and the preventive measures necessary for each worker must be taken. (In this adaptation to individual workers, the Framework Directive links into the later EU policies of "employability," i.e., to promote a high level of employment as a major means for economic development within the union; Walters, 2002a; pp. 54–56.)

Levels of Risk Beyond Specified Regulations

Risk to individuals and levels of risk are related. To ensure the safety and health of sensitive groups and individuals discussed earlier, the employer may well have to assess and control risks beyond adherence to (i.e., are lower than) OELs and other specified national regulations. The limits to how low or improbable OHS risks the employer has to investigate are (partly) defined by national regulations and by the individual cases. However, thorough Risk Assessments have to bear in mind that OELs and other specified measures–limits are temporary regulations on minimum requirements and no guarantee against harm (as is demonstrated in many studies, e.g., in Larsson, 1991, p. 208). National regulations—and the Framework Directive—therefore usually aim at the lowest possible level. Employer organizations have also advised their members the same, for example, when the Swedish Engineering Confederation and the Metal Workers Union jointly set 25% of the OEL as a maximum limit to aim for in engineering plants (Verkstadsföreningen, 1978).

"To ensure the safety and health of workers " (Article 5:1) is also an absolute requirement. Economic relativisations of this duty, such as "so far as is reasonable practicable" (HSE, 1997, p. 46), are common, for example, in British OHS regulations. The UK transposition has been criticized for falling short of the Framework Directive, including the Risk Assessment (e.g., by Bercusson, ibid; Walters, 2002a: 251–255) but the European Court of Justice has not yet tried the issue. However, the limitations in the British employers' duties pertain mainly to the actions they must take to prevent the risks (HSE, 1997). This should not be confused with assessing the risks, which all have to be carefully identified and evaluated, whether they are reasonable practical to prevent or not.

Who Assesses and How

Not Only a Risk-Management Problem

To this broad risk definition, should be added a similarly wide assessment process. Employers unequivocally have to assess all risks. However, the Risk Assessment and its result is not a pure management issue, which is up to them alone. Many works on risk assessment fail to properly understand—or at least communicate—the crucial difference between economic business risks and social OHS risks. They describe OHS as (only) a matter of successful management, without including the political aspect of state regulation and workers' rights.

In doing so, their advice on Risk Assessment may fall short of the Framework Directive, which is clear in this respect. As the major stakeholder in healthy workplaces, workers or their representatives are to be consulted *"on all questions relating to health and safety at work"* (Article 11:1), thus also on the Risk Assessment. Their participation is to be supported by, for example, *"all the necessary information,"* which shall be provided by the employers (Article 10:1), and by various rights for, for example, training and paid time off, as specified in national regulations, and may not involve them in any costs (Article 6.5).

However, participation not only is a worker's right to look after their OHS interest against the business interests of the employer, but also, it is put forward as fundamental to make the employers' OHSM effective and efficient. A thorough assessment has to include workers' experiences of the risks, which cannot fully by identified by experts alone (cf. section under Risk Assessment as Contested Politics: and Gustavsen, 1980). Their practical production competence is also often needed to develop workable preventive measures. The importance of local cooperation between the social partners made Denmark mandate "Workplace Assessment," instead of assessments of individual risks, in their transposal of the Framework Directive (Jensen, 2002). It also motivated Swedish Systematic Work Environment Management (Frick, 2002, p. 221). Worker participation in Risk Assessment (and in the rest of the OHSM) largely follows national industrial relation traditions also in the other EU countries, though in several cases with changes due to the Framework Directive. For example, Italy and France extended such participation rights. The UK did so only after loosing similar cases in the European Court of Justice (ECJ C-382/92 and C-383/92). Yet, in the many nonunionized British workplaces, employers still have much discretion in how to consult their workers (Walters, 2002a, pp. 256–258; see further Boix & Vogel, 1999; and Walters & Frick, 2000, on the motives for and difficulties of participation in Risk Assessment and in OHSM in general).

Competence Is Mandatory

In carrying out the Risk Assessment, the social partners are to be advised by OHS experts. The preamble of the Framework Directive states that "*employers shall be obliged to keep themselves informed of the latest advances in technology and scientific findings . . . concerning the inherent dangers in their undertakings.*" And Article 7 requires employers to either assign internal personnel with enough OHS competence to carry out all the obligations required by the Framework Directive or enlist external expertise with this competence. This article has been especially open to varying national interpretations in its transposition (see Hämäläinen, Husman, Räsänen, Westerholm, & Rantanen, 2001). The so-called preventive services remain a contested issue in several EU states. Sweden transposed Article 7 as late as 2001. Denmark still debates how to comply, whereas France is reluctant to give up its system of company doctors (see the national chapters and the comparison in Walters, 2002a). The UK regulation largely leaves it to employers to decide what competence they may need (Walters, 2002a; 255–256, and 290–293). Recently (May 2003; case C-441/01) the EU court found the Dutch transposal of Article 7 not to be in compliance with the Directive, as it permitted employers to delegate the Risk Assessment to the external consultants of OHS services (i.e., not only to be advised by their competence). However, in whatever way it is organized, the Framework Directive requires employers to have enough OHS competence to identify and evaluate all risks (including organizational ones), which are known by experts to potentially harm workers. But the competence needed thus depends on each workplace's possible risks.

Upstream Assessment for Primary Prevention

The assessment is to guide the prevention, as mandated by Article 6.2. According to this, risks are primarily to be completely avoided and then to be combated at the source. Only as a last resort, shall workers be protected through personal protectives and instructions in safe behavior (among other things, as repeated research has demonstrated "safe behavior" to be much less effective than "safe place"; see, e.g., Gallagher, 1997; Gallagher, Ritter, & Underhill, 2001). The prevention hierarchy may be hard to comply with. Caution—to follow rules of safe behavior in risky situations—remains unavoidable in many production situations. However, whenever behavior control is an important part of the OHSM, the burden of proof—of compliance with the Framework Directive—rests with the employers. They should be able to demonstrate that everything possible has been done to avoid the risks, that is, not to place workers in situations where caution is necessary (on worker behavior versus upstream prevention in OHSM, see Frick & Wren, 2000; Nichols & Tucker, 2000; and Wokutch & VanSandt, 2000).

To support the prescribed prevention, the Risk Assessment must therefore focus upstream, to find the organizational and technological roots of the risks. For example, in ergonomics, macroergonomic risk assessments (and changes) of the work organization may often be more important than microergonomic improvements of workstations and tools to, for example, avoid strenuous tasks (Neumann, 2001). And when "safe behavior" is unavoidable, the Risk Assessment has to search out any detrimental influence on this behavior, for example, by pay systems which encourage cutting corners or by working times, workloads, or noise which may hamper a safe behavior. How the Risk Assessment can identify these and other upstream risks is largely determined by its scope and competence (as discussed earlier).

THE INTERPRETATION AND APPLICATION OF RISK ASSESSMENT AS A STANDARD

Risks Are Poorly Assessed in EU Workplaces

The EU member states have interpreted and transposed the Framework Directive in line with their general political differences (Walters, 2002a.). The UK gives employers much freedom to decide which risks to assess, how workers shall participate in this, and especially what competence they need in the assessment. At the other end, the Nordic model of Denmark, Finland, and Sweden encompasses the broad risk definition of the work environment, is based on local cooperation among the social partners, and emphasizes the need for advice by a holistic OHS competence (despite ongoing debates on how to organize it). The other EU states are often somewhere in between in these aspects of transposing the Directive (with numerous national varieties, due to each country's tradition and structure in OHS policies).

These differences may be larger in theory (law in the book) than in practice (law in action). We know little about how risks are assessed at the EU workplaces. Despite the intended fundamental role of the Framework Directive, very few serious evaluations have been done of its implementation, including of the Risk Assessment. The figures we have are mainly from surveys to employers, that is, of self-reported compliance. Some scattered results—pertaining to the "old" 15 member states between 1996 and 2000—on this are:

- Sweden: 55% of the employers had assessed the risks (AV, 2001, Appendix, p. 68).
- Greece: Very few employers had assessed the risks (Karageorgiou et al., 2000, p. 267).
- Denmark: Some 30% had started, and another 15% finished (Karageorgiou et al., p. 270).
- The Netherlands: Around 25% to 30% had assessed the risks (Karageorgiou et al., p. 273).
- Germany: About half of the employers had assessed the risks (Schaapman, 2002, p. 141).
- UK: Around one third of the employers had assessed the risks (Karageorgiou et al., p. 268).
- Norway (which also must comply with 89/391/EEC): Some 45% of the employers had implemented the OHSM regulation of "internal control," which includes Risk Assessment (Gaupset, 2000, p. 340).
- Spain: 46% to 86% of the employers (depending on their size) had assessed the risks (Walters, 2002b, p. 89).

There are both quantitative and qualitative problems with these data. Self-reported compliance is always likely to be overstated, in this case compared to the labor inspection's evaluations, and possible even more compared to expert studies (as was the case in Sweden; Frick, 2002, p. 228). The widespread lack of competent OHS advice and other survey answers (e.g., "knows the regulation") indicate that employers often do not fully understand what is required of their Risk Assessments. They are therefore even more likely to overstate their compliance. As an example, two thirds of the Swedish local communities rated themselves to comply with the OHSM ordinance (i.e., with the Framework Directive; AV, 2001, p. 14). Most probably they had organized and documented some Risk Assessment. Yet, at the same time, sickness absenteeism was exploding among their personnel, and case studies demonstrated structural obstacles to their OHSM (Larsson, 2000, pp. 209–212).

If one has conducted a Risk Assessment or not thus depends on how this is defined. The Framework Directive is very ambitious. All risks must be assessed in a thorough process. Case studies indicate that few employers fully satisfy these demands. Often, organizational aspects are excluded, worker participation is poor, or the assessments do not go upstream to the roots of the problems, to mention some of the main problems (see Boix & Vogel, 1999, pp. 7–27; Frick & Wren, 2000, pp. 34–42; Walters, 2002a, pp. 285–288). The implementation of the Risk Assessment, intended by the Framework Directive, is therefore probably less, maybe much less, than what the employers have reported.

Risk Assessments are mandatory also in small firms. As mentioned, the European Court of Justice confirmed that all employers, irrespective of size, must have written Assessments. Details on how to conduct and how to document them may be simplified, but all Assessments should anyhow be adapted to the unique local production and work situation, that is, may and should vary according to many other variables than size. Yet, by far the strongest and most pervasive difference in the surveys above is by size. Among firms with fewer than 50 workers, some 15% to 50% report that they have assessed their OHS risks, and even less of the really small ones have done so. Judging from studies of the levels and problems of implementing OHSM in small firms (e.g., Eakin, Lamm, & Limborg, 2000; Frick & Walters, 1998; Walters, 2001, 2002b), their reported compliance is extra likely to be exaggerated.

Why Is the Implementation so Difficult?

Why is it so difficult to implement the Risk Assessment? One reason is that integrated OHSM requires changes in how employers manage their own businesses. On the one hand, even small firms should be able to handle their OHS risks more systematically than they usually do. They know their production. To survive, they also have to assess and handle their other business problems in a systematic manner. On the other, the mandated OHSM was rare to start with. Some large factories have traditions of Safety Management, but these rarely take in the organizational causes (how, e.g., management structures affect accident rates or mental strain) or are supported by a genuine participation. The organizational development needed for a thorough Risk Assessment had to be achieved in and by management structures with limited capacities. All organizations are full of problems, conflicts, failures, and mistakes. Extra demands—in this case, to systematically improve the work environment—are not easy to fulfill.

The interest in the organizational development needed to integrate OHSM properly into the rest of management may also be lacking. Employers-managers may be reluctant to develop their organizations to systematically identify all risks, that is, possibly also new ones, which may well be costly to correct. How much the economic and the occupational health and safety interests are opposing or overlapping is much debated. For example, Dorman (2000) claims that practice demonstrates the conflict to be strong, whereas, for example, the British HSE instead promotes voluntary compliance with the regulated OHSM, as a "business case" of "safety pays" (HSE, 1997). Others see it as depending on the circumstances of production, often with opportunities for OHS competence to increase the joint interests (Frick, 1997; European Agency, 1999). Nevertheless, many managers regard OHS measures more as costs than as investments. With usually tight budgets, they should be less interested to systematically assess the work environment, as they may incur new costs to abate the OHS risks. Ergonomists and others who promote Risk Assessments can rarely avoid the issue of how managers perceive the economic interests in improving the work environment.

The implementation of thorough Risk Assessments (and the rest of the OHSM) at the workplaces is thus obstructed by limited management capacities and more or less severe economic conflicts. But it also is also hampered by changes in the structure of production (as discussed under Only Risks Within the Employment Relation) in which responsible employers are harder to track and large ones are fragmenting into (at least formally) independent small firms. On the other hand, the implementation is promoted by what may be called the OHS infrastructure. This consists of both the resources and the policies of the OHS authorities and of other OHS-actors, such as of the social partners, OHS experts (like health services), and insurance companies. Compared to the broad and deep-going changes needed to implement the Framework Directive and its Risk Assessment, their resources are quite limited (see, further, Walters, 2002a, on how national OHS infrastructures have promoted this implementation).

The Purpose Is More Important Than the Procedure in Risk Assessment

These are some answers of why the implementation has been so poor. However, a more important question is: What to do? How shall ergonomists and other OHS actors apply the EU standard of Risk Assessment? This section does not aim to be a practical guide on how to assess risks at work. As mentioned (under Risk Assessment as Science), there is a large literature of both practical guides and analytical works, oriented to a wide variety of workplace and policy issues. It is also important that applications of the Risk Assessments comply with the varying national regulations (on which there also is ample information and guidance issued by the OHS authorities and others). Here we shall instead discuss some of the principal issues in the application.

First, ergonomists, other OHS practitioners and employers cannot completely rely on the existing guidance material (though much of it is useful). To adhere by its advice may not be enough to secure compliance with the regulations. This material is often sold on a market where employer-managers may prefer messages, which do not interfere too much with how they run their operations. The writers, who commonly produce it, are more skilled in complex risk analysis techniques than in, for example, health effects of organizational factors or on the industrial relations of risk assessments, both of which are covered by the regulations. However, for example, Boix and Vogel's guide is useful not only for unionists (for which it is primarily intended) but also for any other who organizes Risk Assessment as a starting point of a broad workplace development.

Second, how to assess the OHS risks depends much on why it is done. Is it primarily a compliance with the OHS authorities' external demands, or are there also strong internal motives to improve the work environment, as part of the business development? The local answer to this defines the base for the interaction among the concerned actors: management, workers, and often also ergonomists and other OHS experts. It will greatly influence what Risk Assessments one tries to implement. Especially if they look into the production and its problems, OHS experts often have a know-how, which can help managers to combine economic and OHS interests (Oxenburgh, Marlow, & Oxenburgh, 2003).

This does not mean that conflicts of interests disappear with such competence, rather that the implementation of the Risk Assessment must focus as much on its wider setting—why it is done and how its results are going to be used—as on the formal procedures of identifying and evaluating the risks. If the motive is predominantly one of legal compliance to external demands, then ergonomists should note the broad nature of these demands and try to secure that all of them are included in the Risk Assessment. If the Risk Assessment is to support the internal development of work and production, the same broad issues have to be included, but it is easier to discuss and organize the changes in management required for a good implementation of all of the OHSM. However, in both cases, ergonomists and other OHS practitioners have to secure enough competence in the broad organizational and industrial relations issues of the Risk Assessment. This is usually not included in their training.

Usability of OHSM Standards in Small Firms

If and how small firms can use Risk Assessments and other OHSM standards also depends on how they interpret and apply them. As for other employers, the choice is mainly between a legalistic, minimalist, and a more integrated developmental interpretation of the standards. On the one hand, small firms also have to obey the law, and all EU states require fairly extensive Risk Assessments of employers. However, if these requirements are perceived and applied as purely external obligations from the authorities, they will also be regarded as "paperwork", foreign to their informal management. With no internal motivation, the bureaucratic Assessments will have to be implemented in small firms mainly through enforcement. And as these are some 90% of all workplaces, the authorities are hardly able to implement the Risk Assessment (and other OHSM) in small firms through inspections, orders, and sanctions alone.

The alternative is not a purely voluntary compliance. At least some level of integration of Risk Assessment into how they run their businesses is even more important in small firms. The more Risk Assessment is seen as a tool that can (also) support, for example, a more efficient production or improved workplace relations, the more the necessary adaptation of their management will be achievable, that is, acceptable to the owner-managers. However, small firms usually need support to be able use OHSM standards as tools for their internal development from whatever OHS actors there are. These external actors need to help translate the formal regulations into workable methods in the individual firm. This—again—requires competence not only in OHS but also in organizational issues (see Frick & Walters, 1998; Walters, 2001, 2002b).

Ergonomists, who see Risk Assessment as mainly a tool for workplace development, may be important such actors. They, other OHS actors and the few small firm managers who directly try to implement the Risk Assessment can find abundant guidance material, including the checklists, and so forth, supported by the European Agency for Safety and Health at Work. However, for small firms, it is essential that these and other tools are indeed used as development support, not as "tick-in-the-box" formal checklists. The purpose of assessing risks and what comes next must be discussed in order for the tools to become a help to self-help in small firm.

REFERENCES

Ansell, J., & Wharton, F. (Eds.). (1992). *Risk: Analysis, Assessment and Management*. Chichester, John Wiley.

Arbeidsomstandighedenwet. (1999). *Wet van 18 maart 1999, houdende bepalingen ter verbetering van de arbeidsomstandigheden*. The Hague: Staatsblad van het Koninkrijk der Nederlanden.

Ashford, N., & Caldart, C. (1997). *Technology, Law and the Working Environment*. Washington DC: Island Press.

AV. (2001). *Ett aktivt arbetsmiljöarbete*? Rapport 2001:12. Solna: Arbetsmiljöverket.

Beck, U. (1992). *Risk society: Towards a new modernity*. London: Sage.

Bercusson, B. (1996). *European labour law*. London: Butterwick.

Boix, P., & Vogel, L. (1999). *Risk assessment at the workplace—A guide for union action*. Brussels: TUTB.

British Medical Association. (1992). *Living with risk—The British Medical Association Guide*. Chichester: Wiley.

Brodeaur, P. (1974). *Expendable Americans*. New York: Viking Press.

BS 8800. (1996). *Guide to occupational health and safety management systems*. London: British Standard Institution.

Carson, R. (1962). *Silent spring*. Boston: Houghton Mifflin.

Castells, M. (1996). *The rise of network society*. Oxford: Blackwell.

Collinson, D. (1999). Surviving the rigs: Safety and surveillance at north sea oil installations. *Organisational Studies, 20*(4), 579–600.

Colombini, D., Occhipinti, E., & Grieco A. (2002). *Risk assessment and management of repetitive movements and extertions of upper limbs*. Oxford: Elsevier.

Concha, S. (1983). Editorial. *Hazard Prevention, 19*(2):2.

Cox, S., & Cox, T. (1996). *Safety, systems and people*. Oxford: Butterworth-Heinemann.

Cullen, L. (1990). *The public inquiry into the Piper Alpha disaster*. London: Departement of Energy.

Dorman, P. (1996). *Markets and mortality—Economics, dangerous work and the value of the human life*. Cambridge: Cambridge University Press.

Dorman, P. (2000). If safety pays, why don't employers invest in it? In Frick et al. (Eds.), *Systematic occupational health and safety management—Perspectives on an international development*. Oxford: Elsevier.

Dwyer, T. (1991). *Life and death at work—Industrial accidents as a case of socially produced error*. New York: Plenum.

Eakin, J., Lamm, F., & Limborg, H. J. (2000). International perspective on the promotion of health and safety in small workplaces. In Frick et al. (Eds.), *Systematic occupational health and safety management—Perspectives on an international development*. Oxford: Elsevier.

Eckerman, I. (2004). *The Bhopal saga—Causes and consequences of the world's largest industrial disaster*. India: Universities Press Private Ltd, Hyderabad.

European Agency. (1999). Health and safety at work—A question of costs and benefits? *Magazine No. 1, of the European Agency for Safety and Health at Work*. Bilbao. http:/agency.OHSa.eu.int/publications/magazine/en/mag1.html

European Commission. (1996). *Guidance on risk assessment at work*. Luxemburg: Office for Official Publication of the European Communities.

Fallentin, N., Viikari-Juntura, E., Wærsted, M., Kilbom, Å. (2001). Evaluation of physical workload standards and guidelines from a Nordic perspective. *Scandinavian Journal Work Environ Health*, vol. 27, suppl. 2.

Frick, K. (1997). Can managers see any profit in health and safety?—Contradictory views and their penetration into working life. *New Solutions, 7*(4), 32–40.

Frick, K. (2002). Sweden: Occupational health and safety management strategies from 1970-2001. In D. Walters (Ed.), *Regulating health and safety management in the European Union*. Brusels: P.I.E. Peter Lang.

Frick, K. P., Jensen, L., Quinlan, M., & Wilthagen, T. (Eds.). (2000). *Systematic occupational health and safety management—Perspectives on an international development*. Oxford: Elsevier.

Frick, K., & Walters, D. (1998). Worker representation on health and safety in small enterprises: Lessons from a Swedish approach. *International Labour Review, 137*(3), 367–89.

Frick, K., & Wren, J. (2000). Reviewing occupational health and safety management—multiple roots, diverse perspectives and ambiguous outcomes. In Frick et al. (Eds.), *Systematic occupational health and safety management—Perspectives on an international development*. Oxford: Elsevier.

Gallagher, C. (1997). *Planned approaches to health and safety management*. Sydney: National Occupational Health and Safety Commission.

Gallagher, C., Ritter, M., & Underhill, E. (2001). *Occupational health and safety management systems: A review of their effectiveness in securing healthy and safe workplaces*. Sydney: National Occupational Health and Safety Commission.

Gaupset, S. (2000). The Norwegian Internal Control Reform. In Frick et al. (Eds.), *Systematic occupational health and safety management—Perspectives on an international development*. Oxford: Elsevier.

Gunningham, N., & Johnstone, R. (2000). The legal construction of OHS management systems. In Frick et al. (Eds.), *Systematic occupational health and safety management—Perspectives on an international development*. Oxford: Elsevier.

Gustavsen, B. (1980). Improving the work environment: A choice of strategy. *International Labour Review, 119*(3), 271–286.

Hämäläinen, R. M., Husman, K., Räsänen, K., Westerholm, P., & Rantanen, J. (2001). *Survey of the quality and effectiveness of occupational health services in the European Union and Norway and Switzerland*. Helsinki: Finnish Institute of Occupational Health.

Hansson, S. O. (1993). The false promises of risk analysis. *Ratio, 6*, 16–26.

Harms-Ringdahl, L. (2001). *Safety analysis—Principles and practice in occupational safety*. London: Taylor & Francis.

Holmér, I. (2000). Assessment of cold exposure. *International Journal of Circumpolar Health, 60*, 413–421.

Hopkins, A. (2000). *Lessons from Longford—The ESSO gas plant explosion*. Sydney: CCH Australia.

HSE (1992). *Management of health and safety at work*. Sudbury: Health & Safety Executive.

HSE (1994). *Upperlimb disorders: Assessing the risks*. Sudbury: Health & Safety Executive.

HSE (1997). *Successful health and safety management—HSG65*. Sudbury: Health & Safety Executive.

HSE (1998). *Assessment principles for offshore safety cases*. Sudbury: Health & Safety Executive.

HSE (1999). *Five steps to risk assessment*. hse.gov.uk/pubns/indg163.pdf.

IEC (1995). *Dependability management: Risk analysis of technological systems (IEC 300-3-9)*. Geneva: International Electrotechnical Commission.

ISO (1999). *ISO 14121: Safety in machinery—Principles of risk assessment*. Geneva: International Standard Organisation.

Jensen, P. L. (2002). Assessing assessment—The Danish experience of worker participation in risk-assessment. *Economic & Industrial Democracy, 23*(2), 201–227.

Jeynes, J. (2002). *Risk management: 10 Principles*. Oxford: Butterworth-Heinemann.

Johnson, J., & Johansson, G. (Eds.). (1991). *The psychosocial work environment: Work organization democratization and health*. Amityville: Baywood.

Johnstone, R. (1999). Paradigm crossed? The statutory occupational health and safety obligations of the business undertaking. *Australian Journal Labour Law, 12*, 73–112.

Karageorgiou, A., Jensen, P. L., Walters, D., & Wilthagen, T. (2000). Risk assessment in four member states of the European Union. In Frick et al. (Eds.), *Systematic occupational health and safety management—Perspectives on an international development*. Oxford: Elsevier.

Koukoulaki, T., & Boy, S. (2003). *Globalizing technical standards. Impact and challenges for occupational health and safety*. Brussels: TUTB-Saltsa.

Krüger, D., Louhevaara, V., Nielsen, J., & Schneider, T. (1998). *Risk assessment and preventive strategies in cleaning work*. Bremerhaven: Wirtschaftsverlag.

Kumamato, H., & Henley, E. (1996). *Probabilistic risk assessment and management for engineers and scientists*. New York: IEEE Press.

Larsson, T. (1991). *Arbetsmiljöns styrning—"Kristinehamnsmodellen."* Uppsala: SAMU.

Larsson, T. (2000). The diffusion of employer responsibility. In Frick et al. (Eds.), *Systematic occupational health and safety management—perspectives on an international development*. Oxford: Elsevier.

Majone, G. (1985). The international dimension. In H. Otway & M. Peltu (Eds.), *Regulating industrial risks—Science, hazards and public protection*. London: Butterworths.

Marklund, S. (Ed.). (2001). *Worklife and health in Sweden 2000*. Stockholm: National Institute for Working Life.

Markowitz, G., & Rosner, D. (2002). *Deceit and denial: The deadly politics of industrial pollution*. Berkeley: University of California Press.

Mosleh, A., & Bari, R. (Eds.). (1998). *Probabilistic safety assessment and management, PSAM 4*. International Association for Probabilistic Safety Analysis and Management. London: Springer.

Mottel, W., Long, J., & Morrison, D. (1995). *Industrial safety is good business—The DuPont story*. New York: Van Nostrand.

Needleman, C. (2000). OHSA at the crossroad: Conflicting frameworks for regulating OHS in the United States. In Frick et al. (Eds.). *Systematic occupational health and safety management—Perspectives on an international development*. Oxford: Elsevier.

Neumann, P. (2001). *On risk factors for musculoskeletal disorder and their sources in production systems design*. Lund University, Lund: Departement of Design Sciences.

Nichols, T., & Tucker, E. (2000). OHS Management Systems in the UK and Ontario, Canada: A Political Economy Perspective In Frick et al. (Eds.). *Systematic occupational health and safety management—Perspectives on an international development*. Oxford: Elsevier.

Ogden, T. (2003). The 1968 BOHS Chrysotile Asbestos Standard. *Annals of Occupational Hygiene, 47*(1), 3–6.

Olsen, P. B. (1992). *Six cultures of regulation—Labour inspectorates in six European countries*. Copenhagen: Handelshøjskolen.

Otway, H. (1985). Regulation and risk analysis. In H. Otway, & M. Peltu (Eds.). *Regulating industrial risks—Science, hazards and public protection*. London: Butterworths.

Otway, H., & Peltu, M. (Eds.). (1985). *Regulating industrial risks—Science, hazards and public protection*. London: Butterworths.

Oxenburgh, M., Marlow, P., & Oxenburgh, A. (2003). *Increasing productivity and profit through health and safety*. London: Taylor & Francis.

Perrow, C. (1984). *Normal accidents—Living with high-risk technologies*. New York: Basic Books.

Quinlan, M., & Mayhew, C. (2000). Precarious employment, work re-organisation and the fracturing of OHS management. In Frick et al. (Eds.), *Systematic occupational health and safety management—Perspectives on an international development*. Oxford: Elsevier.

Rakel, H. (1996). *Workplace risk assessment: A comparative analysis of regulatory practices in five EU member-states*. Norwich: Environmental Risk Assessment Unit, Norwich University.

Ramazzini, B. (1700). *De Morbis Artificum*. Modena.

Rampal, K., & Sadhra, S. (1999). Basic concepts and developments in health risk assessment and management. In S. Sadhra, & K. Rampal (Eds.), *Occupational health—Risks assessment and management*. Oxford: Blackwell.

Reason, J. (1997). *Managing the risks of organizational accidents*. Aldershot: Ashgate.

Renn, O. (1985). Risk analysis: Scope and limitations. In H. Otway, & M. Peltu (Eds.), *Regulating industrial risks—Science, hazards and public protection*. London: Butterworths.

Rivest, C. (2002). France: From a minimalist transposition to a full scale reform of the OHS system. In D. Walters (Ed.), *Regulating health and safety management in the European Union*. Brussels: P.I.E. Peter Lang.

Roberts-Phelps, G. (1999). *Risk assessment—A Gower health and safety workbook*. Aldershot: Gower.

Ruttenberg, R. (1981a). Why social regulatory policy requires new definitions and techniques for assessing costs and benefits: The case of occupational safety and health. *Labor Studies Journal*, 114–131.

Ruttenberg, R. (1981b). Regulation is the mother of invention. *Working Paper*, (May–June), 42–47.

Sadhra, S., & Rampal, K. (Eds.). (1999). *Occupational health—Risks assessment and management*. Oxford: Blackwell.

Schaapman, M. (2002). Germany: Occupational health and safety discourse and the implementation of the framework directive. In D. Walters (Ed.), *Regulating health and safety management in the European Union*. Brussels: P.I.E. Peter Lang.

Stern, R. (1980). *Introduction to risk assessment. Risk assessment in the welding industry: Part 1.* Copenhagen: The Danish Welding Institute.

Stewart, M., & Melchers, R. (Eds.). (1998). *Integrated risk assessment—Applications and regulations.* Rotterdam: Balkema.

TUTB. (1996). Trade union participation in European standardisation work: TUTB network sounds the alarm. *TUTB Newsletter*, 13–14.

Verkstadsföreningen. (1978). *Se om miljön.* Stockholm: Sveriges Verkstadsförening & Svenska Metallindustriar-betareförbundet.

Vogel, L. (1994). *Prevention at the workplace. An initial review of how the 1989 Community Framework Directive is being implemented.* Brussels: TUTB.

Vogel, L. (1998). *Prevention at the workplace. The impact of the Community Directives on preventive systems in Sweden, Finland, Norway, Austria and Switzerland.* Brussels: TUTB.

Vollmer, G, Giannoni, L., Sokull-Klütgen B. & Kracher, W. (Eds.). (1996). *Risk assessment—Theory and practice.* Luxemburg: Environment Institute, Office for Official Publication of the European Communities.

Walters, D., & Frick, K. (2000). Works, participation and the management of occupational health and safety: Rethinking or conflicting strategies. In Frick et al. (Eds.), *Systematic occupational health and safety management—Perspectives on an international development.* Oxford: Elsevier.

Walters, D. (2001). *Health and Safety in Small Enterprises: European Strategies for Managing Improvement.* Brussels: P.I.E. Peter Lang.

Walters, D. (Ed.). (2002a). *Regulating health and safety management in the European Union.* Brussels: P.I.E. Peter Lang.

Walters, D. (2002b). *Working safely in small enterprises in Europe: Towards a sustainable system for worker partic-ipation and representation.* Brussels: TUTB.

Wharton, F. (1992). Risk management: Basic concepts and general principles. In J. Ansell & F. Wharton (Eds.), *Risk: Analysis, assessment and management.* Chichester: Wiley.

Wokutch, R., & VanSandt, C. (2000). OHS management in the United States and Japan: The DuPont and Toyota Models. In Frick et al. (Eds.), *Systematic occupational health and safety management—Perspectives on an international development.* Oxford: Elsevier.

27

ILO Guidelines on Occupational Safety and Health Management Systems

Daniel Podgórski

Central Institute for Labour Protection—National Research Institute

BEGINNINGS AND DEVELOPMENTS IN STANDARDIZATION OF OSH MANAGEMENT SYSTEMS

First National Standards

It is commonly recognized that the first national standard regarding the occupational safety and health management systems (OSH-MS) was the voluntary British standard BS 8800 (British Standard Institution [BSI], 1996). This standard was developed following the popularity of a systematic and standardized approach to the management systems growing worldwide. Initially, this approach applied to the quality management systems (ISO 9000), and subsequently the environmental management systems (ISO 14000). BS 8800 standard contains guidelines concerning design and implementation of OSH-MS in the way allowing for integration of OSH-MS with the general enterprise management system. The OSH-MS proposed in this standard is based on the continual improvement cycle PDCA (Plan-Do-Check-Act), also known as Deming's cycle. This model is compliant with the management system model applied in the ISO 14001 standard (International Organization for Standardization [ISO], 1996), as well as with the model adopted in 2000 within the ISO 9001 standard regarding quality management (ISO, 2000).

Similar standards intended for voluntary application have been worked out and established also in other countries leading in the area of promotion of the systematic approach to OSH management. One should mention here the Dutch standard NPR 5001 (Nederlands Normalisatie-Institut [NNI], 1996), the Australian guide SAA HB53 (Standards Australia, 1994) intended for application in the construction industry and the joint standard AS/NZS 4804 (Standards Australia/Standards New Zealand [SA/SNZ], 1997/2001) developed by the joint Australia and New Zealand technical committee. These documents constitute the guidelines and may not serve as the basis for management system certification, as it is in the case of ISO 9001 and ISO 14001 standards. The AS/NZS 4804 standard, however, created grounds for elaboration and introduction of the AS/NZS 4801 standard (SA/SNZ, 2001) containing specifications that may be used for the conformity assessment processes regarding such systems, performed by a third party.

Standards and draft standards regarding systematic OSH management have been prepared and published also in Spain where they appeared in a form of six documents (Abad, Mondelo, & Llimona, 2002). The two most important standards within this series include UNE 81900 (Spanish Association of Standardization and Certification [AENOR], 1996a) and UNE 81901 (AENOR, 1996b). Activities aimed at OSH-MS standardization were also started in Poland in 1998 by establishing, within the framework of the Polish Standardization Committee (PKN), the Technical Committee No. 276 for OSH Management. The works of this Committee have resulted in establishing the PN-N-18000 series of standards containing at present three documents: PN-N-18001 (PKN, 1999/2003), PN-N-18002 (PKN, 2000), and PN-N-18004 (PKN, 2001). The first of them contains specification for the OSH-MS and may be used as the basis for certification, whereas the other ones are the practical guidelines supporting occupational risk assessment procedures and OSH-MS implementation in organizations.

In several other countries, a systematic approach to OSH management is popularized on the basis of national law rather than by voluntary standards. Such a situation exists, among others, in the United States, Scandinavian countries, and Japan. In the United States, the system of so-called Voluntary Protection Programs (VPP) has been in force since 1982 under supervision of the Occupational Safety and Health Administration (OSHA, 1982). Under this system, companies implement OSH management programs on the basis of the rules, periodically updated and published by OSHA (OSHA, 2000a). Companies which participate in the VPP, and there are some 900 of them at present, are relieved from routine and programmed OSHA inspections, replaced by periodic audits. Experience generated by the first years of VPP functioning have been used for elaboration of guidelines for OSH-MS implementation in organizations, currently applied for wide popularization and promotion of OSH systematic management in the United States. In 1996, these activities were supported by the American Industrial Hygiene Association (AIHA, 1996), which developed and published guidelines for the OSH-MS on the basis of the quality management system concept contained in the ISO 9000 standards.

In the Scandinavian countries, a systematic approach to OSH management was initiated in the 1990s by establishing mandatory legal provisions regarding so-called "internal control of work environment." Implementation guidelines for such systems were established and published in Norway in 1991 (Kommunaldepartementet, 1991) and in Sweden in 1992 (Swedish National Board of Occupational Safety and Health [SNBOSH], 1992). After a few years of experience, both countries verified regulations governing the internal control of work environment and introduced their new versions in 1997 (Kommunaldepartementet, 1996; SNBOSH, 1997), and in Sweden again in 2001 (Swedish Work Environment Authority, 2001). These guidelines ensure practical implementation, in organizations, of provisions of the EU Framework Directive (89/391) regarding introduction of measures for improvement of employees' health and safety at work (European Union, 1989).

In Japan, OSH-MS guidelines were put in force in the Ordinance of the Minister of Labor of April 30, 1999 (Japanese Ministry Of labor, 1999). Both the structure and the content of provisions in this document are in line with AS/NZS 4804 and BS 8800 standards and other regulating documents in this field. Regardless of their legal provisions status, Japanese guidelines are intended for voluntary application by employers who aim at improved effectiveness of measures related to better work conditions in their enterprises.

Still another situation regarding regulation of the approach to OSH-MS exists in Germany, where the guidelines for implementation of OSH-MS have been established at the regional level. Such solutions are exemplified by the ASCA program, initiated by the Hessian Ministry of Social Matters; Employment and Women (Hessisches Ministerium fuer Frauen, Arbeit und

Socialordnung, 1996), and guidelines for designing, implementation, and integration of OSH-MS in organizations, established by the Bavarian Ministry of Labor, Social Matters, Family, Women, and Health (Bayerisches Staatsministerium fuer Arbeit und Socialordnung, Familie, Frauen und Gesundheit, 1997).

Development of OSH MS Standardization at International Level

In view of the interest in development and establishment of OSH-MS international standards, growing in various countries worldwide, the International Standardization Organization (ISO) analyzed in 1996 the need to initiate OSH-MS standardization process on an international level (Zwetsloot, 2000). Nevertheless, on the basis of discussions held during an international workshop on the need to work out international OSH-MS standards, organized by ISO in Geneva in September 1996, with participation of the International Labor Organization (ILO), as well as after an analysis of results of the voting held among national standardization bodies, the ISO Technical Management Board decided in February 1997 for ILO not to continue further works on standardization of OSH MS specifications. The main reason of this decision consisted in considerable differences in methods and culture of OSH between highly developed and developing countries, expressed in, first of all, different solutions of legal systems covering the OSH issues.

Another attempt by ISO to undertake works related to OSH-MS international standards was made in 1999 due to the proposal of BSI. After the official voting, however, carried out among the standardization bodies—ISO members, this proposal was rejected in April 2000. This situation was caused mainly by the lack of support from developed countries for the concept of certification of OSH-MS conformity with requirements of relevant international standards (Lambert, 2000), by ILO's works, undertaken in the meantime, regarding its guidelines in this field (ILO, 2001a), and by related conviction that ISO would not be an appropriate organization to lay down appropriate requirements for relations between employers and employees, which are the basis of efficient OSH management.

ISO's resignation to undertake activities aimed at development of standards for OSH management generated other international initiatives in this area. The lack of standards which might create a basis for OSH-MS certification encouraged an action by some private consulting companies and certification organizations, operating internationally and specialized so far in certification of quality and environmental management systems. Such organizations, searching for new areas of business development, worked out their own documents containing OSH-MS specifications, and after publishing them, they started to offer OSH-MS certification. Seeing a need to assure a uniform character of the approach to OSH-MS on international level, a dozen or so various certification and standardization institutions, both private and governmental, representing various countries and international certification systems, created a consortium whose objective was to generate a series of documents containing specifications and guidelines for OSH-MS. These activities, carried out under the leadership of BSI, resulted in preparation of OHSAS 18001 (BSI, 1999) and OHSAS 18002 (BSI, 2000) documents. OSH-MS models, adopted within these documents, as well as other provisions are compatible with ISO 9001 and ISO 14001 standards, assuring a possibility for OSH-MS integration with quality management systems and environmental management. Nevertheless, these documents have not been generated within the formal standardization process, therefore they are not recognized as international standards. Although published by BSI, neither are they British standards. While preparing OHSAS 18001, it was assumed that this document would be adopted by some of the institutions participating in the consortium as the basis for their OSH MS certification activities and would replace in this way other documents, hitherto applied by these institutions.

ILO and Its Role in OSH Management Systems International Standardization Process

International Labor Organization—ILO—was set up concurrently with the League of Nations at the Versailles Congress in 1919. Since 1946, ILO has become a specialized agency of the United Nations Organization (UNO). Its objective is to disseminate social justice rules in order to contribute to universal, sustainable peace. ILO acts on the principle of tripartite representation, which is unique among the UNO affiliated agencies, as the ILO Administrative Council consists of representatives of employers and employees organizations as equal partners of the government parties. These three parties are active participants of meetings organized by ILO. Technical secretariat for ILO is provided by the International Labor Office in Geneva.

ILO has established and published for application by Member States a number of commonly respected international conventions and recommendations related to labor and regarding, among others, the freedom of associations, employment, social policies, conditions at work, social security, industrial relations, and labor administration. ILO also carries out advisory activity and provides Member States with technical support through a network of offices and multidisciplinary teams in over 40 countries. Such a support may consist in advise and training in the area of, among others, labor law, employment, development of entrepreneurship, project management, social security, occupational safety, and health and employees education (ILO, 2001b).

In view of the aforementioned competences, empowerment, and scope of activities of ILO, it should have been expected that this organization would be more appropriate to elaborate and popularize on an international scale the OSH management systems standards. This job was undertaken in 1998 in the Department of Working Conditions of the International Labor Office (currently "SafeWork"). A ready draft of ILO guidelines was then subject to verification by international experts and subsequent improvement (ILO, 2001a). Verified draft Guidelines were also opinionated by ILO Member States and in April 2001 served as a basis for discussions of the International Experts Forum, representing employers, employees, and governmental parties.

The final ILO guidelines text was approved on April 27, 2001, and submitted to the ILO Administrative Board which on June 22, 2001, accepted the document for publication as ILO-OSH 2001 (ILO, 2001b). ILO-OSH 2001 guidelines were initially published by ILO in English, French, and Spanish, but in many ILO Member States, guidelines were translated into other languages and published as local versions. An example is provided by the guidelines publication in Polish language made by the Central Institute for Labor Protection (CIOP, 2001).

As it is stated in the introduction to the ILO guidelines, they are addressed to all persons responsible for OSH management. Application of these guidelines is not legally mandatory for ILO Member States or for enterprises located on their territory. Guidelines are not intended to replace provisions of the law or voluntary standards existing in these countries. On the other hand, provisions regarding OSH management contained in this document take into account the main ILO conventions related to OSH, in particular Convention No. 155 regarding occupational safety and health (ILO, 1981) and Convention No. 161 regarding health care service for workers (ILO, 1985).

ILO GUIDELINES' ROLE IN PROMOTING OSH-MS
AT THE NATIONAL LEVEL

Application Levels of ILO Guidelines on OSH MS

ILO-OSH 2001 guidelines are intended for application at two levels: the national and the level of an organization (enterprise). This constitutes the principal difference and advantage of this document as compared with other standards regarding OSH-MS which relate exclusively to the level of organizations. This specific feature of the guidelines is reflected in the document structure, consisting of three parts. The first part contains general objectives of the guidelines; the second one, provisions to be applied at the national level; and the third, guidelines regarding OSH-MS addressing the level of the organization.

National Policy on OSH-MS

Provisions of the ILO guidelines intended to be applied at the national level relate to creation and functioning of national structures responsible for promotion of the systematic approach to OSH management. It is recommended that, as far as possible, such activities should be supported by respective provisions of the national law. These provisions include in particular the following:

a. Nomination of a *competent institution* to formulate and implement national policy concerning establishment and promotion of OSH-MS
b. Formulation of coherent *national policy on OSH-MS*
c. Development of *national and tailored* guidelines concerning voluntary implementation and maintenance of OSH-MS in the organizations

The *competent institution*, mentioned under point a should be nominated in the Member States in agreement with organizations representing interests of employers and employees, as well as with other organizations and institutions competent in the field of OSH. For many countries, government agencies for social affairs and labor law or R&D institutions involved in OSH area are the potential candidates to be nominated as such bodies.

ILO guidelines specify the scope of recommended *national policy on OSH MS*. The most important aspects to be covered by such policy include:

- Promotion OSH-MS as an integral part of the overall management of the organization
- Promotion of voluntary measures directed to systematic activities for OSH improvement
- Avoiding unnecessary bureaucracy, administration, and costs related to OSH-MS
- Support for OSH-MS activities provided by labor inspection and OSH services

National and Tailored Guidelines on OSH-MS

As mentioned under point c, ILO guidelines recommend development and establishing, in Member States, the *national guidelines*, adjusted to the law and practice existing in a specific country, as well as respective *tailored guidelines*. National guidelines should be intended for voluntary application by organizations and should be based on the OSH-MS model provided in the ILO-OSH 2001 document. On the other hand, tailored guidelines should contain the generic elements of the national guidelines, take into consideration the overall ILO guideline objectives, and, first of all, they should reflect specific conditions and needs of the organization or a group of organizations, in particular:

- Their size (small, medium, and large) and organizational structure
- Types of existing hazards and the level of respective occupational risks

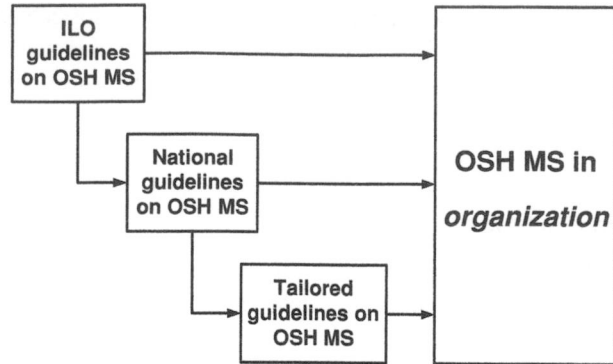

FIG. 27.1. Principles of application of ILO guidelines at the organization level (ILO, 2001b).

The aforementioned provisions of ILO guidelines, relating to the national level, provide possibilities for various solutions in the area of transposition, promotion, and practical application of the OSH-MS concept in various countries and organizations. These provisions indicate in principle three main ways of using the guidelines at the organization level, shown in Figure 27.1 (ILO, 2001b).

Taking into consideration ILO-OSH 2001 recommendations regarding establishing the national guidelines, various actions in this area have been undertaken in various countries. In Germany, for example, transposition of this document was made by publishing of the national guidelines for OSH management systems (Bundesanstalt fuer Arbeitsschutz und Arbeitsmedizin, 2003).

In Poland, on the other hand, it was assumed that the ILO guidelines would be transposed at the national level through verification and amendment of national standards of the PN-N-18001 series. This kind of approach has already resulted in adoption in 2003 of the new version of the PN-N-18001 standard (PKN, 2003) harmonized with the ILO guidelines. Another method of ILO-OSH 2001 transposition at the national level was applied in the Slovak Republic where the National Labour Inspectorate developed and published voluntary guidelines on OSH MS (Narodný Inšpectorat Práce, 2002). The model of OSH-MS adopted in Slovak guidelines is based on the ILO model, whereas the practical recommendations for OSH are in line with the requirements of OHSAS 18001, BS 8800, and the Swedish law on internal control of the work environment.

It should be expected, however, that in many countries the ILO guidelines will not be initially transposed at the national level by establishing new regulations or standards; they will rather be applied directly in their original version or the language version of specific countries.

A subsequent stage in transposition of the ILO guidelines will consist in preparation of respective tailored guidelines in specific countries. In some of them, such documents already exist or are being prepared, in particular in relation to special industries, such as chemical and machine industry. Nevertheless, there are still significant problems with developing a model of systematic approach to OSH management, which would be approved and practically applied by small- and medium-size organizations (SMEs). Some solutions in this area already exist, such as, for example, a guide for OSH-MS implementation in SMEs published in Germany by Länderausschuss für Arbeitsschutz und Sicherheitstechnik (LASI, 2001) and the toolset for occupational risk management in SMEs, worked out by the Institution of Occupational Safety and Health (2002) on the basis of the concept of the Technical Research Center of Finland (VTT).

ILO GUIDELINES MODEL AND PROVISIONS REGARDING OSH-MS IN ORGANIZATIONS

The third part of the ILO guidelines, addressed to organizations, starts with provisions regarding obligations and responsibilities of employers with respect of assuring occupational safety and health, including compliance with the requirements resulting from respective legal regulations in a specific country. Additionally, the employer should provide commitment and leadership in the OSH activities and introduce organizational solutions aimed at implementation of OSH-MS in the organization, in line with specific model. These provisions indicate the OSH-MS model based on the PDCA continual improvement cycle and consisting of five main elements. Graphic presentation of this model is shown on Figure 27.2 (ILO, 2001b).

According to other provisions of the ILO guidelines an employer should aim at OSH-MS integration with the overall management system in the organization, including its subsystems focused on quality management or environmental management for instance. An adopted OSH-MS model facilitates meeting such a requirement as it is philosophically consistent with management system models defined in ISO 9001 and ISO 14001 standards. It is also consistent with OSH MS models adopted in other normative documents, such as PN-N-18001 or OHSAS 18001.

Provisions of the ILO guidelines relating directly to the organization level are presented in this document in the order compliant with the sequence of the main OSH-MS elements shown on Figure 27.2. Their structure and summaries are presented in Table 27.1 (on the basis of ILO, 2001b).

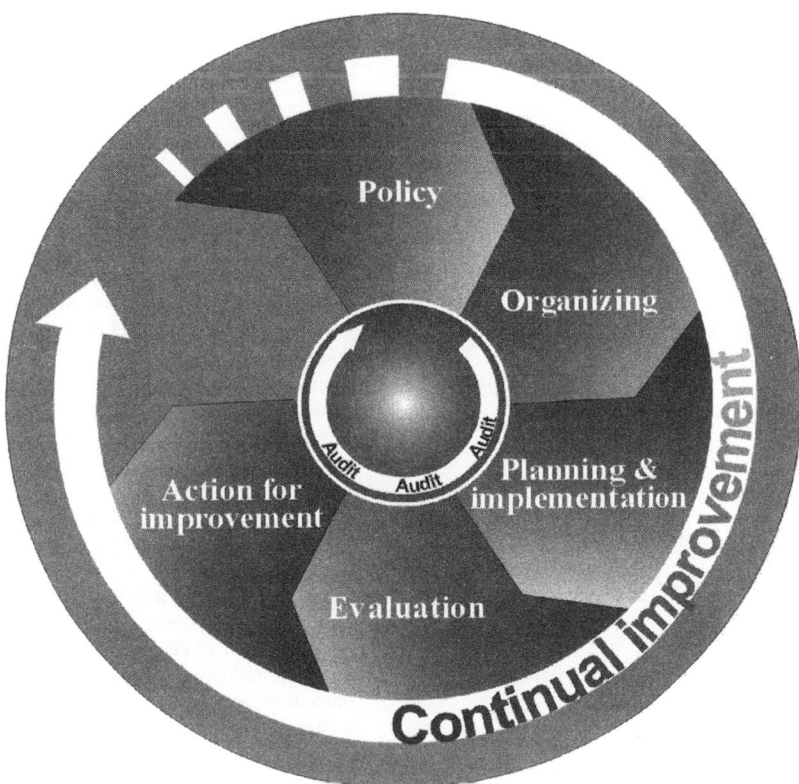

FIG. 27.2.   OSH management system model adopted in the ILO guidelines.

TABLE 27.1
Review of ILO-OSH 2001 Provisions Relating to Organization's Level

Element	Number	Title	Summary
Policy	3.1	Occupational Safety and Health Policy	Requires establishment by the employer of the OSH policy in consultation with workers and their representatives. Defines required content of such a policy and indicates the need to integrate OSH-MS with other systems in the organization.
	3.2	Worker's Participation	Underlines that workers participation is the essential factor ensuring OSH-MS effectiveness. Requires consulting the workers regarding OSH activities and introduction of solutions which encourage participation of workers in OSH-MS, including a safety and health committee.
Organizing	3.3	Responsibilities and Accountability	Indicates the necessity to define responsibilities, duties, and empowerment related to implementation and functioning of OSH-MS. Requires designation of a member of the top management responsible for the OSH-MS implementation and performance as well as promotion of workers participation in this system.
	3.4	Competence and Training	Indicates the requirements related to competence in the field of OSH and introduction of arrangements assuring that all persons will have the competence meeting such requirements. Defines requirements regarding OSH training programs and methods of their implementation.
	3.5	OSH Management System Documentation	Defines requirements regarding establishment and maintenance of OSH-MS documents and records, indicating their content, as well as their identification, reviews, updating, publication, accessibility, and storing.
	3.6	Communication	Requires introduction of organizational arrangements and procedures assuring appropriate receiving and responding to internal and external OSH information as well as their flow among various levels of the organization. Also requires reception and responding to concerns, ideas, and other OSH-related inputs coming from workers.
Planning and Implementation	3.7	Initial Review	Recommends completion of the initial review of OSH management as the basis to create OSH-MS in the organization. Recommends that the review should be carried out in consultation with workers, defines its scope, requires documenting, and indicates using the results to determine the level of reference and decision making relating to the OHS-MS implementation.
	3.8	System Planning, Development, and Implementation	Requires organizational arrangements that support planning the activities related to implementation and maintenance of all the OSH-MS elements. Recommends defining the priorities for activities and quantifiable measures of targets regarding OSH, as well as preparation of plans of tasks to achieve each objective including criteria to verify task implementation and needed resources.

Element	Number	Title	Summary
	3.9	OSH Objectives	Requires establishing measurable objectives in the area of OSH and defines recommendations for their formulation and implementation.
	3.10	Hazard Prevention	This chapter contains important provisions regarding hazard identification and occupational risk management as the key element of the OSH systematic management concept.
	3.10.1	Prevention and Control Measures	Requires arrangements and procedures of ongoing identification of hazards and occupational risk assessment and introduction of adequate preventive and protective measures. Defines required hierarchy of such measures, requires them to comply with national law and good practice, to take into consideration current state of knowledge, including information and reports of labor inspection and other services, and to be subject to periodic reviews and updates.
	3.10.2	Management of Change	Requires assessment of impact of all internal and external changes on OSH, undertaking adequate preventive measures before introduction of such changes providing information and training for workers subject to those changes. Also requires carrying out identification of hazards and assessment of occupational risks before each modification or introduction of new work methods, materials, processes, and machinery.
	3.10.3	Emergency Prevention, Preparedness, and Response	Defines requirements for organizational arrangements in the area of emergency prevention, preparedness, and response including internal communication and coordination, providing first aid, firefighting actions, evacuation, training, and exercises. Requires coordination of such arrangements with external emergency services and communication with such services, authorities, and neighbor organizations.
	3.10.4	Procurement	Requires establishing the procedures to ensure conformity of goods and services purchased by the organization with applicable OSH requirements.
	3.10.5	Contracting	Requires introduction of organizational arrangements assuring contractors' compliance with the same OSH requirements and management rules as those applied by the organization. These arrangements should include selection of contractors, communication, and cooperation methods; rules for registration of accidents; occupational diseases and incidents; OSH training and monitoring contractors for OSH.
Evaluation	3.11	Monitoring and Measurement	Requires establishing monitoring methods for selected OSH aspects including quantitative and qualitative indicators. Indicates the role of proactive and reactive monitoring in OSH MS improvement, in particular in the area of arrangements regarding hazard identification and occupational risk assessment; also defines the main OSH aspects which should be proactively and reactively monitored.

(Continued)

TABLE 27.1
(Continued)

Element	Number	Title	Summary
	3.12	Investigation of Work-Related Injuries, Ill-Health, diseases and Incidents, and Their Impact on Safety and Health Performance	Defines requirements regarding investigations of the causes of work-related injuries, ill-health, diseases and incidents to identify any failures in OSH-MS. Requires the investigations to be carried out by competent persons, with participation of workers, documenting their outcome and taking into consideration the reports of external inspection bodies, communicating results to the safety and health committee, as well as undertaking adequate corrective actions.
	3.13	Audit	Requires carrying out periodic internal audits of OSH-MS according to established policy and program. Indicates that all system elements should be subject to audits and requires implementation of audits by competent persons and consulting the auditing process and results with workers.
	3.14	Management Review	Requires periodic OSH management reviews, carried out by the top management of the organization. Defines objectives for such reviews and required range of reviewed factors. Also requires documenting the reviews and submission of their outcomes to the safety and health committee, workers and their representatives, and persons responsible for the relevant element of OSH-MS.
Action for Improvement	3.15	Preventive and Corrective Action	Requires introduction of organizational arrangements regarding preventive and corrective actions, which should include identification and analysis of the root causes of any nonconformities and initiating, planning, implementation, checking the effectiveness, and documenting preventive and corrective actions.
	3.16	Continual Improvement	Requires introduction of organizational arrangements ensuring continual improvement of all OSH-MS elements and defines range of factors that should be taken into account in such arrangements. Also requires comparing OSH activities and their effects with activities and effects of other organizations.

NEEDS FOR FURTHER ACTIONS IN INTERNATIONAL STANDARDIZATION OF OHS-MS

On the basis of the analysis of ILO-OSH 2001 guidelines content, discussed earlier in this chapter, one may conclude that this document constitutes a unique approach to standardization of requirements for OSH management systems on an international scale. Through a built-in mechanism assuring transposition of its provisions to the national guidelines level, and subsequently to the tailored guidelines level, in many countries, processes have been initiated to create new normative documents adjusted to the needs of such countries, as well as to various sectors and sizes of organizations.

Nevertheless, due to the extent of potential applications, it seems particularly urgent to develop international guidelines regarding systematic OSH management in SMEs. As shown under ILO Guidelines' Role in Promoting OSH-MS at the National Level, such documents may be worked out and established in the developed countries which possess respective competent institutions and long experience in OSH-MS implementation. On the other hand, in the case of other countries, creation of such a document may appear difficult and may require a support in the form of a model that would be commonly approved of at the international level. Such a model does not exist yet, and to work it out would require further research studies regarding its concept, in particular from the point of view of adaptability to the needs of organizations of various sizes (from very small to medium), various sectors typical for such organizations, as well as the needs of various organizations, institutions, and services supporting employers, and acting for the benefit of OSH (e.g., employers organizations, trade unions, insurance institutions, labor inspection, and other government agencies and consulting organizations). In order to ensure useful character and wide application of this model, it should be practically verified by pilot implementations in a large group of organizations representing various sectors and countries, including different levels of legislation, applied management practices, and safety cultures.

Another area where there are needs and potential possibilities for international standardization is the management of programs for ergonomic improvement at workplaces (for simplification called the "ergonomics management" or the "ergonomic program management"). Standardization in this area has been already undertaken first of all in the United States, in the first place to prevent work-related musculoskeletal disorders which cause the biggest social and economic losses (McSweeney, Craig, Congleton, & Miller, 2002). One should mention here the establishment by OSHA (1991) of guidelines for ergonomic program management in meat-processing plants. Further steps in this area included, among others; publishing the guidelines regarding ergonomic programs by the National Institute for Occupational Safety and Health (1997), establishment by OSHA (2000b) of the obligatory standard in this field (subsequently rescinded in 2001), and establishment of the standard regarding ergonomics in nursing homes (OSHA, 2003b).

The aforementioned documents, which define ergonomic improvement management principles, prove the need for further conceptual and standardization works in this area both on the national and on the international level. The OSH-MS model adopted in the ILO-OSH 2001 guidelines and the tripartite representation principle adopted in the process of their creation and establishment may constitute an appropriate model for activities in this new area of international standardization.

REFERENCES

Abad, J., Mondelo, P. R., & Llimona, J. (2002). Towards an international standard on occupational health and safety management. *International Journal of Occupational Safety and Ergonomics (JOSE), 3*, 309–319.

AENOR. (1996a). *Guidelines for the implementation of an occupational safety and health management system* (Standard No. UNE 81900:1996 EX). Madrid, Spain: Spanish Asociation of Standardisation and Certification.

AENOR. (1996b). *Guidelines for the assessment of occupational safety and health management systems. Audit process* (Standard No. UNE 81901:1996 EX). Madrid, Spain: Spanish Asociation of Standardisation and Certification.

AIHA. (1996). *Occupational health and safety management system: An AIHA guidance document.* Fairfax, VA: American Industrial Hygiene Association.

Bayerisches Staatsministerium für Arbeit und Sozialordnung, Familie, Frauen, und Gesundheit. (1997). *Modell zur Entwicklung, Gestaltung, Einführung / Integration eines Managementsystems für Arbeitsschutz und Anlagensicherheit (Occupational Health and Risk Management System).* München, Germany: Bayerisches Staatsministerium für Arbeit und Sozialordnung, Familie, Frauen, und Gesundheit.

BSI. (1996). *Guide to occupational health and safety management systems* (Standard No. BS 8800:1996). London, UK: British Standards Institution.

BSI. (1999). *Occupational health and safety management systems—Specification.* Occupational Health and Safety Assessment Series (Document No. OHSAS 18001:1999). London, UK: British Standards Institution.

BSI. (2000). *Occupational health and safety management systems—Guidelines for the implementation of OHSAS 18001* (Document No. OHSAS 18002:2000). London, UK: British Standards Institution.

Bundesanstalt für Arbeitsschutz und Arbeitsmedizin. (2002). *Leitfaden für Arbeitsschutzmanagementsysteme*. Dortmund, Germany: Bundesanstalt für Arbeitsschutz und Arbeitsmedizin (Federal Institute for Occupational Safety and Health). Retrieved March 31, 2003, from http://www.baua.de/prax/ams/leitfaden_ams.pdf

CIOP. (2001). *Wytyczne do systemów zarządzania bezpieczeństwem i higieną pracy. ILO-OSH 2001* (Polish version of the ILO Guidelines on occupational safety and health management systems. ILO-OSH 2001). Warsaw, Poland: Central Institute for Labour Protection.

European Union. (1989). *Council Directive of 12 June 1989 on the introduction of measures to encourage improvements in the safety and health of workers at work (89/391/EEC)*. Official Journal of the European Communities, No. L 183, 29 June 1989, pp. 1–8.

Hessisches Ministerium für Frauen, Arbeit und Sozialordnung (1996). *ASCA: New directions in state occupational safety*. Wiesbaden, Germany: Hessisches Ministerium für Frauen, Arbeit und Sozialordnung (HMFAS).

ILO. (1981). *Occupational Safety and Health Convention (No. 155)*. Geneva, Switzerland: International Labour Organization.

ILO. (1985). *Occupational Health Services Convention (No. 161)*. Geneva, Switzerland: International Labour Organization.

ILO. (2001a). *Guidelines on Occupational Safety and Health Management Systems (ILO-OSH, 2001)*. Geneva, Switzerland: International Labour Office. Retrieved March 31, 2003, from http://www.ilo.org/public/english/protection/safework/managmnt/index.htm

ILO. (2001b). *Guidelines on Occupational Safety and Health Management Systems* (ILO-OSH 2001). Geneva, Switzerland: International Labour Office.

Institution of Occupational Safety and Health. (2002). SME Risk Management Toolkit. The Grange, UK: Institution of Occupational Safety and Health (IOSH). Retrieved March 31, 2003, from http://www.pk-rh.com/en/index.html

ISO. (1996). *Environmental management systems—Specification with guidance for use* (Standard No. ISO 14001:1996). Geneva, Switzerland: International Organization for Standardization.

ISO. (2000). *Quality management systems—Requirements* (Standard No. ISO 9001:2000). Geneva, Switzerland: International Organization for Standardization.

Japanese Ministry of Labour. (1999). *Guideline for occupational safety and health management systems, ministry of labour notification no. 53*, April 30, 1999, Tokyo, Japan: Ministry of Labour.

Kommunaldepartementet. (1991). *Internkontroll. Forskrift med veiledning* (Internal control. Ordinance with guidelines). Oslo, Norway: Kommunaldepartementet (the former Ministry of Local Government and Labour).

Kommunaldepartementet (1996). Forskrift om systematisk helse-, miljø- og sikkerhetsarbeid i virksomheter m. fl. (Internkontrollforskriften) (Regulation on systematic health, environment and safety activities in enerprises et al., The Internal control regulation). Oslo, Norway: Kommunaldepartementet (the, former, Ministry of Local Government and Labour).

Lambert, J. (2000). *The German position with respect to standardisation of OH&S management systems*. In. D. Podgórski & W. Karwowski (Eds.), *Ergonomics and safety for global business quality and productivity. Proceedings of the Second International Conference ERGON-AXIA 2000, 19–21 May, Warsaw, Poland* (pp. 315–318). Warsaw, Poland: The Central Institute for Labour Protection.

Länderausschuss für Arbeitsschutz und Sicherheitstechnik (LASI). (2001). *Arbeitsschutzmanagementsysteme. Handlungshilfe zur freiwilligen Einführung und Anwendung von Arbeitsschutzmanagementsystemen (AMS) für kleine und mittlere Unternehmen (KMU), LV 22*. Saarbrücken, Germany: Länderausschuss für Arbeitsschutz und Sicherheitstechnik (LASI). Retrieved March 31, 2003, from http://lasi.osha.de/publications/lv/lv22.pdf

McSweeney, K. P., Craig, B. N., Congleton, J. J., & Miller, D. (2002). *Ergonomic program Effectiveness: Ergonomic and medical intervention. International Journal of Occupational Safety and Ergonomics (JOSE), 4*, 433–449.

Narodný Inšpectorat Práce. (2002). Systém riadenia bezpečnosti a ochrany zdravia pri práci. Návod za zavedenie systému (*Occupational safety and health management system. Guideline for system implementation*). Bratislava, Slovak Republic: Narodný Inšpectorat Práce.

National Institute for Occupational Safety and Health. (1997). *Elements of ergonomics programs. a primer based on workplace evaluations of musculoskeletal disorders*. DHHS (NIOSH) Publication No. 97-117. Cincinnati, OH: National Institute for Occupational Safety and Health.

NNI. (1996). *Nederlandese Praktijkrichtlinjn NPR 5001, Model voor een Arbomanagementsysteem* (International version, 1997, Tech. Rep. *NPR 5001: Guide to an Occupational Health and Safety Management System*). Delft, The Netherlands: Nederlands Normalisatie-institut.

OSHA. (1982). *Voluntary protection programs*. Federal Register, 47, 29025. Washington DC: Occupational Safety and Health Administration.

OSHA. (1989). *Safety and health program management guidelines: Issuance of voluntary guidelines*. Federal Register, 54, 3904–3916. Washington DC: Occupational Safety and Health Administration.

OSHA. (1991). *Ergonomics program management guidelines for meatpacking plants*. Washington DC: Occupational Safety and Health Administration, Retrieved March 31, 2003, from http://www.ergoweb.com/resources/reference/guidelines/meatpacking.cfm

OSHA (2000a). *Revision to the voluntary protection programs to provide safe and healthful working conditions.* Federal Register, 65, 45649–45663. Washington DC: Occupational Safety and Health Administration.

OSHA. (2000b). *Final ergonomics program standard.* November 14 (repealed March 8, 2001). Washington DC: Occupational Safety and Health Administration. Retrieved March 31, 2003, from http://www.ergoweb. com/resources/reference/standards/standard.cfm

OSHA (2003a). *Current federal and state-plan-state sites in the VPP as of December 31, 2002.* Washington DC: Occupational Safety and Health Administration. Retrieved March 31, 2003, from http://www.osha.gov/ oshprogs/vpp/alpsites-dec-02.pdf

OSHA (2003b). *Guidelines for Nursing Homes. Ergonomics for the Prevention of Musculoskeletal Disorders.* Washington DC: Occupational Safety and Health Administration. Retrieved March 31, 2003, from http://www.osha.gov/ergonomics/guidelines/nursinghome/final_nh_guidelines.pdf

PKN. (1999). *Systemy zarządzania bezpieczeństwem i higieną pracy—Wymagania* (Standard No. PN-N-18001:1999 *Occupational safety and health management systems—Requirements*). Warsaw, Poland: Polski Komitet Normalizacyjny (Polish Standardization Committee).

PKN. (2000). *Systemy zarządzania bezpieczeństwem i higieną pracy—Ogólne wytyczne do oceny ryzyka zawodowego* (Standard No. PN-N-18002:2000 *Occupational safety and health management systems—General guidelines for assessment of occupational risk*). Warsaw, Poland: Polski Komitet Normalizacyjny (Polish Standardization Committee).

PKN. (2001). *Systemy zarządzania bezpieczeństwem i higieną pracy—Wytyczne* (Standard No. PN-N-18004:2001 *Occupational Safety and Health Management Systems—Guidelines*). Warsaw, Poland: Polski Komitet Normalizacyjny (Polish Standardization Committee).

PKN. (2003). *Systemy zarządzania bezpieczeństwem i higieną pracy—Wymagania* (Standard No. PN-N-18001:2003 *Occupational Safety and Health Management Systems—Requirements*). Warsaw, Poland: Polski Komitet Normalizacyjny (Polish Standardization Committee).

Standards Australia. (1994). *A management system for occupational health, safety and rehabilitation in the construction industry (handbook)* (Standard No. SAA HB53-1994). Homebush, NSW, Australia: Standards Australia.

SA&SNZ. (2001). *Occupational health and safety management systems—Specification with guidance for use* (Standard No. AS/NZS 4801:2001). Homebush, NSW, Australia: Standards Australia and Wellington, New Zealand: Standards New Zealand.

SA&SNZ. (1997/2001). *Occupational health and safety management systems—General guidelines on principles, systems and supporting techniques* (Standards No. AS/NZS 4804:1997 and AS/NSZ 4804:2001). Homebush, NSW, Australia: Standards Australia and Wellington, New Zealand: Standards New Zealand.

SNBOSH. (1992). *Ordinance (AFS 1992:6) Internal Control of the Working Environment.* Statute Book of the Swedish National Board of Occupational Safety and Health, 1992, Solna, Sweden: Swedish National Board of Occupational Safety and Health, Publishing Services.

SNBOSH. (1997). *Ordinance (AFS 1996:6) Internal Control of the Working Environment.* Statute Book of the Swedish National Board of Occupational Safety and Health, 1997, Solna, Sweden: Swedish National Board of Occupational Safety and Health, Publishing Services.

Swedish Work Environment Authority. (2001). *Systematic work environment management (AFS 2001:1).* Work Environment Authority, Solna, Sweden. Retrieved March 31, 2003, from http://www.av.se/english/legislation/ afs/eng0101.pdf

Zwetsloot, I. J. M. (2000). *Developments and debates on OHSM system standardization and certificcation.* In K. Frick, P. L. Jensen, M. Quinlan, & T. Wilthagen (Eds.), *Systematic Occupational Health and Safety Management—Perspectives on an International Development* (pp. 391–412). Oxford, UK: Elsevier Science Ltd.

VII

Safety and Legal Protection Standards

28

Printed Warning Signs, Tags, and Labels: Their Choice, Design, and Expectation of Success

Charles A. Cacha
Ergonix, Inc.

INTRODUCTION

An authority or an owner who has charge of and control of a premises and its physical contents has a moral and legal responsibility toward those individuals who occupy the premises. Similarly, an authority who designed, manufactured, and sold an object has moral and possibly legal responsibility toward the ultimate owner and user of the object. These obligations are usually to protect individuals form physical injury or disease, and in some instances this may apply to protecting personal property as well. Some examples of the responsibility of the authority are (a) the obligation of a landlord to tenants of a dwelling, (b) the obligation of a landlord to a visiting member of the public, (c) the obligation of an employer to an employee, and finally (d) the obligation of a designer and manufacturer of an object to the final owner and user of the object. These obligations may be discharged by removing any hazards present which could cause injury or illness. These hazards may also be permitted to remain, provided the authority or his or her subordinates meticulously and strictly supervises the premises or the object.

In consideration of the use of supervision, it is apparent and commonly understood that immediate constant supervision by the authority is (a) extremely impractical and (b) extremely expensive. It is also commonly agreed that the authority cannot "be every place at once." Thus, a state-of-the-art procedure known as *warning* is unavoidably employed. The warning is very often one of a semantic or ideographic nature and is used to apprise individuals of a particular hazard related to a premises, the contents of the premises, or an object owned by an individual. Usually the warning is of a printed nature and is composed of paints, inks, and so forth, applied to paper, cardboard, plastic, wood, or metal.

OBJECTIVE AND SCOPE

In consideration of the universal and unavoidable use of printed warnings, this chapter deals with the effective use and design of printed warnings so that minimal injuries, illnesses, and

509

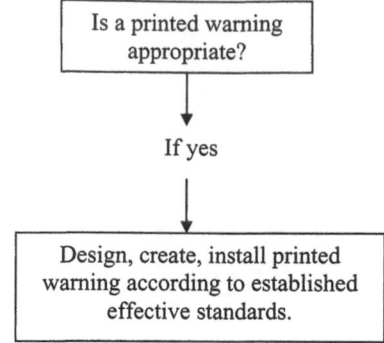

FIG. 28.1.   Sequence in decision making and design of printed warnings.

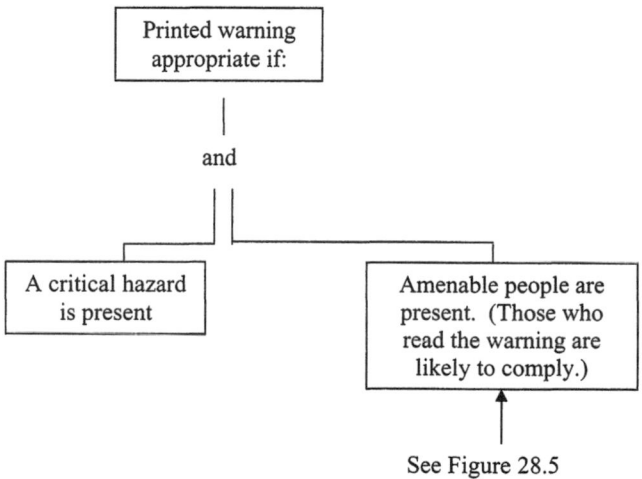

FIG. 28.2.   Decision tree for determining the appropriateness of a printed warning.

property damage occur during exposure to a particular hazard. Printed warnings are generally categorized as (a) signs which appear on walls, equipment, or other relatively large supporting structures; (b) tags which are signs of small dimensions and placed, usually temporarily, on an object; and (c) labels which are small permanent signs placed on an object. All three of these categories are dealt with while adhering to the following discussion format.

1. When is a printed warning appropriate or not appropriate?
2. What is the most effective physical and semantic design of a printed warning? (Figures 28.1 and 28.2).

It is important to note that, in the choice and design of printed warnings, both of the previous questions be employed and that they should be employed in a 1,2 sequence. This sequential application will assure effective meaningful design. It should finally be noted that nonprinted warnings such as those of a visual nature (e.g., lights and blinkers) or of an auditory nature (e.g., buzzers and bells) are not included in this chapter but are covered elsewhere in this book.

Appropriateness of a Printed Warning

Logic indicates that a warning of any kind may be successfully employed only when a proximal critical hazard exists and only when individuals exposed to the critical hazard are likely to comply with the warning. A warning displayed in the absence of a critical hazard is illogical and a waste of resources. Similarly a printed warning in the proximity of a critical hazard has only minimal value if most of those individuals associated with the hazard are not inclined to comply with the warning. Thus, the two primary requirements for the use of a printed warning are (a) the proximate presence of a critical hazard and (b) the presence of individuals who are amenable and likely to comply with the warning.

The Criticality of a Hazard

As was discussed, a printed warning should be used only in the presence of a hazard and particularly in the presence of a critical hazard. A minor hazard that has little consequence need not and should not bear a printed warning because of distractions and confusions that may occur from its presence among many other printed warnings in the environment. Criticality of a hazard is definable by four important major variables: (a) a potential for producing many injuries or illnesses or property damage events (frequency) (b) a potential for producing serious injuries or illnesses or property damage events, (severity) (c) necessity for or inexpendability of the hazard which makes its presence unavoidable (d) an inability to mitigate, contain, or enclose the hazard in order to cancel its harmful effects (Figures 28.3 and 28.4).

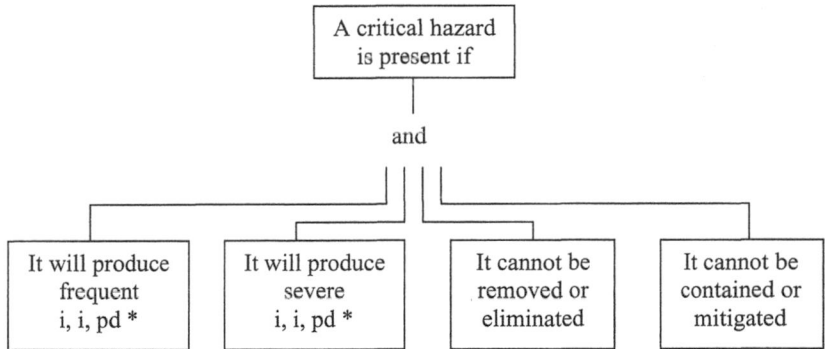

Note. * injury, illness, property damage.

FIG. 28.3. Criteria of a critical hazard.

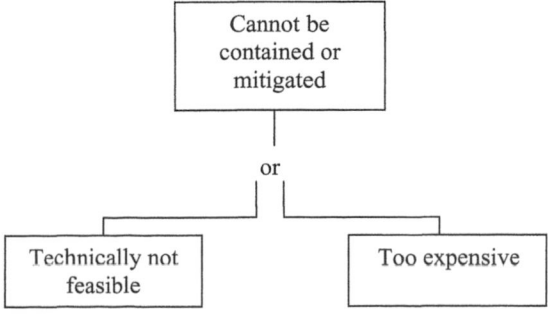

FIG. 28.4. Reason for noncontainment of a hazard.

Frequency and Severity

The concepts of frequency and severity are most often applied in Safety and Risk Management to overall injuries and illnesses experienced by a large organization or subdivisions of a large organization. Procedures for calculating the frequency and severity rates of an organization appear in another section of this book. The concepts of frequency and severity may also apply, however, to an individual hazard to determine whether that hazard is of a critical nature. This analysis may be an empirical *ex post facto* method, wherein observations reveal the number of past injuries and illnesses that have occurred in relationship to the hazard (frequency) or the financial cost of or length of disability time (severity) related to the hazard. Hazard analysis may also be of an *a priori* nature in which competent, experienced person(s) estimate frequency and severity related to a particular hazard. This analysis may then determine the hazard criticality which in turn partially helps decide the appropriateness or inappropriateness of a printed warning. Military Standard 882B provides a method of hazard analysis which may be employed to partially determine the appropriateness of a printed warning. Frequency (synonymous with probability) based on prior incidents or intuitive judgments is categorized as:

A. Frequent: likely to occur frequently
B. Probable: will occur several times in the life of a facility or system
C. Occasional: likely to occur some time in the life of a facility or system
D. Remote: unlikely but possible to occur sometime in the life of a facility or system
E. Improbable: so unlikely it can be assumed that occurrences may not be experienced

Additionally, severity based on prior incidents or intuitive judgments is characterized as.

I. Catastrophic: may cause death or loss of a facility.
II. Critical: may cause severe injury or illness or major property damage.
III. Marginal: may cause minor injury or illness or minor property damage.
IV. Negligible: probably no effect on people or property.

Table 28.1 indicates a matrix method for interfacing frequencies and severities in order to determine the criticality of a particular hazard. This method known as Risk Assessment and

TABLE 28.1
Probability and Severity Matrix

Probability Categories	Severity Categories			
	I Catastrophic	*II* Critical	*III* Marginal	*IV* Negligible
A Frequent	IA[a]	IIA	IIIA	IVA
B Probable	IB	IIB	IIIB	IVB
C Occasional	IC	IIC	IIIC	IVC
D Remote	ID	IID	IIID	IVD
E Improbable	IE	IIE	IIIE	IVE

[a] Interpretations: IA, IB, IC, IIA, IIB, IIIA, Unacceptable Situation (needs immediate emergency action); ID, IIC, IID, IIIB, IIIC, Undesirable Situation (quick management decision needed); IE, IIE, IIID, IIIE, IVA, IVB; Acceptable Situation (management review needed); IVC, IVD, IVE; Acceptable Situation (no management review needed).

Acceptance Coding (RAAC) will not only provide the criticality of a hazard but also suggest appropriate managerial actions when dealing with the hazard. As a general rule, the greater the frequency (probability) of occurrences due to a hazard *plus* the greater the severity of occurrences due to a hazard, the more critical is the hazard and the more important is the need for management to take remedial actions. This action would include the use of printed warnings if no engineering solutions are practicable.

ENGINEERING CONTROLS

Determining the criticality of a hazard also involves an understanding of basic engineering principles which are capable of removing, mitigating, or containing the hazard. This understanding, plus knowledge of the unacceptable or undesirable circumstances related to frequencies and severities, will lead to an ultimate decision to create and install a printed warning or not. An aid to this process is a hierarchical decision tree initially promulgated by William Haddon and further described by Roger L. Brauer (1990) as the Energy Theory. This theory is based on the concept that unwanted events can occur only in the presence of a transfer of energy. This understanding provides a sequential procedure dealing with a hazardous source of energy. The theory and its sequences have been reported in varying formats. A condensed form as created by the author of this chapter follows:

When encountering a potentially injurious source of energy:

1. Eliminate the energy source by removing the energy source from the premises or from any object that people or property are exposed to.
2. If the energy source cannot be eliminated, as above, then retain the source by keeping the source in remote locations so that the effect of its expenditure is diminished to an extent that people and property are not injured or damaged.
3. If the energy source cannot be remotely located as above, then place adequate barriers such as guards, shields, or walls around the source so that an energy expenditure will not cause injury to people or damage to property. Example: machine guards, protective partitions, and so forth.
4. If barriers cannot be placed around the energy source, then place barriers on or around people of property. Example: personal protective equipment such as goggles, steel tip shoes, and so forth.
5. If barriers cannot be placed around people or property then (a) provide a warning which, if heeded, will prevent people or property from contacting the energy expenditure and (b) train, educate, and motivate people exposed to the energy source to comply with the warning.

An Example of Energy Theory Decision Making

A company that manufactures and sells plastic cooking utensils considers manufacturing of a new line of metal cooking utensils. This would require purchasing metal stamping machines which are potentially capable of injuring machine operators.

Ask the following sequential questions:

1. Can the metal stamping operation be excluded from the premises? No. Not if it is essential for business survival that the company open up this new line of business.

2. If this metal stamping operation must be present on premises, can the operation be placed in a remote part of the plant away from employees? No. Machine operators must work in the proximity of the machines in order to successfully manufacture the product.

3. If the stamping machines cannot be isolated and kept away from people, can barrier guards be placed at the point of operation where the stamping occurs? No. Guarding the point of operation will not effectively allow the work piece to be loaded into the machine by the operator.

4. If the stamping operation cannot be guarded, can personal protective equipment for operator's hands be provided? No. Protective gloves would not withstand the forces of the ram in the stamping operation.

5. If personal protective equipment would be ineffective, then can a warning sign be placed in the proximity of the stamping operation which will appraise operators of the hazard and advise them to avoid the hazard? Yes. In addition to placing a warning, training, education, and motivation for those exposed to the hazard should be provided.

As may be concluded from the previous example, a warning as used in safety engineering is a "last resort" and is used only when hierarchical engineering alternatives are not feasible and cannot be technically or economically applied.

ECONOMIC FEASIBILITY

In addition to engineering controls and their technical feasibilities, an additional input into decision making involves economics. Frequently, higher levels of engineering solutions are technically feasible and will give excellent protection to people and property and will consequently eliminate the need for a warning related to a hazard. These engineering solutions, however, must not be economically prohibitive. There are several techniques in Risk Management which will determine the economic viability or inviability a safety engineering solution for controlling a hazard. The following procedure is reported by Harold E. Roland and Brian Moriarty (1983) in their book *System Safety Engineering and Management* and by Charles A. Cacha in his book *Research Design and Statistics for the Safety and Health Professional*.

Case Study: Present Worth of Money Technique.   A safety officer and the treasurer of a corporation are working together as a team to determine the financial feasibility of installing an expensive presence sensing device on all of their metal stamping machines. After researching some sources of information within and outside of the corporation, they determine the following essential facts which are needed to form an opinion.

1. Cost of installation by a competent contractor: $250,000
2. Life expectancy of the stamping machines and thus of the presence sensing device: 30 years
3. Estimated annual savings in workmen's compensation losses and workmen's compensations premiums: $25,000
4. Rate of interest return the corporation expects for investing their excess monetary assets: 8%

The previously cited information is inserted into the following calculation:

Symbols:

> Proposed expenditure (PE) = 250, 000
> life expectancy (L) = 30
> Annual savings (S) = 25, 000
> Value of money (V) = .08
> Present worth of money (PW$_M$)
> Present worth of savings (PW$_S$)

Formulas:

$$PW_M = \frac{(1 + V)^L - 1}{V (1 + V)^L}$$

$$PW_S = PW_M(S)$$

Computation:

$$PW_M = \frac{(1 + .08)^{30} - 1}{.08 (1 + .08)^{30}}$$

$$PW_M = \frac{9.06}{.81} = 11.18$$

$$PW_S = 11.18(25, 000) = \$279, 500$$

$$PE = \$250, 000$$

$$PW_S \text{ (of } \$279, 000) \; > \; PE \text{ (of } \$275, 000)$$

Because Present Worth of Savings is larger than Proposed Expenditure, it becomes financially advantageous and feasible for the corporation to install the presence sensing device. If Present Worth of Savings were substantially less than Proposed Expenditure, corporation management might have reservations about investing in the presence sensing device. Nevertheless, despite the financial disadvantage, the presence sensing device may still be installed for the humanitarian purpose of guaranteeing safety to employees. It is advisable that a procedure such as this be applied in situations where there may be large Safety expenditures. There is one apparent limitation in applying this procedure and that is the question of subjectivity. Methods of predicting Life Expectancy and estimating Annual Savings, which are only partially tempered with limited amounts of research, involve a good deal of judgment. In summary, this method, if it determines there is economic viability for a safety engineering device, will support the utilization of the device and will obviate the need for a printed warning. Similarly, this method if it determines a lack of economic viability for a safety engineering device will indicate that a printed warning coupled with training and education and motivation of people is in order.

COMPLIANCE WITH PRINTED WARNINGS

As discussed earlier, printed warnings are a device of last resort when engineering alternatives cannot be applied to protect people or property from unwanted events that cause injury or damage. Printed warnings, however, are frequently and universally employed and displayed upon premises, equipment, and various objects. Although those in a position of authority are aware of the unavoidability of using these warnings, they should also be aware of the possibilities that those people exposed to these warnings may ignore these warnings even

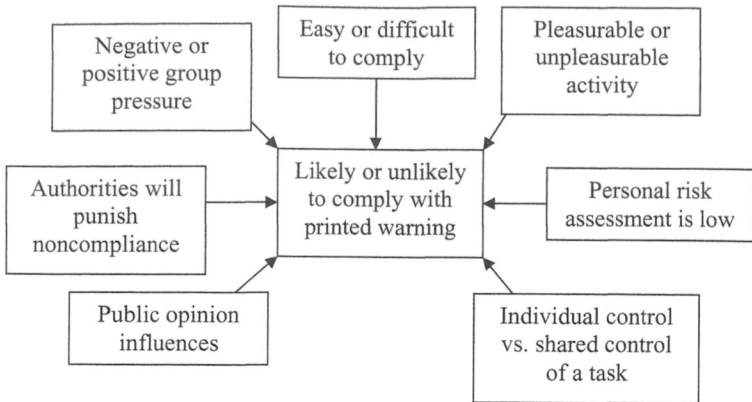

FIG. 28.5.   Factors influencing compliances or noncompliance with printed warnings.

though the warnings are adequately presented and adequately designed. It is important that those authorities understand that there are social psychological variables among people which will prompt compliance or prompt noncompliance with a printed warnings. This understanding, if applied, through competent training, education, and motivation will increase the probabilities that people will comply with a warning and will therefore avoid an unwanted event. Some of the most important sociopsychological variables are listed in the following sections. Many of these variables are described in Volumes I and II of Human Factors Perspectives on Warnings by Laugherty, Wogalter, and Young (1991; Figure 28.5).

Ease or Difficulty in Compliance.   A printed warning may involve an instruction re-quiring an action, an inaction, or an evasion by the recipient who reads the warning. In either case, a warning requiring action, inaction, or evasion which requires minimal physi-cal or mental effort is most likely to be complied with. A warning requiring great physical effort, complex mental processes, and inordinate inputs of time will cause noncompliance (Figure 28.5).

Group Membership.   If the individual reading the warning is part of a group which is also reading the warning, the individual will be influenced by the general responses of the group to the warning. A person, with the exception of the most individualistic of persons, will heed or ignore a warning according to the influence of group pressures ("When in Rome do as the Romans do."). Similarly, an individual, though alone while reading a warning, may still feel obligated to a political, fraternal, or religious group, even though that group is not present while the warning is read.

Sanctions From Authority.   An understanding that punitive actions may be applied by government, or organizational authorities such as employers, may increase the likelihood that a printed warning will be complied with. There may be a range of response in this area depending on how assiduously the authority enforces the warning and how submissive or independent are the attitudes of the recipient of the warning.

Pleasurable Versus Unpleasurable Activity.   A printed warning may particularly suc-ceed if it warns against an unpleasant action. Similarly, the probability of noncompliance with the warning is increased when a warning is implemented against a pleasurable activity, particularly if the activity is of a hedonistic nature.

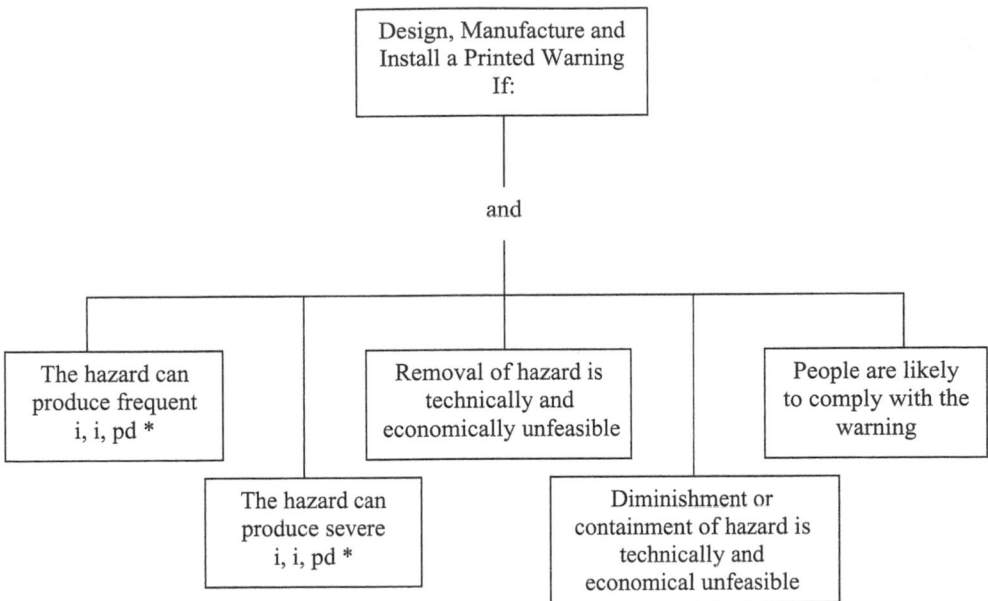

Note. * injury, illness, property damage.

FIG. 28.6.   Requisites for a printed warning.

Personal Risk Assessment.   Design and placement of a written warning by those in authority is predicated upon the authority forming a risk evaluation of the hazard in question. When the warning is finally presented to an individual recipient of the warning, however, the recipient may not agree with the risk evaluation of the authority and may consider the hazard more or less hazardous than the risk evaluation originally promulgated by the authority. If the recipient believes the hazard to be less risky than the warning implies, the recipient may ignore the warning. If the recipient agrees with the risk level implied by the warning or believes the risk level is even higher, the recipient will heed the warning. Judgments by the recipient of risk levels may be based on past personal experiences, an evaluation of immediate circumstance, or current events on a public or local level which report unwanted events (particularly of a catastrophic nature) related to the hazard at hand.

Personal Control.   A recipient when observing a written warning may act individually upon a task or perform a task in conjunction with another individual. When acting alone and with full control of a task, the recipient may ignore a warning because he or she has understanding and confidence in his or her capabilities. When relying fully or partially on the action of other people, the recipient will be more likely to comply to a warning because he or she is not certain of the capabilities of other people. For the final summary for warning design strategy, see Figure 28.6.

STANDARDS FOR PRINTED WARNING DESIGNS

After the authority responsible for a premises, equipment, or a product finally and competently decides to use a printed warning, the authority must design the warning in the most effective manner possible. This effective design will most basically deal with the physical characteristics of the warning as well as with the semantic content of the warning. There are a large number of published standards providing characteristic of an effective warning. The major publications, such as American National Standards Institute (ANSI), International Organization for Standardization (ISO), Underwriters Lab (UL), and Military Standards of Department of Defense (Mil. Std.) are summarized in the following section:

A LISTING OF PROMINENT WARNINGS STANDARDS

ANSI 2535.1-2002: Standard for Safety Color Code

This publication emphasizes the importance of the uniform use of colors on signs, labels, tags, equipment, walkways, and so forth, within an organization or among all organizations. This use will enhance warnings about hazardous conditions. It is also emphasized that any warning regardless of color is no substitute for the reduction or elimination of a hazard. It is also emphasized that excessive use of many colors at a particular location may lead to confusion and minimize the effect of a warning. The standard indicates that there should be adequate illumination on the sign which will allow differentiation of colors. The specifications of various colors are in terms of the Munsell Notation System as described by American Society of Testing Materials (ASTM) Standard ASTM D1535. It should be noted that earlier ANSI standards such as this specification do not appear in the current 2002 standard. Colors have been traditionally used as specified by 2535.1-1998 in the following manner.

Safety Red:	Flammable liquids
	Emergency stop bars on machines
	Emergency stop buttons on machines
	Fire protection equipment and apparatus
	Background for DANGER SIGNS
Safety Orange:	Intermediate level of hazard
	Used on hazardous machine parts
	Used on moveable guards or transmission guards
	Background color for WARNING SIGNS
Safety Yellow:	Physical hazards causing slips, trips, falls, or being caught in between
	Corrosive containers and storage cabinets for flammables
	Background color for CAUTION signs
Safety Green:	Used for emergency egress
	Used for first aid equipment
	Used for safety equipment
Safety Blue:	Used to identify safety information
	Used for mandatory signs related to personal protective equipment

ANSI Z 535.2-1998: Standard for Environmental and Facility Safety Signs

This publication is described as a visual alerting system to aid in potential hazards known to exist in the environment. Major examples of this system are safety signs used in fixed locations in the environment such as industrial facilities, commercial establishments, places of employment, and real estate properties.

The major categories of environmental and facility signs are:

Signal Word	Colors	Use
DANGER	White letters on Safety Red background	Imminent hazardous situation may result in serious injury or death
WARNING	Black letters on Safety Orange background	Potentially hazardous situation may result in serious injury or death
CAUTION	Black letters on Safety Yellow background	Potential hazardous situation may result in minor or moderate injury
NOTICE	White italic letters on Safety Blue background	Statement of policy related to safety
GENERAL Safety	White letters on Safety Green background	General safe practices and procedures

Sign design and layout illustrations in the publication are of two major categories: (a) "portrait" in which the sign's longest dimension is vertical or "landscape" in which the longest dimension is horizontal and (b) the number of panels are three (signal word, message, and symbol and pictorial) or the number of panels are two (signal word and message, but not symbol and pictorial). Generally DANGER, WARNING, CAUTION, and NOTICE signs may make use of these formats (Figure 28.7).

Letter style on signs is required to be sans serif and in upper case for signal words (Figure 28.8). Message panels lettering is a combination of upper- and lower-case sans serif. Letter styles are of a large variety: Arial, Arial Bold, Folio Medium, Franklin Gothic, Helvetica, Helvetica Bold, Meta Bold, News Gothic Bold, Poster Gothic, and Universe.

Legibility is considered to be influenced by the variables of letter height, width, spacing, and stroke width. Size of lettering depends on message length and distance from which the sign can be easily read. Letters must be adequately spaced and not crowded.

A required minimal letter height for signal words is 1 unit of height for every 150 units of safe viewing distance and 1 in 300 for letters in the message panel.

Sign finish requires durable material. Sign placement should be a location that gives sufficient time to take evasive action from the hazard. The sign must be legible, nondistractive, and not hazardous in and of itself, and the sign should not be vulnerable to various motions of objects in the environment which could destroy the sign. Signs should be lit by adequate reliable illumination.

Semantically, the message in the sign should contain in concise words (a) the signal word, (b) the description of the hazard, (c) the consequences related to the hazard, and (d) the related evasive action. The sequence of b, c, and d may vary according to judgment.

Letter characteristics are described in Appendix B and briefly summarized here.

The stroke width/ratio of a black letter on white background should be from 1:6 to 1:8. This ratio of a white letter on black background should be 1:8 to 1:10. The amount of spacing (lead) between lines should be 120% of the height of letters. A table is provided for recommended message letter heights and minimum safe viewing distances (Figure 28.9). This table is based on the following formulas:

Favorable Reading Conditions	Unfavorable Reading Conditions
MSVD[a] (inches)/300	MSVD[a] (feet) × .084

[a] Minimal Safe Visual Distance.

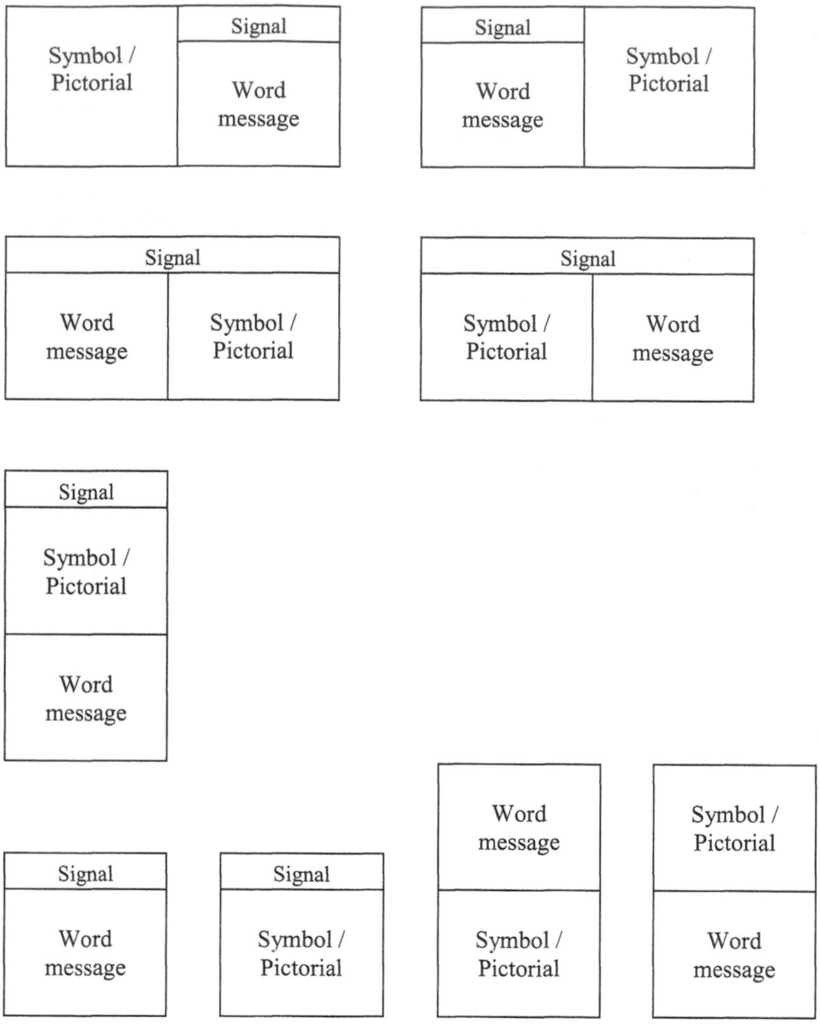

| Symbol /
Pictorial | Signal |
| | Word
message |

| Signal | |
| Word
message | Symbol /
Pictorial |

| Signal | |
| Word
message | Symbol /
Pictorial |

| Signal | |
| Symbol /
Pictorial | Word
message |

| Signal |
| Symbol /
Pictorial |
| Word
message |

| Signal |
| Word
message |

| Signal |
| Symbol /
Pictorial |

| Word
message |
| Symbol /
Pictorial |

| Symbol /
Pictorial |
| Word
message |

FIG. 28.7. Sign layouts.

Sans Serif   Serif

FIG. 28.8. Letter style.

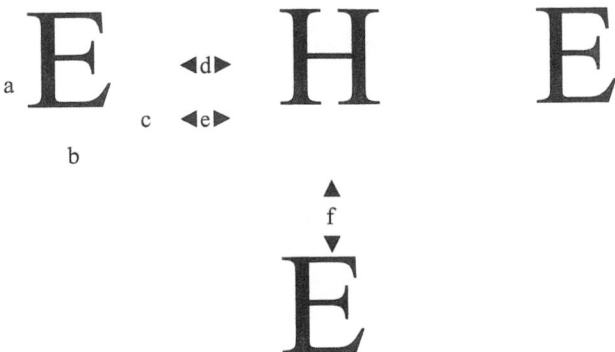

FIG. 28.9. Lettering. a, letter height; b, letter width; c, stroke (width of letter's lines); d, separation between words; e, separation between letters; f, separation between lines (lead).

ANSI Z 535.3-2002: Standard for Criteria for Safety Symbols

This standard recognizes that the recipients of a safety warning may be multicultural and multilingual. This multilingual circumstance will not allow the total success of a written warning because only a portion of the recipient population is fluent in the warning language. Symbols, however, may be comprehensive across all languages and may also provide a supplemental emotional impact on the written warning which will make the warning more effective.

The safety symbol is defined as an image with or without a surrounding geometrical shape which conveys a message without the use of words. The surrounding geometrical shape is defined as the surround shape and is usually a square or a circle.

The major categories of safety symbols are:

Hazard Alert. If a surround shape is used, the symbol should be drawn within a yellow equilateral triangle.
Mandatory action. If a surround shape is used, the image is white within a solid safety blue or safety black circular surround.
Information. If a surround shape is used, it should be placed within a rectangle.
Prohibition. A surround shape is mandatory. The shape is a circle with a 45-degree slash from upper left to lower right. The image is black with safety black or safety red circle and slash (Figure 28.10).

Additional requirements are realistic proportions of figures and avoidance of narrow forms, symmetry of figures vertically and horizontally, direction of focus of the image congruous with the physical layout of the environment, solidly drawn forms rather than an outline form, simplicity of detail with minimal embellishments, adequate symbol size to achieve legibility, and placement of the warning within the field of view and near the hazard.

The standard annexes provide procedures for drawing standardized stick figures and for testing group responses to proposed new safety symbol designs.

ANSI Z 535.4-2002: Standard for Product Safety Signs and Labels

This standard indicates that a product safety sign or label is a sign, label, cord tag, or decal affixed to a product that provides information about that product. A product safety sign or label is either of a permanent nature and cannot be easily removed and warns of a perpetual inherent hazard or of a temporary nature and warns of a temporary hazard related to the product. The temporary sign is removed when the hazard no longer exists.

FIG. 28.10.   Prohibition: Smoking.

The principles of Z 535.2-1998 generally coincide with this standard. There is in this standard additional information related to requirements for letter height of printed material used on a small or larger sign.

Recommended Letter Height for Favorable Reading Conditions

2   feet or less: Viewing Distance (in.)/150

\> 2 to 20 feet: Viewing Distance ((ft.-(2) $\times$ .03 + .16)

over 20 feet: Viewing Distance (ft.)/28.6

Recommended Letter Heights for Unfavorable Reading Conditions

Viewing Distance (ft.) $\times$ .084

ANSI Z 535.5 – 2002: Standard for Safety Tags and Barricade Tapes

This standard is intended to provide warnings about temporary hazards. These warnings are not intended to remain in place as permanent warnings but must be removed when the hazard is neutralized. These warnings, which are usually in the form of attachable tags and barrier tapes, graphically follow the principles found in Z 535.2 and Z 535.4 and color implementation according to Z 535.1. Typically, a tag or tape contains a signal word and a message panel which contains a hazard description and may also contain evasive action. Tags must be hooked as close to the hazard as possible and must not detract from other tags. Tapes should be placed in such a way to give adequate evasion time to the viewer. A table for letter heights is provided. The requirement in this table approximate the letter height requirements of Z 535.2.

ANSI Z 129.1 – 2000: Hazardous Industrial Chemical Precautionary Labeling

This standard provides guidelines for preparation of precautionary information about hazardous chemicals used in industry. This information is in the form of labels placed upon containers.

In order to provide warnings related to immediate and delayed physical and health hazards, a number of label characteristics are expected to be considered:

Identification of the Chemical Product or Critical Components

Signal word: DANGER, WARNING, CAUTION, for immediate hazard
NOTICE and ATTENTION for delayed hazard
Statement of hazard(s)
Precautionary measures
Instructions in case of contact or exposure
Antidotes
Notes to physician
Instruction in case of fire
Instruction in case of spill or leak
Instructions for container handling and storage
References: also additional label, bulletins, and material safety data sheets
Name, address, and telephone number of manufacturer

Underwriters Lab (UL) 969-1995: Marking and Labeling Systems

This standard does not deal with the graphic aspects of a warning; it deals with the physical manufacturer and testing of adhesive attached labels.

ISO (International Organization for Standards) 3864-1984

The stated purpose of this standard is to prescribe safety colors and safety signs in order to prevent accidents and health hazards.

Safety colors required are Red, stop, prohibition, fire fighting equipment; Blue, mandatory action; Yellow, caution, risk of danger; and Green, safe conditions.

Contrasts in color are: Safety Red/White, Safety Blue/White, Safety Yellow/Black, and Safety Green/White.

The geometric shapes of signs are Circle, prohibition or mandatory action; Equilateral Triangle, warning; and Square or Rectangle, information and instructions.

Prohibition signs in addition to having a circular configuration must have a 45° slash from 11 o'clock to 5 o'clock. Warning signs that do not have a suitable image that can be created should instead have an exclamation point.

Effective viewing distance (the greatest distance that a safety sign can be understood) is defined as

$$A > \frac{L^2}{2000},$$

where A is the minimum area of the sign in square meters and L is the distance in meters.

U.S. Federal Legislation: Consumer Product Safety Commission (CPSC)

Various U.S. laws related to labels and warnings exist. 16 CFR Section 1500.121, which was created by the CPSC, appears to provide the most comprehensive design standards for labeling. This section repeats requirements from the Federal Hazardous Substances Act which are:

Signal Word: DANGER, CAUTION, WARNING
Statement of Hazard

Name of Substance
Name and address of manufacturer and supplier
Precautions to follow
Instructions
Special handling and storage
First aid instructions

The section calls for conspicuity, legibility, and contrast in type. Cautionary text must be aligned parallel to the base of the container. The label area must be adequate. A table is included which provides type size required in relationship to the area of the label. Letter height should be no more that three times the letter width.

REFERENCES

Brauer, R. L. (1990). *Safety and health for engineers.* New York: Van Nostrand Reinhold.
Cacha, C. A. (1997). *Research design and statistics for the safety and health professional.* New York: Van Nostrand Reinhold.
Hall, G. (1986). *The failure to warn handbook.* Columbia, MD: Hanrow Press.
Laughery Sr., K. R., Wogalter, M. S., & Young, S. L. (1991). *Human factors perspectives on warnings.* Santa Monica, CA: Human Factors Society.
Roland, H. E., & Moriarty, B. (1983). *System safety engineering and management.* New York: Wiley.
Wogalter, M. S., DeJoy, D. M., & Laughery, K. R. (1999). *Warnings and risk communication.* Philadelphia, PA: Taylor & Francis.

APPENDIX I: DEFINITIONS

Authority. Those in charge of a process, property, or organization who would devise a warning if so required.
Recipient. An individual who reads and is expected to respond to a warning.
Hazard. A physical circumstance that is capable of causing injury, illness, or property damage.
Risk. A degree of probability that a hazard will cause an injury, illness, or property damage.
Frequency. The number of unwanted events caused by a hazard over a specified period of time.
Severity. The degree of injury, illness, or property damage resulting from an unwanted event.
Warning. A stimulus conveying a communication to a recipient that the recipient is exposed to a proximate or long-range hazard.

APPENDIX II: STANDARDS

Z 535.1–2002   Z 129.1–2000
Z 535.2–1998   UL 969–1995
Z 535.3–2002   ISO 864–1994
Z 535.4–2002   16 CFR SECTION 1500.121
Z 535.2–2002

29

Record Keeping and Statistics in Safety and Risk Management

Charles A. Cacha
Ergonix, Inc.

INTRODUCTION

Medium to large-size organizations usually establish a safety or risk management program that is designed to control the incidence of injuries and illnesses to employees. The Safety or Risk Managers who operate this program implement such procedures as safety training, safety motivation, safety inspections, and hazard abatement as integral parts of the program. By necessity, the Safety or Risk Managers are concerned about whether their program is effectively protecting the organization's employees. The criteria for testing the validity and success of a Safety and Risk Management Program has traditionally been the number (frequency) of injuries and illnesses, as well as the severity of injury and illnesses occurring to employees over a given period of time. This chapter describes the traditional statistical method used to validate an existing Safety and Risk Management Program and prognosticate its future success.

SCOPE AND OBJECTIVES

Initially, several forms of record keeping and statistics which have been traditionally used by safety and risk management professionals are described. Second, the currently prescribed method of record keeping and statistics as required by the United States Department of Labor is presented. The traditional and current methods are similar, but distinction is made between them. Finally, some statistical techniques for making comparisons between incident frequencies over time are provided.

Injuries and illnesses which occur within an organization are manifested by incidents known as unwanted events. These unwanted events, which are evidenced by injuries and illnesses, are usually open and obvious and detectable by management and are mutually exclusive of each other and may thus be enumerated as they occur over time. The enumeration of unwanted events over time is subdivided by units of time, generally 1 year, although larger and in particular smaller units such as months may also be used. It is generally agreed that the safer

525

an organization is, the smaller the number of unwanted events that will occur within a given unit of time. This small number also reflects the successful or unsuccessful efforts of the safety or Risk Management Professionals. The number of injuries and illnesses occurring over a unit of time is referred to as frequency. In addition to the concept of frequency, another concept is applied which is referred to as severity. Severity indicates the overall extent of seriousness of injuries across all unwanted events occurring within a unit of time. The seriousness is usually measured temporarily by the length of incapacitance of all employees who are injured or sickened. Formulas for frequency and severity follow.

Frequency

Logically, in order to calculate injury and illness frequency, three variables should be considered: (a) the number of injuries and illnesses which have occurred, (b) the amount of temporal exposure of all workers within the organization to their various job duties, and (c) an interval of time during which the organization had been operating.

The number of injuries and illnesses is generally determined by counting the number of reports generated by supervisors describing injury and illnesses occurrences that require medical aid beyond ordinary perfunctory first aid treatment. This number may apply to the entire organization or to subdivisions of the organization.

The amount of exposure to workers is determined by the total number of hours expended by all workers while working for the organization and performing their required functions over an interval of time. The interval of time is a temporal segment, usually a year, which is uniformly observed in the present and into the future and may be also observed, from existing records, into the past.

The formula for frequency is

$$\text{Frequency} = \frac{\text{Number of Injuries and Illnesses}}{\text{Total Hours Expended by all Employees}}.$$

For example, an organization experienced a total of 21 injuries and illnesses during the course of 1 year. Its employees worked a total of 90,000 hours during that year. Its frequency rate is

$$\frac{21}{90,000} = .00023.$$

Severity

Although the frequency statistic may provide considerable evidence related to Safety or Risk Management efforts, additional data collection may provide even stronger indications. Statistics may be calculated which indicate how critical unwanted events have been. This criticality is measurable by the number of days all workers have lost. This measure is based on the consideration that the more time lost, the more severe the injury (e.g., a worker with a broken leg will in most likelihood lose more time from the job than will a worker with a broken finger):

$$\text{Severity} = \frac{\text{Number of Days Lost by All Workers}}{\text{Total Hours Expended by All Workers}}.$$

For example, an organization experienced a total of 68 days lost from the job among all injured employees during the year. Its employees worked a total of 90,000 hours. The severity rate is

$$\text{Severity Rate} = \frac{68}{90,000} = .00076.$$

Multiplier

The aforementioned two techniques have in the past been exposed to a multiplier in order to create a nondecimalized result and to relate them to realistic circumstances. This multiplier was originally 1,000,000 which is derived from 500 workers working 50 weeks a year at 40 hours per week. This multiplier was recommended by American National Standards Institute (ANSI) Z16.1. The multiplier is currently 200,000, which is derived from 100 workers working 50 weeks a year at 40 hours per week.

The following formulas have been or are currently used:

Prior Method

$$\text{Frequency:} \frac{\text{Number of Injuries and Illnesses}}{\text{Total Hours Worked by All Workers}} \times 1{,}000{,}000.$$

$$\text{Severity:} \frac{\text{Number of Days Lost by All Workers}}{\text{Total Hours Worked by All Workers}} \times 1{,}000{,}000.$$

Present Method

$$\text{Frequency:} \frac{\text{Number of Injuries and Illnesses}}{\text{Total Hours Worked by All Workers}} \times 2{,}000{,}000.$$

$$\text{Severity:} \frac{\text{Number of Days Lost by All Workers}}{\text{Total Hours Worked by All Workers}} \times 2{,}000{,}000.$$

Prior Method

Examples from prior formulas are:

$$\text{Frequency:} \frac{21 \text{ Injuries and Illnesses}}{90{,}000 \text{ hours}} \times 1{,}000{,}000 = 233.$$

$$\text{Severity:} \frac{68 \text{ days lost}}{90{,}000 \text{ hours}} \times 1{,}000{,}000 = 756.$$

Current Method

$$\text{Frequency:} \frac{21 \text{ Injuries and Illnesses}}{90{,}000 \text{ hours}} \times 2{,}000{,}000 = 47.$$

$$\text{Severity:} \frac{68 \text{ days lost}}{90{,}000 \text{ hours}} \times 2{,}000{,}000 = 151.$$

Frequency Related to Severity

These two concepts have dissimilar characteristics, but they are related. Some professionals may argue that frequency has greater importance because all unwanted events have the potential for being a severe event. Other professionals may argue that severity has greater significance because severity reflects a large amount of human suffering and large financial losses to the organization. Attempts have been made to integrate frequency and severity rates into one rate:

$$\text{Severity/Frequency Ratio} = \frac{\text{Severity Rate}}{\text{Frequency Rate}}.$$

An example from prior examples is:

$$\text{Severity/Frequency Ratio: } \frac{151}{47} = 3.2.$$

Whereas frequency and severity rates are used for an entire organization or major subdivisions of an organization, the severity/frequency ratios are often calculated for smaller divisions of the organization. The severity/frequency ratio determined for a small division, if high, is a signal to the safety or risk management department that serious occurrences with severe consequences are happening, and prompt safety engineering practices should be applied for the purpose of hazard abatement.

RECORD-KEEPING STANDARDS

In the past, the appropriate standards were promulgated by the American National Standards Institute (ANSI). These standards follow:

ANSI Z16.1-1967: A Method of Recording and Measurement Work Injury Experience

This standard deals with traditional methods of calculating frequency and severity rates. Rates are based on 1,000,000 worker hours. It also provides schedules of lost days for deaths and permanent injuries which are calculated for severity rates and may be used by private organizations, but has been superceded by OSHA prescribed procedures.

ANSI Z16.2-1962: Method of Recording Basic Facts Relating to the Nature and Occurrence of Work Injuries

This standard deals with procedures for analyzing and categorizing incidents and accidents so that managerial or engineering controls may abate hazards. Procedures are still applicable in current safety and risk management practices. Major categories in analyzing incidents and accidents are:

- Nature of Injury: cuts, lacerations, etc.
- Part of Body Affected: finger, head, etc.
- Source of Injury: machine, tool, vehicle, etc.
- Accident type: struck by, struck against, falls, etc.

ANSI Z16.3-1973: Method of Recording and Measuring the Off-the-Job Disability Accidental Injury Experience of Employees

This method analyzes employees off the job injuries across three categories: transportation, home and public places. The analyses are presented as a percentage of total off-the-job injuries:

$$\frac{\text{Number of Injuries in the Category} \times 100}{\text{Total Number of Off-the-Job Injuries}} = \%$$

An overall frequency rate for off-the-job injuries:

$$\text{Frequency:} \frac{\text{Number of Injuries} \times 200,000}{312 \times \text{Average Number of Employees} \times \text{Number of Months}}.$$

The number 312 is an estimate of the number of hours, monthly, spent off the job but not sleeping.

ANSI Z16-4.-1977: Uniform Record Keeping for Occupational Injuries and Illnesses

This standard tracks the current OSHA record-keeping requirements which will described next.

CURRENT REQUIREMENTS

Presently, the U.S. Occupational Safety and Health Administration (OSHA) requires employees with more than 10 employees to use, among other forms, OSHA Form 300 for recording injuries. OSHA 300 is a log that requires the identity of the employee, job title, date of incident, brief description of injury or illness, part of body affected, and object causing injury. Also required are the indication of whether the worker died, days away from work, and job transfer due to injury or illness. A line entry for each injury or illness is made. An entry (also known as a recording) is made only if injury or illness resulted in death, loss of consciousness, days away from work, restricted work activity, or job transfer or medical treatment beyond first aid. The required method of calculating the frequency rate (referred to by OSHA as the incidence rate) is to add up all line entries on the log, multiply by 200,000, and then divide by the hours worked by all employees.

For example:

$$\text{Incidence Rate} = \frac{\text{Total Injury and Illness Line Entries} \times 200,000}{\text{Hours Worked By All Employees}}.$$

$$\text{Example:} \frac{\text{Total Entries: } 28 \times 200,000}{\text{Hours Worked By All Employees: } 100,000} = 56.$$

Severity Rate (referred to by OSHA as DART) is based on injury or illnesses that caused 2 or more days away from work or which ended in 2 or more days of restricted job activity or job transfer:

$$\text{Example:} \frac{\text{Days Lost Cases (8)} + \text{Restricted Work Cases (2)} \times 200,000}{100,000} = 20.$$

The details for this OSHA record-keeping procedure are explained in 29 CFR Part 1964. Considering issues of legality, this procedure should be considered the future state of the art in record keeping.

There are a number of statistical techniques that may help infer any substantial changes in frequencies or severities in the present or over time. One of these is demonstrated here.

Control Charts

These instruments are used to track and observe changes in frequencies over time. Observing any particularly large beneficial or detrimental change in frequency will aid the safety or risk manager to determine the success or lack of success of his or her safety or risk management program. In order to decide whether a change is large to a point of significance will depend on the application of a statistical technique. The following technique was acquired from *Techniques of Safety Management* by Dan Petersen (1978). A sequence of computational steps follows. It is recommended that at least 20 time intervals be available for study and calculation.

Step 1. Add up all frequencies (at least 20 time intervals) ΣX.
Step 2. Add up all hours worked (at least 20 time intervals) ΣH.
Step 3. Derive a percentage from above: $\frac{\Sigma X}{\Sigma H} = P$.

$$\text{Upper Control Limit (UCL)} = P + 2.576\sqrt{\frac{P(1-P)}{\Sigma H}} \text{ at 99\% Confidence}$$

$$\text{Lower Control Limit (LCL)} = P - 2.576\sqrt{\frac{P(1-P)}{\Sigma H}} \text{ at 99\% Confidence}$$

$$\text{Upper Control Limit (UCL)} = P + 1.96\sqrt{\frac{P(1-P)}{\Sigma H}} \text{ at 95\% Confidence}$$

$$\text{Lower Control Limit (LCL)} = P - 1.96\sqrt{\frac{P(1-P)}{\Sigma H}} \text{ at 95\% Confidence}$$

Judgment will determine the choice of the 99% or 95% confidence level. This technique is usually performed on a monthly basis. Any month's entry on the chart which exceeds the UCL should cause concern to the safety or risk manager who should thereupon apply remedial actions to the physical or managerial environment. Likewise, any month's entry which appears below the LCL should indicate to the safety or risk manager that injury and illness control efforts have been particularly effective and that such efforts should be examined and repeated.

For example: Over the last 20 months, an organization has been record keeping according to OSHA requirements. A total of 60 injuries and illnesses occurred during that time. Also, in the last 20 months, a total 100,000 hours have been worked by employees across the 20 months.

Step 1. Add up all injuries and illnesses over 20 months: 60.
Step 2. Add up all hours worked over 20 months: 100,000.
Step 3. Derive a percentage 60/100,000: .0006.

$$\text{UCL} = .0006 + 2.576\sqrt{\frac{.0006(1-.0006)}{100,000}} = .0008.$$

$$\text{LCL} = .0006 - 2.567\sqrt{\frac{.0006(1-.0006)}{100,000}} = .0004.$$

Enter on the control chart UCL as .0008 × 200,000 and .0004 × 200,000 as LCL. The average is computed by LCL – UCL ÷ 2 × 200,000.

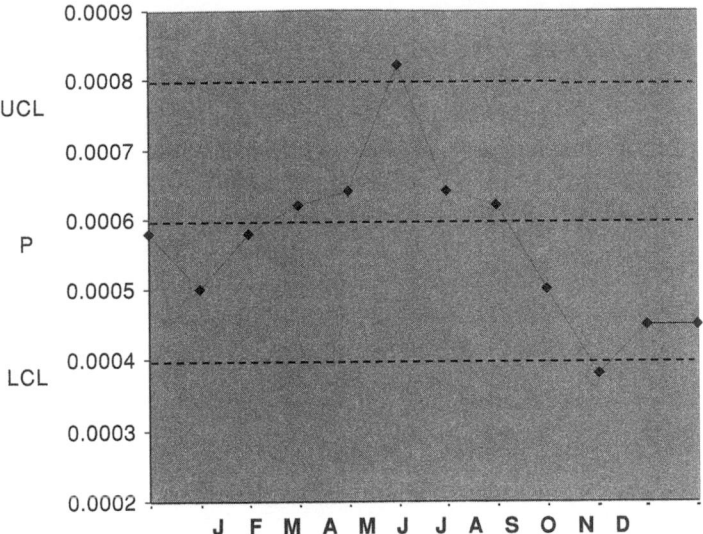

FIG. 29.1. Year 2001 Annual Control Chart Accident Frequencies

See the following figure as an example of future entries. One month (June) demonstrates a particularly undesirable circumstance and another month (October) a particularly desirable circumstance.

REFERENCES

Cacha, C. A. (1997). *Research design and statistics for the safety and health professional.* New York: Van Nostrand Reinhold.
Grimaldi, J. V., & Simonds, R. H. (1975). *Safety management.* Homewood, IL: Richard D. Irwin, Inc.
Petersen, D. (1978). *Techniques of safety management.* New York: McGraw Hill.

APPENDIX I: DEFINITIONS

Frequency. The number of unwanted events caused by a hazard over a specified period of time
Severity. The degree of injury or illnesses resulting from as unwanted event
Record Keeping. A methodical recording of unwanted events and their characteristics as the unwanted events occur
Control Chart. A series of graphic plots representing unwanted events over time

APPENDIX II: STANDARDS

ANSI Z16.1–1967: A method of recording and measuring work injury experience
ANSI Z16.2–1962: Method of recording basic facts related to the nature and occurrence of work injuries
ANSI Z16.3: Method of recording and measuring the off the job disabling accidental injury experience of employees
ANSI Z16.4–1997: Uniform record keeping for occupational injuries and illnesses

30

User Needs in Standards—Older People and People With Disabilities

Anne Ferguson
British Standards Institution

INTRODUCTION

Older people make up an increasing percentage of the world population. A consequence of becoming older is a decrease in various functional abilities, such as hearing, vision, and mobility. Younger people may also have functional impairments, whether from birth or as a result of accident or illness. In a society that recognizes the importance of equal opportunities, it follows that products, services, workplaces, and environments should be accessible to a wide variety of users and thus must be designed to take into account a range of user needs, including those of older people and people with disabilities.

In the standards world, in the past, the needs of people with disabilities have predominantly been considered in the development of specific standards in the area of assistive technology, such as international standards for wheelchairs (ISO, various). Awareness of the need for building accessibility by a wider population has also been covered by various national and international standards, such as *Building Construction—Needs of Disabled People in Buildings—Design Guidelines* (ISO, 1994). However, the needs of people with functional impairments, including older people, have not been adequately addressed in the writing or revision of standards for everyday products and services.

National legislation, such as the Americans with Disabilities Act (1991) in the United States or Disability Discrimination Act (1995) in the United Kingdom (UK), has been instrumental in increasing awareness of needs. Some national standards organizations, such as the Japanese Standards Association (JSA, 2000), have made progress in this area with guidelines on, for example, the usability of consumer products.

The international ISO Guide 71 (2001), *Guidelines for Standards Developers to Address the Needs of Older Persons and Persons with Disabilities*, as its name implies, is intended to help standards writers ensure that *all* standards cater to the widest possible group of users, particularly older people and those with disabilities. It is thus an essential ergonomics standard and useful tool for designers and evaluators of products and services. It is one of several guides produced for standards writers which are also of relevance to designers, ergonomists, and others. These include ISO/IEC Guide 51:1999 *Safety Aspects—Guidelines for Their Inclusion in Standards*, which covers the concept of safety and tolerable risk in fairly general terms, ISO/IEC Guide 50:2002 *Safety Aspects—Guidelines for Child Safety*, which describes a hazard-based approach to ensuring standards take account of child safety, and ISO/IEC Guide 37:1995 *Instructions for Use of Products of Consumer Interest*, which sets out basic criteria for consumer instructions and includes, in an informative annex, a method for assessing instructions, together with "noncomprehensive" compliance and evaluative checklists.

This chapter briefly describes the process which resulted in the Guide 71. A description of the contents follows and then an illustration of how Guide 71 may be used to promote and ensure accessible design in products and services.

DEFINITION OF RELEVANT TERMS

Accessible design—guidelines for the design of consumer products to increase their accessibility to persons with disabilities or who are aging (Trace R & D Center of University of Wisconsin-Madison, 1992)
Alternative format—different presentation or representation intended to make products and services accessible through a different modality or sensory ability (ISO Guide 71)
Assistive technology—equipment, product system, hardware, software or service that is used to increase, maintain, or improve functional capabilities of individuals with disabilities (ISO Guide 71)
Universal design—the design of products and environments to be usable by all people, to the greatest extent possible, without the need for specialized design (North Carolina State University Center for Universal Design)

FULL NAME OF THE STANDARDS (GUIDELINES)

Guidelines for standards developers to address the needs of older persons and persons with disabilities. ISO/IEC Guide 71: 2001

Background

In May 1998, the Japanese Standards Association proposed to the ISO Committee on Consumer Policy [COPOLCO] that guidelines should be drawn up to address the needs of older people and people with disabilities. A policy statement was prepared, by an international group, in a series of meetings in Tokyo, Geneva, Washington DC, Toronto, and Paris, and accepted by the Council of The International Organization for Standardization [ISO] and International Electrotechnical Commission [IEC] in 2000. This statement (ISO/IEC, 2000) draws attention to the commercial benefits, as well as to the gains for Society, in improving access to products, services, and environments. It sets out the three basic principles required to achieve the inclusion of the needs of older persons and people with disabilities in standards production and revision work. These are:

- Universal or Accessible Design
- Consumer representation of older persons and people with disabilities
- Relevant information exchange

The Policy statement describes how these principles may be achieved through raising awareness, in particular, taking account of the subsequently prepared Guide 71, ensuring links between research programs and standardization and increasing the availability of standards, themselves, in accessible formats.

ISO/IEC Guide 71

Guide 71 was drafted by a working group, comprised principally of experts in the fields of disability and human factors, with subsequent input from standards organizations, to ensure a document usable by standards writers. It was published in 2001 and has been widely accepted. For example, in Japan, technical criteria to be incorporated in IT equipment, software, and digital home appliances have been prepared, which are consistent with Guide 71. Also two European standardization bodies, CEN and CENELEC, have adopted and re-published Guide 71 as CEN/CENELEC Guide 6, as the prime means of enacting a mandate by the European Commission (Mandate M/283 Safety and usability of products by people with special needs), which required production of a guidance document in the fields of safety and usability of products by people with special needs, such as elderly and disabled people.

The third European standards body obliged to enact EC Mandate M/283, the European Telecommunication Standards Institute [ETSI], will use Guide 71 as a reference document. The IEC technical committee that deals with domestic electrical appliances (TC 61: Household and Similar Electrical Appliances—Safety), has now included a reference to Guide 71 in its Strategic Policy Statement; it is envisaged that other Technical Committees will also take similar steps.

ISO Guides are normally available, in printed form, in English and French. However, in keeping with the "accessible design" principles of the Guide, an English Braille version can be obtained, at the same price as the printed document, from the ISO Central Secretariat in Geneva. This is the first time an ISO publication has been made available in Braille.

Content

The introduction to Guide 71 recognizes that standards bodies have, for many years, addressed the needs of persons with disabilities in the development of specific standards in the area of assistive technology and accessible building design, but draws attention to the fact that the needs of older people and people with disabilities are not being adequately met, when standards for everyday products or services are written or revised. Thus the Guide lays down a process which can be followed to help to do this.

The very first definition in the Guide is that for ergonomics or human factors, which uses the widely quoted words of Christensen, Topmiller, and Gill (1988): "that branch of science and technology that includes what is known and theorized about human behavioural and biological characteristics that can be validly applied to the specification, design, evaluation, operation and maintenance of products and systems, to enhance safety, and effective and satisfying use by individuals, groups and organizations."

Later the Guide makes clear that ergonomics factors should be addressed when developing solutions to possible conflicts between safety and usability, using the example of child-resistant closures on medicines, which may make the closure more difficult to open for older persons who no longer see as well or have reduced strength or dexterity.

The Guide has four key sections. The first of these, Clause 6 *Developing Standards—Issues to Consider During the Standards Development Process*, suggests a process standards-writers may use to address the needs of older persons and persons with disabilities when drafting a new standard or at each revision of an existing one. This requires definition of the standards project, in terms of factors such as the potential end users, ensuring that the committee is well equipped with relevant background information and experts, in relation to awareness of aging and disability issues, injury data, and so forth, and also making sure that the process of preparing the standard, its review, and eventual publication are all in a form accessible to the group under consideration, that is, older people and people with disabilities.

Clause 7, *Tables of Factors to Consider to Ensure Standards Provide for Accessible Design*, provides seven tables to help standards-writers to identify factors that will affect the use of a product, service, or environment and to consider their significance for persons with different abilities. Each table identifies a clause or section typically found in international standards: These are Information, Packaging, Materials, Installation, User interface, Maintenance, Storage and Disposal, and Built Environments (buildings). Figure 30.1 provides an extract from Table 1 of the Guide, showing how the factors are displayed against Sensory ability. In the Guide, all seven tables have columns for Sensory ability as well as Physical and Cognitive abilities, together with a column for "Allergy." Although not typically recognized as a "disability," allergies can impose limitations on an individual's activities and, in some cases, be life threatening.

Within each table, the user may look up the factors, such as "lighting/glare" or "surface finish," which are particularly significant for users with particular disabilities and then use the remaining key sections, Clauses 8 and 9, to find out more on why particular factors are

Factors to consider in standards clauses on information (labelling, instructions and warnings)	Human abilities				
	9.2 Sensory				
	Seeing 9.2.1	Hearing 9.2.2	Touch 9.2.3	Taste/ smell 9.2.4	Balance 9.2.5
8.2 Alternative format	■	■	■	■	
8.3 Location/ layout	■	■	■		■
8.4 Lighting/glare	■				
8.5 Colour /contrast	■				
8.6 Size/ style of font	■				
8.7 Clear language	■	■			
8.8 Symbols/drawings	■				
8.9 Loudness/pitch		■			
8.10 Slow pace		■			
8.11 Distinctive form	■		■		
8.12 Ease of handling	■				■
8.13 Expiration date marking	■			■	
8.14 Contents labelling	■			■	
8.15 Surface temperature	■		■		
8.16 Accessible routes	■				■

FIG. 30.1.   Extract from Table 1 of Guide 71 on "Factors to consider in clauses in information."

important or, in general terms, what action can be taken. For example, if "ease of handling" was identified as a factor to consider, as it is in column 1, under "seeing" (see Figure 30.1), it could be looked up under Clause 8 *Factors to Consider* to find guidance, spelling out the need to select effective color combinations for presentation of information or consider the size, shape, and mass of a product in relation to ease of handling.

Clause 9, *Detail About Human Abilities and the Consequences of Impairment*, provides a brief definition and description of different abilities, together with information on the effects of aging and the practical implications of impairment. Examples are given of hazards from which older persons and people with disabilities are more at risk because of their functional limitations. So, for example, under "seeing," the section on effects of aging lists changes such as "loss of visual acuity," "reduced field of vision," and "sensitivity to light." Under risk of hazards, "sharp points" and "hot surfaces" appear.

The final section of Guide 71 is a Bibliography, which offers a list of sources that standards-writers can use to investigate more detailed and specific guidance materials with respect to accessible design. Within Europe, a more extensive Bibliography, on this topic, was prepared with funding arising from Mandate 283 (1999), referred to earlier. This covers Healthcare; Personal Care and Protection; Personal Mobility; Housekeeping/Household Equipment; Furnishings; Machinery and Tools, Handling Products and Goods; Communication, Information, and Signalling (ICT); Buildings and Interiors; Outdoor Environment; Traffic and Transportation; Recreation; and general information on accessibility and ergonomics.

Application of Guide 71

The Guide is intended to be used at the point at which a standard is first prepared or revised. Some work has already been undertaken by consumer representatives in the UK, using Guide 71 to assess the effectiveness of existing standards in ensuring access for people with disabilities (BSI-CPC 2003). The review was initially carried out on two parts of the electrical safety standard covering particular requirements for toasters and dishwashers, BS EN 60335 (1990, 1996) and their IEC equivalents and has since been extended to other areas. Use of the tables in Section 7 of Guide 71, together with the products themselves, assisted in identifying the significant factors to consider in relation to different disabilities and possible problems in the current versions of the standards that could be considered in future revisions. For example, a possible change to the dishwasher standard, resulting from the exercise, would be to make explicit that dishwashers should be fitted with an "off" switch to control normal operation, so that users, including those who are partially sighted, could confirm that the machine had stopped. The requested changes have been presented to the UK Technical Committee, and it is hoped that, in due course, the European and international standards will be modified.

Technical committees and working groups are already beginning to take into account Guide 71 in the production of standards and guidance. For example, reference to the Guide by the relevant international working group (ISO/TC 176 SC3 WG10, 2002) helped to ensure that the need for alternative formats for consumer information was made clear in the drafting of guidance on complaints handling. Also, the European working group (CEN TC 224 WG 6: Man-Machine Interface, 2003) compiling guidance on design for accessible card-activated devices, such as public banking and ticket machines, is incorporating definitions and concepts from Guide 71.

The probable consequence of Guide 71, and a range of national legislation in the area of disability, is that more standards that take into account the needs of particular groups, such as older people and people with disabilities, will follow in the future.

REFERENCES

American with Disabilities Act (ADA). (1991). Washington DC, USA.

British Standards Institution (BSI). (1990). *Specification for safety of household and similar electrical appliances. General requirements.* UK: BS 3456-201:1990, EN 60335-1:1988.

British Standards Institution (BSI). (1996). *Specification for safety of household and similar electrical appliances. Particular requirements. Dishwashers.* UK: BS EN 60335-2-5:1996.

British Standards Institution (BSI). (1996). *Specification for safety of household and similar electrical appliances. Particular requirements. Toasters, grills, roasters and similar appliances.* UK: BS EN 60335-2-9:1996.

British Standards Institution Consumer Policy Committee (BSI CPC). (2003). *Personal communication.*

Christensen, J. M., Topmiller, D. A., & Gill, R. T. (1988). Human factors definitions revisited. *Human Factors Society Bulletin, 31,* 7–8.

Disability Discrimination Act (DDA). (1995). UK.

European Commission (EC). (1999). *Mandate to the European Standards Bodies for a guidance document in the field of safety and usability of products by people with special needs (e.g., elderly and disabled).* Brussels, Belgium: M283-EN.

European Committee for Standardization (CEN). (2003). CEN TC 224 WG 6: Man-Machine Interface. Personal communication.

European Committee for Standardization (CEN) and European Committee for Electrotechnical Standardization (CEN-ELEC). (2002). *Guidelines for standards developers to address the needs of older persons and persons with disabilities.* CEN/CENELEC Guide 6.

International Organization for Standardization (ISO). (various). *Wheelchairs.* Switzerland: ISO 7176 series.

International Organization for Standardization (ISO). (1994). *Building construction—Needs of disabled people in buildings—Design guidelines.* Switzerland: ISO/TR 9527:1994.

International Organization for Standardization (ISO) and International Electrotechnical Commission (IEC). (1995). *Instructions for use of products of consumer interest.* Switzerland: ISO/IEC Guide 37:1995.

International Organization for Standardization (ISO). (2002). *Technical aids for disabled persons—Classification.* Switzerland: ISO 9999:2002.

International Organization for Standardization (ISO) and International Electrotechnical Commission (IEC). (1999). *Safety aspects—Guidelines for their inclusion in standards.* Switzerland: ISO/IEC Guide 51:1999.

International Organization for Standardization (ISO). (2000). *Addressing the needs of older persons and people with disabilities in standardization work.* Switzerland: ISO/IEC Policy Statement. 2000.

International Organization for Standardization (ISO) and International Electrotechnical Commission (IEC). (2001). *Guidelines for standards developers to address the needs of older persons and persons with disabilities.* Switzerland: ISO/IEC Guide 71:2001.

International Organization for Standardization (ISO) and International Electrotechnical Commission (IEC). (2002). *Safety aspects—Guidelines for child safety.* Switzerland: ISO/IEC Guide 50:2002.

International Organization for Standardization (ISO). (2002). ISO/TC 176 SC3 WG10 Complaints handling. Personal communication.

Japanese Standards Association (JSA). (2000). *Guidelines for all people including elderly and people with disabilities—Usability of consumer products.* Japan: JIS S 0012:2000.

Japanese Standards Association (JSA). (2000). *Guidelines for all people including elderly and people with disabilities—Marking tactile dots on consumer products.* Japan: JIS S 0011:2000.

Japanese Standards Association (JSA). (2000). *Guidelines for all people including elderly and people with disabilities—Packaging and receptacles.* Japan: JIS S 0021:2000.

VIII

Military Human Factors Standards

VIII

31

The Role of Human Systems Integration Standards in the Modern Department of Defense Acquisition Process

Joe W. McDaniel[1]
Gerald Chaikin[2]

INTRODUCTION

Standardization reform was the cornerstone of acquisition reform. Many believe that all military specifications and standards are gone or that they cannot be used. However, the human engineering standardization documents that were streamlined and consolidated during the process of standards reform were revalidated as important to military acquisition.

Because standards and guidelines should be used with the understanding of when and how they were developed, this chapter discusses the history of the human engineering standardization documents—primarily specifications, standards, handbooks, and data item descriptions—how they evolved into to today's forms and how the currently approved ensemble can best be used.

STANDARDS REFORM

By the early 1990s, the high-tech industry and Department of Defense (DoD) viewed military specifications and standards as pariahs. They were seen as adding cost to government purchases and burdening industry to the point of compromising competitiveness, nationally and internationally. Examples of low-cost, highly capable consumer products include rapidly evolving electronic and computer products. Although capabilities of such products are growing at explosive rates, the human operator and maintainer are not evolving. The military standards

---

[1]Joe W. McDaniel is retired from Human Effectiveness Directorate, Air Force Research Lab, Wright-Patterson AFB, OH. The views presented are those of the authors and do not necessarily represent the views of DoD or its Components.

[2]Gerald Chaikin died 20 Oct 2001. He is best known for his work with the military human factors standards, specifications, and handbooks, first as a civil servant, where for 20 years he chaired the Human Factors Standardization Steering Committee (HFSSC), and later as a contractor supporting the Lead Standardization Activity for Human Factors Standardization at U.S. Army Missile Command, Redstone Arsenal, AL.

for human factors engineering (HFE) and human systems integration (HSI) remain valid and effective.

The Case Against Military Standards

Spearheaded by the electronics industry, a consortium of defense industry groups succeeded in persuading DoD to avoid using military specifications and standards, or if used, to require special permission from high-level officials. How and why did this happen? The state-of-the-art in digital electronics technology has been undergoing a rapid and dramatic evolution. The last quarter of the 20th century saw a progression from single transistors to chips containing 10 transistors, and then 1,000, and then 1 million, and so forth, with no end in sight. This evolution in technology has been so rapid that the military specifications and standards through which the government bought these products had become obsolete and imposed a severe burden on the electronics industry. In many cases, electronics companies required separate assembly lines for their commercial and military products. Moreover, the worldwide market for consumer electronic products now demanded more capability at lower cost.

Back in the heyday of the space program, the government led R&D in microelectronics and was by far the largest customer. Today, the demands of the world marketplace are so vast, that the government is no longer a driving force in the technology. The portion of U.S.-made semiconductors bought by the government declined from about 75% in 1965 to about 1% in 1995. To make matters worse, the military standards for soldered circuit boards required the use of chlorinated fluorocarbons (CFCs), known ozone-depleting chemicals, to clean the flux off the boards. Clearly, the electronic specifications and standards were obsolete, burdensome, and required the use of banned substances.

The Report of the Process Action Team

Responding to complaints from the electronics industry, DoD formed a Process Action Team (PAT) to study the problem and make recommendations. The PAT (1994) reported that the Pentagon *"does not have the ability to subsidize increasingly inefficient defense operations that do not have a self-sustaining market base."* The PAT called for revisions to *"permit reliance on commercial products, practices, and processes."* Describing the problems with military standards, the PAT said, *"The difference between DoD and other major buyers, however, is that the military specifications and standards do not always stop at specifying what is required. Frequently, they also describe how to make a product, indeed, the one acceptable way to make it."* It recommended favoring the use of nongovernment over military specifications and standards.

The PAT recommended that military needs be stated as performance specifications (tell what you want, not how to do it). Stating this would solve the technical obsolescence problem and save $500 million over 2 years. The PAT favored the use of state-of-the-art products and processes, including the best commercial practices and technologies, to achieve lower acquisition costs. The PAT recommended the following test for a military standardization document: Does it impede or facilitate modern manufacturing processes? Does it allow us to do things cheaper?

The DoD has always used nongovernment standards (NGSs). The 1994 DoD Index of Specifications and Standards (DODISS) included over 5,600 NGSs among 49,000 documents. Military standards also incorporated by reference about 5,000 NGSs.

Policy Changes: What the Secretary of Defense Did

The Post-Cold-War era has reduced the inflation-adjusted military acquisition budget by from 40% to 66% (percentage varies according to source), with further reductions possible. DoD officials had said that they cannot carry on business as usual, but that radical changes were needed to match the economic and technical realities of modern times. The DoD had been paying higher prices when lower cost commercial alternatives existed and felt that they must reduce the cost of buying while preserving the defense-unique core capabilities. DoD-unique product and process specifications and standards were often identified as barriers to industry's doing business with DoD. Generally accepting the recommendations of the PAT, (then) Secretary of Defense Dr. William J. Perry issued a policy memorandum on June 29, 1994, that stated:

> Performance specifications shall be used when purchasing new systems, major modifications, upgrades to current systems, and non-developmental and commercial items, for programs in any acquisition category. If it is not practicable to use a performance specification, a non-government standard shall be used. Since there will be cases when military specifications are needed to define an exact design solution because there is no acceptable non-governmental standard or because the use of a performance specification or non-governmental standard is not cost effective, the use of military specifications and standards is authorized as a last resort, with an appropriate waiver.

The June 29, 1994, Perry Memorandum introduced the following policy changes:

- Priority for use:
  1. Performance specifications
  2. Nongovernment standards
  3. Military specifications and standards as a last resort, with an appropriate waiver
- Waivers must be approved by the Milestone Decision Authority (usually the service secretary).
- Encourage contractors are to propose nongovernment standards and industry-wide practices.
- Standards cited by other standards are automatically downgraded to nonbinding guidance documents.
- Deactivate management and manufacturing specifications and standards.
- Identify and remove obsolete specifications.
- Replace military standards with NGSs.
- Reducing direct government oversight.
- Identify and reduce or eliminate toxic pollutants.

To promote interoperability among the services, the DoD is advocating using open systems specifications and standards. Some have questioned whether this is contrary to the policy that restricts the use of detailed military specifications and standards. However, Specification and Standards Reform favors performance specifications and NGSs in solicitations and requires a waiver for using detailed specifications and some kinds of military standards (e.g., interface standards can be used without a waiver, but a design criteria standard requires a waiver before it can be used in a solicitation). This preference for performance specifications and NGSs, presumably, encourages contractors to propose specifications, standards, and products that are used in the private sector. This preference paves the way for contractor proposals using industry solutions consistent with the open systems approach, that is, widely accepted, standard products, available from multiple suppliers at competitive prices. Thus the "Open Systems" approach and Standards Reform are very consistent as to purpose and effect.

The military HFE community tried unsuccessfully for years to persuade the Defense Standards Improvements Council (DSIC) to re-designate key HFE standards from Design Criteria Standards to Interface Standards because Design Criteria Standards cannot be cited as contractually binding without a high-level waiver. In general, only the following standardization documents may be used without a waiver:

- Performance specifications (identified by the "MIL-PRF-" designation)
- Guide specifications (e.g., JSSG 2010 Aircrew Systems)
- Commercial item descriptions (CIDs)
- Interface standards (e.g., MIL-STD-1787B *Department of Defense Interface Standard: Aircraft Display Symbology*)
- Standard practices (e.g., MIL-STD-882, System Safety Program Requirements)
- Military handbooks[3]
- Nongovernment standards

As a result of MILSPEC Reform, the number of military specifications and standards was reduced by 38%, from 45,531 in 1994 to 28,326 in 1999. As a part of this streamlining initiative, the number of HFE standardization documents was reduced from 21 to 11. Only six of these are standards[4] (Chaikin, 1998). The major impact was that most of the retained HFE standardization documents lost their influence by re-designating them as non-binding guidance documents (handbooks) or as design criteria standards that require a waiver.

HISTORY OF THE MILITARY HUMAN FACTORS ENGINEERING STANDARDS

HFE design criteria standards began as responses to accidents resulting from human error. Currently, human error is still the leading category of causes of all accidents. So-called "lessons learned" are merely frequently occurring errors that we hope not to repeat. Jehan (1994) discussed the rationale for having military standards and how they exist because of bona fide requirements:

> Not created for economic reasons, they exist because, as history has shown, they were required to reduce combat risk. Simply put, they represent dollars paid now to save lives later.

World War II provided the disastrous accidents that motivated what we know today as human factors engineering. The rush to build war matériel resulted in many "horror stories." Military pilots were required to fly different types of aircraft, and in those days, there was no standard control arrangement in cockpits. For example, one twin-engine bomber had the engine controls stacked vertically, with primer on top, mixture between, and throttle on the bottom. A twin-engine cargo plane also had them arranged vertically, but the order was different: throttle, primer, and mixture. Another twin-engine cargo plane had a horizontal arrangement with yet another order: mixture, throttle, and primer running from left to right. Planes crashed because pilots erroneously reverted to previous behavior patterns and operated the wrong control. The human engineering solutions included standardizing a single arrangement for engine controls and using distinct shapes for the control handles (shape coding). These standards all but eliminated this type of accident.

---

[3] If a military specification or standard is cited in a contract, it becomes a binding part of that contract. However, specifications and standards can be cited for guidance only, in which case they are not binding documents. Military handbooks are never binding.

[4] The August 1998 issue of HFAC Highlights tabulates these standards, as well as other HFAC documents and (then) current key points-of-contact for the HFAC (Human Factors Standardization Area) Program, other US Government human factors standards organizations, US NGS Committee Chairs, and related information.

An important additional reason for development of HFE standards emerged in 1958 and is best expressed when one contractor human factors manager pressured the Army to provide such design criteria, using the following reasoning: *"Testing our deliverables and identifying human factors problems without giving us your requirements and guidelines beforehand doesn't help anyone and is wasteful. If you would provide design criteria standards before our system is designed, we will know what you will test (among other things) and can design to comply with your needs."* The Army started its HFE design criteria standards efforts as a direct result of this expressed industry need.

Beginning with World War II, as the military began to realize the importance of acquiring systems that could be operated and maintained effectively, efficiently, cheaply, and safely, each of the services[5] generated a host of specs and standards covering specific systems, subsystems, and classes of systems (e.g., ballistic missiles, and ground vehicles). As these proliferating documents became more numerous and costly to maintain, each of the services began to consolidate into more general-purpose specs and standards.

But it was the missile and space programs of the late 1950s and early 1960s that provided the impetus to elevate organizational human factors standards to military standards. Perhaps the most visible U.S. ballistic missile and space programs (before NASA was formed in Oct 1958) were run by the Air Force Ballistic Missile Division in Inglewood, CA, and the Army Ballistic Missile Agency (ABMA) of the Army Ordnance Missile Command (AOMC) at Redstone Arsenal, AL. Later AOMC became the Army Missile Command (MICOM).[6] In 1967, MICOM was selected as DoD's Lead Standardization Activity (LSA) for the Human Factors (HFAC) standardization to consolidate the principal service-peculiar human engineering specifications and standards into one triservice specification and one triservice standard.

Working together over the last 40 years, human factors engineers from the three services, industry, and technical societies jointly developed a small set of consensus-type military standards that embody accumulated HFE knowledge. The two key documents were a military specification providing human engineering program requirements and guidelines (MIL-H-46855) and a military standard providing human engineering design criteria (MIL-STD-1472).

The first true human factors military standard was AFBM[7] 57-8A *Human Engineering Design Standards for Missile System Equipment* (November 1, 1958) that superseded a policy exhibit 57-8 dated August 1, 1957. This standard had the following major sections:

General requirements
Visual displays
Controls
Physical characteristics (components)
Ambient environment
Workplace Characteristics (anthropometry)
Hazards and Safety

The material in AFBM 57-8A was drawn from a number of technical reports, many of which eventually became chapters in the *Joint Services Human Engineering Guide to Equipment*

---

[5]The commonly used phrase "triservice" began when the National Security Act of 1947 became law on July 26, 1947, it created the Department of the Air Force as a third service. Prior to 1947, the Air Force was the Army Air Forces (AAF) and prior to March 1942, the Army Air Corps.

[6]On October 1, 1997, major components of the U.S. Army Missile Command and the U.S. Army Aviation and Troop Command (ATCOM) formed the U.S. Army Aviation and Missile Command (AMCOM) at Redstone Arsenal.

[7]AFBM stands for Air Force Ballistic Missile.

Design[8] With some minor changes, AFBM Exhibit 57-8A, was reformatted as a MIL-STD and released as MIL-STD-803 (USAF, November 5, 1959) *Human Engineering Criteria for Aircraft, Missile, and Space Systems, Ground Support Equipment*. MIL-STD-803 then evolved into a three volume set: MIL-STD-803A-1 ((January 27, 1964) *Human Engineering Design Criteria for Aerospace System **Ground Equipment***), MIL-STD-803A-2 (December 1, 1964) *Human Engineering Design Criteria for Aerospace System **Facilities and Facility Equipment***, and MIL-STD-803A-3 (May 1967) *Human Engineering Design Criteria for Aerospace **Vehicles and Vehicle Equipment***.

In March 1960, the Army approved ABMA XPD-844, *PERSHING Weapon System Human Factors Engineering Criteria*. In October 1961, this was updated and expanded to include all missile systems as ABMA-STD-434, *Weapon System Human Factors Engineering Criteria*. Typical source documents for ABMA-STD-434 were the same as those used for MIL-STD-803. The Army's MIL-STD-1248, *Missile Systems Human Factors Engineering Criteria* (January 20, 1964) was essentially a MIL-STD-formatted version of ABMA-STD-434A.

The MIL-STD-803A series, together with MIL-STD-1248, were the seminal documents for the original triservice MIL-STD-1472 (February 9, 1968) *Human Engineering Design Criteria for **Military Systems, Equipment, and Facilities***. MIL-STD-1472 has survived the test of time, remains the key U.S. standard for military human engineering, and, for other than software ergonomics design, may still be the most cited human engineering standard in the technical literature, worldwide.

A more detailed history of MIL-STD-1472 from 1957 through 1978 (prestandard predecessor documents and sources, development of first standards, conversions into military standards, consolidations of service military standards, technical changes in each version and revision, and approaches of the three services) can be found in Chaikin (1978).

THE GENERAL FAMILY OF DoD HUMAN FACTORS DOCUMENTS

An important outcome of the standardization reform initiative of the late 1990s was the cancellation of most of the single-service standards and the consolidation of their materials in a few DoD standards and handbooks. Because of the criticality of aircraft design, there continue to be two primary categories of human factors documents: *general* (MIL-STD-1472 and related handbooks) and *aircraft* (JSSG 2010 and related handbooks). The general family of DoD human factors documents includes:

MIL-STD-1472F (August 23, 1999). *Department of Defense Design Criteria Standard Human Engineering*. Design data and information were removed from MIL-STD-1472E and inserted in MIL-HDBK-759. Some material from the cancelled MIL-STD-1801 User/ Computer Interface (USAF) was added to 1472.

MIL-STD-1474D (Notice, August 1, 1997). *Department of Defense Design Criteria Standard: Noise Limits*. Implementing the policies of standardization reform, this document was updated as a triservice design criteria standard. MIL-STD-1474 was first issued March 1, 1973 as an Army standard on noise limits, based on HEL STD S-1-63C. Since then it's been extensively revised and expanded. As a result of recent consolidations, MIL-STD-1474D now serves as the *DoD Design Criteria Standard on Noise Limits* that is used by all the services.

---

[8]This was later published as *Human Engineering Guide to Equipment Design*, (Morgan, Cook, Chapanis, & Lund, eds., McGraw-Hill Book Co., Inc., New York, 1963). Popularly called "the HEGED," it was widely used as a textbook.

MIL-HDBK-46855A (May 17, 1999). *Human Engineering Program, Process, and Procedures.* This handbook was extensively updated to include MIL-F-46855 and DoD-HDBK-763 *Human Engineering Procedures Guide.* The superseded DoD-HDBK-763 (canceled on July 31, 1998) covered human engineering methods *and* tools. MIL-HDBK-46855A adopted or revised only those traditional methods in DoD-HDBK-763 that have remained stable over time; the section does not describe currently available automated human engineering tools, which have rapidly evolving names and features, but refers the reader to the DSSM (Directory of Design Support Methods) on the MATRIS Web site (http://dtica.dtic.mil/ddsm/).

MIL-HDBK-1908B (August 1999). *Department of Defense Handbook: Definitions of Human Factors Terms.* This handbook (previously a standard, but converted to a handbook in accordance with standardization reform) is the single source of definitions for all documents in the HFAC standardization area. This avoids conflicting definitions of the same terms in human factors documents as each is developed or revised.

MIL-HDBK-759C, Notice 2 (March 31, 1998). *Department of Defense Handbook: Human Engineering Design Guidelines.* This is a companion to MIL-STD-1472 and provides design data and extended guidelines. It includes data removed from MIL-STD-1472E.

DoD-HDBK-743A (February 1991). *Anthropometry of US Military Personnel.* This contains statistics from about 40 military surveys,[9] including the 1988 Army ANthropometric SURvey (ANSUR) of 1,774 men and 2,208 women with more than 132 measures.

Design Criteria Standard Human Engineering—MIL-STD-1472F

Specific HFE design criteria are found in MIL-STD-1472F (August 23, 1999) *Department of Defense Design Criteria Standard Human Engineering.* MIL-STD-1472 has incorporated, by reference, ANSI/HFS 100 on Visual Display Terminal (VDT) Workstations, and defers to JSSG-2010 on issues relating to aircraft crew stations, including aircraft passenger accommodation. Because JSSG-2010 does not address aircraft maintainability, MIL-STD-1472 is the appropriate guidance on design for maintainer issues for all systems, including aircraft.

The purpose of MIL-STD-1472 is to achieve mission success, system effectiveness, simplicity, efficiency, reliability, and safety of system operation, training, and maintenance. It contains a mix of requirements and guidelines to facilitate achieving required human performance and ensuring that design is compatible with human characteristics of operators and maintainers. The standard is divided into the following major sections:

Control/display integration	Design for the maintainer
Visual displays	Design for remote handling
Audio displays	Small systems (portable)
Controls	Operational and maintenance
Labeling	ground/shipboard vehicles
Physical accommodation	Hazards and safety
Workplace design	User–computer interface
Environment	VDT Workstations (ANSI/HFS 100)

MIL-STD-1472 provides time-tested design limits and guidance for systems, equipment, and facilities that warfighters, other operators, and maintainers can use effectively.

---

[9]Digital files from recent surveys are available from Human Systems Integration Information Analysis Center [HSIIAC].

A frequently cited MANPRINT[10] success story is how the T-800 engine in the Army's Comanche helicopter can be maintained with a nine-piece tool kit. This is far from a new idea, rather applying guidance that has been in the standard since the original MIL-STD-1472 (February 9, 1968),[11]

> 5.9.10.1 General The number and diversity of fasteners used shall be minimized commensurate with stress, bonding, pressurization, shielding, thermal, and safety requirements.

The services revised MIL-STD-1472 many years ago to incorporate provisions to ensure that women would be able to operate and maintain military systems, equipment, and facilities. Such design limits (e.g., dimensions and forces) were adjusted for compatibility with size, strength, and other characteristics of the female military population. Accordingly, MIL-STD-1472 is the primary technical tool used by DoD to ensure that women in the services are not inappropriately excluded from opportunities merely as a result of design-induced incompatibilities with the systems to which they might otherwise be assigned. In other words, MIL-STD-1472 can be considered to be a highly significant and effective Equal Employment Opportunity (EEO) tool.

Because of the increasing complexity of new military systems, deficiencies in the human–system interface are frequently cited in accidents. To increase the emphasis on HFE, DoD acquisition policy, circa 1993, specifically named MIL-STD-46855 and MIL-STD-1472 as "Key Standards."

Human Engineering Program, Process, and Procedures—MIL-HDBK-46855A

The other key document was MIL-H-46855 (February 16, 1968) *Human Engineering Requirements for Military Systems, Equipment and Facilities*, a military specification. This specification defined the requirements for applying human engineering to the development and acquisition of military systems. It covered the tasks to be performed by contractors in conducting a human engineering effort, including:

Defining and allocating system functions
Equipment selection
Analysis of tasks
Preliminary system and subsystem design
Studies, experiments, and laboratory tests (mockups, simulation, etc.)
Equipment detail design drawings
Work environment, crew stations, and facilities design
Human engineering in performance and design specifications
Equipment procedure development
Human engineering in test and evaluation
Failure analysis

MIL-H-46855 was originally a consolidation of one Army, two Navy, and one Air Force specifications—conducted simultaneously with the MIL-STD-1472 consolidation effort that

---

[10]MANPRINT is the Army's MANpower and PeRsonnel INTegration (MANPRINT) program.
[11]This requirement was also in the November 5, 1959 MIL-STD-803.

was completed in February 1968. On May 26, 1994, pursuant to a re-definition of the term, *standard*, MIL-H-46855B was revised and converted to a military standard, MIL-STD-46855. On January 31, 1996, as part of standardization reform, MIL-STD-46855 was redesignated as MIL-HDBK-46855, *Human Engineering **Guidelines** for Military Systems, Equipment, and Facilities*. Because MIL-HDBK-46855 and its companion guidelines, DoD-HDBK-763,[12] were now both handbooks, it was decided to consolidate them into a new handbook. MIL-HDBK-46855A *Human Engineering Program, Process, and Procedures*, the surviving document guides DoD and contractor program managers and practitioners regarding analysis, design, and test aspects of the human engineering program.

Key selections from MIL-HDBK-46855A

4. PROGRAM TASKS
   4.2 Detailed guidelines
      4.2.1 Analysis [Definition and allocation of system functions]
      4.2.2 HE in design and development
      4.2.3 HE in test and evaluation
5. THE SIGNIFICANCE OF HE FOR PROGRAM ACQUISITION
   5.1 HE support in system acquisition
      5.1.1 Total system approach
      5.1.2 HE and Human Systems Integration (HSI)
      5.1.5 Manpower, personnel, and training interactions and implications
      5.1.6 Scope of HE concerns
         5.1.6.1 Operators and maintainers
         5.1.6.2 Nonhardware issues
   5.2 HE activity areas
   5.3 The value of HE
6. HE PROCEDURES FOR DoD ORGANIZATIONS
   6.2 Application of HE during system acquisition
   6.3 Program planning, budgeting, and scheduling
      6.3.2 Work breakdown structure (WBS)
   6.4 Coordination [Participation in IPTs]
   6.5 Preparation of the Request for Proposal (RFP)
   6.6 Proposal evaluation
   6.7 Contractor monitoring
7. HE PROCEDURES FOR CONTRACTORS
   7.1 HE design standards and guidelines
   7.2 Program organization and management
   7.3 Application of HE during system acquisition
   7.4 General contractor considerations
8. HE METHODS AND TOOLS
   8.1 Methods and tools section overview
   8.2 HE during analysis efforts
   8.3 HE analysis methods
   8.4 HE during design and development
   8.5 HE design and development methods
   8.6 HE during test and evaluation
   8.7 HE T&E methods

---

[12] See MIL-HDBK-46855A summary at the beginning of this section.

Over the years, human factors researchers and practitioners have developed many powerful methods to aid in HE work. Section 8 (previous) provides information regarding a number of the methods that can be applied by HE practitioners during system acquisition. The focus of Section 8 is on HE methods that are stable over time. Automated tools, however, are not included because they typically have rapidly evolving names and features. Therefore, descriptions of currently available HE tools can, instead, be found at the Web site for the Manpower and Training Information System (MATRIS) Office (http://dticam. dtic.mil/). The MATRIS homepage lists MIL-HDBK-46855A tools that implement many of the methods that continue to appear in the handbook. The basic description and point of contact information for the tools are available from this Web site. Additional data collected regarding the tools are available for DoD HFE TAG[13] members and approved DoD contractors from the Standardization office at the U.S. Army Aviation and Missile Command, Redstone Arsenal, AL.

All prehandbook versions of MIL-HDBK-46855 also contained Data Item Descriptions (DIDs) that specified the deliverable documentation of the HFE work done on a program. In the past, calling out MIL-STD-46855 in a contract and citing one or more tailored DIDs constituted the formal HFE reporting requirements for an acquisition program. The HFAC DIDs are:

Human Engineering Program Plan
Human Engineering Progress Report
Human Engineering Dynamic Simulation Plan
Human Engineering Test Plan
Human Engineering Test Report
Human Engineering System Analysis Report
Human Engineering Design Approach Document—Operator
Human Engineering Design Approach Document—Maintainer
Critical Task Analysis Report

Downgrading MIL-STD-46855 to a handbook meant that the DIDs, which were correlated with the provisions of MIL-STD-46855, became stand-alone documents that are now rarely, if ever, contractually cited.

HUMAN SYSTEMS INTEGRATION IN AIRCRAFT

When standardization reform placed all specifications and standards in jeopardy, the military aviation community, led by the Joint Aeronautical Commanders Group,[14] reorganized and completely replaced the entire system of specifications and standards with a new system of Joint Service Specification Guides. These JSSGs cover all aspects of military aviation systems, not just human systems. However, designating "Crew Systems" as one of the 10 top-level domains gives HSI unprecedented visibility in the aviation development. The following list shows the architecture of the JSSG system.

---

[13]Department of Defense Human Factors Engineering Technical Advisory Group [http://dticam.dtic.mil/hftag /index.html].

[14]The JACG is comprised of senior military and civilian representatives from the Army, Navy, Air Force, Marine Corps, Coast Guard, DLA, NASA, and FAA. The JACG's charter is to develop and continuously improve joint processes and procedures that will facilitate the design, development, and acquisition of aviation systems that are identical (to the maximum extent possible) or common, and that maximize interoperability.

Joint Service Specification Guides Master Index	Approved
JSSG-2000B Air System	September 21, 2004
JSSG-2001B Air Vehicle	April 30, 2004
JSSG-2002 Training	Incomplete
JSSG-2003 Support Systems	Incomplete
JSSG-2004 Weapons	Incomplete
JSSG-2005 Avionics	October 30, 1998
JSSG-2006 Structures	October 30, 1998
JSSG-2007A Engines	January 29, 2004
JSSG-2008 Air Vehicle Control & Management	October 1, 2003
JSSG-2009 Air Vehicle Subsystems	October 30, 1998
JSSG-2010 Crew Systems	October 30, 1998

The concept of the "specification guides" evolved out of the Air Force's MIL-PRIME initiative. These documents have two major parts: One is a draft specification (e.g., JSSG 2010) with key numbers and requirements replaced by blanks. The second part is a set of fourteen handbooks (e.g., JSSG 2010-1 through JSSG 2010-14) that discuss the issues for filling in the blanks. The actual filling in of the blanks can be a joint decision of military and contractors. Once filled in, these guide specs become a binding part of the contract. The JSSG, then, avoided the problems of getting a waiver by not being standard, yet becomes contractually binding in the final form. Because some of the data in the JSSG series are restricted, it was decided to limit the distribution of all JSSGs to DoD and DoD contractors.[15]

Joint Service Specification Guides Crew Systems, JSSG-2010

JSSG-2010 summarizes a unified process for applying the required disciplines to the development, integration, test, deployment and support of military aircraft crew systems. This document supports a human-centered crew station approach to the acquisition process, where the platform is designed around the human and human-generated requirements for performance as the driving force. JSSG-2010 has 14 accompanying handbooks as follows:

- JSSG-2010-1 Systems engineering guidance for the *design of crew stations* in fixed and rotary wing aircraft.
- JSSG-2010-2 Guidance for the *development requirements and verifications* for crew systems.
- JSSG-2010-3 Guidance for the criteria to *optimize cockpit/crew station/ cabin designs* without hindering the development of new, improved systems, including fixed and rotary wing.
- JSSG-2010-4 Guidance for the design and verification of *aircrew alerting systems*.
- JSSG-2010-5 Guidance for the development requirements and verifications for *interior and exterior airborne lighting* equipment, including specific requirements for interior lighting compatible with type I or II and class A or B night-vision-imaging systems).
- JSSG-2010-6 Guidance for the design and test information for *sustenance and waste management systems* for the support of aircrew and passengers.

---

[15]Portions of JSSG-2000 are restricted to government employees and government contractors. Those currently restricted are JSSG-2005, JSSG-2006, JSSG-2008, and JSSG-2010. When these are updated, they may be available to the public. Qualified users can order JSSG-2000 by regular mail at ASC/ENOI; 2530 Loop Road West; Wright-Patterson AFB OH; 45433-7107 or email at Engineering, Standards@wpafb.af.mil. Those currently available to the public are JSSG-200b, JSSG-2001B, JSSG-2007A, and JSSG-2009.

- JSSG-2010-7 Guidance for the development requirements and verifications for *occupant crash protection* and for crash protective aspects of seating, restraint, and crewstation and passenger/troop station design.
- JSSG-2010-8 Rationale, guidance, lessons learned, and instructions for the *Energetic Systems*[explosive actuators] section.
- JSSG-2010-9 Guidance for the development requirements and verifications for *aircrew personal protective equipment*.
- JSSG-2010-10 Guidance for the development requirements and verifications for an *aircraft oxygen system* and its components.
- JSSG-2010-11 Guidance for the development requirements and verifications for *aircraft emergency escape systems*.
- JSSG-2010-12 Guidance for the development requirements and verifications for *deployable aerodynamic decelerator* (DAD) system or subsystem. [Parachutes are DADs]
- JSSG-2010-13 Guidance for the development requirements and verifications for an *airborne survival and flotation system* and its components. This includes many provisions for emergency egress, life support, descent, and land and water survival for extended time periods until recovery.
- JSSG-2010-14 Guidance for the performance, development, compatibility, manufacturability, and supportability requirements and verification procedures for an *aircraft windshield/canopy system* and its components.

The JSSG-2010 is incomplete in two areas: (a) aircraft maintainability that is covered in the general-purpose MIL-STD-1472 and (b) aircraft symbology that is covered by MIL-STD-1787C (January 5, 2001) *Department of Defense Interface Standard: Aircraft Display Symbology*. MIL-STD-1787B was originally planned to be one of the handbooks included in JSSG 2010, but was approved as an interface standard (that may be cited without a waiver) before the JSSG series was finished. Because an interface standard has more authority, it was decided to leave it as a stand-alone document.

RELATION TO OSHA STANDARDS

The HFE practitioner should be familiar with both OSHA (Occupational Safety and Health Administration) and military HFE-related standards on SOH (Safety and Occupational Health). The policy is to apply the "more stringent" of those available, insofar as practicable.

Application of OSHA Versus DoD Safety and Health Standards

Because the OSHA statutes were not written to apply to military systems, there is often some confusion about whether or when OSHA standards apply. As clarification, DoD policy, according to DoDI 6055.1, *DoD Safety and Occupational Health (SOH) Program*, states that OSHA standards apply to non-military-unique operations and workplaces, and in certain circumstances apply to military-unique systems if they are more stringent than military standards. In fact, military standards are usually more stringent because their objective is ensuring performance rather than merely avoiding injury as in OSHA standards.

Non-military Unique System Coverage by OSHA

DoD Instruction 6055.1 (May 6, 1996) states that DoD Components shall comply with the standards promulgated by the OSHA in all non-military-unique DoD operations and workplaces (office, maintenance shops, and other nonfrontline activities). This applies regardless of whether work is performed by military or civilian personnel. However, DoD Components may develop and apply standards that are alternate or supplemental to such OSHA standards, and DoD standards may need to be more stringent than OSHA ones if the military situation warrants. DoD Components shall apply OSHA and other non-DoD regulatory safety and health standards to military-unique equipment, systems, operations, or workplaces, in whole or in part, insofar as practicable and when supported by good science. According to DoD Instruction 6055.1, if a

> DoD Component determines that compliance in a non-military unique work environment with an OSHA standard is not feasible, a proposed alternate standard shall be developed and submitted after consultation with other DoD Components and with affected employees or their representatives. *For example, OSHA health standards designed to protect personnel from 8 hour exposures to hazardous chemicals may not be applicable for 24 hr exposures, or for multiple exposures and various modes of entry into the body during military operations and deployment situations.* When military design, specifications, or deployment requirements render these standards unfeasible or inappropriate, or when no standard exists for such military application, DoD Components shall develop, publish, and follow special military standards, rules, or regulations prescribing SOH measures, dosimetry, and acceptable exposure levels. Acceptable exposure measures and limits shall be derived from use of the risk management process described elsewhere in this Instruction.

CONTINUING ISSUES FOR MILITARY STANDARDS

There are many fundamental differences between the HFE military standards and those specifications and standards that have been attacked as wasteful. There are at least seven issues distinguishing HFE military standards from other military standards and commercial standards:

- Obsolescence
- Specifying a solution
- Targeted systems and subsystems
- COTS/NDI (commercial off-the-shelf/nondevelopmental items)
- Performance standards
- Commercial standards
- Work breakdown structure

Issue of Obsolescence

Industry's primary reason for attacking military specifications and standards was their use of obsolete technology and manufacturing processes. Such specifications and standards were probably valid at the time they were promulgated, but became obsolete as technology evolved.

Some would argue that the HFE design standards are not up-to-date because the latest technologies are not covered. Those seeking a quick solution to high-technology human–system interfaces are often disappointed at not finding standard answers. ***In this regard, it is important to understand that such standards are not intended to provide solutions; they are limits on design***. Moreover, if the HFE standards did attempt to provide design criteria or preferred practices for rapidly evolving technology issues, they would be open to the same criticism of the electronics and manufacturing standards. An HFE design standard should properly provide criteria for which there is common agreement. This means that the technology has settled down to the point where a consensus can be reached on a needed human engineering design provision. If there is no consensus on design limits or process issues, a standard is premature. So when we say that the HFE military standards are current, we mean that all its provisions reflect current consensus.

Even older HFE standards are not obsolete because, although certain technologies have evolved rapidly, the human has not. Sensors have ever increasing resolution and spectral range, but the human still has two eyes, with no evidence that visual perception is any better now than it was in the past. The speed of computers has progressed so much that an ordinary desktop PC now outperforms the early super computers. Computer memory has increased from kilobytes to gigabytes. Yet the human brain cannot be said to have any greater memory or new capability. The only noticeable change—average human stature—that increased about 2% in the first half of the 20th century (attributed to better health and nutrition) appears to have leveled off.

Unlike the rapidly evolving digital electronics, the capabilities and limitations of humans exhibit negligible evolution, and the design principles for the human–system interface are not obsolete. The HFE standards describe design limits to ensure that the system will be effectively, efficiently, safely, and cheaply operable and maintainable by its intended users (both men and women), irrespective of whether yesterday's or cutting-edge technology is involved. The HFE standards community has always kept their standards and DIDs up-to-date and consistent with current DoD policy and needs. Many years ago, when DoD policy called for reducing the number of DIDs, the HFE DIDs were reduced from more than 30 to 10. When prior acquisition reform initiatives emphasized streamlining, the HFE community revised its process standard to bring it into compliance as the first such document to add comprehensive tailoring guides tied to each of the acquisition phases to avoid excessive requirements.

Issue of Specifying a Solution

MIL-STD-1472 does not specify any solutions; it provides time-tested design limits as requirements or guidelines. These represent performance standards in the sense that most of its criteria are human performance-driven. Failing to meet these minimum standards will cause performance to be degraded. By specifying performance-based design limits for routine elements of the human system interface, the designer (a) avoids repeating past mistakes, (b) devotes more effort to the new human systems issues (see Figure 31.1), and (c) is provided the flexibility to be innovative within relatively liberal design limits that reflect consensus of the technical community. Some military-unique performance-driven design limits include weightlift maxima for the military population, label size and color for low light operations, and control guarding options to prevent inadvertent actuation under certain conditions. In addition to not specifying solutions, the military's Human Factors Engineering standards do not prescribe materials or manufacturing techniques. MIL-STD-1472 does not specify solutions.

Issue of Targeted Systems and Subsystems

Although the military's HFE standards deal with the broad spectrum of design issues, they contain unique human performance requirements peculiar to the military and the military operating environment that include worldwide operations and war fighting. On the other hand, most voluntary standards apply to the commercial marketplace and, justifiably, have different priorities, for example, aesthetics, use by a wider range population, use by an untrained population, sales appeal, and use in benign environments.

HFE military standards frequently address mission environments that are unique, or near unique to the military or the battlefield. For example, design must accommodate operation

APPLYING LABOR TO DESIGN HUMAN INTERFACE

Case 1 - With   Case 2 - Without

HFE Standards   HFE Standards

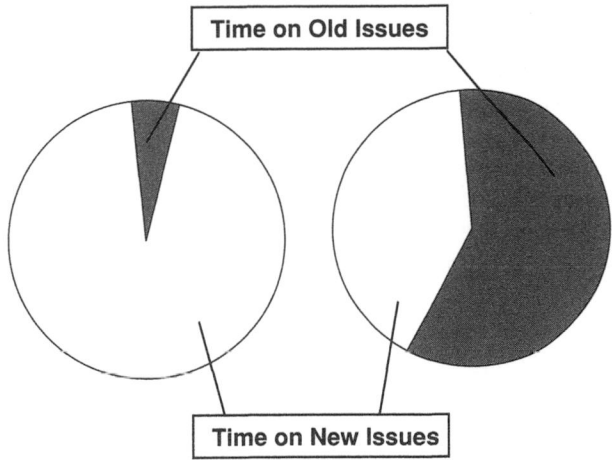

FIG. 31.1. Using HFE standards to design performance into the routine aspects of the system leaves more engineering labor to apply to the new design issues. Without HFE standards, every detail of every aspect of the human-system interface must be researched, designed, and tested. Much time is spent "re-inventing the wheel." In most programs, the HFE budget is fixed. Using accepted military standards frees up labor for solving new issues.

and maintenance by military personnel wearing protective equipment and clothing, such as chemical and biological protective gear that retains body heat, reduces body mobility, and aggravates accessibility and operability of equipment. Commercial products very seldom have a need to deal with such issues.

Issues of COTS/NDI (Commercial Off-the-Shelf/Nondevelopmental Items)

The market for consumer electronics has increased so rapidly that it is driving new technology. Seeing the advanced capability available to the public at small costs has made the military envious. Systems and equipment developed for the military is relatively more costly because of the smaller quantities purchased, the longer development period, and frequently required military-peculiar features such as ruggedness and operability in climatic extremes. A modern weapon system might take more than 10 years to develop and field. As a result, computers embedded in new aircraft may use 10-year-old technology.

It might help clarify the proper role of nonmilitary standard products if the acquisition system recognized that there are two distinct military functions that must be equipped: (a) the peacetime military plus DoD civilian personnel and (b) the combat military. The "peacetime military and DoD civilian personnel" is by far the largest group and is made up of the military members performing non-combat duties at their home base and the civilian workforce of the DoD (approximately one third of the DoD is made up of civilians). The work performed by this category of DoD personnel is similar in most respects to the work performed by business and industry. In this role, it is both reasonable and necessary that these workers have the same commercial-grade equipment that is found in business and industry. A military office already has ordinary commercial furniture, carpet, telephone, desktop computer, fax machine, and copier.

When we talk about military systems and equipment, we are referring to systems and equipment to be used by the "combat military—those military men and women who are deployed somewhere in the world and performing or training for the traditional military mission, either fighting a war or keeping the peace. Consumer products are rarely suitable in a military combat environment. If you do not believe this, all you have to do is get out the manual and read the safety instructions and the warranty.

Here is a specific example: A new cordless phone intended for home and office use has an encryption feature that prevents its radio transmissions from being monitored. Such a phone would seemingly have utility for the military. It is state-of-the-art technology, mass produced, and inexpensive. However, the instruction manual reveals five fatal flaws that would cause this product to fail miserably if used by the military in the field. First, the product may not be used near water or liquid of any kind. If the unit is exposed to rain or water, it must be unplugged immediately and returned to the manufacturer. This limitation may make it unattractive to the Army and Marines, or troops deployed in a tent in the rain. Second, the unit should not be operated in a hot environment, so it must be kept out of the sun and the desert. Third, one must not let anything rest on the power cord, nor should anyone walk on it. In other words, the power cord is very delicate. Fourth, this item may cause interference with other radio or television equipment. Fifth, the small, low-contrast labels cannot be read reliably in other than a well-lit area with good, unobstructed vision (i.e., incompatible with certain types of eye protection). It cannot be used at all with chemical defense gear on. When wearing gloves one cannot use the keypad.

Clearly, this commercial product was designed for use in an air-conditioned home or office, and will not function reliably in any other environment. For peacetime military and civilian use, this product would provide acceptable service. However, despite its economical price and ready availability, it would fail to perform in a combat military environment.

Traditionally, products for combat use are not designed for eye-appeal. A military jet has neither simulated wood-grain instrument panel nor carpeted floor, and yet it costs a fortune. With some obvious exceptions, such as dress uniforms, made-for-the-military emphasizes functionality, not appearance. On the other hand, the private consumer demands that products have a stylish appearance, with functionality being of secondary importance. The military human factors engineer cares about how well a product works, but not how it looks. Consumer products must look attractive if they are to compete in the open marketplace. However, shiny trim on commercial products, while offering eye appeal, may reflect light and disclose location to the enemy.

There have been occasions when the military has attempted to use consumer or commercial equipment in a war, usually with poor results. For example, in the Gulf War, to aid communication, many of the deploying troops took their office Fax machines with them. In a very short time, the fine desert sand had found it way into every nook and cranny, and these marvels of modern technology literally ground to a halt. The hero of the day turned out to be the Air Force, who had acquired a few Fax machines that had been hardened according to military standards. These machines continued to work, despite the sand, and were heavily used, because they were shared with the other services whose COTS fax machines had quickly failed.

We tend to forget that equipment designed for use in air-conditioned offices will fail when operated in many other places in the world where our military must deploy. Taking commercial and consumer products designed for office or even factory use to a cold, hot, wet, or dusty region of the world can quickly convert high tech equipment into expensive door stops. Buying an inexpensive item that does not meet the need is never a bargain. DoD must consider both the life-cycle costs and the environment of use. This is the "military-unique" environment the policymakers were referring to when talking of exceptions to COTS.

Issues With Developing HFE Performance Standards

The Standards Process Action Team said: *"Because of the uniqueness of military require-ments, it is unlikely that the DoD will ever be able to rely completely on performance-based specifications."* According to Lowell (1994), the DSIC had subsequently developed a common waiver process for military specifications and standards. There is a precedence where HFE and system safety engineering have been set aside in an exempt category, as illustrated by the following quote from a House of Representatives Report (1976):

> Content of Performance Standards. A performance standard established for a device under the proposed legislation must provide reasonable assurance of a device's safe and effective perfor-mance. Although use of the term "performance standard" reflects a preference for standards which allow the fullest use of technological alternatives, the Committee does not intend the term to be construed as excluding design-related requirements, as it is when it is used in the engineering community. *Design-related requirements that are necessary to provide reasonable assurance of safe and effective performance or that improve device safety and effectiveness by reducing the likelihood of human error should be included in a performance standard.*

This quotation tells how even if performance standards are used, they must be supplemented with specific design-related requirements to assure safe and effective performance. This is precisely the essence of the military standards for HFE.

According to Lowell (1994), the DSIC had approved the following definition of performance specifications:

> A performance specification . . . states requirements in terms of the required results with criteria for verifying compliance, but without stating the methods for achieving the required results. A performance specification defines the functional requirements for the item, the environment in which it must operate, and interface and interchangeability characteristics.

The military's HFE standards do not dictate materials or manufacturing techniques. Ar-guably, they are performance standards, because they provide design criteria that (a) allow users to operate and maintain with efficiency and with minimal error and (b) because most of the criteria are human performance-based. For example, we know from studying human performance that as keys of a keyboard get smaller and closer together, keying errors and time will increase exponentially. MIL-STD-1472 describes the size and separation of keys on a keyboard that human can operate with minimal errors. For barehanded use, the minimum spacing for keys on a keyboard is 16.4 mm (0.675 inch) center to center, with 19 mm (0.75 inch) preferred. (Note the human performance basis for this provision.) The typical computer key-board meets the preferred separation, so the standard does not limit access to computer key-boards and typewriters. This standard key separation is easy to understand, easy to apply, consistent with 50 years of competent research, and compliance can be economically veri-fied in a few seconds with an ordinary ruler. It guarantees adequate performance on military equipment. In arriving at this standard, there was no opposition from industry because industry was interested in optimizing speed and accuracy.

To convert this key separation requirement to a true "performance standard," we must replace it with a statement of our performance goal. For example, a performance standard would not specify a key separation, but have words to the effect that "the key size and spacing must be such that keying errors should not exceed 'X'%, and the average time to correctly reach and operate push buttons should not exceed 'Y'-tenths of a second, when operated by all members of target audience under all conditions of intended use." Unlike the previous standard, this is not straightforward in its application. What would a contractor have to do to comply with such a performance requirement? To start, the contractor must acquire data on the relationships described. As a minimum, this would involve a literature search by someone skilled in that technology. If relevant data were not found, the contractor may have to perform an expensive and time-consuming research study to gather the needed data.

After the product is delivered, verification by the government that the goal has been met may also require a very expensive and time-consuming test and evaluation. A real example illustrates the difficulty. One manufacturer makes a compact keyboard with keys 12 mm (0.48 inch) between centers. Designed for the less demanding consumer market, the manufacturer would also like to sell it to the military. After all, it is economical commercial off-the-shelf (COTS) equipment. It fails to meet the minimum separation in the old HFE standard, but what about the new performance standard? Reputable research has shown that the close key spacing causes many times more keying errors than does the preferred spacing. However, in addition to competent studies, the research literature also has studies with inadequate design or sample size that failed to find conclusive results. Because the literature contains studies with conflicting results, the contractor not only is free to choose one that supports the offered design, but also has an economic incentive to do so.

Next, the government's program manager will be confronted by the company's self-serving arguments together with the military's human engineer pointing out the problems with the questionable research. The program manager has better things to do than adjudicate disputes over sample sizes in unfamiliar research studies. The contractor's arguments for low cost and saving of valuable control panel space will likely carry more weight than debatable research results. Sadly, none of these players will ever know if years later the troops in the field had problems with this design.

In an alternate scenario, the contractor's strategy might be not to worry about research at all and take a chance on passing the government's final acceptance test. At the end of the program, when the product is delivered for test and evaluation, the key spacing problem will again surface as a concern, and warrant a test study be performed. Contentious issues with this study include such factors as hand size of the test subjects, number of subjects, the test procedure, and the definition of what the performance specification meant by "correctly reach and operate." So, in this case, technical practitioners' comparing test measurements with design limits have been supplanted by lawyers and contract specialists' arguing about words. If a competent evaluation were performed, the results would show that the keyboard did not meet the spec. However, the odds are greatly against the study being performed. The program manager is now faced with an even more difficult problem, because correcting the design at this late date will be prohibitively costly and further delay a program that is already behind schedule. With the program facing further delays and cost increases, the original requirement will come under fire. Why did the government specify a maximum of 1% errors, anyway? Isn't 2% good enough? The program manager may not want to reject the entire system on the basis a test that showed only 2% errors. What difference does an extra 1% make anyway?

Although performance standards for high-technology items will undoubtedly save money and increase opportunities for innovation where cutting-edge technology is involved, applying them to the human–system interface may cause the contractors unnecessary work, and may give the military error-prone designs, and increase costs of test and evaluation. It is likely that, rather than heading off problematic designs early, when they can be dealt with inexpensively, design issues will be postponed until late in the program when changes are difficult and prohibitively expensive. Experience has shown that industry and DoD can agree on reasonable HFE design standards, provided the decision is made outside the context of a specific program. Once the program begins, schedules, existing designs, and profit incentives tend to cloud the issues and make resolution expensive and contentious.

Issues With Converting to Commercial (Nongovernment Standards)

Standards reform gave priority to using voluntary standards and converting military specs and standards to NGSs, as applicable. Unfortunately, when military standards are converted to commercial standards, their scope is broadened to include nonmilitary systems and equipment and nonmilitary populations.[16] Because there are more non-military systems, expanding the scope dilutes the amount of time and effort that goes toward updating the standards.

Currently, there are no comprehensive, general HFE NGSs.[17] There are some point-designs, such as the video display terminals covered by ANSI/HFS 100, but the DoD is the only customer whose interest includes virtually all products. It is for this reason that the military has led and still leads the development of HFE design technology. The military has been a consistent user of HFE technology for the past 50 years because the consequences of failing to do so can be catastrophic. One of the reasons stated for emphasizing NGSs is to take advantage of the best consensus standards industry has to offer without burdening them. Industry supports the military's HFE standards because it has participated in their development (as voting participants) and kept them reasonable. Today, many nondefense companies use applicable provisions of MIL-STD-1472 voluntarily on their commercial products because it is recognized as the best available.

There are two fundamental reasons why the military should have its own HFE standards. First, the mission and weapons functions are unique to the military. The military should retain control of performance requirements for all equipment the troops take to the field in a military action. These requirements are almost always life critical, with mission performance and system safety at stake. Certainly, these standards do not have to be applied to the everyday equipment used by military and DoD civilian personnel in performance of noncombat duties. Second, the military needs an integrated HFE standard, not a large number of piecemeal standards. The mixing and matching from a set of hundreds or thousands of commercial standards not only is inefficient for HFE requirements but also will likely lead to omissions of important considerations. When the military considers commercial off-the-shelf equipment, it should always be tested to determine whether it is compatible with the military environment. When

---

[16]Military populations have body height and weight limitations, and for all practical purposes, age restrictions that make their physical characteristics differ from the general civilian population.

[17]Voluntary standards themselves are copyrighted products that must be purchased and cannot be reproduced. When commercial standards organizations considered converting MIL-STD-1472 to a voluntary standard, they planned to split the comprehensive document into many single-topic standards. Anybody in government and industry would have to purchase up to 16 separate documents to get the same material. In contrast, hard copies of military standards and handbooks are available for a modest page charge, may be reproduced without charge, and are instantly available in electronic format at http://www.dodssp.daps.mil/. Reinforcing the concern regarding the need to purchase or otherwise access an unreasonably large number of HFE NGS is identification of 364 HFE NGS under 36 topical categories by the DoD HFE TAG in September 2002. Compare having to search a large portion of these 364 HFE NGS versus using a single standard, MIL-STD-1472F with its 17 NGS cited as referenced documents (TS/I, 1997).

modified commercial off-the-shelf equipment is developed, it should be consistent with the military's human–system interface standards.

Reduced military budgets make it increasingly difficult for DoD HFE practitioners and researchers to participate in updates of commercial standards and guidelines. Reduced budgets drive the military to sacrifice all but strictly mission-essential activities, and updating of commercial standards and guidelines are not seen as mission-essential.

Obtaining Nongovernment Standards

After October 1, 2000, the DODSSP ceased to provide DoD users NGSs. Instead, each organization will use government purchase cards to acquire NGSs as needed. The NSSN[18] Web site will refer all users to a designated source. Many commercial resellers provide ready access to both government and nongovernment standards through subscriptions or individual document sales. Commercial resellers often also add other value (e.g., improved indexing, document summaries, and full-text search capability), and some have extensive collections of historical documents. The Navy's surface ship community now uses a commercialized version of 1472 tailored for ships for the Navy and Coast Guard. ASTM F1166-95a *Standard Practice for Human Engineering Design for Marine Systems, Equipment and Facilities* is copyrighted, and can be ordered for $60.00 per copy. MIL-STD-1472 is not copyrighted, and unlimited numbers can be reproduced or downloaded from the Web free of charge.

Government Participation in Nongovernment Standards Bodies (NGSB)

To reduce to a minimum the reliance by agencies on government-unique standards, the Office of Management and Budget has published Revised OMB Circular A-119[19] that directs government agencies to use voluntary consensus standards in lieu of government standards except where inconsistent with law or otherwise impractical. It also provides guidance for agencies participating in voluntary consensus standards bodies and describes procedures for satisfying the reporting requirements. This policy encourages DoD employees to participate as "equal partners" with private-sector and other government agency employees on technical committees of NGSBs. Such participation ensures proper consideration of DoD requirements, enhances the technical knowledge of DoD personnel, and allows DoD to contribute the considerable technical capabilities of its employees in the development of "world-class" national standards.

Although government personnel are encouraged to participate in the development of NGSs, it is costly to do so. One should never forget that many voluntary standards are typically written to "the least common denominator" so that all the standard-writing participants' products are acceptable. This has both positive and negative implications.

Travel costs can be considerable. Moreover, some NGSBs charge government personnel a fee[20] to attend each standard committee meeting. Some government personnel sense a conflict

---

[18]NSSN originally stood for National Standards Systems Network, but now prefers National Resource for Global Standards; http://www.nssn.org/index.html. The NSSN contains over 250,000 references to standards from more than 600 developers worldwide. These have been grouped into six categories: 200,000 Approved Industry Standards; 15,000 Approved International Standards; 46,000 Approved U.S. Government Standards; 10,000 Industry Standards Under Development; 3,000 International Standards Under Development; and 4,000 U.S. Government Standards Under Development

[19]Revised OMB Circular A-119, *Federal Participation in the Development and Use of Voluntary Consensus Standards and in Conformity Assessment Activities, 10 Feb 1998.* This circular establishes policies on Federal use and development of voluntary consensus standards and on conformity assessment activities.

[20]One of the major NGS organizations, for example, charges government personnel a $200 fee to attend each NGS development meeting.

of interest when paying a fee to donate the taxpayer's labor to develop NGSs that the NGSBs then sell back to the government, its contractors, and the public for a profit. These can be reasons why the number of DoD participants on NGSB committees has dropped dramatically from over 2,200 DoD participants in 1994 to fewer than 500 in 1999.[21]

The Problems With Nongovernment Standards (NGS)

When examining a list of NGSs in the human factors domain, the two most striking characteristics are the spotty coverage and the duplication in isolated popular areas. NGSs tend not to be comprehensive, but focus on specific products, product lines, or product components. An example of overlapping standards is the group: agricultural equipment (with seven standards), earthmoving equipment (with ten standards), off-road equipment (three standards), and graders. The overlap among these is obvious. In addition to numerous standards on telecommunications, there are six human factors standards relating to telephones. With the repetition and overlap of NGSs, just selecting the one applicable for a military program would be a laborious task. It is also an expensive task, because you first must buy all the NGS on your topic before you can determine which one, if any, is applicable to the design.

Human Engineering—Principles and Practices (HEB1)

The G-45 (Human Factors) Committee of the Government Electronics and Information Technology Association (GEIA), with the support of the Human Factors Standardization SubTAG, prepared a Human Engineering—Principles and Practices Bulletin. The Bulletin and its annexes provide guidance to the application of human engineering principles and practices in the analysis, design, development, testing, fielding, support, and training for military and commercial systems, equipment, and facilities. As an industry (nongovernment) document, the use of this document is consistent with DoD's acquisition reform and could be applied in DoD solicitations. The Electronic Industries Alliance (EIA) published this bulletin in June 2002.[22]

HEB1 resulted from the need to present to the Program Office a succinct human engineering management approach that would explain human engineering requirements for both government and industry systems, equipment and facilities. HEB1 is a 21-page document based on Section 4 of MIL-HDBK-46855A (May 1999), an update of MIL-H-46855B, Rev 2. It has been edited to address both government and industry needs, and to include a list of, and links to, current Data Item Descriptions (DIDs) developed at the DoD and by the Federal Aviation Administration (FAA). The bulletin is a Human Engineering (HE) Best Practices document developed by both DoD and industrial HE practitioners. It also includes a list of acronyms used in the document and the terms they represent and a list of documents that give additional information.

With the acquisition policy of the DoD discouraging the use of military and government standards in favor of industry practices and standards, HEB1 may prove to be a valuable document. As an "engineering bulletin," HEB1 is not a standard, but recommended practices. When the DoD's MIL-H-46855 was converted to MIL-HDBK-46855A in May 1999, access to the Data Item Descriptions (DIDs) was lost, because they cannot be invoked by a handbook. As indicated in the following, the FAA currently uses some of the DIDs. By recommending and providing links to the DIDs listed here, HEB1 returns them from limbo to a prominent place in the acquisition community.

---

[21] According to Under Secretary of Defense for Acquisition and Technology J. S. Gansler, October 14, 1999.
[22] See Global Engineering: http://global.ihs.com/.

- Human Engineering Simulation Concept [DI-HFAC-80742B and FAA-HF-005]
- Human Engineering Design Approach Document-Operator [DI-HFAC-80746A and FAA-HF-002]
- Human Engineering Design Approach Document—Maintainer [DI-HFAC-80747B and FAA-HF-003]
- Noise Measurement Report (NMR) [DI-HFAC-80938A]
- Critical Task Analysis Report [DI-HFAC-81399 and FAA-HF-004]
- Human Engineering Program Plan [FAA-HF-001]

THE DEFENSE STANDARDIZATION PROGRAM

Department of Defense Instruction 4120.24 (June 18, 1998) *Defense Standardization Program* (DSP) establishes the DSP under the Defense Logistics Agency (DLA). It is DoD policy to promote standardization of materiel, facilities, and engineering practices to improve military operational readiness, reduce total ownership costs, and reduce acquisition cycle time. These objectives are accomplished by a single, integrated DSP and a uniform series of specifications, standards, and related documents. Specific implementation and guidance is found in the current issue of DoD 4120.24-M *Defense Standardization Program (DSP) Policies and Procedures (March 2000).*[23]

Resources

The DSP Web site (http://www.dsp.dla.mil/) provides ready access to current and recently obsolete[24] military specifications and standards. The DSP mission is to identify, influence, develop, manage, and provide access to standardization processes, products, and services for the warfighter, the acquisition community, and the logistics community to promote interoperability, reduce total ownership costs, and sustain readiness.

Search tools allow users to locate and view Defense specifications, standards, handbooks, other documents listed in the DoD Index of Specifications and Standards (DODISS), and data item descriptions (DIDs). As of March 1, 2000, the ASSIST database is the official source for all active DIDs. Earlier versions of most recently revised DIDs are also available.

National Standards Systems Network (NSSN)[18]

The NSSN is a national resource for global standards that indexes documents of over 600 standards-developing organizations. There is no charge, nor do users need to register. NSSN is a service of the American National Standards Network (ANSI). Users perform searches by document number or by key words within the document title or description. Once a document is located, the NSSN index describes where to obtain it. The 33 organizations with significant HFE standards or guidelines are shown in the "Index of Nongovernment Standards on Human Engineering Design Criteria and Program Requirements/Guidelines" at http://dtica.dtic.mil/hftag/product.html.

Defense Technical Information Center (DTIC)

The DTIC lets users search the Public Scientific and Technical Information Network and retrieve copies of unclassified, unrestricted technical papers.

---

[23]DoD directives, instructions, regulations, manuals, etc. are available at http://www.dtic.mil/whs/directives/.

[24]The version of military standards and handbooks placed on the original contract remain in force despite subsequent updates during the life of the program.

The DoD Single Stock Point (DODSSP)

All interested parties can request copies of Defense specifications and standards, Federal specifications and standards used by DoD, and other DoD standardization documents from the DODSSP in Philadelphia, PA. The DODSSP maintains the official repository of all DoD standardization documents and publishes the DoD Index of Specifications and Standards (DoDISS)at http:/assist.daps.dla.mil/quick search. Registered users can query the ASSIST database and download most document images as Adobe PDF files.

OTHER GOVERNMENT HFE STANDARDS

NASA Standards

The principal human engineering standard used by NASA is NASA-STD-3000, Man-Systems Integration Standards (MSIS). This family of standards provides specific user information to ensure proper integration of the man–system interface requirements with those of other aerospace disciplines. These man–system interface requirements apply to launch, entry, on-orbit, and extraterrestrial space environments. This document is intended for use by design engineers, systems engineers, maintainability engineers, operations analysts, human factors specialists, and others engaged in the definition and development of manned space projects or programs. Concise design considerations, design requirements, and design examples are provided. Requirements specified are applicable to all U.S. manned spaceflight programs.

In addition to the basic document, additional volumes of the MSIS are created and maintained which specifically address the human factors and crew interface needs for that program. As specialized volumes of this type are updated and revised, the information gathered for them is also evaluated for possible inclusion in the basic MSIS volume. To date, there are three volumes planned, with four already published and released:

Vol. I, Man-Systems Integration Standards, first published in 1987 and last updated as Rev. B in June, 1995 [http://msis.jsc.nasa.gov/]

Vol. II, Man-Systems Integration Standards—Appendices, first published in 1987 and last updated in 1995, at the same time as, and to the same revision letter as Vol. I [http://msis.jsc.nasa.gov/]

Vol. III, Man-Systems Integration Standards—Design Handbook (the data in this volume coincides with Rev. A, of Vol. I)

Vol. IV, Space Station Freedom Man-Systems Integration Standards, a subset of Vol. I, published in 1987 (Inactive)

Federal Aviation Administration (FAA) standards

The FAA HF-SFD-001 (June 2003) Human Factors Design Standard (HFDS) provides reference information to assist in the selection, analysis, design, development, and evaluation of new and modified FAA systems and equipment. This guide covers a broad range of human factors topics that pertain to automation, maintenance, human interfaces, workplace design, documentation, system security, safety, the environment, and anthropometry. This document also includes extensive human–computer interface guidance. The HFDG draws heavily from human factors information developed by DoD, NASA, and DOE. This document (Wagner, Birt, Snyder, & Duncanson, 1996) is available to the public through the National Technical Information Service (NTIS), Springfield, VA 22161 and online at http://hf.tc.faa.gov/hfds.

CONCLUSIONS

A comparison of the HFE standards against the recommendations of the PAT reveal that they met most of the stated goals as is because (a) they were jointly developed by government and industry, (b) they have been kept up-to-date for the issues they cover, (c) and the HFE design criteria are based on human performance. HFE standards do not require dual manufacturing processes because they are not manufacturing standards, do not define hardware, and are not obsolete. They save the government money by reducing need for design studies, tests, and evaluation. They lower life-cycle costs by providing solutions early in design, not during T&E when repair is most costly. They are military unique, embodying descriptions of the capabilities and limitations of military personnel (somewhat different from civilian populations) and their personal protective equipment (very different than civilian counterparts).

There are no nongovernment standard alternatives. Indeed, the superiority of the HFE military standards causes them to be used for civilian applications (dual use). Military and industry jointly developed and updated these standards. They are coordinated through industry groups, professional societies, and other standards organizations.

Standards should not be written for rapidly evolving technology, such as is now ongoing in the electronics technology area. If standards are to be efficient and effective, they should be based on consensus of both the buyer and the seller. The HFE standards do not have the serious deficiencies that are addressed by the acquisition reform.

Although not the cause of the problems that drove standardization reform, HFE standards were flushed out with the rest. Although HFE standards are important to the HFE profession, they are probably too small to merit special consideration from DoD management in these busy times.

Reforms and fine-tuning of acquisition are always needed. The HFE/HSI community must continue to keep up with the changing policies and comply with the new directives in a way that is efficient, effective, and in the long term best interest of the military.

So far, the largest effect on HFE of deemphasizing use of military standards to date has been severe cuts in the size of HFE staffs in defense industries (McDaniel, 1996). Part of these cuts result from the general decline in military spending, but the HFE staffs have taken disproportionately large cuts. Anticipating less emphasis on HFE, some companies have significantly cut back on HFE staff. Government HFEs cannot take up the slack because their numbers have also been reduced. As a result, there are fewer HFE people in industry to do the work, and fewer HFE people in the government to write the performance specs and test the finished products. This trend may have a chilling effect on the profession as a whole.

The prognosis for the future of military standards is favorable. On March 29, 2005 the Under Secretary of Defense for Acquisition, Technology and Logistics signed Policy Memo 05-3 titled "Elimination of Waivers to Cite Military Specifications and Standards in Solicitations and Contracts." This memo allows program managers to use and cite MIL-STD-1472 and MIL-STD-1474 as contractually binding requirements for the first time in over 10 years. This memo was not widely advertised, but is known to those in the human factors community. The pendulum has swung the other way, and future acquisitions will benefit for it.

REFERENCES

Chaikin, G. (1978). Human engineering design criteria—The value of obsolete standards and guides. *Proceedings of the Human Factors Society—22nd Annual Meeting—1978*, 409–415.
Chaikin, G. (Ed.). (1998). *HFAC highlights August 1998*. Redstone Arsenal, AL: US Army Aviation and Missile Command.

DOT, Department of Transportation. (1995). *Human factors in the design and evaluation of air traffic control systems, April 1995*. Cambridge, MA: Federal Aviation Administration, John A. Volpe National Transportation Systems Center.

House of Representatives. Report. (1976). *Report No. 94-853, Medical Device Amendments of 1976* (p. 26) Washington, DC.

Jehan, H. I., Jr. (1994). MIL-SPECS and MIL-STDS no more? DoD changes prioritizing policy. In *Program Manager, July-August*, (pp. 8–10). Ft Belvoir, VA: Defense Systems Management College Press.

Lowell, S. C. (1994). Effects of specs and standards reform on HFE. Unpublished oral presentation to the Human Factors Standardization Steering Committee, November 1, 1994, Orlando FL.

McDaniel, J. W. (1995a). Demise of military standards may affect ergonomics. In A. Bittner (Ed.), *Advances in industrial ergonomics and safety VII* (pp. 811–818), Bristol, PA: Taylor & Francis.

McDaniel, J. W. (1995b). Obsolete accounting model hinders crew system integration. *CSERIAC Gateway, VI*(3), 1–4.

McDaniel, J. W. (1996). Demise of military standards may affect ergonomics. In A. Mital (Ed.), *International Journal of Industrial Ergonomics* (Vol. 18(5-6), pp. 339–348). Amsterdam, The Netherlands: Elsevier Science BV.

Perry, W. P. (1994). *Policy memorandum on military specifications and standards*. Washington, DC: Office of the Secretary of Defense.

Process Action Team. (PAT). (1994). *Report of the process action team on military specifications and standards*. Washington, DC: Office of the Under Secretary of Defense for Acquisition Technology.

Technical Society/Industry Subgroup (TS/I). (1997). *Index of non-government standards on human engineering design criteria and program requirements/guidelines*. San Antonio, TX: DoD Human Factors Engineering Technical Advisory Group.

Wagner, D., Birt, J., Snyder, M., & Duncanson, J. (1996). *FAA human factors design guide (HFDG) for acquisition of commercial off-the-shelf subsystems, non-developmental items, and developmental systems*. Atlantic City, NJ: FAA William J. Hughes Technical Center.

IX

Sources of Human Factors and Ergonomics Standards

32

Sources and Bibliography of Selected Human Factors and Ergonomics Standards

David Rodrick
Waldemar Karwowski
University of Louisville

INTRODUCTION

International Organization for Standardization (ISO) defines standard as "a documented agreement containing technical specifications or other precise criteria, to be used consistently as rules, guidelines, or definitions of characteristics, to ensure that materials, products, processes and services are fit for the purpose served by those making reference to the standard" (ISO, 2004). Over 50 years of research and practice in human factors and ergonomics discipline clearly demonstrated that consideration of workers as "human-being" in designing work and production systems results in beneficial outcomes. The objective of this chapter is to identify selected human factors and ergonomics standards developed by the international, national, and local bodies and provide selected bibliography to assist researchers and practitioners in human factors and ergonomics domain.

The listing of the standards is reasonably current as of 2004. The standards that were selected contained the terms *human factors* or *ergonomics* in their titles. Those standards, the title of which did not contain these terms, were not included. The bibliography section of the chapter also contains an up-to-date compilation of books and journal articles that reflect theoretical views and empirical research on existing human factors and ergonomics standards.

SOURCES OF SELECTED HF/E STANDARDS AND GUIDELINES

Ergonomics: General Guiding Principles

TC 159/SC 1 Ergonomic guiding principles, International Organization for Standardization (http://www.iso.org/iso/en/CatalogueListPage.CatalogueList?COMMID=3906&scopelist=)

ISO 6385:2004 Ergonomic principles in the design of work systems

ISO 10075:1991 Ergonomic principles related to mental work-load—General terms and definitions

ISO 10075-2:1996 Ergonomic principles related to mental workload—Part 2: Design principles

ISO 10075-3:2004 Ergonomic principles related to mental workload—Part 3: Principles and requirements concerning methods for measuring and assessing mental workload

Anthropometry and Biomechanics

NASA RP-1024 Anthropometric Source Book (http://msis.jsc.nasa.gov/volume2/Appx_ a_Bibli.htm)

Guidelines for Using Anthropometric Data in Product Design (ISBN 0-945289-23-5), Human Factors and Ergonomics Society, Santa Monica, CA. (http://www.hfes.org/publications/anthropometryguide.html)

ANSI B11 Technical Report: Ergonomic Guidelines for the Design, Installation and Use of Machine Tools (Reported in http://www.ergoweb.com/resources/reference/guidelines/ansib11.cfm)

TC 159/SC 3 Anthropometry and biomechanics, International Organization for Standardization (http://www.iso.org/iso/en/CatalogueListPage.CatalogueList?COMMID= 3904&scopelist=)

> ISO 7250:1996 Basic human body measurements for technological design
>
> ISO 11226:2000 Ergonomics—Evaluation of static working postures
>
> ISO 11228-1:2003 Ergonomics—Manual handling—Part 1: Lifting and carrying
>
> ISO 14738:2002 Safety of machinery—Anthropometric requirements for the design of workstations at machinery
>
> ISO 14738:2002/Cor 1:2003
>
> ISO 15534-1:2000 Ergonomic design for the safety of machinery—Part 1: Principles for determining the dimensions required for openings for whole-body access into machinery
>
> ISO 15534-2:2000 Ergonomic design for the safety of machinery—Part 2: Principles for determining the dimensions required for access openings
>
> ISO 15534-3:2000 Ergonomic design for the safety of machinery—Part 3: Anthropometric data
>
> ISO 15535:2003 General requirements for establishing anthropometric databases
>
> ISO/TS 20646-1:2004 Ergonomic procedures for the improvement of local muscular workloads—Part 1: Guidelines for reducing local muscular workloads

JIS Z8500 Basic Human Body Measurements for Technological Design. (2002). Japanese Industrial Standard (http://www.jsa.or.jp/default_english.asp)

DIN 33402-2 Human Body Dimensions—Values. (1986). Deutsches Institut für Normung (http://www2.din.de/, http://www.techstreet.com/)

SAE J833 Human Physical Dimensions. (1989). Society of Automotive Engineers (http://www.sae.org/servlets/index)

EN ISO 7250:1997 Basic human body measurements for technological design (ISO 7250:1996), European Committee for Standardization (CEN) (http://www.cenorm.be/cenorm/index.htm)

Clothing

ISO 13688 Protective Clothing—General Requirements. (1998). International Organization for Standardization (http://www.iso.ch/iso/en/prods-services/ISOstore/store.html)

ASTM F1154 Standard Practices for Qualitatively Evaluation the Comfort, Fit, Function, and Integrity of Chemical-Protective Suit Ensembles. (1999). American Society for Testing and Materials (http://www.astm.org/cgi-bin/SoftCart.exe/index.shtml?E+mystore)

ISO 9920 Ergonomics of the Thermal Environment—Estimation of the Thermal Insulation and Evaporative Resistance of a Clothing Ensemble. (1995). International Organization for Standardization (http://www.iso.ch/iso/en/prods-services/ISOstore/store.html)

Collision Avoidance

SAE ARP 4153 Human Interface Criteria for Collision Avoidance Systems in Transport Aircraft. (1988). Society of Automotive Engineers (http://www.sae.org/servlets/index)

Communication

SAE ARP 4791 Human Engineering Recommendations for Data Link Systems. (1996). Society of Automotive Engineers (http://www.sae.org/servlets/index)

ETSI ETR 070 The Multiple Index Approach (MIA) for the Evaluation of Pictograms. (1993). European Telecommunications Standardization Institute (http://www.etsi.org/, http://www.techstreet.com/)

Telecommunication

ETSI ETR 029 Access to Telecommunications for People with Special Needs: Recommendations for Improving and Adapting Telecommunication Terminals and Services for People with Impairments. (1998). European Telecommunications Standardization Institute (http://www.etsi.org/, http://www.techstreet.com/)

ETSI ETR 068 European Standardization Situation of Telecommunications Facilities for People with Special Needs. (1998). European Telecommunications Standardization Institute (http://www.etsi.org/, http://www.techstreet.com/)

ETSI ETR 170 Generic User Control Procedures for Telecommunication Terminals and Services. (1995). European Telecommunications Standardization Institute (http://www.etsi.org/, http://www.techstreet.com/)

ETSI ETR 095 Guide for Usability Evaluations of Telecommunications Systems and Services. (1993). European Telecommunications Standardization Institute (http://www.etsi.org/, http://www.techstreet.com/)

ETSI ETR 160 Human Factors Aspects of Multimedia Telecommunications. (1995). European Telecommunications Standardization Institute (http://www.etsi.org/, http://www.techstreet.com/)

ETSI ETR 039 Human Factors Standards for Telecommunications Applications. (1992). European Telecommunications Standardization Institute (http://www.etsi.org/, http://www.techstreet.com/)

ETSI ETR 165 Recommendation for a Tactile Identifier on Machine Readable Cards for Telecommunication Terminals. (1995). European Telecommunications Standardization Institute (http://www.etsi.org/, http://www.techstreet.com/)

ETSI ETR 167 User Instructions for Public Telecommunications Services; Design Guidelines. (1995). European Telecommunications Standardization Institute (http://www.etsi.org/, http://www.techstreet.com/)

Telephones

ETSI ETR 051 Human Usability Checklist for Telephones: Basic Requirements (1992). European Telecommunications Standardization Institute (http://www.etsi.org/, http://www.techstreet.com/)

ETSI ETR 096 Human Factors Guidelines for the Design of Minimum Phone Based User Interface to Computer Services. (1993). European Telecommunications Standardization Institute (http://www.etsi.org/, http://www.techstreet.com/)

ETSI ETR 166 Evaluation of Telephones for People with Special Needs: An Evaluation Method. (1995). European Telecommunications Standardization Institute (http://www.etsi.org/, http://www.techstreet.com/)

ETSI ETR 187 Recommendation of Characteristics of Telephone Services Tones When Locally Generated in Telephony Terminals. (1995). European Telecommunications Standardization Institute (http://www.etsi.org/, http://www.techstreet.com/)

Videophones

ETSI ETS 300 375 Pictograms for Point-to-Point Video Telephony. (1994). European Telecommunications Standardization Institute (http://www.etsi.org/, http://www.techstreet.com/)

ETSI ETR 175 User Procedures for Multipoint Video Telephony. (1995). European Telecommunications Standardization Institute (http://www.etsi.org/, http://www.techstreet.com/)

Control Rooms

ISO 11064-1 Ergonomic Design of Control Centres—Part 1: Principles for the Design of Control Centers. (2000). International Organization for Standardization (http://www.iso.ch/iso/en/prods-services/ISOstore/store.html)

ISO 11064-2 Ergonomic Design of Control Centers—Part 2: Principles for the Arrangement of Control Suites. (2000). International Organization for Standardization (http://www.iso.ch/ iso/en/prods-services/ISOstore/store.html)

ISO 11064-3 Ergonomic Design of Control Centers—Part 3: Control Room Layout. (1999). International Organization for Standardization (http://www.iso.ch/iso/en/prods-services/ISOstore/store.html)

ISA RP 60.3 Human Engineering for Control Centers. (1985). Instrumentation, Systems, and Automation Society (http://www.isa.org)

BS 7517 Nuclear Power Plants—Control Rooms—Operator Controls. (1995). British Standard (http://www.bsi-global.com/index.xalter)

IEC 61227 Nuclear Power Plants—Control Rooms—Operator Controls. (1993). International Electrotechnical Commission (http://www.iec.ch/)

IEC 60964 Design for Control Rooms of Nuclear Power Plants. (1989). International Electrotechnical Commission (http://www.iec.ch/)

Aircraft Controls and Displays

SAE ARP 4102 Flight Deck Panels, Controls, and Displays. (1988). Society of Automotive Engineers (http://www.sae.org/servlets/index)

SAE ARP 4102/7 Flight Deck Panels, Controls, and Displays, Part 7: Electronic Display Symbology for EADI/PFD. (1993). Society of Automotive Engineers (http://www. sae.org/servlets/index)

SAE ARP 4102/8 Flight Deck Panels, Controls, and Displays, Part 8: Flight Deck Head-Up Displays. (1998). Society of Automotive Engineers (http://www.sae.org/servlets/index)

SAE ARD 50016 Head-Up Display Human Factors Issues. (1998). Society of Automotive Engineers (http://www.sae.org/servlets/index)

SAE ARP 4032 Human Engineering Considerations in the Application of Color to Electronic Aircraft Displays. (1988). Society of Automotive Engineers (http://www.sae.org/servlets/index)

Control and Display Design

ISO 9355-1 Ergonomic Requirements for the Design of Displays and Control Actuators—Part 1: Human Interactions with Displays and Control Actuators. (1999). International Organization for Standardization (http://www.iso.ch/iso/en/prods-services/ ISOstore/store.html)

ISO 9355-2 Ergonomic Requirements for the Design of Signals and Control Actuators—Part 2: Displays. (1999). International Organization for Standardization (http://www.iso.ch/iso/en/prods-services/ISOstore/store.html)

SAE J107 Operator Controls and Displays on Motorcycles. (1996). Society of Automotive Engineers (http://www.sae.org/servlets/index)

SAE J680 Location and Operation of Instruments and Controls in Motor Truck Cabs. (1988). Society of Automotive Engineers (http://www.sae.org/servlets/index)

Office Ergonomics

BSR/HFES100 Human Factors Engineering of Computer Workstations (Draft standard for trial use). (2002). Human Factors and Ergonomics Society, Santa Monica, CA. [NATIONAL] (http://www.hfes.org/publications/HFES100.html)

ISO 9241 Usability Standards, International Organization for Standardization. Brief description reported in http://www.ergoweb.com/resources/reference/guidelines/iso 9241.cfm

Part 1: General Introduction contains general information about the standard and provides an overview of each of the parts.

Part 2: Task Requirements discusses the enhancement of user interface efficiency and the well-being of users by applying practical ergonomic knowledge to the design of VDT work tasks.

Part 3: Display Requirements specifies requirements for visual displays and their images.

Part 4: Keyboard Requirements specifies the characteristics that determine the effectiveness in accepting keystrokes from a user.

Part 5: Workstation Requirements specifies the design characteristics of workplaces in which VDTs are used.

Part 6: Environmental Requirements specifies characteristics of the working environment in which VDTs are used.

Part 7: Display requirements with reflections describe how to maintain usable and acceptable VDT image quality by evaluating the reflection properties of a screen and the image quality of the screen over a range of typical office lighting conditions.

Part 8: Requirements for displayed color states specifications for display color images, color measurement metrics, and visual perception tests.

Part 9: Requirements for non-keyboard-input devices specifies requirements for the design and usability of input devices other than keyboards.

Part 10: Dialogue Principles specifies a set of high-level dialogue design principles for command languages, direct manipulation, and form-based entries.

Part 11: Guidance on Usability explains the way in which the user, equipment, task, and environment should be described—as Part of the total system—and how usability can be specified and evaluated.

Part 12: Presentation of Information specifies requirements for the coding and formatting of information on computer screens.

Part 13: User Guidance specifies requirements and attributes to be considered in the design and evaluation of the software user interfaces.

Part 14: Menu Dialogues provides conditional requirements and recommendations for menus in user–computer dialogues.

Part 15: Command Dialogues provides conditional recommendations for common languages.

Part 16: Direct Manipulation Dialogues provides guidance on the design of manipulation dialogues in which the user directly acts upon object or object representations (icons) to be manipulated.

Part 14: Menu Dialogues

Part 15: Command Dialogues

Part 16: Direct Manipulation Dialogues

Part 17: Form-Filling Dialogues

BIFMA G1-2002 Ergonomics Guideline for VDT (Visual Display Terminal) Furniture Used in Office Work Spaces, The Business and Institutional Furniture Manufacturers Association [LOCAL/ORGANIZATIONAL] (http://www.bifma.org/standards/index.html)

CSA Z412-00 Guideline on Office Ergonomics, The Canadian Standards Association (CSA). [NATIONAL] (http://www.csa-intl.org/onlinestore)

ACGIH 9331 Ergonomics in Computerized Offices. (1992). American Conference of Governmental Industrial Hygienists (http://www.acgih.org/home.htm)

ACGIH 99-036 Visual Ergonomics in the Workplace. (1998). American Conference of Governmental Industrial Hygienists (http://www.acgih.org/home.htm)

Human–System Interaction

TC 159/SC 4 Ergonomics of human-system interaction, International Organization for Standardization (http://www.iso.org/iso/en/CatalogueListPage.CatalogueList?COMMID=3916&scopelist=)

ISO 1503:1977 Geometrical orientation and directions of movements

ISO 9241-1:1997 Ergonomic requirements for office work with visual display terminals (VDTs)—Part 1: General introduction

ISO 9241-1:1997/Amd 1: 2001

ISO 9241-2:1992 Ergonomic requirements for office work with visual display terminals (VDTs)—Part 2: Guidance on task requirements

ISO 9241-3:1992 Ergonomic requirements for office work with visual display terminals (VDTs)—Part 3: Visual display requirements

ISO 9241-3:1992/Amd 1: 2000.

ISO 9241-4:1998 Ergonomic requirements for office work with visual display terminals (VDTs)—Part 4: Keyboard requirements

ISO 9241-4:1998/Cor 1:2000

ISO 9241-5:1998 Ergonomic requirements for office work with visual display terminals (VDTs)—Part 5: Workstation layout and postural requirements

ISO 9241-6:1999 Ergonomic requirements for office work with visual display terminals (VDTs)—Part 6: Guidance on the work environment

ISO 9241-7:1998 Ergonomic requirements for office work with visual display terminals (VDTs)—Part 7: Requirements for display with reflections

ISO 9241-8:1997 Ergonomic requirements for office work with visual display terminals (VDTs)—Part 8: Requirements for displayed colors

ISO 9241-9:2000 Ergonomic requirements for office work with visual display terminals (VDTs)—Part 9: Requirements for non-keyboard input devices

ISO 9241-10:1996 Ergonomic requirements for office work with visual display terminals (VDTs)—Part 10: Dialogue principles

ISO 9241-11:1998 Ergonomic requirements for office work with visual display terminals (VDTs)—Part 11: Guidance on usability

ISO 9241-12:1998 Ergonomic requirements for office work with visual display terminals (VDTs)—Part 12: Presentation of information

ISO 9241-13:1998 Ergonomic requirements for office work with visual display terminals (VDTs)—Part 13: User guidance

ISO 9241-14:1997 Ergonomic requirements for office work with visual display terminals (VDTs)—Part 14: Menu dialogues

ISO 9241-15:1997 Ergonomic requirements for office work with visual display terminals (VDTs)—Part 15: Command dialogues

ISO 9241-16:1999 Ergonomic requirements for office work with visual display terminals (VDTs)—Part 16: Direct manipulation dialogues

ISO 9241-17:1998 Ergonomic requirements for office work with visual display terminals (VDTs)—Part 17: Form filling dialogues

ISO 9355-1:1999 Ergonomic requirements for the design of displays and control actuators—Part 1: Human interactions with displays and control actuators

ISO 9355-2:1999 Ergonomic requirements for the design of displays and control actuators—Part 2: Displays

ISO 11064-1:2000 Ergonomic design of control centers—Part 1: Principles for the design of control centers

ISO 11064-2:2000 Ergonomic design of control centers—Part 2: Principles for the arrangement of control suites

ISO 11064-3:1999 Ergonomic design of control centers—Part 3: Control room layout

ISO 11064-3:1999/Cor 1:2002

ISO 11064-4:2004 Ergonomic design of control centers—Part 4: Layout and dimensions of workstations

ISO 13406-1:1999 Ergonomic requirements for work with visual displays based on flat panels—Part 1: Introduction

ISO 13406-2:2001 Ergonomic requirements for work with visual displays based on flat panels—Part 2: Ergonomic requirements for flat panel displays

ISO 13407:1999 Human-centered design processes for interactive systems

ISO 14915-1:2002 Software ergonomics for multimedia user interfaces—Part 1: Design principles and framework

ISO 14915-2:2003 Software ergonomics for multimedia user interfaces—Part 2: Multimedia navigation and control

ISO 14915-3:2002 Software ergonomics for multimedia user interfaces—Part 3: Media selection and combination

ISO/TS 16071:2003 Ergonomics of human-system interaction—Guidance on accessibility for human-computer interfaces

ISO/TR 16982:2002 Ergonomics of human-system interaction—Usability methods supporting human-centered design

ISO/PAS 18152:2003 Ergonomics of human-system interaction—Specification for the process assessment of human-system issues

ISO/TR 18529:2000 Ergonomics—Ergonomics of human-system interaction—Human—centered lifecycle process descriptions

CEN/TC 122 Ergonomics, European Committee for Standardization (http://www.cenorm.be/cenorm/index.htm):

EN ISO 9241-1:1997 Ergonomic requirements for office work with visual display terminals (VDTs)—Part 1: General introduction (ISO 9241-1:1997)

EN ISO 9241-2:1993 Ergonomic requirements for office work with visual display terminals (VDTs)—Part 2: Guidance on task requirements (ISO 9241-2:1992)

EN ISO 9241-3:1993 Ergonomic requirements for office work with visual display terminals (VDTs)—Part 3: Visual display requirements (ISO 9241-3:1992)

EN ISO 9241-4:1998 Ergonomic requirements for office work with visual display terminals (VDTs)—Part 4: Keyboard requirements (ISO 9241-4:1998)

EN ISO 9241-5:1999 Ergonomic requirements for office work with visual display terminals (VDTs)—Part 5: Workstation layout and postural requirements (ISO 9241-5:1998)

EN ISO 9241-6:1999 Ergonomic requirements for office work with visual display terminals (VDTs)—Part 6: Guidance on the work environment (ISO 9241-6:1999)

EN ISO 9241-7:1998 Ergonomic requirements for office work with visual display terminals (VDTs)—Part 7: Requirements for display with reflections (ISO 9241-7:1998)

EN ISO 9241-8:1997 Ergonomic requirements for office work with visual display terminals (VDTs)—Part 8: Requirements for displayed colours (ISO 9241-8:1997)

EN ISO 9241-9:2000 Ergonomic requirements for office work with visual display terminals (VDTs)—Part 9: Requirements for non-keyboard input devices (ISO 9241-9:2000)

EN ISO 9241-10:1996 Ergonomic requirements for office work with visual display terminals (VDTs)—Part 10: Dialogue principles (ISO 9241-10:1996)

Physical Environment

TC 159/SC 5 Ergonomics of the physical environment, International Organization for Standardization (http://www.iso.org/iso/en/CatalogueListPage.CatalogueList?COMMID=3916&scopelist=)

ISO 7243:1989 Hot environments—Estimation of the heat stress on working man, based on the WBGT-index (wet bulb globe temperature)

ISO 7726:1998 Ergonomics of the thermal environment—Instruments for measuring physical quantities

ISO 7730:1994 Moderate thermal environments—Determination of the PMV and PPD indices and specification of the conditions for thermal comfort

ISO 7731:2003 Ergonomics—Danger signals for public and work areas—Auditory danger signals

ISO 7933:2004 Ergonomics of the thermal environment—Analytical determination and interpretation of heat stress using calculation of the predicted heat strain

ISO 8996:2004 Ergonomics of the thermal environment—Determination of metabolic rate

ISO 9886:2004 Ergonomics—Evaluation of thermal strain by physiological measurements

ISO 9920:1995 Ergonomics of the thermal environment—Estimation of the thermal insulation and evaporative resistance of a clothing ensemble

ISO 9921:2003 Ergonomics—Assessment of speech communication

ISO 10551:1995 Ergonomics of the thermal environment—Assessment of the influence of the thermal environment using subjective judgment scales

ISO/TR 11079:1993 Evaluation of cold environments—Determination of required clothing insulation (IREC)

ISO 11399:1995 Ergonomics of the thermal environment—Principles and application of relevant International Standards

ISO 11428:1996 Ergonomics—Visual danger signals—General requirements, design and testing

ISO 11429:1996 Ergonomics—System of auditory and visual danger and information signals

ISO 12894:2001 Ergonomics of the thermal environment—Medical supervision of individuals exposed to extreme hot or cold environments

ISO 13731:2001 Ergonomics of the thermal environment—Vocabulary and symbols

ISO/TS 13732-2:2001 Ergonomics of the thermal environment—Methods for the assessment of human responses to contact with surfaces—Part 2: Human contact with surfaces at moderate temperature

ISO 15265:2004 Ergonomics of the thermal environment—Risk assessment strategy for the prevention of stress or discomfort in thermal working conditions

ISO/TR 19358:2002 Ergonomics—Construction and application of tests for speech technology

CEN/TC 122 Ergonomics, European Committee for Standardization (http://www.cenorm.be/cenorm/index.htm):

EN ISO 7726:2001 Ergonomics of the thermal environment—Instruments for measuring physical quantities (ISO 7726:1998)

EN ISO 7730:1995 Moderate thermal environments—Determination of the PMV and PPE indices and specification of the conditions for thermal comfort (ISO 7730: 1994)

Information Technology/Software Engineering

ICS 35 Information technology. Office machines http://www.iso.org/iso/en/CatalogueListPage.CatalogueList?ICS1=35&scopelist

ISO/IEC 11581-1:2000 Information technology—User system interfaces and symbols—Icon symbols and functions Part 1: Icons—General

ISO/IEC 11581-2:2000 Information technology—User system interfaces and symbols—Icon symbols and functions Part 2: Object icons

ISO/IEC 11581-3:2000 Information technology—User system interfaces and symbols—Icon symbols and functions Part 3: Pointer icons

ISO/IEC 11581-5:2004 Information technology—User system interfaces and symbols—Icon symbols and functions Part 5: Tool icons

ISO/IEC 11581-6:1999 Information technology—User system interfaces and symbols—Icon symbols and functions Part 6: Action icons

NASA-STD-8719.13A Software Safety Standard. (1997). National Aeronautics and Space Administration (http://satc.gsfc.nasa.gov/assure/nss8719_13.html)

JTC 1 / SC 7 Software and system engineering (http://www.iso.org/iso/en/stdsdevelopment/tc/tclist/TechnicalCommitteeDetailPage.TechnicalCommitteeDetail?COMMID=40)

ISO/IEC 9126-1:2001 Software engineering—Product quality—Part 1: Quality model

ISO/IEC TR 9126-2:2003 Software engineering—Product quality—Part 2: External metrics

ISO/IEC TR 9126-3:2003 Software engineering—Product quality—Part 3: Internal metrics

ISO/IEC TR 9126-4:2004 Software engineering—Product quality—Part 4: Quality in use metrics

Quality and Environmental Management Standards

ISO 9000. Quality management and related standard series: (http://www.iso.org/iso/en/iso 9000-14000/iso9000/iso9000index.html)

ISO 9000:2000. Quality management systems—Fundamentals and vocabulary

ISO 9001:2000. Quality management systems—Requirements

ISO 9004:2000. Quality management systems—Guidelines for performance improvements

ISO 19011. Guidelines on Quality and/or Environmental Management Systems Auditing (currently under development)

ISO 10005:1995. Quality management—Guidelines for quality plans

ISO 10006:1997. Quality management—Guidelines to quality in project management

ISO 10007:1995. Quality management—Guidelines for configuration management

ISO/DIS 10012. Quality assurance requirements for measuring equipment—Part 1: Metrological confirmation system for measuring equipment

ISO 10012-2:1997. Quality assurance for measuring equipment—Part 2: Guidelines for control of measurement of processes

ISO 10013:1995. Guidelines for developing quality manuals

ISO/TR 10014:1998. Guidelines for managing the economics of quality

ISO 10015:1999. Quality management—Guidelines for training

ISO/TS 16949:1999. Quality systems—Automotive suppliers

ISO 14000 Environment Management Standard series: (http://www.iso.org/iso/en/iso9000-14000/iso14000/iso14000index.html)

ISO 14040 series: Life cycle assessment

ISO 14062: Design for environment

ISO 14020 series: Environmental labels and declarations

ISO 14063: Environmental communication

ISO 19011: Environmental management systems auditing

Occupational Safety and Health Standards

ILO-OSH 2001—Principles for occupational safety and health management system, International Labour Organization (http://www.ilo.org/public/english/protection/safework/managmnt/guide.htm)

AFMA Voluntary Ergonomics Guideline for the Furniture Manufacturing Industry, American Furniture Manufactures Association (AFMA, 2003) (http://www.afma4u.org)

OSHA Voluntary Ergonomics Standards (http://www.osha.gov/pls/publications/pubindex.list)

1. Nursing Home Guideline (issued on March 13, 2003)
2. Draft Guideline for Poultry Processing (issued on June 3, 2003)
3. Guideline for the Retail Grocery Industry (issued on May 28, 2004)

AIHA ASC Z10 Occupational Health Safety Systems, American Industrial Hygiene Association (http://www.aiha.org/ANSICommittees/html/z10committee.htm)

ASC Z-365 Management of Work-Related Musculoskeletal Disorders (MSD). (2002, final draft). (http://www.nsc.org/ehc/z365/finldrft.htm)

ACGIH 99-049 Ergonomics and Safety in Hand Tool Design. (1999). American Conference of Governmental Industrial Hygienists (http://www.acgih.org/home.htm)

Machinery Safety

CEN/TC 122 Ergonomics, European Committee for Standardization (http://www.cenorm.be/cenorm/index.htm):

EN 457:1992. Safety of machinery—Auditory danger signals—General requirements, design and testing (ISO 7731:1986 modified)

EN 547-1:1996. Safety of machinery—Human body measurements—Part 1: Principles for determining the dimensions required for openings for whole body access into machinery

EN 547-2:1996. Safety of machinery—Human body measurements—Part 2: Principles for determining the dimensions required for access openings

EN 547-3:1996. Safety of machinery—Human body measurements—Part 3: Anthropometric data

EN-563 1994. Safety of machinery—Temperature of touchable surfaces—Ergonomics data to establish temperature limit values for hot surfaces

EN 614-1:1995. Safety of machinery—Ergonomic design principles—Part 1: Terminology and general principles

EN 614-2:2000. Safety of machinery—Ergonomic design principles—Part 2: Interactions between the design of machinery and work tasks

EN 842:1996. Safety of machinery—Visual danger signals—General design requirements, design and testing

EN 894-1:1997. Safety of machinery—Ergonomics requirements for the design of displays and control actuators—Part 1: General principles for human interactions with displays and control actuators

EN 894-2:1997. Safety of machinery—Ergonomics requirements for the design of displays and control actuators—Part 2: Displays

EN 894-3:2000. Safety of machinery—Ergonomics requirements for the design of displays and control actuators—Part 3: Control actuators

EN 981:1997. Safety of machinery—System of auditory and visual danger and information signals

EN 1005-1:2001. Safety of machinery—Human physical performance—Part 1: Terms and definitions

EN 1005-2. Safety of machinery—Human physical performance—Part 2: Manual handling of machinery and component parts of machinery

EN 1005-3:2002. Safety of machinery—Human physical performance—Part 3: Recommended force limits for machinery operation

Transportation

TC 22/SC 13 Ergonomics applicable to road vehicles (http://www.iso.org/iso/en/CatalogueList Page. CatalogueList?COMMID=869&scopelist=CATALOGUE)

ISO 2575:2004. Road vehicles—Symbols for controls, indicators and tell-tales

ISO 3409:1975. Passenger cars—Lateral spacing of foot controls

ISO 3958:1996. Passenger cars—Driver hand-control reach

ISO 4040:2001. Road vehicles—Location of hand controls, indicators and tell-tales in motor vehicles

ISO 6549:1999. Road vehicles—Procedure for H- and R-point determination

ISO/TR 9511:1991. Road vehicles—Driver hand-control reach—In-vehicle checking procedure

ISO/TS 12104:2003. Road vehicles—Gearshift patterns—Manual transmissions with power-assisted gear change and automatic transmissions with manual-gearshift mode

ISO 12214:2002. Road vehicles—Direction-of-motion stereotypes for automotive hand controls

ISO 15005:2002. Road vehicles—Ergonomic aspects of transport information and control systems—Dialogue management principles and compliance procedures

ISO 15006:2004. Road vehicles—Ergonomic aspects of transport information and control systems—Specifications and compliance procedures for in-vehicle auditory presentation

ISO 15007-1:2002. Road vehicles—Measurement of driver visual behaviour with respect to transport information and control systems—Part 1: Definitions and parameters

ISO/TS 15007-2:2001. Road vehicles—Measurement of driver visual behaviour with respect to transport information and control systems—Part 2: Equipment and procedures

ISO 15008:2003. Road vehicles—Ergonomic aspects of transport information and control systems—Specifications and compliance procedures for in-vehicle visual presentation

ISO/TS 16951:2004. Road vehicles—Ergonomic aspects of transport information and control systems (TICS)—Procedures for determining priority of on-board messages presented to drivers

ISO 17287:2003. Road vehicles—Ergonomic aspects of transport information and control systems—Procedure for assessing suitability for use while driving

TC 20/SC 14 Space systems and operations (http://www.iso.org/iso/en/CatalogueListPage. CatalogueList?COMMID=739&scopelist=CATALOGUE)

ISO 14620-1:2002. Space systems—Safety requirements—Part 1: System safety

ISO 14620-2:2000. Space systems—Safety requirements—Part 2: Launch site operations

ISO 17399:2003. Space systems—Man-systems integration

ISO 17666:2003. Space systems—Risk management

TC 23/SC 3 Tractor Safety and comfort of the operator (http://www.iso.org/iso/en/Catalogue ListPage. CatalogueList?COMMID=938&scopelist=CATALOGUE)

ISO 3463:1989. Wheeled tractors for agriculture and forestry—Protective structures—Dynamic test method and acceptance conditions

ISO 3463:1989/Amd 1:1998

ISO 3776:1989. Tractors for agriculture—Seat belt anchorages

ISO 4254-1:1989. Tractors and machinery for agriculture and forestry—Technical means for ensuring safety—Part 1: General

ISO 4254-1:1989/Amd 1:1998

ISO 4254-2:1986. Tractors and machinery for agriculture and forestry—Technical means for providing safety—Part 2: Anhydrous ammonia applicators

ISO 4254-5:1992. Tractors and machinery for agriculture and forestry—Technical means for ensuring safety—Part 5: Power-driven soil-working equipment

ISO 4254-9:1992. Tractors and machinery for agriculture and forestry—Technical means for ensuring safety—Part 9: Equipment for sowing, planting and distributing fertilizers

ISO 5700:1989. Wheeled tractors for agriculture and forestry—Protective structures—Static test method and acceptance conditions

ISO 5700:1989/Amd 1:1998

ISO 12140:1998. Agricultural machinery—Agricultural trailers and trailed equipment—Drawbar jacks

ISO/TS 15077:2002. Tractors and self-propelled machinery for agriculture and forestry—Operator controls—Actuating forces, displacement, location and method of operation

TC 23/SC 14 Tractor Operator controls, operator symbols and other displays, operator manuals (http://www.iso.org/iso/en/CatalogueListPage. CatalogueList?COMMID=991&scopelist=CATALOGUE)

ISO 3600:1996. Tractors, machinery for agriculture and forestry, powered lawn and garden equipment—Operator's manuals—Content and presentation

ISO 3767-1:1998. Tractors, machinery for agriculture and forestry, powered lawn and garden equipment—Symbols for operator controls and other displays—Part 1: Common symbols

ISO 3767-2:1991. Tractors, machinery for agriculture and forestry, powered lawn and garden equipment—Symbols for operator controls and other displays—Part 2: Symbols for agricultural tractors and machinery

ISO 3767-2:1991/Amd 1:1995

ISO 3767-2:1991/Amd 2:1998. Additional symbols

ISO 3767-2:1991/Amd 3:2000

ISO 3767-3:1995. Tractors, machinery for agriculture and forestry, powered lawn and garden equipment—Symbols for operator controls and other displays—Part 3: Symbols for powered lawn and garden equipment

ISO 3767-4:1993. Tractors, machinery for agriculture and forestry, powered lawn and garden equipment—Symbols for operator controls and other displays—Part 4: Symbols for forestry machinery

ISO 3767-4:1993/Amd 1:2000. Additional symbols

ISO 3767-5:1992. Tractors, machinery for agriculture and forestry, powered lawn and garden equipment—Symbols for operator controls and other displays—Part 5: Symbols for manual portable forestry machinery

ISO 3767-5:1992/Amd 1:2001. Revised and additional symbols

ISO 11684:1995. Tractors, machinery for agriculture and forestry, powered lawn and garden equipment—Safety signs and hazard pictorials—General principles

TC 110/SC 2 Safety of powered industrial trucks (http://www.iso.org/iso/en/CatalogueList Page. CatalogueList?COMMID=3090&scopelist=CATALOGUE)

ISO 3287:1999. Powered industrial trucks—Symbols for operator controls and other displays

ISO 15870:2000. Powered industrial trucks—Safety signs and hazard pictorials—General principles

Manual Materials Handling

ISO 780:1997 Packaging—Pictorial marking for handling of goods (http://www. iso.org/iso/en/CatalogueList Page. CatalogueList?COMMID=3293&scopelist=CATALOGUE)

ISO 11228-1:2003. Manual handling—Part 1: Lifting and carrying (http://www.iso.org/iso/en/CatalogueDetailPage. CatalogueDetail?CSNUMBER=26520&ICS1=13&ICS2=180&ICS3=)

ISO/CD 11228-2 Manual handling—Part 2: Pushing and Pulling (Draft—not available)

ISO/CD 11228-3 Manual Handling—Part 3: Handling, at High Repetition, of Low Loads (Draft—not available)

ISO/TS 20646-1:2004. Ergonomic procedures for the improvement of local muscular workloads—Part 1: Guidelines for reducing local muscular workloads (http://www.iso.org/iso/en/CatalogueDetailPage. CatalogueDetail? CSNUMBER=35501&ICS1=13&ICS2=180&ICS3=)

NASA-STD-8719.9 Standard for Lifting Devices and Equipment. (2002). National Aeronautics and Space Administration (http://www.hq.nasa.gov/office/codeq/doctree/87199.pdf)

Guidelines for Elderly/Disabled Users

ANSI A117.1 Guidelines for Accessible and Usable Buildings and Facilities. (1998). American National Standards Institute (http://www.ansi.org/)

BS 4467 Guide to Dimensions in Designing for Elderly People. (1991). British Standard (http://www.bsi-global.com/index.xalter)

Ground Vehicle Standards

SAE J2364: Navigation and Route Guidance Function Accessibility While Driving, Society of Automotive Engineers (http://www.sae.org/servlets/productDetail?PROD_TYP=STD&PROD_CD=J2364_200408)

SAE J2365: Calculation of the Time to Complete In-Vehicle Navigation and Route Guidance Tasks, Society of Automotive Engineers (http://www.sae.org/servlets/productDetail?PROD_TYP=STD&PROD_CD=J2365_200205)

SAE J1050 Describing and Measuring the Driver's Field of View. (1994). Society of Automotive Engineers (http://www.sae.org/servlets/index)

SAE J941 Motor Vehicle Driver's Eye Locations. (1997). Society of Automotive Engineers (http://www.sae.org/servlets/index, http://www.ansi.org/)

ISO 7397-1 Verification of Driver's Direct Field of View—Part 1: Vehicle Positioning for Static Measurement. (1993). International Organization for Standardization (http://www.iso.ch/iso/en/prods-services/ISOstore/store.html)

ISO 7397-2 Verification of Driver's Direct Field of View—Part 2: Test Method. (1993). International Organization for Standardization (http://www.iso.ch/iso/en/prods-services/ISOstore/store.html)

Furniture Standards

ISO 5970 Chairs and Tables for Educational Institutions—Functional Sizes (1979), International Organization for Standardization (http://www.iso.ch/iso/en/prods-services/ISOstore/store.html)

BS 3044 Guide to Ergonomics Principles in the Design and Selection of Office Furniture. (1990). British Standard (http://www.bsi-global.com/index.xalter)

Human Error

ACGIH 9658 Human Error Reduction and Safety Management. (1996). American Conference of Governmental Industrial Hygienists (http://www.acgih.org/home.htm)

AICHE G15 Guidelines for Preventing Human Error in Process Safety. (1994). American Institute of Chemical Engineers (http://www.aiche.org/)

API 770 A Manager's Guide to Reducing Human Errors: Improving Human Performance in the Process. (2001). American Petroleum Institute (http://api-ec.api.org/newsplashpage/index.cfm/)

Guidelines for Medical Devices

AAMI HE48 Human Factors Engineering Guidelines and Preferred Practices for the Design of Medical Devices. (1993). Association for the Advancement of Medical Instrumentation (http://www.aami.org/)

ANSI/AAMI HE74 Human Factors Design Process for Medical Devices. (2001). Association for the Advancement of Medical Instrumentation (http://www.aami.org/)

Application Guidelines for Nuclear Power

ANSI/IEEE STD1023 Guide for the Application of Human Factors Engineering to Systems, Equipment, and Facilities of Nuclear Power Generating Stations. (1988). American National Standards Institute (http://www.ansi.org/)

Guidelines for Human Performance and Reliability

AIAA G-035 Guide to Human Performance Measurements. (2000). American Institute of Aeronautics and Astronautics (http://www.aiaa.org/)

ACGIH 9651 Evaluation of Human Work, A Practical Ergonomics Methodology. (1995). American Conference of Governmental Industrial Hygienists (http://www.acgih.org/home.htm)

IEEE 1082 Guide for Incorporating Human Action Reliability for Nuclear Power Generating Stations. (1997). Institute of Electrical and Electronics Engineers (http://www.ieee.org/portal/index.jsp)

Robot Design and Safety

ANSI/RIA R15.02-1 Industrial Robots and Robot Systems—Hand-Held Robot Control Pendants—Human Engineering Design Criteria (1990). American National Standards Institute (http://www.ansi.org/, http://www.roboticsonline.com/store/)

ANSI/RIA R15.06 Industrial Robots and Robot Systems—Safety Requirements. (1999). American National Standards Institute (http://www.ansi.org/, http://www.roboticsonline.com/store/)

User Interface Standards/Guidelines

IBM Web Design Guidelines (http://www-3.ibm.com/ibm/easy/eou_ext.nsf/publish/572)

IBM User Interface Architecture Administrative Guidelines (http://www-3.ibm.com/ibm/easy/eou_ext.nsf/publish/1392/$File/ibm_uia.pdf)

IBM OOBE Usability Guidelines (http://www-3.ibm.com/ibm/easy/eou_ext.nsf/publish/577)

GNOME Foundation Human Interface Guidelines. (2004). (http://developer.gnome.org/projects/gup/hig/2.0/)

ESD-TR-86-278 Guidelines for Designing User Interface Software. (1986). The MITRE Corporation, Bedford, MA (http://hcibib.org/sam/)

Apple Human Interface Guidelines (http://developer.apple.com/documentation/User Experience/Conceptual/OSXHIGuidelines/)

GUI Standard by Human Factors International, Inc. (Local/Organizational) (http://www.humanfactors.com/downloads/GUIbooklet.asp)

Web Site Design Standards/Guidelines

36 CFR Part 1194 Electronic and Information Technology Accessibility Standards. (2000). Architectural and Transportation Barriers Compliance Board (http://www.access-board.gov/sec508/508standards.htm)

Research-Based Web Design and Usability Guidelines. (2003). The U.S. Department of Health and Human Services (HHS) and National Cancer Institute (http://usability.gov/pdfs/guidelines.html)

Web Style Guide, 2nd edition (http://www.webstyleguide.com/index.html)

Telstra Online Standards (http://www.telstra.com.au/standards/standards/standards_all.cfm)

W3C User Agent Accessibility Guidelines. (2002). (http://www.w3.org/TR/UAAG10/)

Internet Standard by Human Factors International, Inc. (Local/Organizational) (http://www.humanfactors.com/downloads/Intranetbooklet.asp)

U.S. Department of Defense Standards

DOD-HDBK-743A Anthropometry of U.S. Military Personnel. (1991). (http: assist.daps.dla.mil/docimages/0000/40/29/54083.pd0)

U.S. Department of Transportation Standards—Federal Aviation Administration

HF-STD-001 Human Factors Design Standard. (2003). Department of Transportation, Federal Aviation Administration (http://www.hf.faa.gov/docs/508/docs/wjhtc/hfds.zip)

DOT-VNTSC-FAA-95-3 Human Factors in the Design and Evaluation of Air Traffic Control Systems. (1995). Department of Transportation, Federal Aviation Administration (http://www.hf.faa.gov/docs/volpehndk.zip)

FAA-HF-001 Human Engineering Program Plan. (1999). Department of Transportation, Federal Aviation Administration (http://www.hf.faa.gov/docs/did_001.htm)

FAA-HF-002 Human Engineering Design Approach Document—Operator. (1999). Department of Transportation, Federal Aviation Administration (http://www.hf.faa.gov/docs/did_002.htm)

FAA-HF-003 Human Engineering Design Approach Document—Maintainer (1999). Department of Transportation, Federal Aviation Administration (http://www.hf.faa.gov/docs/did_003.htm)

FAA-HF-004 Critical Task Analysis Report (2000). Department of Transportation, Federal Aviation Administration (http://hfetag.dtic.mil/docs-hfs/faa-hf-004_critical_task_ analysis_report.doc)

FAA-HF-005 Human Engineering Simulation Concept. (2000). Department of Transportation, Federal Aviation Administration (http://hfetag.dtic.mil/docs-hfs/faa-hf-005_human-engineering_simulation.doc)

U.S. Department of Transportation Standards—Federal Highway Agency

FHWA-JPO-99-042 Preliminary Human Factors Guidelines for Traffic Management Centers. (1999). Department of Transportation, Federal Highway Agency, (http://plan2op.fhwa.dot.gov/pdfs/pdf2/edl10303.pdf)

FHWA-RD-98-057 Human Factors Design Guidelines for Advanced Traveler Information Systems (ATIS) and Commercial Vehicle Operations (CVO). (1998). Department of Transportation, Federal Highway Agency (http://www.fhwa.dot.gov/tfhrc/safety/pubs/atis/index.html)

FHWA-RD-01-051 Guidelines and Recommendations to Accommodate Older Drivers and Pedestrians. (2001). Department of Transportation, Federal Highway Agency (http://www.tfhrc.gov/humanfac/01105/cover.htm)

FHWA-RD-01-103 Highway Design Handbook for Older Drivers and Pedestrians. (2001). Department of Transportation, Federal Highway Agency (http://www.tfhrc.gov/humanfac/01103/coverfront.htm)

U.S. Department of Energy Standard

DOE-HDBK-1140-2001 Human Factors/Ergonomics Handbook for the Design for Ease of Maintenance. (2001). Department of energy (http://tis.eh.doe.gov/techstds/standard/hdbk1140/hdbk11402001_part1.pdf)

Military Standards and Related Documentations

MIL-STD-882D Standard Practice for System Safety. (2000). (http://assist.daps.dla.mil/docimages/0001/95/78/std882d.pd8)

MIL-STD-1472F Human Engineering. (1999). (http://assist.daps.dla.mil/docimages/0001/87/31/milstd14.pd1)

MIL-STD-1474D Noise Limits. (1997). (http://assist.daps.dla.mil/docimages/0000/31/59/1474d.pd1)

MIL-STD-1477C Symbols for Army Systems Displays. (1996). (http://assist.daps.dla.mil/docimages/0000/42/03/69268.pd9)

MIL-STD-1787C Aircraft Display Symbology. (2001). (not available in public domain)

MIL-HDBK-759C Human Engineering Design Guidelines. (1995). (http://assist.daps.dla.mil/docimages/0000/40/04/mh759c.pd8)

MIL-HDBK-767 Design Guidance for Interior Noise Reduction in Light-Armored Tracked Vehicles. (1993). (http://assist.daps.dla.mil/docimages/0000/13/24/767.pd1)

MIL-HDBK-1473A Color and Marking of Army Materiel. (1997). (http://assist.daps.dla.mil/docimages/0000/85/40/hdbk1473.pd6)

MIL-HDBK-1908B Definitions of Human Factors Terms. (1999). (http://assist.daps.dla.mil/docimages/0001/81/33/1908hdbk.pd9)

MIL-HDBK-46855 Human Engineering Requirements for Military Systems Equipment and Facilities (not available in public domain)

National Aeronautics and Space Administration (NASA) Man-System Integration Standard

NASA-STD-3000B Man-Systems Integration Standards. (1995). National Aeronautics and Space Administration (http://msis.jsc.nasa.gov)

Federal Accessibility Standard

FED-STD-795 Uniform Federal Accessibility Standards. (1988). (http://assist.daps.dla.mil/docimages/0000/46/05/53835.pd5)

REFERENCES

Akoumianakis, D., Stephanidis, C. (1997). Supporting user-adapted interface design: The USE-IT system. *Interacting with Computers, 9*, 73–104.

Albin, T. J. (2004). Board of Standards Review/Human Factor and Ergonomics Society 100—Human Factors Engineering of Computer Workstations—Draft Standard for Trial Use. In W. Karwowski (Ed.), *2005, Handbook of human factors and ergonomics standards and guidelines*. Hillsdale, NJ: Lawrence Erlbaum Publishers.

American Furniture Manufactures Association. (2003). *Voluntary Ergonomics Guideline for the furniture manufacturing industry*. High Point, NC: AFMA.

Anshel, J. (1998). *Visual ergonomics in the workplace*. Taylor & Francis: London, UK.

Babakri, K. A., Bennett, R. A., and Franchetti, M. (2003). Critical factors for implementing ISO 14001 standard. in United States industrial companies. *Journal of Cleaner Production, 11*, 749–752.

Babakri, K. A., Bennett, R. A., Rao, S., & Franchetti, M. (2004). Recycling performance of firms before and after adoption of the ISO 14001 standard. *Journal of Cleaner Production, 12*, 633–637.

Baleani, M., Cristofolini, L., & Viceconi, M. (1999). Endurance testing of hip prostheses: A comparison between the load fixed in ISO 7206 standard and the physiological loads. *Clinical Biomechanics, 14*, 339–345.

Barre, F., & Lopez, J. (2001). On a 3D extension of the MOTIF method (ISO 12085). *International Journal of Machine Tools & Manufacture, 41*, 1873–1880.

Bastien, J. M. C., Scapin, D. L., & Leulier, C. (1999). The ergonomic criteria and the ISO/DIS 9241-10 dialogue principles: A pilot comparison in an evaluation task. *Interacting with Computers, 11*, 299–322.

Berlage, T. (1995). OSF/Motif as a user interface standard. *Computer Standards and Interface, 17*, 99–106.

Besuijen, K., & Spendelink, G. P. J. (1998). Standardizing visual display quality. *Displays, 19,* 67–76.

Button, K., Clarke, A., Palubinskas, G., Stough, R., & Thibault, M. (2004). Conforming with ICAO safety oversight standards. *Journal of Air Transport Management, 10*(4), 249–255.

Cakir, A., & Dzida, W. (1997). International ergonomic HCI standards. In M. Helander, T. K. Landauer, & P. Prabhu (Eds.). *Handbook of human-computer interaction* (pp. 407–420). The Netherlands: Elsevier.

Calonius, O., & Saikko, V. (2002). Slide track analysis of eight contemporary hip simulator designs. *Journal of Biomechanics, 35,* 1439–1450.

Carson, B. E. Sr., Alper, M., Barrett, C. B., & Brink, K. (2004). ISO 9001:2000—A quality management system (QMS) to make your IVF center better. *Fertility and Sterility, 82*(Suppl. 2), S190–S191.

CEN. (2004). European Standardization Committee website. http://www.cenorm.be/cenorm/index.htm

Chapanis, A. (1996). *Human factors in systems engineering.* New York: Wiley.

Cho, D. S., Kim, J. H., Choi, T. M., Kim, B. H., & Manvell, D. (2004). Highway traffic noise prediction using method fully compliant with ISO 9613: Comparison with measurements. *Applied Acoustics, 65,* 883–892.

Chord-Auger, S., de Bouchony, E. T., Moll, M. C., Boudart, D., & Follea, G. (in press). Satisfaction survey in general hospital personnel involved in blood transfusion: implementation of the ISO 9001: 2000 standard, *Transfusion Clinique et Biologique.*

Department of Defense. (2002). Index of non-government standards on human engineering design criteria and program requirements/guidelines, http://hfetag.dtic.mil/docs/index_ngs.doc

Department of Industrial Relation. (2004). California Department of Industrial Relation homepage. http://www.dir.ca.gov/

Dickinson, C. E. (1995). Proposed manual handling international and European Standards. *Applied Ergonomics, 26*(4), 265–270.

Dowlatshahi, S., & Urias, C. (2004). An empirical study of ISO certification in the maquiladora industry. *International Journal of Production Economics, 88,* 291–306.

Dul, J., de Vlaming, P. M., & Munnik, M. J. (1996). A review of ISO and CEN standards on ergonomics. *International Journal of Industrial Ergonomics, 17*(3), 291–297.

Dul, J., de Vries, H., Verschoof, S., Eveleens, W., & Feilzer, A. (2004). Combining economic and social goals in the design of production systems by using ergonomics standards. *Computers & Industrial Engineering, 47*(2–3), 207–222.

Dzida, W. (1995). Standards for user-interfaces. *Computer Standards & Interfaces, 17*(1), 89–97.

Dzida, W. (1997). International user-interface standardization. In A. B. Tucker (Ed.)., *The computer science and engineering handbook* (pp. 1474–1493). Boca Raton, FL: CRC Press.

Earthy, J., Jones, B. S., & Bevan, N. (2001). The improvement of human-centred processes—Facing the challenge and reaping the benefit of ISO 13407. *International Journal of Human-Computer Studies, 55,* 553–585.

Eibl, M. (in press). International Standards of Interface Design. In W. Karwowski (Ed.), *2005. Handbook of Human Factors and Ergonomics Standards and Guidelines.* Hillsdale, NJ: Lawrence Erlbaum Associates.

Emam, K. E., & Birk, A. (2000). Validating the ISO/IEC 15504 measures of software development process capability. *The Journal of Systems and Software, 51,* 119–149.

Emam, K. E., & Garro, I. (2000). Estimating the extent of standards use: The case of ISO/IEC 15504. *The Journal of Systems and Software, 53,* 137–143.

Emam, K. E., & Jung, H-W. (2001). An empirical evaluation of the ISO/IEC 15504 assessment model. *The Journal of Systems and Software, 59,* 23–41.

Escanciano, C., Fernandez, E., & Vazquez, C. (2002). Linking the firm's technological status and ISO 9000 certification: Results of an empirical study. *Technovision, 22,* 509–515.

Ghisellini, A., & Thurston, D. L. (in press). Decision traps in ISO 14001 implementation process: Case study results from Illinois certified companies. *Journal of Cleaner Production.*

Gingele, J., Childe, S. J., & Miles, M. E. (2002). A modeling technique for re-engineering business processes controlled by ISO 9001. *Computers in Industry, 49,* 235–251.

Gordon, C. C., Churchill, T., Clauser, T.E., Bradtmiller, B., McConville, J.T., Tebbets, I., et al. (1989). *1988 Anthropometric survey of U.S. Army personnel: Summary statistics interim report* (Tech. Rep. NATICK/TR-89-027).

Grieco, A., Occhipinti, E., Colombini, D., & Molteni, G. (1997). Manual handling of loads: The point of view of experts involved in the application of EC Directives 90/269. *Ergonomics, 40*(10), 1035–1056.

Griefahn, B., & Brode, P. (1999). The significance of lateral whole-body vibrations related to separately and simultaneously applied vertical motions: A validation study of ISO 2631. *Applied Ergonomics, 30,* 505–513.

Griefahn, B. (2000). Limits of and possibilities to improve the IREQ cold stress model (ISO/TR 11079): A validation study in the field. *Applied Ergonomics, 31,* 423–431.

Harker, S. (1995). The development of ergonomics standards for software. *Applied Ergonomics, 26*(4), 275–279.

Hiyassat, M. A. S. (2000). Applying the ISO standards to a construction company: A case study. *International Journal of Project Management, 18,* 275–280.

Hoyle, D. (2001). *ISO 9000: Quality systems handbook*. Oxford: Butterworth Heinemann.

Human Factors and Ergonomics Society. (2002). Board of Standards Review/Human Factors and Ergonomics Society 100—*Human factors engineering of computer workstations*—Draft Standard for Trial Use. Human Factors and Ergonomics Society, Santa Monica, CA.

ILO. (2004). International Labor Organization Web site. http://www.ilo.org/public/english/index.htm

ILO-OSH. (2001). *Guidelines on occupational safety and health management systems. ILO-OSH 2001*. Geneva, International Labour Office. http://www.ilo.org/public/english/protection/safework/managmnt/guide.htm

Ishitake, T., Miyazaki, Y., Noguchi, R., Ando, H., & Matoba, T. (2002). Evaluation of frequency weighting (ISO 2631-1) for acute effects of whole-body vibration on gastric motility. *Journal of Sound and Vibration, 253*(1), 31–36.

ISO (2004). International Standardization Organization Web site. http://www.iso.org/iso/en/ISOOnline.openerpage

Jabir, & Moore, J. W. (1998). A search for fundamental principles of software engineering, Computer Standards and Interfaces, 19, 155–160.

Jung, H-W., & Hunter, R. (2001). The relationship between ISO/IEC 15504 process capability levels, ISO 9001 certification and organization size: an empirical study. *The Journal of Systems and Software, 59*, 43–55.

Kampmann, B., & Piekarski, C. (2000). The evaluation of workplaces subjected to heat stress: Can ISO 7933. (1989) adequately describe heat strain in industrial workplaces? *Applied Ergonomics, 31*, 59–71.

Karltun, J., Axelsson, J., & Eklund, J. (1998). Working conditions and effects of ISO 9000 in six furniture-making companies: implementation and processes. *Applied Ergonomics, 29*(4), 225–232.

Karwowski, W. (Ed.). (in press). *Handbook of human factors and ergonomics standards and guidelines*, Hillsdale, NJ: Lawrence Erlbaum Associates.

Kenny, D. (2001). ISO and CEN documents on quality in medical laboratories. *Clinica Chimica Acta, 309*, 121–125.

Kosanke, K., & Nell, J. G. (1999). Standardization in ISO for enterprise engineering and integration. *Computers in Industry, 40*, 311–319.

Lindermeier, R. (1994). Quality assessment of software prototypes. *Reliability Engineering & System Safety, 43*(1), 87–94.

Lindfors, M. (1998). Accuracy and repeatability of the ISO 9241–7 test method. *Displays, 19*, 3–16.

Long, J. (1996). Specifying relations between research and the design of human-computer interactions. *International Journal of Human-Computer Studies, 44*, 875–920.

MacDonald, J. P. (in press). Strategic sustainable development using the ISO 14001 standard. *Journal of Cleaner Production*.

MATRIS. (2004). Directory of Design Support Methods (DSSM). http://dtica.dtic.mil/ddsm/

McDaniel, J. W. (1996). The demise of military standards may affect ergonomics. *International Journal of Industrial Ergonomics, 18*(5–6), 339–348.

Nachreiner, F. (1995). Standards for ergonomics principles relating to the design of work systems and to mental workload. *Applied Ergonomics, 26*(4), 259–263.

Nanthavanij, S. (2000). Developing national ergonomics standards for Thai industry. *International Journal of Industrial Ergonomics, 25*(6), 699–707.

Occupational Safety and Health Administration (OSHA). (2000). *Ergonomics Program Rule* (Federal Register) 65(220).

Olesen, B. W., & Parsons, K. C. (2002). Introduction to thermal comfort standards and to the proposed new version of EN ISO 7730. *Energy and Buildings, 34*(6), 537–548.

Olesen, B. W. (1995). International standards and the ergonomics of the thermal environment. *Applied Ergonomics, 26*(4), 293–302.

OSHA. (2004). Occupational Safety and Health Administration website. http://www.osha-slc.gov

Parsons, K. (1995). Ergonomics and international standards. *Applied Ergonomics, 26*(4), 237–238.

Parsons, K. C. (1995). Ergonomics of the physical environment: International ergonomics standards concerning speech communication, danger signals, lighting, vibration and surface temperatures. *Applied Ergonomics, 26*(4), 281–292.

Parsons, K. C. (1995). Ergonomics and international standards: introduction, brief review of standards for anthropometry and control room design and useful information. *Applied Ergonomics, 26*(4), 239–247.

Parsons, K. C. (2000). Environmental ergonomics: A review of principles, methods and models. *Applied Ergonomics, 31*(6), 581–594.

Parsons, K. C., Shackel, B., & Metz, B.. (1995). Ergonomics and international standards: History, organizational structure and method of development. *Applied Ergonomics, 26*(4), 249–258.

Public Law 101-336. (1990). Americans with Disabilities Act. Public Law 336 of the 101st Congress, enacted July 26, 1990.

Ragothaman, S., & Korte, L. (1999). The ISO 9000 international quality registration: An empirical analysis of implications for business firms. *International Journal of Applied Quality Management, 2*(1), 59–73.

Raines, S. S. (2002). Implementing ISO 14001—an international survey assessing the benefits of certification. *Corporate Environment Strategy, 9*(4), 418–426.

Reed, P., Holdaway, K., Isensee, S., Buie, E., Fox, J., Williams, J., Lund, A. (1999). User interface guidelines and standards: Progress, issues, and prospects. *Interacting with Computers, 12*(2), 119–142.

Reed, P., Holdaway, K., Isensee, S., Buie, E., Fox, J., Williams. J., et al. (1999). User interface guidelines and standards: Progress, issues, and prospects. *Interacting with Computers, 12*(2), 119–142.

Rosenthal, I., Ignatowski, A. J., & Kirchsteiger, C. K. (2002). A generic standard for the risk assessment process: discussion on a proposal made by the program committee of the ER-JRC workshop on 'Promotion of Technical Harmonization of Risk-base Decision Making.' *Safety Science, 40*, 75–103.

Saito, S. Piccoli, B., Smith, M. J., Sotoyama, M., Sweitzer, G., Villanueva, M. B. G., et al. (2000). Ergonomic guidelines for using notebook personal computers. *Industrial Health, 38*, 421–434.

Seabrook, K. A. (2001). International Standards Update: Occupational Safety and Health Management Systems. In *Proceedings of the American Society of Safety Engineers' 2001 Professional Development Conference*, Anaheim, CA.

Sherehiy, B., Karwowski, W., & Rodrick, D. (in press). Human factors and ergonomics standards, In G. Salvendy (Ed.), *Handbook of human factors and ergonomics*, New York: Wiley.

Smith, W. J. (1996). *ISO and ANSI ergonomic standards for computer products.* Prentice Hall: Upper Saddle River, NJ.

Spivak, S. M., & Brenner, F. C. (2001). *Standardization essentials: Principles and practice.* New York: Dekker.

Staccini, P., Joubert, M., Quaranta, J-F., Fieschi, M. (in press). Mapping care processes within a hospital: From theory to a web-based proposal merging enterprise modeling and ISO normative principles. *Medical Informatics.*

Stevenson, T. H., & Barnes, F. C. (2001). Fourteen years of ISO 9000: Impact, criticisms, costs, and benefits. *Business Horizons*, May-June, 45–51.

Stewart, T.. (1995). Ergonomics standards concerning human-system interaction: Visual displays, controls and environmental requirements. *Applied Ergonomics, 26*(4), 271–274.

Stuart-Buttle, C. (in press). Overview of International Standards and Guideliness. In W. Karwowski (Ed.), *Handbook of human factors and ergonomics standards and guidelines.* Hillsdale, NJ: Lawrence Erlbaum Publishers.

Umezu, N., Nakano, Y., Sakai, T., Yoshitake, R., Herlitschke, W., & Kubota, S. (1998). Specular and diffuse reflection measurement feasibility study of ISO 9241 Part 7. Method. *Displays, 19*, 17–25.

Vikari-Juntura, E. R. A. (1997). The scientific basis for making guidelines and standards to prevent work-related musculoskeletal disorders. *Ergonomics, 40*(10), 1097–1117.

Walker, A. J. (1998). Improving the quality of ISO 9001 audits in the field of software. *Information and Software Technology, 40*, 865–869.

Wegner, E. (1995). Quality of software packages: The forthcoming international standard. *Computer Standards & Interfaces, 17*, 115–120.

Welzel, D., & Hausen, H-L. (1995). A method for software evaluation: contribution of the European project SCOPE to international standards. *Computer Standards & Interfaces, 17*, 121–129.

Wettig, J. (2002). New developments in standardization in the past 15 years—product versus process related standards. *Safety Science, 40*(1–4), 51–56.

Zuo, L., & Nayfeh, S. A. (2003). Low order continuous-time filters for approximation of the ISO 2631-1 human vibration sensitivity weightings. *Journal of Sound and Vibration, 265*, 459–465.

Author Index

602 AUTHOR INDEX

Wood, D. J., 151, *156*
Woolrych, A., 354, *360*
Workers' Compensation Board of British Columbia, 85,
  88, 90, 97, 98, 100, 101, *108*
Wren, J., 485, 487, *490*
Wright, M. S., 431, *440*
Wulff, I. A., 60, *77*

Y

Yamaoka, T., 372, *379*
Yamazaki, K., 372, *379*
Yanagida, K., 372, *379*
Yang, L., 158, *165*
Yang, M., 158, *166*
Yoblick, D. A., 408, *408*

Yoshitake, R., 33, *46, 589*
Young, S. L., 516, *524*

Z

Zehner, G. F., 180, 181, *196*
Zeng, G., 163, *166*
Zhai, S., 32, *46*
Zhao, H., *166*
Zheng, T., 158, *166*
Zheng, Z., 163, *166*
Zhou, B., 158, *166*
Zimolong, B., 416, 419, 431, *440*
Zuo, L., *589*
Zwetsloot, I. J. M., 495, *505*

Subject Index

Milton Keynes UK
Ingram Content Group UK Ltd.
UKHW052029071024
449327UK00027B/2495